英汉
化工设计施工
图解词汇

◉ 中国寰球工程有限公司 编

化学工业出版社
·北京·

本书以炼油、化工、石油化工、煤化工、LNG 等行业词汇为主要内容，广泛收集了工艺、系统、环保、安全、设备、蒸汽动力系统、转动机器、管道、仪表、电气、分析、实验室、焊接、无损检验、土建、施工机具等工程建设中常用的词汇。全书共分 15 部分，约 20000 条词汇，内容按专业分类，以图形示例，英中文对照编排，书末附有词条索引，便于检索。

本书可供化工设计、施工、科研、教学等领域的科技工作者使用。

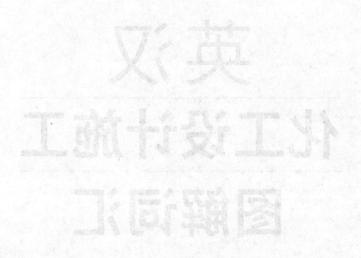

图书在版编目（CIP）数据

英汉化工设计施工图解词汇/中国寰球工程有限公司编．—北京：化学工业出版社，2018.11
ISBN 978-7-122-32991-2

Ⅰ.①英… Ⅱ.①中… Ⅲ.①化工设计-词汇-图解-英、汉 Ⅳ.①TQ02-61

中国版本图书馆 CIP 数据核字（2018）第 207851 号

责任编辑：刘　哲　潘新文　　　　　　　　　　　　装帧设计：韩　飞
责任校对：王　静

出版发行：化学工业出版社（北京市东城区青年湖南街 13 号　邮政编码 100011）
印　　装：北京新华印刷有限公司
787mm×1092mm　1/16　印张 58¾　字数 1427 千字　2019 年 3 月北京第 1 版第 1 次印刷

购书咨询：010-64518888　　售后服务：010-64518899
网　　址：http://www.cip.com.cn

凡购买本书，如有缺损质量问题，本社销售中心负责调换。

定　　价：368.00 元　　　　　　　　　　　　　　　　　　版权所有　违者必究

编委会名单

名誉主任： 张永久

主　　任： 赵　敏

编委会委员： 胡　健　刘　博　赵振月　弓普站　魏　毅　郑建华　林海涛　姚桂华

参编人员（以汉语拼音排序）：

安晓霞　白新泽　包光磊　陈凤秋　陈红捷　陈　晖
陈　萍　崔晓爽　崔迎春　丁聚庆　丁宇慧　杜　飞
杜　剑　樊　清　冯延忠　冯志华　高全乐　郭德庆
郭慧波　韩　景　贺　丁　侯越峰　胡　健　胡　平
简　明　蒋　宇　金晓晨　鞠林青　雷霁霞　李芳玲
李改云　李光宇　李广良　李锦辉　李立三　李　敏
李生彬　李文堂　李文忠　林　畅　林　珩　林晶虹
林贤莉　刘灿刚　刘广宇　刘建宾　刘小顿　刘晓波
路聿轩　马明燕　彭振河　邵　晨　沈　辉　舒小芹
宋　磊　隋建伟　孙长庚　谭　薇　唐东江　唐　峰
唐　硕　田红霞　万　茜　汪　洋　王　琥　王　炜
王喜全　王小鹏　王欣月　王　秀　王雪梅　王　颖
王　勇　王　昭　文　涛　吴德娟　吴　帆　武海坤
息　宁　肖　燕　谢　旸　邢桂坤　邢　睿　杨思思
杨伟冲　杨亚芝　叶　凌　叶　宁　应兴荣　游　宇
岳　巍　张焕照　张　瑾　张　婧　张　俊　张　可
张　鹏　张世忱　张树青　张旭辉　张　怡　张　芫
赵栓柱　赵文瑾　赵　诤　周　晖　朱　卉　左文耀

前言

由原化工工业部化学工业设计公司（现中国寰球工程有限公司北京分公司前身）和原北京石油化工总厂基本建设指挥部合作编写的《英汉化学工程图解词汇》一书，采用图形示例、英汉对照的独特编写方式，为广大石油化工战线的科技人员和工程技术人员翻译、阅读国外技术资料，提供了一本极为实用的工具书。自1985年第一版、2003年第二版出版以来，深受广大读者的欢迎。

2017年，中国寰球工程有限公司北京分公司与化学工业出版社商定，拟在2003年第二版的基础上，由中国寰球工程有限公司北京分公司组织各专业具有丰富工程设计及施工经验的技术人员，重新修编《英汉化学工程图解词汇》并更名为《英汉化工设计施工图解词汇》，删减或修改过时词汇，增收近年国内外新兴化工工艺流程相关的各专业词汇，以适应更多工程技术人员和广大读者的需求，力求做到与时俱进。

本书以炼油、化工、石油化工、煤化工、LNG等行业词汇为主要内容，收集了工艺、系统、环保、安全、设备、管道、仪表、电气、分析、建筑、结构、施工机械及检验等主要专业工程建设中的常用词汇，重点收集了一批在一般词典中不易收载的专业性较强的词汇和缩写词，共计2万余条，分成15部分。为保证本书图文的正确性，且有利于与国际通用资料接轨，书中的图例均选自与国外合作的大型工程项目的资料和厂商文件，并逐一进行核订。词汇的中文译名，以我国现行的国家标准、行业标准的命名为依据，无上述标准的，则以专业手册或习惯叫法为依据，有些词条的译名，不是机械地按英文直译，而是按我国工程上的实际使用名称译出。

由于图解词汇涉及的专业繁多，加之我们的水平有限，面对浩瀚的工程资料，虽然力求收集全面，但收集工作难免有偏颇之处，书中的词汇与译名也还会有不妥之处，敬请广大读者批评指正，提出宝贵意见，以便修订时改进。

中国寰球工程有限公司北京分公司
2018年9月

目 录

1 工艺及公用工程 Process and Utilities ··· 1
　1.1 工艺流程图图例说明 Process Flow Diagram/Legent Identification ·············· 1
　　1.1.1 物料代号 Line Service(Fluid)Identification ·· 1
　　1.1.2 在仪表符号中字母标志的意义 Meanings of Identification Letters
　　　　 in Instrument Symbols ··· 2
　　1.1.3 管道图例 Piping Symbols ·· 3
　　1.1.4 绘图示例——氨库装置 Illustrative Drawing—Ammonia Storage Facility ··· 5
　1.2 工艺装置典型工艺流程图 Typical Process Unit Flow Diagram ····················· 6
　　1.2.1 炼厂流程框图,润滑油型 Refining Block Diagram,Lube Type ················· 6
　　1.2.2 炼化一体化流程框图,炼油部分 Refining Chemical Complex Block
　　　　 Diagram,Refining Part ·· 7
　　1.2.3 炼化一体化流程框图,化工部分 Refining Chemical Complex Block
　　　　 Diagram,Chemical Part ·· 9
　　1.2.4 合成氨装置工艺流程图 Process Flow Diagram for Ammonia Plant ········· 11
　　1.2.5 尿素装置工艺流程图 Process Flow Diagram for Urea Plant ·················· 13
　　1.2.6 乙烯装置工艺流程图 Process Flow Diagram For Ethylene Plant ············ 15
　　1.2.7 线性低密度聚乙烯装置工艺流程图 Process Flow Diagram for LLDPE
　　　　 Plant ·· 16
　　1.2.8 管式法低密度聚乙烯/乙烯-醋酸乙烯共聚物装置工艺流程图 Process Flow
　　　　 Diagram for L-Tubular DPE/EVA Plant ·· 17
　　1.2.9 聚丙烯装置工艺流程图 Process Flow Diagram for Polypropylene
　　　　 Plant ·· 18
　　1.2.10 聚氯乙烯装置工艺流程图 Process Flow Diagram for PVC (Polyvinyl
　　　　　Chloride) Plant ·· 20
　　1.2.11 环氧乙烷/乙二醇装置工艺流程图 Process Flow Diagram for Ethylene
　　　　　Oxide/Ethylene Glycol Plant ··· 21
　　1.2.12 合成气制乙二醇装置工艺流程图 Process Flow Diagram for Syngas
　　　　　to Ethylene Glycol Plant ·· 23
　　1.2.13 天然气液化装置工艺流程图 Process Flow Diagram for Natural Gas
　　　　　Liquefaction Plant ·· 25
　　1.2.14 液化天然气接收站工艺流程图 Process Flow Diagram for Liquefied

    Natural Gas (LNG) Receiving Terminal ………………………… 27
  1.2.15 MTO装置工艺流程图 Process Flow Diagram for Methanol to Olefin (MTO) …………………………………………………………………… 29
  1.2.16 MTP工艺流程图 Process Flow Diagram for Methanol to Propylene …… 30
  1.2.17 丙烷脱氢装置工艺流程图 Process Flow Diagram for Propane Dehydrogenation (PDH) Plant ………………………………………… 31
  1.2.18 煤气化工艺流程图 Process Flow Diagram for Coal Gasification ………… 32
  1.2.19 煤制油工艺流程图 Process Flow Diagram for Coal to Oil Plant ………… 34
  1.2.20 煤制甲醇工艺流程图 Process Flow Diagram for Coal to Methanol ……… 36
  1.2.21 低温甲醇洗工艺流程图 Process Flow Diagram for Rectisol ………… 37
  1.2.22 精对苯二甲酸装置（PTA）工艺流程图 Process Flow Diagram for PTA Plant ……………………………………………………………………… 38
  1.2.23 苯乙烯/环氧丙烷（SM/PO）装置工艺流程图 Process Flow Diagram for SM/PO Plant …………………………………………………………… 38
  1.2.24 聚醚多元醇装置工艺流程图 Process Flow Diagram for Polyether Polyol Plant ……………………………………………………………………… 39
1.3 公用工程典型工艺流程图 Typical Utility Flow Diagram ……………………… 40
  1.3.1 空分空压装置工艺流程图 Process Flow Diagram for Air Separation Unit and Compressed Air "Station" ………………………………………… 40
   1.3.1.1 深冷空分装置工艺流程图 Process Flow Diagram for Cryogenic Air Separation Unit ……………………………………………………… 40
   1.3.1.2 变压吸附制氮装置工艺流程图 Process Flow Diagram for PSA Nitrogen Generation Unit ……………………………………………… 42
   1.3.1.3 真空变压吸附制氧装置工艺流程图 Process Flow Diagram for VPSA Oxygen Generation Unit ………………………………………………… 42
   1.3.1.4 膜分离制氮装置工艺流程图 Process Flow Diagram for Membrane Module Nitrogen Generation Unit ……………………………………… 42
   1.3.1.5 空压站工艺流程图 Process Flow Diagram for Compressed Air Station ………………………………………………………………… 43
  1.3.2 循环冷却水工艺流程图 Process Flow Diagram for Recirculating Cooling Water System …………………………………………………………… 43
  1.3.3 净水场工艺流程图 Process Flow Diagram for Raw Water Treatment Plant ……………………………………………………………………… 44
  1.3.4 储运系统工艺流程图 Process Flow Diagram for Storage and Transportation System …………………………………………………………… 44
  1.3.5 离子交换除盐水系统工艺流程图 Process Flow Diagram for Demineralized Water System (Ion Exchanger System) ………………… 45
  1.3.6 污水处理工艺流程图 Process Flow Diagram for Waste Water Treatment System ……………………………………………………………………… 47
  1.3.7 含盐废水处理装置流程图 Process Flow Diagram for Salty Waste Water Treatment Unit ………………………………………………………… 49

1.3.8 浓盐水蒸发系统流程图　Process Flow Diagram for Brine Water Evaporation System ………………………………………………………………… 51

1.3.9 浓盐水结晶系统流程图　Process Flow Diagram for Brine Water Crystallization System …………………………………………………………… 51

1.3.10 反渗透除盐水工艺流程图　Process Flow Diagram for Demineralized Water System（Reverse Osmosis System） ……………………………… 52

1.3.11 锅炉房系统工艺流程图　Process Flow Diagram for Boiler House ………… 53

 1.3.11.1 锅炉房给水系统工艺流程图　Process Flow Diagram for Feed Water System of Boiler House ……………………………………………… 53

 1.3.11.2 锅炉房锅炉给水系统工艺流程图　Process Flow Diagram for Boiler Feed Water System of Boiler House ……………………………… 54

 1.3.11.3 锅炉房锅炉燃烧系统工艺流程图　Process Flow Diagram for Boiler Combustion System of Boiler House ……………………………… 55

1.3.12 热电站系统工艺流程图　Process Flow Diagram for Thermal and Power Station ……………………………………………………………………… 56

 1.3.12.1 热电站汽机系统工艺流程图　Process Flow Diagram for Steam Turbine System of Thermal and Power Station ………………………… 56

 1.3.12.2 汽轮机空冷系统工艺流程图　Process Flow Diagram for Steam Turbine Air-Cooling Steam Condenser System ………………………… 57

 1.3.12.3 热电站锅炉系统工艺流程图　Process Flow Diagram for Boiler System of Thermal and Power Station ………………………………………… 58

 1.3.12.4 热电站锅炉燃烧系统工艺流程图　Process Flow Diagram for Boiler Combustion System of Thermal and Power Station ……………… 58

 1.3.12.5 氨法烟气脱硫工艺流程图　Process Flow Diagram for Flue Gas Desulfurization（FGD）System（Ammonia） ……………………… 59

 1.3.12.6 石灰石/石灰-石膏法烟气脱硫工艺流程图　Process Flow Diagram for Flue Gas Desulfurization（FGD）System（Limestone/Lime-Gypsum） ……………………………………………………………… 60

 1.3.12.7 烟气脱硝 SCR 系统流程图　Process Flow Diagram for Flue Gas Selective Catalytic Reduction（SCR）Denitrification System ……… 60

 1.3.12.8 尿素 SNCR 脱硝系统流程图　Process Flow Diagram for Flue Gas Urea Selective Non-Catalytic Reduction（SNCR）Denitrification System ………………………………………………………………… 61

 1.3.12.9 余热 ORC 发电系统工艺流程图　Process Flow Diagram for Waste Heat Organic Rankiee Cycle（ORC）Power Generation System ……… 62

1.3.13 物料处理系统典型工艺流程图　Typical Process Flow Diagram for Material Handling System ……………………………………………………… 62

 1.3.13.1 破碎筛分系统工艺流程图　Process Flow Diagram for Crushing and Screening System ………………………………………………… 62

 1.3.13.2 添加剂计量系统工艺流程图　Process Flow Diagram for Additive Metering System ……………………………………………………… 63

  1.3.13.3 挤压造粒系统工艺流程图 Process Flow Diagram for Extrusion Pelletizing System ……… 63

  1.3.13.4 掺混及储存系统工艺流程图 Process Flow Diagram for Blending and Storing System ……… 64

  1.3.13.5 包装码垛系统工艺流程图 Process Flow Diagram for Packing and Palletizing System ……… 64

  1.3.13.6 煤储运系统工艺流程图 Process Flow Diagram for Coal Storage and Transport System ……… 65

  1.3.13.7 磨煤干燥装置（CMD）工艺流程图 Process Flow Diagram for Coal Milling and Drying System ……… 66

1.4 工艺流程模拟计算 Process Flowsheet Simulation ……… 67

 1.4.1 工艺流程图 Process Flowsheet ……… 67

 1.4.2 热力学模型 Thermodynamic Model ……… 68

 1.4.3 模型修正 Model Modify ……… 69

 1.4.4 用户输入 User Input ……… 70

 1.4.5 石油馏分数据 Petroleum Component Data ……… 71

 1.4.6 蒸馏塔 Distillation Column ……… 72

 1.4.7 尺寸设计和校核 Sizing and Rating ……… 73

 1.4.8 收敛数据 Convergence Data ……… 74

 1.4.9 优化 Optimization ……… 75

 1.4.10 流程收敛 Flowsheet Convergence ……… 76

 1.4.11 用户定义 User Interface ……… 77

 1.4.12 结果汇总 Result Summary ……… 78

 1.4.13 设置选项 Setup Option ……… 79

 1.4.14 用户定制模拟程序 Customizing Simulation Engine ……… 80

 1.4.15 塔内件 Column Internals ……… 81

 1.4.16 负荷性能图 Hydraulic Plots ……… 82

 1.4.17 流动场分析 Computational Fluid Dynamics ……… 83

2 环境保护 Environmental Protection ……… 84

2.1 化工企业排放的主要污染物 Pollutants Discharged by Chemical Industrial Enterprise ……… 84

 2.1.1 废气污染物 Waste Gas Pollutants ……… 84

 2.1.2 废水污染物 Waste Water Pollutants ……… 84

 2.1.3 固体废弃物 Waste Solid ……… 85

 2.1.4 其他 Others ……… 85

2.2 环境质量 Environmental Quality ……… 85

 2.2.1 环境空气质量 Ambient Environmental Quality ……… 85

 2.2.2 地表水环境质量 Surface Water Environmental Quality ……… 86

 2.2.3 声环境质量 Noise Environmental Quality ……… 86

 2.2.4 其他 Others ……… 86

2.3 化工企业三废污染控制 Chemical Industrial Enterprise Wastes Pollution

  Control ·· 86
  2.3.1 大气污染控制 Air Pollution Control ·· 86
  2.3.2 工艺废水污染控制 Process Wastewater Pollution Control ································ 86
  2.3.3 固体废物污染控制 Waste Solid Pollution Control ·· 88
  2.3.4 噪声污染控制 Noise Pollution Control ·· 88
 2.4 大气污染物扩散 Air Pollutant Dispersion ·· 88
  2.4.1 烟羽形态 The Shapes of Stack Plumes ·· 88
  2.4.2 局地环流对大气污染的影响 Effects of Atmospheric Circulation in
    Localized Areas on Air Pollution ·· 89
   2.4.2.1 海陆风对沿岸地带空气污染的影响 Effect of Ocean Wind and Land
     Wind on Air Pollution in Coast-Land ·· 89
   2.4.2.2 热岛效应 Heat Island Effect ·· 89
 2.5 固体废弃物处理 Waste Solid Treatment ·· 89
  2.5.1 安全填埋场 Security Landfill Disposal Site ·· 89
  2.5.2 焚烧厂 Incineration Plant ·· 90
 2.6 污染物治理措施 Pollution Control Facility ·· 91
  2.6.1 废气治理措施 Waste Gas Treatment ·· 91
   2.6.1.1 VOCs 吸收工艺流程图 VOCs Absorption Process Flow
     Diagram ·· 91
   2.6.1.2 蓄热式氧化器工艺流程简图 Simplified Process Flow Diagram of
     RTO ·· 91
   2.6.1.3 废气流化床吸附工程示意图 Fluidized Bed Adsorption of Waste
     Gas ·· 91
   2.6.1.4 废气催化氧化处理工艺 Waste Gas Catalyzed Oxidation Process
     Diagram ·· 92
   2.6.1.5 袋式除尘器 Bag Dust collector ·· 92
   2.6.1.6 湿式除尘器 Wet Collector ·· 93
   2.6.1.7 氨法脱硫工艺 Process Flow Diagram Of De-SO_2 With Ammonia ············ 94
   2.6.1.8 氨法选择性催化氧化还原工艺流程图 Process Flow Diagram of SCR
     with Ammonia ·· 94
   2.6.1.9 尿素选择性非催化氧化还原工艺流程图 Process Flow Diagram of
     SNCR with Urea ·· 95
  2.6.2 多效蒸发系统 Multi-effect Evaporation System ·· 95
  2.6.3 重点污染防治区防渗结构图示 Special Important Pollution Prevention
    and Control Area Penetration Prevention Structure Diagram ···························· 96
  2.6.4 噪声控制措施 Noise Control Facility ·· 96

3 消防、安全和职业卫生 Fire Fighting, Safety and Occupational Health ········ 97
 3.1 消防系统流程图 Flow Diagram of Fire Fighting System ································ 97
  3.1.1 水喷淋典型流程图 Typical Flow Diagram of Water Sprinkler System ··········· 97
  3.1.2 泡沫消防典型流程图 Typical Flow Diagram of Foam Extinguishing
    System ·· 97

3.1.3 气体灭火系统典型流程图 Typical Flow Diagram of Gas Extinguishing System ······ 98
3.1.4 干粉灭火系统典型流程图 Typical Flow Diagram of Dry Chemical Extinguishing System ······ 98
3.2 消防设备 Fire Fighting Equipment ······ 99
  3.2.1 消防车 Fire Truck ······ 99
    3.2.1.1 泡沫消防车 Foam Fire Truck ······ 99
    3.2.1.2 干粉消防车 Dry Powder Fire Truck ······ 99
    3.2.1.3 水罐消防车 Water Fire Truck ······ 100
    3.2.1.4 直臂云梯车 Straight Arm and Scaling Ladder Truck ······ 100
  3.2.2 消防泵 Fire Pump ······ 101
  3.2.3 泡沫灭火设备 Foam Extinguishing Equipment ······ 102
    3.2.3.1 泡沫液储罐 Foam Storage Tank ······ 102
    3.2.3.2 混合器及比例混合系统 Proportioners and Proportioning Systems ······ 103
    3.2.3.3 泡沫炮 Foam Monitor ······ 104
    3.2.3.4 拖车式泡沫比例混合装置 Proportioning Foam Trailer ······ 105
    3.2.3.5 泡沫产生器 Foam Generator/Maker ······ 105
    3.2.3.6 泡沫喷头 Foam Nozzle ······ 106
    3.2.3.7 泡沫枪 Foam Nozzle ······ 106
  3.2.4 消火栓 Fire Hydrant ······ 107
  3.2.5 消防水泵接合器 Fire Pump Connection ······ 108
  3.2.6 消防水炮 Fire Water Monitor ······ 108
  3.2.7 电控消防水炮 Electric Monitor ······ 109
  3.2.8 消防接口和闷盖 Fire Quick Coupling and Cap ······ 109
    3.2.8.1 管牙接口 Quick Coupling ······ 109
    3.2.8.2 肘接口 Elbow Coupling ······ 109
    3.2.8.3 进水口闷盖 Water Inlet Cap ······ 110
    3.2.8.4 出水口闷盖 Water Outlet Cap ······ 110
  3.2.9 灭火器 Fire Extinguisher ······ 110
    3.2.9.1 推车式干粉灭火器 Wheel Dry Powder Extinguisher ······ 110
    3.2.9.2 手提式干粉灭火器 Portable Dry Powder Extinguisher ······ 111
    3.2.9.3 手提式二氧化碳灭火器 Portable Carbon Dioxide Extinguisher ······ 111
    3.2.9.4 手提式泡沫灭火器 Portable Foam Extinguisher ······ 111
  3.2.10 消防水带 Fire Hose ······ 111
  3.2.11 灭火剂 Fire Extinguishing Agent ······ 112
    3.2.11.1 清水 Water ······ 112
    3.2.11.2 泡沫液 Foam Concentrate (or Concentrate) ······ 112
    3.2.11.3 干粉 Dry Powder ······ 112
    3.2.11.4 卤代烷 Halogenated Agent ······ 112
    3.2.11.5 二氧化碳 Carbon Dioxide (CO_2) ······ 112

   3.2.11.6 氮气 Nitrogen（N_2）……112
 3.3 安全设备 Safety Equipment ……112
  3.3.1 呼吸器 Breathing Apparatus ……112
  3.3.2 防护服 Chemical Suit ……112
  3.3.3 其他安全设备 Other Safety Equipment ……113
   3.3.3.1 防护手套 Gloves ……113
   3.3.3.2 口罩 Gauze Mask ……113
   3.3.3.3 面罩 Mask ……113
   3.3.3.4 安全帽及防护屏 Safety Helmet and Face Shields ……113
   3.3.3.5 耳罩及耳塞 Earmuffs and Earplugs ……113
   3.3.3.6 安全带和安全绳 Safety Belt and Safety Ropes ……113
   3.3.3.7 防护眼镜 Protection Spectacles ……114
   3.3.3.8 防护鞋 Protection Shose ……114
   3.3.3.9 急救箱 First Aid Box ……114
   3.3.3.10 灭火毯 Fire Blanket ……114
   3.3.3.11 事故淋浴洗眼器 Safety Shower and Eye Washer ……114
   3.3.3.12 便携式事故淋浴洗眼器 Portable Emergency Eye Washer ……115

4 成套设备 Package Equipment ……116
 4.1 空分设备 Air Separation Unit ……116
  4.1.1 冷箱 Cold Box ……116
  4.1.2 变压吸附制氮装置/真空变压吸附制氧装置 PSA Nitrogen Generation Unit/VPSA Oxygen Generation Unit ……117
  4.1.3 膜分离制氮装置 Membrane Module Nitrogen Generation Unit ……117
 4.2 制冷机 Refrigerator ……118
  4.2.1 制冷装置 Refrigeration Unit ……118
  4.2.2 辅助设备 Auxiliary Equipments ……118
 4.3 低温冷箱 Cryogenic Cold Box ……120
  4.3.1 CO分离 CO Separation ……120
  4.3.2 液氮洗 Nitrogen Wash ……121
 4.4 冷却塔 Cooling Towers ……121
  4.4.1 自然通风冷却塔 Natural Draft Cooling Tower ……121
  4.4.2 横流冷却塔 Crossflow Cooling Tower ……122
  4.4.3 逆流冷却塔 Counterflow Cooling Tower ……122
 4.5 过滤器 Filter ……123
  4.5.1 压力过滤器 Pressure Filter ……123
  4.5.2 重力无阀过滤器 Gravity Valveless Filter ……123
 4.6 双室浮动阳（阴）离子交换器 Double Chamber Floating Cation (Anion) Exchanger ……124
 4.7 逆流再生阳（阴）离子交换器 Countercurrent Regeneration Cation (Anion) Exchanger ……125
 4.8 顺流再生阳（阴）离子交换器 Concurrent Regeneration Cation (Anion)

    Exchanger ··· 125
 4.9 混合离子交换器 Mixed Ion Exchanger ··· 126
 4.10 反渗透膜元件 Reverse Osmosis Membrane ·· 126
 4.11 脱气塔 Degassing Tower ·· 127
 4.12 微孔曝气器 Fine Bubble Diffusion Aerator ··· 127
 4.13 转刷曝气机 Brush Aerator ·· 128
 4.14 表面曝气机 Surface Aerator ·· 128
 4.15 火炬 Flare Stacks ·· 129
  4.15.1 火炬系统 Flare System ··· 129
  4.15.2 火炬头 Flare Tip ·· 131
  4.15.3 地面火炬 Multi-point Ground Flare ··· 132
 4.16 输送机和提升机 Conveyor and Elevator ·· 133
  4.16.1 带式输送机 Belt (Band) Conveyor ·· 133
   4.16.1.1 带式输送机示意图 Belt Conveyor Sketch ··························· 133
   4.16.1.2 带式输送机系统 Belt Conveyor Systems ··························· 133
   4.16.1.3 带式输送机部件 Belt (Band) Conveyor Components ··········· 135
   4.16.1.4 拼装式皮带输送机和移动式皮带输送机 Pre-built Sectional Belt Conveyor and Mobile Belt Conveyor ······································· 138
   4.16.1.5 裙状挡板大倾角带式输送机 Skirt Side Plate Inclined Belt Conveyor ··· 139
   4.16.1.6 双带提升机 Twin Riser ··· 140
   4.16.1.7 中心距超过180m的单轮传动带式输送机用的张紧装置 Take-up Unit for Single Drum Drive Belt Conveyor Exceeding 180 Meters Centres ··· 140
   4.16.1.8 双轮传动带式输送机的张紧装置 Take-up Unit for Dual Drum Dive Belt Conveyor ··· 141
  4.16.2 垂直提升(输送)机 Vertical Elevator (Conveyor) ····················· 141
   4.16.2.1 斗式提升机 Bucket Elevator ··· 141
   4.16.2.2 垂直提升机和箱类提升机 Vertical Rising Conveyor and Case Elevator ··· 142
  4.16.3 埋刮板式输送机 En Masse Conveyor ······································· 143
  4.16.4 螺旋输送机 Screw Conveyor ·· 143
  4.16.5 辊子输送机 Roller Conveyor ·· 144
  4.16.6 吊挂式链输送机和振动输送机 Chain Trolley Conveyor and Vibrating Conveyor ··· 145
   4.16.6.1 吊挂式链输送机 Chain Trolley Conveyor ························· 145
   4.16.6.2 吊挂式链输送机部件 Unit of Chain Trolley Conveyor ······· 145
   4.16.6.3 振动输送机 Vibrating Conveyor ····································· 147
  4.16.7 气力输送机系统 Pneumatic Conveying System ··························· 147
   4.16.7.1 正压气力输送系统 Pressure Pneumatic Conveying System ···· 147
   4.16.7.2 负压气力输送系统 Vacuum Pneumatic Conveying System ····· 147

4.16.7.3　正负压相结合的系统　Pressure with Vacuum Pneumatic Conveying System ………… 148
　　　4.16.7.4　封闭环形气力输送系统　Closed Pneumatic Conveying System …… 148
　　　4.16.7.5　密相气力输送系统　Dense Phase Pneumatic Conveying System …… 148
　　　4.16.7.6　稀相气力输送系统　Dilute Phase Pneumatic Conveying System …… 149
4.17　破碎和筛分设备　Crushing and Screening Equipment ………… 149
　4.17.1　颚式冲击破碎机　Impact Jaw Crusher ………… 149
　4.17.2　颚式破碎机　Jaw Crusher ………… 150
　4.17.3　回转球形破碎机　Gyrasphere Crusher ………… 151
　4.17.4　盘式回转破碎机　Tray Type Gyratory Crusher ………… 152
　4.17.5　液压锥形破碎机和冲击破碎机（叶片破碎机）　Hydraulic Cone Crusher and Impact Crusher（Impeller Breaker）………… 154
　4.17.6　辊子粉碎机　Roller Mill ………… 155
　4.17.7　双轴锤式破碎机　Double Shaft Hammer Mill ………… 156
　4.17.8　磨机　Mill ………… 157
　　　4.17.8.1　球磨机　Ball Mill ………… 158
　　　4.17.8.2　中速磨　Medium Speed Mill ………… 158
　　　4.17.8.3　风扇磨　Fan Mill ………… 158
　4.17.9　干燥粉磨机　Dryer-Pulveriser ………… 159
　4.17.10　返混设备布置　Layout of Back Mixing Equipment ………… 160
　4.17.11　振动筛　Vibrating Screen ………… 161
　　　4.17.11.1　往复式振动筛　Reciprocating Vibrating Screen ………… 161
　　　4.17.11.2　圆振筛　Round Separator ………… 162
　4.17.12　分级机　Classifiers ………… 163
4.18　料仓、料斗与阀门　Storage Bin, Hopper and Valve ………… 164
　4.18.1　筒仓　Silo ………… 164
　4.18.2　重力式掺混仓　Gravity Blending Silo ………… 164
　4.18.3　料斗与溜槽　Hopper and Chute ………… 165
　4.18.4　气动粉体插板阀　Pneumatic Slide Valve ………… 165
　4.18.5　手动粉体插板阀　Manual Slide Valve ………… 166
　4.18.6　粉体重力换向阀　Gravity Flow Diverter ………… 166
　4.18.7　气动翻板换向阀　Pneumatic Flap Diverter ………… 167
　4.18.8　气动转子换向阀　Pneumatic Rotary Diverter ………… 168
　4.18.9　粉体旋转给料阀　Rotary (Feeder) Valve ………… 168
4.19　给料、称重、包装及码垛设备　Feeding, Weighing, Bagging and Palletizing Machine ………… 169
　4.19.1　振动给料机　Vibrating Feeder ………… 169
　4.19.2　电振动给料机　Electric Vibrating Feeder ………… 170
　4.19.3　板式给料机　Apron Feeder ………… 170
　4.19.4　（粉末）均匀自动给料机　Smooth Auto-Feeder ………… 171
　4.19.5　带式计量秤　Dosing Belt Weigher ………… 172

4.19.6 定量给料秤 Constant Feed Weigher …… 173
4.19.7 失重式给料机 Loss-in-Weight Feeder …… 173
4.19.8 自动装袋系统 Auto Bagging System …… 173
4.19.9 自动码垛系统 Automatic Palletizing System …… 174
4.19.10 托盘式缠绕机 Pallet Stretch Wrapping Machine …… 175
4.19.11 热缩膜垛盘打包机 Shrink Packing System …… 175
4.20 仓储设备 Storage Equipment …… 176
 4.20.1 堆取料机 Stacker and Reclaimer …… 176
  4.20.1.1 斗轮堆取料机 Bucket Wheel Stacker and Reclaimer …… 176
  4.20.1.2 圆形堆场堆取料机 Stacker and Reclaimer for Circular Store …… 176
  4.20.1.3 侧式悬臂堆料机 Stacker for Longitudinal Store …… 177
  4.20.1.4 门式取料机 Portal Reclaimer …… 177
 4.20.2 火车装车机 Train Carriage Loader …… 178
4.21 LNG 成套设备 LNG Package Equipment …… 178
 4.21.1 装卸臂 Loading (Unloading) Arm …… 178
 4.21.2 装车撬 Truck Loading Skid …… 179
 4.21.3 装车臂 Truck Loading Arm …… 179
 4.21.4 浸没燃烧式气化器 Submerged Combustion Vaporizer …… 180
4.22 造粒机 Granulator …… 180
 4.22.1 钢带造粒机 Steel Belt Granulator …… 180
 4.22.2 转筒造粒机 Rotating Drum Granulator …… 181

5 蒸汽动力系统 Steam and Power System …… 182
5.1 锅炉 Boiler …… 182
 5.1.1 燃煤锅炉 Coal Fired Boiler …… 182
  5.1.1.1 火管锅炉及水管锅炉的基本形式 Basic Patterns for Fire and Water Tube Boiler …… 182
  5.1.1.2 椭圆管板换热器 Ellipsoidal Shell and Tube Heat Exchanger …… 183
  5.1.1.3 水冷管夹套换热器 Cooling Tubes and Jacket Heat Exchanger …… 184
  5.1.1.4 蒸汽净化及锅筒内件 Steam Purification and Drum Internals …… 185
  5.1.1.5 过热器 Superheater …… 190
  5.1.1.6 减温器 Attemperator …… 191
  5.1.1.7 空气预热器 Air Preheater …… 192
  5.1.1.8 炉排 Grates …… 193
  5.1.1.9 下饲炉排 Underfeed Stoker …… 194
  5.1.1.10 省煤器 Economizer …… 195
  5.1.1.11 喷燃器，燃烧器 Burners …… 196
  5.1.1.12 抛煤机 Spreader Feeders …… 198
  5.1.1.13 磨煤机 Pulverizers …… 199
 5.1.2 燃油（气）锅炉 Oil-Fired Boiler and Gas-Fired Boiler …… 200
  5.1.2.1 燃油（气）锅炉的结构 Configure of Oil-Fired Boiler and Gas-Fired Boiler …… 200

  5.1.2.2　燃油蒸汽锅炉系统工艺流程图　Process Flow Diagram for Oil-Fired Steam Boiler System ·· 200

  5.1.2.3　燃气蒸汽锅炉系统工艺流程图　Process Flow Diagram for Gas-Fired Steam Boiler System ·· 201

 5.1.3　循环流化床锅炉　Circulation Fluidized Bed Boiler ······························ 202

 5.1.4　特种锅炉　Special Boiler ··· 202

 5.1.5　热回收设备　Heat Recovery Equipment ··· 203

  5.1.5.1　废液焚烧炉　Waste Liquid Incinerator ··· 203

  5.1.5.2　余热锅炉　Waste Heat Boiler ·· 204

  5.1.5.3　垃圾焚烧炉　Garbage Incinerator ·· 205

  5.1.5.4　垃圾焚烧发电厂工艺流程图　Process Flow Diagram for Garbage Incineration Power Plant ·· 206

5.2　汽轮机　Steam Turbine ·· 207

 5.2.1　汽轮机循环　Steam Turbine Cycles ··· 207

 5.2.2　冲动式汽轮机　Impulse Turbine ·· 208

 5.2.3　单级汽轮机　Single Stage Steam Turbine ··· 209

 5.2.4　汽轮机轴封　Turbine Glands and Gland Sealings ······························ 210

 5.2.5　超速脱扣装置（保安器）　Overspeed Tripping Device ······················ 211

 5.2.6　汽轮机的润滑　Lubrication of Steam Turbine ···································· 212

 5.2.7　汽轮机供汽方式　Methods of Steam Supply to a Turbine ················ 213

 5.2.8　汽轮机调速器　Turbine Governor ·· 214

 5.2.9　汽轮机调速器及调速　Governors and Governing of Steam Turbine ··· 216

 5.2.10　汽轮机的安装　Installation of Steam Turbine ··································· 218

5.3　燃气轮机　Gas Turbine ·· 219

 5.3.1　燃气轮机结构　Constitution of Gas Turbine ······································ 219

 5.3.2　燃气轮机发电机组　Gas Turbine Power Plant ···································· 219

 5.3.3　燃气轮机简单循环发电厂　Gas Turbine Simple Cycle Power Plant ··· 220

 5.3.4　燃气轮机-蒸汽轮机联合循环发电厂　Gas Turbine-Steam Turbine Combined Cycle Power Plant ·· 221

6　设备　Equipment ··· 222

6.1　设备设计常用术语　Terms and Definitions ·· 222

 6.1.1　一般术语　General Terms ··· 222

 6.1.2　设备类型术语　Types of Equipment ·· 222

 6.1.3　压力容器分类术语　Classification for Categories of Pressure Vessels ······ 222

6.2　塔　Column ··· 223

 6.2.1　板式塔　Plate Column ·· 223

  6.2.1.1　液流型式　Liquid-Flow Patterns ·· 225

  6.2.1.2　塔盘型式　Types of Trays ·· 226

  6.2.1.3　塔盘的支承　Supports of Tray ··· 233

  6.2.1.4　塔底结构及重（再）沸器　Bottom Structures and Reboiler ············ 234

  6.2.1.5　进料和抽出　Feed and Draw-off ·· 235

- 6.2.2 填料塔 Packed Column ⋯⋯ 237
  - 6.2.2.1 填料型式 Types of Packing ⋯⋯ 239
  - 6.2.2.2 液体分布器和再分布器 Liquid Distributors and Redistributors ⋯⋯ 241
  - 6.2.2.3 填料压板和支承板 Packing Hold-Down Plates and Support Plates ⋯⋯ 243
- 6.2.3 楼梯（梯子）和平台 Stair and Platform ⋯⋯ 244
- 6.2.4 塔附件 Tower Attachments ⋯⋯ 245
- 6.2.5 塔结构示例 Column Examples ⋯⋯ 249
  - 6.2.5.1 CO_2 吸收塔 CO_2 Absorber ⋯⋯ 249
  - 6.2.5.2 再生塔/CO_2 汽提塔 Regenerator/CO_2 Stripper ⋯⋯ 250

6.3 反应设备 Reactor ⋯⋯ 251
- 6.3.1 氨合成塔 Ammonia Converter ⋯⋯ 251
- 6.3.2 尿素合成塔 Urea Reactor ⋯⋯ 253
- 6.3.3 环管反应器 Loop Reactor ⋯⋯ 255
- 6.3.4 氧氯化反应器 Oxychlorination Reactor ⋯⋯ 256
- 6.3.5 聚合釜 Polymerizer ⋯⋯ 257
- 6.3.6 隔膜电解槽 Diaphragm Cells ⋯⋯ 258
- 6.3.7 变换炉 Shift Converter ⋯⋯ 259

6.4 造粒设备 Prilling Equipment ⋯⋯ 260
- 6.4.1 （造粒塔）总图及造粒喷头组装图 General Assembly and Prill-Spray Assembly ⋯⋯ 260
- 6.4.2 造粒塔扒料机 Prill Tower Reclaimer ⋯⋯ 261

6.5 换热设备 Heat Exchanger ⋯⋯ 263
- 6.5.1 换热器的类型 Types of Heat Exchangers ⋯⋯ 263
- 6.5.2 管壳式换热器 Tubular Heat Exchanger ⋯⋯ 266
- 6.5.3 套管式换热器和刮面式换热器 Double-Pipe Heat Exchanger and Scraped-Surface Exchanger ⋯⋯ 277
- 6.5.4 套管式纵向翅片换热器 Double Pipe Longitudinal Finned Exchanger ⋯⋯ 278
- 6.5.5 板式换热器 Plate Exchangers ⋯⋯ 279
- 6.5.6 蒸汽表面冷凝器，凝汽器 Steam Surface Condensers ⋯⋯ 280
- 6.5.7 蒸发器 Evaporators ⋯⋯ 281
- 6.5.8 空冷器、空气冷却器 Air-Cooled Heat Exchangers ⋯⋯ 282

6.6 储罐 Storage Tanks ⋯⋯ 288
- 6.6.1 储罐类型 Types of Storage Tanks ⋯⋯ 288
- 6.6.2 固定顶储罐 Fixed-Roof Tanks ⋯⋯ 289
  - 6.6.2.1 储罐零部件名称 Nomenclature of Tank Parts ⋯⋯ 289
  - 6.6.2.2 罐顶结构 Roof Structure ⋯⋯ 292
- 6.6.3 内浮顶罐 Internal Floating Roof Tank ⋯⋯ 295
- 6.6.4 外浮顶罐 External Floating Roof Tank ⋯⋯ 296
  - 6.6.4.1 外浮顶罐零部件名称 Nomenclature of External Floating Roof Tank Parts ⋯⋯ 296
  - 6.6.4.2 外浮顶型式 External Floating Roof Types ⋯⋯ 298

  6.6.4.3 外浮顶罐的密封型式　Seal Types of External Floating Roof Tank ……………… 300

 6.6.5 球形储罐及其他低压大型储罐　Spherical Tanks and Other Large Low-Pressure Storage Tanks …………………………………………………………………… 301

 6.6.6 低温储罐　Refrigerated Storage Tanks ……………………………………………… 302

6.7 衬里设备　Lining Equipment ………………………………………………………………… 303

 6.7.1 橡胶衬里设备　Rubber Lining Equipment …………………………………………… 303

 6.7.2 搪玻璃设备　Glass-Lined Equipment ………………………………………………… 304

 6.7.3 衬砖设备　Brick Lining Equipment …………………………………………………… 306

6.8 非金属设备　Non-metal Equipment ………………………………………………………… 307

 6.8.1 石墨设备　Graphite Equipment ………………………………………………………… 307

  6.8.1.1 石墨设备的类型　Type of Graphite Equipment ………………………………… 307

  6.8.1.2 石墨设备零件名称　Components Nomenclature of Graphite Equipment …… 308

 6.8.2 玻璃钢设备　FRP（Fibre Reinforce Plastics）Equipment ………………………… 309

  6.8.2.1 玻璃钢设备的类型　Type of FRP Equipment …………………………………… 309

  6.8.2.2 玻璃钢管连接型式　Type of FRP Pipe Joint …………………………………… 309

6.9 混合设备　Mixing Equipment ………………………………………………………………… 310

 6.9.1 搅拌器型式　Types of Agitator ………………………………………………………… 310

 6.9.2 混合（搅拌）槽　Mixing Tanks ……………………………………………………… 312

 6.9.3 管道混合器　Line Mixers (Flow Mixers) …………………………………………… 313

 6.9.4 静止混合器　Static Mixers …………………………………………………………… 314

 6.9.5 双螺杆连续混合机　Double Screw Continuous Mixer ……………………………… 315

 6.9.6 膏状物料及黏性物料混（拌）合设备　Paste and Viscous-Material Mixing Equipments ……………………………………………………………………………… 316

 6.9.7 固体混合机械　Solids Mixing Machines …………………………………………… 318

6.10 萃取器　Extractors …………………………………………………………………………… 319

 6.10.1 浸提设备　Leaching Equipments …………………………………………………… 319

 6.10.2 连续萃取设备，连续抽提设备　Continuous Contact (Differential Contact) Equipments ……………………………………………………………………………… 320

6.11 旋风分离器、澄清器、过滤器和离心机　Cyclone, Decanter, Filter and Centrifuger …………………………………………………………………………………… 322

 6.11.1 旋风分离器　Cyclone Separators …………………………………………………… 322

 6.11.2 沉降罐、澄清器　Gravity Settlers (Decanters) …………………………………… 325

 6.11.3 气体洗涤器　Gas Scrubbers ………………………………………………………… 326

 6.11.4 过滤机　Filters ………………………………………………………………………… 328

  6.11.4.1 压滤机　Pressure Filters ……………………………………………………… 328

  6.11.4.2 叶滤机　Pressure Leaf Filters ………………………………………………… 329

  6.11.4.3 袋式过滤器　Bag Filers ………………………………………………………… 330

  6.11.4.4 转鼓真空过滤机　Rotary-Drum Vacuum Filter ……………………………… 331

 6.11.5 离心式分离机　Centrifugal Separator ……………………………………………… 332

6.11.5.1 离心机　Centrifuges ……………………………………………… 332
6.11.5.2 双鼓真空离心过滤机　Double-Bowl Vacuum Centrifuge ……… 334
6.11.5.3 静止叶片型离心式分离器　Stationary Vane Type Centrifugal Separators …………………………………………………………… 335
6.12 干燥器　Dryers …………………………………………………………… 336
  6.12.1 直接干燥器　Direct Dryers ………………………………………… 336
  6.12.2 间接干燥器　Indirect Dryers ……………………………………… 338
  6.12.3 喷雾干燥器　Spray Dryers ………………………………………… 339
    6.12.3.1 喷雾干燥装置　Spray Dryer Installation ……………………… 339
    6.12.3.2 雾化喷头、喷雾嘴、雾化器　Spray Nozzles（Atomizers）…… 340
  6.12.4 气流（气动）输送干燥器　Pneumatic Conveyor Dryers ………… 341
6.13 其他静设备及部件　Other Equipments and Components ……………… 342
  6.13.1 石油炼制中的流化过程　Fluidization Processes in Petroleum Refinery ……………………………………………………………… 342
    6.13.1.1 流态化　Fluidization …………………………………………… 343
    6.13.1.2 流化床分布器　Distributors for Fluidized Bed ……………… 344
  6.13.2 壳体和封头　Shells and Heads …………………………………… 345
  6.13.3 破沫器　Demister …………………………………………………… 348
    6.13.3.1 破沫器及其应用　Demister and Its Applications …………… 348
    6.13.3.2 破沫网的安装和纤维除雾气　Installation of Mesh and Fiber Mist Eliminator …………………………………………………… 349
  6.13.4 设备支座　Supports of Equipments ……………………………… 350
  6.13.5 密封型式　Types of Seal ………………………………………… 352
  6.13.6 非圆形截面容器　Vessels of Noncircular Cross Section ……… 353
  6.13.7 其他容器及零部件名称　Nomenclature of Other Vessels and Components ………………………………………………………… 355
6.14 耐火衬里结构　Refractory Lining Constructions ……………………… 358
  6.14.1 陶瓷纤维衬里结构　Ceramic Fiber Lining Construction ……… 358
    6.14.1.1 陶瓷纤维毯结构　Ceramic Fiber Blanket Layered Construction … 358
    6.14.1.2 陶瓷纤维模块结构　Ceramic Fiber Modular Construction … 359
  6.14.2 砖炉衬结构　Fire Brick Lining Construction …………………… 360
  6.14.3 浇注料炉衬结构　Castable Lining Construction ………………… 362
6.15 炉用零部件　Furnace Components ……………………………………… 362
  6.15.1 炉管、管件及联箱（集合管）　Tubes, Fitting and Header (Manifold) ……………………………………………………………… 362
  6.15.2 燃烧器（烧嘴）和附属设备　Burners and Accesseries ………… 364
  6.15.3 管板及管架　Tube Sheet and Tube Supports …………………… 365
  6.15.4 其他炉用附件　Furnace Accessories ……………………………… 366
6.16 炉型及结构　Types and Structure of Furnace ………………………… 368
  6.16.1 管式加热炉　Tubular Heater ……………………………………… 368
  6.16.2 典型加热炉结构　Typical Structure of Heater ………………… 370

6.16.3 转化炉 Reformers ………………………………………………………… 371
 6.16.3.1 一段转化炉/转化炉 Primary Reformer …………………………… 371
 6.16.3.2 二段转化炉 Secondary Reformer ………………………………… 373
 6.16.3.3 换热式转化炉 Reforming Exchanger …………………………… 374
6.16.4 裂解炉 Cracking Furnace …………………………………………… 375
6.16.5 气化炉 Gasifier ……………………………………………………… 376
 6.16.5.1 粉煤加压气化炉 Pulverized-coal Pressurized Gasifier ………… 376
 6.16.5.2 水煤浆气化炉 Coal-water Slurry Gasifier ……………………… 377
 6.16.5.3 加压固定床煤气化炉 Pressurized Fixed-bed Coal Gasifier …… 378
 6.16.5.4 流化床煤气化炉 Fluidized-bed Coal Gasifier …………………… 379
6.16.6 余热回收和废热锅炉 Waste Heat Recovery and Waste Heat Boiler … 380
 6.16.6.1 余热回收 Waste Heat Recovery ………………………………… 380
 6.16.6.2 CO 燃烧废热锅炉 CO Firing Waste Heat Boiler ……………… 381
 6.16.6.3 第一废热锅炉 Primary Waste Heat Boiler ……………………… 382
 6.16.6.4 第二废热锅炉 Secondary Waste Heat Boiler …………………… 383
6.16.7 热载体加热炉 Heat Carrier Heater ………………………………… 384
 6.16.7.1 热载体加热炉系统 Heat Carrier Heater System ……………… 384
 6.16.7.2 水（油）浴炉 Water (Oil) Bath Heater ………………………… 384
 6.16.7.3 热载体加热炉 Heat Carrier Heater ……………………………… 385

7 转动机器 Rotary Machine ……………………………………………………… 386
 7.1 泵 Pump …………………………………………………………………… 386
  7.1.1 各种形式的泵 Various Types of Pumps ……………………………… 386
  7.1.2 离心泵 Centrifugal Pump …………………………………………… 391
  7.1.3 轴流泵 Axial Flow Pump …………………………………………… 395
  7.1.4 管道泵 Inline Pump ………………………………………………… 396
  7.1.5 混流泵 Mixed Flow Pump …………………………………………… 397
  7.1.6 斜流泵 Diagonal Flow Pump ………………………………………… 398
  7.1.7 螺杆泵 Screw Pump ………………………………………………… 399
   7.1.7.1 单螺杆泵 Progressive Cavity Pump …………………………… 399
   7.1.7.2 双螺杆泵 Twin Screw Pump …………………………………… 400
   7.1.7.3 三螺杆泵 Three-Spindle Screw Pump ………………………… 400
  7.1.8 计量泵 Metering Pump ……………………………………………… 401
  7.1.9 真空泵 Vacuum Pump ……………………………………………… 402
   7.1.9.1 水环真空泵 Liquid Ring Vacuum Pump ……………………… 402
   7.1.9.2 旋片式真空泵 Vane Vacuum Pump …………………………… 402
   7.1.9.3 往复式真空泵 Piston Vacuum Pump …………………………… 404
  7.1.10 气动隔膜泵 Air Operated Pump …………………………………… 405
  7.1.11 往复泵 Reciprocating Pump ………………………………………… 406
   7.1.11.1 活塞式往复泵 Reciprocating Piston Pump …………………… 406
   7.1.11.2 双作用蒸汽往复泵 Duplex Acting Steam-Driven Reciprocating Pump ……………………………………………………………… 406

7.1.11.3 双作用活塞式往复泵（皮碗式） Double Action Reciprocating Pump (Bucket Type) ······ 407
7.1.12 喷射泵 Jet Pumps ······ 409
7.1.13 喷射装置 Ejector Units ······ 410
7.1.14 皮托管泵 Pitot Tube Pump ······ 412
7.2 压缩机、鼓风机和风机 Compressors, Blowers and Fans ······ 412
  7.2.1 螺杆压缩机 Screw Compressor ······ 412
    7.2.1.1 无油螺杆压缩机 Dry Screw Compressor ······ 412
    7.2.1.2 喷油螺杆压缩机 Oil-injected Screw Compressor ······ 413
  7.2.2 往复式压缩机 Reciprocating Compressor ······ 414
    7.2.2.1 活塞式压缩机 Piston Compressor ······ 416
    7.2.2.2 迷宫式压缩机 Labyrinth Compressor ······ 417
    7.2.2.3 隔膜式压缩机 Diaphragm Compressor ······ 418
  7.2.3 低密度聚乙烯（超）高压压缩机 High Pressure Compressor for Low Density Polythylene Process ······ 419
    7.2.3.1 高压气缸和中心型组合阀 High-Pressure Cylinder and Central Valve ······ 421
    7.2.3.2 卸荷阀及其他阀 Unloading Valve and Other Valves ······ 422
  7.2.4 离心压缩机 Centrifugal Compressor ······ 424
    7.2.4.1 高速压缩机 High Speed Compressor ······ 424
    7.2.4.2 水平剖分式离心压缩机 Horizontally Split Centrifugal Compressor ······ 425
    7.2.4.3 径向剖分式离心压缩机 Radially Split Centrifugal Compressor ······ 426
    7.2.4.4 整体齿轮式压缩机 Integrally Geared Centrifugal Compressor ······ 427
    7.2.4.5 轴流压缩机 Axial-flow Compressor ······ 428
  7.2.5 鼓风机 Blowers ······ 428
  7.2.6 风机 Fans ······ 430
  7.2.7 典型的空气压缩机装置 Typical Air Compressor Unit ······ 432
7.3 搅拌器 Mixer ······ 433
7.4 驱动机 Driver ······ 434
  7.4.1 蒸汽轮机 Steam Turbine ······ 434
  7.4.2 蒸汽轮机的分类与结构 Classification and Structure of Steam Turbines ······ 435
  7.4.3 液力透平 Hydraulic Turbine ······ 438
  7.4.4 燃气轮机 Gas Turbine ······ 438
  7.4.5 烟气轮机 Flue Gas Turbine ······ 439
  7.4.6 膨胀机/再压缩机 Expander/Compressor ······ 439
  7.4.7 柴油机 Diesel Engine ······ 440
7.5 转动设备辅机 Rotating Equipment Auxiliaries ······ 441
  7.5.1 液力耦合器 Fluid Coupling ······ 441
  7.5.2 齿轮箱 Gear ······ 441

7.5.3 透平冷凝系统　Turbine Condensing System ………………………… 442
8　管道工程　Piping Engineering …………………………………………… 443
　8.1　设备及管道布置　Equipment and Piping Layout …………………… 443
　　8.1.1　设备布置　Equipment Layout ……………………………………… 443
　　8.1.2　管道布置图，配管图　Piping Layout Drawing …………………… 444
　　　8.1.2.1　管道平面布置　Piping Plan Layout ………………………… 444
　　　8.1.2.2　立面图　Section Drawing …………………………………… 445
　　　8.1.2.3　管道布置——装置配管　Piping Layout—Installation Piping … 446
　　8.1.3　管道组装图（轴测图，管段图）　Erection Diagram（Pipe Line
　　　　　　Isometric Diagram） ………………………………………… 447
　　8.1.4　伴热及夹套　Tracing Lines and Jacket ………………………… 448
　　　8.1.4.1　蒸汽伴热管　Steam Tracing Lines ………………………… 448
　　　8.1.4.2　热水伴热管　Hot Water Tracing Lines …………………… 449
　　　8.1.4.3　夹套管　Jacket Lines ………………………………………… 449
　　　8.1.4.4　电伴热　Electric Heat Tracing ……………………………… 450
　　8.1.5　急救冲洗和洗眼站　Safety Shower and Eyewash Station ……… 451
　8.2　管道材料　Piping Material …………………………………………… 452
　　8.2.1　管子，弯管　Pipe, Pipe Bends …………………………………… 452
　　8.2.2　管件　Pipe Fitting ………………………………………………… 453
　　　8.2.2.1　钢焊接管件　Steel-Welding Fittings ………………………… 453
　　　8.2.2.2　螺纹管件　Threaded Fittings ………………………………… 454
　　　8.2.2.3　支管台与承插管件　Outlet and Socket Welding Fittings …… 455
　　　8.2.2.4　法兰管件　Flanged Fittings …………………………………… 456
　　　8.2.2.5　塑料压接管接头　Plastics Compression Joints ……………… 457
　　8.2.3　法兰、法兰密封面　Flanges, Flange Facings …………………… 458
　　8.2.4　管螺纹　Pipe Threaded …………………………………………… 459
　　8.2.5　垫片　Gasket ……………………………………………………… 460
　　8.2.6　管道用紧固件　Piping Fastener ………………………………… 461
　　8.2.7　阀门　Valve ………………………………………………………… 461
　　　8.2.7.1　阀杆与阀盖结构　Valve Stem and Bonnet Structure ……… 461
　　　8.2.7.2　阀门　Valves …………………………………………………… 463
　　　8.2.7.3　闸阀　Gate Valve ……………………………………………… 471
　　　8.2.7.4　波纹管密封闸阀　Bellows Sealed Gate Valve ……………… 473
　　　8.2.7.5　截止阀　Globe Valve …………………………………………… 474
　　　8.2.7.6　止回阀　Check Valve …………………………………………… 475
　　　8.2.7.7　球阀　Ball Valve ……………………………………………… 476
　　　8.2.7.8　旋塞阀　Plug Valve …………………………………………… 477
　　　8.2.7.9　蝶阀　Butterfly Valve ………………………………………… 478
　　　8.2.7.10　弹簧安全泄压阀　Spring Safety-Relief Valve ……………… 480
　　　8.2.7.11　液面控制浮球阀　Pilot Operated Ball Float Valve ………… 481
　　　8.2.7.12　阀门操纵机构　Valve Operating Mechanisms ……………… 482

8.2.7.13 蒸汽疏水阀（器）和空气疏水阀（器） Steam Traps and Air Traps ……………………………………………………………………………… 484
8.2.7.14 热膨胀阀 Thermo Expansion Valve …………………………………… 486
8.2.8 管道特殊件 Pipe Specialty …………………………………………………… 487
  8.2.8.1 过滤器，阻火器 Strainer，Flame Arrester ………………………… 487
  8.2.8.2 分离器，消声器，取样冷却器 Separator, Silencer, Sample Cooler ……………………………………………………………………… 488
  8.2.8.3 视镜 Sight Glass ……………………………………………………… 489
8.2.9 管道特殊元件 Pipe Special Element ………………………………………… 490
8.2.10 填料 Packings ………………………………………………………………… 491
8.2.11 管子端部连接 End Connection in Tubing and Pipe ……………………… 492
8.3 管道机械 Piping Mechanics …………………………………………………………… 493
  8.3.1 金属软管 Metal Flexible Hose ……………………………………………… 493
  8.3.2 金属补偿器（膨胀节） Metal Expansion Joint …………………………… 493
    8.3.2.1 填函式伸缩节 Packed Slip Expansion Joint ………………………… 493
    8.3.2.2 波纹补偿器（膨胀节） Bellows Expansion Joint ………………… 494
  8.3.3 管道支吊架 Pipe Hangers and Pipe Supports …………………………… 495
    8.3.3.1 管道支架 Pipe Supports ……………………………………………… 495
    8.3.3.2 管道吊架 Pipe Hangers ……………………………………………… 496
    8.3.3.3 弹簧架 Spring Supports ……………………………………………… 497
    8.3.3.4 管架零部件 Attachments of Piping Supports ……………………… 497
8.4 地下管道 Underground Piping ………………………………………………………… 498
  8.4.1 地下管平面布置图 Underground Piping Arrangement Drawing ………… 498
    8.4.1.1 地下管道图例 Underground Piping Legend ………………………… 498
    8.4.1.2 地下管道管线代号 Underground Pipe Line Code ………………… 499
  8.4.2 地下管常见附属设施及构筑物 Common Facility and Structure of U/G Piping ………………………………………………………………………… 499
8.5 绝热 Insulation ………………………………………………………………………… 500
  8.5.1 管道绝热 Piping Insulation ………………………………………………… 500
  8.5.2 立式容器的外部隔热 External Thermal Insulation for Vertical Vessel ………………………………………………………………………… 502
8.6 工程图常用缩写词（按字母顺序排列） Abbreviations for Use on Drawing (in Alphabetical Order) ……………………………………………………………… 503
8.7 防腐 Anticorrosion ……………………………………………………………………… 518
  8.7.1 内防护 Internal Anticorrosive Coating …………………………………… 518
  8.7.2 三层聚乙烯涂层 Three Layer Polyethylene Coating ……………………… 519
  8.7.3 表面处理 Surface Preparation Vocabulary ………………………………… 520
  8.7.4 涂料 Paint …………………………………………………………………… 520
  8.7.5 腐蚀 Corrosion ……………………………………………………………… 521
  8.7.6 其他常用词汇 Others ……………………………………………………… 521
8.8 管道应力分析 Piping Stress Analysis ………………………………………………… 522

- 8.8.1 应变　Strain 522
- 8.8.2 应力　Stress 523
- 8.8.3 管道应力　Piping Stress due to Forces and Moments 524
- 8.8.4 管道应力分析　Piping Stress Analysis 525
- 8.8.5 管道应力分析输入条件　Input of Piping Stress Analysis 526
- 8.8.6 管道柔性　Piping Flexibility 527
- 8.8.7 管道动态效应　Dynamic Effects on Piping 527
- 8.8.8 疲劳曲线　Fatigue Curves 528

8.9 三维模型　3D Model 528
- 8.9.1 三维管道设计软件专用词汇　Vocabulary for 3D Piping Design Software 528
  - 8.9.1.1 Smart Plant 3D管道模型分层管理词汇　Hierarchical Management Vocabulary for Smart Plant 3D Piping Model 528
  - 8.9.1.2 PDMS模型分层管理词汇　Hierarchical Management Vocabulary for PDMS Model 529
  - 8.9.1.3 PDMS数据库类型词汇　Database Type Vocabulary for PDMS 529
  - 8.9.1.4 PDS模块　Module for PDMS 530
  - 8.9.1.5 建立设备模板　Create Equipment Template 530
  - 8.9.1.6 建立基本体　Create Primitives 531
- 8.9.2 Smart Plant Review软件专用词汇　Specific Vocabulary for Smart Plant Review 531
  - 8.9.2.1 按钮　Button 531
  - 8.9.2.2 菜单　Menu 533

9 工业自动化仪表　Process Measurement and Control Instrument 537
9.1 温度计　Thermometer 537
- 9.1.1 测温元件　Thermometer Element 537
- 9.1.2 温度变送器　Temperature Transmitter 539
- 9.1.3 辐射高温计　Radiation Pyrometer 540
- 9.1.4 双金属温度计　Bimetal Thermometer 541
- 9.1.5 光学高温计　Radiation Pyrometer 542

9.2 压力测量仪表　Pressure Instruments 543
- 9.2.1 电接点压力表和压力开关　Pressure Gauge with Electric Contact and Pressure Switch 543
- 9.2.2 压力表　Pressure Gauge 544
- 9.2.3 压力变送器　Pressure Transmitter 546
- 9.2.4 基地式压力指示调节器　Field Installed Type of Pressure Indicating and Controller 547
- 9.2.5 压力表的安装　Installation of Pressure Gage 548

9.3 流量仪表　Flow Meter 549
- 9.3.1 科氏力质量流量计　Coriolis Mass Flow Meters 549
- 9.3.2 一次流量元件　Primary Flow Element 550

9.3.3 流量测量用差压变送器 Difference Pressure Transmitter for Flow Measurement …… 552
9.3.4 流速式流量计 Fluid Velocity Meter …… 553
  9.3.4.1 旋涡流量计 Vortex Flow Meter …… 553
  9.3.4.2 涡轮流量计 Turbine Meter …… 554
9.3.5 容积式流量计 Positive Displacement Type Flow Meters …… 555
9.3.6 可变面积式流量计 Variable Area Type Flow Meters …… 557
9.3.7 电磁流量计 Electro-magnetic Flow Meters …… 558
  9.3.7.1 电磁流量计及测量原理 Electro-magnetic Flow Meters and Principle of Measurement …… 558
  9.3.7.2 电磁流量计电极 Electro-magnetic Flow Meters Electrode …… 559
  9.3.7.3 电磁流量计变送器 Electro-magnetic Flow Meters Transmitter …… 559
9.3.8 热式质量流量计 Thermal Mass Flow Meters …… 560
9.3.9 差压流量计的安装 Installation of Head Meters …… 561
9.4 液位测量仪表 Level Meter …… 563
9.4.1 直读式就地液位指示仪 Locally Mounted Direct Reading Level Gages …… 563
9.4.2 液位测量用差压变送器 Difference Pressure Transmitter for Liquid Level Measurement …… 564
9.4.3 浮筒式液位计 Displacement Type Level Meter …… 565
9.4.4 伽马射线液位计 Gamma Radiometric Level Meter …… 566
9.4.5 雷达液位计 Radar Level Meter …… 567
9.4.6 伺服液位计 Servo Level Gauge …… 569
9.4.7 液位调节器及液位开关 Level Controllers and Switches …… 570
9.4.8 气动液位调节器 Pneumatic Liquid Level Controller …… 571
9.5 过程分析仪表 Process Analyzer …… 572
9.5.1 过程气相色谱仪 Process Gas Chromatograph …… 572
9.5.2 非色散红外线气体分析仪 Non-dispersive Infra-red Gas Analyzer (NDIR) …… 573
9.5.3 氧分析器 Oxygen Analyzer …… 573
  9.5.3.1 氧化锆分析仪 Zirconium Oxide Oxygen Analyzer …… 573
  9.5.3.2 顺磁式氧分析仪 Paramagnetic Oxygen Analyzer …… 574
  9.5.3.3 电化学式氧分析仪 Electrochemical Oxygen Analyzer …… 574
  9.5.3.4 溶解氧分析仪 Dissolved Oxygen Analyzer …… 574
9.5.4 热导式分析仪 Thermal Conductivity Analyzer …… 575
9.5.5 pH 分析仪 pH Analyzer …… 575
9.5.6 电导仪 Conductivity Analyzer …… 575
9.6 控制阀 Control Valve …… 576
9.6.1 气动执行机构 Pneumatic Actuator …… 577
9.6.2 电动执行机构 Electric Actuator …… 577
9.6.3 控制阀型式 Control Valve Type …… 578
  9.6.3.1 直通单座控制阀 Single-Ported Globe Control Valve …… 578

- 9.6.3.2 直通双座控制阀 Double-Ported Globe Control Valve …… 578
- 9.6.3.3 角形控制阀 Angle Control Valve …… 579
- 9.6.3.4 三通控制阀 Three-Way Control Valve …… 579
- 9.6.3.5 控制蝶阀 Butterfly Control Valve …… 579
- 9.6.3.6 偏心旋转控制阀 Rotary Control Valve …… 580
- 9.6.3.7 V形控制球阀 V-Ball Control Valve …… 580
- 9.6.3.8 控制球阀 Ball Control Valve …… 580
- 9.6.4 控制阀上阀盖 Control Valve Bonnet …… 581
  - 9.6.4.1 波纹管密封型上阀盖 Bonnet with Guardian Bellows Seal …… 581
  - 9.6.4.2 低温型上阀盖 Cryogenic Bonnet …… 581
- 9.6.5 自力式调节阀 Self-Operated Regulator …… 582
  - 9.6.5.1 自力式温度调节阀 Temperature Regulator …… 582
  - 9.6.5.2 自力式流量调节阀 Flow Regulator …… 582
  - 9.6.5.3 自力式差压调节阀 Differential Pressure Regulator …… 583
  - 9.6.5.4 自力式压力调节阀 Pressure Regulator …… 583
- 9.6.6 控制阀附件 Control Valve Accessory …… 584
  - 9.6.6.1 电/气转换器 I/P Converter (E/P Converter) …… 584
  - 9.6.6.2 电/气阀门定位器 I/P Positioner …… 584
  - 9.6.6.3 数字式阀门定位器 Digital Positioner …… 585
  - 9.6.6.4 现场总线式阀门定位器 Fieldbus Positioner …… 585
  - 9.6.6.5 限位开关 Limit Switch …… 586
  - 9.6.6.6 电磁阀 Solenoid Valve …… 587
  - 9.6.6.7 阀位变送器 Position Transmitter …… 587
  - 9.6.6.8 气锁阀 Pneumatic Lock-up Valve …… 588
  - 9.6.6.9 空气过滤减压器 Air Pressure Reducing Station …… 588
  - 9.6.6.10 一体化过滤减压阀 Integrated Filter Regulator …… 589
  - 9.6.6.11 气动放大器（增速继动器）Pneumatic Volume Booster (Booster Relay) …… 589
  - 9.6.6.12 快速排放阀 Quick Exhaust Valve …… 590
  - 9.6.6.13 止回阀 Check Valve …… 590
- 9.6.7 安全泄压阀 Safety Relief Valve …… 591
  - 9.6.7.1 弹簧式安全泄压阀 Spring Loaded Safety Relief Valve …… 591
  - 9.6.7.2 带导阀的安全泄压阀 Pilot Operated Safety Relief Valve …… 592
- 9.6.8 爆破片 Burst Disc …… 592
- 9.7 盘装仪表 Panel Mounting Instrument …… 593
  - 9.7.1 安全栅 Safety Barriers …… 593
    - 9.7.1.1 用于变送器的安全栅 Safety Barriers for Transmitter …… 593
    - 9.7.1.2 用于数字（开关）输入的安全栅 Safety Barriers for Digital (Switch) Inputs …… 593
    - 9.7.1.3 用于数字（开关）输出的安全栅 Safety Barriers for Digital (Switch) Outputs …… 594

9.7.2　隔离器　Isolator ………… 594
  9.7.2.1　用于变送器和控制器的隔离器　Isolator for Transmitter and Controller ………… 594
  9.7.2.2　用于电磁阀/报警驱动器的隔离器　Isolator for Solenoid/Alarm Driver ………… 594
  9.7.2.3　用于开关/趋近检测器的隔离器　Isolator for Switch/Proximity Detector ………… 595

9.8　特殊仪表　Special Instrument ………… 595
  9.8.1　速度传感器　Speed Sensor ………… 595
  9.8.2　轴位移变送器　Transmitter for Thrust Position ………… 595
  9.8.3　称重系统　Weighing System ………… 596
    9.8.3.1　负荷传感器　Load Cells ………… 596
    9.8.3.2　称重系统　Weighing System ………… 596

9.9　火灾自动报警系统　Automatic Fire Alarm System ………… 597
  9.9.1　火灾自动报警系统　Automatic Fire Alarm System ………… 597
  9.9.2　火灾报警控制盘　Fire Alarm Control Panel ………… 598
  9.9.3　激光感烟探测器　Laser Smoke Detector ………… 599

9.10　工业电视监视系统　Industry Television Surveillance System ………… 600
  9.10.1　工业电视监视系统　Industry TV Surveillance System ………… 600
  9.10.2　矩阵控制器　Matrix Switcher ………… 601

9.11　数字控制系统　Digital Control System ………… 602
  9.11.1　分散控制系统　Distributed Control System ………… 602
  9.11.2　现场总线控制系统　Fieldbus Control System（FCS） ………… 603
  9.11.3　工厂资源管理系统　Plant Resource Management System（PMR） ………… 604
  9.11.4　可编程控制器　Programmable Logic Controller（PLC） ………… 604
  9.11.5　紧急停车系统　Emergency Shutdown System（ESD） ………… 605

9.12　仪表盘，柜和操作台　Instrument Panels, Cabinets and Consoles ………… 606

10　电气工程　Electrical Engineering ………… 608

10.1　旋转电机　Electrical Rotating Machine ………… 608
  10.1.1　直流电动机　Direct-Current Motor ………… 608
  10.1.2　直流发电机　Direct-Current Generator ………… 609
  10.1.3　AC换向电机　AC Commutator Machine ………… 610
  10.1.4　同步电动机的无刷励磁　Brushless Excitation of Synchronous Motor ………… 610
  10.1.5　交流电动机　Alternating-Current Motor ………… 611
  10.1.6　滑环式感应电动机　Slip-Ring Type Induction Motor ………… 612
  10.1.7　无刷同步电动机　Brushless Synchronous Motor ………… 614
  10.1.8　线槽和绕组　Slots and Windings ………… 615
  10.1.9　柴油发电机组　Diesel Generator Set ………… 616
    10.1.9.1　柴油发电机组侧视图　Diesel Generator Set Side View ………… 616
    10.1.9.2　就地控制屏　Local Control Panel ………… 616
    10.1.9.3　柴油机左前视图　Diesel Engine Front and Left Side View ………… 617

10.1.9.4 柴油机右后视图 Diesel Engine Rear and Right Side View …… 617
10.1.9.5 发电机 Power Generator …… 618
10.2 变压器 Transformer …… 619
10.2.1 油浸式变压器 Oil Immersed Transformer …… 619
10.2.1.1 油浸式变压器正视图 Oil Immersed Transformer Front View …… 619
10.2.1.2 油浸式变压器侧视图 Oil Immersed Transformer Side View …… 620
10.2.1.3 油浸式变压器外视图 Oil Immersed Transformer Exterior …… 621
10.2.2 油浸式密闭变压器 Sealed Oil Immersed Transformer …… 622
10.2.3 树脂绝缘干式变压器 Cast Resin Transformer …… 623
10.2.4 箱式变电站 Transformer Substation …… 624
10.3 整流器和电池 Rectifier and Battery …… 625
10.3.1 整流器 Rectifier …… 625
10.3.2 原电池 Primary Batteries …… 626
10.3.3 阀控式密封铅酸蓄电池 Stationary Lead Acid Valve Regulated Sealed Batteries …… 629
10.3.4 交流不间断电源 AC Uninterrupted Power Supply …… 630
10.3.5 直流不间断电源 DC Uninterrupted Power Supply …… 631
10.4 高中压开关装置 High and Middle Voltage Switchgear …… 631
10.4.1 110kV 六氟化硫气体封闭式组合电器 110kV SF_6 Gas-Insulated Switchgear …… 631
10.4.2 110kV 空气绝缘高压组合电器 110kV Prefabricated Air Insulated Switchgear …… 632
10.4.3 中压六氟化硫气体绝缘开关装置 Medium Voltage SF_6 Gas-Insulated Switchgear …… 632
10.4.4 金属封闭式开关装置 Metal-Enclosed Switchgear …… 633
10.4.5 中压金属铠装开关装置 Middle Voltage Metal-Clad Switchgear …… 634
10.4.6 环网柜 Ring Switchgear …… 635
10.4.6.1 六氟化硫气体绝缘环网柜 SF_6 Gas Insulated Ring Switchgear …… 635
10.4.6.2 金属封闭环网柜 Metal Clad Ring Switchgear …… 636
10.4.7 高压户外六氟化硫断路器 High Voltage Outdoor SF_6 Circuit Breakers …… 636
10.4.8 中压六氟化硫断路器 Medium Voltage SF_6 Circuit Breakers …… 637
10.4.9 户内真空断路器 Indoor Vacuum Circuit Breaker …… 638
10.4.10 避雷器 Lightning Arrester …… 638
10.4.11 真空接触器 Vacuum Contactor …… 639
10.4.12 高压熔断器 HV Fuses …… 640
10.4.12.1 高压管式熔断器 HV Fuse Tubes …… 640
10.4.12.2 高压跌开式熔断器 HV Falling Type Fuse …… 640
10.4.13 补偿电容器成套装置 Compensation Capacitors Cabinet …… 641
10.4.14 操动机构 Operating Device …… 641
10.4.14.1 直动操动机构 Direct Operating Device …… 641

10.4.14.2 转向操动机构 Swivel Operating Device ……………………………… 642
10.4.14.3 弹簧储能操动机构 Spring-Stored Energy Operating Device ……… 642
10.5 低压开关装置 Low Voltage Switchgear ………………………………………… 643
　10.5.1 低压开关柜 Low Voltage Switchgear …………………………………… 643
　10.5.2 框架式空气断路器 Air Circuit Breaker ………………………………… 644
　　10.5.2.1 框架式空气断路器前面板 Air Circuit Breaker Front Face ……… 644
　　10.5.2.2 框架式空气断路器分解图 Air Circuit Breaker Explode Drawing … 645
　10.5.3 低压熔断器 LV Fuses …………………………………………………… 646
　　10.5.3.1 螺旋式熔断器 Screw Base Fuse ……………………………………… 646
　　10.5.3.2 无填料封闭管式熔断器 Sealed Tube Fuse without Filling ……… 646
　　10.5.3.3 有填料快速熔断器 Filled Fast Fuse ………………………………… 647
　　10.5.3.4 插入式熔断器 Plug-in Fuse ………………………………………… 647
　10.5.4 微型断路器及附件 Miniature Circuit Breaker and Accessories ……… 648
　　10.5.4.1 微型断路器附件 Miniature Circuit Breaker Accessories ………… 648
　　10.5.4.2 微型断路器 Miniature Circuit Breaker …………………………… 649
　10.5.5 塑壳断路器 Moulded Case Circuit Breaker …………………………… 649
　　10.5.5.1 塑壳断路器 Moulded Case Circuit Breaker（MCCB） …………… 649
　　10.5.5.2 塑壳断路器模块化系统 MCCB Modularized System ……………… 650
　10.5.6 真空负荷开关 Vacuum Load Switch …………………………………… 651
　10.5.7 控制开关 Control Switch ………………………………………………… 652
　　10.5.7.1 按钮开关 Pushbutton Switches ……………………………………… 653
　　10.5.7.2 凸轮旋转开关 Cam Switches ………………………………………… 654
　10.5.8 变频调速装置 Variable Velocity Variable Frequency Device ………… 655
　10.5.9 电动机启动器 Motor Starter …………………………………………… 656
　　10.5.9.1 手动启动器 Manual Starter ………………………………………… 656
　　10.5.9.2 自动启动器 Automatic Starter ……………………………………… 656
10.6 控制及保护器件 Control and Protection Devices ……………………………… 657
　10.6.1 电磁机构和器件 Electromagnetic Mechanism and Devices …………… 657
　10.6.2 电磁继电器 Electromagnetic Relays …………………………………… 658
　10.6.3 固态继电器 Solid State Relays …………………………………………… 659
10.7 电气防爆设备 Explosion Protected Apparatus ………………………………… 660
　10.7.1 防爆安装单元 Explosion Protected Installation Control Units ……… 660
　10.7.2 防爆照明灯具 Explosion Protected Luminaries ……………………… 661
　10.7.3 防爆控制设备 Explosion Protected Control Equipment ……………… 662
10.8 供电系统 Power Supply System …………………………………………………… 662
　10.8.1 预装式变电所 Electrical House（E-HOUSE） ………………………… 662
　10.8.2 热电厂 Thermal Power Plant …………………………………………… 663
　10.8.3 变电所屋内配电装置 Indoor Installations of Electric Substation …… 664
　10.8.4 节能发电厂 Energy Saving Power Plant ……………………………… 665
　10.8.5 铁塔及电杆 Towers and Poles …………………………………………… 667
　10.8.6 静电除尘器 Electrostatic Precipitator ………………………………… 668

10.9 电线、电缆及附件 Wires, Cables and Accessories ………………… 669
  10.9.1 电缆、电线结构 Wire and Cables Structure ………………… 669
    10.9.1.1 一般结构 General Structure ………………… 669
    10.9.1.2 防火/阻燃电缆结构 Fire-Proof/Flame Retardant Cable Structure ………………… 669
    10.9.1.3 铝绞线/钢芯铝绞线结构 Al Twisted Wire/ ACSR Wire ………………… 669
  10.9.2 电缆、电线产品 Wire and Cables Products ………………… 670
  10.9.3 电缆附件 Accessories for Cables ………………… 671
    10.9.3.1 45kV 以下中、低压电缆附件 Up to 45kV Accessories for Low and Medium Voltage Cables ………………… 671
    10.9.3.2 高压电缆附件 Accessories for High Voltage Cables ………………… 672
    10.9.3.3 电缆梯架 Cable Ladder ………………… 673
10.10 灯泡 Lamps ………………… 674
10.11 电伴热系统 Electrical Heat Tracing System ………………… 675
  10.11.1 电伴热产品 Electrical Heat Tracing Products ………………… 675
    10.11.1.1 自调控电伴热带 Self-Regulating Heat Tracing Cable ………………… 675
    10.11.1.2 限功率伴热带 Power Limiting Heat Tracing Cable ………………… 675
    10.11.1.3 并联恒功率伴热带 Parallel Constant Watt Cable ………………… 675
    10.11.1.4 串联恒功率伴热带 Series Constant Watt Cable ………………… 676
    10.11.1.5 矿物绝缘伴热带 Mineral Insulated Cable ………………… 676
    10.11.1.6 罐和容器的加热板 Tank and Vessel Heating Panel ………………… 676
    10.11.1.7 仓斗伴热板 Hopper Heating Module ………………… 677
  10.11.2 典型电伴热系统 Typical Heat Tracing System ………………… 677
  10.11.3 电伴热控制和监测 Heat Tracing Control and Monitoring ………………… 678
10.12 阴极保护 Cathodic Protection ………………… 679
  10.12.1 阴极保护测试站和接线箱 Cathodic Protection Test Stations and Junction Box ………………… 679
  10.12.2 长效参比电极 Permanent Reference Cell ………………… 679
  10.12.3 阳极 Anodes ………………… 680
  10.12.4 填料 Backfill ………………… 681
  10.12.5 整流器 Rectifier ………………… 681
  10.12.6 绝缘材料 Insulation Material ………………… 682
  10.12.7 接地电池 Grounding Cells ………………… 682
  10.12.8 阴极保护系统 Cathodic Protection Systems ………………… 683
    10.12.8.1 深阳极系统 Deep Anode System ………………… 683
    10.12.8.2 水罐系统 Water Storage Tank System ………………… 683
    10.12.8.3 地上储罐罐底保护 Protection for Above Ground Storage Tank Bottoms ………………… 684
    10.12.8.4 地下储罐保护 Protection for Underground Storage Tanks ………………… 684
10.13 接地防雷系统 Grounding and Lightning System ………………… 685
  10.13.1 接地产品 Grounding Products ………………… 685

10.13.2	典型防雷系统 Typical Lightning System	686
10.14	监测控制和数据采集 Supervise Control and Data Acquiring (SCADA)	687
10.15	通信系统 Telecommunication System	688
10.15.1	用户电话交换机 PABX	688
10.15.2	结构化布线系统 Structured Cabling System	689
10.15.3	扩音对讲系统 Paging System	690
10.15.4	有线电视系统 Cable TV System	690

11 实验室 Laboratory — 691

11.1	实验室常用仪器 General Apparatus and Instruments	691
11.1.1	玻璃器皿 Glassware	691
11.1.2	实验室主要分析仪器 General Laboratory Equipment	691
11.2	化学试剂和药品 Chemicals and Reagents	692
11.2.1	无机试剂 Inorganic Reagent	692
11.2.2	有机试剂 Organic Reagent	693
11.2.3	指示剂 Indicator	694
11.3	实验室常用词汇 Laboratory Vocabulary	694
11.4	主要分析仪器示例图 Legend for Main Laboratory Equipment	695
11.4.1	分析天平 Analytical Balance	695
11.4.2	电极 Electrode	695
11.4.2.1	玻璃电极 Glass Electrode	695
11.4.2.2	甘汞电极 Calomel Electrode	696
11.4.3	色谱仪类 Chromatography	696
11.4.3.1	气相色谱法原理图 Gas Chromatograph Schematic Diagram	696
11.4.3.2	气相色谱仪外观图 Gas Chromatograph Outline	697
11.4.3.3	气相色谱进样模式 Gas Chromatograph Injection Mode	697
11.4.3.4	液相色谱示意图 Liquid Chromatograph Schematic Diagram	698
11.4.4	光谱类仪器 Spectrometric Instruments	698
11.4.4.1	紫外-可见分光光度计光路图 UV-Visible Spectrophotometer Optical Schematic Diagram	698
11.4.4.2	傅立叶变换红外光谱仪原理图 Fourier Transform Infrared Spectrometer Schematic Diagram	699
11.4.4.3	原子吸收分光光度计示意图 Atomic Absorption Spectrophotometer Schematic Diagram	699
11.4.5	电化学类仪器 Electrochemistry Instruments	700
11.4.5.1	卡尔费休水分滴定仪 Karl Fisher Auto Titrator	700
11.4.5.2	卡尔费休水分滴定仪原理图 Karl Fisher Auto Titrator Schematic Diagram	700
11.4.5.3	pH 计 pH Meter	701
11.4.6	水质分析仪器 Apparatus for Water Analysis	701
11.4.6.1	生物耗氧量测定仪 Apparatus for Biological Oxygen Command Determination	701

11.4.6.2 化学耗氧量测定仪　Apparatus for Chemical Oxygen Command Determination ………………………… 702

12　土建工程　Civil Engineering ……………………………………………… 703
12.1　建筑　Architectural ……………………………………………………… 703
12.1.1　平面图、立面图、剖面图　Plan, Elevation and Section ………… 703
12.1.2　楼梯、电梯　Staircase and Lift …………………………………… 704
12.1.3　卫生间　Toilet ………………………………………………………… 704
12.1.4　门的形式　Type of Doors ………………………………………… 705
12.1.5　窗的形式　Type of Windows …………………………………… 706
12.1.6　窗的组成　Components of Windows …………………………… 706
12.1.7　屋顶的形式　Type of Roofs ……………………………………… 707
12.1.8　钢结构厂房　Steel Structure House …………………………… 708
12.1.9　钢结构厂房节点　Detail of Steel Structure House …………… 709
12.1.10　薄壳屋顶　Shell Roofs …………………………………………… 711
12.2　结构　Construction ……………………………………………………… 712
12.2.1　设计条件　Design Information …………………………………… 712
12.2.2　材料　Material ………………………………………………………… 713
　12.2.2.1　混凝土　Concrete ……………………………………………… 713
　12.2.2.2　钢筋　Reinforcement ………………………………………… 713
　12.2.2.3　结构钢　Structural Steel ……………………………………… 713
12.2.3　荷载条件　Load Condition ………………………………………… 714
　12.2.3.1　几何图形　Geometry ………………………………………… 714
　12.2.3.2　荷载种类　Type of Loads …………………………………… 714
　12.2.3.3　施加载荷　Imposed Loads ……………………………………… 714
12.2.4　结构计算　Structure Calculation ………………………………… 715
　12.2.4.1　受力及受力条件　Force and Application ………………… 715
　12.2.4.2　参数　Parameter ………………………………………………… 715
　12.2.4.3　支承、变形和稳定　Support, Deformation and Stability …… 716
　12.2.4.4　框架　Frames …………………………………………………… 716
12.2.5　土壤与基础　Soil and Foundation ………………………………… 717
　12.2.5.1　土壤符号　Symbols for Soil ………………………………… 717
　12.2.5.2　土层剖面图　Subsoil Profile ………………………………… 717
　12.2.5.3　基础类型　Type of Foundation ……………………………… 718
　12.2.5.4　基础剖面　Section of Foundation ………………………… 718
　12.2.5.5　桩的形式　Type of Pile ……………………………………… 719
　12.2.5.6　设备基础　Foundation for Equipment ……………………… 720
　12.2.5.7　大型设备的锚固　Anchorage of Heavy Machine ………… 721
12.2.6　多层工业厂房　Multistory Industrial Buildings ………………… 723
12.2.7　钢筋混凝土结构　Reinforced Concrete Construction ………… 724
12.2.8　工业构筑物　Industrial Structures ……………………………… 724
　12.2.8.1　烟囱的形式　Types of Chimneys ………………………… 724

12.2.8.2 管支架 Pipe Supports ········ 725
12.2.8.3 排气筒 Vent Stacks ········ 726
12.2.8.4 水塔的形式 Types of Water Towers ········ 727
12.2.8.5 冷却塔 Cooling Tower ········ 727
12.2.8.6 通廊和栈桥 Galleries and Trestles ········ 728
12.2.8.7 筒仓和储斗 Silos and Bunkers ········ 728
12.2.8.8 低温液化气体储罐 Tank for the Storage of Refrigerated Liquefied Gases ········ 729
12.2.9 钢结构 Steel Construction ········ 730
12.2.9.1 钢构件的连接 Connection of Steel Members ········ 731
12.2.9.2 钢结构连接详图 Structural Steel Connection Details ········ 732
12.2.9.3 钢栏杆 Steel Balustrade ········ 733
12.2.9.4 钢扶梯和爬梯 Steel Stairs and Ladders ········ 734

12.3 暖通 Heating Ventilation and Air Conditioning ········ 735
12.3.1 采暖系统与散热器 Heating System and Radiator ········ 735
12.3.1.1 散热器 Radiator ········ 735
12.3.1.2 暖风机 Unit Heater ········ 736
12.3.1.3 自动及手动排气阀 Automatic and Manual Air Vent-Valve ········ 737
12.3.1.4 热水分/集水器和蒸汽分汽缸 Supply/Return Header for Hot Water and Supply Header for Steam ········ 738
12.3.1.5 开式膨胀水箱 Open Type Expansion Tank ········ 739
12.3.2 通风与空调设备 Ventilation and Air Conditioning Equipment ········ 740
12.3.2.1 离心风机及减震台座 Centrifugal Fan and Shock Absorption Support ········ 740
12.3.2.2 轴流风机 Axial Fan ········ 741
12.3.2.3 混流风机 Mixed Flow Fan ········ 741
12.3.2.4 屋顶风机 Roof Fan ········ 742
12.3.2.5 风机箱 Cabinet Fan ········ 743
12.3.2.6 组合式空气处理机 Modular Air Handling Unit ········ 744
12.3.2.7 整体式水冷空调机 Water-Cooled Packaged Air Conditioner ········ 745
12.3.2.8 分体式风冷空调机 Air-Cooled Split Air Conditioner ········ 746
12.3.2.9 屋顶式空调机 Roof Air Conditioning Unit ········ 747
12.3.3 空调制冷与末端设备 Refrigerating and Terminal Equipment ········ 747
12.3.3.1 离心式冷水机组 Centrifugal-Type Water Chiller ········ 747
12.3.3.2 直燃式溴化锂吸收式冷水机组 Direct-Fired Lithium-Bromide Absorption Water Chiller ········ 748
12.3.3.3 蒸汽型溴化锂吸收式制冷机 Steam Type Lithium-Bromide Absorption Water Chiller ········ 749
12.3.3.4 热水型溴化锂吸收式制冷机 Hot Water Type Lithium-Bromide Absorption Water Chiller ········ 749
12.3.3.5 螺杆式冷水机组 Screw-Type Water Chiller ········ 750

12.3.3.6　新风净化机组　Fresh Air Unit ·· 751
12.3.3.7　多联机空调　VRV System ·· 751
12.3.3.8　新风换气机　Fresh-Air Exchanger ·· 752
12.3.3.9　风机盘管　Fan Coil ·· 752
12.3.4　通风管道及部件　Duct and Fitments ·· 753
12.3.4.1　风管　Duct ·· 753
12.3.4.2　风管弯头　Elbow for Duct ·· 753
12.3.4.3　风管三通　Tee for Duct ·· 754
12.3.4.4　风管异径　Reducer for Duct ·· 755
12.3.4.5　手动对开多叶调节阀　Manual Opposed Multi-Blade Regulating
　　　　　Damper ·· 755
12.3.4.6　手柄式蝶阀　Hand Grip Type Butterfly Damper ····························· 756
12.3.4.7　拉链式蝶阀　Dragline Type Butterfly Damper ······························· 756
12.3.4.8　风管止回阀　Check Damper ··· 757
12.3.4.9　变风量文丘里阀　Variable Air Volume Venturi Valve ····················· 758
12.3.4.10　抗爆阀　Blast Valve ·· 758
12.3.4.11　通风空调系统送回风口　Supply Air Outlet and Return Air Intake
　　　　　 for Ventilation and Air Conditioning System ································ 759
12.3.4.12　防火阀　Fire Damper ·· 761
12.3.4.13　排气罩　Exhaust Hood ·· 761
12.3.4.14　风管支吊架　Bracket Support and Hanger Frame for Duct ··············· 762
12.3.4.15　风管保温　Insulation of Duct ·· 762
12.3.4.16　管式消声器　Cell Type Attenuator ··· 763
12.3.4.17　片式消声器　Plate Type Attenuator ·· 763

12.4　总图运输　General Plot Plan and Transportation ···································· 764
12.4.1　总图　General Plot plan ·· 764
12.4.1.1　总图　Overall Plot Plan ··· 764
12.4.1.2　风玫瑰　Wind Rose ·· 765
12.4.1.3　地形图　Topographic Map ·· 765
12.4.2　道路　Road ··· 766
12.4.2.1　城市型道路　City Road ·· 766
12.4.2.2　公路型道路　Highway ··· 766
12.4.2.3　道路结构　Road Structure ·· 767
12.4.3　铁路　Railway ·· 768
12.4.4　厂区竖向　Plant Vertical Arrangement ·· 768
12.4.5　土方工程　Earth Work ··· 769
12.4.6　绿化　Greening ·· 769
12.4.7　其他构筑物　Other Structure ··· 770
12.4.7.1　护坡　Slope Protection ·· 770
12.4.7.2　急流槽　Chute ··· 770
12.4.7.3　跌水　Drop Water ··· 770

  12.4.7.4 挡土墙 Retaining Wall …… 771
  12.4.7.5 涵洞 Culvert …… 771
13 焊接 Welding …… 772
 13.1 焊接符号 Welding Symbols …… 772
 13.2 金属焊接 Welding of Metals …… 773
 13.3 保护式电弧焊原理 Principles of Shielded Arc Welding …… 775
 13.4 焊接位置、接头形式及焊接形式 Welding Positions, Types of Joints and Welding …… 776
 13.5 坡口 Grooves …… 777
 13.6 坡口详图 Detail of Grooves …… 778
 13.7 填角焊、角焊 Fillet Welding …… 779
 13.8 焊接缺陷 Defects of Welding …… 780
 13.9 焊接方法及设备 Type of Welding and Welding Machine …… 781
  13.9.1 自动埋弧焊 Automatic Submerged Arc Welding …… 781
  13.9.2 金属极气体保护焊焊枪 Electrode Guns of Gas Metal-Arc Welding …… 782
  13.9.3 气体保护电弧焊 Gas-Shielded Arc Welding …… 783
  13.9.4 金属极惰性气体保护焊 Gas Metal-Arc Welding …… 784
  13.9.5 普通电渣焊 Conventional Electroslag Welding …… 785
  13.9.6 熔嘴电渣焊 Electroslag Welding by Consumable Guide Tube …… 786
  13.9.7 电气焊 Electrogas Welding …… 787
  13.9.8 管状焊丝电弧焊 Flux-Cored Arc Welding …… 789
  13.9.9 气焊设备 Gas Welding Equipment …… 790
  13.9.10 移动式乙炔发生器 Portable Acetylene Generators …… 791
  13.9.11 乙炔发生器的基本形式 Basic Types of Acetylene Generators …… 792
 13.10 塑料焊接 Welding of Plastics …… 793
 13.11 焊接术语 Welding Term …… 795
 13.12 熔焊术语 Fusion Welding Term …… 795
 13.13 钎焊术语 Soldering Term …… 795
 13.14 焊接材料术语 Welding Material Term …… 795
14 无损检验 Non-destructive Testing …… 796
 14.1 无损检验方法 Non-destructive Testing Method …… 796
 14.2 无损检验设备 Non-destructive Testing Apparatus …… 798
  14.2.1 探孔镜及显微镜 Borescope and Microscope …… 798
  14.2.2 X射线发生器及其线路 X-Ray Tube and Its Circuit …… 799
  14.2.3 射线照相及电子照相 Photoradiography and Electroradiography …… 800
  14.2.4 轻便X射线机及透度计 Mobile X-Ray Unit and Penetrometer …… 801
  14.2.5 超声波探伤方法及探头 Ultrasonic Test Methods and Search Units …… 802
  14.2.6 超声波发射探头及接收探头 Ultrasonic Transducer and Refraction …… 803
  14.2.7 配管焊缝的超声波探伤 Ultrasonic Testing of Weld in Piping …… 804
  14.2.8 成套式渗透探伤仪 Self-Contained Penetrant Inspection Unit …… 805
 14.3 磁化法 Methods of Magnetization …… 806
 14.4 超声波检测术语 Ultrasonic Testing Term …… 806

14.5 射线检测术语　Radiation Testing Term ……… 807
14.6 渗透检测术语　Penetrant Testing Term ……… 807
14.7 磁粉表面检测术语　Magnetization Surface Testing Term ……… 807
14.8 涡流检测术语　Eddy Current Testing Term ……… 807

15 施工设备和工具　Construction Equipment and Tool ……… 808

15.1 起重机械　Hoisting Machinery ……… 808

15.1.1 麻绳、绳结、吊索和吊具　Hemp Ropes, Knots, Sling and Hardware ……… 808
  15.1.1.1 麻绳和绳结　Hemp Ropes and Knots ……… 808
  15.1.1.2 吊索　Sling ……… 809
  15.1.1.3 吊具　Hardware ……… 811
15.1.2 起重机构造与吊车手语　Crane Structure and Sign Language ……… 815
15.1.3 各类起重机　Types of Crane ……… 817
15.1.4 手动葫芦和千斤顶　Manual Hoist and Lifting Jack ……… 818
15.1.5 电动吊篮　Electromotion Basket ……… 820
15.1.6 手动吊篮　Manual Operation Basket ……… 821
15.1.7 液压汽车起重机　Hydraulic Pressure Mobile Cranes ……… 822
15.1.8 塔吊　Tower Crane ……… 824
15.1.9 桥式起重机　Overhead Crane ……… 824
15.1.10 门式起重机　Gantry Crane ……… 825
15.1.11 手动式屋面移动吊机　Manual Operating Roof Suspension Hoist Machine ……… 825
15.1.12 电动式屋面移动吊机　Electrocution Roof Suspension Hoist Machine ……… 825
15.1.13 叉车和附属配件　Forklift and Accessory ……… 826

15.2 工具　Tools ……… 827

15.2.1 木工工具　Wood Working Tools ……… 827
15.2.2 扳手　Wrenches ……… 828
15.2.3 活扳手及管钳　Adjustable Wrenches and Pipe Wrenches ……… 830
15.2.4 钳和剪钳　Pliers and Nippers ……… 831
15.2.5 刀具的柄部、套节及套筒　Shakes, Sockets and Sleeves ……… 832
15.2.6 丝锥　Taps ……… 833
15.2.7 铰刀　Reamers ……… 834
15.2.8 麻花钻　Twist Drills ……… 835
15.2.9 检测规　Inspection Gages ……… 836
15.2.10 量具　Measuring Tools ……… 837

15.3 混凝土搅拌设备　Concrete Mixer Equipment ……… 838

15.3.1 工业用搅拌机　Industrial Mixer ……… 838
15.3.2 现场用搅拌机　On-site Mixer ……… 839
15.3.3 混凝土用车　Concrete Mixer Trucks ……… 839
15.3.4 预搅拌混凝土站　Ready-mix Concrete Batch Plant ……… 840
15.3.5 商品混凝土搅拌站　Commodity Concrete Mixing Station ……… 841
15.3.6 汽车式混凝土搅拌运输车　Concrete Truck Mixer ……… 842

- 15.4 土方设备 Earthwork Equipment ……… 843
  - 15.4.1 挖掘机 Excavator ……… 843
  - 15.4.2 推土机 Bulldozer ……… 845
  - 15.4.3 装载机 Loader ……… 846
  - 15.4.4 挖掘装载机（两头忙） Backhoe ……… 847
  - 15.4.5 平地机 Grader ……… 848
  - 15.4.6 滑移转向装载机（机器猫） Skid Steer Loader (Bobcat) ……… 848
- 15.5 地基处理机械 Foundation Treatment Machinery ……… 849
  - 15.5.1 步履式打桩机 Walking Piling Rig ……… 849
  - 15.5.2 螺旋钻打桩机 Auger Piling Rig ……… 850
  - 15.5.3 碾压机 Compactor Rollers ……… 851
  - 15.5.4 水平定向钻孔机 Horizontal Directional Drilling Machinery ……… 853
  - 15.5.5 钻机 Drilling Machine ……… 854
  - 15.5.6 强夯机械 Rammer Machinery ……… 856
- 15.6 高处作业平台和工程车辆 Height Platform and Construction Vehicle ……… 857
  - 15.6.1 叉车和附属配件 Forklift and Attachments ……… 857
  - 15.6.2 混凝土泵车 Concrete Boom Truck ……… 858
  - 15.6.3 液压升降平台 Hydraulic Pressure Lift Platform ……… 859
  - 15.6.4 升降车 Man Lift ……… 860
  - 15.6.5 工程车辆 Construction Vehicle ……… 862
    - 15.6.5.1 公路型运输车 On-highway Trucks ……… 862
    - 15.6.5.2 越野型运输车 Off-highway Trucks ……… 864
    - 15.6.5.3 拖车 Trailer ……… 864
- 15.7 滑模施工 Slipform Construction ……… 865
- 15.8 大模板爬模施工 Large Formwork Climbing Construction ……… 868
- 15.9 钢筋加工机械 Process Machinery for Steel Bar ……… 869
  - 15.9.1 钢筋切断器 Steel Bar Cutter ……… 869
  - 15.9.2 钢筋拔丝机 Steel Bar Stretcher ……… 871
  - 15.9.3 钢筋调直机 Steel Bar Straightening Machines ……… 874
  - 15.9.4 钢筋弯曲机 Steel Bar Bender ……… 875
  - 15.9.5 钢筋笼滚焊机 Rotary Welding Machinery for Steel Cage ……… 877
  - 15.9.6 数控钢筋切锯套丝机 NC Thread Machinery for Steel Sawing ……… 878
  - 15.9.7 数控钢筋弯曲机 CNC Rebar Bending Machinery ……… 878
- 15.10 脚手架 Scaffolding ……… 879
  - 15.10.1 脚手架构件 Scaffolding Component ……… 879
  - 15.10.2 脚手架形式 Types of Scaffolding ……… 882
- 15.11 测绘 Surveying ……… 884
  - 15.11.1 经纬仪 Theodolite ……… 884
  - 15.11.2 光电测距仪 Photoelectricity Rangefinder ……… 885
  - 15.11.3 水准仪 Optical Level ……… 886
  - 15.11.4 全站仪 Total Station ……… 887
  - 15.11.5 其他测绘工具 Other Surveying Tools ……… 887

索引 ……… 889

1 工艺及公用工程 Process and Utilities

1.1 工艺流程图图例说明 Process Flow Diagram/Legent Identification

1.1.1 物料代号 Line Service（Fluid） Identification

1　A　air　空气
2　AC　acid　酸
3　AG　acid gas　酸性气体
4　BD　blow down　排污
5　BFW　boiler feed water　锅炉给水
6　BOG　boil off gas　驰放气
7　BR　brine　盐水
8　CA　caustic　碱
9　CAT　catalyst　催化剂
10　CBD　cold blowdown　冷排放
11　CF　chemical feed　化学品进料
12　CFL　cold flare　冷火炬
13　CG　caustic gas　碱性气体
14　CH　chemical　化学品
15　CO_2　carbon dioxide　二氧化碳
16　CS　chemical sewer　化学污水
17　CW　cooling water　冷却水
18　DA　decoking air　清焦空气
19　DI　deionized water　脱离子水
20　DM　demineralized water　脱盐水
21　DR　drain　排液，排水
22　DS　dilution steam　稀释蒸汽
23　FG　fuel gas　燃料气
24　FL　flue gas　烟气
25　FO　fuel oil　燃料油
26　FS　fused salt　熔盐
27　FW　fire fighting water　消防水
28　GO　gland oil　填料油
29　H_2　hydrogen　氢气
30　HBD　hot blowdown　热排放
31　HC　high pressure condensate　高压蒸汽凝液
32　HFL　hot flare　热火炬
33　HS　high pressure steam　高压蒸汽

34　HW　hot water　热水
35　IA　instrument air　仪表空气，仪表风
36　IG　inert gas　惰性气体
37　IW　industrial water　生产水
38　LC　low pressure condensate　低压蒸汽凝液
39　LLS　low low pressure steam　低低压蒸汽
40　LNG　liquified natural gas　液化天然气
41　LO　lube oil　润滑油
42　LPG　liquified petroleum gas　液化石油气
43　LS　low pressure steam　低压蒸汽
44　MC　medium pressure condensate　中压蒸汽凝液
45　MS　medium pressure steam　中压蒸汽
46　N　nitrogen　氮气
47　NG　natural gas　天然气
48　NH　ammonia　氨
49　O　oxygen　氧气
50　OG　off gas　尾气
51　OW　oily water　含油水
52　OWS　oily water　含油污水排放
53　PA　plant air　工厂空气，工厂风
54　PF　process flush　工艺冲洗流体
55　PG　process gas　工艺气体
56　PL　process liquid　工艺液体
57　PO　pan oil　盘油
58　PS　process steam　工艺蒸汽
59　PW　process water　工艺水
60　QO　Quench oil　急冷油
61　QW　Quench watch　急冷水
62　RW　raw water　原水
63　SC　steam condensate　蒸汽冷凝水
64　SG　synthesis gas　合成气
65　SO　seal oil　密封油

66	SS	superheated steam	过热蒸汽，super high pressure steam 超高压蒸汽	69	VE	vacuum exhaust	真空排气
67	ST	auxiliary steam	辅助蒸汽	70	VG	vent gas	放空气
68	TW	treated water	处理水	71	WA	warm cooling water	温冷却水
				72	WO	wash oil	洗油

1.1.2 在仪表符号中字母标志的意义 Meanings of Identification Letters in Instrument Symbols

(1) First letter is indicated as a measured variable, if is used in combination with a modifying letter D (differential), F or f (fraction), the combination shall be treated as a first-letter entity. 第一个字母表示所检测的变量，如第一位字母与修饰字母 D（差）、F 或 f（比）连在一起用时，则这两个组合文字应作为第一位字母的整体来对待。

　　A——analysis　分析（成分）
　　C——conductivity　电导率
　　D——density　密度
　　E——position　位置
　　F——flow　流量
　　Ff——flow fraction　流量比
　　H——hand　手动
　　I——current　电流
　　J——power　功率
　　L——level　液面
　　P——pressure　压力
　　PD（pd）——pressure differential　压差
　　Q——quantity　数量
　　S——speed　速度
　　T——temperature　温度
　　TD（Td）——temperature differential　温差
　　U——multivariable　多变量
　　V——vibration　振动
　　W——weight　重量

(2) Succeeding letters are indicated as the variable function of instrument. 后继字母表示仪表的不同功能。

　　A——alarm　报警
　　C——control　控制，调节
　　E——primary element　主要元件
　　LL——low low value　低低值
　　H——high value　高值
　　HH——high high value　高高值
　　I——indicate　指示
　　L——light　灯
　　L——low value　低值
　　R——record　记录
　　S——switch　开关
　　T——transmit　变速

1.1.3 管道图例 Piping Symbols

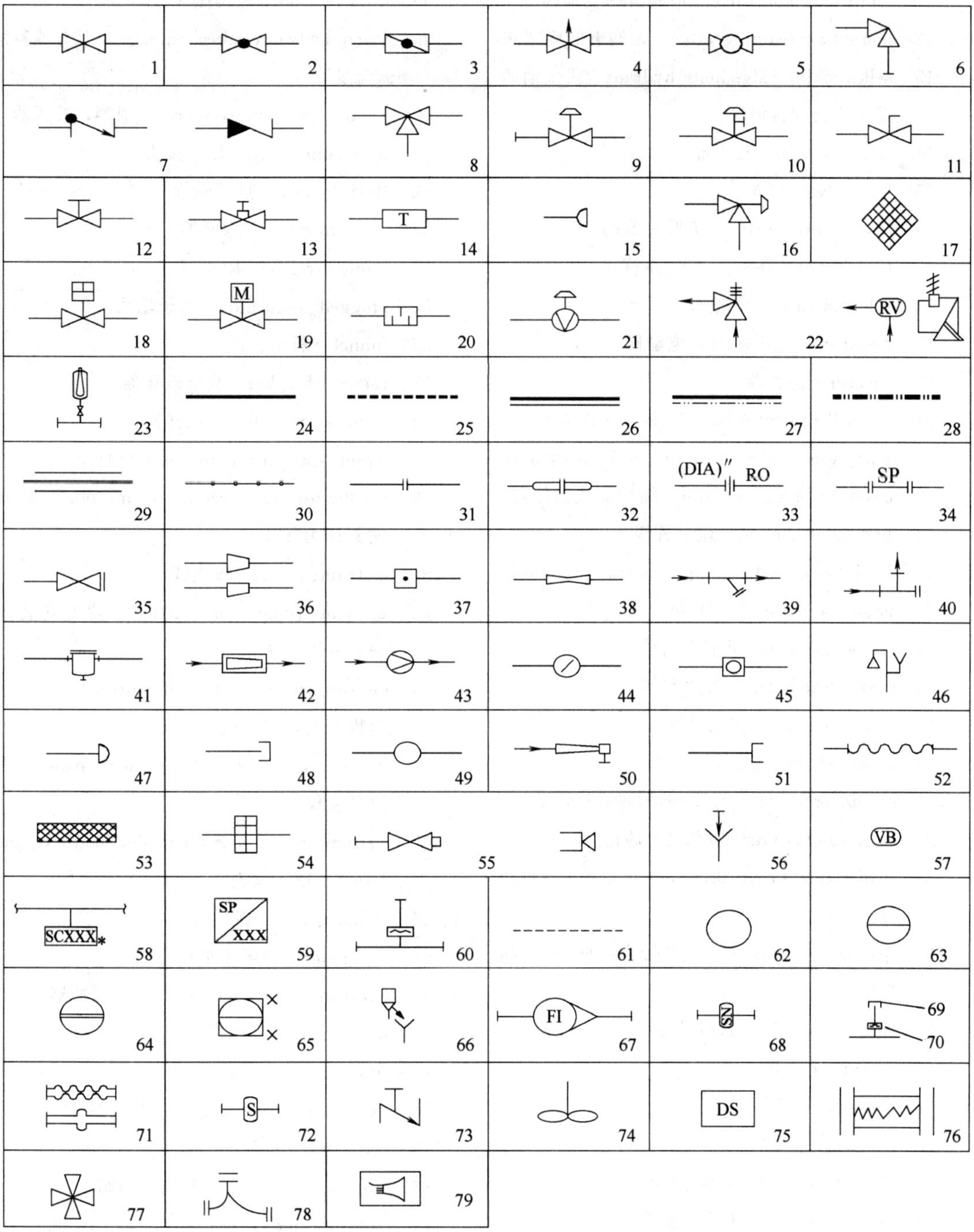

1　gate valve　闸阀
2　globe valve　截止阀
3　butterfly valve　蝶阀
4　needle valve　针型阀
5　ball valve　球阀
6　angle valve　角阀
7　check valve　止回阀，单向阀
8　three way valve　三通阀

9	control valve (air diaphram)	调节阀（气动薄膜）
10	control valve with handwheel	带手轮的调节阀
11	lubricated plug valve	带润滑的旋塞阀
12	non-lubricated plug valve	无润滑的旋塞阀
13	self draining valve kelly hydrant	凯氏消防龙头式自动排泄阀
14	trap	疏水阀，疏水器
15	foot valve	底阀
16	angle control valve	角形调节阀
17	bird/insect screen	防鸟/虫网
18	solenoid valve	电磁阀
19	motor operated valve	电动阀
20	silencer	消音器
21	vee ball control valve	V形落球调节阀
22	safety valve and relief valve	安全阀与泄放阀
23	desurger or accumulator	缓冲器或蓄能器
24	line above ground	地上管线
25	line below ground (underground line)	地下管线
26	steam traced line	蒸汽伴热管线
27	electric traced line	电伴热管线
28	process battery limit	生产界区
29	jacketed pipe line	夹套管线
30	perforated pipe	多孔管
31	line size orifice run	与管线同径的孔板管段
32	increased orifice run	比管径大的孔板管段
33	restriction orifice and diameter	节流孔板及直径（尺寸）
34	spool piece removable	（两端带法兰的）可拆卸短管
35	blinded connection	盲板封头
36	reducer	异径管
37	pitot tube	皮托管
38	flow tube	流量管
39	"Y" type strainer	Y型过滤器
40	"T" type strainer	T型过滤器
41	basket type strainer	篮式过滤器
42	puritan hat strainer	帽型过滤器
43	cone (conical) type strainer	锥型过滤器
44	sight flow indicator	观察式流量指示器
45	sight glass	视镜
46	shower and eye washer	喷淋器和洗眼器
47	weldcap	焊帽
48	screw type cap	螺纹管帽
49	flow quantity or displacement meter	质量或容积式流量计
50	ejector, eductor, booster	喷射器，升压器
51	hose connection	软管接头
52	flexible hose	挠性软管
53	demisting pad	除雾器
54	flame arrestor	阻火器
55	plugged connection	塞子堵头
56	funnel	漏斗
57	vacuum breaker	真空破坏器
58	sample connection	取样器
59	piping speciality item	特种管件
60	diaphragm seal pressure connection	隔膜密封压力接头
61	instrument line	仪表线
62	local instrument or function	就地安装仪表，一次仪表
63	instrument or function in control building	盘装仪表，二次仪表
64	instrument and function on local panel	就地盘装仪表
65	instrument and function in distributed control system	DCS仪表
66	exhaust head	排气头
67	rotameter	转子流量计
68	pulsation dampener	脉动流体缓冲器
69	weather cap	防护罩
70	rupture disc	爆破片
71	expansion joint	膨胀节，补偿器
72	separator	分离器
73	stop check valve	切断式止回阀
74	agitator/mixer	搅拌器
75	desuper heater	减温器
76	static mixer	静态混合器
77	4-way valve	四通阀
78	swing elbow	转换弯头
79	spray nozzle	喷嘴

1.1.4 绘图示例——氨库装置 Illustrative Drawing—Ammonia Storage Facility

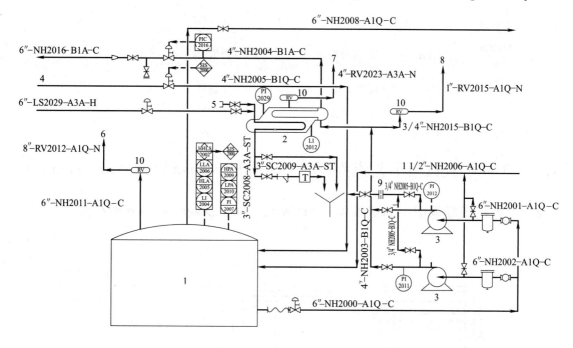

1　ammonia storage tank　氨储罐
2　ammonia heater　氨加热器
3　ammonia transfer pumps　氨输送泵
4　liquid ammonia　液氨
5　hose connection　软管接头
6　to atmosphere after treatment　处理后排向大气
7　to atmosphere at safe location　在安全地点排向大气
8　to ammonia flare　去氨火炬排放
9　RO restriction orifice　限流孔板
10　RV relief valve　泄放阀

管道编号　物料代号详见 1.1.1 节。

6″-NH1008-B1Q-C	6″——管径	NH——氨
	1008——管道编号	B1Q——管道等级　C——保冷
6″-LS2027-A3A-H	6″——管径	LS——低压蒸汽
	2027——管道编号	A3A——管道等级　H——热保温
1″-RV2015-A1Q-N	1″——管径	RV——泄放阀排放
	2015——管道编号	A1Q——管道等级　N——无保温

仪表代号　仪表符号中的字母标志意义详见 1.1.2 节。

$\dfrac{LLA}{2006}$　　LLA——低液位报警　　2006——仪表编号

$\dfrac{HLA}{2005}$　　HLA——高液位报警　　2005——仪表编号

$\dfrac{LI}{2004}$　　LI——液位指示　　2004——仪表编号

$\dfrac{HPA}{2009}$　　HPA——高压报警　　2009——仪表编号

$\dfrac{LPA}{2010}$　　LPA——低压报警　　2010——仪表编号

$\dfrac{PI}{2007}$　　PI——压力指示　　　　　2007——仪表编号

$\dfrac{PIC}{2016}$　　PIC——压力指示控制　　2016——仪表编号

$\dfrac{HHLS}{2002}$　　HHLS——高高液位联锁　　2002——仪表编号

$\dfrac{SIS}{2000}$　　SIS——安全仪表系统　　　2000——联锁编号

管道图例　详见 1.1.3 节。

1.2 工艺装置典型工艺流程图　Typical Process Unit Flow Diagram

1.2.1 炼厂流程框图，润滑油型　Refining Block Diagram, Lube Type

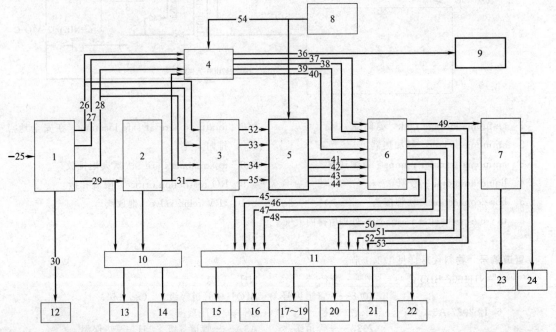

1　vacuum distillation unit（VDU）　减压蒸馏装置
2　propane deasphalting unit（DAS）　丙烷脱沥青装置
3　solvent refining unit　溶剂精制装置
4　hydrotreating（hydrocracking）unit　加氢处理（加氢裂化）装置
5　hydrofinishing unit　加氢补充精制装置
6　solvent dewaxing unit　溶剂脱蜡装置
7　wax fractionation and treating unit　蜡脱油及精制装置
8　hydrogen plant　制氢装置
9　gasoline and fuel oil scheme　汽油和燃料系统
10　asphalt plant　沥青装置
11　continuous product lube blending and additive compounding unit　成品润滑油及添加剂连续调和装置
12　fuel oil blending stock　燃料油调和油料
13　straight asphalt　渣油沥青
14　blown asphalt　氧化沥青
15　transformer oil　变压器油
16　spindle oil　锭子油
17　dynamo oil　发电机油
18　refrigerator oil　冷冻机油
19　turbine oil　透平油
20　machine oil　机械油
21　motor oil　车用机油

22	cylinder oil 气缸油			处理油
23	soft wax 软蜡		41	light hydrofinished oil 轻加氢补充精制油
24	hard wax 硬蜡		42	medium hydrofinished oil 中加氢补充精制油
25	topped crude 拔头油		43	heavy hydrofinished oil 重加氢补充精制油
26	light vacuum distillate 减压轻馏分油		44	high viscosity hydrofinished oil 高黏度补充精制油
27	medium vacuum distillate 减压中馏分油			
28	heavy vacuum distillate 减压重馏分油		45	light lube oil (hydrotreated) 轻质润滑油（加氢精制）
29	vacuum residue (VR) 减压渣油			
30	vacuum gas oil (VGO) 减压瓦斯油		46	medium lube oil (hydrotreated) 中质润滑油（加氢精制）
31	deasphalted oil 脱沥青油			
32	light solvent refined oil 轻溶剂精制油		47	heavy lube oil (hydrotreated) 重质润滑油（加氢精制）
33	medium solvent refined oil 中溶剂精制油			
34	heavy solvent refined oil 重溶剂精制油		48	high viscosity lube oil (hydrotreated) 高黏度润滑油（加氢精制）
35	high viscosity solvent refined oil 高黏度溶剂精制油			
			49	slack wax 蜡膏
36	light distillate 轻馏分油		50	high viscosity lube oil 高黏度润滑油
37	light hydrotreated oil 轻加氢处理油		51	heavy lube oil 重质润滑油
38	medium hydrotreated oil 中加氢处理油		52	medium lube oil 中质润滑油
39	heavy hydrotreated oil 重加氢处理油		53	light lube oil 轻质润滑油
40	high viscosity hydrotreated oil 高黏度加氢		54	H_2 氢气

1.2.2 炼化一体化流程框图，炼油部分　Refining Chemical Complex Block Diagram, Refining Part

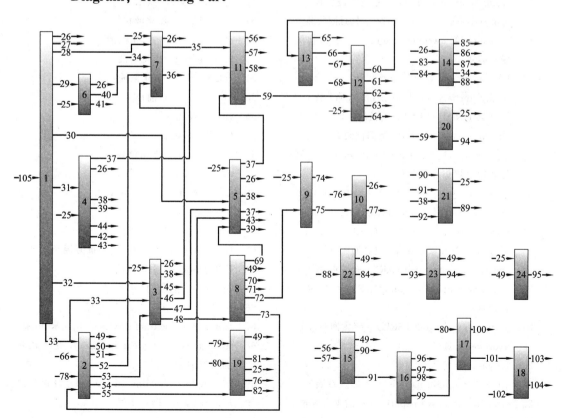

#		#	
1	crude distillation and vacuum distillation unit (CDU&VDU) 常减压蒸馏装置	30	straight run diesel 直馏柴油
2	delayed coking unit (DCU) 延迟焦化装置	31	vacuum gas oil (VGO) 减压蜡油
3	residue hydrodesulfurization unit (RDS) 渣油加氢装置	32	heavy vacuum gasoil (HVGO) 减压重蜡油
4	gasoil hydrocracking unit (HCK) 蜡油加氢裂化装置	33	vacuum residue (VR) 减压渣油
5	diesel hydrocracking unit (DHCU) 柴油加氢裂化装置	34	heavy naphtha 重石脑油
6	kerosene hydrotreating unit (KHT) 煤油加氢装置	35	hydrogenated heavy naphtha 预处理重石脑油
7	naphtha hydrotreating unit (NHT) 石脑油加氢装置	36	hydrogenated C_5 预处理轻石脑油
		37	hydrocracker heavy naphtha 加裂重石脑油
8	residue fluid catalytic cracking unit (RFCC) 渣油催化裂化装置	38	CLPS vapor 低分气
		39	hydrocracker diesel 加裂柴油
9	gasoline hydrotreating unit (GHT) 汽油加氢装置	40	naphtha 石脑油
		41	hydrogenated kerosene 加氢煤油
10	FCC naphtha etherificating unit 轻汽油醚化装置	42	hydrocracker light naphtha 加裂轻石脑油
		43	hydrocracker kerosene 加裂煤油
11	continuous catalytic reformer 连续重整装置	44	hydrocracker wax 加裂尾油
12	aromatic complex 芳烃联合装置	45	hydrogenated light naphtha 加氢轻石脑油
13	aromatic extraction unit 芳烃抽提装置	46	hydrogenated heavy naphtha 加氢重石脑油
14	light ends recovery unit (LER) 轻烃回收装置	47	hydrogenated diesel 加氢柴油
15	off gas/LPG sweetening unit 干气/液化气脱硫装置	48	hydrogenated residue 加氢重油
		49	acid gas 酸性气
16	LPG spllitter 液化气分离装置	50	coker off gas 焦化干气
17	methyl tertiary butyl ether unit (MTBE) 甲基叔丁基醚 (MTBE) 装置	51	coker LPG 焦化液化气
		52	coker gasoline 焦化汽油
18	alkylation unit (ALK) 烷基化装置	53	coker gasoil 焦化蜡油
19	H_2 unit (coal coke partial oxidation) 制氢装置（煤焦部分氧化）	54	coker diesel 焦化柴油
		55	green coke 石油焦
		56	LPG 液化气
20	pressure swing adsorption (PSA for CCR H_2) 变压吸附（重整氢提浓）	57	C_5 oil 戊烷油
		58	CCR hydrogen 重整氢
		59	C_{6+} reformate C_{6+} 重整汽油
21	pressure swing adsorption (PSA for H_2 rich off gas) 变压吸附（富氢气体回收氢气）	60	C_7 reformate C_7 重整汽油
		61	benzene 苯
22	amine regeneration unit (AR) 胺液再生装置	62	para-xylene (PX) PX
23	sour water stripping unit (SWS) 酸性水汽提装置	63	C_{11+} heavy aromatics C_{11+} 重芳烃
		64	CCR gasoline 重整汽油
		65	raffinate 抽余油
24	surfur recovery unit (SRU) 硫黄回收装置	66	C_6/C_7 aromatics C_6/C_7 芳烃
25	hydrogen 氢气	67	beznene/toulene from ethylene 化工苯/甲苯
26	off gas 干气	68	hydrogenated ethylene gasoline 化工乙烯加氢汽油
27	straight run light naphtha 直馏轻石脑油		
28	straight run heavy naphtha 直馏重石脑油	69	FCC light circulating oil (LCO) 催化柴油
		70	FCC off gas 催化干气
29	straight run kerosene 直馏煤油	71	FCC LPG 催化液化气

72 FCC naphtha 催化汽油
73 slurry 催化油浆
74 hydrogenated heavy FCC naphtha 加氢重汽油
75 FCC light naphtha 催化轻汽油
76 methanol 甲醇
77 etherificating light gasoline 醚化轻汽油
78 styrene tar oil 化工苯乙烯焦油
79 coal 煤
80 oxygen 氧气
81 syngas (fuel gas) 合成气（燃料气）
82 ammonia 合成氨
83 stright, hydrogenated and hydrocracker light naphtha 直馏、加氢和加裂轻石脑油
84 lean amine 贫胺液
85 sweetened off gas 脱硫干气
86 sweetened LPG 脱硫液化气
87 sweetened LPG 脱硫液化气
88 rich amine 富胺液
89 PSA purge gas PSA 尾气
90 propylene dehydrogenation hydrogen 化工丙烷脱氢氢气
91 styrene dehydrogenation tail gas 化工苯乙烯脱氢尾气
92 hydrogen rich gas from LER 加氢轻烃富氢干气
93 sour water 酸性水
94 stripped water 汽提水
95 sulfur 硫黄
96 ethane 乙烷
97 propylene 丙烯
98 propane 丙烷
99 C₄ mixture 混合 C₄
100 methyl tert-butyl ether (MTBE) MTBE
101 MTBE LPG MTBE 液化气
102 isobutane 异丁烷
103 alkylate 烷基化油
104 normal butane 正丁烷
105 crude 原油

1.2.3 炼化一体化流程框图，化工部分 Refining Chemical Complex Block Diagram, Chemical Part

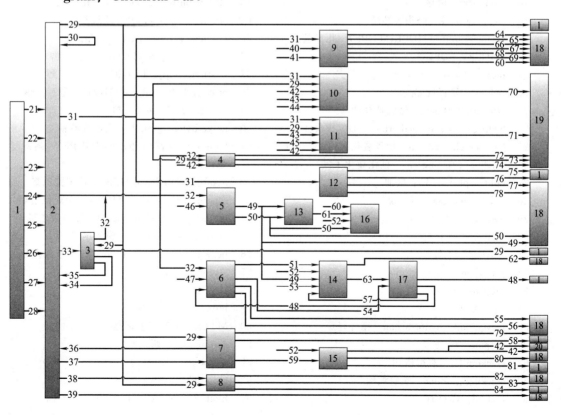

#	English	中文
1	refining units	炼油装置
2	ethylene unit	乙烯装置
3	propane dehydrogenation (PDH) unit	丙烷脱氢装置
4	polypropylene (PP) unit	聚丙烯装置
5	phenol/acetone unit	苯酚/丙酮装置
6	acrylonitrile (AN) unit	丙烯腈装置
7	butadiene extraction unit	丁二烯抽提装置
8	cracked gasoline hydrogenation unit	裂解汽油加氢装置
9	ethylene glycol (EG) unit	乙二醇装置
10	full density polyethylene (FDPE) unit	全密度聚乙烯装置
11	high density polyethylene (HDPE) unit	高密度聚乙烯装置
12	styrene (ethyl benzene) unit	苯乙烯（含乙苯）装置
13	bisphenol A (BPA) unit	双酚A装置
14	methyl methacrylate (MMA) unit	甲基丙烯酸甲酯装置
15	MTBE/1-butene unit	MTBE/1-丁烯装置
16	Polycarbonate (PC) unit	聚碳装置
17	sulfuric acid unit	硫酸装置
18	tank farm	罐区
19	shipping dock	码头装卸站
20	chemical units (HDPE, FDPE)	化工装置（HDPE，FDPE）
21	butane	丁烷
22	rich C_5	富正构C_5
23	aromatic raffinate	芳烃抽余油
24	hydrogenation topped crude	加氢拔头油
25	hydrocracker tail oil	加氢裂化尾油
26	catalytic cracking gas	催化裂化干气
27	light naphtha	轻石脑油
28	rich ethane gas	富乙烷气
29	hydrogen	氢气
30	fuel gas	燃料气
31	ethylene	乙烯
32	propylene	丙烯
33	propane	丙烷
34	C_4	碳四
35	recycle light hydrocarbon	循环轻烃
36	propine	丙炔
37	cracking C_4	裂解碳四
38	cracking gasoline	裂解汽油
39	fuel oil	燃料油
40	oxygen	氧气
41	steam	水蒸气
42	1-ethylene	1-丁烯
43	1-hexene	1-己烯
44	isopentane	异戊烷
45	isobutane	异丁烷
46	benzene	苯
47	ammonia	液氨
48	sulfuric acid	硫酸
49	acetone	丙酮
50	phenol	苯酚
51	formonitrile	氢氰酸
52	methanol	甲醇
53	neutralizer	中和剂
54	ammonium sulphate concentrate	硫铵浓缩液
55	acetonitrile	乙腈
56	acrylonitrile	丙烯腈
57	oleum	发烟硫酸
58	hydrogenation C_4	加氢碳四
59	raffinate C_4	抽余碳四
60	carbon dioxide	二氧化碳
61	bisphenol A (BPA)	双酚A
62	methyl methacrylate (MMA)	甲基丙烯酸甲酯
63	alkylation waste acid	烷基化废酸
64	ethylene oxide (EO)	环氧乙烷
65	ethylene glycol (EG)	一乙二醇
66	diethylene glycol (DEG)	二乙二醇
67	triethylene glycol (TEG)	三乙二醇
68	heavy ethylene glycol (HEG)	重乙二醇
69	waste glycol	废乙二醇
70	polyethylene granule	聚乙烯粒料
71	high density polyethylene (HDPE)	高密度聚乙烯
72	homopolyethylene (HPE)	均聚聚乙烯
73	random polyethylene (RPE)	无规聚乙烯
74	impact polyethylene (IPE)	抗冲聚乙烯
75	dehydrogenation gas	脱氢尾气
76	benzene/toluene mixture	苯/甲苯混合物
77	styrene	苯乙烯
78	tar	焦油
79	butadiene	丁二烯
80	MTBE	甲基叔丁基醚
81	residual C_4	剩余C_4
82	C_5	C_5
83	C_9+	C_9+
84	hydrogenation $C_6 \sim C_8$	加氢$C_6 \sim C_8$

1.2.4 合成氨装置工艺流程图 Process Flow Diagram for Ammonia Plant

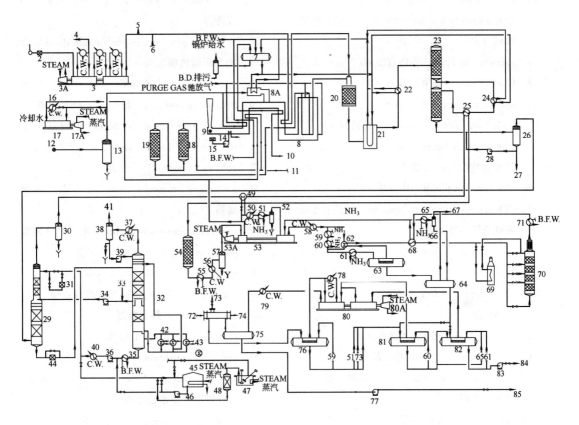

1	process air 工艺空气	16	feed gas compressor kickback cooler 原料气压缩机回流冷却器
2	filter 过滤器		
3，3A	process air compressor 工艺空气压缩机	17	feed gas compressor 原料气压缩机
4	process air compressor turbine 工艺空气压缩机透平	17A	feed gas compressor turbine 原料气压缩机透平
4	instrument air to dryers 仪表空气去干燥器	18	hydrogenator (or hydrogenation reactor) 加氢反应器
5	vent 放空		
6	steam 蒸汽	19	ZnO guard chamber (desulfurizer) 氧化锌脱硫槽
7	steam drum 汽包		
8，8A	primary reformer 一段转化炉	20	secondary reformer 二段转化炉
9	flue gas 引风机	21	primary waste heat boiler 第一废热锅炉
10	super-heated steam 过热蒸汽	22	secondary waste heat boiler 第二废热锅炉
11	process steam 工艺蒸汽	23	shift converter 变换炉
12	natural gas feed and fuel 原料和燃料天然气	24	primary shift effluent waste heat boiler 一段变换出口废热锅炉
13	feed gas compressor suction drum 原料气压缩机吸气罐		
		25	methanator feed heater 甲烷化进料加热器
14	combustion air fan 鼓风机	26	separator 分离器
15	filter 过滤器	27	process condensate 工艺冷凝液

11

28	condensate pump 冷凝液泵		甲烷化炉出口锅炉给水加热器
29	CO$_2$ absorber 二氧化碳吸收塔	56	methanator effluent cooler 甲烷化炉出口冷却器
30	CO$_2$ absorber knockout drum 二氧化碳吸收塔分离罐	57	synthesis gas compressor suction drum 合成气压缩机吸入罐
31	side stream filter 旁路过滤器	58	water cooler 水冷却器
32	CO$_2$ stripper 二氧化碳解吸塔	59	first stage ammonia chiller 一级氨冷器
33	anti-foam agent 消泡剂	60	second stage ammonia chiller 二级氨冷器
34	semi-lean solution pump 半贫液泵	61	third stage ammonia chiller 三级氨冷器
35	lean solution boiler feed water exchanger 贫液锅炉给水换热器	62	cold exchanger 冷交换器
36	lean solution pump 贫液泵	63	secondary ammonia separator 第二氨分离器
37	CO$_2$ stripper overhead condenser 二氧化碳解吸塔塔顶冷凝器	64	primary ammonia separator 第一氨分离器
38	CO$_2$ stripper reflux drum 二氧化碳解吸塔回流罐	65	purge gas chiller 驰放气冷却器
		66	purge gas separator 驰放气分离罐
39	CO$_2$ stripper reflux pump 二氧化碳解吸塔回流泵	67	purge gas to fuel 驰放气去燃料系统
		68	hot exchange 热交换器
40	lean solution cooler 贫液水冷器	69	start-up heater 开工加热炉
41	CO$_2$ product 二氧化碳产品	70	ammonia synthesis converter 氨合成塔
42	CO$_2$ stripper gas reboiler 二氧化碳解吸塔气体再沸器	71	ammonia converter boiler feed water exchanger 合成塔锅炉给水换热器
43	CO$_2$ stripper steam reboiler 二氧化碳解吸塔蒸汽再沸器	72	ammonia refrigerant 氨冷冻剂
		73	flash gas chiller 闪蒸气氨冷器
44	hydraulic turbine 水力透平	74	refrigerant out 冷冻剂出口
45	solution storage tank 溶液储槽	75	ammonia receiver 氨受槽
46	solution make-up pump 脱碳溶液补充泵	76	first stage ammonia flash drum 一级氨闪蒸槽
47	solution sump 脱碳溶液地下槽	77	hot ammonia product pump 热氨产品泵
48	solution filter 溶液过滤器	78	ammonia refrigerant compressor inter stage cooler 氨压缩机段间冷却器
49	synthesis gas methanator feed exchanger 合成气/甲烷化进料换热器	79	ammonia condenser 氨冷凝器
50	synthesis gas compressor inter stage cooler 合成气压缩机段间冷却器	80	ammonia compressor 氨压缩机
51	ammonia chiller 氨冷却器	80A	ammonia compressor turbine 氨压缩机透平
52	synthesis gas compressor inter stage separator 合成器压缩机段间分离器	81	second stage ammonia flash drum 二级氨闪蒸槽
53	synthesis gas compressor 合成气压缩机	82	third ammonia flash drum 三级氨闪蒸槽
53A	synthesis gas compressor turbine 合成器压缩机透平	83	cold ammonia product pump 冷氨产品泵
54	methanator 甲烷化炉	84	to atmospheric ammonia storage tank 去常压氨储槽
55	methanator effluent boiler feed water heater	85	ammonia product 氨产品

1.2.5 尿素装置工艺流程图　Process Flow Diagram for Urea Plant

#	English	中文
1	liquid ammonia	液氨
2	carbon dioxide	二氧化碳
3	HP ammonia pumps	高压氨泵
4	CO_2 compressor	CO_2压缩机
5	stripper	汽提塔
6	HP carbamate condenser	高压甲铵冷凝器
7	urea reactor	尿素合成塔
8	HP scrubber	高压洗涤塔
9	carbamate ejector	甲铵喷射器
10	hydrogen converter	脱氢反应器
11	LP decomposer separator	低压分解器分离器
12	LP decomposer	低压分解器
13	vacuum preconcentrator separator	真空预浓缩分离器
14	vacuum preconcentrator	真空预浓缩器
15	urea solution tank	尿液储槽
16	urea solution pumps	尿素溶液泵
17	1st vacuum concentrator	一段真空浓缩器
18	1st vacuum concentrator separator	一段真空浓缩器分离器
19	2nd vacuum concentrator	二段真空浓缩器
20	2nd vacuum concentrator separator	二段真空浓缩器分离器
21	urea melt pumps	熔融尿素泵
22	prilling tower	造粒塔
23	urea product	尿素产品
24	LP carbamate condenser	低压甲铵冷凝器
25	LP carbamate condenser seperator	低压甲铵冷凝器分离器
26	HP carbamate pumps	高压甲铵泵
27	4bar① absorber	4bar 吸收塔
28	1st absorber	第一吸收塔
29	2nd absorber	第二吸收塔
30	process condensate tank	工艺冷凝液槽
31	process condensate pumps	工艺冷凝液泵
32	absorber feed cooler	吸收塔给料冷却器
33	1st vacuum system	一段真空系统
34	2nd vacuum system	二段真空系统
35	desorber feed pumps	解吸塔给料泵
36	desorber preheater	解吸塔预热器
37	desorber	解吸塔
38	hydrolyzer feed pumps	水解器给料泵
39	hydrolyzer	水解器
40	hydrolyzer preheater	水解器预热器
41	reflux condenser	回流冷凝器
42	reflux pumps	回流泵
43	2nd absorber circulation cooler	第二吸收塔循环冷却器
44	2nd absorber circulation pumps	第二吸收塔循环泵
45	stack	放空筒
46	prilling buckets	造粒喷头
47	scraper	刮料机
48	purified process condensate	净化工艺冷凝液

① $1\,\text{bar}=10^5\,\text{Pa}$。

1.2.6 乙烯装置工艺流程图 Process Flow Diagram For Ethylene Plant

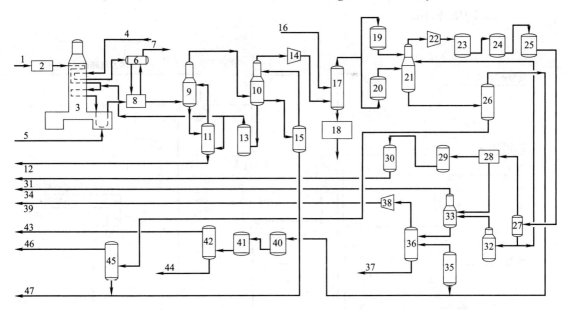

1	raw material 原料	25	cracked gas secondary dryer 裂解气第二干燥器
2	raw material preheat unit 原料预热	26	low pressure depropanizer 低压脱丙烷塔
3	pyrolysis furnace 裂解炉	27	high pressure depropanizer reflux drum 高压脱丙烷回流罐
4	boiler feed water 锅炉给水	28	cold separation and cold box 冷分离和冷箱
5	fuel gas 燃料气	29	methanator reactor 甲烷化反应器
6	steam drum 汽包	30	hydrogen dryer 氢气干燥器
7	super high pressure steam 超高压蒸汽	31	hydrogen product 氢气产品
8	quench fitting 急冷器	32	demethanizer prefractionator 预切割塔
9	quench oil tower 急冷油塔	33	demethanizer 脱甲烷塔
10	quench water tower 急冷水塔	34	fuel gas 燃料气（甲烷氢）
11	fuel oil stripper 燃料油汽提塔	35	deethanizer 脱乙烷塔
12	fuel oil 燃料油	36	ethylene tower 乙烯塔
13	dilution steam generator 稀释蒸汽发生器	37	ethane (to pyrolysis furnace) 循环乙烷（去裂解炉）
14	cracked gas compressor stage 1st to 4th 裂解气压缩机 1~4 段	38	ethylene compressor 乙烯压缩机
15	gasoline stripper 汽油汽提塔	39	ethylene product 乙烯产品
16	caustic 碱液	40	C_3 arsine removal bed C_3加氢脱砷保护床
17	caustic tower 碱洗塔	41	C_3 hydrogenation reactor C_3加氢反应器
18	spent caustic oxidation unit 废碱氧化单元	42	propylene tower 丙烯塔
19	cracked gas drier 裂解气干燥器	43	propylene product 丙烯产品
20	liquid hydrocarbon dryer 液烃干燥器	44	propylene to pyrolysis tower 丙烷去裂解炉
21	high pressure depropanizer 高压脱丙烷塔	45	debutanizer 脱丁烷塔
22	cracked gas compressor 5th stage 裂解气压缩机5段	46	mixed C_4 product 混合 C_4 产品
23	C_2 arsine removal bed C_2加氢脱砷保护床	47	pyrolysis gasoline product 裂解汽油
24	C_2 hydrogenation reactor C_2加氢反应器		

1.2.7 线性低密度聚乙烯装置工艺流程图 Process Flow Diagram for LLDPE Plant

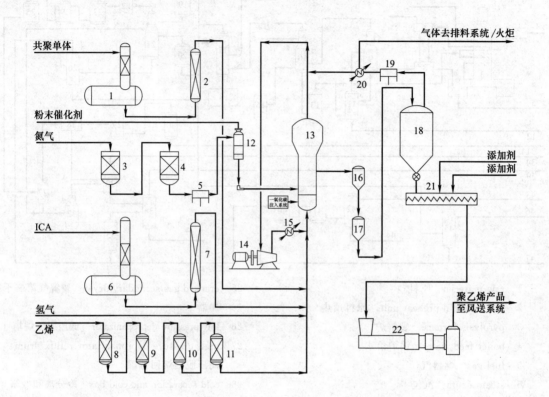

1	comonomer stripper 共聚单体脱气塔	12	catalyst feeding vessel 催化剂加料器
2	comonomer dryer 共聚单体干燥器	13	reactor 反应器
3	nitrogen deoxygen drum 氮气脱氧罐	14	recycle gas compressor 循环气压缩机
4	nitrogen dryer 氮气干燥器	15	recycle gas cooler 循环气冷却器
5	nitrogen compressor 氮气压缩机	16	product chamber drum 产品出料罐
6	ICA stripper ICA脱气塔	17	product blow drum 产品吹出罐
7	ICA dryer ICA干燥塔	18	product purge bin 产品脱气仓
8	ethylene CO removal drum 乙烯脱CO罐	19	discharge gas compressor 排放气压缩机
9	ethylene deoxygen drum 乙烯脱氧罐	20	discharge gas cooler 冷却器
10	ethylene dryer 乙烯干燥罐	21	resin additive feeder 树脂添加剂给料器
11	ethylene CO_2 removal drum 乙烯脱CO_2罐	22	extruder 混炼机

1.2.8 管式法低密度聚乙烯/乙烯-醋酸乙烯共聚物装置工艺流程图　Process Flow Diagram for L-Tubular DPE/EVA Plant

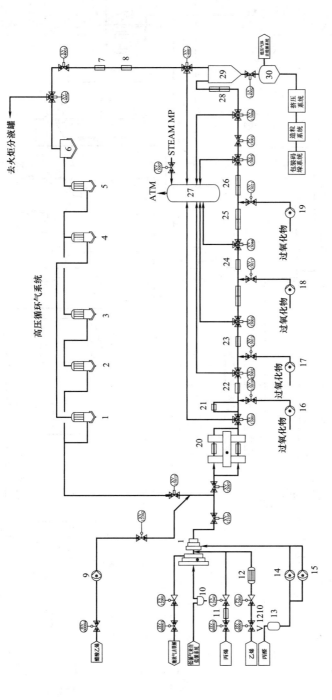

1~5	high pressure recycle gas cooler	高压循环气冷却器
6	product separator	产品分离器
7, 8	product cooler	产品冷却器
9	VA feeding pump	醋酸乙烯供给泵
10	compressor inlet separator	压缩机入口分离器
11	propylene vaporizer	丙烯蒸发器
12	ethylene heater	乙烯加热器
13	modifier buffer tank	调整剂缓冲罐
14, 15	modifier dosing pump	调整剂注入泵
16~19	initiator dosing pump	引发剂注入泵
20	primary compressor	一次压缩机
21	second compressor	二次压缩机
22	reactor preheater	反应器预热器
23	first section reactor	一段反应器
24	second section reactor	二段反应器
25	third section reactor	三段反应器
26	fourth section reactor	四段反应器
27	discharge vessel	排放罐
28	reactor subcooler	反应器后冷器
29	high pressure product separator	高压产品分离器
30	low pressure product separator	低压产品分离器

1.2.9 聚丙烯装置工艺流程图　Process Flow Diagram for Polypropylene Plant

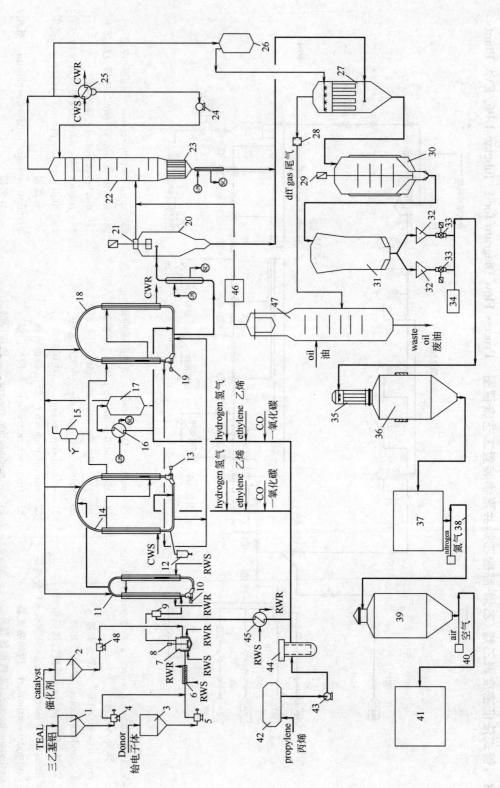

1. TEAL storage drum 三乙基储铝罐
2. catalyst dispersion drum 催化剂分散罐
3. donor storage drum 给电子体储罐
4. TEAL metering pump 三乙基铝计量泵
5. donor metering pump 给电子体计量泵
6. co-catalyst intermediate cooler 助催化剂中间冷却器
7. catalyst precontacting pot 催化剂预接触罐
8. precontacting pot agitator 预接触罐搅拌器
9. on-line mixer 在线混合器
10. prepolymerization reactor circulation pump 预聚合循环泵
11. prepolymerization reactor 预聚合反应器
12. first loop reactor sampling pot 第一环管反应器取样罐
13. slurry circulation pump 浆料循环泵
14. first loop polymerization reactor 第一环管反应器
15. expansion tank of reactor cooling circulating water 反应器循环水膨胀罐
16. propylene vaporizer 丙烯气化器
17. reactor surge drum 反应器缓冲罐
18. second loop polymerization reactor 第二环管反应器
19. slurry circulation pump 浆料循环泵
20. flash drum 闪蒸罐
21. dynamic separator 动力分离器
22. recycle propylene scrubber 循环丙烯洗涤塔
23. column reboiler 塔再沸器
24. column reflux pump 塔回流泵
25. recycle propylene condenser 循环丙烯冷凝器
26. blow back gas drum 返回气罐
27. recycle gas filter 循环气体过滤器
28. recycle gas guard filter 循环气保护过滤器
29. steamer agitator 汽蒸器搅拌器
30. steamer 汽蒸器
31. dryer 干燥器
32. PP powder surges bin 聚丙烯粉料缓冲料斗
33. rotary valve 旋转阀
34. PP powder conveying system 聚丙烯粉传输系统
35. bagging filter 袋式过滤器
36. PP surge silo 聚丙烯缓冲料仓
37. extrusion unit 挤压单元
38. pellets conveying system 粒料传输系统
39. blending silos 掺混料仓
40. product pellets conveying system 产品粒料传输系统
41. bagging system 包装单元
42. propylene feed drum 丙烯进料罐
43. propylene feed pump 丙烯进料泵
44. propylene feed guard filter 丙烯进料保护过滤器
45. prepolymerizer feed cooler 预聚合进料冷却器
46. recycle propylene compressor 循环丙烯压缩机
47. low pressure propylene scrubber 低压丙烯洗涤塔
48. catalyst feed metering pump 催化剂进料计量泵

CWR: cooling water return 冷却水回水
CWS: cooling water supply 冷却水供水
LPS: low pressure steam 低压蒸汽
SC: steam condensate 蒸汽凝液
RWS: refrigerate water supply 冷冻水给水
RWR: refrigerated water return 冷冻水回水

1.2.10 聚氯乙烯装置工艺流程图 Process Flow Diagram for PVC (Polyvinyl Chloride) Plant

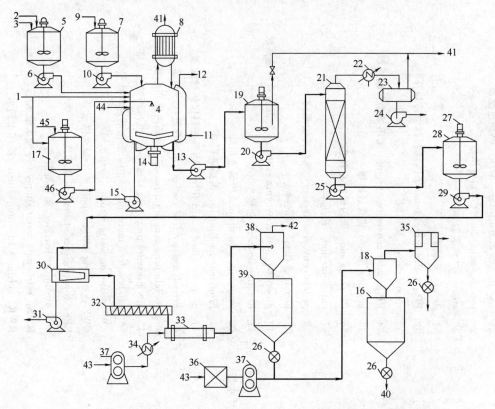

1	demineralizing water 脱盐水	25	transfer pump 传输泵
2	stabilizer 稳定剂	26	rotary valve 旋转阀
3	antioxidant 抗氧剂	27	agitator 搅拌器
4	autoclave 高压反应釜	28	centrifugal feeding tank 离心进料罐
5	stabilizer tank 稳定剂罐	29	centrifugal feeding pump 离心进料泵
6	stabilizer pump 稳定剂泵	30	centrifugal separator 离心分离器
7	catalyst drum 催化剂罐	31	waste water pump 废水泵
8	reflux condenser 回流冷凝器	32	spiral transfer 螺旋给料器
9	catalyst 催化剂	33	rotary dryer 旋转干燥器
10	catalyst pump 催化剂泵	34	air heater 空气加热器
11	refrigerated water supply 冷冻水供水	35	bagging filter 袋式过滤器
12	refrigerated water return 冷冻水回水	36	air-filter 空气过滤器
13	slurry stansfer pump 浆料输送泵	37	air-blower 鼓风机
14	reactor agitator 反应器搅拌器	38	cyclone 旋风分离器
15	waste water pump 废水泵	39	powder storage silo 粉料料仓
16	storage silo 料仓	40	PVC (polyvinyl chloride) product 聚氯乙烯产品
17	granulating agent drum 悬浮剂罐		
18	feeder hopper 进料料斗	41	to recycle tank 去回收气柜
19	slurry tank 浆料罐	42	vent to air 放空
20	stripper column feed pump 汽提塔进料泵	43	air 空气
21	stripper column 汽提塔	44	vinyl chloride monomer 氯乙烯单体
22	condenser 冷凝器	45	granulating agent 悬浮剂
23	vacuum tank 真空罐	46	granulating agent pump 悬浮剂泵
24	vacuum pump 真空泵		

1.2.11 环氧乙烷/乙二醇装置工艺流程图　Process Flow Diagram for Ethylene Oxide/Ethylene Glycol Plant

1	sulfur guard bed	脱硫床
2	oxygen mixing station	氧气混合站
3	EO reactor	环氧乙烷反应器
4	wash tower	洗涤塔
5	recycle compressor	循环气压缩机
6	EO reactor steam drum	高压汽包
7	EO reactor gas cooler steam drum	中压汽包
8	CO_2 stripping column	二氧化碳解吸塔
9	EO stripping column	环氧乙烷解吸塔
10	reclaim compressor	尾气压缩机
11	acid scrubber	酸洗塔
12	EO reabsorber	环氧乙烷再吸收塔
13	glycol feed stripper	乙二醇反应器进料汽提塔
14	EO purification column	环氧乙烷精制塔
15	glycol reactor	乙二醇反应器
16	first effect evaporator	一效蒸发塔
17	second effect evaporator	二效蒸发塔
18	third effect evaporator	三效蒸发塔
19	fourth effect evaporator	四效蒸发塔
20	fifth effect evaporator	五效蒸发塔
21	sixth effect evaporator	六效蒸发塔
22	vacuum effect evaporator	真空效蒸发塔
23	aldehyde stripper	脱醛塔
24	cycle water treating unit	循环水处理单元
25	drying column	脱水塔
26	MEG column	一乙二醇塔
27	MEG splitter	一乙二醇分离塔
28	DEG column	二乙二醇塔
29	TEG column	三乙二醇塔
30	MEG post-treatment resin bed	一乙二醇精制树脂床
…		
…		
41	ethylene	乙烯
42	oxygen	氧气
43	ethyl chloride	一氯乙烷
44	vent to waste heat boiler	放空至废热锅炉
45	BFW boiler feed water	锅炉给水
46	steam	蒸汽
47	process steam	工艺蒸汽
48	methane	甲烷
49	carbon dioxide	二氧化碳
50	DMW demineralized water	脱盐水
51	EO product	环氧乙烷产品
52	acetaldehyde purge	含醛 EO
53	acetaldehyde	乙醛
54	to vacuum ejector system	去真空喷射系统
55	to drying column hotwell	去脱水塔热水井
56	MEG product	一乙二醇产品
57	DEG product	二乙二醇产品
58	TEG product	三乙二醇产品
59	PEG	重乙二醇
60	Cooling water	循环冷却水

1.2.12 合成气制乙二醇装置工艺流程图　Process Flow Diagram for Syngas to Ethylene Glycol Plant

1	liquid ammonia storage vessel	液氨储罐
2	liquid ammonia vaporizer	液氨汽化器
3	ammoxidation furnace	氨氧化炉
4	ammoxidation steam drum	氨氧化汽包
5	NO reclaiming column	NO 回收塔
6	residual gas compressor	尾气压缩机
7	MN reclaiming column	MN 回收塔
8	Pressure Swing Adsorption (PSA)	变压吸附
9	dehydrator	脱水塔
10	esterification column	酯化反应塔
11	methanol reclaiming column	甲醇回收塔
12	nitric acid concentrator	硝酸提浓塔
13	carbonylation steam drum	羰化汽包
14	carbonylation reactor	羰化反应器
15	carbonylation products KO drum	羰化产物汽液分离器
16	methanol scrubber	甲醇洗涤塔
17	carbonylation recycle compressor	羰化循环压缩机
18	methanol rectification column	甲醇精馏塔
19	DMO rectification column	DMO 精馏塔
20	DMO hydrogenation reactor	DMO 加氢反应器
21	hydrogenation steam drum	加氢汽包
22	recycle hydrogen compressor	循环氢压缩机
23	reclaiming hydrogen compressor	回收氢压缩机
24	membrane separation	膜分离
25	high pressure KO drum	高压汽液分离器
26	low pressure KO drum	低压汽液分离器
27	methanol separating column	甲醇分离塔
28	light ends rectification column	轻组分精馏塔
29	EG rectification column	乙二醇精馏塔
30	heat exchanger	换热器
31	liquid ammonia	液氨
32	oxygen	氧气
33	methanol	甲醇
34	acid waste water	酸性废水
35	cooling water	循环冷却水
36	carbon monoxide	一氧化碳
37	nitrogen	氮气
38	fuel gas	燃料气
39	hydrogen	氢气
40	boiler feed water (BFW)	锅炉给水
41	steam	蒸汽
42	DMC	碳酸二甲酯
43	to vacuum system	去真空系统
44	mixed alcohol esters	混合醇酯
45	qualified ethylene glycol	合格乙二醇
46	polymer grade ethylene glycol	优等品乙二醇
47	heavy glycols	重质二元醇

1.2.13 天然气液化装置工艺流程图 Process Flow Diagram for Natural Gas Liquefaction Plant

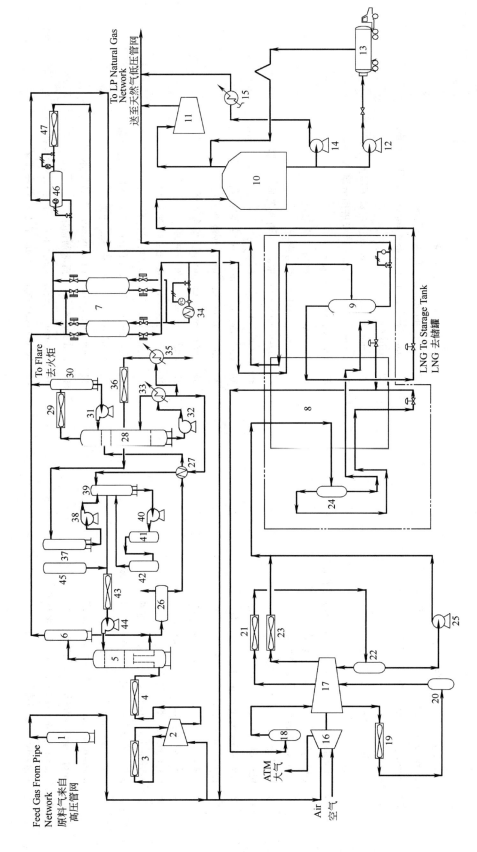

1	natural gas filter	天然气过滤器
2	natural gas compressor	天然气压缩机
3	interstage cooler	中间冷却器
4, 23	after cooler	后冷却器
5	CO_2 absorber	CO_2 脱除塔
6	separator	分离器
7	molecular sieve dehydrators	分子筛脱水塔
8	plate fin heat exchanger	板翅式换热器
9	heavies separator	重烃分离器
10	LNG storage tank	LNG 储罐
11	boil-off gas compressor	BOG 压缩机
12	LNG loading pump	LNG 装车泵
13	LNG tank truck	LNG 槽车
14	LNG send out pump	LNG 送出泵
15	LNG evaporizer	LNG 蒸发器
16	gas turbine	燃气透平
17	refrigerant compressor	冷剂压缩机
18	refrigerant suction drum	冷剂入口缓冲罐
19	interstage cooler I	中间冷却器 I
20	interstage separator I	中间分离罐 I
21	interstage cooler II	中间冷却器 II
22	interstage separator II	中间分离罐 II
24	separator	分离罐
25	refrigerant pump	冷剂泵
26	amine flash tank	胺液闪蒸罐
27	amine rich/lean heat exchanger	贫富胺换热器
28	stripper	解吸塔
29	stripper reflux condenser	解吸塔回流液冷凝器
30	stripper reflux drum	解吸塔回流罐
31	reflux pump	回流泵
32	stripper bottom pump	塔釜泵
33	amine reboiler	胺再沸器
34	regeneration gas heater	再生气体加热器
35	reclaimer	复活器
36	reclaimer condenser	复活器冷凝器
37	amine make-up tank	胺液补充罐
38	amine make-up pump	胺液补充泵
39	amine drum	胺液缓冲罐
40	amine filtration pump	胺液过滤泵
41	lean amine filter	贫胺过滤器
42	lean amine charcoal filter	贫胺活性炭过滤器
43	lean amine cooler	贫胺冷却器
44	lean amine circulation pump	贫胺循环泵
45	anti-foam tank	消泡剂储罐
46	regeneration gas separator	再生气体分离器
47	regeneration gas cooler	再生气体冷却器

1.2.14 液化天然气接收站工艺流程图　Process Flow Diagram for Liquefied Natural Gas (LNG) Receiving Terminal

1　LNG carrier　液化天然气运输船
2　LNG unloading arm　液化天然气卸船臂（可以为 4 台或 3 台或 2 台，视计算定）
3　vapour return arm　气相返回臂
4　LNG storage tank　液化天然气储罐
5　LP pump/in-tank pump　低压泵/罐内泵
6　recondenser　再冷凝器
7　BOG compressor K.O. drum　蒸发气分液罐
8　BOG compressor　蒸发气压缩机
9　HP pump/sendout pump　高压泵/外输泵
10　ORV (open rack vaporizer)　开架式气化器
11　other vaporizer such as IFV (intermediate fluid vaporizer), STV (shell and tube vaporizer), AAV (ambient air vaporizer)　其他形式的气化器，如中间介质气化器、管壳式气化器、空温式气化器
12　SCV (submerged combustion vaporizer)　浸没燃烧式气化器
13　metering station　计量站
14　pig launcher　清管器发射器
15　truck loading arm　槽车装车橇
16　LNG truck　液化天然气汽车
17　jetty nitrogen buffer drum　码头氮气缓冲罐（非必需，视计算定）
18　jetty drain drum　码头排净罐（非必需，视计算定）
19　jetty drain drum heater　码头排净罐电加热器（非必需，视计算定）
20　vapor return blower　回流鼓风机
21　BOG compressor K.O. drum drain drum　蒸发气压缩机分液罐排净罐
22　fuel gas heater　燃料气电加热器
23　truck loading area drain drum　槽车站排净罐
24　flare K.O. drum　火炬分液罐
25　flare K.O. drum heater　火炬分液罐电加热器
26　flare　火炬（高架火炬或地面火炬）

1.2.15 MTO 装置工艺流程图 Process Flow Diagram for Methanol to Olefin (MTO)

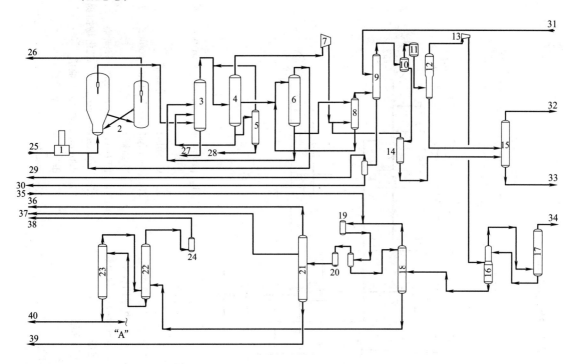

1 heater 加热炉
2 MTO Reactor/Regenerator system MTO 反应再生系统
3 quenchtower 急冷塔
4 water wash tower 水洗塔
5 water stripper 水汽提塔
6 oxygenate stripper 氧化物汽提塔
7 product gas compressor 1st—3st 产品气压缩机 1~3 段
8 methanol washing tower 甲醇洗涤塔
9 caustic wash tower 碱洗塔
10 product gas knockout drum 产品气分离罐
11 product gas dryer 产品气干燥器
12 depropanizer 脱丙烷塔
13 MTO Product gas compressor 4st MTO 产品气压缩机四段
14 condensate stripper 凝液汽提塔
15 debutanizer 脱丁烷塔
16 pre-cutting tower 预切割塔
17 oil absorption tower 油吸收塔
18 deethanizer 脱乙烷塔
19 acetylene hydrogenation reactor 乙炔加氢反应器
20 ethylene dryer 乙烯干燥器
21 ethylene tower 乙烯塔
22 2# propylene tower 2#丙烯塔
23 1# propylene tower 1#丙烯塔
24 propylene guard bed 丙烯产品保护床
25 methanol 甲醇
26 flue gas 烟气
27 waste water 废水
28 stripped water 汽提水
29 waste gas 废气
30 spent caustic 废碱
31 BFW 锅炉给水
32 C_4 Distillate C_4 馏分
33 C_{5+} Distillate C_{5+} 馏分
34 fuel gas 燃料气
35 hydrogen 氢气
36 non condensable gas 不凝气
37 ethylene 乙烯产品
38 propylene 丙烯产品
39 ethane 乙烷
40 propane 丙烷

1.2.16 MTP 工艺流程图　Process Flow Diagram for Methanol to Propylene

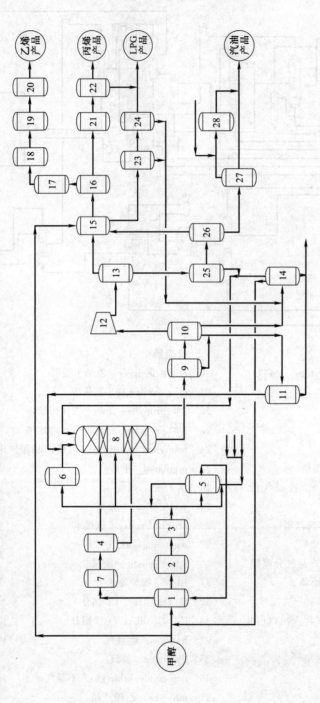

1	methanol preheater	甲醇预热器
2	methanol superheater	甲醇过热器
3	methanol superheater/DME partial condenser	甲醇过热器/DME 分凝器
4	DME separator	DME 分离罐
5	DME reactor	DME 反应器
6	fired heater	加热炉
7	DME reactor effluent cooler	DME 反应器产物冷却器
8	MTP reactor	MTP 反应器
9	pre-quench column	预急冷塔
10	quench column	急冷塔
11	process steam boiler	工艺蒸汽发生器
12	HC compressor	烃压缩机
13	separator stage 4	四段出口分离罐
14	MeOH recovery column	甲醇回收塔
15	depropanizer (DME absorber)	脱丙烷塔
16	deethanizer	脱乙烷塔
17	CO_2 scrubber	CO_2 洗涤塔
18	C_2 dryer	C_2 干燥器
19	demethanizer	脱甲烷塔
20	C_2 splitter	C_2 分离塔
21	propylene product guard bed	丙烯产品保护床
22	C_3 splitter	C_3 分离塔
23	depropanizer bottoms cooler	脱丙烷塔底冷却器
24	LPG separator	LPG 分离罐
25	oxygenate extractor	氧化物抽提塔
26	debutanizer	脱丁烷塔
27	dehexanizer	脱己烷塔
28	gasoline stabilizer column	汽油稳定塔

1.2.17 丙烷脱氢装置工艺流程图 Process Flow Diagram for Propane Dehydrogenation (PDH) Plant

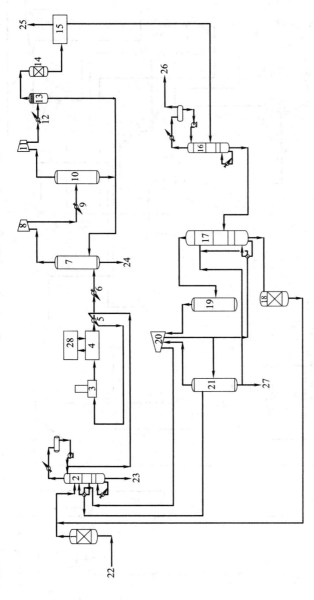

1	feed pretreated system (if any)	原料预处理系统（如有）
2	depropanizer	脱丙烷塔
3	heater	加热炉
4	reactor	反应器
5	feed/effluent heat exchanger	进出料换热器
6	reactor effluent cooler	产品气冷却器
7	reactor effluent compressor 1st suction drum	产品气压缩机一段吸入罐
8	reactor effluent compressor 1st stage	产品气压缩机一段
9	first stage aftercooler	后冷却器
10	second stage suction drum	二段吸入罐
11	reactor effluent compressor 2nd stage	产品气压缩机二段
12	reactor effluent compressor discharge cooler	产品气压缩机出口冷却器
13	reactor effluent compressor discharge drum	产品气压缩机排出罐
14	reactor effluent drier	产品气干燥器
15	separation system	分离系统
16	deethanizer	脱乙烷塔
17	C_3 splitter	丙烯塔
18	C_3 Hydrogenation reactor	C_3 加氢反应器
19	heat pump compressor first suction drum	热泵压缩机一段吸入罐
20	heat pump compressor	热泵压缩机
21	heat pump compressor second suction drum	热泵压缩机二段吸入罐
22	fresh feed	新鲜原料
23	C_{4+} fraction	C_4 以上馏分
24	waste water	废水
25	net gas	干气
26	offgas	尾气
27	poly grade propylene	聚合级丙烯
28	regeneration system	再生系统

1.2.18 煤气化工艺流程图 Process Flow Diagram for Coal Gasification

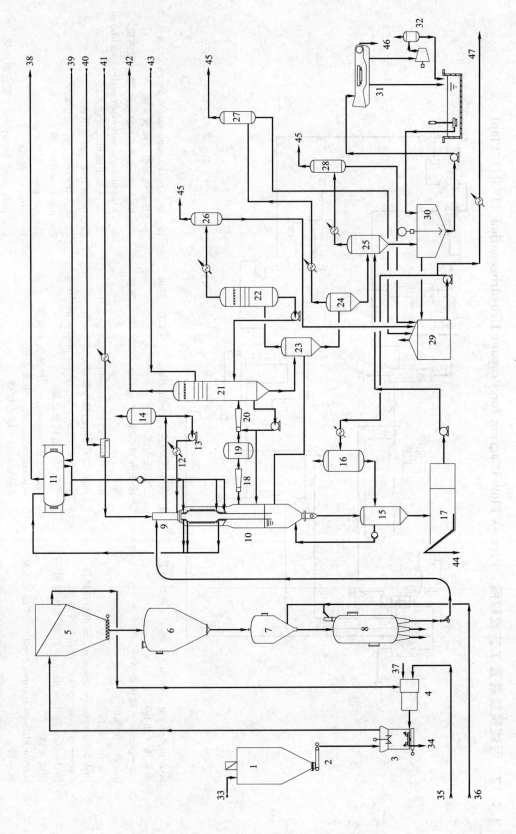

1	crushed coal bunker	原料煤储仓
2	gravimetric coal feeder	称重给煤机
3	coal mill	磨煤机
4	inert gas generator	惰性气体发生器
5	pulverized coal bag filter	粉煤袋式过滤器
6	pulverized coal storage	粉煤储罐
7	lock hopper	粉煤锁斗
8	coal feed vessel	粉煤给料罐
9	coal burner	粉煤烧嘴
10	gasifier	气化炉
11	steam drum	汽包
12	burner cooling cycle cooler	烧嘴冷却循环水冷却器
13	burner cooling cycle pump	烧嘴冷却循环水泵
14	burner cooling cycle drum	烧嘴冷却循环水罐
15	slag hopper	渣锁斗
16	flushing vessel	冲洗水罐
17	drag chain	捞渣机
18	#1 venturi scrubber	一级文丘里洗涤器
19	venturi K/O drum	文丘里气液分离罐
20	#2 venturi scrubber	二级文丘里洗涤器
21	scrubber	洗涤塔
22	saturation tower	增湿塔
23	high pressure flash drum	高压闪蒸罐
24	low pressure flash drum	低压闪蒸罐
25	vacuum flash drum	真空闪蒸罐
26	sour gas K/O drum	酸性气冷凝液分离罐
27	LP flash overhead separator	低压闪蒸罐冷凝液分离罐
28	vacuum flash overhead separator	真空闪蒸罐冷凝液分离罐
29	cycle water vessel	循环水槽
30	clarifier	澄清槽
31	vacuum belt filter	真空带式过滤机
32	vacuum K/O drum	气液分离罐
33	raw coal	原煤
34	foreign matter	杂质
35	fuel gas	燃料气
36	HP CO$_2$	高压二氧化碳
37	air	空气
38	steam	蒸汽
39	boiler feed water	锅炉给水
40	MP superheated steam	中压过热蒸汽
41	oxygen	氧气
42	raw gas	粗合成气
43	shift condensate	冷凝液
44	slag	渣
45	sour gas	酸性气
46	filter cake	滤饼
47	waste water	废水

1.2.19 煤制油工艺流程图 Process Flow Diagram for Coal to Oil Plant

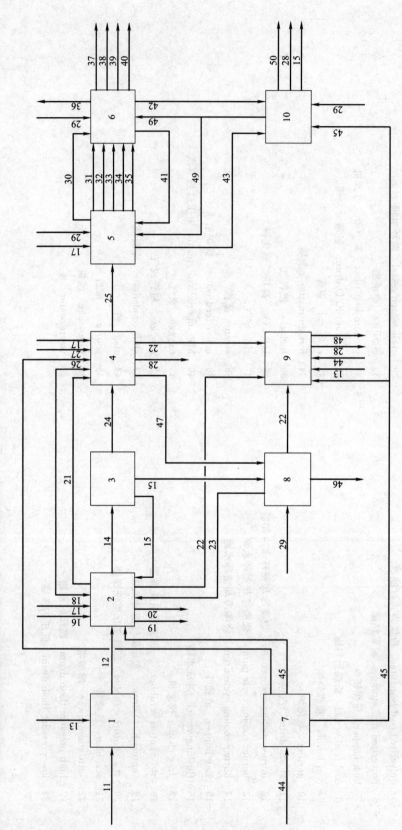

注1:(气化工艺为干煤粉气化工艺)
Note1: (Dry Coal Powder Gasification Process)

注2:(若气化工艺为水煤浆气化工艺,则无物流号13、16、18、21)
Note2: (If Gasification Process Is Slurry Gasification, There Is No Stream No.13、16、18、21)

1	coal milling dry unit or slurry preparing unit	磨煤干燥装置或水煤浆制备装置
2	coal gasification unit	煤气化装置
3	CO shift unit	一氧化碳变换装置
4	rectisol unit	低温甲醇洗装置
5	F-T oil synthesis	F-T 合成装置
6	petroleum refinery unit	油品加工装置
7	air separation unit	空分装置
8	waste water stripper unit	酸水汽提装置
9	sulfur recovery unit	硫黄回收装置
10	tail-gas hydrogen production unit	尾气制氢装置
11	raw coal	原煤
12	pulv. coal or slurry	煤粉或水煤浆
13	fuel gas	燃料气
14	raw syntheis gas	粗合成气
15	process condensate	工艺冷凝液
16	process steam	工艺蒸汽
17	demineralized water	脱盐水
18	CO_2 for conveying coal	输煤二氧化碳
19	slag	排渣
20	filter cake	滤饼
21	vent CO_2	排放二氧化碳
22	sour gas	酸气
23	stripping condensate	汽提凝液
24	CO shift gas	变换气
25	purification gas	净化气
26	N_2	氮气
27	boiler water	锅炉水
28	vent gas	排气
29	steam	蒸汽
30	compressed condensate	压缩凝液
31	naphtha	石脑油
32	heavy oil	重油
33	wax	蜡
34	F-T synthetic water	F-T 合成水
35	decarbonization tail gas	脱碳尾气
36	synthetic water	合成水
37	diesel oil	柴油
38	naphtha product	产品石脑油
39	LPG	液化石油气
40	mixed alcohols	混醇
41	reduced heavy diesel	还原重油
42	oil wash dry gas	油洗干气
43	reduced tail gas	还原尾气
44	air	空气
45	oxygen	氧气
46	waste water	废水
47	sour water	酸水
48	sulfur	硫黄
49	hydrogen	氢气
50	PSA tail gas	PSA 尾气

1.2.20 煤制甲醇工艺流程图　Process Flow Diagram for Coal to Methanol

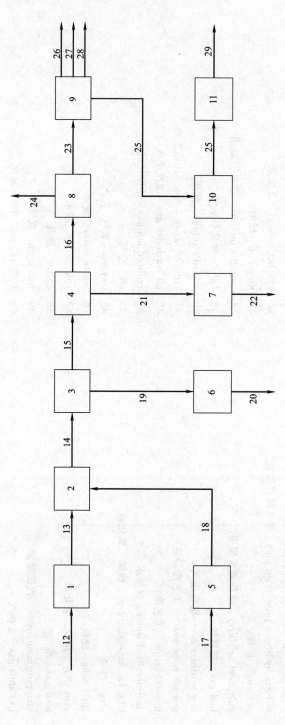

1　coal milling dry (CMD) unit (or slurry preparation unit)　磨煤干燥（CMD）单元或煤浆制备单元
2　coal gasification unit　煤气化单元
3　CO shift unit　一氧化碳变换单元
4　rectisol unit　低温甲醇洗单元
5　air separation unit　空分单元
6　sour water stripping unit　酸水汽提单元
7　sulfur recovery unit　硫黄回收单元
8　methanol synthesis unit　甲醇合成单元
9　methanol distillation unit　甲醇精馏单元
10　intermediate tank farm　中间罐区
11　methanol storage　甲醇罐区
12　raw coal　原煤
13　pulverized coal (or slurry)　煤粉或水煤浆
14　raw gas　粗煤气
15　shifted gas　变换气
16　synthesis gas　合成气
17　air　空气
18　oxygen　氧气
19　process condensate　工艺冷凝液
20　waste water　废水
21　acid gas　酸性气
22　sulfur　硫黄
23　crude methanol　粗甲醇
24　purge gas　驰放气
25　pure methanol　精甲醇
26　fuel gas　燃料气
27　side draw　侧线采出液
28　process water　工艺水
29　methanol product　产品甲醇

1.2.21 低温甲醇洗工艺流程图 Process Flow Diagram for Rectisol

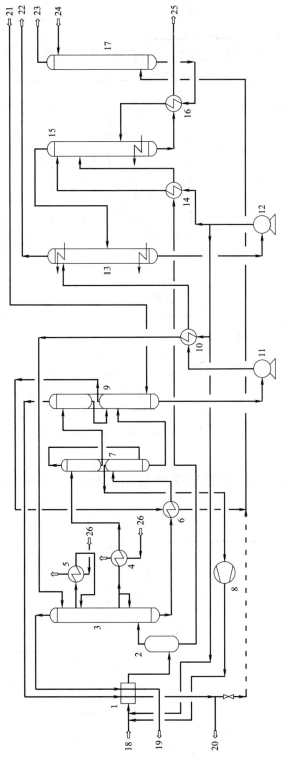

1	raw gas cooler	原料气冷却器	
2	K.O. drum	分水罐	
3	absorber	吸收塔	
4	H₂S-free methanol cooler	无硫甲醇冷却器	
5	stage cooler	段间冷却器	
6	H₂S-laden methanol cooler	富硫甲醇冷却器	
7	flash column	闪蒸塔	
8	recycle gas compressor	循环气压缩机	
9	reabsorbor	解吸塔	
10	lean/Laden MeOH exchanger	贫/富甲醇
11	hot regenerator feed pump	热再生塔进料泵
12	MeOH/water seperator feed pump	甲醇/水分离塔进料泵
13	hot regenerator	热再生塔
14	MeOH/H₂O separator feed cooler	甲醇水分离塔进料冷却器
15	MeOH/H₂O seperator	甲醇/水分离塔
16	waste water	污水换热器
17	offgas scrubber	尾气洗涤塔
18	raw gas	原料气
19	product gas	产品气
20	CO₂ product	二氧化碳产品
21	LP N₂	低压氮气
22	sour gas	酸性气
23	offgas	尾气
24	demin. water	脱盐水
25	waste water	污水
26	refrigerant	冷剂

1.2.22 精对苯二甲酸装置（PTA）工艺流程图 Process Flow Diagram for PTA Plant

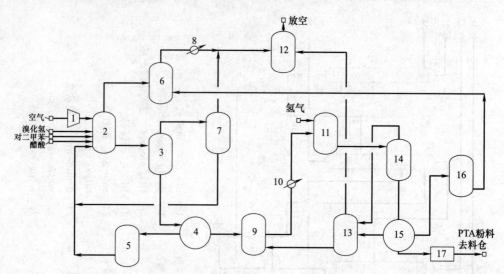

1 air compressor 空气压缩机
2 TA reactor 对苯二甲酸氧化反应器
3 TA crystallizer 粗对苯二甲酸结晶器
4 TA solid separator 粗对苯二甲酸固体分离器
5 TA ML drum 粗对苯二甲酸母液罐
6 TA dehydration tower 粗对苯二甲酸脱水塔
7 absorber 尾气吸收塔
8 steam boiler 蒸汽发生器
9 PTA feed mixing drum 精对苯二甲酸进料混合器
10 PTA heat exchanger 精对苯二甲酸反应器进料加热器
11 PTA reactor 对苯二甲酸精制反应器
12 vent stack 放空烟囱
13 bromine scrubber 溴洗涤塔
14 PTA crystallizer 精对苯二甲酸结晶器
15 PTA solid separator 精对苯二甲酸结晶器
16 PTA ML drum 精对苯二甲酸母液罐
17 PTA dryer 精对苯二甲酸产品干燥器

1.2.23 苯乙烯/环氧丙烷（SM/PO）装置工艺流程图 Process Flow Diagram for SM/PO Plant

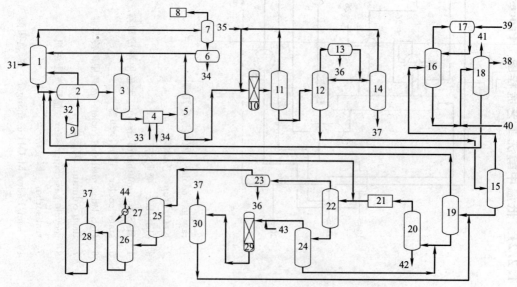

1	EB preheater 乙苯预热器	23	water separator 水分离器
2	EBHP reactor 乙苯过氧化氢反应器	24	MPK column MPK 塔
3	pre-concentrator 预浓缩器	25	EB/SM column EB/SM 塔
4	WAS section WAS 段	26	styrene purification column SM 精制塔
5	EBHP concentrator EBHP 浓缩器	27	styrene purification column condenser SM 精制塔冷凝器
6	separator 分离罐	28	heavy ends column 重组分塔
7	EB absorber 乙苯吸收塔	29	MPK hydrogenation reactor MPK 加氢反应器
8	furnace 加热炉		
9	air compressor 空气压缩机	30	de-gassing vessel 脱气罐
10	epoxidation reactor 环氧化反应器	31	EB 乙苯
11	first propylene removal column 一段丙烯脱除塔	32	air 空气
12	second propylene removal column 二段丙烯脱除塔	33	caustic 碱
		34	waste caustic 废碱
13	condensate collector 凝液收集罐	35	propylene 丙烯
14	propylene purification column 丙烯精制塔	36	waste water 废水
15	crude PO column 粗 PO 塔	37	fuel 燃料
16	PO drying column PO 干燥塔	38	epoxy propane product 环氧丙烷产品
17	coalescer 聚结器	39	azeotropic agent 共沸剂
18	PO purification column PO 精制塔	40	extractant 萃取剂
19	EB recovery column 乙苯塔	41	crude PO bleeding 粗 PO 排放
20	MPC column MPC 塔	42	heavy ends 重组分
21	styrene reactor 苯乙烯反应器	43	hydrogen 氢气
22	crude SM column 粗苯乙烯塔	44	styrene product 苯乙烯产品

1.2.24 聚醚多元醇装置工艺流程图 Process Flow Diagram for Polyether Polyol Plant

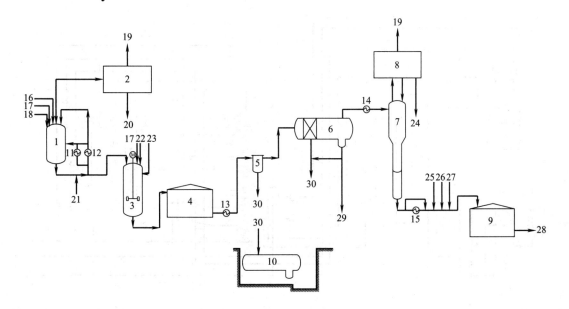

1　polymerization reactor　聚合反应器
2　reactor vacuum unit　反应器真空单元
3　neutralizer　中和器
4　coalescer buffer tank　聚结器缓冲罐
5　coalescer feed filter　聚结器进料过滤器
6　coalescer　聚结器
7　polyol stripper/dryer　多元醇汽提/干燥塔
8　polyol stripper/dryer vacuum unit　多元醇汽提/干燥塔真空单元
9　polyether polyol rundown tank　聚醚多元醇成品罐
10　drain vessel　废液罐
11　reactor heater　反应器加热器
12　reactor cooler　反应器冷却器
13　coalescer feed heater　聚结器进料加热器
14　stripper/dryer feed preheater　汽提/干燥塔进料预热器
15　polyether polyol product cooler　聚醚多元醇产品冷却器
16　glycerine　甘油
17　KOH　氢氧化钾
18　base polyol　基础多元醇
19　off gas　废气
20　PO contaminated liquid　含环氧丙烷的废液
21　EO/PO　环氧乙烷/环氧丙烷
22　H_3PO_4　磷酸
23　demineralized water　脱盐水
24　process condensate　工艺凝液
25　anti-oxidant 1　抗氧剂1
26　anti-oxidant 2　抗氧剂2
27　anti-oxidant 3　抗氧剂3
28　polyether polyol　聚醚多元醇
29　brine effluent　含盐废水
30　effluent bleeding　废液排放

1.3　公用工程典型工艺流程图　Typical Utility Flow Diagram

1.3.1　空分空压装置工艺流程图　Process Flow Diagram for Air Separation Unit and Compressed Air "Station"

1.3.1.1　深冷空分装置工艺流程图　Process Flow Diagram for Cryogenic Air Separation Unit

（1）深冷空分内压缩工艺流程图　Internal Compression Process Flow Diagram for Cryogenic Air Separation Unit

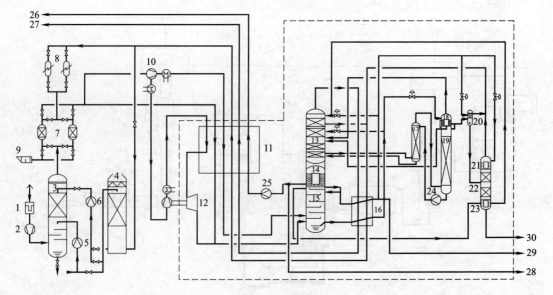

1	air filter 空气过滤器	16	subcooler 过冷器
2	air compressor 空气压缩机	17	crude argon column Ⅰ 粗氩塔Ⅰ
3	air water tower 空冷塔	18	crude argon condenser 粗氩冷凝蒸发器
4	nitrogen water tower 水冷塔	19	crude argon column Ⅱ 粗氩塔Ⅱ
5	cooling water pump 冷却水泵	20	crude argon liquefier 粗氩液化器
6	chilling water pump 冷冻水泵	21	pure argon condenser 精氩冷凝器
7	molecular sieve purifiers 分子筛纯化器	22	pure argon column 精氩塔
8	steam/electic regeneration heater 蒸汽/电再生加热器	23	pure argon vaporizer 精氩蒸发器
9	sliencer 消声器	24	recycle liquid argon pump 循环液氩泵
10	air booster 空气增压机	25	liquid oxygen pump 液氧泵
11	main heat exchanger 主换热器	26	O₂ 氧气
12	booster-turbexpander unit 增压透平膨胀机	27	N₂ 氮气
13	upper column 上塔	28	LOX（liquid oxygen） 液氧
14	main condenser 主冷凝蒸发器	29	LIN（liquid nitrogen） 液氮
15	lower column 下塔	30	LAR（liquid argon） 液氩

（2）深冷空分外压缩工艺流程图 External Compression Process Flow Diagram for Cryogenic Air Separation Unit

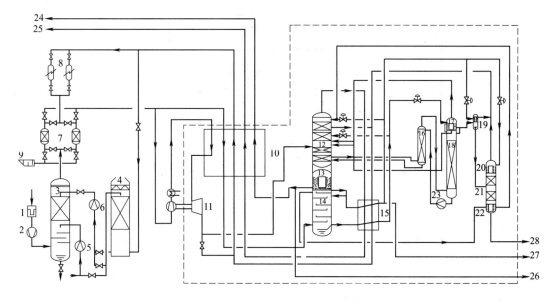

1	air filter 空气过滤器	15	subcooler 过冷器
2	air compressor 空气压缩机	16	crude argon column Ⅰ 粗氩塔Ⅰ
3	air water tower 空冷塔	17	crude argon condenser 粗氩冷凝蒸发器
4	nitrogen water tower 水冷塔	18	crude argon column Ⅱ 粗氩塔Ⅱ
5	cooling water pump 冷却水泵	19	crude argon liquefier 粗氩液化器
6	chilling water pump 冷冻水泵	20	pure argon condenser 精氩冷凝器
7	molecular sieve purifiers 分子筛纯化器	21	pure argon column 精氩塔
8	steam/electic regeneration heater 蒸汽/电再生加热器	22	pure argon vaporizer 精氩蒸发器
9	sliencer 消声器	23	recycle liquid argon pump 循环液氩泵
10	main heat exchanger 主换热器	24	O₂ 氧气
11	booster-turbexpander unit 增压透平膨胀机	25	N₂ 氮气
12	upper column 上塔	26	LOX（liquid oxygen） 液氧
13	main condenser 主冷凝蒸发器	27	LIN（liquid nitrogen） 液氮
14	lower column 下塔	28	LAR（liquid argon） 液氩

41

1.3.1.2 变压吸附制氮装置工艺流程图 Process Flow Diagram for PSA Nitrogen Generation Unit

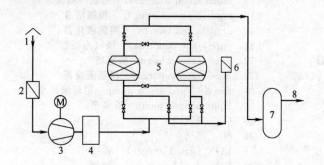

1 air 空气
2 air filter 空气过滤器
3 air compressor 空气压缩机
4 air dryer 空气干燥器
5 PSA nitrogen generator 变压吸附制氮装置
6 silencer 消声器
7 nitrogen vessel 氮气罐
8 nitrogen 氮气

1.3.1.3 真空变压吸附制氧装置工艺流程图 Process Flow Diagram for VPSA Oxygen Generation Unit

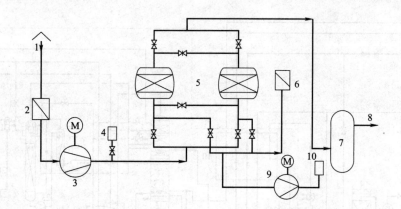

1 air 空气
2 air filter 空气过滤器
3 blower 鼓风机
4 sliencer 消音器
5 VPSA oxygen generator 变压吸附制氧装置
6 blow-off silencer 放空消音器
7 oxygen vessel 氧气罐
8 oxygen 氧气
9 vacuum blower 真空风机
10 waste nitrogen 污氮气

1.3.1.4 膜分离制氮装置工艺流程图 Process Flow Diagram for Membrane Module Nitrogen Generation Unit

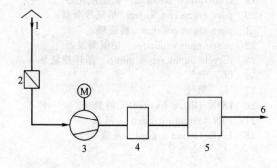

1 air 空气
2 air filter 空气过滤器
3 air compressor 空气压缩机
4 air dryer 空气干燥器
5 membrane module 膜组
6 nitrogen 氮气

1.3.1.5 空压站工艺流程图 Process Flow Diagram for Compressed Air Station

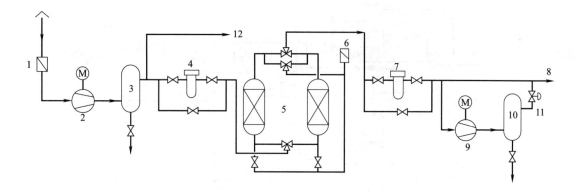

1	air filter 空气过滤器	7	after filter 后置过滤器
2	air compressor 空气压缩机	8	instrument air 仪表空气
3	air buffer vessel 空气缓冲罐	9	instrument air booster 仪表空气增压机
4	pre filter 前置过滤器	10	instrument air vessel 仪表空气储罐
5	air dryer 空气干燥器	11	control valve 调节阀
6	silencer 消声器	12	plant air 工厂空气

1.3.2 循环冷却水工艺流程图 Process Flow Diagram for Recirculating Cooling Water System

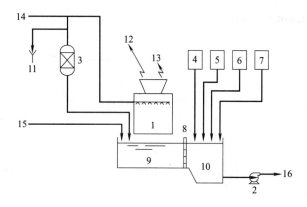

1	cooling tower 冷却塔	9	cooling water basin 冷却水池
2	cooling water pump 冷却水泵	10	suction sump 水泵吸水池
3	side-stream filter 旁滤器	11	cooling water blow down 冷却水排污
4	biocide dosing system 杀菌剂投加系统	12	evaporation loss 蒸发损失
5	dispersant dosing system 阻垢剂投加系统	13	drift loss 风吹损失
6	corrosion inhibitor dosing system 缓蚀剂投加系统	14	cooling water return 冷却水回水
7	acid dosing system 加酸系统	15	make-up water 补充水
8	screen 格栅	16	cooling water supply 冷却水给水

1.3.3 净水场工艺流程图 Process Flow Diagram for Raw Water Treatment Plant

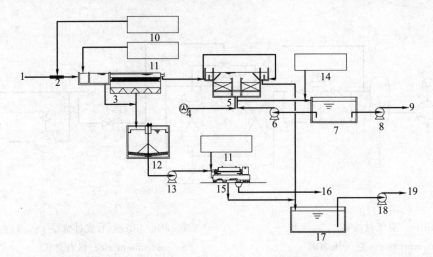

1	raw water 原水		11	coagulant dosing package 助凝剂
2	static mixer 静态混合器		12	sludge thickener 污泥浓缩池
3	reaction and sedimentation basin 沉淀池		13	sludge feed pump 污泥提升泵
4	backwash blower 反洗风机		14	disinfectant dosing system 消毒剂
5	sand filter 滤池		15	frame filter 板框压滤机
6	sand filter backwash pump 反洗水泵		16	sludge shipping 污泥外送
7	service water basin 清水池		17	waste water basin 污水池
8	service water pump 送水泵		18	backwash wastewater pump 反洗废水提升泵
9	to user 去用户		19	waste water treatment 去排水系统
10	flocculant dosing package 絮凝剂			

1.3.4 储运系统工艺流程图 Process Flow Diagram for Storage and Transportation System

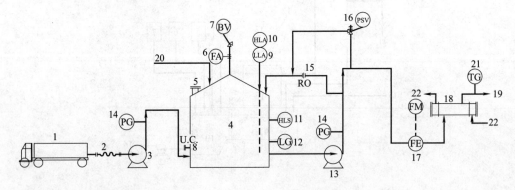

1	tank truck 汽车槽车		12	level gauge 液位计
2	hose connection 软管接头		13	transfer pump 输送泵
3	unloading pump 卸料泵		14	pressure gauge 压力表
4	storage tank 储罐		15	restriction orifice 限流孔板
5	man hole 人孔		16	pressure safety valve 安全阀
6	flare arrestor 阻火器		17	flow elements, flow meter 流量元件,流量计
7	breath valve 呼吸阀		18	heater 加热器
8	utility connection 公用工程接口		19	to users 去用户
9	low level alarm 低液位报警		20	nitrogen 氮气
10	high level alarm 高液位报警		21	temperature gauge 温度计
11	high level switch 高液位开关		22	heat medium 加热介质

1.3.5 离子交换除盐水系统工艺流程图　Process Flow Diagram for Demineralized Water System (Ion Exchanger System)

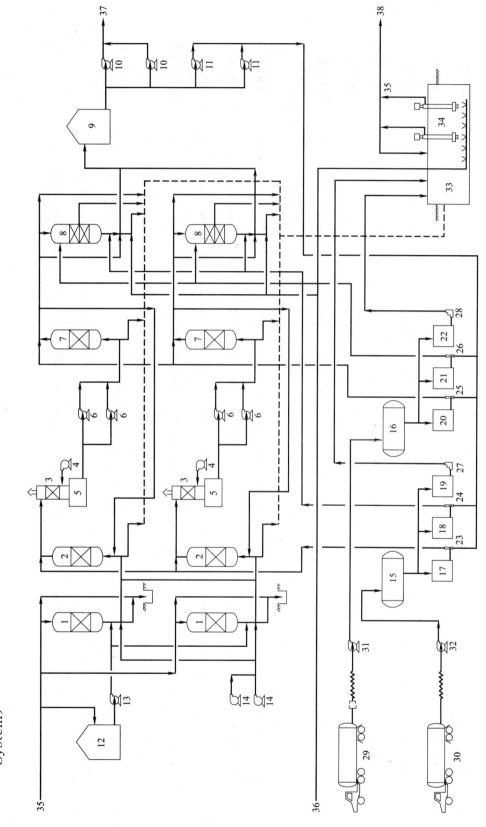

1	high efficiency filter 高效过滤器	17	HCl metering tank (for cation exchanger) 盐酸计量箱（用于阴离子交换器）	25	NaOH injector (for anion exchanger) 氢氧化钠喷射器（用于阴离子交换器）
2	cation exchanger 阳离子交换器	18	HCl metering tank (for mixed ion exchanger) 盐酸计量箱（用于混合离子交换器）	26	NaOH injector (for mixed ion exchanger) 氢氧化钠喷射器（用于混合离子交换器）
3	degassing tower 除二氧化碳器	19	HCl metering tank (for neutralization water basin) 盐酸计量箱（用于中和水池）	27	HCl metering pump 盐酸计量泵
4	degassing tower blower 脱气塔风机	20	NaOH metering tank (for anion exchanger) 氢氧化钠计量箱（用于阴离子交换器）	28	NaOH metering pump 氢氧化钠计量泵
5	middle water tank 中间水箱	21	NaOH metering tank (for mixed ion exchanger) 氢氧化钠计量箱（用于混合离子交换器）	29	NaOH tank truck 氢氧化钠槽车
6	middle water pump 中间水泵	22	NaOH metering tank (for neutralization water basin) 氢氧化钠计量箱（用于中和水池）	30	HCl tank truck 盐酸槽车
7	anion exchanger 阴离子交换器	23	HCl injector (for cation exchanger) 盐酸喷射器（用于阴离子交换器）	31	NaOH unloading pump 氢氧化钠卸车泵
8	mixed ion exchanger 混合离子交换器	24	HCl injector (for mixed ion exchanger) 盐酸喷射器（用于混凝土离子交换器）	32	HCl unloading pump 盐酸卸车泵
9	demineralized water tank 脱盐水箱			33	neutralization water basin 中和水池
10	demineralized water pump 脱盐水泵			34	neutralization water pump 中和水泵
11	regenerating water pump 再生水泵			35	fresh water 新鲜水
12	backwash water tank 反洗水箱			36	plant air 工厂空气
13	backwash water pump 反洗水泵			37	user 用户
14	backwash blower 反洗风机			38	drain 排放
15	HCl storage tank 盐酸储罐				
16	NaOH storage tank 氢氧化钠储罐				

1.3.6 污水处理工艺流程图 Process Flow Diagram for Waste Water Treatment System

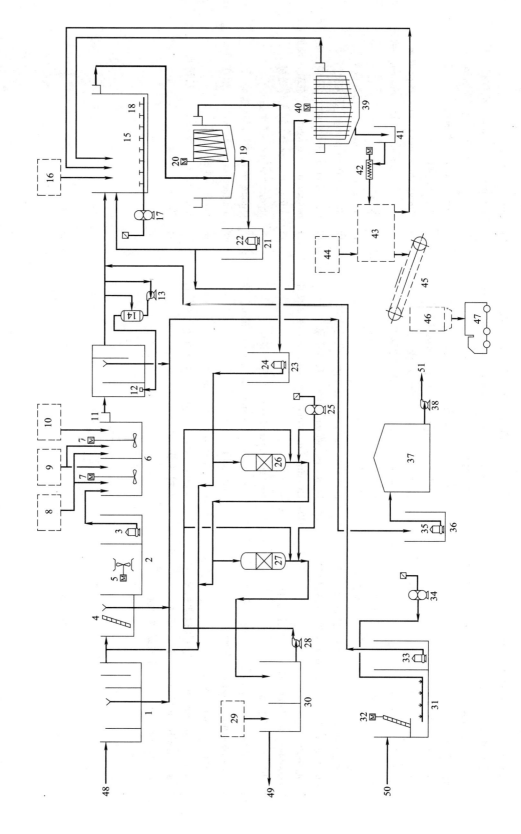

1	oil interceptor (API/CPI)	隔油池
2	equalization basin	均衡池
3	production waste water lifting pump	生产废水提升泵
4	bar screen	格栅
5	equalization basin mixer	均衡池搅拌器
6	neutralization basin	中和池
7	neutralization basin mixers	中和池搅拌器
8	sulphuric acid feed system	硫酸投加系统
9	caustic soda feed system	氢氧化钠投加系统
10	polyelectrolyte feed system	聚合电解质投加系统
11	dissolved air flotation basin (DAF)	气浮池
12	diffuser	扩散器
13	flotation recycle pump	气浮循环水泵
14	air dissolving drum	溶气罐
15	aeration basin	曝气池
16	nutrient feed system	营养物投加系统
17	aeration basin blower	曝气池风机
18	air diffusion system	空气扩散系统
19	secondary sedimentation basin	二沉池
20	secondary sedimentation basin sludge scraper	二沉池刮泥机
21	sludge pit	污泥池
22	sludge recirculation pump	污泥循环泵
23	clarified water pit	澄清池
24	clarified water pump	澄清水泵
25	backwash blower	反洗风机
26	sand filter	砂滤器
27	activated carbon filter	活性炭过滤器
28	backwash water pump	反洗水泵
29	chlorine liquid injection system	液氯投加系统
30	treated water basin	处理后水池
31	sanitary waste water equalization basin	生活污水均衡池
32	moving screen	机械格栅
33	sanitary waste water lifting pump	生活污水提升泵
34	blower	风机
35	slop oil lifting pump	废油提升泵
36	slop oil collecting pit	废油池
37	slop oil tank	废油罐
38	slop oil pump	废油泵
39	sludge thickener basin	污泥浓缩池
40	thickener basin sludge scraper	浓缩池刮泥机
41	sludge collecting pit	污泥收集池
42	sludge lifting pump	污泥提升泵
43	belt filter press system	带式压缩系统
44	coagulant feed system	混凝剂投加系统
45	sludge transport	污泥传送
46	sludge hopper	污泥漏斗
47	sludge truck	污泥运输车
48	production waste water	生产污水
49	drain	排放
50	sanitary waste water	生活污水
51	slop oil	废油

1.3.7 含盐废水处理装置流程图 Process Flow Diagram for Salty Waste Water Treatment Unit

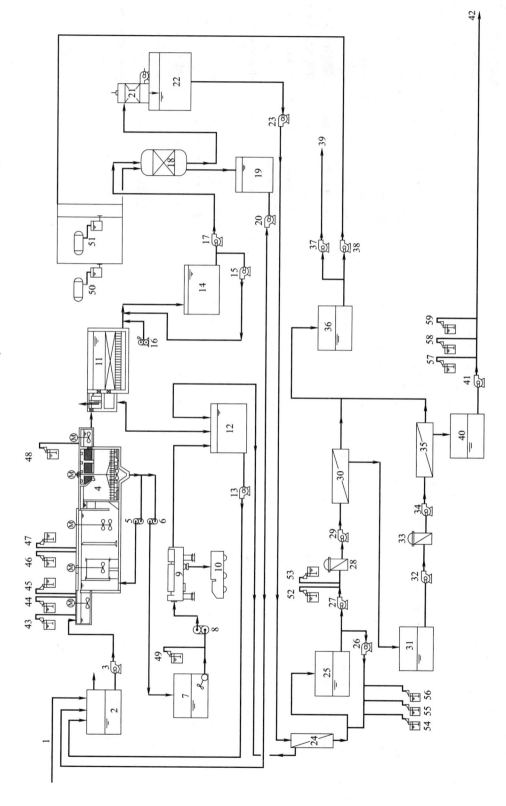

#	English	Chinese	#	English	Chinese
1	raw water	原水	24	ultrafilter	超滤
2	raw water basin	原水调节池	25	UF water basin	超滤水池
3	raw water pump	原水泵	26	UF backwashing pump	超滤反洗泵
4	clarifier	高密度澄清池	27	RO feed water pump	RO 给水泵
5	sludge circulating pump	污泥循环泵	28	5μm cartridge filter	5μm 保安过滤器
6	sludge discharge pump	污泥排放泵	29	high pressure pump	高压泵
7	sludge buffer basin	污泥缓冲池	30	RO system	反渗透系统
8	sludge feed pump	污泥给料泵	31	RO concentrated water basin	RO 浓水池
9	sludge dehydrator	污泥脱水机	32	RO concentrated water pump	RO 浓水泵
10	sludge truck	污泥运输车	33	con. water 5μm cartridge filter	浓水 5μm 保安过滤器
11	sand filter	滤池	34	con. water high pressure pump	浓水 RO 高压泵
12	wastewater collecting basin	废水收集池	35	con. RO system	浓水 RO 系统
13	wastewater collecting pump	废水收集泵	36	reused water tank	回用水箱
14	filtered water basin	过滤水池	37	reused water pump	回用水泵
15	filter backwashing pump	滤池反洗水泵	38	softener regeneration pump	软化器再生泵
16	filter backwashing blower	滤池反洗风机	39	to user	去用户
17	softener feed water pump	软化给水泵	40	brine water basin	浓盐水池
18	softener	软化器	41	brine pump	浓盐水泵
19	neutralization water basin	中和水池	42	to evaporation system	去蒸发系统
20	neutralization water pump	中和水泵	43	$Ca(OH)_2$ dosing system	石灰加药系统
21	decarbonayor	脱碳器	44	Na_2CO_3 dosing system	纯碱加药系统
22	softened water basin	软化水池	45	MgO dosing system	镁剂加药系统
23	UF feedwater pump	超滤给水泵	46	coagulant dosing system	絮凝剂加药系统
			47	PAM dosing system	PAM 加药系统
			48	H_2SO_4 dosing system	加药系统
			49	PAM dosing system for sludge treatment	污泥 PAM 加药系统
			50	softener acid regeneration system	软化器酸再生系统
			51	softener alkali regeneration system	软化器碱再生系统
			52	$NaHSO_3$ dosing system	亚硫酸氢钠加药系统
			53	antiscalant dosing system	阻垢剂加药系统
			54	UF backwashing	超滤反洗酸系统
			55	UF backwashing NaOH dosing system	反洗加碱系统
			56	UF backwashing NaClO dosing system	反洗加次氯酸钠系统
			57	H_2SO_4 dosing system	硫酸加药系统
			58	Evaporator antiscalant dosing system	阻垢剂加药系统 蒸发器
			59	Evaporator antifoamer dosing system	消泡剂加药系统 蒸发器

1.3.8 浓盐水蒸发系统流程图 Process Flow Diagram for Brine Water Evaporation System

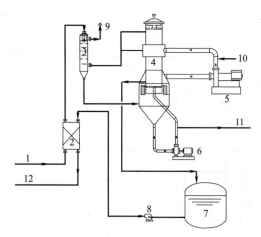

1　brine water　浓盐水
2　brine pre-heat exchanger　盐水预换热器
3　degasifier　脱气器
4　MVR evaporator　MVR 蒸发器
5　steam compressor　蒸汽压缩机
6　evaporator circulating pump　蒸发器循环泵
7　condensate water tank　冷凝液水箱
8　condensate water pump　冷凝液水泵
9　ambience emission　排入大气
10　steam　蒸汽
11　to crystallization system　去结晶系统
12　to reused water tank　去回用水池

1.3.9 浓盐水结晶系统流程图 Process Flow Diagram for Brine Water Crystallization System

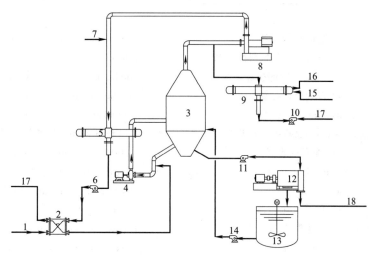

1　concentrated brine from evaporator system　来自蒸发系统浓缩液
2　feed pre-heat exchanger　进料预换热器
3　crystallizer　结晶器
4　crystallizer circulating pump　结晶器循环泵
5　crystallizer forced circulating heat exchanger　结晶器强制循环换热器
6　condensate water pump　冷凝水泵
7　steam　蒸汽
8　crystallizer steam compressor　结晶器蒸汽压缩机
9　condensator　冷凝器
10　condensate water pump　冷凝水泵
11　salt centrifuge feed pump　盐脱水机给料泵
12　salt centrifuge　盐离心脱水机
13　centrate tank　滤液罐
14　centrate pump　滤液泵
15　cooling water supply　冷却水给水
16　cooling water return　冷却水回水
17　condensate for reuse　冷凝水回用
18　to salt dryer　去干燥系统

1.3.10 反渗透除盐水工艺流程图 Process Flow Diagram for Demineralized Water System (Reverse Osmosis System)

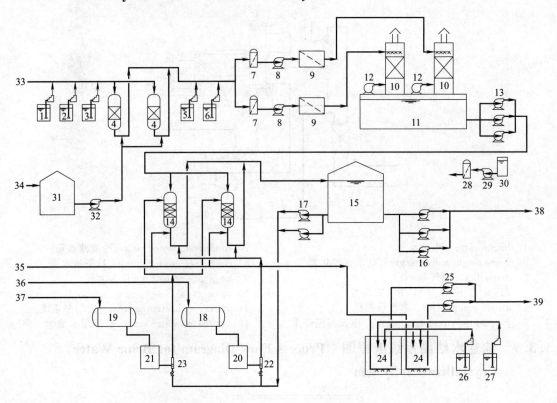

1 HCl feed system 盐酸投加系统
2 flocculant aid feed system 助凝剂投加系统
3 coagulant feed system 混凝剂投加系统
4 multimedia filter 多介质过滤器
5 antiscalant feed system 阻垢剂投加系统
6 NaHSO₃ feed system 亚硫酸氢钠投加系统
7 5 micron filter 5μ保安过滤器
8 high pressure pump 高压水泵
9 RO system 反渗透系统
10 degassing tower 除二氧化碳器
11 middle water tank 中间水箱
12 degassing tower blower 脱气塔风机
13 middle water pump 中间水泵
14 mixed ion exchanger 混合离子交换器
15 demineralized water tank 脱盐水箱
16 demineralized water pump 脱盐水泵
17 regenerating water pump 再生水泵
18 acid storage tank 酸储罐
19 NaOH storage tank 氢氧化钠储罐
20 acid metering tank 酸计量箱
21 NaOH metering tank 氢氧化钠计量箱
22 acid injector 酸喷射器
23 alkalic injector 碱喷射器
24 neutralization water basin 中和水池
25 neutralization water pump 中和水泵
26 acid feed system 酸投加系统
27 NaOH feed system 氢氧化钠投加系统
28 micron filter 清洗保安过滤器
29 cleaning water pump 清洗水泵
30 cleaning water tank 清洗水箱
31 backwash water tank 反洗水箱
32 backwash water pump 反洗水泵
33 raw water 原水
34 make-up water 补充水
35 plant air 工厂空气
36 acid 酸
37 NaOH 氢氧化钠
38 user 用户
39 drain 排放

1.3.11 锅炉房系统工艺流程图 Process Flow Diagram for Boiler House

1.3.11.1 锅炉房给水系统工艺流程图 Process Flow Diagram for Feed Water System of Boiler House

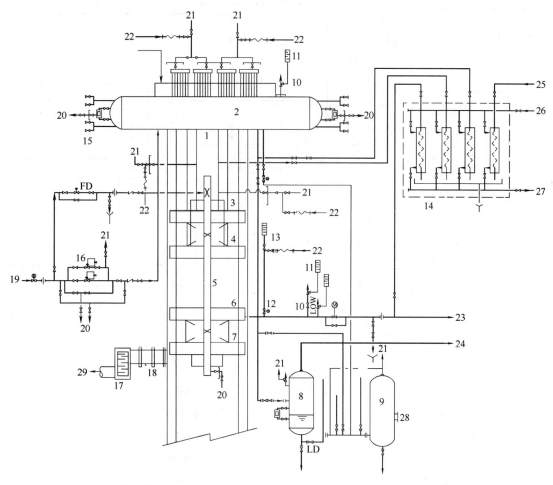

1   steam boiler   蒸汽锅炉
2   steam drum   上汽包
3   low temperature superheater inlet header   低温过热器进口联箱
4   low temperature superheater outlet header   低温过热器出口联箱
5   de-superheater   减温器
6   high temperature superheater inlet header   高温过热器进口联箱
7   high temperature superheater outlet header   高温过热器出口联箱
8   continuous blowdown tank（flash tank）   连排扩容器
9   intermittent blowdown tank（blow tank）   定排扩容器
10   safety valve   安全阀
11   silencer   消音器
12   vent valve   放空阀
13   vent silencer   放空消音器
14   sampling cooler   取样冷却器
15   level gauge   液位计
16   control valve   控制阀
17   air pre-heater   空气预热器
18   sampling flue gas   烟气取样
19   boiler feed water   锅炉给水
20   drain   排净
21   vent   放空
22   nitrogen   氮气
23   main steam   主蒸汽
24   flash steam   闪蒸汽
25   boiler feed water sample   锅炉给水取样
26   cooling water supply   冷却水给水
27   cooling water return   冷却水回水
28   manhole   人孔
29   flue gases to stack   烟道气去烟囱

1.3.11.2 锅炉房锅炉给水系统工艺流程图 Process Flow Diagram for Boiler Feed Water System of Boiler House

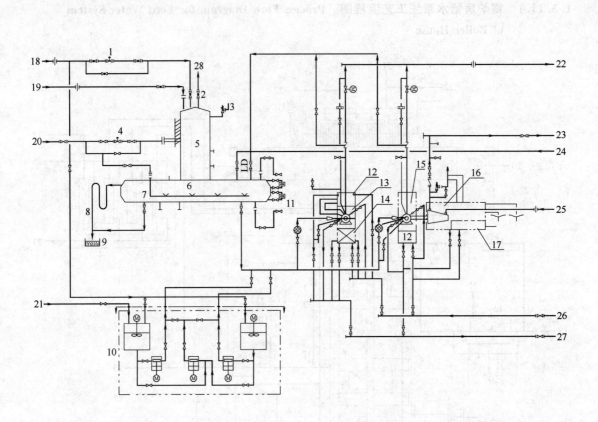

1	feed water control valve 给水控制阀	15	boiler feed water pump driven by turbine 汽机驱动锅炉给水泵
2	vent valve 放空阀	16	steam turbine 汽轮机
3	safety valve 安全阀	17	steam turbine local panel 汽机就地盘
4	steam control valve 蒸汽控制阀	18	demineralized water 脱盐水
5	deaerator 除氧器	19	return condensate 返回凝液
6	deaerator tank 除氧水箱	20	steam 蒸汽
7	startup heating line 启动加热管	21	ammonia water 氨水
8	water sealed device 水封装置	22	boiler feed water 锅炉给水
9	cooling pond 冷却水池	23	LPS (light pressure steam) 低压蒸汽
10	chemical injection unit 加药装置	24	flash steam 排污闪蒸汽
11	level gauge 液位计	25	MPS (medium pressure steam) 中压蒸汽
12	oil tank 滑油箱	26	cooling water supply 冷却水给水
13	boiler feed water pump driven by motor 电机驱动锅炉给水泵	27	cooling water return 冷却水回水
14	motor 电机	28	vent 放空

54

1.3.11.3 锅炉房锅炉燃烧系统工艺流程图 Process Flow Diagram for Boiler Combustion System of Boiler House

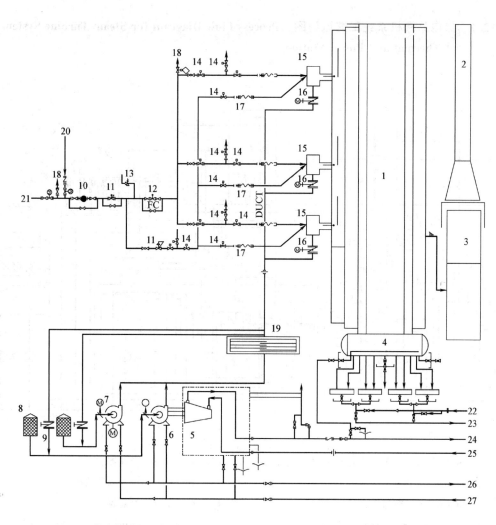

1	steam boiler 蒸汽锅炉		15	burner 燃烧器
2	stack 烟囱		16	damper 风门
3	air pre-heater 空气预热器		17	pilot igniter 点火器
4	water drum 下汽包		18	vent 放空
5	steam turbine for fresh fan 风机驱动用汽机		19	preheater 预热器
6	fresh fan driven by turbine 风机（汽机驱动）		20	nitrogen 氮气
7	fresh fan driven by motor 风机（电机驱动）		21	fuel gas 燃料气
8	air inlet filter silencer 空气过滤器消音器		22	boiler feed water 锅炉给水
9	air duct damper 风门		23	blowdown 排污水
10	strainer 过滤器		24	LPS 低压蒸汽
11	pressure regulating valve 调压器		25	MPS 中压蒸汽
12	control valve 控制阀（调节阀）		26	cooling water supply 冷却水给水
13	safety valve 安全阀		27	cooling water return 冷却水回水
14	solenoid valve 电磁阀			

55

1.3.12 热电站系统工艺流程图 Process Flow Diagram for Thermal and Power Station

1.3.12.1 热电站汽机系统工艺流程图 Process Flow Diagram for Steam Turbine System of Thermal and Power Station

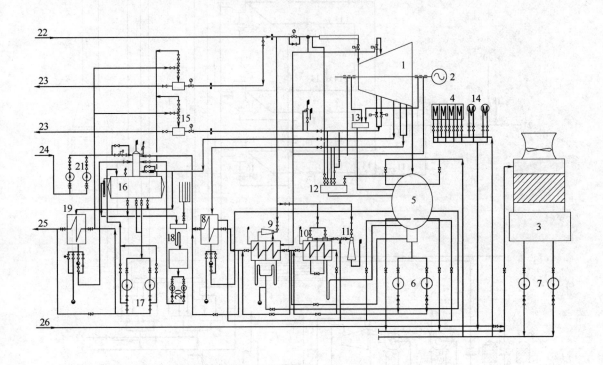

1	steam turbine （蒸）汽轮机	14	oil cooler 冷油器
2	generator 发电机	15	desuperheater 减温减压器
3	cooling tower 冷却塔	16	deaerator 除氧器
4	generator cooler 发电机冷却器	17	boiler feed water pump 锅炉给水泵
5	steam condenser 蒸汽冷凝器	18	condensate flash tank 疏水扩容器
6	condensate pump 凝结水泵	19	H.P. heater 高压加热器
7	circulating water pump 循环水泵	20	condesate return pump 疏水泵
8	L.P. heater 低压加热器	21	demineralized water pump 脱盐水泵
9	nozzle 喷射器	22	from boiler 来自锅炉
10	condensate heater 凝结水加热器	23	to outside piperack 去外管廊
11	nozzle for startup 启动喷射器	24	from demineralized water tank 来自脱盐水罐
12	pipe line condensate flash tank 管道凝结水闪蒸罐	25	to boiler 去锅炉
13	turbine condensate flash tank 透平凝结水闪蒸罐	26	raw water 原水

1.3.12.2 汽轮机空冷系统工艺流程图 Process Flow Diagram for Steam Turbine Air-Cooling Steam Condenser System

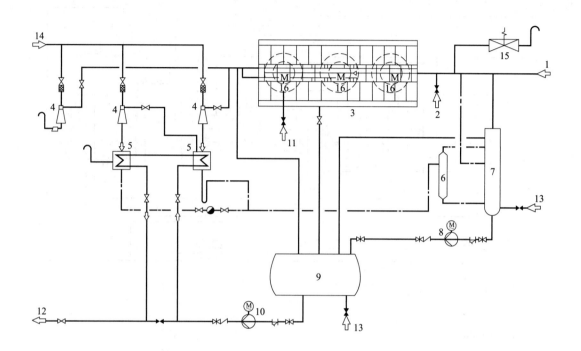

1	exhaust steam from turbine 汽轮机排汽	10	condensate water pump 凝结水泵
2	by-pass steam 旁路蒸汽	11	cleaning water 清洗水
3	air-cooled steam condenser 空冷凝气器	12	condensate water to deaerator or refining system 凝结水去往除氧器或精制系统
4	steam ejector 射汽抽气器		
5	condenser 抽气凝汽器	13	demineralized water 脱盐水
6	drainage flash tank 疏水膨胀箱	14	steam from header 蒸汽来自母管
7	condensate collecting tank 集液箱	15	exhaust steam safety valve 排汽安全阀
8	drainage pump 疏水泵	16	air fan 风扇
9	hot well 热井		

57

1.3.12.3 热电站锅炉系统工艺流程图 Process Flow Diagram for Boiler System of Thermal and Power Station

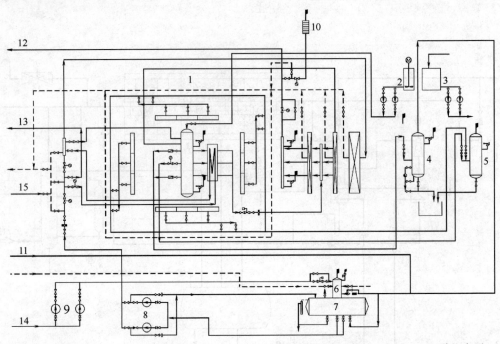

1	boiler 锅炉	9	demineralized water pump 脱盐水泵
2	chemical dosing system 加药系统	10	silencer 消音器
3	low level water tank 低位水箱	11	from HP heater 来自高压分气缸
4	continuous blowdown tank 连续排污扩容器	12	to steam turbine 去汽轮机（透平）
5	periodic blowdown tank 定期排污扩容器	13	to condensate flash tank 去凝液闪蒸罐
6	deaerator 除氧器	14	from demineralized water tank 来自脱盐水罐
7	deaerated water tank 除氧水箱		
8	boiler feed water pump 锅炉给水泵	15	chemical water 化学水

1.3.12.4 热电站锅炉燃烧系统工艺流程图 Process Flow Diagram for Boiler Combustion System of Thermal and Power Station

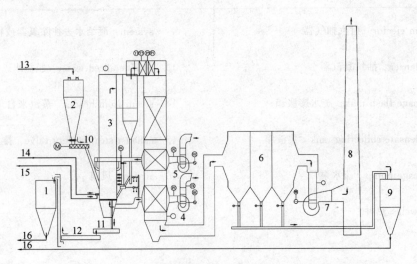

58

1	slag silo 渣仓	9	fly-ash silo 灰仓
2	coal silo 煤仓	10	coal conveyor 输煤机
3	CFB (circulating fluidized bed) boiler 循环流化床锅炉	11	slag cooler 冷渣器
		12	slag conveyor 输渣机
4	secondary air fan 二次风机	13	from coal convey system 来自输煤系统
5	forced draft fan 鼓风机	14	startup oil 开车油
6	electrostatic precipitator 静电除尘器	15	startup oil return 开车油返回
7	induced fan 引风机	16	truck 卡车
8	stack 烟囱		

1.3.12.5 氨法烟气脱硫工艺流程图 Process Flow Diagram for Flue Gas Desulfurization (FGD) System (Ammonia)

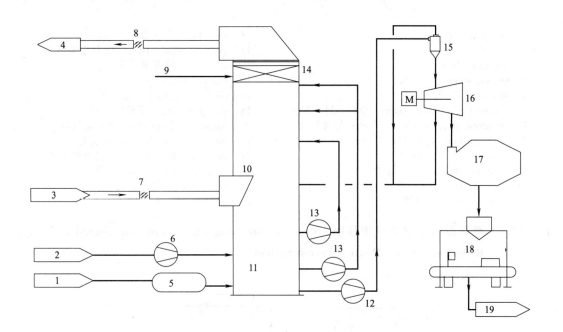

1	ammonia 氨	11	oxidation pond 氧化池
2	air 空气	12	bleed pump 产出泵
3	flue gas from draft fan 烟气来自引风机	13	circulating pump 循环泵
4	flue gas to stack 烟气去往烟囱	14	demister 除雾器
5	ammonia solution tank 氨罐	15	cyclone 旋流器
6	oxidation fan 氧化风机	16	centrifuge 离心机
7	entrance damper 入口挡板	17	drier 干燥机
8	exit damper 出口挡板	18	packager 包装机
9	flushing water 冲洗水	19	ammonium sulfate 成品硫铵
10	absorber 吸收塔		

59

1.3.12.6 石灰石/石灰-石膏法烟气脱硫工艺流程图 Process Flow Diagram for Flue Gas Desulfurization（FGD）System（Limestone/Lime-Gypsum）

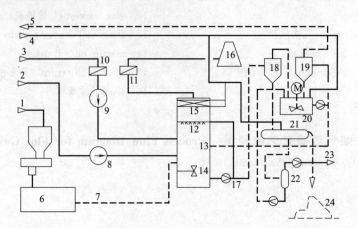

1	limestone 石灰石	13	absorber 吸收塔
2	air 空气	14	absorber slurry pond 吸收塔浆池
3	flue gas 烟气	15	demister 除雾器
4	process water 工艺水	16	stack 烟囱
5	waste water 废水	17	circulating pump 循环泵
6	absorbent preparation system 吸收剂制备系统	18	slurry cyclone 浆液旋流器
7	limestone slurry pipeline 石灰石加浆管线	19	waste water cyclone 废水旋流器
8	oxidation fan 氧化风机	20	filtered water tank 过滤水箱
9	booster fan 增压风机	21	vacuum belt filter 真空皮带脱水机
10	entrance damper 进口挡板	22	vacuum system 真空系统
11	exit damper 出口挡板	23	exhaust air 排出空气
12	spray levels 喷淋器	24	gypsum 石膏

1.3.12.7 烟气脱硝 SCR 系统流程图 Process Flow Diagram for Flue Gas Selective Catalytic Reduction（SCR）Denitrification System

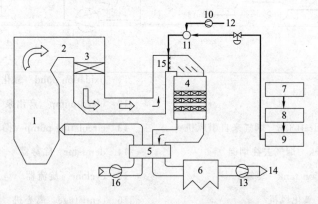

1	boiler 锅炉	9	ammonia preparation system 氨制备系统
2	flue gas 烟气	10	dilution air fan 稀释风机
3	economizer 省煤器	11	ammonia air mixer 烟气混合器
4	SCR reactor SCR 反应器	12	air 空气
5	air preheater 空气预热器	13	induced draft fan 引风机
6	dust catcher 除尘器	14	flue gas exit 烟气出口
7	liquid ammonia unloading system 氨卸料系统	15	ammonia inject grid（AIG） 氨喷射格栅
8	liquid ammonia storage system 氨存储系统	16	forced draft fan 送风机

1.3.12.8 尿素 SNCR 脱硝系统流程图 Process Flow Diagram for Flue Gas Urea Selective Non-Catalytic Reduction (SNCR) Denitrification System

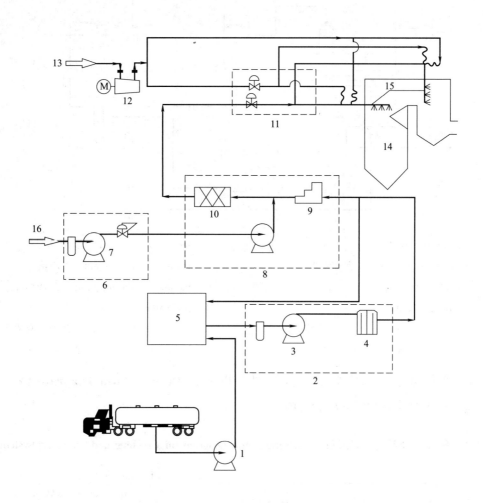

1　urea unloading pump　尿素卸载泵
2　supply and circulating module　供应循环模块
3　circulating pump　循环泵
4　electric heater　电加热器
5　urea storage tank　尿素存储罐
6　dilution water pressure control module　稀释水压力控制模块
7　water pump　水泵
8　metering module　计量模块
9　metering device　计量装置
10　mixer　混合器
11　distribution module　均分模块
12　air compressor　空压机
13　air　空气
14　boiler　锅炉
15　injection device　注入器
16　water　水

1.3.12.9 余热 ORC 发电系统工艺流程图 Process Flow Diagram for Waste Heat Organic Rankiee Cycle (ORC) Power Generation System

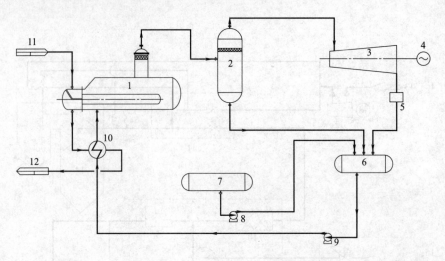

1　evaporator　蒸发器
2　gas-liquid separator　气液分离器
3　expander　膨胀机
4　generator　发电机
5　evaporative condenser　蒸发式冷凝器
6　condensate tank　凝液罐
7　medium storage tank　工质储存罐
8　medium delivery pump　工质输送泵
9　medium circulating pump　工质循环泵
10　preheater　预热器
11　hot source from B.L.　热源来自界区
12　hot source to B.L.　热源去往界区

1.3.13 物料处理系统典型工艺流程图 Typical Process Flow Diagram for Material Handling System

1.3.13.1 破碎筛分系统工艺流程图 Process Flow Diagram for Crushing and Screening System

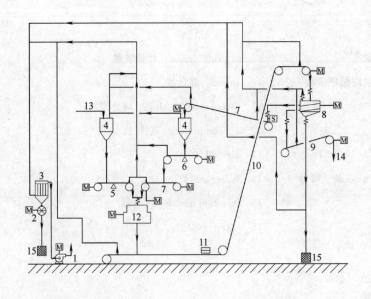

1　fan　通风机
2　rotary valve　旋转阀
3　dust collector　除尘器
4　bin　料仓
5, 6　weighing feeder　称重给料机
7, 9　belt conveyor　带式输送机
8　inclined vibrating screen　倾斜式振动筛
10　big slop belt conveyor　大倾角带式输送机
11　deironer　除铁器
12　crusher　破碎机
13　material　原料
14　product　产品
15　collector　收集箱

1.3.13.2 添加剂计量系统工艺流程图 Process Flow Diagram for Additive Metering System

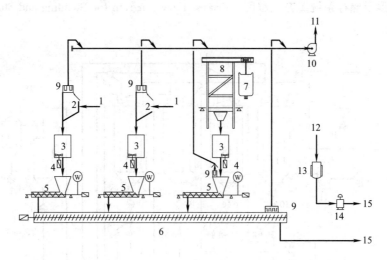

1	solid additive 固体添加剂		9	bag filter 袋式过滤器
2	bag dump station 倒袋站		10	fan 风机
3	solid additive surge bin 固体添加剂缓冲罐		11	vent 排放气
4	agitator 搅拌器		12	liquid additive 液体添加剂
5	loss-in-weight feeder 失重式给料机		13	liquid additive drum 液体添加剂罐
6	screw conveyor 螺旋输送机		14	metering pump 计量泵
7	big bag solid additive 大袋固体添加剂		15	downstream 下游
8	big bag unloading station 大袋卸料站			

1.3.13.3 挤压造粒系统工艺流程图 Process Flow Diagram for Extrusion Pelletizing System

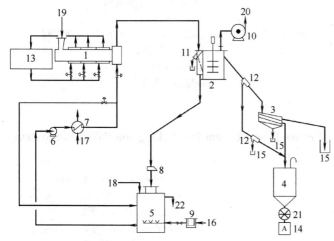

1	extruder 挤压造粒机		7	cutting water cooler 切粒水冷却器
2	centrifugal dryer 离心干燥器		8	cutting water filter 切粒水过滤器
3	pellets screen 筛分机		9	low-pressure steam filter 低压蒸汽过滤器
4	pneumatic transport hopper 气流输送系统缓冲料仓		10	wet air fan 湿空气抽风机
			11	pellets and water separator 粒水分离器
5	cutting water heater 切粒水箱		12	three-way valve 三通阀
6	cutting water pumps 切粒水泵		13	cooling/heating system 冷却/加热系统
			14	to pneumatic system 去风送系统
			15	off-size collector 不合格品收集器
			16	low pressure steam 低压蒸汽
			17	cooling water 冷却水
			18	DW (demineralized water) 脱盐水
			19	powder material 粉料
			20	vent 排放气
			21	rotary feeder 旋转给料器
			22	product crumb 产品碎屑

1.3.13.4 掺混及储存系统工艺流程图 Process Flow Diagram for Blending and Storing System

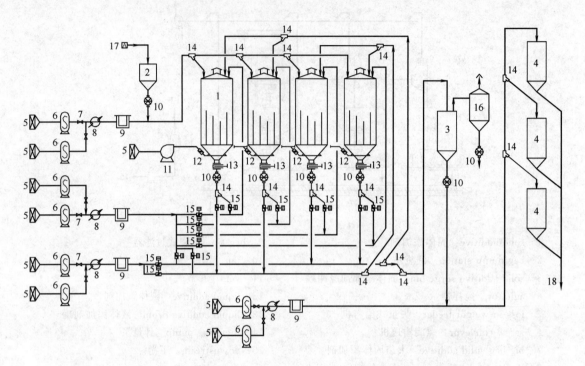

1	pellets homogenizing storage silo 颗粒均化储仓	10	rotary feeder 旋转给料器
2	pneumatic feeding hopper 气流输送系统喂料斗	11	silo venting blower 料仓吹扫鼓风机
3	elutriator 淘析器	12	butterfly valve 蝶阀
4	pellets storage silo 粒料储仓	13	slide gate valve 插板阀
5	intake filter 吸入口过滤器	14	diverter 换向阀
6	pellets pneumatic transport blower 粒料气流输送鼓风机	15	pneumatic butterfly valve 气动蝶阀
7	globe valve 截止阀	16	cyclone 旋风分离器
8	discharge cooler 空冷器	17	product pellets 粒料
9	filter 过滤器	18	to packing 去包装

1.3.13.5 包装码垛系统工艺流程图 Process Flow Diagram for Packing and Palletizing System

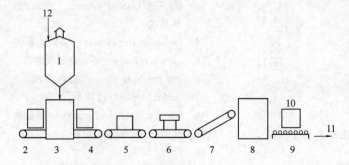

1	packing hopper 包装料斗	3	bagging machine 包装机
2	automatic bag placer 自动上袋机	4	check weigher & bag printing device 质量复检器和打印机
		5	metal detector 金属探测器
		6	bag rejector 剔袋器
		7	belt conveyor 带式输送机
		8	palletizing unit 码垛机
		9	conveyor 输送机
		10	shrink hood 收缩膜罩袋机
		11	to warehouse 去仓库
		12	product material 产品物料

1.3.13.6 煤储运系统工艺流程图　Process Flow Diagram for Coal Storage and Transport

(1) 筒仓储存　Bin Storage System

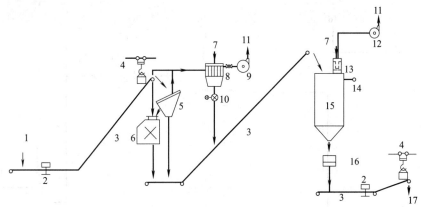

1	raw coal 原煤		10	rotary valve 旋转给料阀
2	electric belt scale 电子皮带秤		11	vent 排气
3	belt conveyor 带式输送机		12	filter fan 过滤器风机
4	magnetic iron separator 电磁除铁器		13	raw coal bin filter 原煤仓过滤器
5	vibrating screen 振动筛		14	level indicator 料位计
6	ring hammer crusher 环锤式破碎机		15	raw coal bin 原煤仓
7	impulse purge air 脉冲式吹扫空气		16	feeder 给料机
8	filter 除尘器		17	downstream 下游
9	filter fan 除尘器风机			

(2) 圆形堆料场储存　Circular Stacking Yard Storage

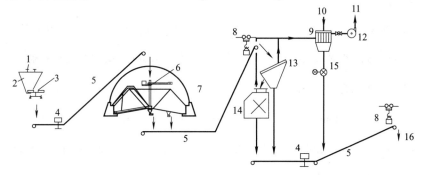

1	railway 火车轨道		9	filter 除尘器
2	coal discharge chute under the rail 轨下卸煤槽		10	impulse purge air 脉冲式吹扫空气
3	coal wheel feeder 叶轮给煤机		11	vent 排气
4	electric belt scale 电子皮带秤		12	filter fan 除尘器风机
5	belt conveyor 带式输送机		13	vibrating screen 振动筛
6	stacker and reclaimer 堆取料机		14	ring hammer crusher 环锤式破碎机
7	circular stacking yard 圆形堆场		15	feeder 给料机
8	magnetic iron separator 电磁除铁器		16	to downstream 去下游

1.3.13.7 磨煤干燥装置（CMD）工艺流程图　Process Flow Diagram for Coal Milling and Drying System

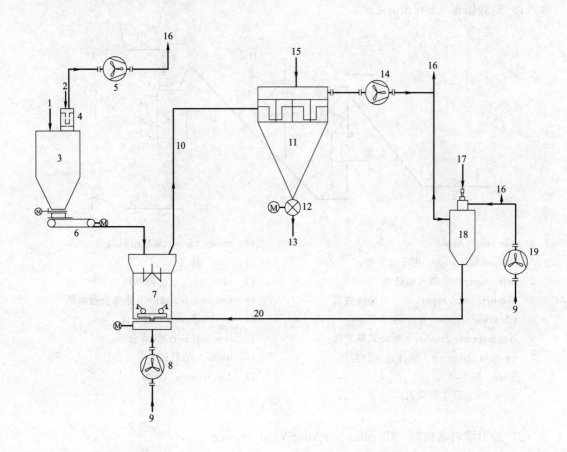

1	raw coal　原煤	11	pulverized coal collector　煤粉收集器
2	impulse purge air　脉冲式吹扫空气	12	rotary feeder　旋转给料器
3	raw coal bin　原煤仓	13	pulverized coal　煤粉
4	raw coal bin filter　原煤仓过滤器	14	circulation fan　循环风机
5	raw coal bin filter fan　原煤仓过滤器风机	15	impulse purge nitrogen　脉冲式吹扫氮气
6	gravimetric coal feeder　称重给煤机	16	vent　排气
7	coal mill　磨煤机	17	fuel gas　燃料气
8	coal mill seal fan　磨煤机密封风机	18	hot gas generator　热风炉
9	air　空气	19	combustion air fan　燃烧空气风机
10	pulverized coal＋inert gas　煤粉＋惰性气体	20	hot inert gas　热惰性气体

1.4 工艺流程模拟计算 Process Flowsheet Simulation

1.4.1 工艺流程图 Process Flowsheet

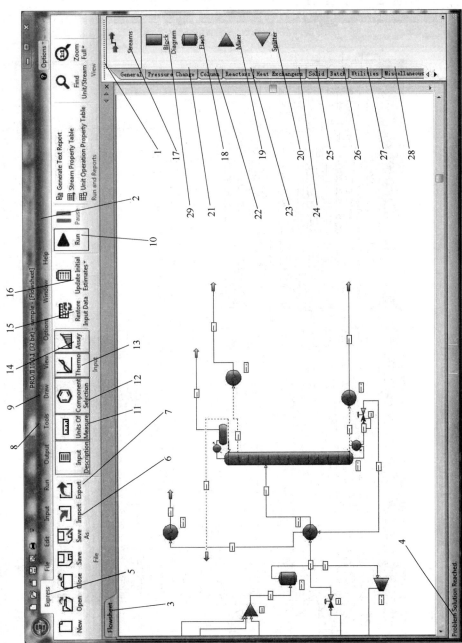

1　Unit Operation Palette　操作模块
2　Menu Bar　菜单条
3　Flowsheet　流程图
4　Status　状态
5　Express　快捷
6　Import　导入
7　Export　导出
8　Tools　工具
9　Draw　绘图
10　Run　运行
11　Units of Measure　计量单位
12　Component Selection　组分选择
13　Thermo　热力学模型
14　Assay　石油馏分分析
15　Restore Input Data　重置输入
16　Update Initial Estimates　更新初始估计值
17　Stream　流股
18　Flash　闪蒸
19　Mixer　混合器
20　Splitter　流股分支
21　Pressure Change　压力变化
22　Column　塔器
23　Reactors　反应器
24　Heat Exchanger　换热器
25　Solid　固体
26　Batch　间歇单元
27　Utilities　公用程序
28　Miscellaneous　其他
29　Block Diagram　模块图

1.4.2 热力学模型 Thermodynamic Model

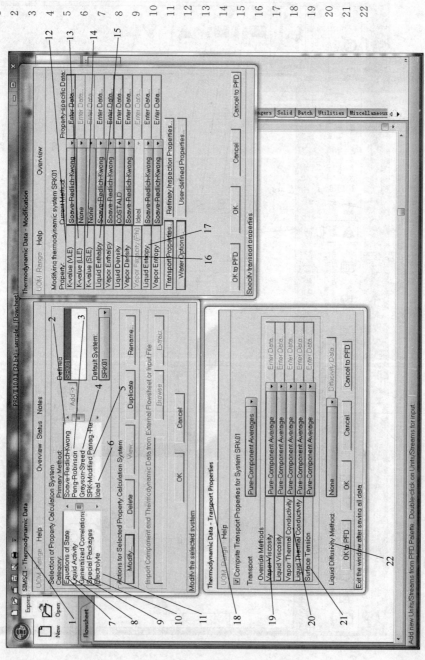

1 Thermodynamic Data 热力学数据
2 Soave-Redlich-Kwong SRK 状态方程
3 Peng-Robinson PR 状态方程
4 Grayson-Streed GS 通用关联式
5 SRK-Modified SRK 修正方程
6 Ideal 理想气体法
7 Equations of State 状态方程
8 Liquid Activity 液相活度
9 Generalized Correlations 通用关联式
10 Special Packages 专用包
11 Electrolyte 电解质
12 K-value (VLE) K 值 (气液平衡)
13 K-value (LLE) K 值 (液液平衡)
14 K-value (SLE) K 值 (固液平衡)
15 Enthalpy 焓
16 Density 密度
17 Entropy 熵
18 Transport Properties 传递性质
19 Viscosity 黏度
20 Thermal Conductivity 热导率
21 Surface Tension 表面张力
22 Diffusivity Method 扩散系数法

1.4.3 模型修正 Model Modify

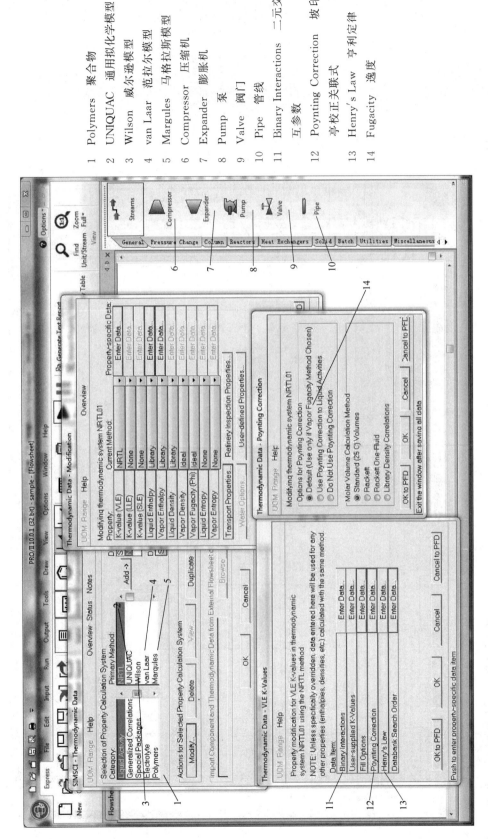

1	Polymers	聚合物
2	UNIQUAC	通用拟化学模型
3	Wilson	威尔逊模型
4	van Laar	范拉尔模型
5	Margules	马格拉斯模型
6	Compressor	压缩机
7	Expander	膨胀机
8	Pump	泵
9	Valve	阀门
10	Pipe	管线
11	Binary Interactions	二元交互参数
12	Poynting Correction	坡印亭校正关联式
13	Henry's Law	亨利定律
14	Fugacity	逸度

1.4.4 用户输入 User Input

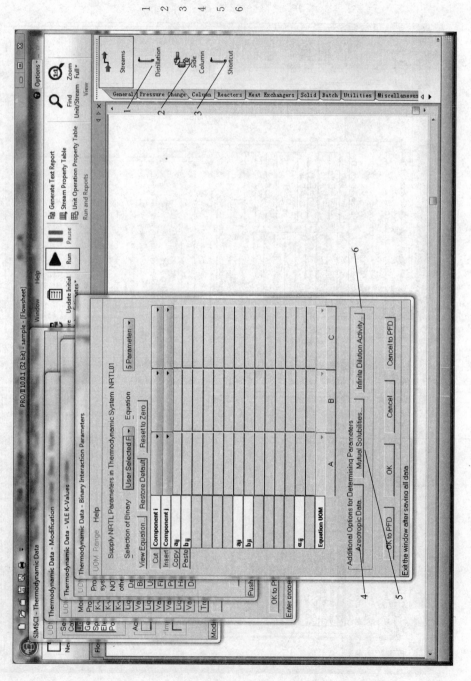

1 Distillation 蒸馏
2 Side Column 侧线塔
3 Shortcut 简捷塔
4 Azeotropic Data 恒沸点数据
5 Mutual Solubilities 互溶性
6 Infinite Dilution Activity 无限稀释活度

1.4.5 石油馏分数据 Petroleum Component Data

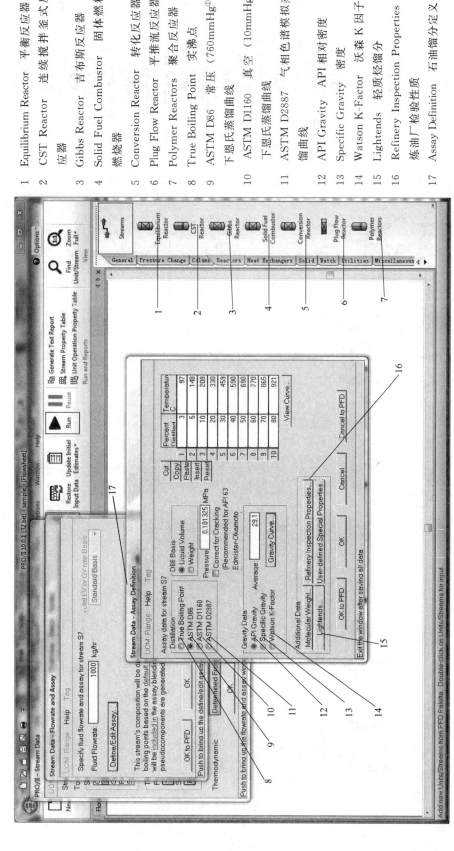

1　Equilibrium Reactor　平衡反应器
2　CST Reactor　连续搅拌釜式反应器
3　Gibbs Reactor　吉布斯反应器
4　Solid Fuel Combustor　固体燃料燃烧器
5　Conversion Reactor　转化反应器
6　Plug Flow Reactor　平推流反应器
7　Polymer Reactors　聚合反应器
8　True Boiling Point　真沸点
9　ASTM D86　常压（760mmHg[①]）下恩氏蒸馏曲线
10　ASTM D1160　真空（10mmHg）下恩氏蒸馏曲线
11　ASTM D2887　气相色谱模拟蒸馏曲线
12　API Gravity　API 相对密度
13　Specific Gravity　密度
14　Watson K-Factor　沃森 K 因子
15　Lightends　轻质轻馏分
16　Refinery Inspection Properties　炼油厂检验性质
17　Assay Definition　石油馏分定义

① $1\text{mmHg}=1.013\times10^5\text{Pa}$。

1.4.6 蒸馏塔 Distillation Column

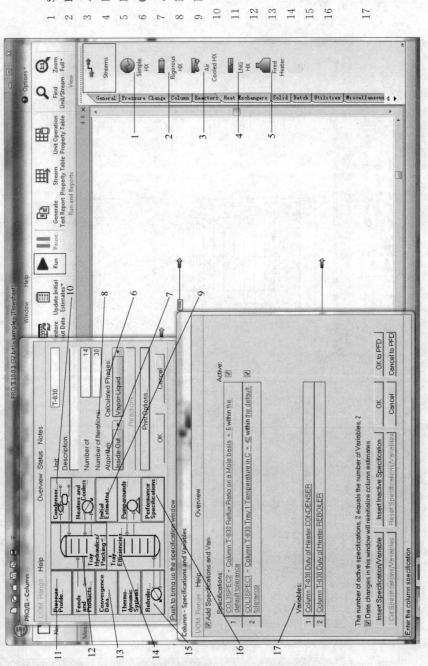

1　Simple HX　换热器简捷计算
2　Rigorous HX　换热器严格计算
3　Air Cooled HX　空冷器
4　LNG HX　多股流换热器
5　Fired Heater　加热炉
6　Calculated Phases　计算相态
7　Algorithm　算法
8　Number of Iterations　迭代次数
9　Initial Estimates　初值估计
10　Condenser　冷凝器
11　Pressure Profile　压力剖面
12　Tray Hydraulics　塔板水力学
13　Convergence　收敛
14　Tray Efficiencies　板效率
15　Reboiler　再沸器
16　Performance Specifications　性能规定
17　Variables　变量

1.4.7 尺寸设计和校核 Sizing and Rating

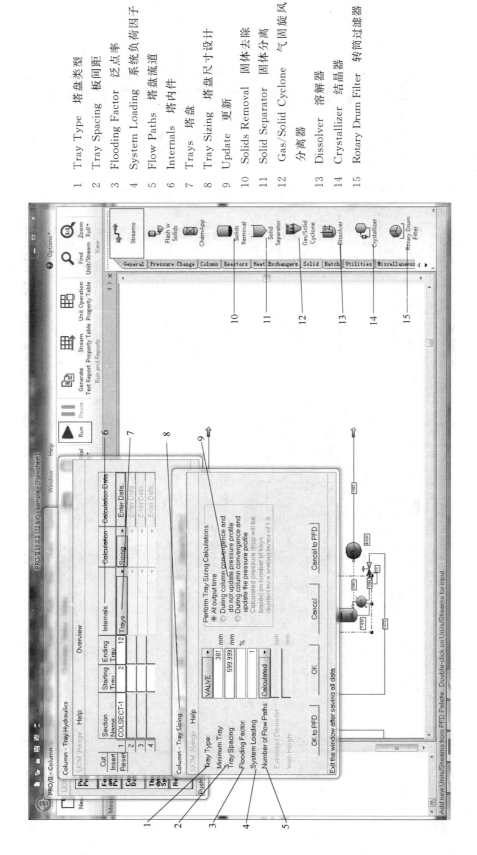

1　Tray Type　塔盘类型
2　Tray Spacing　板间距
3　Flooding Factor　泛点率
4　System Loading　系统负荷因子
5　Flow Paths　塔盘流道
6　Internals　塔内件
7　Trays　塔盘
8　Tray Sizing　塔盘尺寸设计
9　Update　更新
10　Solids Removal　固体去除
11　Solid Separator　固体分离
12　Gas/Solid Cyclone　气固旋风分离器
13　Dissolver　溶解器
14　Crystallizer　结晶器
15　Rotary Drum Filter　转筒过滤器

1.4.8 收敛数据 Convergence Data

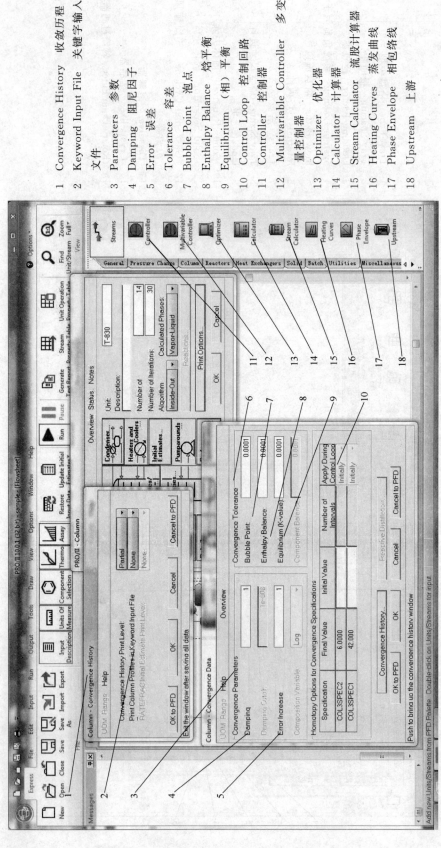

1. Convergence History 收敛历程
2. Keyword Input File 关键字输入文件
3. Parameters 参数
4. Damping 阻尼因子
5. Error 误差
6. Tolerance 容差
7. Bubble Point 泡点
8. Enthalpy Balance 焓平衡
9. Equilibrium （相）平衡
10. Control Loop 控制回路
11. Controller 控制器
12. Multivariable Controller 多变量控制器
13. Optimizer 优化器
14. Calculator 计算器
15. Stream Calculator 流股计算器
16. Heating Curves 蒸发曲线
17. Phase Envelope 相包络线
18. Upstream 上游

1.4.9 优化 Optimization

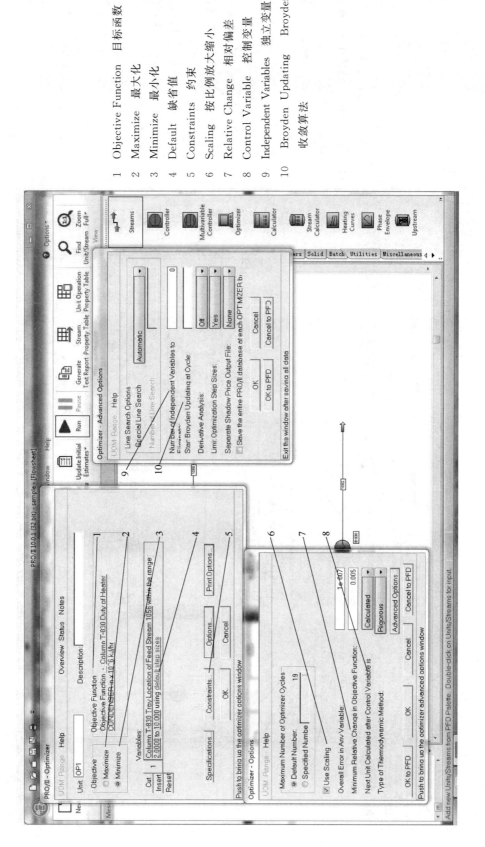

1 Objective Function 目标函数
2 Maximize 最大化
3 Minimize 最小化
4 Default 缺省值
5 Constraints 约束
6 Scaling 按比例放大缩小
7 Relative Change 相对偏差
8 Control Variable 控制变量
9 Independent Variables 独立变量
10 Broyden Updating Broyden 收敛算法

1.4.10 流程收敛 Flowsheet Convergence

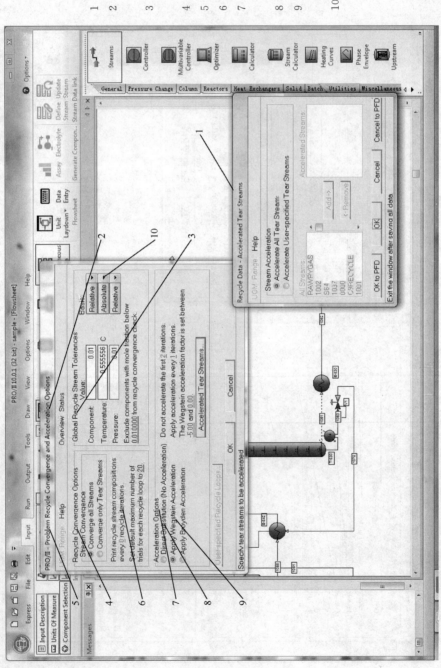

1　Tear Streams　切断流股
2　Acceleration　迭代计算加速收敛法
3　Global Recycle Stream　全局循环流股
4　Recycle Convergence　循环收敛
5　Problem　题目
6　recycle iterations　循环迭代
7　Acceleration Options　加速收敛方法选择
8　Direct Substitution　直接迭代
9　Wegstein Acceleration　威格斯坦加速收敛法
10　Absolute　绝对（答差）

1.4.11 用户定义 User Interface

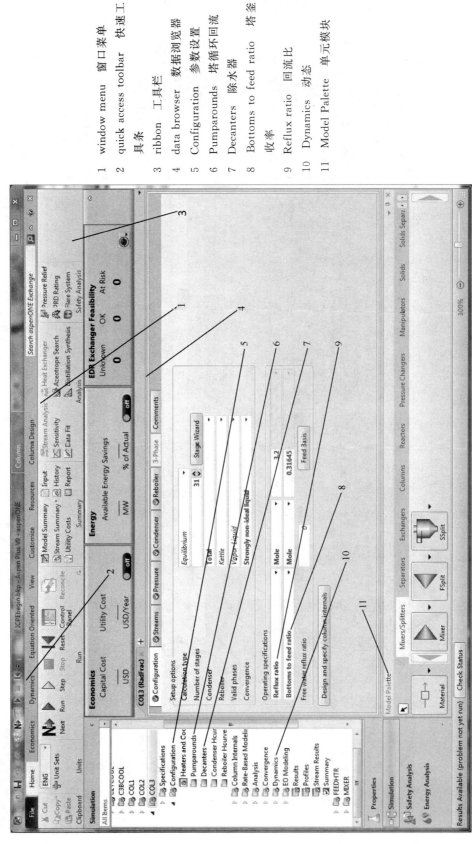

1　window menu　窗口菜单
2　quick access toolbar　快速工具条
3　ribbon　工具栏
4　data browser　数据浏览器
5　Configuration　参数设置
6　Pumparounds　塔循环回流
7　Decanters　除水器
8　Bottoms to feed ratio　塔釜收率
9　Reflux ratio　回流比
10　Dynamics　动态
11　Model Palette　单元模块

1.4.12 结果汇总 Result Summary

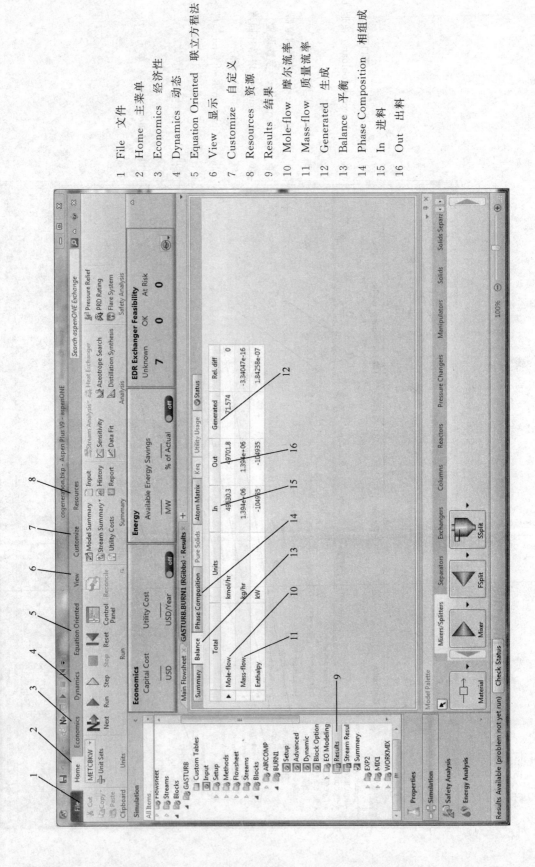

1　File　文件
2　Home　主菜单
3　Economics　经济性
4　Dynamics　动态
5　Equation Oriented　联立方程法
6　View　显示
7　Customize　自定义
8　Resources　资源
9　Results　结果
10　Mole-flow　摩尔流率
11　Mass-flow　质量流率
12　Generated　生成
13　Balance　平衡
14　Phase Composition　相组成
15　In　进料
16　Out　出料

1.4.13 设置选项 Setup Option

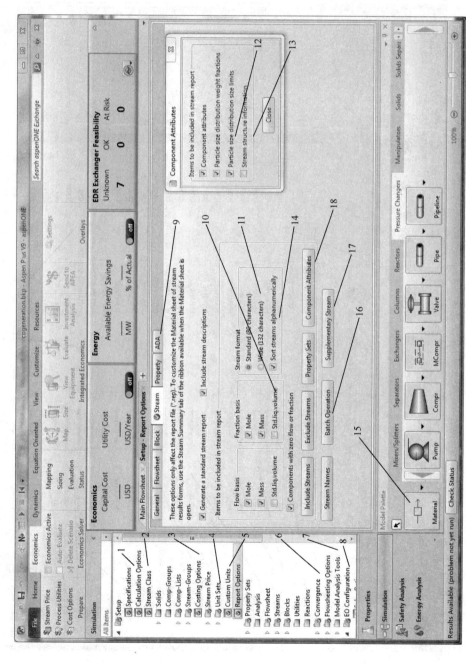

1 Calculation Options 计算选项
2 Stream Class 物流分类
3 Costing Options 费用选项
4 Unit Sets 单位集
5 Custom Units 自定义单位
6 Convergence 收敛
7 Model Analysis Tools 模型分析工具
8 EO Configuration 联立方程法参数设置
9 ADA (assay data analysis) 试验数据分析
10 Standard liquid volume 标准液体体积
11 Wide 宽幅
12 Particle size distribution 固体粒径分布
13 Stream structure 流股结构
14 Sort streams alphanumerically 按照字母数字顺序归类流股
15 Material stream 物质流
16 Batch Operation 间歇操作
17 Supplementary Stream 补充流股说明
18 Component Attributes 组分属性

1.4.14 用户定制模拟程序 Customizing Simulation Engine

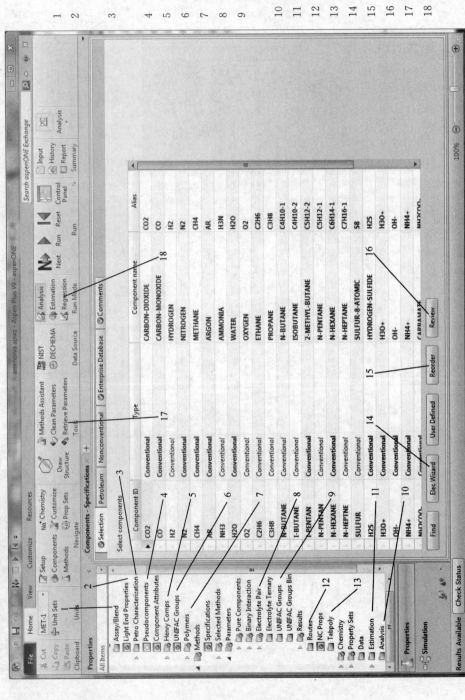

1 Assay/Blend 油品/调和
2 Light End Properties 轻馏分性质
3 Petro Characterization 油品特性
4 Pseudocomponents 虚拟组分
5 Henry Comps 亨利组分
6 UNIFAC Groups 基团
7 Polymers 聚合物
8 Electrolyte Pair 电解质离子对
9 Electrolyte Ternary 三元电解质子组
10 Analysis 分析
11 Estimation 估计
12 Routes 调用子程序
13 Tabpoly 表列多项式
14 Elec Wizard 电解质向导
15 Reorder 重排序
16 Review 检查
17 Retrieve 检索
18 Regression 回归

1.4.15 塔内件 Column Internals

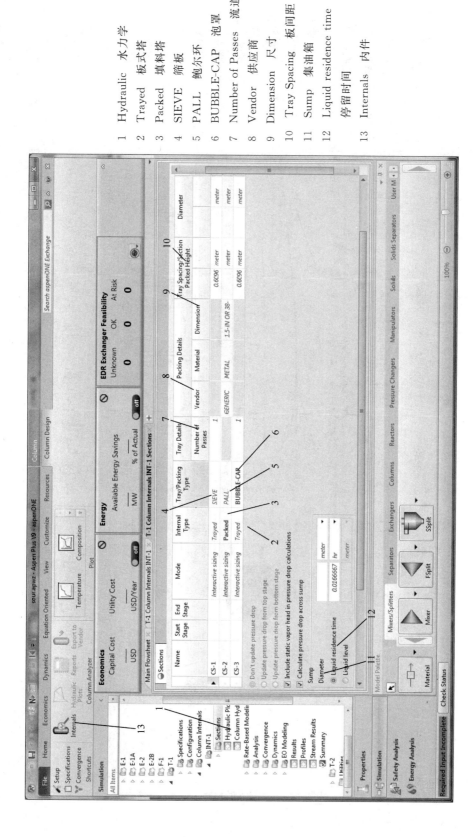

1 Hydraulic 水力学
2 Trayed 板式塔
3 Packed 填料塔
4 SIEVE 筛板
5 PALL 鲍尔环
6 BUBBLE-CAP 泡罩
7 Number of Passes 流道数
8 Vendor 供应商
9 Dimension 尺寸
10 Tray Spacing 板间距
11 Sump 集油箱
12 Liquid residence time 液体停留时间
13 Internals 内件

1.4.16 负荷性能图 Hydraulic Plots

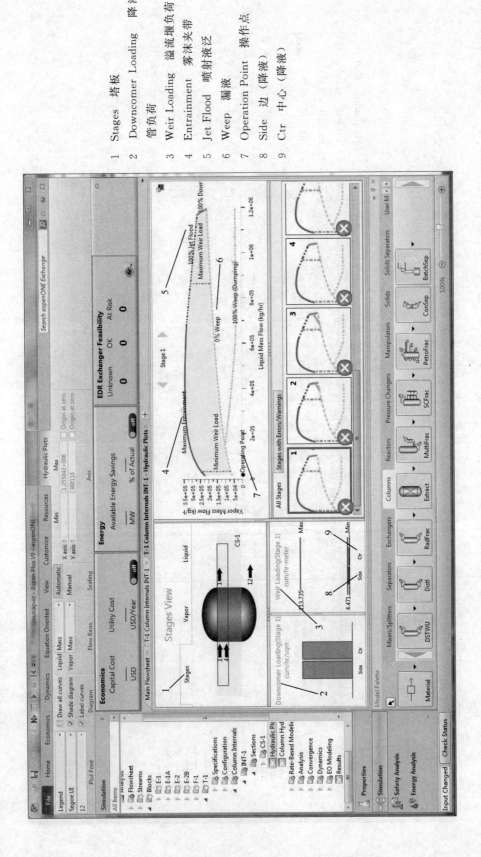

1　Stages　塔板
2　Downcomer Loading　降液管负荷
3　Weir Loading　溢流堰负荷
4　Entrainment　雾沫夹带
5　Jet Flood　喷射液泛
6　Weep　漏液
7　Operation Point　操作点
8　Side　边（降液）
9　Ctr　中心（降液）

1.4.17 流动场分析 Computational Fluid Dynamics

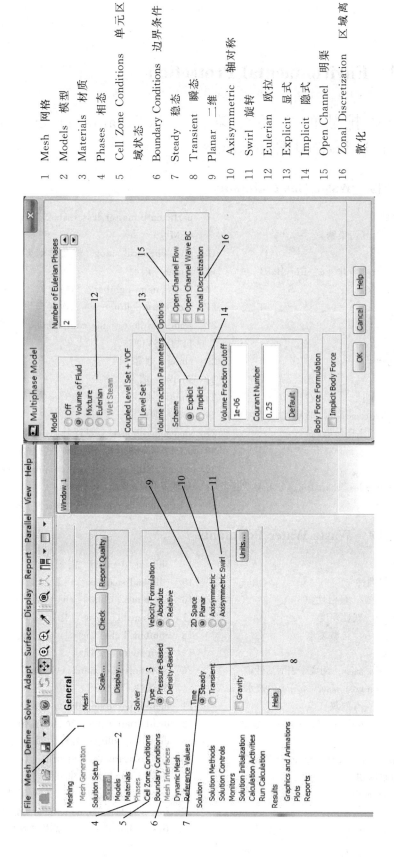

1　Mesh　网格
2　Models　模型
3　Materials　材质
4　Phases　相态
5　Cell Zone Conditions　单元区域状态
6　Boundary Conditions　边界条件
7　Steady　稳态
8　Transient　瞬态
9　Planar　三维
10　Axisymmetric　轴对称
11　Swirl　旋转
12　Eulerian　欧拉
13　Explicit　显式
14　Implicit　隐式
15　Open Channel　明渠
16　Zonal Discretization　区域离散化

2 环境保护 Environmental Protection

2.1 化工企业排放的主要污染物 Pollutants Discharged by Chemical Industrial Enterprise

2.1.1 废气污染物 Waste Gas Pollutants

1 aerosol 气溶胶
2 air pollutants 大气污染物
3 ash 灰
4 benzo [α] pyrene 苯并[α]芘（BaP）
5 dioxin 二噁英
6 droplet 液滴
7 dust 粉尘
8 fly ash 飞灰
9 fluoride 氟化物
10 fume 烟尘
11 grit 尘粒
12 non-methane hydrocarbon 非甲烷总烃
13 odor pollutants 恶臭污染物
14 ozone 臭氧
15 particle 颗粒
16 particulate matter less than $10\mu m$ 可吸入颗粒物（PM_{10}）
17 particulate matter less than $2.5\mu m$ 细颗粒物（$PM_{2.5}$）
18 photochemical smog 光化学烟雾
19 plume 烟羽
20 Ringelmann number 林格曼数
21 secondary pollutants 二次污染物
22 semi-volatile organic compounds 半挥发性有机物（SVOC）
23 smog 烟雾
24 smoke 烟
25 smuts 烟炱
26 soot 烟粒
27 suspended matter 悬浮物质
28 total suspended particulate 总悬浮颗粒物
29 total volatile organic compounds 总挥发性有机物（TVOC）
30 volatile organic compounds 挥发性有机物（VOCs）

2.1.2 废水污染物 Waste Water Pollutants

1 acidity 酸度
2 aggressivity 侵蚀性
3 albuminoid nitrogen 蛋白性氮
4 alkaline hardness 碱性硬度
5 alkaline wastewater 含碱废水
6 alkalinity 碱度
7 anionic surface active agent 阴离子表面活性剂
8 aromatic hydrocarbon wastewater 含苯系物废水
9 available chlorine 有效氯
10 biochemical oxygen demand（BOD） 生化需氧量
11 carbon adsorption chloroform extraction（CCE） 炭吸附-氯仿萃取物
12 cationic surface active agent 阳离子表面活性剂
13 chemical oxygen demand（COD） 化学需氧量
14 chlorine demand 需氯量
15 colloidal suspension 胶态悬浮物
16 combined available chlorine 化合有效氯
17 combined chlorine 化合氯
18 combined chlorine residual 化合余氯
19 corrosivity 腐蚀
20 dichromate demand（COD_{Cr}） 重铬酸盐需氧量
21 dichromate oxidizability 重铬酸盐氧化性
22 dichromate value 重铬酸盐值
23 dissolved solids 溶解性固体
24 eutrophication 富营养化
25 flush water 地面冲洗水
26 free available chlorine 游离有效氯
27 free available chlorine residual 游离有效余氯
28 free carbon dioxide 游离二氧化碳

29	free chlorine residual 游离余氯		46	sludge volume index (SVI) 污泥容积指数
30	hardness 硬度		47	solids 固体
31	Kjeldahl nitrogen 凯氏氮		48	sour water 含硫含氨酸性水
32	living wastewater 生活污水		49	specific conductance 电导率
33	methyl red end Point alkalinity 甲基红碱度		50	surface active agent 表面活性剂
34	nitrogen compounds 氮的化合物		51	surfactant 表面活化剂
35	non-alkaline hardness 非碱性硬度		52	temporary hardness 暂时硬度
36	non-ionic surface active agent 非离子型表面活性剂		53	total ammonia 总氨
			54	total available chlorine 总有效氯
37	permanent hardness 永久性硬度		55	total carbon dioxide 总二氧化碳
38	permanganate demand (COD$_{Mn}$) 高锰酸盐需氧量		56	total organic carbon (TOC) 总有机碳
39	permanganate oxidizability 高锰酸盐氧化性		57	total oxygen demand (TOD) 总需氧量
40	permanganate value 高锰酸盐值		58	total organic nitrogen 总有机氮
41	phenolphthalein end Point alkalinity 酚酞碱度		59	total residual chlorine 总余氯
42	polluted rainwater 污染雨水		60	total solids 总固体
43	process wastewater 工艺废水		61	turbidity 浊度
44	residual chlorine 余氯		62	unionized free ammonia 游离氨
45	salinity 盐度		63	volatile organic liquid 挥发性有机液体

2.1.3 固体废弃物 Waste Solid

1	domestic waste solid 生活垃圾		3	leachate 渗滤液
2	hazardous wastes 危险废物		4	waste solid 固体废弃物

2.1.4 其他 Others

1	approach zero discharging for wastewater 废水近零排放		9	non-organized emission 无组织排放
2	background noise 背景噪声		10	online monitoring 在线监测
3	concentration at battery limits 厂界浓度		11	organized emission 有组织排放
			12	pollutant emission amount 污染物排放量
4	discharge outlet 排放口		13	pollutant emission concentration 污染物排放浓度
5	emission height 排放高度		14	railway noise 铁路噪声
6	flue gas volume 烟气排放量		15	stack 排气筒
7	manual monitoring 手动监测		16	super-low emission for flue gas 烟气超低排放
8	noise 噪声		17	total amount control factors 总量控制指标

2.2 环境质量 Environmental Quality

2.2.1 环境空气质量 Ambient Environmental Quality

1	Air Pollution Index (API) 空气污染指数		8	monthly mean 月平均
2	air quality 空气质量		9	PM$_{10}$, Particular matter less than 10μm 可吸入颗粒物
3	Air Quality Index (AQI) 空气质量指数			
4	ambient air 环境空气		10	quarterly average 季平均
5	daily average concentration 日平均浓度		11	standard condition 标准状态
6	hourly average concentration 一小时平均浓度		12	TSP, Total Suspended Particular 总悬浮颗粒物
7	limited value 限值		13	yearly average concentration 年平均浓度

2.2.2 地表水环境质量 Surface Water Environmental Quality

1　potable water　饮用水
2　surface water　地表水
3　water body　水体
4　water quality　水质
5　water quality assessment　水质评价
6　water quality monitoring　水质监测
7　water sources　水源

2.2.3 声环境质量 Noise Environmental Quality

1　absorbing coefficient　吸收系数
2　burst noise　突发噪声
3　day-time equivalent sound level　昼间等效声级
4　decay　衰减
5　decibel　分贝
6　direction factor　方向性因子
7　equal A noise level　等效 A 声级
8　maximum sound level　最大声级
9　night-time equivalent sound level　夜间等效声级
10　noise　噪声
11　noise-sensitive building　噪声敏感建筑物
12　noise level　声级
13　noise power level　声功率级
14　noise pressure level　声压级
15　percentile sound level　累积百分声级
16　weigh equal continuous perceive noise level（WECPNL）　计权等效连续感觉噪声级

2.2.4 其他 Others

1　environmental baseline value　环境背景值
2　environmental capacity　环境容量
3　environmental carrying capacity　环境承载力
4　key pollution source under national monitoring program　国控重点污染源
5　industrial pollution source　工业污染源
6　pollution discharge report and registration　排污申报登记
7　pollutant discharge permit　排污许可证
8　three simultaneousness system of construction projects and proposes solution　三同时制度
9　total amount control of pollutants　污染物排放总量控制
10　double among reduction　倍量削减

2.3 化工企业三废污染控制 Chemical Industrial Enterprise Wastes Pollution Control

2.3.1 大气污染控制 Air Pollution Control

1　abatement　抑制
2　chimney effect　烟囱效应
3　cut-off point　截止点
4　dry adiabatic lapse rate　干绝热递减率
5　effective chimney height　有效烟囱高度
6　elutriation　水平淘洗法
7　emission　排放
8　emission flux　排放通量
9　emission rate　排放速率
10　equivalent diameter　当量直径
11　ground level concentration　地面浓度
12　immission　接受
13　immission dose　接受剂量
14　immission flux　接受通量
15　immission rate　接受速率
16　impinger　冲击式采样器
17　isokinetic sampling　等速采样
18　lapse rate　递减率
19　micrometeorology　微气象学
20　monitoring　监测
21　photochemical reaction　光化学反应
22　probe　探头
23　rain out　雨洗效应
24　retention efficiency　保留效率
25　scavenging　自净
26　scrubbing　涤气法
27　source　排放源
28　transmission　迁移
29　wash out　洗脱

2.3.2 工艺废水污染控制 Process Wastewater Pollution Control

1　abiotic degradation（non-biological degradation）　非生物降解
2　activated sludge　活性污泥
3　activated sludge treatment　活性污泥处理
4　aeration　曝气
5　aerobic　需氧的、好氧的（细菌）

6	aerobic sludge digestion 需氧污泥消化	64	non-point source 非点污染源
7	algae 藻类	65	odor threshold 嗅阈
8	ammonification 氨化作用	66	oxidation pond 氧化塘
9	anaerobic 厌氧的（细菌）	67	oxidation reduction potential (redox potential ORP) 氧化还原电位
10	anaerobic sludge digestion 污泥厌氧消化		
11	antagonism 拮抗作用	68	oxygen sag curve 氧垂曲线
12	bacteria 细菌	69	ozonation 臭氧处理
13	bacteria bed 细菌滤床	70	percolating filter 渗滤池
14	bacteriological sample 细菌样品	71	permeability 渗透性
15	bacteriophage 噬菌体	72	pH equilibrium pH 平衡
16	balancing tank 均衡池	73	photo autorophic bacteria 光能自养菌
17	benthic region 水底区	74	physicochemical treatment 物理化学处理
18	bioaccumulation 生物积累	75	polyelectrolytes 聚合电解质
19	biological filter 生物滤池	76	prechlorination 预氯化
20	centrifuging 离心	77	preliminary treatment (of sewage) 预处理（污水）
21	chemical coagulation 化学混凝		
22	chemical treatment 化学处理	78	primary degradation 初级降解
23	coliform organisms 大肠杆菌	79	primary production 初级生产力
24	compartmentalization 相迁移	80	primary treatment (of sewage) 初级处理（污水）
25	conductivity (electrical conductivity) 电导率	81	pulse dose 脉冲剂量
		82	raw sludge 生污泥
26	de-aeration 除空气	83	readily biodegradable substances 易生物降解物质
27	dechlorination 脱氯	84	regeneration (ion exchange) 离子交换材料的再生
28	degasification 除气		
29	deionifiction 去离子	85	respiration 呼吸
30	demineralization 去矿化（脱矿质）	86	returned activated sludge 回流活性污泥
31	denitrification 反硝化（脱硝）	87	reverse osmosis 反渗透
32	deoxygenation 除氧	88	salmonis (fish) 鲑类（鱼）
33	desalination 脱盐	89	scale deposit 水垢
34	dewatering 脱水	90	screen 滤筛
35	digestion 消化	91	secondary treatment 二级处理
36	disinfection 消毒	92	sedimentation 沉降
37	distillation 蒸馏	93	self-purification 自净
38	effluent 流出物	94	septic tank 化粪池
39	electrodialysis 电渗析	95	service reservoir 配水池
40	Escherichia coli (E. coli) 大肠埃希氏菌	96	settled sewage 沉淀后的污水
41	filter run 过滤周期	97	sewage 污水
42	filtration 过滤	98	sewage fungi 污水真菌
43	floatation 浮选	99	sink 污汇点
44	floc 絮凝物	100	sludge 污泥
45	flocculation 凝聚	101	sludge cake 污泥饼
46	flocculation aid 助凝剂	102	sludge conditioning 污泥调节
47	fluoridation 氟化	103	sludge thickening 污泥浓缩
48	freshwater limit 淡水限界	104	softening 软化
49	gathering ground 集水区	105	sparging 鼓泡
50	grey water 灰水	106	spray aeration 喷雾曝气
51	half-life period 半衰期	107	sterilization 灭菌
52	hard detergent 硬洗涤剂	108	sulfur bacteria 硫细菌
53	ion exchange 离子交换	109	superchlorination 过氯化
54	ion-exchange material 离子交换材料	110	surface loading rate 表面负荷率
55	lysimeter 测渗计	111	swallow hole 落水洞
56	mesphilic micro-organisms 嗜温微生物	112	synergism 协同作用
57	metanmnion 斜温层	113	thermalwater 热泉水
58	methaemoglobinaemia 高铁血红蛋白血症	114	thermophilic digestion (conditioning) 高温消化（调节）
59	micronutrient 微量营养素	115	thickening 增稠
60	migration 迁移	116	trace element (analytical) 痕量元素（分析方面）
61	mixed bed (ion exchange) 离子交换混合床	117	trace element (essential) 微量元素（必需的）
62	nitrification 硝化	118	ultimate biodegradation 最终生物降解
63	nitrogen cycle bacteria 氮循环菌	119	ultra-filtration 超滤

2.3.3 固体废物污染控制 Waste Solid Pollution Control

1 dispose 处置
2 incineration 焚烧
3 landfill 填埋
4 leachate 滤液

2.3.4 噪声污染控制 Noise Pollution Control

1 damp 阻尼
2 damper 阻尼器
3 deadening 隔声材料
4 enclosed 屏蔽
5 impulsive noise 脉冲噪声
6 interleaver 衬垫
7 isolator 隔声装置
8 muffler 消声器
9 octave band sound pressure level 倍频带声压级
10 silencer 消声器
11 sound absorption 吸声
12 sound insulation 隔声
13 transmission coefficient 透射系数
14 vibration isolation 隔振

2.4 大气污染物扩散 Air Pollutant Dispersion

2.4.1 烟羽形态 The Shapes of Stack Plumes

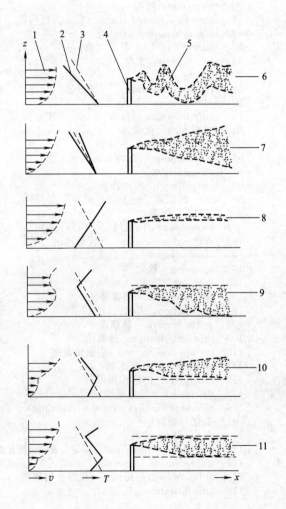

1 velocity profile 风速廓线
2 ambient lapse rate 环境温度递减率
3 dry adiabatic lapse rate 干绝热递减率
4 stack 烟囱,排气管
5 plume shape 烟羽形态
6 looping, strong instability 翻卷型,强不稳定
7 coning, near neutral stability 锥型,近中性
8 fanning, surface inversion 平展型,地面大气逆温
9 fumigation, aloft inversion 熏烟型,高处逆温
10 lofting, inversion below stack 上升型,排放口下逆温
11 trapping, inversion below and above stack height 封闭型,排放口上下两侧逆温

2.4.2 局地环流对大气污染的影响 Effects of Atmospheric Circulation in Localized Areas on Air Pollution

2.4.2.1 海陆风对沿岸地带空气污染的影响 Effect of Ocean Wind and Land Wind on Air Pollution in Coast-Land

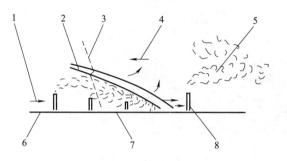

1　cool ocean wind　海洋冷风
2　inversion layer　逆温层
3　dry adiabatic lapse rate　干绝热递减率
4　warm land wind　陆地暖风
5　stack plume　烟羽
6　sea　海面
7　land　陆地
8　stack　烟囱

2.4.2.2 热岛效应 Heat Island Effect

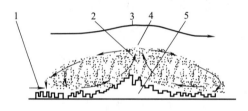

1　cool air of the countryside　乡村冷空气
2　warm air　热空气
3　free air flow　自由空气流
4　dust dome　尘幕
5　city heat island　城市热岛

2.5 固体废弃物处理 Waste Solid Treatment

2.5.1 安全填埋场 Security Landfill Disposal Site

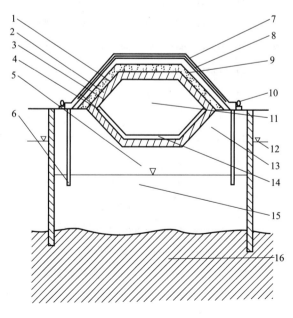

1　secondary organic lining　二层有机衬里
2　clay　黏土
3　first organic lining　一层有机衬里
4　sandy soil　砂
5　lower groundwater level　低地下水位
6　well point　井点
7　vegetation　植被
8　soil　土壤
9　sand and gravel　砂和砾面
10　pump　泵
11　waste solid　固体废弃物
12　normal groundwater level　正常地下水位
13　leachate　浸出液
14　collecting pipe　收集管
15　leachable soil　可渗透土壤
16　unleachable soil　不可渗透土壤

2.5.2 焚烧厂 Incineration Plant

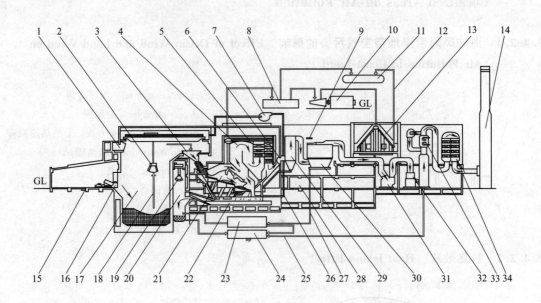

1	refuse crane control room 垃圾吊车控制室	16	refuse bunker gate 垃圾储槽口
2	refuse crane 垃圾吊车	17	refuse bunker 垃圾储槽
3	refuse hopper 垃圾料斗	18	refuse flow 垃圾流
4	gas flow 废气流	19	ash crane 灰烬吊车
5	boiler 锅炉	20	ash bunker 储灰槽
6	economizer 省煤器	21	incinerator 焚烧炉
7	induced draft fan 引风机	22	ash conveyer 出灰机
8	high-pressure steam receiver 高压蒸汽收集器	23	fly ash treatment system 飞灰处理系统
9	slaked lime 消石灰	24	waste water treatment system 污水处理系统
10	steam turbine generator 蒸汽发电机	25	waste water flow 污水流
11	condensate tank 冷凝槽	26	air flow 空气流
12	steam and condensate flow 蒸汽或冷凝液流	27	ash flow 灰流
13	turbine exhaust condenser 涡轮尾气冷凝器	28	central control room 中央控制室
		29	quenching chamber 急冷室
		30	filter-type dust collector 滤式除尘器
14	stack 烟囱	31	forced draft fan 排风机
15	plat form 平台	32	exhaust gas washer 尾气洗涤器
		33	steam reheater 蒸汽再热器
		34	deNO$_x$ catalyst facility 催化剂脱氮塔

2.6 污染物治理措施 Pollution Control Facility

2.6.1 废气治理措施 Waste Gas Treatment

2.6.1.1 VOCs 吸收工艺流程图 VOCs Absorption Process Flow Diagram

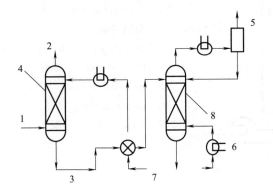

1　VOCs　挥发性有机物
2　clean gas　净化气
3　absorbent　吸收剂
4　absorber　吸收塔
5　knock-out　气液分离器
6　reboiler　再沸器
7　condensate　冷凝液
8　stripping tower　汽提塔

2.6.1.2 蓄热式氧化器工艺流程简图 Simplified Process Flow Diagram of RTO

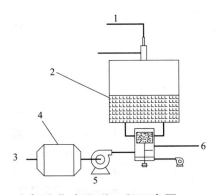

1　fuel gas　燃料气
2　RTO reactor　RTO 反应器
3　organic waste gas　有机废气
4　pretreat system　预处理系统
5　secondary air blower　二次风机
6　off gas　外排气体

2.6.1.3 废气流化床吸附工程示意图 Fluidized Bed Adsorption of Waste Gas

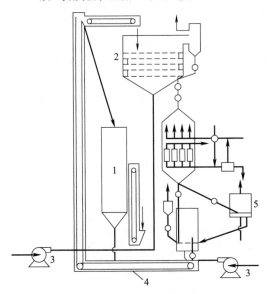

1　hopper　料斗
2　multi-level fluidized bed absorber
　　多层流化床吸附器
3　air blower　风机
4　belt conveyor　皮带传送机
5　regeneration tower　再生塔

2.6.1.4 废气催化氧化处理工艺 Waste Gas Catalyzed Oxidation Process Diagram

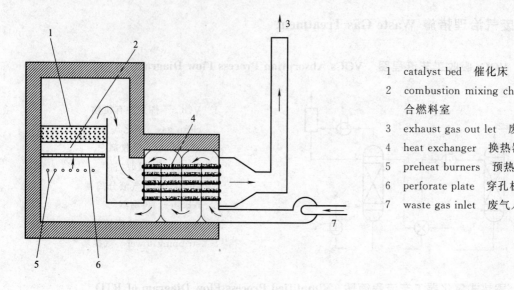

1 catalyst bed 催化床
2 combustion mixing chamber 混合燃料室
3 exhaust gas out let 废气排放口
4 heat exchanger 换热器
5 preheat burners 预热燃烧器
6 perforate plate 穿孔板
7 waste gas inlet 废气入口

2.6.1.5 袋式除尘器 Bag Dust collector

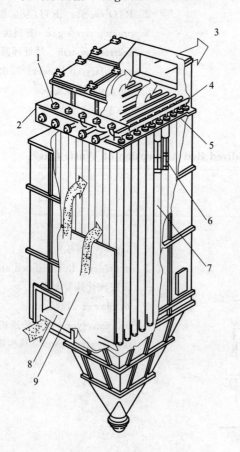

1 diaphragm valve 隔膜阀
2 pulse air manifold 脉冲气体集流箱
3 clean gas 净化排气
4 pulse pipe 脉冲管
5 tube sheet 管板
6 cage 集箱
7 filter bag 过滤袋
8 dirty gas inlet 含尘气体入口
9 inlet baffle 进气栅板

2.6.1.6 湿式除尘器 Wet Collector

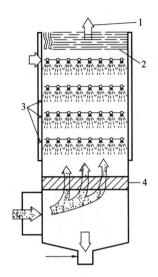

spray tower 直接喷淋式除尘器

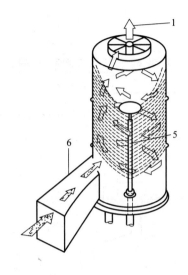

cyclone spray tower 旋风喷淋式除尘器

1 clean gas 净化后气体
2 mist eliminator 除雾器
3 sprays 喷淋
4 gas distributor plate 气体分布盘
5 spray manifold 喷水减温器
6 tangential inlet 切线进口

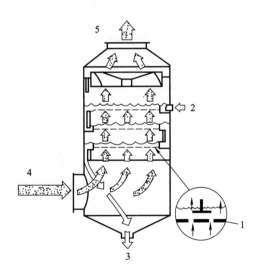

impingement scrubber 冲击洗涤式除尘器

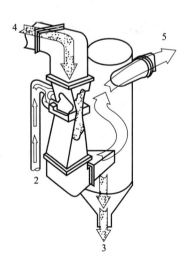

venturi scrubber 文丘里除尘器

1 impingement baffle plate 冲击盘
2 water in 水入口
3 waste out 水出口
4 dirty gas in 废气入口
5 clean gas out 净化气出口

93

2.6.1.7 氨法脱硫工艺 Process Flow Diagram Of De-SO₂ With Ammonia

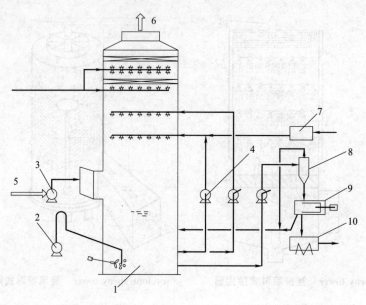

1　desulfurization tower　脱硫塔
2　roots blower　罗茨风机
3　pressure blower　增压风机
4　circulating pump　循环泵
5　dirty gas in　废气入口
6　clean gas out　净化后烟气
7　ammonia storage tank　氨原料储罐
8　hydroclone　水力旋流器
9　ammonium sulfatecentrifuge　硫铵离心机
10　drying system　干燥系统

2.6.1.8 氨法选择性催化氧化还原工艺流程图 Process Flow Diagram of SCR with Ammonia

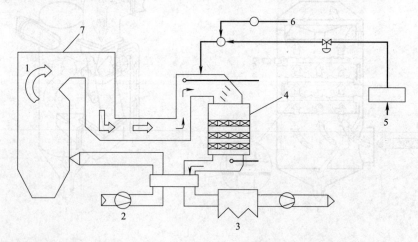

1　锅炉烟气　boiler fume
2　风机　air blower
3　除尘器　dust filter
4　SCR反应器　SCR reactor
5　氨入口　ammonia inlet
6　注入空气　air inlet
7　锅炉　boiler

2.6.1.9 尿素选择性非催化氧化还原工艺流程图 Process Flow Diagram of SNCR with Urea

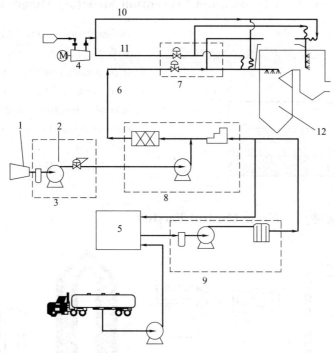

1　water　水
2　water pump　水泵
3　dilution water pressure control module　稀释水压力控制模块
4　air compressor　空压机
5　50% urea　50%尿素
6　10% urea　10%尿素
7　split module　均分模块
8　metering module　计量模块
9　supply/recycle module　供应/循环模块
10　cool air　冷却空气
11　gasified air　气化空气
12　boiler　锅炉

2.6.2 多效蒸发系统 Multi-effect Evaporation System

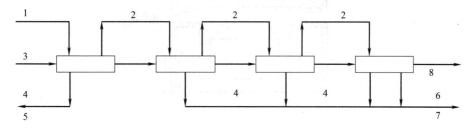

1　steam　蒸汽
2　secondary steam　二次蒸汽
3　RO water　反渗透浓水
4　condensate　冷凝水
5　back to the boiler　回锅炉
6　reuse　回用
7　to replenish water　去循环水补充水
8　evaporation of the residue　蒸发残液

2.6.3 重点污染防治区防渗结构图示 Special Important Pollution Prevention and Control Area Penetration Prevention Structure Diagram

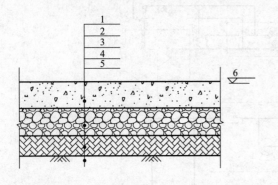

1 cement-based impermeable coating 水泥基防渗涂层
2 impermeable concrete 抗渗混凝土
3 sand and gravel pavement 砂卵石铺砌基层
4 natural gravel base 天然沙砾基层
5 secondary field flat soil (compacted) 二次场平土（夯实）
6 site elevation 场地标高

2.6.4 噪声控制措施 Noise Control Facility

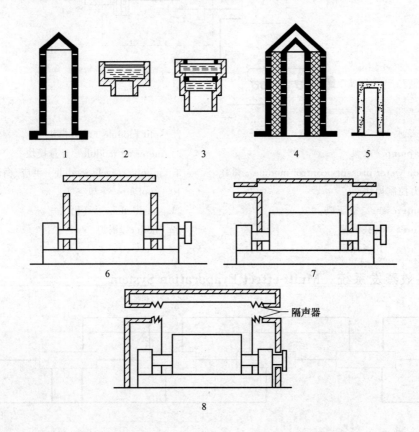

1 small hole diffuser 小孔消声器
2 primary metal mesh diffuser 一级金属丝网扩散消声器
3 secondary metal mesh diffuser 二级金属丝网扩散消声器
4 small hole metal mesh diffuser 小孔丝网扩散消声器
5 cooper powder diffuser 铜粒粉末扩散消声器
6 sound insulation cover 隔声罩
7 sound insulation cover with diffuser 带消声器的隔声罩
8 sound insulation cover with diffuser and vibration isolation 带隔振设施和消声器的隔声罩

3 消防、安全和职业卫生　Fire Fighting, Safety and Occupational Health

3.1 消防系统流程图　Flow Diagram of Fire Fighting System

3.1.1 水喷淋典型流程图　Typical Flow Diagram of Water Sprinkler System

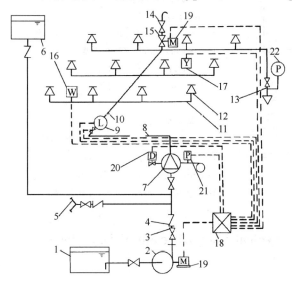

1　water basin　水池
2　water pump　水泵
3　gate valve　闸阀
4　check valve　止回阀
5　water pump connection　水泵接合器
6　fire water drum　消防水箱
7　alarm valve set　报警阀组
8　feed mains　配水干管
9　flow indication meter　水流指示器
10　cross mains　配水管
11　branch line　配水支管
12　sprinkler　喷头
13　end water testing device　末端试水装置
14　rapid vent valve　快速排气阀
15　motor valve　电动阀
16　heat detector　感温探测器
17　smoke detector　感烟探测器
18　alarm control panel　报警控制器
19　driven motor　驱动电机
20　solenoid valve　电磁阀
21　pressure switch　压力开关
22　pressure gauge　压力表

3.1.2 泡沫消防典型流程图　Typical Flow Diagram of Foam Extinguishing System

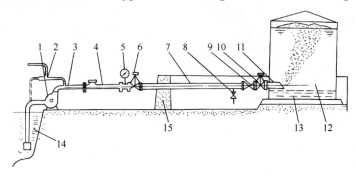

1　fire water pump　消防泵
2　air foam proportioner　空气泡沫比例混合器
3　foam solution pipe　混合液管道
4　high back pressure foam maker　高背压泡沫产生器
5　back pressure gauge　背压表
6　back pressure regulating valve　背压调节阀
7　foam pipe　泡沫管道
8　foam sampling valve or drain valve　泡沫取样阀或放水阀
9　check valve　止回阀
10　global valve　截止阀
11　foam inlet pipe　泡沫注入管
12　oil tank　油罐
13　water layer　水层
14　water basin　水池
15　dike　防火堤

97

3.1.3 气体灭火系统典型流程图 Typical Flow Diagram of Gas Extinguishing System

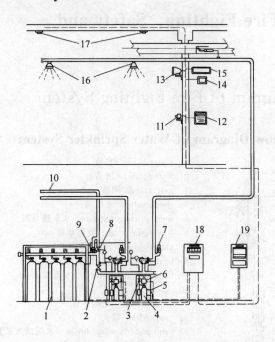

1　carbon dioxide cylinder　二氧化碳储气瓶
2　global valve on collection pipe　集流管截止阀
3　global valve on start-up pipe　启动管截止阀
4　actuation unit　启动装置
5　pressure switch　压力开关
6　solenoid valve　电磁阀
7　peculiar smell detection unit　异味检测装置
8　vent valve　放空阀
9　relief valve　安全阀
10　vent pipe　放空管
11　gas detector　气体探测器
12　operation box　操作箱
13　alarm horn　声讯报警器
14　indoor hazardous concentration indicator　室内危险浓度指示器
15　relief indication light　释放指示灯
16　nozzle　喷头
17　fire detectors　火灾探测器
18　system control panel　系统控制盘
19　gas alarm panel　气体报警盘

3.1.4 干粉灭火系统典型流程图 Typical Flow Diagram of Dry Chemical Extinguishing System

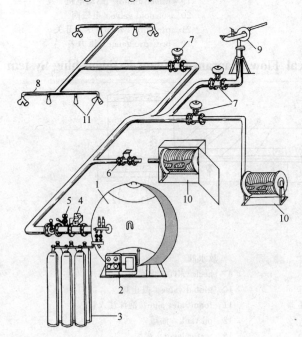

1　powder container　干粉罐
2　control panel　控制盘
3　gas cylinder　气瓶
4　main powder valve　主阀
5　test connection　检测口
6　directional valve for hose line with extinguishing pistol　选择阀（与带灭火枪的软管相连）
7　directional valve　选择阀
8　pipe system　管网系统
9　powder monitor　干粉炮
10　hose reel with extinguishing pistol　带灭火枪的软管卷盘
11　extinguishing nozzle　喷头

3.2 消防设备 Fire Fighting Equipment

3.2.1 消防车 Fire Truck

3.2.1.1 泡沫消防车 Foam Fire Truck

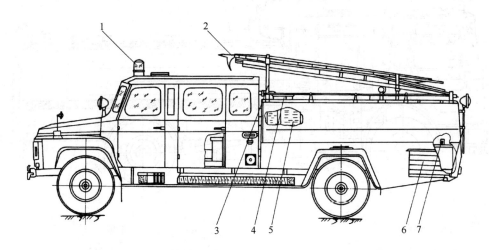

1　alarm bell　警灯
2　foam solution handline　泡沫混合液管枪
3　ladder assembly　梯架总成
4　foam concentrate inlet　泡沫液加注口
5　foam concentrate/water drum　泡沫液/水罐
6　apparatus box　器材箱
7　appurtenances and tools　附件及工具

3.2.1.2 干粉消防车 Dry Powder Fire Truck

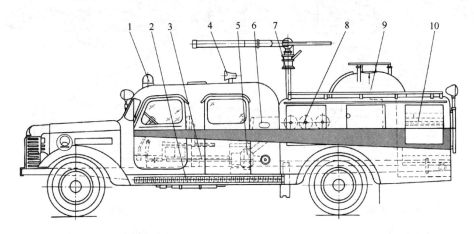

1　alarm bell　警灯
2　water suction tube　吸水管
3　pump rotation shaft　水泵传动轴
4　siren　警笛
5　water pump　水泵
6　operation handle bar　操纵手柄
7　dry powder monitor　干粉炮
8　nitrogen cylinder　氮气钢瓶
9　dry powder drum　干粉罐
10　apparatus box　器材箱

99

3.2.1.3 水罐消防车　Water Fire Truck

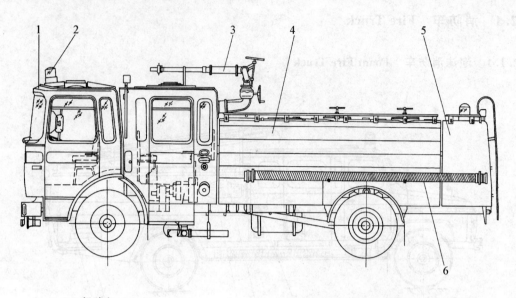

1　cab　驾驶室
2　alarm bell　警灯
3　water monitor　水炮
4　water drum　水罐
5　apparatus box　器材箱
6　water suction tube　吸水管

3.2.1.4 直臂云梯车　Straight Arm and Scaling Ladder Truck

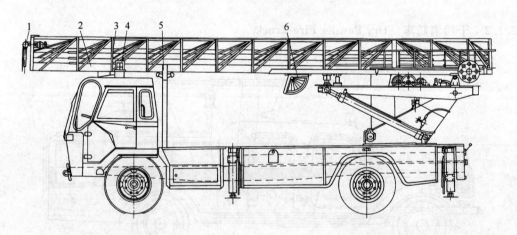

1　fire nozzle　水枪
2　scaling ladder　云梯
3　siren　警笛
4　alarm bell　警灯
5　support pillar　支承桁架
6　angle panel　角度盘

3.2.2 消防泵 Fire Pump

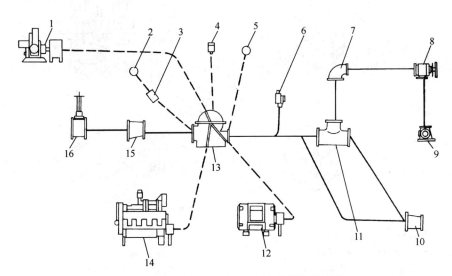

horizontal fire pump system 卧式消防泵系统

1　horizontal steam turbine　卧式蒸汽透平
2　compound type suction gauge　复合式入口压力表
3　gauge protector　压力表保护装置
4　automatic air release valve　自动放空阀
5　discharge gauge　出口压力表
6　casing relief valve　壳体放空阀
7　elbow　弯头
8　relief valve　安全阀
9　enclosed cone　角阀
10　reducer　异径管
11　tee　三通
12　horizontal electric motor　卧式电机
13　pump body　泵体
14　diesel engine　柴油发电机
15　eccentric suction reducer　入口偏心异径管
16　valve　阀门

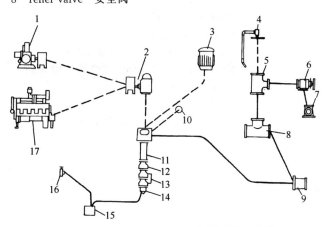

vertical fire pump system 立式消防泵系统

1　horizontal steam turbine　卧式蒸汽透平
2　right angle gear　直角齿轮
3　vertical electric motor　立式电机
4　automatic air release valve　自动放空阀
5，8　tee　三通
6　relief valve　安全阀
7　enclosed cone　角阀
9　reducer　异径管
10　discharge gauge　出口压力表
11　column　筒体
12　discharge bowl　出口段
13　intermediate bowls　中间段
14　bell suction　斗形吸入口
15　basket strainer　篮形过滤器
16　level gauge　液位计
17　diesel engine　柴油发电机

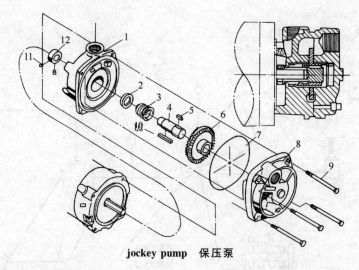

jockey pump 保压泵

1 motor bracket 电机支架
2 seal stationary seat 密封静止件
3 seal rotating element 密封转动件
4 shaft sleeve 轴套
5 key/impeller 键/叶轮
6 impeller 叶轮
7 casing 壳体
8 cover 盖
9 thru bolt 长螺栓
10 key, sleeve drive 键
11 set screw 紧定螺钉
12 lock collar 锁紧环

3.2.3 泡沫灭火设备 Foam Extinguishing Equipment

3.2.3.1 泡沫液储罐 Foam Storage Tank

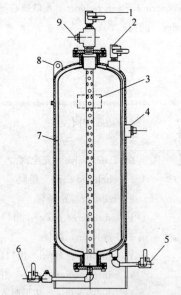

1 bladder vent/fill valve 胆囊放空/加液阀
2 tank shell vent valve 储罐放空阀
3 nameplate 铭牌
4 water inlet 入水口
5 tank shell drain valve 储罐放净阀
6 bladder drain/fill valve 胆囊放净/加液阀
7 bladder 胆囊
8 lifting lug 吊耳
9 concentrate outlet 泡沫液出口

vertical bladder tank 立式隔膜型储罐

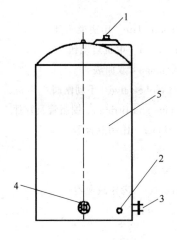

1　vent　放空口
2　drain　放净口
3　pump suction connection　泵吸入接口
4　pump return connection　泵回路接口
5　foam tank　泡沫液罐

atmospheric foam tank　常压泡沫储罐

3.2.3.2　混合器及比例混合系统　Proportioners and Proportioning Systems

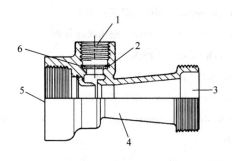

1　foam concentrate inlet　泡沫液进口
2　retaining ring　护圈
3　foam solution discharge　泡沫混合液出口
4　body　本体
5　water inlet　进水口
6　metering orifice　计量孔板

line propoertioner　管线式比例混合器

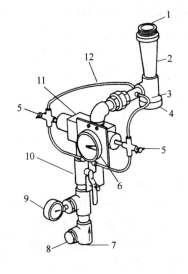

1　foam solution discharge　泡沫混合液出口
2　proportioner　比例混合器
3　foam concentrate sensing line　泡沫传感线
4　water inlet　进水口
5　drain cock valve　放净旋塞阀
6　duplex gauge　双向表
7　foam concentrate inlet　泡沫液进口
8　check valve　止回阀
9　pressure gauge　压力表
10　foam concentrate valve　泡沫液阀
11　spool valve　柱式阀
12　water sensing line　水传感线

**in-line balanced pressure proportioner
平衡压力式比例混合器**

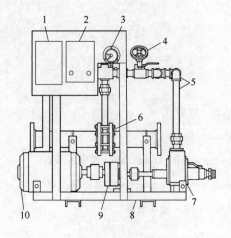

1 instruction plate 用法说明盘
2 motor starter 电机启动钮
3 duplex gauge 双向表
4 manual globe valve 手动闸阀
5 brass pipe and fittings 铜制管及管件
6 proportioner 比例混合器
7 pump 泵
8 skid 滑板
9 gear reducer 齿轮减速器
10 motor 电机

balanced pressure pump proportioning system
平衡压力泵比例混合系统

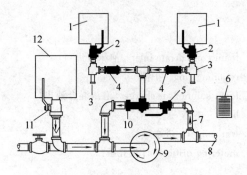

1 foam liquid tank 泡沫液罐
2 foam supply valve 泡沫液供应阀
3 foam metering valve 泡沫液计量阀
4 check valve 止回阀
5 water by-pass valve 水旁路阀
6 instruction plate 用法说明盘
7 inlet from discharge side of pump 泵出口侧进液口
8 solution discharge 泡沫混合液出口
9 pump 泵
10 eductor 环泵比例混合器
11 optional water suction valve 吸水阀（可选）
12 water tank 水罐

around the pump proportioning system
环泵式比例混合系统

3.2.3.3 泡沫炮 Foam Monitor

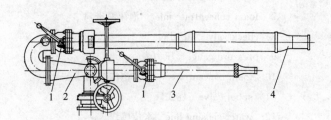

1 ball valve 球阀
2 rotation body 转塔
3 water nozzle 喷射水炮筒
4 foam nozzle 喷射泡沫炮筒

foam/water monitor 泡沫/水两用炮

3.2.3.4 拖车式泡沫比例混合装置 Proportioning Foam Trailer

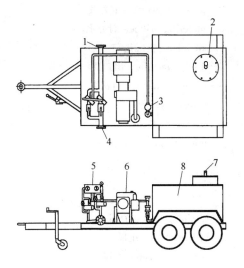

1 water inlet flange 进水口法兰
2 expansion dome and fill opening 膨胀顶和加液口
3 pressure control valve 压力控制阀
4 foam solution discharge flange 泡沫混合液出口法兰
5 proportioning system 比例混合系统
6 diesel driven concentrate pump 柴油驱动的泡沫液泵
7 pressure/vacuum valve 压力/真空阀
8 foam concentrate tank 泡沫液罐

3.2.3.5 泡沫产生器 Foam Generator/Maker

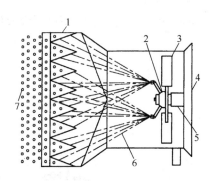

high expansion foam generator 高倍数泡沫产生器

1 foam screen 泡沫筛
2 water reaction motor 水力电机
3 blower fan 风机
4 air flow 空气流股
5 foam solution inlet 泡沫混合液进口
6 foam solution spray 泡沫混合液喷雾
7 foam 泡沫

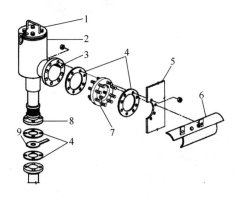

low expansion foam chamber 低倍数泡沫产生器

1 hinged inspection hatch 带铰链的检查盖
2 chamber body 产生器本体
3 outlet flange 出口法兰
4 gasket 垫片
5 tank wall 罐壁
6 split deflector 导流板
7 mounting pad 装配垫
8 inlet flange 进口法兰
9 orifice 孔板

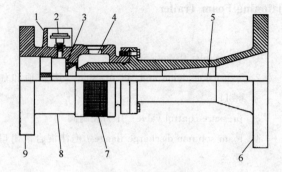

1	protective shroud　保护罩
2	pressure gauge　压力表
3	metering orifice　计量孔板
4	air inlet　空气进口
5	discharge tube　出口管
6	outlet flange　出口法兰
7	screen　筛子
8	body　本体
9	inlet flange　进口法兰

high back-pressure foam maker
高背压泡沫产生器

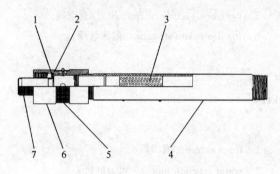

1	retaining ring　护圈
2	orifice plate　孔板
3	aeration screen　鼓风筛
4	barrel　桶体
5	air inlet screen　空气入口筛
6	body　本体
7	foam solution strainer　泡沫混合液过滤器

floating roof foam maker
浮顶泡沫产生器

3.2.3.6　泡沫喷头　Foam Nozzle

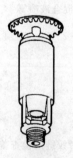 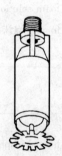 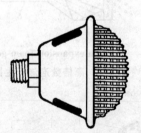

3.2.3.7　泡沫枪　Foam Nozzle

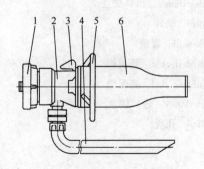

1	quick coupling　管牙接口
2	body　枪体
3	on/off grip　启闭柄
4	suction pipe　吸液管
5	hand wheel　手轮
6	nozzle body　枪筒

3.2.4 消火栓 Fire Hydrant

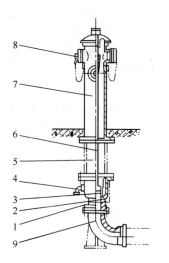

1 valve body 阀体
2 valve seat 阀座
3 valve disc 阀瓣
4 drain valve 排水阀
5 flange connection pipe 法兰接管
6 plug 阀杆
7 body 本体
8 hose connection 水龙带接口
9 water inlet bend 进水弯管

aboveground outdoor hydrant 地上式室外消火栓

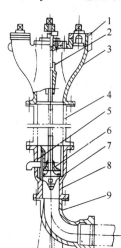

1 hose connection 水龙带接口
2 plug 阀杆
3 body 本体
4 flange connection pipe 法兰接管
5 drain valve 排水阀
6 valve disc 阀瓣
7 valve seat 阀座
8 valve body 阀体
9 water inlet bend 进水弯管

underground outdoor hydrant 地下式室外消火栓

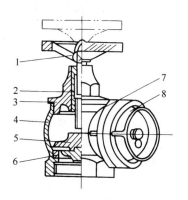

1 handwheel 手轮
2 plug 阀杆
3 bonnet 阀盖
4 body 本体
5 valve disc 阀瓣
6 valve seat 阀座
7 hose connection 水龙带接口
8 cap 闷盖

indoor hydrant 室内消火栓

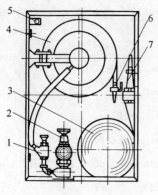

1 quick connection 快速接口
2 indoor hydrant 内消火栓
3 rubber-lined hose 衬胶水带
4 hose reel 软管卷盘
5 control button 控制按钮
6 small size solid stream nozzle 小口径直流水枪
7 big size solid stream nozzle 大口径直流水枪

hydrant cabinet 消火栓箱

3.2.5 消防水泵接合器 Fire Pump Connection

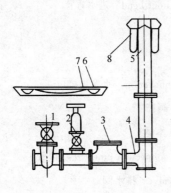

1 gate valve 闸阀
2 safety valve 安全阀
3 check valve 止回阀
4 bend 弯管
5 body 本体
6 well hatch seat 井盖座
7 well hatch 井盖
8 WSK type of fixed connection WSK 型固定接口

3.2.6 消防水炮 Fire Water Monitor

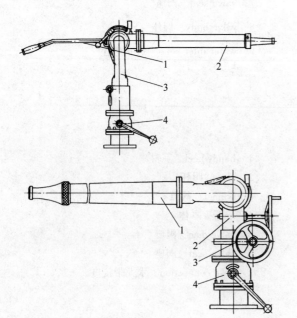

1 hand grip 手柄
2 nozzle 炮筒
3 rotation body 转塔
4 ball valve 球阀

1 nozzle 炮筒
2 rotation body 转塔
3 handwheel 手轮
4 ball valve 球阀

3.2.7 电控消防水炮 Electric Monitor

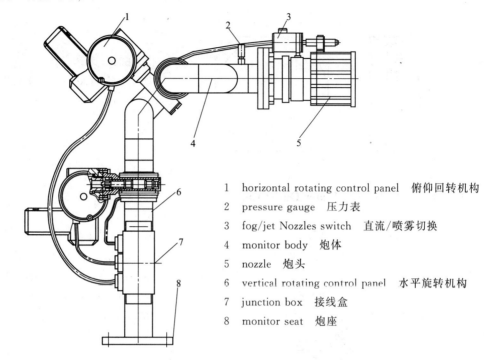

1 horizontal rotating control panel 俯仰回转机构
2 pressure gauge 压力表
3 fog/jet Nozzles switch 直流/喷雾切换
4 monitor body 炮体
5 nozzle 炮头
6 vertical rotating control panel 水平旋转机构
7 junction box 接线盒
8 monitor seat 炮座

3.2.8 消防接口和闷盖 Fire Quick Coupling and Cap

3.2.8.1 管牙接口 Quick Coupling

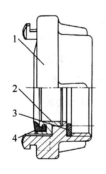

1 body 本体
2 sealing ring base 密封圈座
3 plane sealing ring 平面密封圈
4 sealing ring 密封圈

3.2.8.2 肘接口 Elbow Coupling

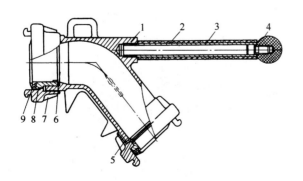

1 elbow coupling body 肘接口本体
2 handle grip shroud 手把外套
3 handle grip assembly 手把部件
4 handle grip ball 手柄球
5 quick coupling 管牙接口
6 plane gasket ring 平面垫圈
7 hose connection 水带接口
8 sealing ring base 密合圈座
9 sealing ring 密合圈

109

3.2.8.3 进水口闷盖 Water Inlet Cap

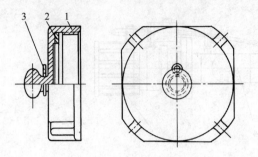

1 body 本体
2 plane gasket ring 平面垫圈
3 double hole ring 双眼圈

3.2.8.4 出水口闷盖 Water Outlet Cap

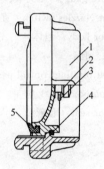

1 body 本体
2 plane gasket ring 平面垫圈
3 double hole ring 双眼圈
4 retaining ring 护圈
5 sealing gasket 密封垫

3.2.9 灭火器 Fire Extinguisher

3.2.9.1 推车式干粉灭火器 Wheel Dry Powder Extinguisher

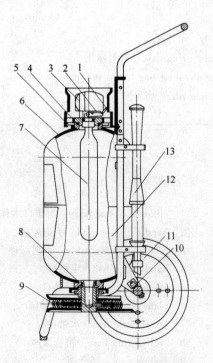

1 lifting bail for gas inlet 进气压杆提环
2 gas inlet tube 进气压管
3 pressure gauge 压力表
4 sealing gasket 密封胶圈
5 protection shroud 护罩
6 body 桶体
7 gas cartridge 蓄气瓶
8 sealing gasket of powder outlet 出粉口密封胶圈
9 powder outlet tube 出粉管
10 wheel shaft 轮轴
11 wheel 车轮
12 carriage 车架
13 dry powder nozzle 喷粉枪

3.2.9.2 手提式干粉灭火器 Portable Dry Powder Extinguisher

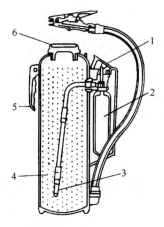

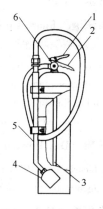

cartridge-operated dry powder extinguisher
外置式干粉灭火器

1　safety pin　保险销
2　gas cartridge　储气瓶
3　gas tube　进气管
4　dry powder　干粉
5　carrying handle　提把
6　cap　桶帽

stored-pressure dry powder extinguisher
内置式干粉灭火器

1　discharge lever　压把
2　carrying handle　提把
3　siphon tube　虹吸管
4　nozzle　喷枪
5　wand applicator　出粉管
6　pressure gauge　压力表

3.2.9.3 手提式二氧化碳灭火器 Portable Carbon Dioxide Extinguisher

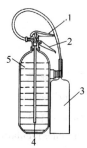

1　discharge lever　压把
2　carrying handle　提把
3　discharge horn　喷管
4　siphon tube　虹吸管
5　carbon dioxide in a fluid state　液态二氧化碳

3.2.9.4 手提式泡沫灭火器 Portable Foam Extinguisher

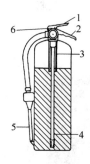

1　discharge lever　压把
2　carrying handle　提把
3　anti-overfill tube　防溢管
4　siphon tube　虹吸管
5　air aspirating foam nozzle　空气吸入式泡沫喷嘴
6　pressure gauge　压力表

3.2.10 消防水带 Fire Hose

1　hose made of hemp, flax or jute　麻质消防水带
2　hose made of cotton　棉质消防水带
3　rubber-lined hose　衬胶消防水带

111

3.2.11 灭火剂 Fire Extinguishing Agent

3.2.11.1 清水 Water

3.2.11.2 泡沫液 Foam Concentrate（or Concentrate）

1 protein concentrate　蛋白泡沫液
2 fluoroprotein concentrate　氟蛋白泡沫液
3 aqueous film-forming foam(AFFF)concentrate　水成膜泡沫液
4 film-forming fluoroprotein(FFFP)concentrate　成膜氟蛋白泡沫液
5 alcohol-resistant concentrate　抗溶性泡沫液

3.2.11.3 干粉 Dry Powder

1 sodium bicarbonate　碳酸氢钠
2 potassium bicarbonate　碳酸氢钾
3 ammonium phosphate　磷酸铵盐

3.2.11.4 卤代烷 Halogenated Agent

1 Halon 1211 (i.e. bromochlorodifluoromethane)　卤代烷1211（即二氟一氯一溴甲烷）
2 Halon 1301 (i.e. bromotrifluoromethane)　卤代烷1301（即三氟一溴甲烷）

3.2.11.5 二氧化碳 Carbon Dioxide（CO_2）

3.2.11.6 氮气 Nitrogen（N_2）

3.3 安全设备 Safety Equipment

3.3.1 呼吸器 Breathing Apparatus

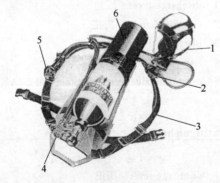

1 face-piece　面罩
2 first stage regulator　一级调节器
3 backpack　背架
4 second stage regulator　二级调节器
5 pressure gauge　压力表
6 cylinder　气瓶

3.3.2 防护服 Chemical Suit

3.3.3 其他安全设备 Other Safety Equipment

3.3.3.1 防护手套 Gloves

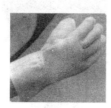

3.3.3.2 口罩 Gauze Mask

3.3.3.3 面罩 Mask

3.3.3.4 安全帽及防护屏 Safety Helmet and Face Shields

3.3.3.5 耳罩及耳塞 Earmuffs and Earplugs

3.3.3.6 安全带和安全绳 Safety Belt and Safety Ropes

3.3.3.7 防护眼镜 Protection Spectacles

3.3.3.8 防护鞋 Protection Shoes

3.3.3.9 急救箱 First Aid Box

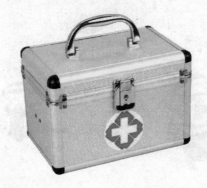

3.3.3.10 灭火毯 Fire Blanket

3.3.3.11 事故淋浴洗眼器 Safety Shower and Eye Washer

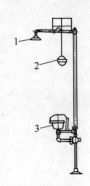
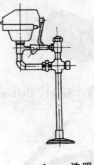

1　shower nozzle　淋浴喷头
2　pull ring　拉环
3　eye washer　洗眼器

eye washer　洗眼器　　safety shower　淋浴器

3.3.3.12 便携式事故淋浴洗眼器 Portable Emergency Eye Washer

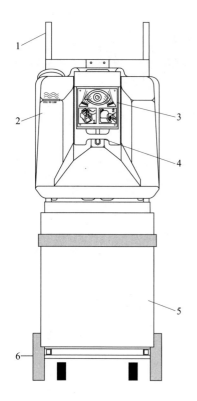

1 bracket 支架
2 water tank 水箱
3 eye wash nozzle 洗眼喷头
4 eye wash switch 洗眼开关
5 waste water can 废水槽
6 wheel 轮子

4 成套设备 Package Equipment

4.1 空分设备 Air Separation Unit

4.1.1 冷箱 Cold Box

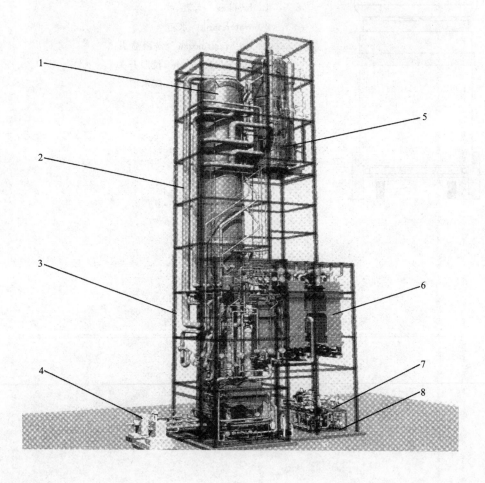

1 rectification column 精馏塔
2 piping 管道
3 frame 框架
4 cryogenic pump 低温泵
5 crude argon column 粗氩塔
6 main heater exchanger 主换热器
7 expander 膨胀机
8 base plate 底座

4.1.2 变压吸附制氮装置/真空变压吸附制氧装置 PSA Nitrogen Generation Unit/VPSA Oxygen Generation Unit

1　adsorber vessel　吸附器
2　piping　管件
3　air inlet　空气进口
4　nitrogen outlet　氮气出口
5　manhole　人孔
6　control panel　控制盘
7　silencer　消声器

4.1.3 膜分离制氮装置 Membrane Module Nitrogen Generation Unit

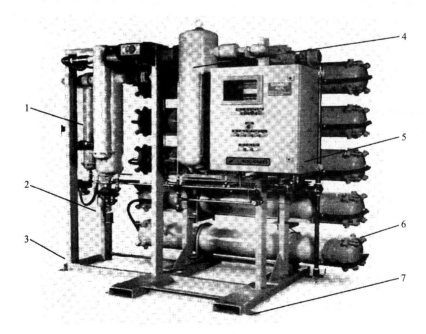

1　filter　过滤器
2　trap　疏水阀
3　support　支架
4　distribution　分配器
5　control panel　控制盘
6　membrane module　膜组件
7　baseplate　底盘

4.2 制冷机 Refrigerator

4.2.1 制冷装置 Refrigeration Unit

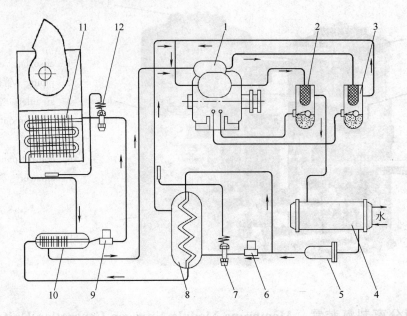

1 compressor 压缩机
2，3 oil separator 油分离器
4 cooler 冷凝器
5 dryer and filter 干燥过滤器
6 electromagnetic valve 电磁阀
7 expansion valve 膨胀阀
8 intercooler 中间冷却器
9 electromagnetic valve 电磁阀
10 regenerator 回热器
11 evaporator 蒸发器
12 expansion valve 膨胀阀

4.2.2 辅助设备 Auxiliary Equipments

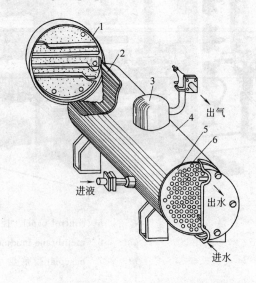

horizontal tubular evaporator 管壳卧式蒸发器

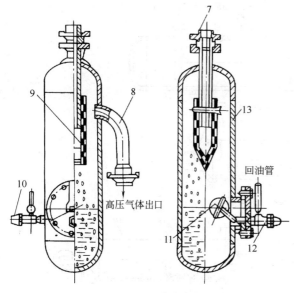

1 cover 端盖
2 still tube 蒸发管
3 collection chamber 集气室
4 casing 外壳
5 tube sheet 管板
6 rubber ring 橡皮垫圈
7 inlet 进口
8 outlet 出口
9 filter gauze 滤网
10 hand oil feed-back valve 手动回油阀
11 float valve 浮球阀
12 oil feed-back valve 回油阀
13 casing 筒体

oil separator 油分离器

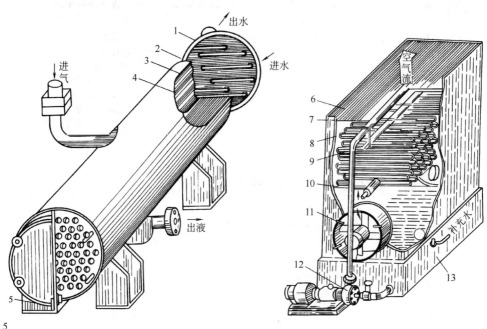

horizontal tubular condenser 卧式管壳式冷凝器

evaporation condenser 蒸发式冷凝器

1 cover 端盖
2 rubber ring 橡皮圈
3 tube sheet 管板
4,8 cold finger 冷凝管
5 drain 排净口
6 water retain grill 挡水栅

7 nozzle 喷嘴
9 gas inlet 进气管
10 fluid outlet 出液管
11 fan 风机
12 water pump 水泵
13 float 浮子

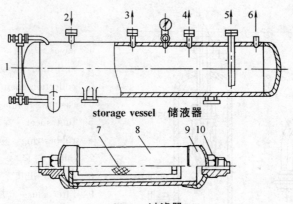

storage vessel 储液器

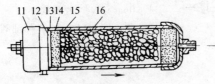

filter 过滤器

filter & dryer 过滤干燥器   **filter (with dessicant) 过滤器（加装干燥剂）**

1   liquid level indicator   液位计
2   inlet   进液口
3   balance pipe   平衡管
4   safety valve   安全阀
5   outlet   出液口
6   vent   放气口
7,18   filter gauze   滤网
8   casing   外壳

9,11,17   cover   端盖
10   pipe joint   管接头
12   wire fence   铁丝布
13   copper wire fence   铜丝布
14   cotton wire fence   纱布
15   medicinal cotton   药棉
16   anhydrous $CaCl_2$ 无水氯化钙
19   dessicant   干燥剂

4.3   低温冷箱   Cryogenic Cold Box

4.3.1   CO 分离   CO Separation

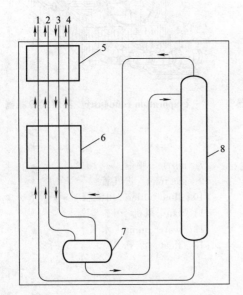

1   一氧化碳产品气   CO product gas
2   粗氢气   raw hydrogen gas
3   原料气   feed gas
4   闪蒸气   flash gas
5   冷却器1   cooler 1
6   冷却器2   cooler 2
7   分离器   seperator
8   气提塔   stripping column

4.3.2 液氮洗 Nitrogen Wash

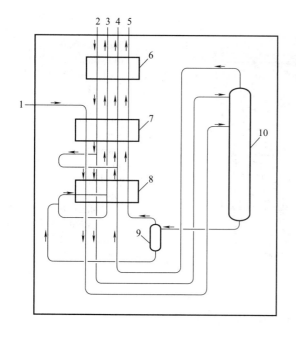

1　原料气　feed gas
2　高压氮气　high pressure nitrogen gas
3　尾气　tail gas
4　氨合成气　ammonia syngas
5　闪蒸气　flash gas
6　预冷器　precooler
7　冷却器1　cooler 1
8　冷却器2　cooler 2
9　闪蒸罐　flash drum
10　氮洗塔　nitrogen wash column

4.4 冷却塔 Cooling Towers

4.4.1 自然通风冷却塔 Natural Draft Cooling Tower

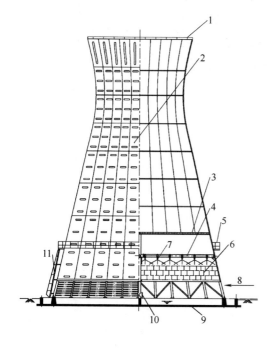

1　lightning arrester　避雷设施
2　casing　围护结构
3　drift eliminator　收水器
4　water distribution system　配水系统
5　stairway and handrail　走道及扶手
6　fill　填料
7　branch arm and spray nozzle assembly　支管和喷水器
8　air inlet　空气进口
9　cooling water basin　冷却水池
10　water inlet　进水管
11　ladder　爬梯

4.4.2 横流冷却塔 Crossflow Cooling Tower

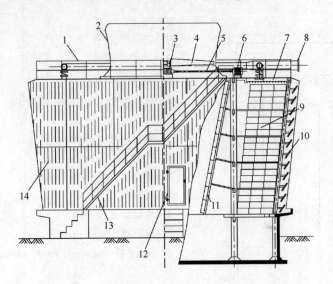

1 handrail 扶手
2 fan cylinder 风筒
3 speed reducer 减速器
4 fan 风机
5 drive shaft 驱动轴
6 motor 电机
7 water distribution system 配水系统
8 water inlet 进水管
9 fill 填料
10 louver 百叶窗
11 drift eliminator 收水器
12 access door 检修门
13 stairway 梯子
14 casing 围护结构

4.4.3 逆流冷却塔 Counterflow Cooling Tower

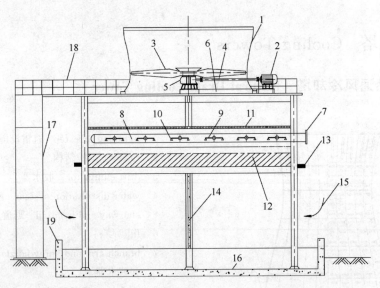

1 fan cylinder 风筒
2 motor 电机
3 fan 风机
4 drive shaft 驱动轴
5 speed reducer 减速器
6 fan deck 风机平台
7 water inlet 进水管
8 water distribution system 配水系统
9 branch arm 配水支管
10 spray nozzle assembly 喷头
11 drift eliminator 收水器
12 fill 填料
13 air inlet guide 进风导流板
14 windwall 隔风墙
15 air inlet 进风口
16 cooling water basin 冷却水池
17 stairway 梯子
18 handrail 扶手
19 curb wall 池壁

4.5 过滤器 Filter

4.5.1 压力过滤器 Pressure Filter

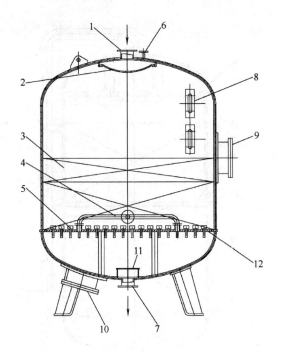

1 water inlet 进水口
2 water inlet device 进水装置
3 filter media 滤料
4 air inlet device 进气装置
5 filter nozzle 滤帽
6 vent 排气
7 water outlet 出水口
8 sight glass 视镜
9,10 manhole 人孔
11 water outlet device 出水装置
12 jet tray 滤帽安装板

4.5.2 重力无阀过滤器 Gravity Valveless Filter

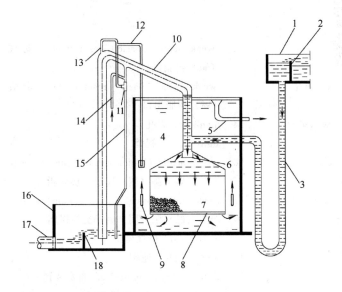

1 inlet distribution tank 配水箱
2 distribution weir 分配堰
3 water inlet pipe 进水管
4 clean water tank 清水箱
5 water outlet 出水管
6 water inlet device 进水装置
7 filter media 滤料
8 supporting layer 承托层
9 connection pipe 连通管
10 siphon pipe 虹吸管
11 ejector 喷射器
12 siphon break pipe 虹吸破坏管
13 air suction pipe 抽气管
14 forced wash pipe 强制冲洗管
15 auxiliary siphon pipe 辅助虹吸管
16 drainage tank 排水箱
17 drainage pipe 排水管
18 seal weir 水封堰

4.6 双室浮动阳（阴）离子交换器 Double Chamber Floating Cation (Anion) Exchanger

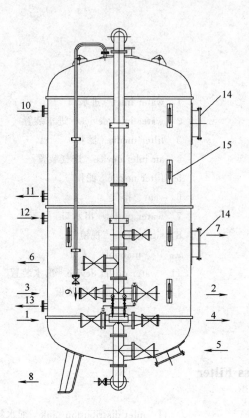
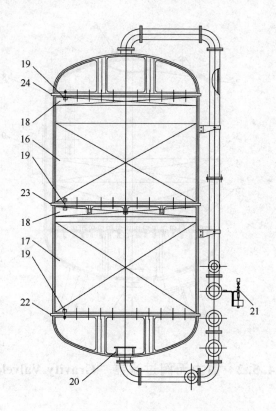

1	feed water inlet 进水口	12	weak acidic (basic) resin inlet 弱酸（碱）树脂进口
2	product water outlet 出水口	13	weak acidic (basic) resin outlet 弱酸（碱）树脂出口
3	acidic (basic) solution inlet 酸（碱）再生液进口	14	manhole 人孔
4	acidic (basic) waste water outlet 酸（碱）废水出口	15	sight glass 视镜
5	recirculating wash inlet (for cation exchanger) 循环洗进水口（用于阳离子交换器）	16	strong acidic (basic) cation (anion) resin 强酸（碱）阳（阴）离子交换树脂
6	recirculating wash outlet (for anion exchanger) 循环洗出水口（用于阳离子交换器）	17	weak acidic (basic) cation (anion) resin 弱酸（碱）阳（阴）离子交换树脂
7	forward-wash water outlet 正洗水出口	18	inert resin 惰性树脂
8	drain 排净	19	filter nozzle 滤帽
9	vent 放空	20	water distribution device 布水装置
10	strong acidic (basic) resin inlet 强酸（碱）树脂进口	21	sampling device 取样装置
		22	lower jet tray 下层滤帽安装板
11	strong acidic (basic) resin outlet 强酸（碱）树脂出口	23	intermediate jet tray 中间滤帽安装板
		24	upper jet tray 上层滤帽安装板

4.7 逆流再生阳（阴）离子交换器 Countercurrent Regeneration Cation (Anion) Exchanger

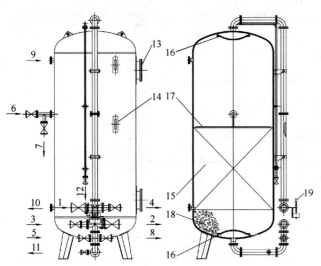

5 acidic (basic) solution inlet 酸（碱）再生液进口
6 backwash water inlet (brief) 小反洗水进口
7 forward-wash outlet (brief) 小正洗水出口
8 forward-wash outlet 正洗水出口
9 resin inlet 树脂进口
10 resin outlet 树脂出口
11 drain 排净
12 vent 放空
13 manhole 人孔
14 sight glass 视镜
15 cation (anion) resin 阳（阴）离子交换树脂
16 water distribution device 布水装置
17 middle water distribution device 中间布水装置
18 sand supporting layer 石英砂承托层
19 sampling device 取样装置

1 feed water inlet 进水口
2 product water outlet 出水口
3 backwash water inlet 反洗水进口
4 backwash water outlet 反洗水出口

4.8 顺流再生阳（阴）离子交换器 Concurrent Regeneration Cation (Anion) Exchanger

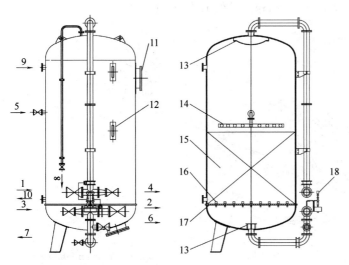

4 backwash water outlet 反洗水出口
5 acidic (basic) solution inlet 再生液进口
6 forward-wash water outlet 正洗水出口
7 drain 排净
8 vent 放空
9 resin inlet 树脂进口
10 resin outlet 树脂出口
11 manhole 人孔
12 sight glass 视镜
13 water distribution device 布水装置
14 acidic (basic) solution distribution device 布酸（碱）装置
15 cation (anion) resin 阳（阴）离子交换树脂
16 filter-nozzle 滤帽
17 jet tray 滤帽安装板
18 sampling device 取样装置

1 feed water inlet 进水口
2 product water outlet 出水口
3 backwash water inlet 反洗水进口

4.9 混合离子交换器 Mixed Ion Exchanger

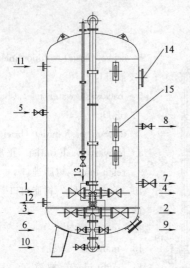
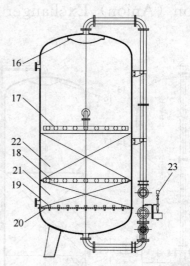

1 feed water inlet 进水口
2 product water outlet 出水口
3 backwash water inlet 反洗水进口
4 backwash water outlet 反洗水出口
5 basic solution inlet 碱液进口
6 acidic solution inlet 酸液进口
7 middle drain 中间排水
8 upper drain 上部排水
9 forward-wash water outlet 正洗排水口
10 compressed air inlet 压缩空气进口
11 resin inlet 树脂进口
12 resin outlet 树脂出口
13 vent 放空
14 manhole 人孔
15 sight glass 视镜
16 upper water distribution device 上布水装置
17 basic solution distribution device 进碱装置
18 middle drain device 中间排水装置
19 filter-nozzle 滤帽
20 jet tray 滤帽安装板
21 strong acidic cation resin 强酸阳离子交换树脂
22 strong basic anion resin 强碱阴离子交换树脂
23 sampling device 取样装置

4.10 反渗透膜元件 Reverse Osmosis Membrane

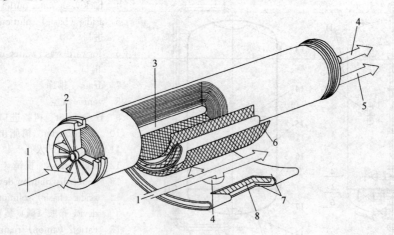

1 feed water 进水
2 brine seal 密封垫圈
3 product tube 集水管
4 permeate 透过水
5 concentrate 浓缩水
6 mesh spacer 网状隔层
7 reverse osmosis membrane 反渗透膜
8 permeate carrier 导流层

4.11 脱气塔 Degassing Tower

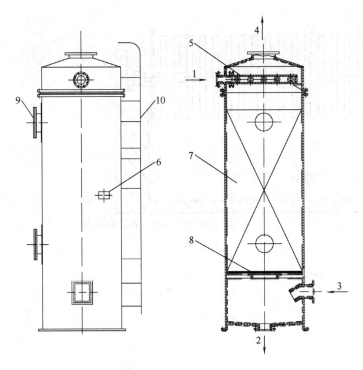

1 feed water inlet 进水口
2 product water outlet 出水口
3 air inlet 进风口
4 exhaust outlet 排气口
5 water distribution device 进水装置
6 nameplate 铭牌
7 filling 填料
8 jet tray 塔板
9 manhole 人孔
10 ladder 梯子

4.12 微孔曝气器 Fine Bubble Diffusion Aerator

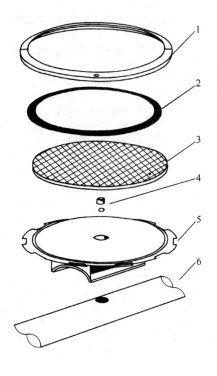

1 screw-on ring 螺盖
2 retaining ring 密封垫圈
3 membrane disc 布气膜片
4 non-return valve 止回阀
5 diffuser base 布气托盘
6 air distribution pipe 布气管

4.13 转刷曝气机　Brush Aerator

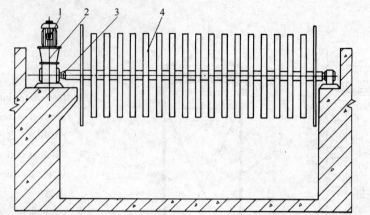

1　electric motor　电动机
2　speed reducer　减速器
3　flexible coupling　柔性联轴器
4　brush　转刷

4.14 表面曝气机　Surface Aerator

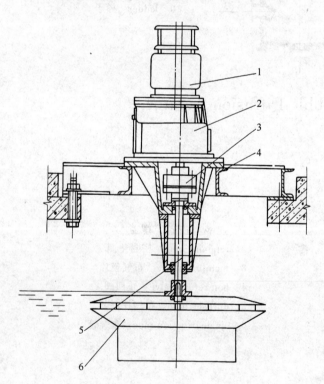

1　electric motor　电动机
2　speed reducer　减速器
3　coupling　联轴器
4　support　机座
5　shaft　主轴
6　impeller　叶轮

4.15 火炬 Flare Stacks

4.15.1 火炬系统 Flare System

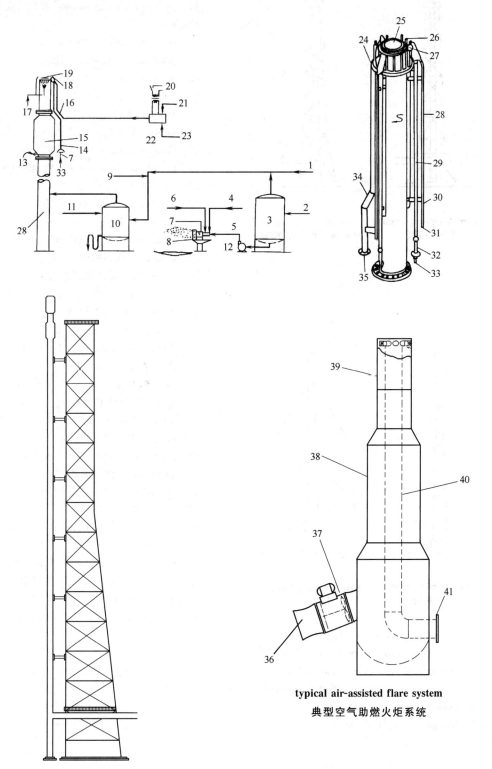

typical air-assisted flare system
典型空气助燃火炬系统

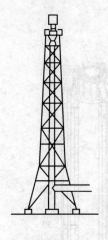
derrick-support structure 塔架式结构

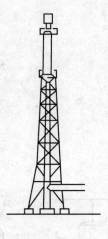
half derrick-support structure 半塔架式结构

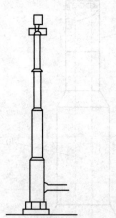
self-supported structure 自支撑式机构

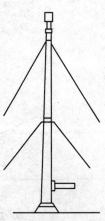
guyed-supported structure 拉线式机构

1 dry gas from process 工艺干气
2 mixed flow from process 工艺混合物料
3 liquid-vapor separator 气液分离器
4 atomizing air or steam 雾化空气或蒸汽
5 liquid 液体
6 fuel gas 燃料气
7 atmospheric air 常压空气
8 liquid field flare 燃液地面火炬
9 methane purge 甲烷吹扫气
10 seal drum 密封罐
11 purge or heating steam 吹扫气或热蒸汽
12 pump 泵
13 purge gas inlet 吹扫气入口
14 pilot tubes 长明灯导管
15 molecular seal 分子密封器
16 igniter tubes 引火管
17 steam 蒸汽
18 pilots 长明灯
19 flare 火炬头
20 battery or plant power 电池或工厂电源
21 air 空气
22 ignition system 点火系统
23 methane 甲烷
24 steam manifold 蒸汽集气管
25 flare tip 火炬喷嘴
26 steam tip 无烟火炬蒸汽喷嘴
27 ignitor tip 点燃器顶部的引火器喷嘴
28 flare stack 火炬筒体
29 pilot gas tube 长明灯供气管
30 flame front tube for pilot ignition 引火器点火用的火焰管
31 gas and flame from ignition system 从点火系统来的气体和火焰
32 pilot fuel-air mixer assembly 长明灯进气口-空气混合器（组件）
33 pilot gas supply 长明灯进气口
34 steam line 供气管线
35 steam connection 蒸汽接口
36 inlet bell 入口喇叭口
37 vane-axial low pressure air blower 低压轴流鼓风机
38 low-pressure air riser 低压空气上升管
39 stainless steel flare tip 不锈钢火炬头
40 gas riser 气体上升管
41 flanged inlet for gases to be flared 火炬气入口法兰

4.15.2 火炬头 Flare Tip

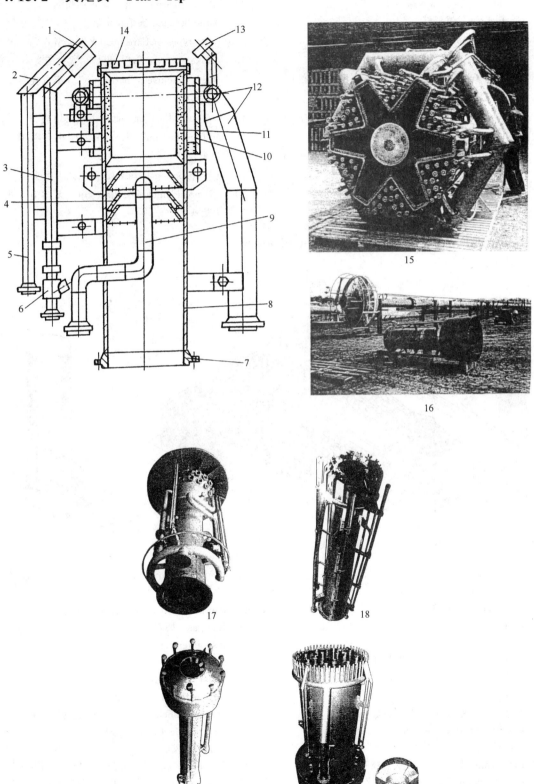

1 pilots 长明灯
2 igniter tips 点火器
3 fuel gas 燃料气
4 dynamic seal 动态密封器
5 igniter tubes 引火管
6 pilot fuel-air mixer assembly 长明灯燃料-空气混合器(组件)
7 flange 法兰
8 flare 火炬头(本体)
9 centric steam tip assembly 中心蒸汽喷嘴组件
10 wind proofing cover 防风罩
11 lining 耐火衬里
12 steam manifold 蒸汽集气管
13 "plum-blossom" tip 梅花形喷嘴
14 flame holder 火焰稳定器
(15~20 types of flare tip 火炬头型式)
15 smokeless "star tip" 无烟星式火炬头
16 steam smokeless "barrel tip" 蒸汽无烟筒式火炬头
17 low noise flare tip 低噪声式火炬头
18 "plum-blossom" flare tip 梅花形喷嘴火炬头
19 wind proofing flare tip 防风式火炬头
20 low noise smokeless flare tip 无烟消音式火炬头

4.15.3 地面火炬 Multi-point Ground Flare

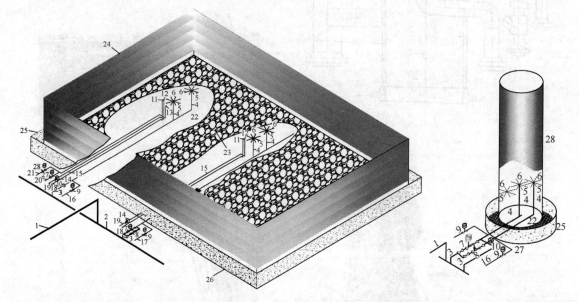

1 flare gas header 可燃气总管
2 distribute maidle 分配管
3 runner 分支管
4 riser 竖管
5 tip 燃烧器
6 flare gas orifices 可燃气体喷射孔
7 stage valve 分级控制阀
8 emergency bypass 紧急旁路
9 purge gas valve 吹扫气阀
10 fire arrester 阻火器
11,12 pilots 长明灯
13 steam ejector 消烟蒸汽喷嘴
14 manual igniter 爆燃式点火器
15 ignater tube 引火管
16 continues purge nitrogen or fuel gas 连续吹扫氮气燃料气
17 end purge nitrogen 后吹扫氮气
18 fuel gas 燃料气
19 instrument air 仪表空气
20 assist steam 消烟蒸汽
21 assist air 助燃空气
22 combustion area 燃烧区
23 gravel or shelter plate 碎石或防辐射板
24 radiate shield fence 围栏
25 wall 挡风墙
26 open multi-point ground flare 开放式地面火炬
27 close multi-point ground flare 封闭式地面火炬
28 stack 炉膛

4.16 输送机和提升机　Conveyor and Elevator

4.16.1 带式输送机　Belt（Band）Conveyor

4.16.1.1 带式输送机示意图　Belt Conveyor Sketch

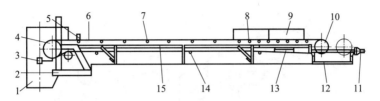

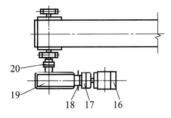

1	head chute　头部漏斗		11	take up device　张紧装置
2	head frame　头架		12	tail frame　尾架
3	head scraper　头部清扫器		13	scraper　清扫器
4	drive pulley　传动滚筒		14	return idlers　回程托辊
5	safety device　安全保护装置		15	stringer　中间架
6	belt　输送带		16	motor　电机
7	carrying idlers　承载托辊		17	hydraulic coupling　液力耦合器
8	impact idlers　缓冲托辊		18	break　制动器
9	loading skirt　导料槽		19	gearbox　减速器
10	tail pulley　尾部滚筒		20	coupling　联轴器

4.16.1.2 带式输送机系统　Belt Conveyor Systems

single pulley drive　单轮驱动

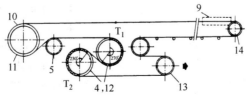

double or dual pulley drive　双轮驱动

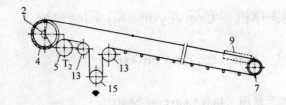

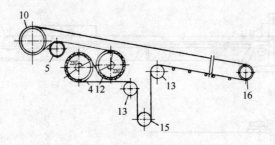

inclined conveyor 倾斜式输送机

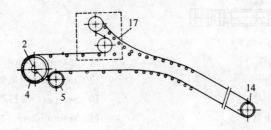

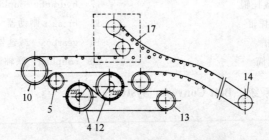

tripper conveyor 自动倾斜卸料输送机

1	high tension 高张力端		10	head pulley 头轮
2	220° angle of wrap 220°包角		11	head or discharge pulley 头部或卸料轮
3	scraper 清扫器		12	440° angle of wrap 440°包角
4	driving pulley 驱动轮		13	bend pulley 改向滚筒
5	snub pulley 增面滚筒		14	low tension 低张力端
6	plough 扫除器		15	gravity take-up 重力张紧装置
7	tail pulley 尾轮		16	take-up pulley 张紧轮
8	screw take-up 张紧螺钉		17	tripper 倾卸器
9	loading section 装载段			

4.16.1.3 带式输送机部件　Belt (Band) Conveyor Components

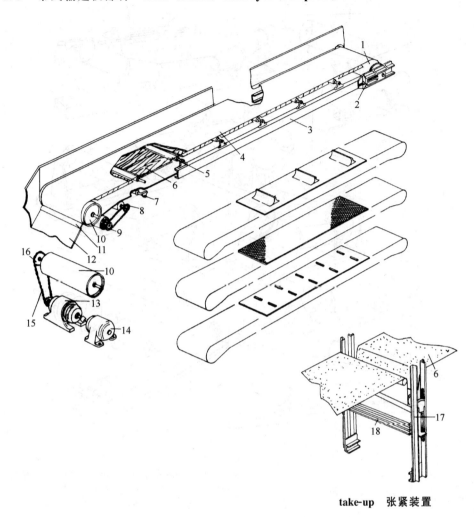

take-up　张紧装置

1	tail pulley　尾轮		10	driving pulley　驱动轮
2	take-up　张紧装置		11	scraper　刮板
3	stringer　中间架		12	chute　溜槽
4	conveyor belt　输送带		13	reduction gear　减速齿轮箱
5	carrying idler　承重托辊		14	motor　电动机
6	belt　（输送）带		15	transmission chain　传动链条
7	return idler　回程托辊		16	sprocket　链轮
8	snub pulley　防滑增面托辊		17	take-up frame　张紧装置架
9	dust sweeper　清扫器		18	counter weight　张紧重锤

135

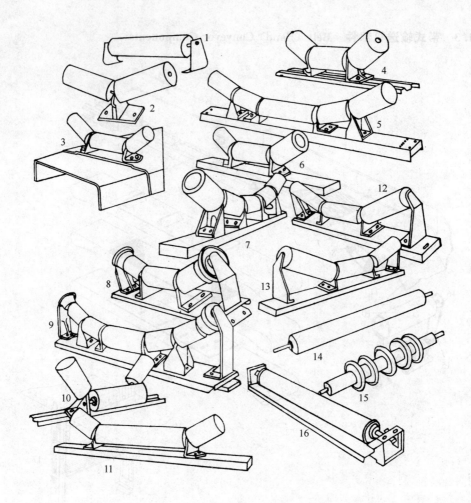

1	for a flat belt　平皮带用	9	steering, five roller　导向式，五辊
2	two rollers　双辊	10	deep trough　深槽式
3	on inverted trough frame　装在槽形架上	11	transition idler　槽角过渡用
4	closed end on idler-board, three-roller　装在惰轮底板上的闭端式，三辊	12	end supported, roller bearing　端部支承式，滚珠轴承
5	closed end on idler-board, five-roller　装在惰轮底板上的闭端式，五辊	13	end supported, ball bearing　端部支承式，球轴承
6	cushion, three-roller　减震式，三辊	14	flat idler　平托辊
7	cushion, five-roller　减震式，五辊	15	disc return idler　梳形回程托辊
8	steering, three-roller　导向式，三辊	16	steering return　导向回程（惰轮）

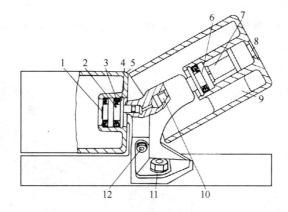

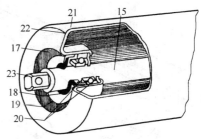

drop-in idler roll 落入式托辊

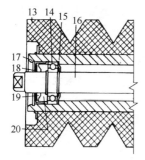

section of a closed end idler 闭端式托辊截面图

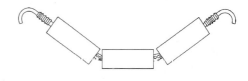

suspended carrying idler 悬挂式运载托辊

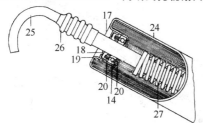

drop in impact idler roll 落入式耐震托辊

1	shoulder on spindle 轴肩		15	inner grease retainer 润滑油内护圈
2	grease-hole 润滑脂孔		16	shaft 轴
3	grease-labyrinth 润滑脂迷宫（密封）		17	bores 孔
4	spring ring 弹簧圈		18	contact seal 接触密封圈
5	round 倒圆角		19	seal retained by an outer steel shield 装在外钢罩内的密封圈
6	shoulders 肩部		20	grease chambers 润滑脂室
7	lubricant 润滑剂		21	shell thickness 辊体厚度
8	permanent plug 永久性堵头		22	all-welded steel shell 全焊接式钢辊体
9	equidistant ribs 等距肋		23	shaft ends 轴端
10	spindle 心轴		24	all-welded steel roll 全焊接钢辊体
11	brackets 支架		25	zinc plated sliding hook bar 镀锌滑动挂钩棒
12	oiler 油杯		26	flexible bellows 挠性波纹管
13	fire resistant synthetic rubber cushions 耐火合成橡胶垫		27	heavy duty spring 重型弹簧
14	ball bearing 球轴承，滚珠轴承			

4.16.1.4 拼装式皮带输送机和移动式皮带输送机 Pre-built Sectional Belt Conveyor and Mobile Belt Conveyor

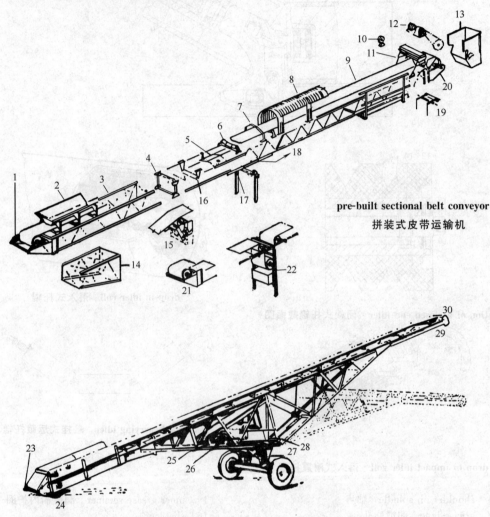

pre-built sectional belt conveyor 拼装式皮带运输机

mobile belt conveyor 移动式皮带运输机

1	screw take-up foot terminal 尾端螺旋张紧装置	16	return belt idler 下托辊
2	loading hopper 给料斗	17	bent 托架
3	truss section 桁架段	18	corbel connection 连接托架
4	lateral frame 横向构架	19	belt wiper 清带刷
5	decking 遮板	20	head end 前端
6	troughed belt idler 槽形带托辊	21	fixed end 固定式尾端
7	wind cover fixing device 防风罩固定装置	22	gravity take-up 重力张紧装置
8	hood 罩壳	23	trailer bar 牵引杆
9	conveyor belt 输送带	24	tail pully 尾滚轮
10	backstop 制动器	25	belt tensioning drum 皮带张紧滚筒
11	drive support 驱动装置支架	26	belt driving drum 皮带驱动滚筒
12	drive 驱动装置	27	axle frame 轮架
13	discharge chute 卸料槽斗	28	oil jack for luffing 转向液压千斤顶
14	transition section 过渡段	29	belt cleaner 清带器
15	knuckle joint 折向节点，转向节	30	head pully 头滚轮

4.16.1.5 裙状挡板大倾角带式输送机 Skirt Side Plate Inclined Belt Conveyor

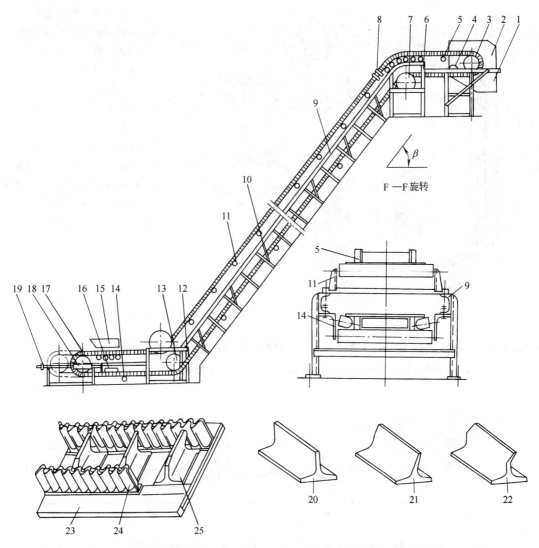

1	discharge chute 卸料漏斗		13	band pulley 改向滚筒
2	head hood 头部护罩		14	returning idler 下托辊
3	driving pulley 传动滚筒		15	loading channel 导料槽
4	beating cleaner 拍打清扫器		16	safety cleaner 空段清扫器
5	skirt side plate 挡边带		17	tension pulley 尾部滚筒
6	conveyor structure for arch section 凸弧段机架		18	tension device 拉紧装置
7	belt pressing wheel 压带轮		19	end structure 尾架
8	steering roller 挡辊		20	type T T型
9	main structure 中间机架		21	type C C型
10	leg of main structure 中间架支腿		22	type TC TC型
11	conveying idler 上托辊		23	base belt 基带
12	conveyor structure for up turning section 凹弧段机架		24	skirt side belt 挡边
			25	seperater 隔板

4.16.1.6 双带提升机 Twin Riser

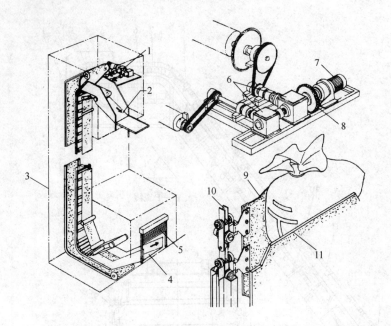

1　drive unit　传动装置
2　unloading chute　卸料槽
3　shaft　机壳
4　hatch　入口
5　fire door　消防门
6　reverse turn gear　反向转动齿轮箱
7　motor　电动机
8　torque limiter　力矩限制器
9　expansion cover belt　张紧带
10　main chain　主链
11　main belt　主带

4.16.1.7 中心距超过180m的单轮传动带式输送机用的张紧装置 Take-up Unit for Single Drum Drive Belt Conveyor Exceeding 180 Meters Centres

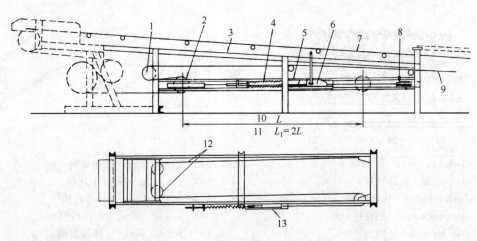

1　fixed pulley　固定轮
2　pulley in sliding carriage　滑动架内的带轮
3　deck plates over loop and jib　（输送）带环上方的悬臂面板
4　tension springs　张紧弹簧
5　belt tension scale　（输送）带的张紧计
6　tensioning unit　张紧装置
7　carrying belt　运输带
8　rope sheave　绳轮
9　return belt　回程带
10　carriage travel L　滑动架行程 L
11　belt storage $L_1=2L$　带的备用长度 $L_1=2L$
12　rope & pulley system　绳和滑轮系统
13　tension unit located on this side only　张紧装置，仅装在（图示）一侧

140

4.16.1.8 双轮传动带式输送机的张紧装置　Take-up Unit for Dual Drum Dive Belt Conveyor

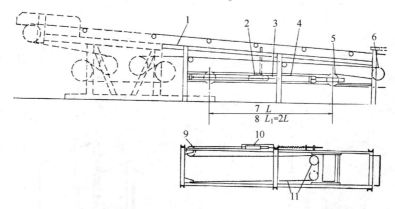

1　deck plates over loop and jib　（输送）带环上方的悬臂面板
2　tensioning unit on far side　张紧装置（在对面一侧）
3　belt tension scale on far side　（输送）带的张紧计（在对面一侧）
4　tension springs on far side　张紧弹簧（在对面一侧）
5　pulley in sliding carriage　滑动架内的带轮
6　fixed pulley　固定轮
7　carriage travel L　滑动架行程 L
8　belt storage $L_1 = 2L$　（输送）带的备用长度 $L_1 = 2L$
9　rope sheave　绳轮
10　tension unit located on this side only　张紧装置，仅在（图示）一侧
11　rope and pulley system　绳和滑轮系统

4.16.2　垂直提升（输送）机　Vertical Elevator(Conveyor)

4.16.2.1　斗式提升机　Bucket Elevator

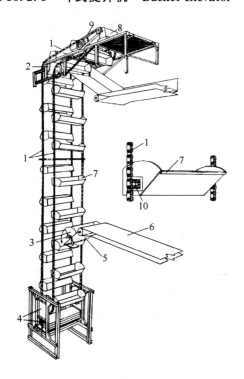

1　main chain　主链
2　sprocket　链轮
3　synchronizing damper　同步挡板
4　take-up weight　张紧重块
5　chute　滑板
6　band conveyor　带式输送机
7　bucket　斗
8　motor and speed reduction gear　电机和减速齿轮
9　transmission V-belt　V形传动带
10　attachment　连接件

141

4.16.2.2 垂直提升机和箱类提升机 Vertical Rising Conveyor and Case Elevator

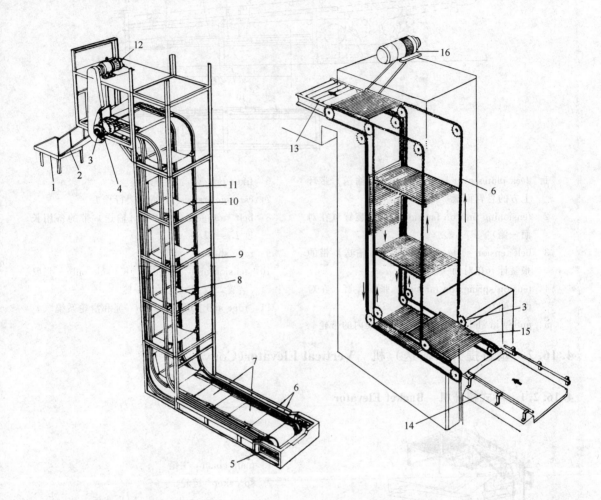

1 table 平台

2 chute 滑槽

3 sprocket 链轮

4 transmission chain 传动链

5 take-up screw 张紧螺杆

6 main chain 主链

7 loading position 装载位置

8 band rack 带式货台

9 frame 机架

10 attachment 连接件

11 guide rail 导轨

12 motor with reduction gear 带减速齿轮箱的电动机

13 unloader 卸料辊

14 loader 装载器

15 slat 条板

16 motor 电动机

4.16.3 埋刮板式输送机　En Masse Conveyor

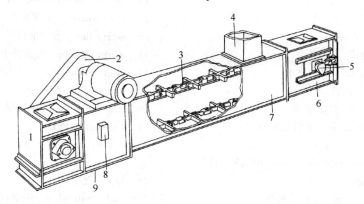

1　head section　头部
2　driving unit　驱动装置
3　en masse chain　刮板链条
4　inlet　加料口
5　chain breaking detector　断链指示器
6　end section　尾部
7　middle section　中间段
8　blocking detector　堵料探测口
9　outlet　卸料口

4.16.4 螺旋输送机　Screw Conveyor

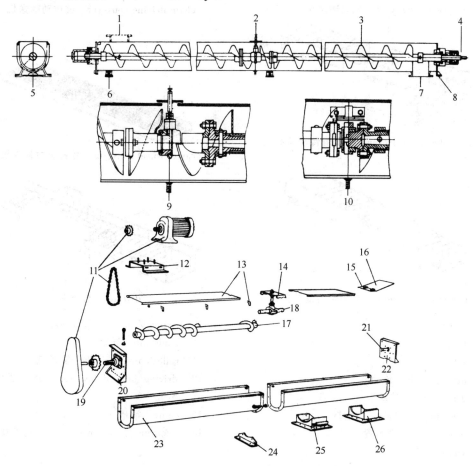

1　inlet　加料口
2　sleeve bearing　套筒轴承
3　helix　螺线
4　drive end　驱动端
5　trough　料槽
6　support sliding　滑动支承
7，26　outlet　出料口
8　support fixed　固定支承
9　lubricated bearing　油润滑轴承
10　dry bearing　干式轴承
11　motor, chain drive and guard　电动机、链传动和防护罩
12　motor platform　电动机底板
13　cover　盖
14　intermediate bearing　中间轴承
15　microswitch　微动开关
16　safety flap　安全活板
17　worm on pipe　装在管上的螺旋板
18　intermediate shaft　中间轴
19　drive shaft　传动轴
20　box end drive　箱端驱动装置
21　tail shaft　尾轴
22　tail bearing　尾部轴承
23　casing　壳体
24　foot　支座
25　outlet with slide　带滑板的出料口

4.16.5　辊子输送机　Roller Conveyor

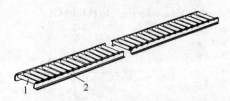

gravity conveyor　重力输送机

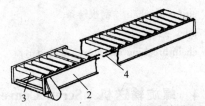

chain driving conveyor　链传动输送机

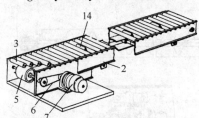

flat band driving conveyor　平皮带传动输送机

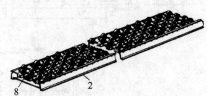

gravity wheel conveyor　轮式重力输送机

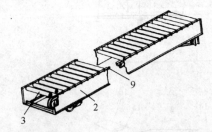

Vband driving conveyor　V形带传动输送机

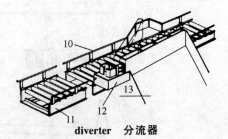

diverter　分流器

1　idler roller　惰辊
2　frame　机架
3　roller　辊子
4　driving chain　传动链
5　driving belt　传动带
6　transmission chain　传动链
7　drive unit　传动装置
8　wheel　轮
9　driving V belt　驱动V形带
10　guide rail　导轨
11　roller conveyor　辊子输送机
12　driving unit of diverter　分流器的传动装置
13　chute　溜槽
14　automatic pressure regulator　自动压力调整器

4.16.6 吊挂式链输送机和振动输送机 Chain Trolley Conveyor and Vibrating Conveyor

4.16.6.1 吊挂式链输送机 Chain Trolley Conveyor

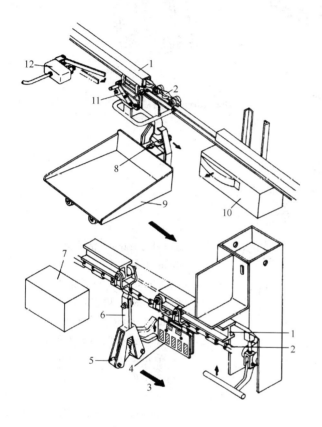

1 rail 轨道
2 chain 链
3 retractable bag release mechanism 伸缩式料袋释放机构
4 magnetic code memory 磁代码存储器
5 bag carrier 料袋运载架
6 hanger 吊钩
7 sensing device 传感装置
8 trigger hook 碰钩
9 tray 盘
10 discharge device 卸载装置
11 escort memory device 跟踪存储装置
12 code reading device 读码装置

4.16.6.2 吊挂式链输送机部件 Unit of Chain Trolley Conveyor

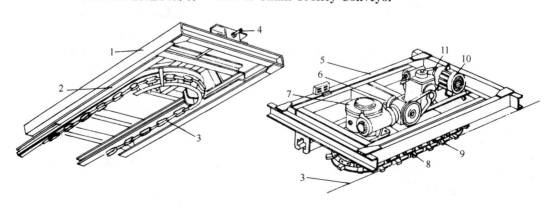

drive unit（caterpillar drive system） 传动装置（履带传动系统）

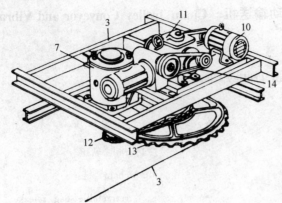

drive unit（sprocket drive system）
传动装置（链轮传动系统）

bag carrier 料袋运载架

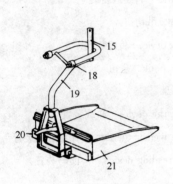

tray（single tilting）
盘（单向倾卸）

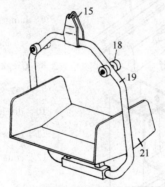

tray（double tilting）
盘（双向倾卸）

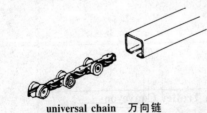

universal chain　万向链

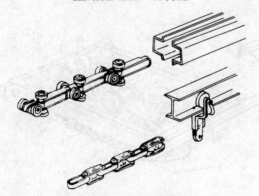

rivetless chain　无铆钉链

1	frame	架
2	movable frame	活动架
3	main chain	主链
4	tension screw	拉紧螺钉
5	fixed frame	固定架
6	floating frame	浮动架
7	reduction gear	减速齿轮箱
8	caterpillar chain	履带链
9	dog	卡齿
10	motor	电动机
11	speed change gear	变速齿轮箱
12	transmission chain	传动链
13	driving sprocket	驱动链轮
14	transmission V belt	V形传动带
15	hanger	吊架
16	lever	触杆
17	gripping tooth	夹紧齿轮
18	guide roller	导辊
19	hang pipe	吊管
20	trigger hook	碰钩
21	tray	盘

4.16.6.3 振动输送机　Vibrating Conveyor

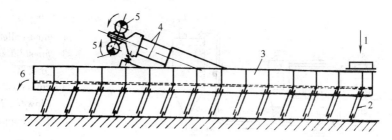

1　inlet　物料进口
2　leaf-spring guides　板簧导向装置
3　trough　槽体
4　transmission springs　传动弹簧
5　two identical out-of balance motors　两台对称的偏心电机
6　outlet　物料出口

4.16.7　气力输送机系统　Pneumatic Conveying System

4.16.7.1　正压气力输送系统　Pressure Pneumatic Conveying System

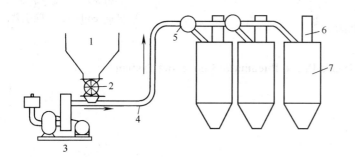

1　feed hopper　进料斗
2　rotary valve　旋转阀
3　blower　鼓风机
4　pipe　输送管
5　diverter valve　换向阀
6　dust-collector　除尘器
7　storage silo　储料仓

4.16.7.2　负压气力输送系统　Vacuum Pneumatic Conveying System

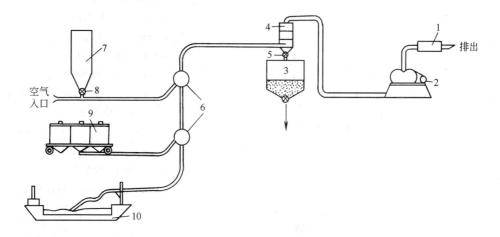

1　silencer　消音器
2　blower　鼓风机
3　silo　料仓
4　dust-collector　除尘器
5　discharge rotary valve　卸料旋转阀
6　diverter valve　换向阀
7　feed hopper　加料斗
8　charge rotary valve　加料旋转阀
9　train hopper wagon　铁路漏斗车
10　cabin　船舱

4.16.7.3 正负压相结合的系统 Pressure with Vacuum Pneumatic Conveying System

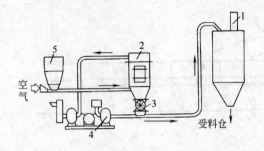

1 dust-collector 除尘器
2 air-solid seperator 气固分离器
3 rotary valve 旋转阀
4 blower 鼓风机
5 feed hopper 加料斗

4.16.7.4 封闭环形气力输送系统 Closed Pneumatic Conveying System

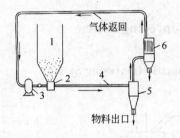

1 feed hopper 进料斗
2 charging unit 供料器
3 blower 鼓风机
4 pipe 输送管
5 cyclone 旋风分离器
6 dust-collector 除尘器

4.16.7.5 密相气力输送系统 Dense Phase Pneumatic Conveying System

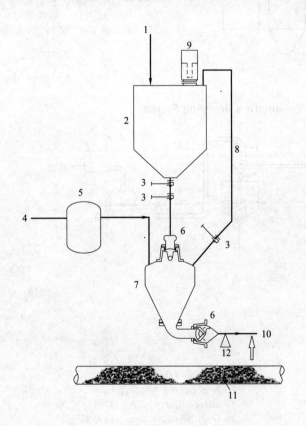

1 incoming material 进料
2 feed hopper 进料料斗
3 slide valve 插板阀
4 air inlet 气体入口
5 surge drum 缓冲罐
6 dome valve 圆顶阀
7 dense phase conveyor 密相发送罐
8 vent line 排气线
9 feed hopper filter 进料料斗过滤器
10 downstream 下游
11 plug flow 栓塞流
12 damping support 减震管支座

4.16.7.6 稀相气力输送系统 Dilute Phase Pneumatic Conveying System

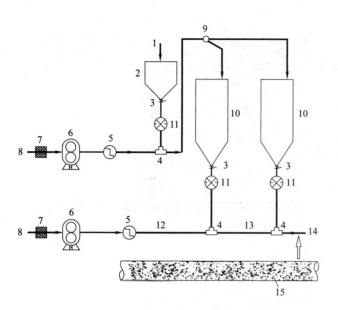

1 incoming material 进料
2 feed hopper 进料料斗
3 slide valve 插板阀
4 feeding shoe 下料三通
5 heat exchanger 换热器
6 blower 风机
7 air inlet filter 进气过滤器
8 air inlet 气体入口
9 diverter valve 换向阀
10 silo 料仓
11 rotary valve 旋转阀
12 conveying air pipeline 输送气管线
13 shot-peened pipeline 喷砂管线
14 downstream 下游
15 dispersion flow 悬浮分散流

4.17 破碎和筛分设备 Crushing and Screening Equipment

4.17.1 颚式冲击破碎机 Impact Jaw Crusher

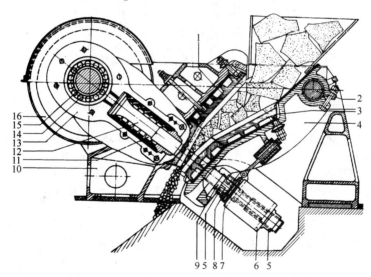

1 stationary jaw 固定颚板
2 swing jaw axle 摆动颚轴
3 swing jaw plate 摆动颚板
4 swing jaw 摆动颚
5 gap adjusting nuts 间隙调整螺母
6 cross member 横梁
7 toggle seat 肘座
8 wedge for adjusting nut 调整螺母的楔
9 toggle plate 肘板
10 crusher frame 破碎机架
11 overload protection spring 超负荷保护弹簧
12 pitman 连杆
13 spring housing 弹簧座体
14 self-aligning roller bearing 自定心滚柱轴承
15 eccentric shaft 偏心轴
16 flywheel 飞轮

4.17.2 颚式破碎机 Jaw Crusher

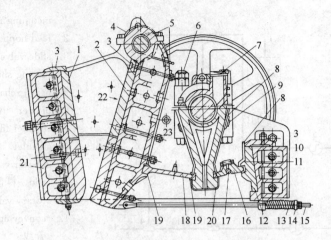

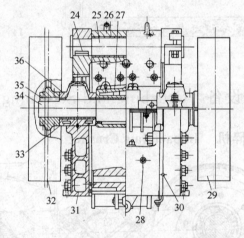

1	fixed jaw plate-manganese steel 固定颚板-锰钢	18	front toggle 前肘
2	swing jaw plate 摆动颚板	19	toggle seat 肘座
3	retainer 压板	20	pitman 连杆
4	swing jaw 摆动颚	21	front frame 前机架
5	alemite grease fittings and pipe 压力润滑器和管子	22	check plate 挡板
		23	frame rod 机架拉杆
6	pitman cap bolts 连杆盖螺栓	24	key 键
7	pitman grease well 连杆润滑脂池	25	swing jaw shaft 摆动颚轴
8	oil pipe 油管	26	bushings 衬套
9	oil pipe wedges 油管楔铁	27	pitman cap 连杆盖
10	rear frame 后机架	28	toggle block bolts 肘块螺栓
11	shims 垫片	29	flywheel 飞轮
12	toggle block 肘块	30	rear frame dowels 后机架销钉
13	tension rod spring 拉杆弹簧	31	side section of frame 机架侧剖面
14	washer 拉杆弹簧垫圈	32	driving sheave 传动带轮
15	tension rod 拉杆	33	pitman shaft bearing base 连杆轴轴承底座
16	rear toggle 后肘	34	pitman shaft 连杆轴
17	toggle hinge pin 肘铰接销	35	sheave key 带轮键
		36	bearing cap 轴承盖

4.17.3 回转球形破碎机　Gyrasphere Crusher

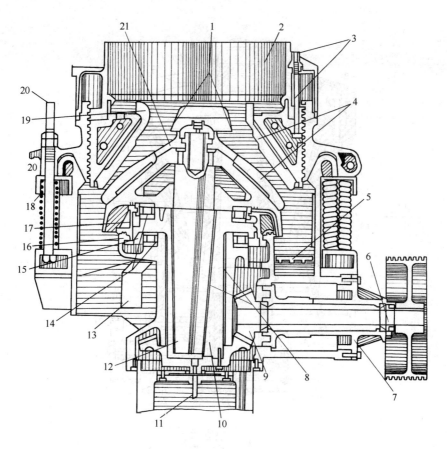

1　large unobstructed feed opening　敞通的大加料口
2　heavy steel hopper　重型钢料斗
3　discharge opening adjusting mechanism　出料口调节机构
4　manganese steel crushing members　锰钢破碎板
5　replaceable countershaft box shield　可更换的主轴箱护盖
6　heavy countershaft　加重传动轴
7　countershaft box removable as a unit　可整体拆卸的传动轴箱
8　large diameter long sleeve eccentric bearings　大直径长套筒偏心轴承
9　cut steel drive gears　铣齿钢驱动齿轮
10　large heavy eccentric　重型大偏心轴
11　main oil supply　主供油口
12　large diameter heavy duty main shaft　大直径重型主轴
13　frame arm shields　机架臂护板
14　heavy duty roller bearings　重型滚柱轴承
15　cam and lever　凸轮和杠杆
16　piston ring and labyrinth seals　活塞环和迷宫式密封
17　rotary seal　回转密封件
18　spring relief　弹簧保险
19　gun lock　扣紧枪机
20　supports for adjusting mechanism　调节机构立柱
21　spherical shaped crushing head　球形破碎头

4.17.4 盘式回转破碎机　Tray Type Gyratory Crusher

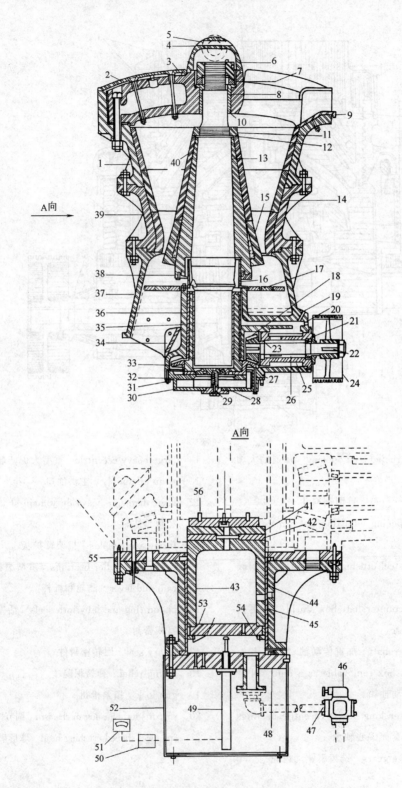

#	English	中文
1	top shell	上部机体
2	spider	多幅架
3	spider shields	多幅架护罩
4	spider cap	多幅架帽
5	eyebolt for main shaft	主轴环首螺栓
6	suspension nut	悬挂螺母
7	wearing ring	耐磨环
8	suspension bushing	悬挂轴衬
9	hopper wearing plates	加料斗耐磨板
10	suspension oil seal	悬挂轴封
11	head nut	轴头（肩）螺母
12	head nut locking pin	轴头螺母锁紧销
13	main shaft	主轴
14	concave ring	凹面圈（衬板）
15	lower head mantle	头部下衬板
16	dust seal bonnet	防尘套
17	rib shields	肋护板
18	driving pinion	传动小齿轮
19	oil fill and inspection plug	注油和检查丝堵
20	cleanout plug	清理丝堵
21	countershaft seal retainer	传动轴密封保持盖
22	key for driving sheave	传动带轮键
23	countershaft Timken bearings	传动轴铬镍钼耐热钢轴承
24	driving sheave	传动带轮
25	countershaft housing	传动轴套筒
26	countershaft	传动轴
27	oil level pipe	油位管
28	oil reservoir drain plug	油檐排放塞
29	oil inlet body	进油口体
30，55	bottom plate	底板
31	bottom plate bolts	底板螺栓
32	eccentric wearing ring	耐磨偏心环
33	driving gear	驱动齿轮
34	gear case shields	齿轮箱罩
35	eccentric sleeve	偏心套
36	eccentric bushing	偏心（套）轴衬
37	bottom shell	下部机体
38	dust seal ring retainer	防尘圈保持环
39	intermediate shell	中部机体
40	upper head mantle	头部上衬板
41	center wearing ring	中间耐磨环
42	piston wearing ring	活塞耐磨环
43	piston	活塞
44	upper piston bushing	活塞上部衬套
45	lower piston bushing	活塞下部衬套
46	relief valve	安全阀
47	adapter	接头
48	bottom cover	底盖
49	transmitter	变送器
50	controller	控制器
51	position indicator	位置指示器
52	bottom shield	底罩
53	packing	填料
54	seal retainer	密封保持器
56	wearing ring dowel	耐磨环定位销

4.17.5 液压锥形破碎机和冲击破碎机（叶片破碎机） Hydraulic Cone Crusher and Impact Crusher (Impeller Breaker)

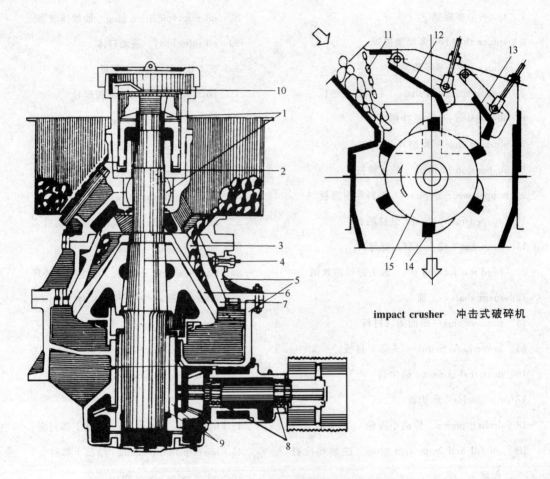

impact crusher 冲击式破碎机

hydraulic cone crusher 液压锥形破碎机

1	upper main bearing 上主轴承		9	eccentric sleeve 偏心套筒
2	hydraulic cylinder 液压缸		10	selsyn generator 同步发动机
3	crushing chamber 破碎室		11	chain curtain 链帘
4	main shaft 主轴		12	impact plate No. 1 1号冲击板
5	frame 固定锥壳		13	impact plate No. 2 2号冲击板
6	mantle 回转衬板		14	striking blades 撞击片
7	bowl liner 锥壳衬板		15	rotor 转子
8	counter shaft bearing 传动轴轴承			

4.17.6 辊子粉碎机　Roller Mill

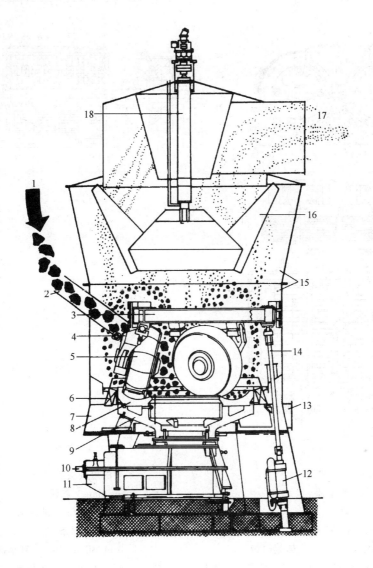

1	raw material　原料	11	speed reducer　减速器
2	feed spout　加料流槽	12	hydraulic loading cylinder　承重液压缸
3	roller support　辊子支架	13	hot gas intake port　热气体吸入口
4	roller housing　辊子壳	14	loading rod　承重杆
5	grinding roller　碾磨辊	15	totally enclosed housing　全封闭外壳
6	ported air ring　排气环	16	classifier blade　分级器叶片
7	gas plenum　气体强制通风	17	product discharge port　成品出料口
8	wearing ring　耐磨环	18	classifier shaft (drive by variable speed drive)　分级器轴（变速传动）
9	rotating grinding table　转动碾磨盘		
10	high speed shaft　高速轴		

4.17.7 双轴锤式破碎机　Double Shaft Hammer Mill

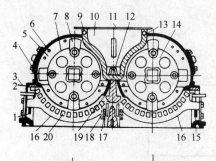
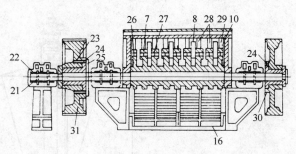
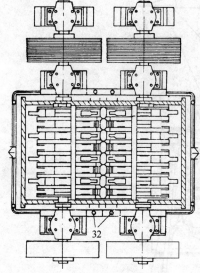

1	support bar　支承扁钢		条架的可调支承扁钢
2	locking bars　锁紧棒	18	cover plate　盖板
3	flat cover　平盖	19	breaking plates　破碎板
4	inspection door　检查门	20	grid bars　栅条
5	side liners for casing　壳体侧衬板	21	bearing oiling rings　轴承润滑环
6	bolts for casing side liners　壳体侧衬板螺栓	22	bearing bushes in halves　对开轴瓦
7	hammer carriers, regular　锤盘——常规用的	23	shearing pin bushes　剪切销套
8	hammers　锤	24	shearing pins　剪切销
9	tie bolts　拉杆螺栓	25	covers　盖
10	end plate　端板	26	end hammer carriers　端部锤盘
11	feed opening　进料口	27	stop plate　挡板
12	anvil block　锤砧块	28	bars for receiving basket　受料框条板
13	grate basket　栅筐	29	bars end for receiving basket　受料框端部条板
14	rotor　转子		
15	door　门	30	flywheel　飞轮
16	grid frames　栅条架	31	driving wheel　驱动轮
17	adjustable support bar for grid frames　栅	32	adjusting bolts　调整螺栓

156

4.17.8 磨机 Mill

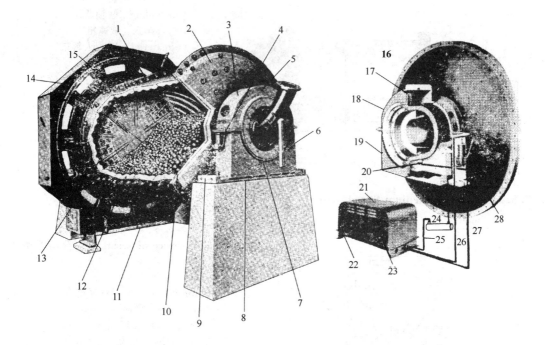

1 heavy duty shell 加重机壳
2 conical head and trunnion（cast integral） 锥形头盖和轴颈（铸成整体）
3 water cooled bearing 水冷轴承
4 triple protected bearing lubrication 三重保护轴承润滑
5 oversize trunnion bearing surfaces 加大尺寸的轴承承压面
6 accurate alignment of mill on bearings 在轴承上球磨机精密对准直线
7 oil tight bearing seal 油密封的轴承密封
8 thick machined steel soleplate 机加工的厚钢底板
9 adjusting jack screw for bearing position 调整轴承位置用的螺旋顶重器
10 head liners 头盖衬里
11 dust and water-tight grinding chamber 防尘防水的碾磨室
12 liners 衬里
13 diaphragm grate 孔栅板
14 totally enclosed gear guards 全封闭齿轮护罩
15 spur or single helical gears 正齿轮或斜齿轮
16 automatic lubrication 自动润滑
17 oil bucket 油斗
18 bearing cap 轴承罩
19 bearing base 轴承座
20 bearing insert 轴承嵌片
21 protecting cover 保护盖
22 high pressure gage 高压表
23 low pressure gage 低压表
24 heater 加热器
25 suction 吸入
26 low pressure 低压
27 high pressure 高压
28 auxiliary manual high pressure starting pump 手动高压辅助启动泵

4.17.8.1 球磨机 Ball Mill

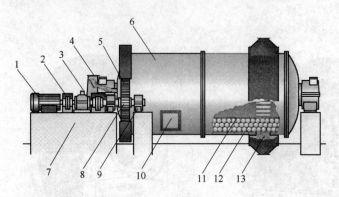

1 motor 电机
2 coupling 联轴器
3 gearbox 减速箱
4 material inlet 入料口
5 large gear 大齿轮
6 shell 壳体
7 support 支座
8 small gear 小齿轮
9 bearing pedestal 轴承座
10 access hole 检修口
11 mill ball 研磨球
12 liner plate 衬板
13 material outlet 出料口

4.17.8.2 中速磨 Medium Speed Mill

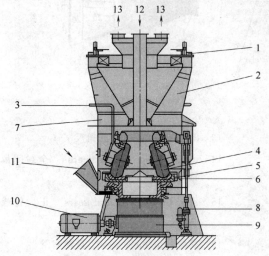

1 adjust device of separator 分离器调节装置
2 separator 分离器
3 seal air system 密封气系统
4 grinding roll 磨辊
5 nozzle ring 喷嘴环
6 grinding table 磨盘
7 shell 壳体
8 hydraulic push pod system 液压压杆系统
9 gearbox 减速箱
10 mail motor 主电机
11 primary air inlet 一次风入口
12 coal inlet 进煤
13 coal fines outlet 出粉口

4.17.8.3 风扇磨 Fan Mill

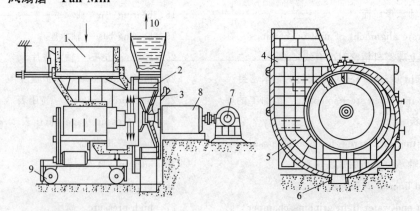

1 loading chute 给料槽
2 impact wheel 冲击轮
3 front beater 前置锤
4 discharge chute 排料管道
5 volute 蜗壳区
6 pit 凹坑
7 motor 电机
8 gearbox 减速箱
9 maintenance bogie 检修小车
10 coal fines outlet 出粉口

4.17.9 干燥粉磨机 Dryer-Pulveriser

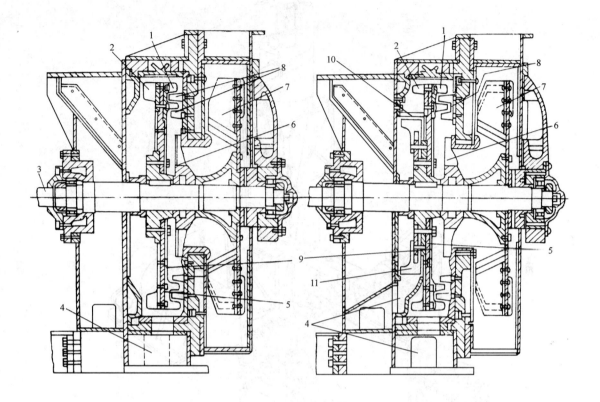

1　second effect peg segments　第二级磨柱

2　first effect hammers　第一级磨锤

3　rotor shaft　转子轴

4　tramp metal pocket　金属夹杂物槽

5　rotor disc　转子盘

6　ejectors　分离臂

7　internal fan　内风扇叶

8　stator pegs　定子柱

9　second effect rotor wear plate　第二级转子的耐磨板

10　hammer screen　锤栅

11　swing hammers　摆动锤

4.17.10 返混设备布置 Layout of Back Mixing Equipment

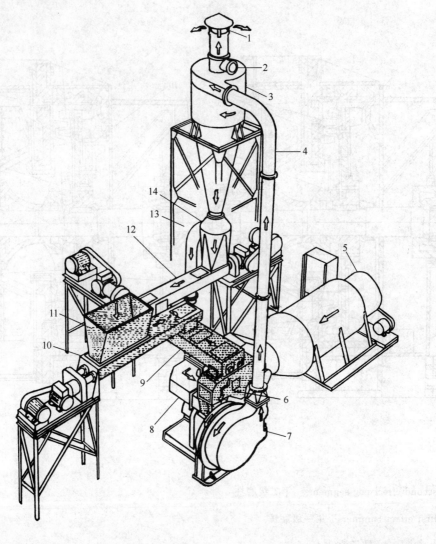

1 vent to atmosphere or to secondary dust collecting equipment 通入大气或二级集尘设备的排气口
2 ducts to secondary collectors 至二级收集器的管道
3 cyclone collector 旋风收集器
4 finished product to cyclone 成品至旋风收集器
5 stove for supplying hot air 供应热空气的加热炉
6 magnetic separator 磁力分离器
7 dryer-pulveriser 干燥-粉磨机
8 hot gas duct 热气道
9 twin paddle backmixer 双桨返混器
10 raw wet material feeder 湿原料给料器
11 raw wet material inlet hopper 湿原料入口斗
12 pulverised dry material feeder 干粉磨料给料器
13 finished product chute 成品溜槽
14 distributor 分配器

4.17.11 振动筛　Vibrating Screen

4.17.11.1 往复式振动筛　Reciprocating Vibrating Screen

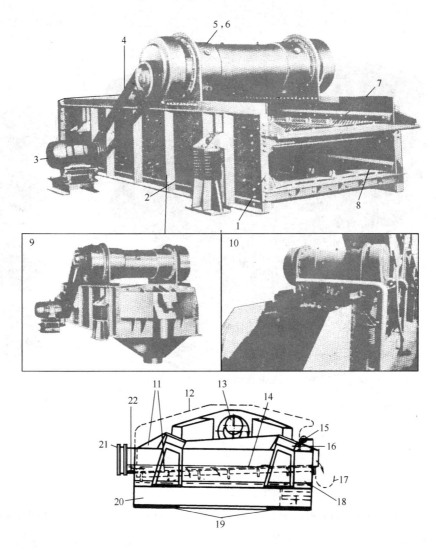

1	frame　筛架		增强塑料制的罩
2	spring pad　弹簧垫	13	vibrator　振动器
3	motor　电动机	14	screen plate　筛板
4	V-belt　V带	15	shower pipe　喷水管
5	eccentric shaft　偏心轴	16	rubber supports　橡皮垫
6	counter weight　配重	17	separated tailing trough　分离余料溜槽
7	coarse screen　粗筛网	18	vat　受料槽
8	fine screen　细筛网	19	damping rubber plate　减震橡胶板
9	totally enclosed screen　全封闭筛	20	concrete plate　混凝土板
10	washing screen　冲洗筛	21	flexible inlet flange sealing　挠性进口法兰密封
11	screen cradle　筛摇架		
12	hood made of fibreglass plastic　玻璃纤维	22	inlet sealing　进口密封

161

4.17.11.2 圆振筛 Round Separator

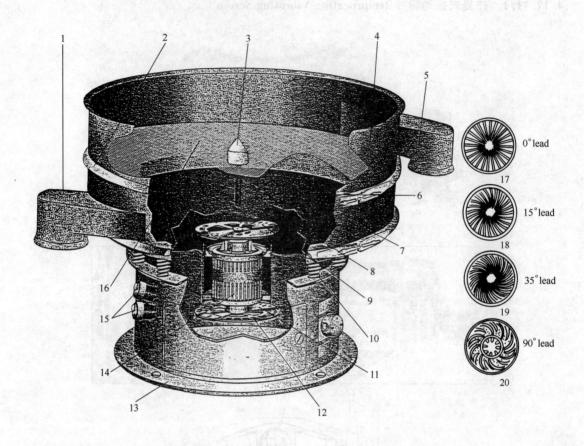

1	undersize discharge 筛下物料出口	11	bottom force wheel 下部压紧轮
2	quick-release clamp ring 快拆夹持环	12	lead angle adjustment 垂锤夹角调整器
3	center-tie-down 中心张紧	13	base 基座
4	upper frame 上部框架	14	motion generator 振动发生器
5	oversize discharge 筛上物料出口	15	automatic re-greaser 滑脂自动注油器
6	discharge dome 排料室	16	screen cloth 筛网
7	lower frame 下部框架	17	0° lead 0°重锤夹角(物料方向图)
8	platform 承料平台	18	15° lead 15°重锤夹角(物料方向图)
9	top force wheel 上部压紧轮	19	35° lead 35°重锤夹角(物料方向图)
10	springs 弹簧	20	90° lead 90°重锤夹角(物料方向图)

4.17.12 分级机 Classifiers

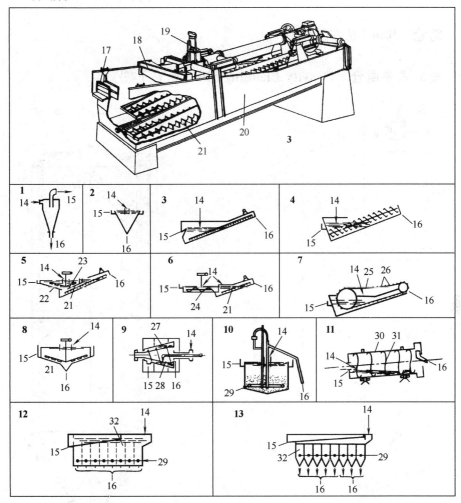

1. liquid cyclone 水力旋流（分级）器
2. cone classifier 圆锥分级机
3. rake classifier 耙式分级机
4. spiral classifier 螺旋分级机
5. bowl classifier 浮槽分级机
6. bowl desiltor 浮槽脱泥机
7. drag classifier 刮板式分级机
8. hydroseparator 水力分选机
9. solid bowl centrifuge 卧式离心分选机
10. D-O siphonsizer 多尔-奥利弗虹吸分级机
11. countercurrent classifier 逆流分级机
12. jet sizer 多室式水力分级机
13. supersorter 多室式水力分级机
14. feed 进料
15. overflow product 溢出产品
16. sand product 尾矿产品
17. overflow weir 溢流堰
18. feed launder 进料槽
19. lifting mechanism 提升机构
20. sloping bottom tank 斜底槽
21. rake 耙
22. shallow bowl 浅槽
23. revolving plow 转动犁
24. rotating blade 转动刮板
25. chain 链
26. cross flight 横刮板
27. solid-shell conical-shaped bowl 锥形实壳体转筒
28. helical-screw conveyor 螺旋输送器
29. hydraulic water 加压水
30. rotating cylindrical drum 圆柱滚筒
31. spiral flight 螺旋提升片
32. classification pocket 分级室

4.18 料仓、料斗与阀门 Storage Bin, Hopper and Valve

4.18.1 筒仓 Silo（见左下图）

4.18.2 重力式掺混仓 Gravity Blending Silo（见右下图）

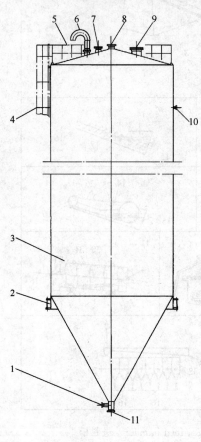
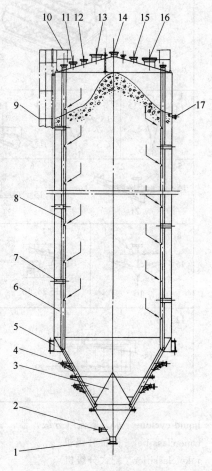

1　level min. port　低料位计接口
2　skirt　裙座
3　shell　壳体
4　straight ladder　直爬梯
5　guard rail　护栏
6　vent pipe　排空管
7　spare port　备用接口
8　inlet　进料口
9　manhole　人孔
10　level max. port　高料位计接口
11　outlet　出料口

1　outlet　出料口
2　level min. port　低料位计接口
3　counter-cone　反向锥
4　air inlet　进气口
5　skirt　裙座
6　shell　壳体
7　supporting ribbon　支承条板
8　blending pipe　掺混管
9　straight ladder　直爬梯
10　guard rail　护栏
11　spare port　备用口
12　inlet　进料口
13　outlet aspiration　吸气口
14　inlet　进气口
15　level indication port　料位指示接口
16　manhole　人孔
17　level max. port　高料位计接口

4.18.3 料斗与溜槽 Hopper and Chute

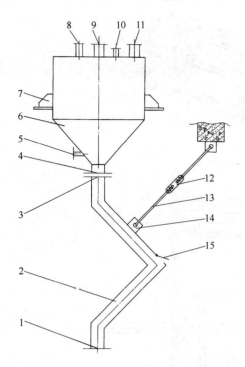

1　outlet　溜槽出口
2　chute shell　溜槽壳体
3　inlet　溜槽进口
4　outlet　料斗出料口
5　hand hole　手孔
6　hopper shell　料斗壳体
7　support　支座
8　spare　备用口
9　inlet　料斗进料口
10　level detector port　料位计接口
11　manhole　人孔
12　adjuster　调节器
13　hanger　吊杆
14　lug　吊耳
15　inspection door　检查门

4.18.4 气动粉体插板阀 Pneumatic Slide Valve

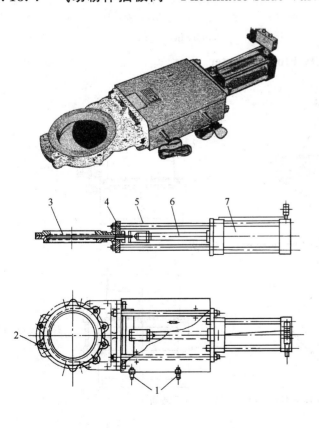

1　limit switch　限位（行程）开关
2　connecting flange　连接法兰
3　slide plate　插（滑）板
4　valve base　阀座
5　link case　连杆箱
6　link　连杆
7　cylinder　气缸

165

4.18.5 手动粉体插板阀　Manual Slide Valve

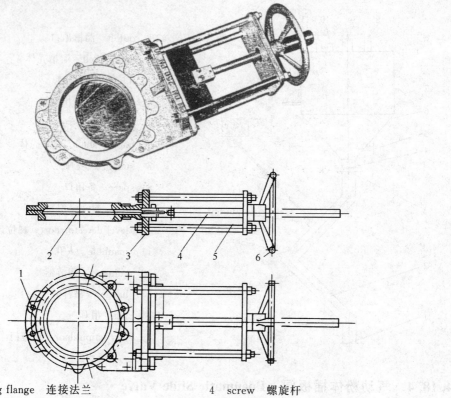

1　connecting flange　连接法兰
2　slide plate　插（滑）板
3　valve base　阀座
4　screw　螺旋杆
5　supporting rod　支承杆
6　hand wheel　手轮

4.18.6 粉体重力换向阀　Gravity Flow Diverter

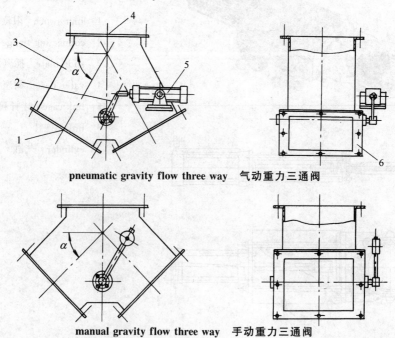

pneumatic gravity flow three way　气动重力三通阀

manual gravity flow three way　手动重力三通阀

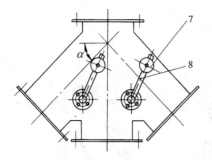

manual gravity flow three way 手动重力三通阀

1 outlet 出料口
2 rod (connected with baffle) 连杆（与挡板相接）
3 valve shell 阀壳体
4 inlet 进料口
5 cylinder 气缸
6 connecting flange 连接法兰
7 balance weight 平衡重锤
8 operation handle (connected with baffle) 操纵杆（与挡板相接）

4.18.7　气动翻板换向阀　Pneumatic Flap Diverter

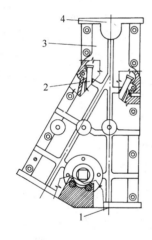
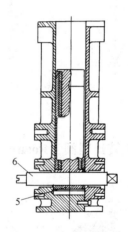
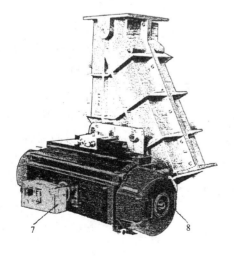

1 outlet 出料口
2 flap 翻（挡）板
3 valve case 阀体
4 inlet 进料口
5 bushing 轴套
6 shaft 轴
7 solenoid valve 电磁阀
8 cylinder 气缸

167

4.18.8 气动转子换向阀 Pneumatic Rotary Diverter

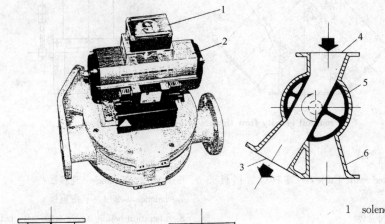

1 solenoid valve 电磁阀
2 cylinder 气缸
3 outlet 出料口
4 inlet 进料口
5 rotor (plug) 转子
6 valve case 阀体
7 end cover 端盖
8 driving shaft 驱动轴

4.18.9 粉体旋转给料阀 Rotary (Feeder) Valve

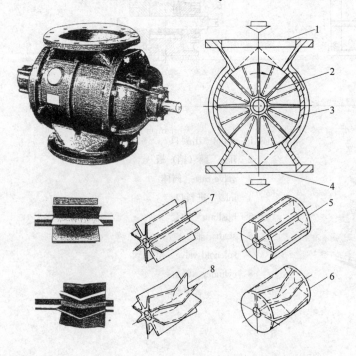

1 inlet 进料口
2 rotor 转子
3 case 壳体
4 outlet 出料口
5 closed rotor (straight blade) 闭式转子（直叶片）
6 closed rotor ("V" type blade) 闭式转子（"V"形叶片）
7 open rotor (straight blade) 开式转子（直叶片）
8 open rotor ("V" type blade) 开式转子（"V"形叶片）

4.19 给料、称重、包装及码垛设备　Feeding, Weighing, Bagging and Palletizing Machine

4.19.1 振动给料机　Vibrating Feeder

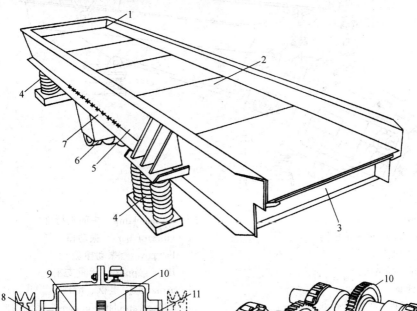

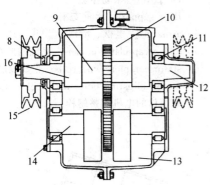

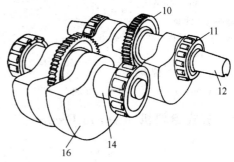

1　AR steel liners　耐磨钢衬板

2　integrally welded body　整体焊接机身

3　heavy-duty pan and grizzly　加重料盘和铁栅

4　coil springs　螺旋弹簧

5　side channel members　槽钢侧板

6　self-contained low-head machanism　整体安装的低振幅机构

7　heavy steel beams　重型钢梁

8　oil-tight mechanism　油封结构

9　husky shaft　刚性轴

10　hardened steel gears　淬火齿轮

11　precision high capacity straight roller bearings　高精度重型滚柱轴承

12　double ended shaft　两端外伸的轴

13　continuous splash lubrication　连续溅油润滑

14　short bearing centers　短的轴承中心距

15　deep groove sheave　深槽带轮

16　accurately calibrated counterweight　精密校准过的平衡重

4.19.2 电振动给料机 Electric Vibrating Feeder

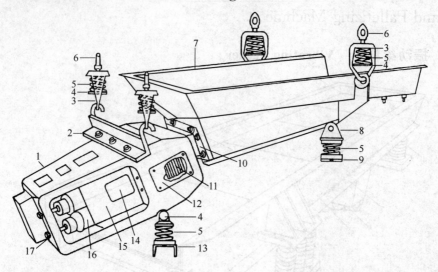

1	main frame 主机架	10	centre clamp 中间夹持件
2	suspension angle 吊重角钢	11	vibrator bars 振动棒
3	shackle U形吊环	12	bar cover 振动棒盖
4	spring cap 弹簧盖	13	floor support 承重支座
5	spring 弹簧	14	armature 电枢、衔铁
6	eye nut 环首（羊眼）螺母	15	coils encapsulated in pairs 密闭的成对线圈
7	tray 料盘	16	stator complete with coils 带线圈的定子
8	spring cup 弹簧座	17	motor adjusting bolts with lock nuts 带锁紧螺母的电机调整螺栓
9	spring base support 弹簧底支座		

4.19.3 板式给料机 Apron Feeder

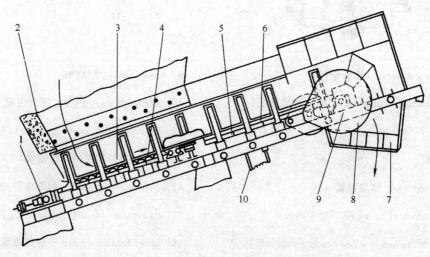

1	take-up station 张紧装置	6	chain 链
2	concrete feed hopper 混凝土加料斗	7	outlet 出口
3	lamellae 板条	8	gear guard 齿轮罩
4	supporting rollers 支承滚	9	drive station 传动装置
5	rail 轨道	10	scraper 刮板

4.19.4 （粉末）均匀自动给料机　Smooth Auto-Feeder

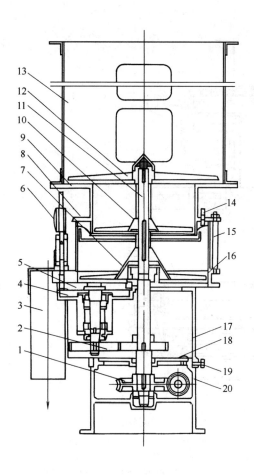
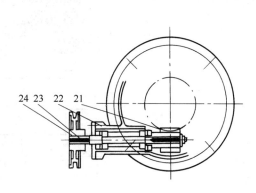

1	worm wheel	蜗轮	13	upper vessel	上部料槽
2	spur gear	正齿轮	14	knob for gate	料门调节柄
3	shoot	出料管	15	lower vessel	下部料槽
4	feed table case	给料盘壳体	16	bottom plate	底板
5	feed table	给料盘	17	upper case	上部料箱
6	feed meter	给料计量装置	18	worm case cover	蜗轮箱盖
7	lower agitator	下搅动器	19	set bolt	紧定螺栓
8	gate	料门	20	worm case	蜗轮箱壳体
9	partition plate	分配板	21	worm	蜗杆
10	upper agitator	上搅动器	22	bearing case	轴承壳
11	vertical shaft	立轴	23	worm shaft	蜗轮轴
12	agitator wing	搅动器翼	24	V-pulley	V形带轮

4.19.5 带式计量秤 Dosing Belt Weigher

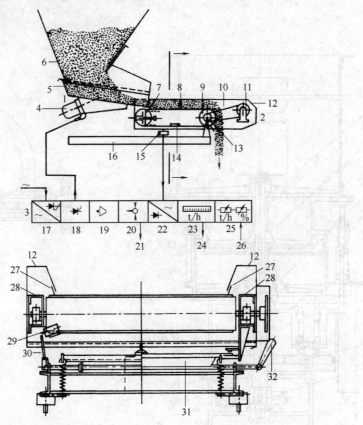

1 discharging feeder with electromagnetic drive 电磁传动给料器
2 weigh belt 称量带
3 electronic weighing equipment 电子称量装置
4 electromagnetic vibrator 电磁振动器
5 trough of discharging feeder 给料器槽
6 bunker with skirt plates 带边板的料仓
7 tension drum 张紧轮
8 conveyor belt of polyester with edge protection 带护边的聚酯输送带
9 drive drum 驱动轮
10 chain drive 链条传动
11 synchronous motor 同步电动机
12 protective and guide plates 保护和导向板
13 pivot for the frame 座架心轴
14 interior and exterior strippers 内侧和外侧的清扫器
15 loading pick-up 载重传感器
16 frame for feeder and weigh belt 给料器和称量带的构架
17 power supply unit 供电装置
18 connection unit for the electromagnetic vibrator 电磁振动器的连接装置
19 control amplifier with matching unit 带耦合装置的控制放大器
20 set-value/actual-value comparison 调定值/真值比较器
21 connection for supplementary unit for fault signalling 出错信号辅助装置的接口
22 power supply for loading pick-up and processing of actual value 供载负传感器和真值处理用电源
23 actual-value indicator 真值指示器
24 connections for output equipment 输出设备的接口
25 set-value adjustment 调定值调整器
26 connection for set-value input equipment 调定值输入设备的接口
27 seal 密封
28 bearing frame 轴承座架
29 centering rollers 对中滚轮
30 limit sensors for monitoring straight running of the belt 控制皮带直线运动的限位传感器
31 calibrating weight 校准重量
32 selector lever for recalibration 供再校准用的选择杆

4.19.6 定量给料秤 Constant Feed Weigher

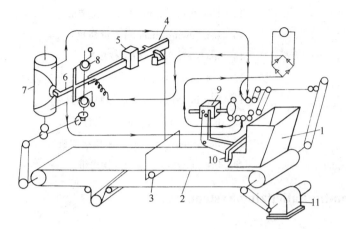

1. supply hopper 供料斗
2. weigh-checking conveyor belt 称重输送带
3. weigh-carrier 称重辊
4. scale beam 秤梁
5. counterpoise 砝码
6. roller finger 滚子触杆
7. rotary switch 旋转开关
8. clamp cam 压紧凸轮
9. magnet clutch 磁力离合器
10. regulating gate 调节门
11. vari-speed motor 变速电动机

4.19.7 失重式给料机 Loss-in-Weight Feeder

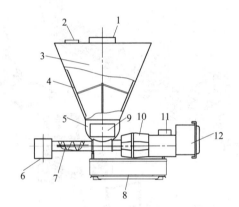

1. inlet 进料口
2. vent 排气口
3. hopper 料斗
4. rotating scraper 旋转刮板
5. trough 进料槽
6. outlet 出料口
7. screw 螺杆
8. weigher 计量秤
9. vibrating plate 振动板
10. gearbox 减速箱
11. motor 电动机
12. control box 控制箱

4.19.8 自动装袋系统 Auto Bagging System

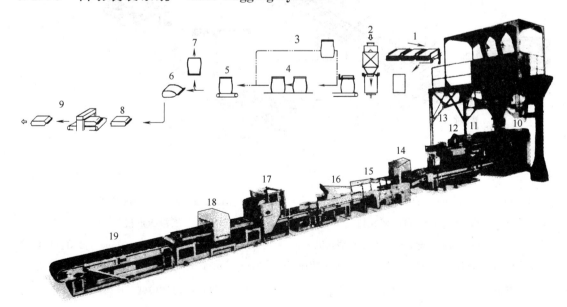

173

1　bag feeding　送袋
2　bag filling　装袋
3　bag sealing　封袋
4　bag sewing　缝袋
5　mass checking　校核质量
6　bag inverting　翻袋
7　discharging　剔除
8　forming　整形
9　metal detecting　探测金属夹杂物
10　automatic bagging machine with weighing scale　自动称量装袋机
11　heat sealing machine　热封口机
12　bag inserting machine　夹袋机
13　bag sewing machine　缝袋机
14　mass checker　质量校核计
15　automatic sorting device　自动分类装置
16　bag inversion device　翻袋装置
17　bag forming machine　袋整形机
18　metal detector　夹杂金属检测机
19　lift conveyor　提升式输送带

4.19.9　自动码垛系统　Automatic Palletizing System

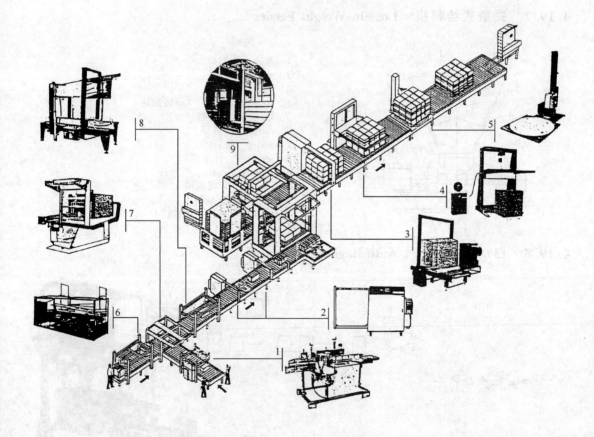

1　automatic labeler　自动贴标机
2　automatic strapper　自动捆扎机
3　automatic vertical strapper　自动垂直捆扎机
4　automatic horizontal strapper　自动水平捆扎机
5　wrapping machine　缠绕机
6　cartoner　纸箱成形机
7　automatic packing machine　自动包装机
8　automatic sealing machine　自动封箱机
9　automatic palletizer　自动码垛机

4.19.10 托盘式缠绕机 Pallet Stretch Wrapping Machine

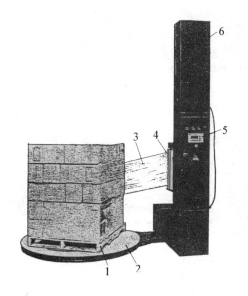

1　pallet　托盘
2　turnable platform　可转动平台
3　stretch film　伸缩薄膜
4　film coil holder　薄膜卷轴架
5　control panel　操作盘
6　support frame　支承机架

4.19.11 热缩膜垛盘打包机 Shrink Packing System

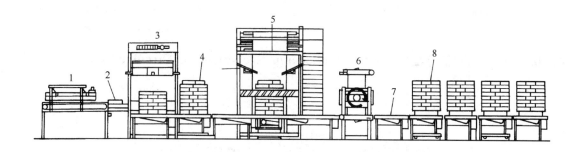

1　bag flattener　袋子整平机
2　flattened bag　整平后的成品袋
3　automatic stacker　自动堆垛机
4　stacked bags　成品垛
5　combi shrink system with profiling equipment　带有整形设备的热缩膜打包系统
6　stack turner　成品翻转机
7　roller conveyor　辊子输送机
8　packaged stack product　打包后的成品垛

175

4.20 仓储设备 Storage Equipment

4.20.1 堆取料机 Stacker and Reclaimer

4.20.1.1 斗轮堆取料机 Bucket Wheel Stacker and Reclaimer

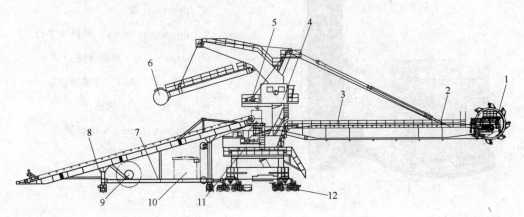

1 bucket wheel 斗轮
2 stacking and reclaiming boom 堆取料臂
3 boom conveyor 臂上皮带
4 driving cab 驾驶室
5 luffing mechanism 俯仰机构
6 counter weight 配重
7 tripper 尾车
8 yard conveyor 料场皮带
9 cable reel 电缆卷盘
10 electric room 电气室
11 slewing mechanism 回转机构
12 travelling mechanism 行走机构

4.20.1.2 圆形堆场堆取料机 Stacker and Reclaimer for Circular Store

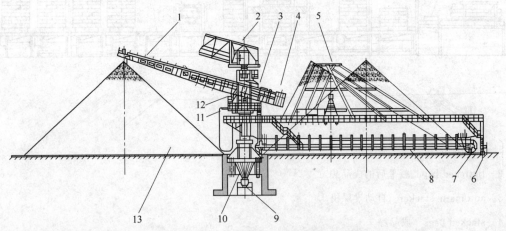

1 boom 悬臂
2 incoming belt conveyor 进料皮带
3 central column 中心柱
4 boom counterweight 悬臂配重
5 raking harrow 料耙
6 travel bogie 行走机构
7 scraper chain tension unit 链条张紧机构
8 scraper 刮板
9 outgoing belt conveyor 取料带
10 outgoing chute 卸料溜槽
11 slewing unit 旋转机构
12 luffing unit 俯仰机构
13 stock pile 料堆

4.20.1.3 侧式悬臂堆料机 Stacker for Longitudinal Store

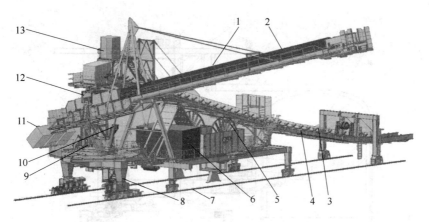

1 boom 悬臂
2 boom conveyor 悬臂带
3 tripper car conveyor 尾车带
4 tripper car 尾车
5 cable reel 电缆卷盘
6 control/motor cabinet room 机柜间
7 rail 轨道
8 travel bogie 行走机构
9 slewing unit 旋转机构
10 luffing unit 俯仰机构
11 boom counterweight 悬臂配重
12 feeding chute 进料溜槽
13 operator cabin 司机间

4.20.1.4 门式取料机 Portal Reclaimer

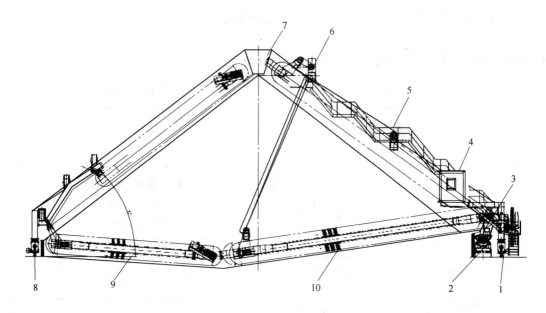

1 main travelling gear 主行走装置
2 reclaiming conveyor 取料带
3 chute 溜槽
4 driving cabinet 驾驶室/配电室
5 hoisting mechanism 起升装置
6 hoisting wheel 起升导向轮
7 portal 门架
8 auxiliary traveling gear 副行走装置
9 auxiliary boom 副耙臂
10 main boom 主耙臂

4.20.2 火车装车机 Train Carriage Loader

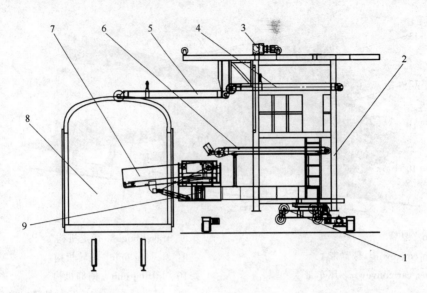

1	travelling mechanism 行走机构	6	transferring conveyor 过渡输送机
2	travelling frame 行走钢架	7	telescopic conveyor 变幅输送机
3	top telescopic mechanism 顶棚变幅机构	8	train carriage 火车车厢
4	top transferring conveyor 顶棚过渡输送机	9	slewing and telescopic mechanism 旋转变幅机构
5	top telescopic conveyor 顶棚变幅输送机		

4.21 LNG 成套设备 LNG Package Equipment

4.21.1 装卸臂 Loading（Unloading）Arm

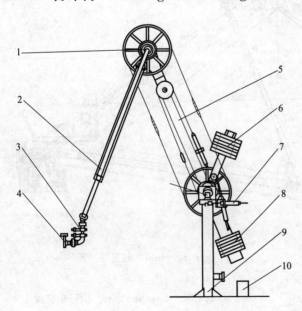

1　swivel joint　旋转接头
2　outboard arm　外臂
3　emergency release system（ERS）紧急脱离系统
4　quick connect\disconnect couple（QC-DC）快速连接系统
5　inboard arm　内臂
6　secondary counterweight　副配重
7　slewing drive cylinder　回转驱动缸
8　primary counterweight　主配重
9　base riser　基座
10　hydraulic power unit（HPU）液压单元

4.21.2 装车撬 Truck Loading Skid

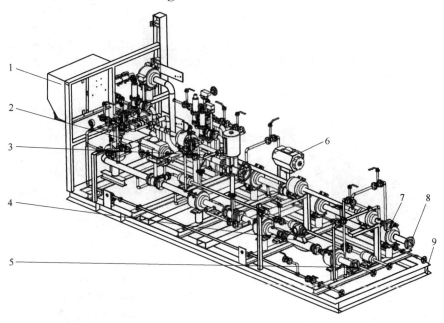

1 local control panel 就地控制盘
2 LNG shut-off valve 液相切断阀
3 flow control valve 流量控制阀
4 flowmeter 流量计
5 LNG line 液相管线
6 return gas shut-off valve 回气切断阀
7 pressure release line 压力卸放管线
8 return gas line 回气管线
9 skid 撬块

4.21.3 装车臂 Truck Loading Arm

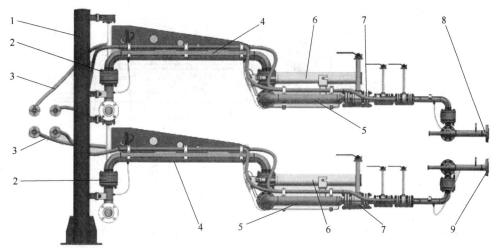

1 post 立柱
2 swivel joint 旋转接头
3 N_2 purge line 氮气吹扫管线
4 inboard arm 内臂
5 outboard arm 外臂
6 cylinder 气缸
7 shut-off valve 切断阀
8 LNG line 液相管线
9 NG line 气相管线

4.21.4 浸没燃烧式气化器 Submerged Combustion Vaporizer

1	stack 烟囱		6	dosing tank 加药罐
2	burner 燃烧器		7	tube bundle 换热盘管
3	air duct 风管		8	cooling water pump 冷却水泵
4	blower 风机		9	control panel 控制盘
5	fuel gas system 燃料气系统		10	water sink 水槽

4.22 造粒机 Granulator

4.22.1 钢带造粒机 Steel Belt Granulator

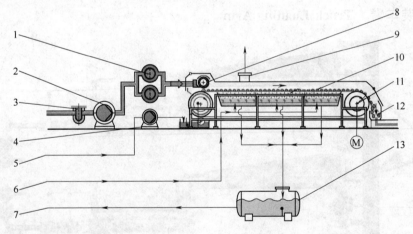

1	fine filter 过滤器		8	rotorform 滴落机
2	product pump 产品泵		9	exhaust vent 排放口
3	pre-filter 预过滤器		10	steel belt cooler 钢带冷却机
4	release agent pump 脱膜剂泵		11	main drum 主动滚筒
5	agent inlet line 脱模剂入口管线		12	outlet chute 出口溜槽
6	cooling water inlet line 冷却水入口管线		13	water collection tank 冷却水收集罐
7	cooling water outlet line 冷却水出口管线			

4.22.2 转筒造粒机　Rotating Drum Granulator

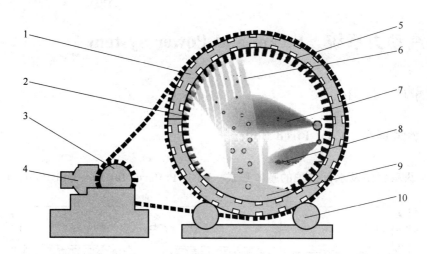

1　rotating drum　转鼓
2　flights　抄板
3　chain-driven system　链条驱动系统
4　drive unit　驱动单元
5　inner drum　壳体
6　product nuclei particles　物料核粒子
7　product spray　物料喷淋
8　cooling spray　冷却喷淋
9　product granules　物料颗粒
10　support idler　支承托棍

5 蒸汽动力系统 Steam and Power System

5.1 锅炉 Boiler

5.1.1 燃煤锅炉 Coal Fired Boiler

5.1.1.1 火管锅炉及水管锅炉的基本形式 Basic Patterns for Fire and Water Tube Boiler

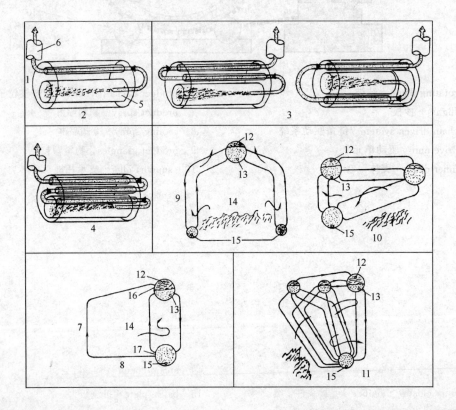

1 basic gas-flow patterns for fire-tube boiler 火管锅炉烟气流的基本形式
2 two-pass 双程
3 three-pass 三程
4 four-pass 四程
5 combustion chamber 燃烧室
6 stack 烟囱
7 circulation in water tube boiler 水管锅炉中的（汽水）循环
8 two-drum D type boiler with water tube furnace 具有水管炉膛的 D 型双锅筒锅炉
9 three drum A boiler A 型三锅筒锅炉
10 three drum, low-head boiler 低型三锅筒锅炉
11 four-drum boiler 四锅筒锅炉
12 steam out 蒸汽出口
13 feedwater 给水
14 furnace 炉膛
15 blow down 排污口
16 steam drum 汽包，汽鼓
17 mud drum 泥包，泥鼓

5.1.1.2 椭圆管板换热器 Ellipsoidal Shell and Tube Heat Exchanger

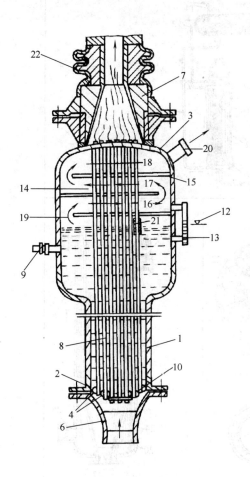
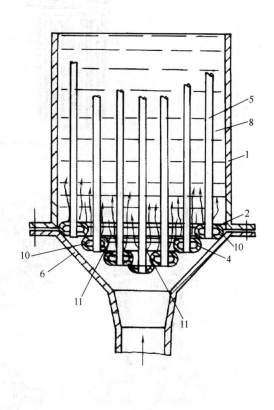

1 cylindrical pressure jacket 圆筒形压力壳体
2 lower closure member 下封头件
3 upper closure member 上封头件
4 annular chamber 环行室，（椭圆）集流管
5 heat-exchange tube 换热管
6 lower header chamber 下部气体进口分配室
7 upper header chamber 上部气体出口集流室
8 interior 内腔，储水室
9 water inlet 水进口
10 oversize aperture 大于管径的孔
11 aperture 侧孔
12 water level 水位
13 water level gauge 水位计
14 steam chamber 蒸汽空间
15 baffle plate 折流板
16 sub compartment 下层蒸汽空间
17 successive compartment 中层蒸汽空间
18 upper most compartment 上层蒸汽空间
19 opening of the deflection baffle 折流板切口
20 steam outlet 蒸汽出口
21 anticorrosive short tube 防腐短管
22 expansion joint 膨胀节

5.1.1.3 水冷管夹套换热器 Cooling Tubes and Jacket Heat Exchanger

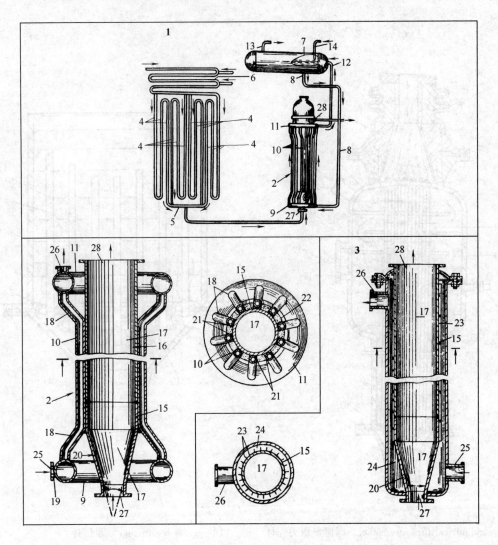

1	line （系统）管线	16	inner surface 内表面
2	heat exchanger 换热器	17	unobstructed passage 直通通道
3	convection preheat section 对流预热段	18	curved portion 弯管
4	coils 蛇管，盘管	19	connecting conduit 接管
5	outlet manifold 出口总管	20	unobstructed inlet diffuser section 直通进口扩散段
6	steam line 蒸汽系统，蒸汽管线		
7	steam drum 汽包，汽鼓	21	weld 焊缝
8	down comer 下降管	22	heat transfer material 导热材料
9，11	torus 环形集汽管	23	cooling fins 冷却翅片
10	water cooling tubes 水冷管	24	coolant jacket 冷却剂夹套
12	riser 上升管	25	coolant inlet 冷却剂进口
13	hot steam line 饱和蒸汽管	26	coolant outlet 冷却剂出口
14	feed water line 给水管	27	hot gas inlet 热气进口
15	concentric pipe 同心管	28	gas outlet 气体出口

5.1.1.4 蒸汽净化及锅筒内件 Steam Purification and Drum Internals

蒸汽净化及锅筒内件（1） Steam Purification and Drum Internals（Ⅰ）

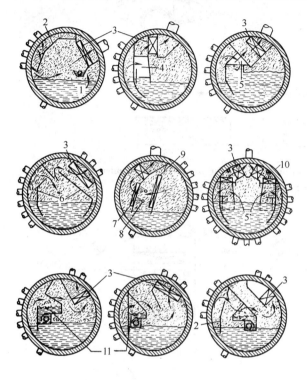

typical arrangements of boiler drum internals in high pressure boiler 高压锅炉锅筒内件的典型配置

1　feedwater tray　给水槽
2　deflector plate　折流板
3　scrubber　洗汽器，汽水分离器
4　cyclone separator　旋风分离器
5　reversing hood　转向罩
6　baffle　挡板
7　washer　洗汽器
8　spray　喷雾器
9　primary separator　一次分离器
10　spray washer　喷雾洗汽器
11　reversing hood washer　转向罩洗汽器
12　to superheater　至过热器
13　dryer　干燥器
14　dry drum　干锅筒
15　drain　排水管
16　riser　上升管
17　downcomer　下降管

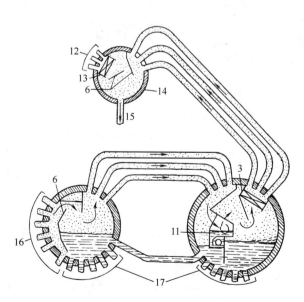

dry drum arrangement　干锅筒配置

蒸汽净化及锅筒内件（2） Steam Purification and Drum Internals（Ⅱ）

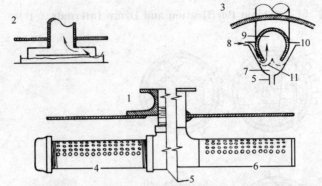

typical drum outlet primary steam separator for saturated steam boiler
饱和蒸汽锅炉典型的锅筒出口一次蒸汽分离器

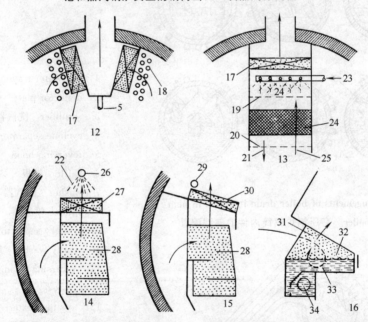

boiler drum internals—steam washer type   锅炉汽包内件——蒸汽清洗型式

1 dry pipe steam outlet 干汽管蒸汽出口
2 pan type separator 盘式分离器
3 steam outlet with internal separator 设有内分离器的蒸汽出口
4 screwed 螺纹连接管
5 drain 排水管
6 welded 焊接管
7 water chamber 水室
8 steam slot 蒸汽槽道
9 dry pipe 干汽管
10 special dry pipe 特殊干汽管

11 perforated sheet 孔板
12 condensing steam washer 凝结洗汽器
13 wire-mesh steam washer 钢丝网洗汽器
14 spray type steam washer 喷淋式洗汽器
15 wetted-scrubber steam washer 湿式洗汽器
16 hood type steam washer 罩式洗汽器
17 corrugated secondary scrubber 波形板二次洗汽器
18 condensing coil cooled by feedwater 给水冷却的冷凝蛇管

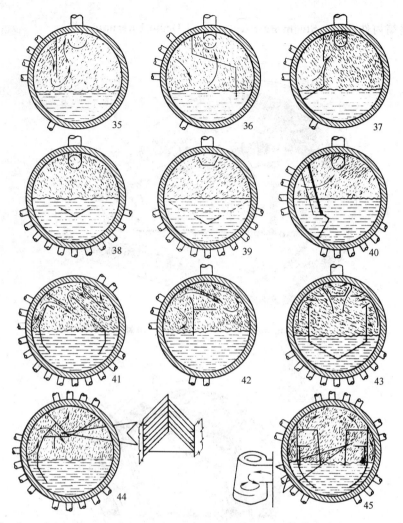

boiler drum internals——primary steam separator method 一次蒸汽分离法用的锅筒内件

19	perforated tray (upper) 上孔板	34	feed water distributor 给水分配器
20	perforated tray (lower) 下孔板	35	deflector (baffle plate) 折流板
21	wash water drainage 清洗水排水	36	offset deflector (baffle plate) 迂回折流板
22	spray 喷淋	37	slotted deflector (baffle plate) 缝隙式折流板
23	wash water feed pipe 清洗水给水管	38	V baffle V 形挡板
24	stainless steel wire mesh 不锈钢丝网	39	perforated plate and V baffle 多孔 V 形挡板
25	steam flow 蒸汽流	40	angle iron deflected baffle 角铁折流板
26	spray nozzle 喷水嘴	41，42	hydraulic barrage baffle 水力折流板
27	primary scrubber 一次洗汽器	43	compartment baffle 室形挡板
28	cyclone separator 旋风分离器	44	triangular reversing hood barrage baffle 三角形转向罩折流板
29	wash water distributor 清洗配水器	45	cyclone type primary separator 旋风式一次分离器
30	inclined scrubber 倾斜式清洗器		
31	reversing hood 转向罩		
32	notched weir plate 缺口堰板		
33	water feed box 给水箱		

蒸汽净化及锅筒内件（3） Steam Purification and Drum Internals（Ⅲ）

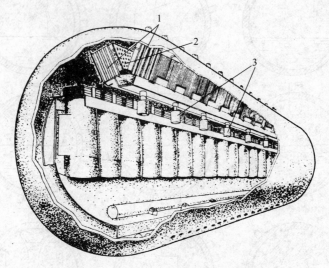

single row arrangement of drum internals 锅筒内件单排配置

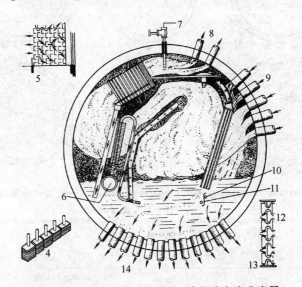

condenser type of steam purifier 冷凝式汽水分离器

1 corrugated scrubber 波形板洗汽器
2 perforated distribution plate 配汽多孔板
3 drain port 疏水口
4 condenser arrangement 冷凝器的配置
5 section thru dryer 干燥器的剖面
6 feed water 给水
7 steam sampling 蒸汽取样
8 steam to superheater 至过热器的蒸汽
9 steam carrying tube 带汽管
10 continuous blowdown 连续排污
11 chemical feed 加药，加化学剂
12 flow 流向
13 section thru primary separator 一次分离器的剖面
14 downflow tube 下降管

蒸汽净化及锅筒内件（4） Steam Purification and Drum Internals（Ⅳ）

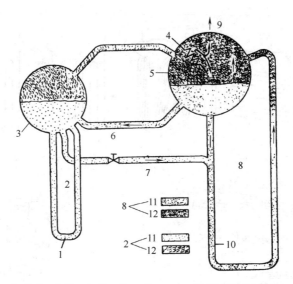

dual circulation boiler flow diagram 双循环锅炉流程图

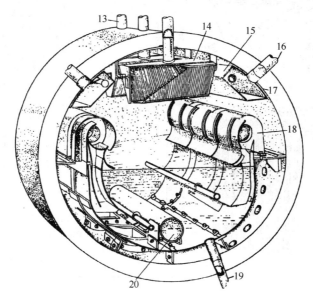

dual circulation boiler drum internals 双循环锅炉锅筒内件

1 boiler tube 锅炉管
2 secondary section 次循环段
3 drum 锅筒
4，17，20 feed water 给水
5 condenser compartment 凝汽器区
6 blow down to secondary section 排至次循环段
7 recirculating line 再循环管
8 primary section 主循环段
9 steam generator output 蒸汽发生器输出
10 waterwall tube 水冷壁管
11 water 水
12 steam 蒸汽
13 superheater supply tube 过热器供汽管
14 unit chevron drier 折纹干燥器单元
15 condensing compartment 凝汽室
16 steam from secondary section 来自次循环段的蒸汽
18 horizontal separator 卧式分离器
19 downtake tube 下降管

5.1.1.5 过热器 Superheater

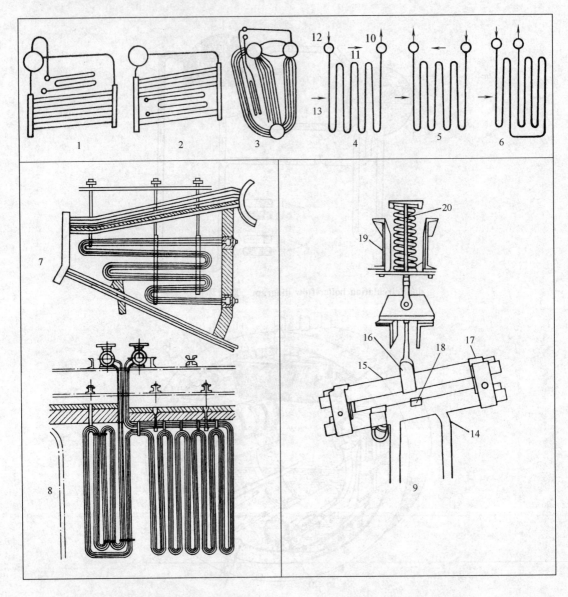

1 overdeck superheater 炉管上方过热器
2 interdeck superheater 炉管中间过热器
3 interbank superheater 炉管束间过热器
4 parallel flow superheater 顺流式过热器
5 counterflow superheater 逆流式过热器
6 mixed superheater 混流式过热器
7 horizontal superheater 卧式过热器
8 vertical superheater 立式过热器
9 superheater support and tube clamp 过热器吊架及管卡
10 steam outlet 蒸汽出口
11 steam 蒸汽
12 steam inlet 蒸汽进口
13 gas 烟气
14 superheater element 过热器元件
15 top row of boiler tube 顶排锅炉管
16 roof tube support 顶管吊架
17 alloy clamp 合金管卡
18 alloy support 合金支架
19 structural steel frame 钢架
20 spring loaded hanger 弹簧吊架

5.1.1.6 减温器 Attemperator

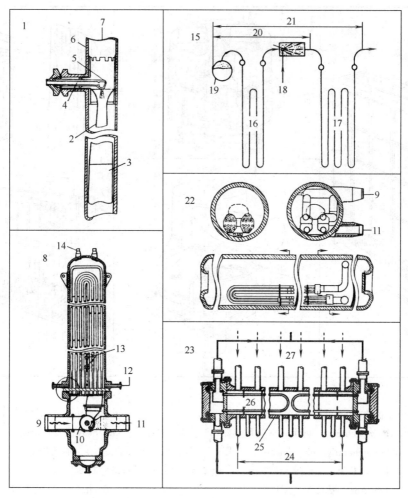

1	spray attemperator 喷水减温器	15	direct contact (spray) type attemperator 直接喷水减温器
2	Venturi mixing and thermal sleeve section 文丘里式混合及热保护套管段	16	primary superheater 第一级过热器
3	thermal sleeve 热保护套管	17	secondary superheater 第二级过热器
4	water 水	18	feed water 给水
5	spray nozzle 喷嘴	19	boiler drum 锅筒,汽包、汽鼓
6	steam line 蒸汽管	20	single-stage superheater 单级过热器
7	steam flow 蒸汽流	21	two-stage superheater 两级过热器
8	shell type surface attemperator 管壳式表面减温器	22	drum type surface attemperator 筒式表面减温器
9	steam inlet 蒸汽进口	23	condenser type surface attemperator 冷凝式表面减温器
10	rotary valve 回转阀	24	superheater element 过热器元件
11	steam outlet 蒸汽出口	25	superheater inlet header 过热器进口联箱
12	blowdown 排污	26	loop tube 环行管
13	water inlet 进水口	27	saturated steam connection 饱和蒸汽连接管
14	steam/water outlet 汽/水出口		

191

5.1.1.7 空气预热器 Air Preheater

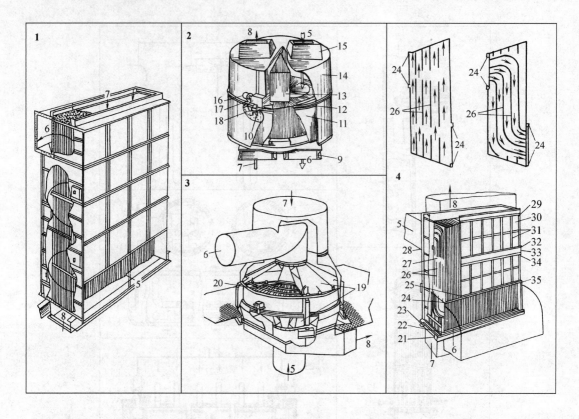

1	tubular air preheater 管式空气预热器	19	rotating air hood 旋转空气室（罩）
2	regenerative air preheater 再生式空气预热器	20	stationary heating surface 固定受热面
3	Rothemuhle regenerative air preheater 风罩型再生式空气预热器	21	stiffening bar 加强杆
		22	stop bar 制动杆
4	plate air preheater 板式空气预热器	23	gas inlet channel 烟气进口的槽钢
5	air inlet 空气进口	24	element lug 传热元件吊耳
6	air outlet 空气出口	25	sealing strip 密封带
7	gas inlet 烟气进口	26	spacer bar 间隔杆
8	gas outlet 烟气出口	27	element 传热元件
9	metal plate 金属板	28	air inlet channel for duct connection 进气管连接用的槽钢
10	gear 齿轮		
11	rotor 转体	29	gas outlet channel 烟气出口的槽钢
12	external gear 外齿轮	30	heater casing 加热器外壳
13	diaphragm 隔板	31	stiffening angle 加强角铁
14	casing 外壳	32	clamp 压板
15	connection 连接件	33	cleanout door 清洗门
16	motor 电动机	34	hinge 折叶
17	gear box 齿轮箱	35	air outlet channel for duct connection 出气管连接用的槽钢
18	support 支架		

5.1.1.8 炉排 Grates

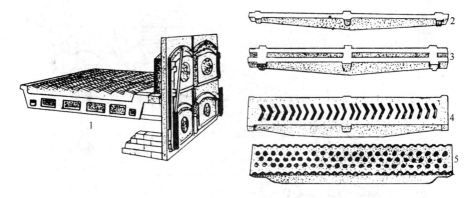

hand-fired grate 手烧炉排

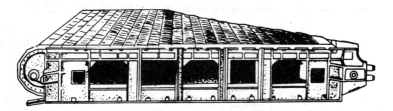

traveling-grate stoker 链条炉排

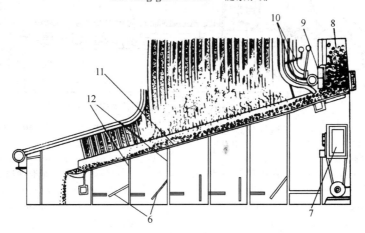

vibrating-grate stoker 振动炉排

1	shaking grate 摇动炉排	7	vibration generator 振动器
2	No. 9 grate, 1in. wide 9号炉排，1in[①]宽	8	coal hopper 煤斗
3	No. 8 grate, 3in. wide 8号炉排，3in宽	9	coal gate 煤闸门
4	tupper grate for coal, 6in. wide 烧煤炉排，6英寸宽	10	overfire-air nozzle 二次风喷嘴
5	sawdust grate 木屑炉排	11	grate tuyere block 炉排风口段
6	air-control damper 空调挡板	12	flexing plate 挠性板

① 1in=25.4mm

5.1.1.9 下饲炉排 Underfeed Stoker

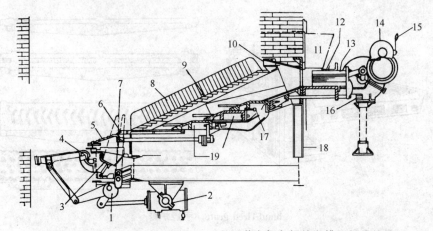

multiple-retort stoker with dumping grate 装有卸灰板的多槽下饲式炉排

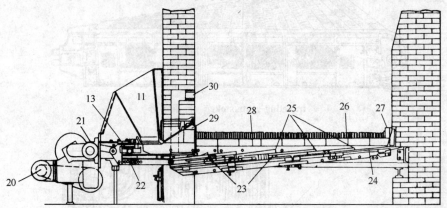

single-retort underfeed stoker 单槽下饲式炉排

1　dump plate locking device　卸灰板锁定装置
2　steam cylinder for operating ash dump plate　操作卸灰板的汽缸
3　extension grate cleanout　炉排延伸段清除口
4　steam-operated dump plate　汽动卸灰板
5　reciprocating extension　往复式延长炉排
6　renewable tuyere and coal plate nose　可更换的风口及煤板头部
7　dump plate operating lever　卸灰板操作杆
8　side wall air back　侧壁空气靠板
9　air distribution tuyere　空气分配口
10　renewable ram box cap extension　可更换的柱塞箱帽延长部分
11　coal hopper　煤斗
12　ram box　柱塞箱
13　main feeding ram　主给煤柱塞
14　two speed spur gear power box　两速正齿轮动力箱
15　speed control lever　调速杆
16　adjuster for fuel distributing system　燃料分配系统的调整器
17　stroke adjuster for individual distributing pusher　单独分配推进器的炉排行程调整装置
18　individual retort supporting column　单煤槽支承柱
19　distributing pusher　分配推进器
20　motor　电动机
21　mechanical drive　机械驱动装置
22　adjustable secondary ram driving arm　可调第二柱塞驱动臂
23　adjustable driving block　可调驱动件
24　retort bottom　煤槽底
25　secondary ram plate　第二柱塞板
26　grate　炉排
27　compensator　补偿器
28　retort side　煤槽侧面
29　retort throat　煤槽喉口
30　secondary air port　二次风口

5.1.1.10 省煤器 Economizer

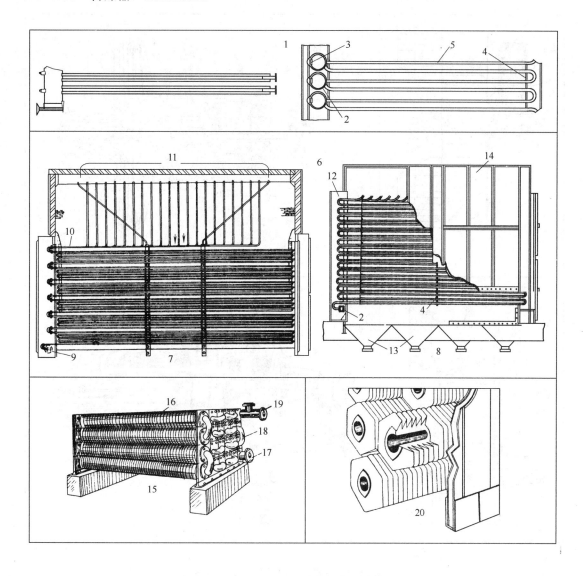

1 construction of economizer 省煤器的结构
2 header 集管箱,联箱
3 hand hole 手孔
4 tube support plate 管支承板
5 economizer tube 省煤器管
6 return-bend economizer 回弯管省煤器
7 continuous-tube type economizer 蛇形管式省煤器
8 loop-tube type economizer 环(蛇)形管式省煤器
9 economizer inlet 省煤器进口
10 gas flow 烟气流向
11 economizer outlet to boiler drum 通至锅炉锅筒的省煤器出口
12 removable steel panel 可拆卸的锅炉板
13 soot hopper 灰斗
14 casing 外壳
15 cast-iron economizer 铸铁省煤器
16 gill 肋片
17 water inlet 进水口
18 return bend 回弯头
19 water outlet 出水口
20 cast-iron gilled tube 铸铁肋片管

5.1.1.11 喷燃器，燃烧器 Burners

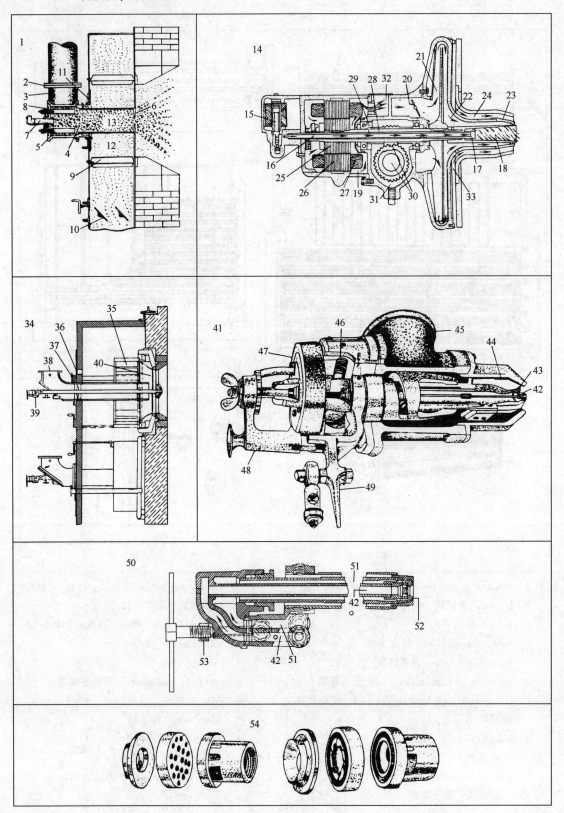

1　flare-type pulverized coal burner　扩口式煤粉喷燃器
2　burner pipe flange　喷燃器接口法兰
3　inlet chamber　进口室
4　nozzle　喷嘴
5　adjustable vane　可调导向叶片
6　coal spreader　煤粉分布器
7　spreader carrying tube　分布器支承管
8　spreader carrying tube locking screw　分布器支承管锁紧螺钉
9　secondary air vane　二次风叶片
10　secondary air plenum chamber damper　二次风风室挡板
11　opening for inserting lighting torch　点火棒插入口
12　secondary air　二次风
13　coal and primary air　煤粉及一次风
14　rotary oil burner with integral pump　带内装泵的旋转式油喷燃器
15　oil metering valve　油计量阀
16　fuel oil tube　燃料油管
17　oil distribution head　油管喷头
18　rotary atomizing cup　雾化转杯
19　cooling air passage around motor　电动机周围的冷却空气道
20　adjustable primary air shutter　一次风调风门
21　primary air fan rotor　一次风扇的转子
22　primary air passage to burner　至喷燃器的一次风道
23　angular-vaned air nozzle　角叶片空气喷嘴
24　passage for induced air to coal burner front plate and sleeve　到煤粉喷燃器前板及套管的引风道
25　motor rotor　电动机转子
26　motor stator　电动机定子
27　motor stator winding　电动机定子绕组
28　motor shaft for driving oil pump and rotating cup　驱动油泵及转杯的电动机轴
29　ball bearing　滚珠轴承，球轴承
30　oil pump driving gear　油泵传动齿轮
31　splash feed lubricating oil reservoir　飞溅润滑油池
32　motor leads junction box　电动机接线盒
33　clearance　间隙
34　multifuel burner　多燃料复合喷燃器
35　adjustable secondary air register　二次风调节器
36　observation and lighting door　观察及点火孔
37　gas inlet　燃料器进口
38　pulverized coal inlet　煤粉进口
39　oil atomizer　燃油雾化器
40　gas ring　配气环
41　proportioning air-atomizing oil burner　定比空气雾化式油喷燃器
42　oil　燃料油
43　primary atomizing air　一次雾化空气
44　secondary atomizing air　二次雾化空气
45　air inlet　空气进口
46　oil inlet　燃油进口
47　micro-metering oil valve　微调测油阀
48　quick-disconnect oil valve lever　速断油阀杆
49　single control　单调节杆
50　steam atomizing oil burner　蒸汽雾化式油喷燃器
51　steam　蒸汽
52　sprayer　喷雾器
53　quick-detachable coupling　速卸联轴器
54　sprayer plate and orifice assembly　喷雾器板及喷嘴组件

5.1.1.12 抛煤机 Spreader Feeders

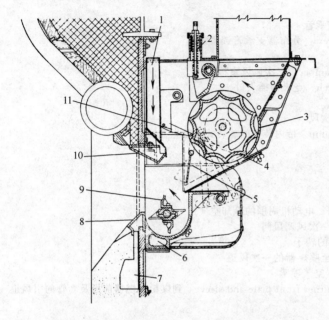

spreader stoker drum-type coal feeder and rotary distribution mechanism 鼓轮式给煤及旋转抛煤（分配）机构，抛煤机炉排

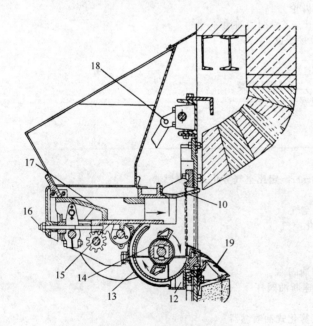

spreader stoker reciprocating-type coal feeder and rotary distribution mechanism 往复式给煤及旋转抛煤机构，抛煤机炉排

1 air duct for deflecting tuyere 折向风口风道
2 release-apron spring assembly 释放拖板（调节煤量用）弹簧组件
3 feed drum 给煤鼓轮
4 adjusting rod for trajectory plate 落煤挡板调整杆
5 trajectory plate 落煤调整杆
6 water jacket 冷却水夹套
7 underfeeder air duct 下饲空气道
8 air swept cut off plate 冲刷空气截止板
9 distributor blade 抛煤转子叶片
10 deflector tuyere 折向风口
11 pocket wiper assembly 刮槽器组件
12 air duct to rotor housing 至抛煤转子外壳的风道
13 rotor housing 抛煤转子外壳
14 rotor 抛煤转子
15 spilling plate 落煤板
16 spilling plate adjusting screw 落煤板调整螺栓
17 feed plate 给煤板，滑煤板
18 control shaft 调节轴
19 apron tuyere 风口

5.1.1.13 磨煤机 Pulverizers

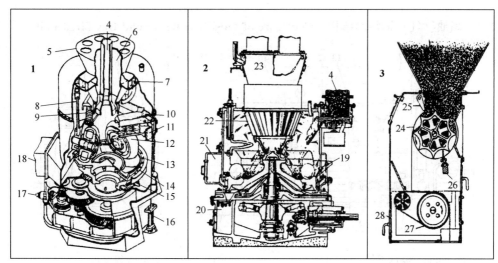

1 M.P.S. (Milldiun pulverizering speed) type pulverizers 中速磨煤机
2 ball-and-race type pulverizer mill 钢球座圈式磨煤机
3 pulverizer feeder with magnetic separator 装有磁铁分离器的磨煤给煤机
4 raw coal inlet 原煤进口
5 discharge turret 煤粉出口
6 pulverized coal to burner 至燃烧器的煤粉
7 classifier 粗粉分离器
8 seal air pipe 密封空气管
9 spring 弹簧
10 spring frame 弹簧框架
11 pressure frame 辊轴压架
12 stationary tire 固定辗轮
13 segmented grinding ring 扇形磨盘
14 throat 喉道
15 pyrites plow (plough) 黄铁矿刮板
16 loading cylinder 装载汽缸
17 gear drive shaft 齿轮主动轴
18 air inlet 空气进口
19 heated grinding element 加热的磨件
20 pyrites trap 黄铁矿收集器
21 hot primary air 一次热空气
22 rotating classifier 旋转式粗粉分离器
23 fines discharge 煤粉出口
24 rotating drum 滚筒,旋转鼓轮
25 spring loaded apron 弹簧加压挡煤板
26 wiper 刮煤板
27 magnet pulley 磁性轮
28 inspection door 观察门

5.1.2 燃油（气）锅炉　Oil-Fired Boiler and Gas-Fired Boiler

5.1.2.1 燃油（气）锅炉的结构　Configure of Oil-Fired Boiler and Gas-Fired Boiler

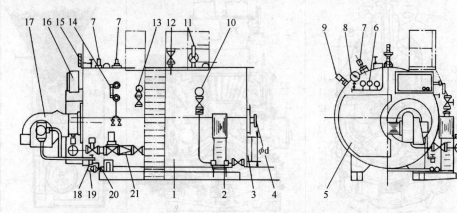

1	steam boiler　蒸汽锅炉	11	safety valve　安全阀
2	feed water pump with motor　给水泵带电动机	12	main steam valve　主汽阀
3	flue interface　烟道接口	13	level blow-down valve　液面排污阀
4	peep hole　观火孔	14	water level gauge　水位表
5	front back-flame chamber　前回烟室	15	electronic water level indicator　电子水位显示器
6	pressure regulator　压力控制器	16	control box　电控箱
7	electronic water level regulator　电子水位控制器	17	burner　燃烧器
8	pressure gauge　压力表	18	oil filter　油过滤器
9	water admittance detector　水导电率探头	19	blow-down valve　排污阀
10	water intake valve　进水阀	20	fast blow-down valve　快速排污阀
		21	fuel gas valves train　燃气阀组

5.1.2.2 燃油蒸汽锅炉系统工艺流程图　Process Flow Diagram for Oil-Fired Steam Boiler System

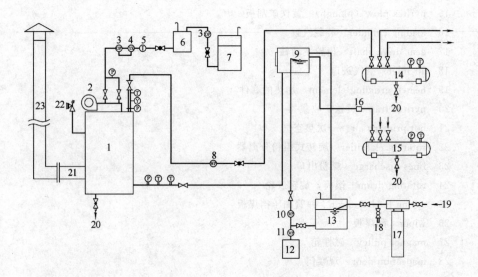

1	boiler	锅炉
2	burner	燃烧器
3	oil pump	油泵
4	oil-water seperator	油水分离器
5	oil filter	油过滤器
6	indoor oil tank	室内油箱
7	oil storage tank	储油罐
8	circulating water pump	循环水泵
9	surge water tank	缓冲水箱
10	make up water pump	补水泵
11	dosing pump	加药泵
12	dosing drum	加药桶
13	soft water tank	软水箱
14	steam header	分汽缸
15	water header	集水器
16	blowdown tank	排污箱
17	water softening device	软水器
18	softened water sampling point	软水取样口
19	water supply	供应水
20	blow-down point	排污口
21	economizer	省煤器
22	safety valve	安全阀
23	stack	烟囱

5.1.2.3 燃气蒸汽锅炉系统工艺流程图　Process Flow Diagram for Gas-Fired Steam Boiler System

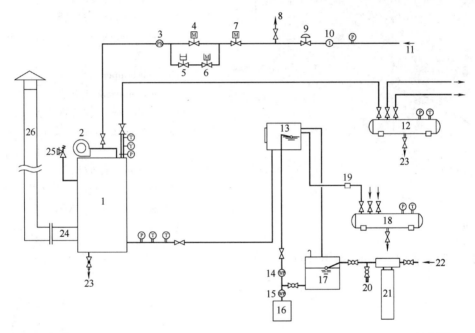

1	boiler	锅炉
2	burner	燃烧器
3	pressure controller	压力控制器
4	steady pressure control valve	稳压控制阀
5	regulating valve	调压阀
6	ignition valve	点火阀
7	emergency shut down valve	事故切断阀
8	vent	放空
9	pressure regulator	调压器
10	fuel gas filter	燃料气过滤器
11	fuel gas supply	燃料气供应
12	steam header	分汽缸
13	surge water tank	缓冲水箱
14	make up water pump	补水泵
15	dosing pump	加药泵
16	dosing drum	加药桶
17	soft water tank	软水箱
18	water header	集水器
19	blowdown tank	排污箱
20	softened water sampling point	软水取样口
21	water softening device	软水器
22	water supply	供应水
23	blow-down point	排污口
24	economizer	省煤器
25	safety relief valve	安全阀
26	stack	烟囱

5.1.3 循环流化床锅炉　Circulation Fluidized Bed Boiler

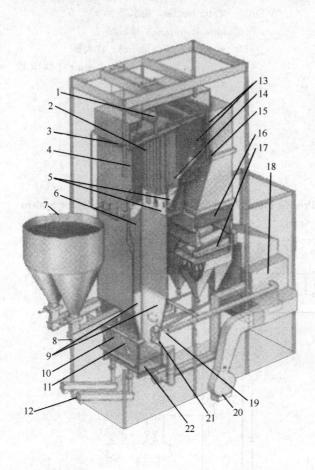

1　steam drum　汽包
2　internal channel separator　内槽型分离器
3,5,9　water-cooled refractory　水冷耐火层
4　evaporator plate　蒸发屏
6　segregation　分隔
7　coal hopper　煤斗
8　gravity feed stoker　重力给煤机
10　secondary air nozzle　二次风喷嘴
11　feeder chute　给煤槽
12　quencher　冷渣器
13　superheater　过热器
14　external channel separator　外槽型分离器
15　ash hopper　飞灰斗
16　economizer　省煤器
17　multiclone　多管旋风分离器
18　tubular preheater　管式空气预热器
19　recirculation system　再循环系统
20　blower　鼓风机
21　above-bed burner　床上燃烧器
22　primary air　一次风

5.1.4 特种锅炉　Special Boiler

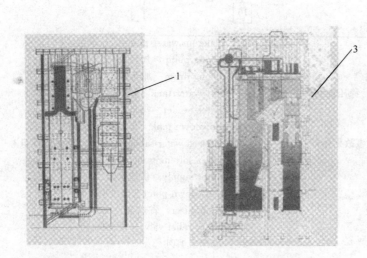
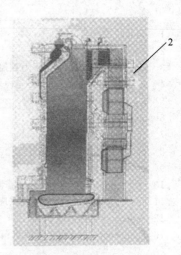

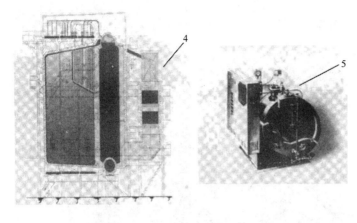

1 chemical recovery boiler 碱回收锅炉
2 bagasse-fired boiler 甘蔗渣锅炉
3 vertical cyclone boiler 立式旋风锅炉
4 carbon monoxide boiler 一氧化碳锅炉
5 thermal storage electric boiler 蓄热型电加热锅炉

5.1.5 热回收设备 Heat Recovery Equipment

5.1.5.1 废液焚烧炉 Waste Liquid Incinerator

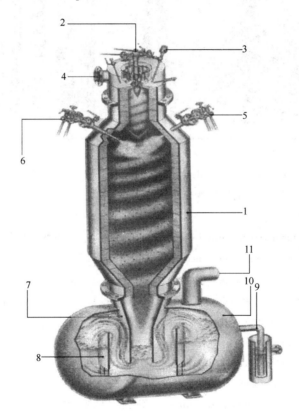

1 waste liquid incinerator 废液焚烧炉
2 fuel nozzle 燃料喷嘴
3 flame detector 火焰探测器
4 combustion air inlet 燃烧空气入口
5 waste liquid nozzle 1 废液喷嘴 1
6 waste liquid nozzle 2 废液喷嘴 2
7 downcomer tube 下降管
8 weir 堰板
9 overflow pipe 溢流管
10 quencher 冷却器
11 exit gas outlet 烟气出口

5.1.5.2 余热锅炉 Waste Heat Boiler

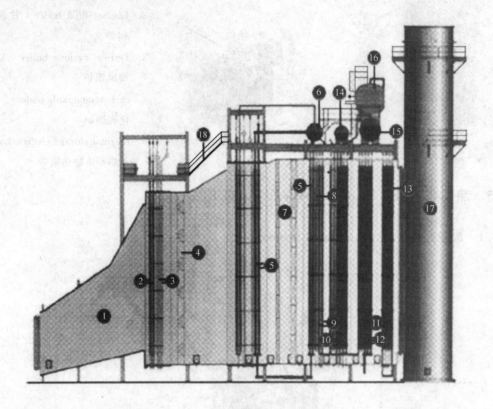

1 high pressure flue gas inlet and stainless steel flue　高压烟气入口及不锈钢烟道

2 high pressure superheater　高压过热器

3 reheater　再热器

4 burner and filter for make-up natural gas and diesel oil　天然气及柴油补燃器和过滤装置

5 high pressure boiler section and downcomer　高压锅炉区及下降管

6 high pressure steam drum　高压汽包

7 CO transformer and catalyst reducing device　一氧化碳转换器及催化剂减量装置

8 middle pressure superheater　中压过热器

9 high pressure compression section　高压降压区

10 middle pressure boiler section and downcomer　中压锅炉区及下降管

11 middle pressure compression section　中压降压区

12 low pressure boiler section and downcomer　低压锅炉区及下降管

13 flue gas condenser (feed water preheater)　烟气冷凝器（给水预热器）

14 middle pressure steam drum　中压汽包

15 low pressure steam drum (a device connecting water header in it)　低压汽包（内设与集水器相连装置）

16 water header　集水器

17 stack　烟囱

18 operating platform and ladder　操作平台及楼梯

5.1.5.3 垃圾焚烧炉 Garbage Incinerator

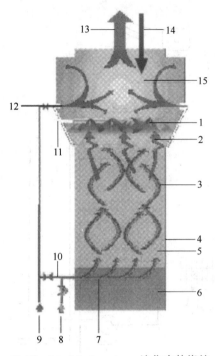

1 main flame 主火焰
2 gas air mixture 空气燃气混合气
3 pyrolysis gas 垃圾分解气
4 fluidized bed 流化床
5 calibrated quartz sand 校准石英砂
6 insulating layer 保温层
7 gas-air distributor 燃气空气分配器
8 fuel gas 燃料气
9 air supply 空气送入
10 primary air 一次空气
11 pilot burner 点火燃烧器
12 secondary air supply 二次空气
13 flue gas 烟气
14 waste being injected 垃圾注入
15 post combustion chamber 后烧室
16 gas chamber 气化室
17 blower for gas chamber 气化室用鼓风机
18 exhaust pipe 排气筒
19 burning chamber 燃烧室
20 burning burner 燃烧用喷嘴
21 blower for burning chamber 燃烧室用鼓风机
22 charging port 投入口
23 burner for ignition 点火用喷嘴

fluidized bed incinerator 流化床焚烧炉

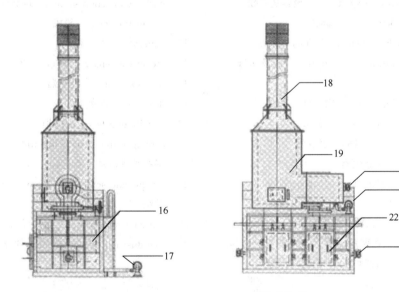

gasifying incinerator 气化焚烧炉

5.1.5.4 垃圾焚烧发电厂工艺流程图 Process Flow Diagram for Garbage Incineration Power Plant

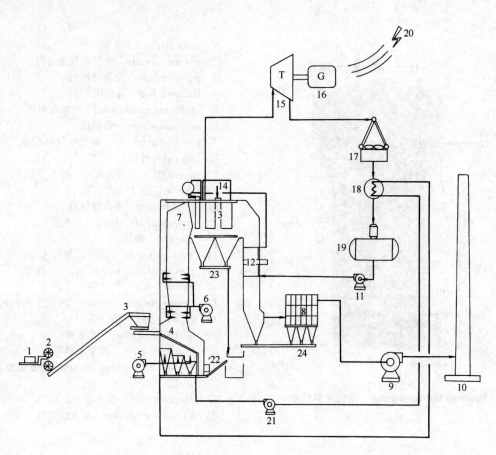

1	garbage conveying system 垃圾输送系统	14	garbage injection system 垃圾注入系统
2	garbage cracker 垃圾破碎机	15	steam turbine 汽轮机
3	chamber feeding system 炉膛进料系统	16	generator 发电机
4	chamber and multi-stage grate 炉膛及多级炉排	17	steam condenser 蒸汽冷凝器
5	primary air injected 一次风注入	18	heat exchanger 换热器
6	secondary air injected 二次风注入	19	deaerator 除氧器
7	boiler/evaporator 锅炉/蒸发器	20	distribution box 配电柜
8	bag filter 袋式除尘器	21	cooling water system for multi-stage grate 多级炉排冷却水系统
9	induced fan 引风机	22	residue and screening system 灰渣及筛选系统
10	stack 烟囱		
11	boiler feed pump 锅炉给水泵	23	ash collection system 锅炉灰收集系统
12	economizer 省煤器	24	flue dust collection system 飞灰收集系统
13	superheater 过热器		

5.2 汽轮机 Steam Turbine

5.2.1 汽轮机循环 Steam Turbine Cycles

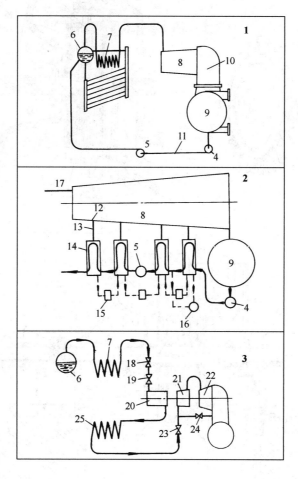

1 Rankine cycle 朗肯循环
2 regenerative feed-heating cycle 回热循环
3 reheating cycle 再热循环
4 condensate extraction pump 凝结水泵
5 boiler feed pump 锅炉给水泵
6 boiler 锅炉
7 superheater 过热器
8 steam turbine 汽轮机
9 condenser 凝汽器
10 isentropic expansion 等熵膨胀
11 adiabatic compression 绝热压缩
12 intermediate stage 中间级
13 extracted steam 抽汽
14 feed heater 给水加热器
15 steam trap 蒸汽疏水器
16 water pump 水泵
17 fresh steam 新蒸汽
18 emergency stop valve 紧急切断阀
19 governor valve 调速器
20 H.P. cylinder 高压汽缸
21 M.P. cylinder 中压汽缸
22 L.P. cylinder 低压汽缸
23 interceptor valve 截断阀
24 dumping valve 排汽阀
25 reheater 再热器

207

5.2.2 冲动式汽轮机 Impulse Turbine

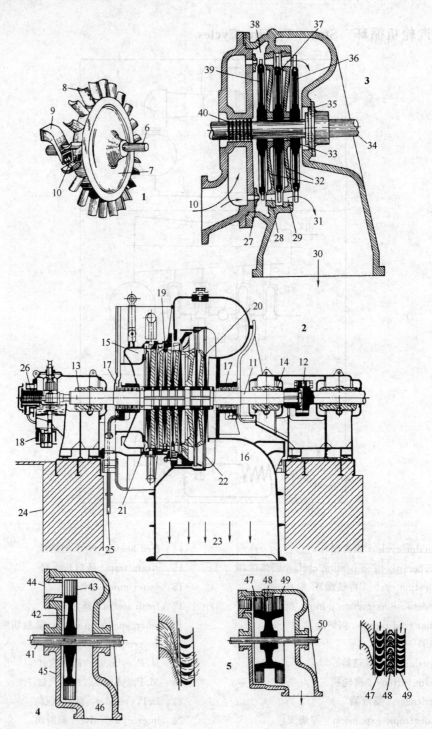

1. simple impulse turbine 简单冲动式汽轮机
2. multistage impulse turbine 多级冲动式汽轮机
3. three pressure stage impulse turbine 三压力级冲动式汽轮机
4. single-stage impulse turbine 单级冲动式汽轮机
5. single-stage impulse turbine with two velocity

stages 复速单级冲动式汽轮机
6,41 shaft 轴
7,42 disc 叶轮
8 blade 叶片
9,44 nozzle 喷嘴
10 steam 蒸汽
11 shaft with seven discs 带7个叶轮的轴
12 coupling 联轴器
13 front journal bearing 前轴颈轴承
14 rear journal bearing 后轴颈轴承
15 annular chamber 环形进汽室
16,46 exhaust pipe 排气管
17 shaft gland packing 轴封
18 gear oil pump 齿轮油泵
19 turbine casing 汽轮机汽缸
20,32 diaphragm 隔板
21 first stage nozzle 一级喷嘴
22 last stage blade 末级叶片
23 to condenser 至凝汽室
24 foundation of turbine 汽轮机基础
25 drain pipe 疏水管

26 collar-type thrust bearing 环式止推轴承
27 first stage nozzle 第一级喷嘴
28 second stage nozzle 第二级喷嘴
29 third stage nozzle 第三级喷嘴
30 back pressure 背压
31 exhaust steam chamber 排气室
33 diaphragm labyrinth packing 隔板迷宫式轴封
34,50 shaft 主轴
35 separate packing casing 分离式轴封箱
36 third stage disc 第三级叶轮
37 second stage disc 第二级叶轮
38 steam chamber 汽室
39 first stage disc 第一级叶轮
40 labyrinth packing 迷宫式轴封
43 moving blade 动叶片
45 stator 机体
47 first row of moving blades 第一排动叶
48 guide blades 导叶
49 second row of moving blades 第二排动叶

5.2.3 单级汽轮机 Single Stage Steam Turbine

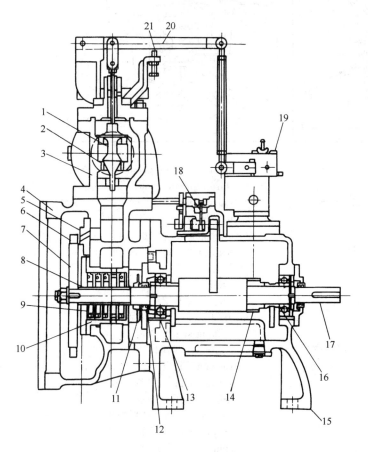

1 governor valve 调速阀
2 governor valve seats 调速阀座
3 steam chest 汽室
4 turbine casing 汽轮机汽缸
5 nozzle ring 喷嘴环
6 blades 叶片
7 turbine wheel 叶轮
8 shaft sleeve 轴套
9 carbon rings 石墨环,炭精环
10 packing case 轴封箱
11 slinger 甩水器
12 oil baffle 油挡
13 thrust bearing 止推轴承
14 worm gear 蜗轮(蜗杆)传动装置
15 bearing bracket 支架
16 line bearing 轴承
17 turbine shaft 汽轮机轴
18 pilot shaft 错油阀,导阀
19 Woodward governor 伍德瓦德调速器
20 governor lever 调速杠杆
21 automatic start adjust bolt 自动启动调节螺栓

5.2.4 汽轮机轴封 Turbine Glands and Gland Sealings

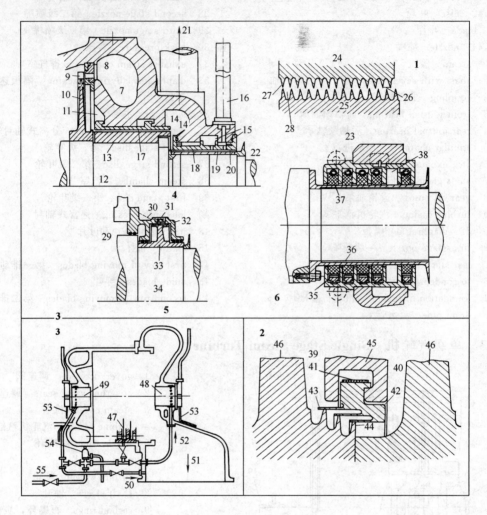

1	internal gland	内轴封
2	external gland	外轴封
3	steam connection of external gland	外轴封蒸汽连接
4	labyrinth gland	迷宫式轴封
5	hydraulic gland (water seal)	水封
6	carbon ring gland	碳精轴封
7	distribution box	配汽室
8	nozzle	喷嘴
9	blade	叶片
10, 46	disc	叶轮
11	first stage wheel chamber	第一级叶轮汽室
12	rotor	转子
13	sleeve	轴套
14	steam space	蒸汽扩容室
15	pocket	汽穴
16	vapor pipe	放气管
17	first separate gland	第一分轴封
18	second separate gland	第二分轴封
19	third separate gland	第三分轴封
20	fourth separate gland	第四分轴封
21	to exhaust space	至排汽室
22	atmosphere	大气
23	high pressure turbine gland	高压汽轮机轴封
24	cylinder	汽缸
25	rotor	转子
26	axial pitch	轴向间距
27	packing strip	梳齿
28	clearance	间隙
29	vacuum	真空
30	steam supply	供汽
31	water supply	供水
32	circular casing	环形外壳
33	paddle wheel	桨轮
34	spindle	汽轮机轴

35	coach spring 托簧		47	intermediate stage of turbine 汽轮机中间级
36	steel sleeve 钢制轴套		48	low pressure gland 低压轴封
37	carbon ring 碳精环		49	high pressure gland 高压轴封
38	leak-off pocket 外漏汽穴		50	to condenser 至凝汽器
39	low pressure side 低压侧		51	exhaust branch 排汽管
40	high pressure side 高压侧		52	steam to low pressure gland 至低压轴封的蒸汽
41	plate spring 板簧		53	open drain to sump 排水至下水道
42	packing ring 轴封环		54	relief valve 安全泄液阀
43	project strip 外伸带环		55	live steam supply 供新蒸汽
44	fine (projecting ridge) 细突缘			
45	diaphragm 隔板			

5.2.5 超速脱扣装置（保安器） Overspeed Tripping Device

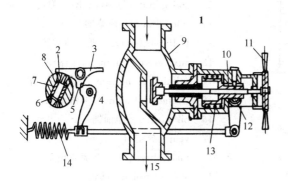

1　overspeed trip with lever control　杠杆式超速保安器
2　eccentric pin　偏心销
3　trips lever　安全杆
4　lever　杠杆
5　interlock　咬口
6　astatic regulator　调整用重块
7，13　spring　弹簧
8　turbine shaft　汽轮机轴
9　stop valve　切断阀，主汽阀
10　sleeve　套筒
11　handwheel　手轮
12　segment　掣子
14　helical spring　螺旋弹簧
15　steam to turbine　蒸汽至汽轮机

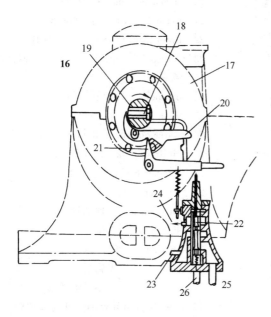

16　overspeed tripping relay with hydraulic control　油压控制超速脱扣继动器
17　steam turbine casing　汽轮机汽缸
18　limiting speed pin　限速销
19　turbine shaft　汽轮机轴
20　left arm of lever　拉杆
21　lock　咬口
22　servomotor piston　伺服马达活塞
23　oil under pressure　压力油
24　oil to lubricating system　油至润滑系统
25　oil to regulating system　油至调节系统
26　drain oil　排油

5.2.6 汽轮机的润滑 Lubrication of Steam Turbine

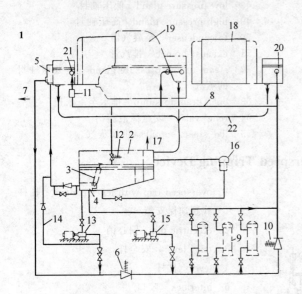

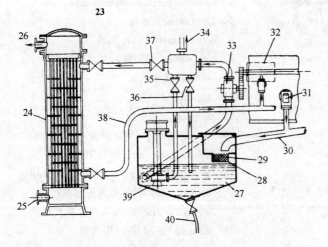

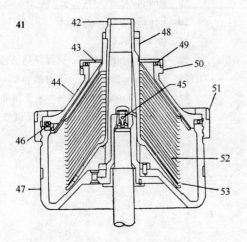

1 lubrication system for Turbo-generator 汽轮发电机组润滑系统
2 drain from bearings 轴承回油
3 oil strainer 滤油器
4 clean oil chamber 净油室
5 main oil pump 主油泵
6 pressure sustaining valve 保压阀
7 oil to relay 至继动器的油
8 oil to bearings 至轴承的油
9 oil coolers 冷油器组
10 spring-loaded bypass valve 弹簧旁通阀
11 fine strainer 精滤器
12 L.P. relief valve 低压泄放阀
13 auxiliary oil pump 辅助油泵
14 priming connection 启动油管路
15 flushing oil pump 冲洗油泵
16 oil purifier 净油器
17 vent 放空口
18 main alternator 交流发电机
19 steam turbine 汽轮机
20 thrust bearing 推力轴承
21 journal bearing 轴颈轴承
22 drain from bearing 轴承回油
23 forced lubrication system 强制润滑系统
24 oil cooler 冷油器
25 water inlet 水入口
26 water outlet 水出口
27 oil drain tank 回油箱
28 strainer basket 过滤框
29 strainer 过滤网
30 oil drain pipe 回油管
31 sight glass 视窗
32 bearing 轴承
33 gear pump 齿轮泵
34 oil to relay 至继动器的油
35 normal return valve 正常回油阀
36 relief valve 泄压阀
37 reducing valve 减压阀
38 oil to bearings 至轴承的油

39	auxiliary oil pump 辅助油泵		47	bowl shell 鼓壳
40	drain pipe 放油管		48	top disc 顶盘
41	bowl of centrifugal purifier 离心分离机转鼓		49	gravity rubber ring 橡胶重力密封圈
42	distributor 分配管		50	gravity lock ring 重力锁紧环
43	gravity disc 重力盘		51	main lock ring 主锁紧环
44	top shell 顶壳		52	intermediate discs 中间盘
45	conveyor 输送器		53	bottom disc 底盘
46	main rubber ring 主橡胶圈			

5.2.7 汽轮机供汽方式 Methods of Steam Supply to a Turbine

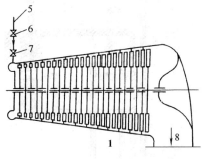
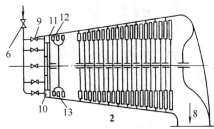
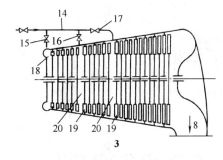
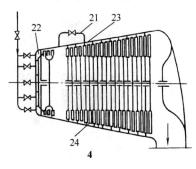

1 throttle governing 节流调节
2 nozzle control governing 喷嘴调节
3 external bypass governing 外旁路调节
4 internal bypass governing 内旁路调节
5 fresh steam 新蒸汽
6 stop valve 主汽阀，切断阀
7 throttle valve (governor valve) 节流阀，调速阀
8 exhaust pipe 排汽管
9 regulating valve 调节阀
10 nozzle box 喷嘴室
11 nozzle groups 喷嘴组件
12 first stage blade 第一级叶片
13 guide blade 导向叶片
14 external blade 外旁路
15 main throttle valve 主节流阀
16 first bypass valve 第一旁通阀
17 second bypass valve 第二旁通阀
18 first stage 第一级
19 intermediate stage 中间级
20 bypass chamber 旁路汽室
21 internal bypass 内旁通
22 steam chest 蒸汽室
23 fourth stage nozzle 第四级喷嘴
24 steam belt 蒸汽通道

213

5.2.8 汽轮机调速器 Turbine Governor

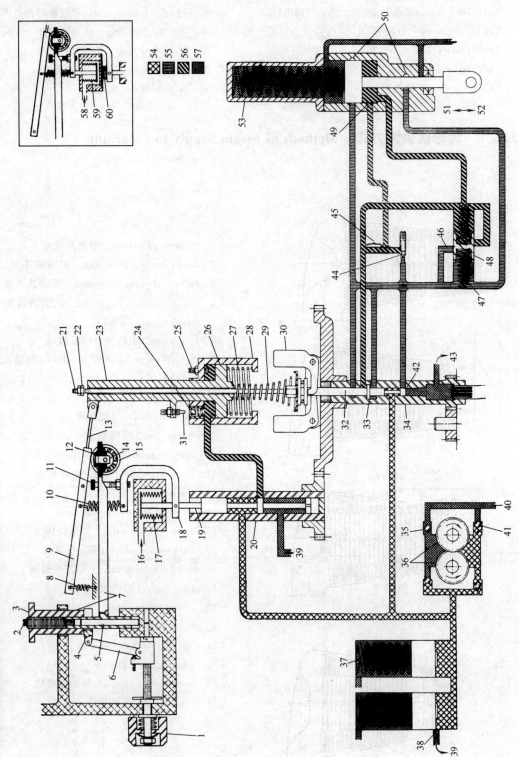

1 manual speed adjusting knob 手动调速旋钮
2 high speed stop adjusting setscrew 高速停止定位螺钉
3 speed adjusting nut 调速螺母
4 stop collar 止动环
5 high speed stop pin 高速停止销
6、18 link 连杆
7 speed setting assembly 速度整定（调定）件
8 loading spring 负载弹簧
9 restoring lever 复位杠杆
10 restoring spring 复位弹簧
11 low speed adjusting screw 低速调节螺钉
12 knurled nut 滚花螺母
13 bimetal strip 双金属条
14 adjustable pivot bracket 可变支点托架
15 range ADJ 调整范围
16、58 control air pressure 控制气压
17、59 bellows 波纹管
19 speed setting plunger 速度调定柱塞
20 speed setting bushing 速度调定衬套
21 shutdown rod 停机杆
22 shutdown nuts 停机螺母
23 piston rod 活塞杆
24 maximum speed limiting valve 高速限制阀
25 piston stop setscrew 活塞停止定位螺钉
26 speed setting piston 转速调定活塞
27 piston spring 活塞弹簧
28 speed setting cylinder 转速调定汽缸
29 speeder spring 调速弹簧
30 flyweight 飞锤
31 limiting valve adjusting screw 限制阀调节螺钉
32 pilot valve plunger 错油（导）阀柱塞
33 compensating land 补偿挡圈
34 rotating pilot valve bushing 旋转式错油（导）阀衬套
35 check valve 止回阀
36 pump gears 齿轮泵
37 accumulator 蓄力器
38 relief bypass 泄放旁路
39、43 to sump 去储槽
40 from sump 来自储槽
41 check valve 止回阀
42 control land 控制柱塞
44 compensation needle valve 补偿针阀
45 compensation cutoff 补偿截止线路
46 bypass port 旁通口
47 buffer spring 缓冲弹簧
48 buffer piston 缓冲活塞
49 compensation cutoff port 补偿截止孔
50 seal grooves 密封槽
51 increase 增速
52 decrease 减速
53 spring loaded power cylinder assembly 弹簧动力缸组件
54 pump oil pressure 泵油压
55 intermediate oil pressure 中间油压
56 trapped oil & power cylinder oil pressure 截留油和动力缸油压
57 sump oil pressure 储槽油压
60 bellows spring 波纹管弹簧

5.2.9 汽轮机调速器及调速 Governors and Governing of Steam Turbine

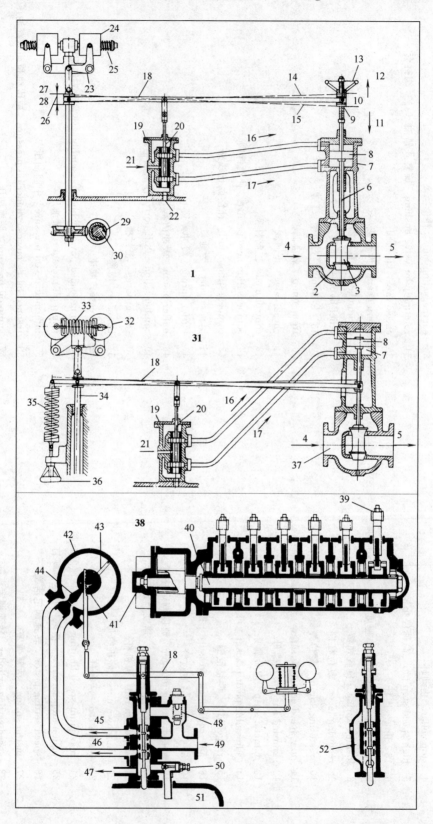

1　oil relay governor gear　油继动器调速装置

2　double seat balanced type vale　双座平衡型阀门

3　throttle valve　节流阀

4　fresh steam　新蒸汽

5　steam to turbine　至汽轮机的蒸汽

6　regulate valve spindle　调节阀阀杆

7，42　servo motor　伺服马达

8　piston　活塞

9　screwed sleeve　螺纹套筒

10　lift　升程

11　to raise speed　提高转速

12　to lower speed　降低转速

13　speeder handwheel　调速手轮

14　full load　满负荷

15　no load　无负荷

16　to close valve　关阀门

17　to open valve　开阀门

18　floating lever　浮动杠杆

19　oil relay　油继动器

20　pilot valve　错油阀，导阀

21　oil inlet　油入口

22　oil drain　油出口

23　governor　调速器

24　rotating weight　飞锤

25　spring　弹簧

26　sleeve travel　套筒行程

27　fast　加速

28　slow　减速

29　worm　蜗杆

30　turbine spindle　汽轮机主轴

31　alternative form of speeder gear　调速装置的另一种形式

32　fly ball　飞锤

33　governor spring　调速器弹簧

34　governor axis　调速器轴

35　tension spring　拉簧

36　speed adjustive handwheel　调速手轮

37　steam　蒸汽

38　diagrammatic arrangement of governor gear　调速装置布置图

39　nozzle valves　喷嘴阀组

40　cam shaft　凸轮轴

41　rack and pinion　齿条及小齿轮

43　moving blade　动叶片

44　fixed partition　固定隔板

45　oil flow to open steam valve　开蒸汽阀的油路

46　oil flow to close steam valve　关蒸汽阀的油路

47　oil exhaust to bearings　排至轴承的油

48　high pressure by pass valve　高压旁通阀

49　oil from pump　油泵来油

50　low pressure by pass valve　低压旁通阀

51　oil reservoir　储油箱

52　oil exhaust passage　排油通道

5.2.10 汽轮机的安装 Installation of Steam Turbine

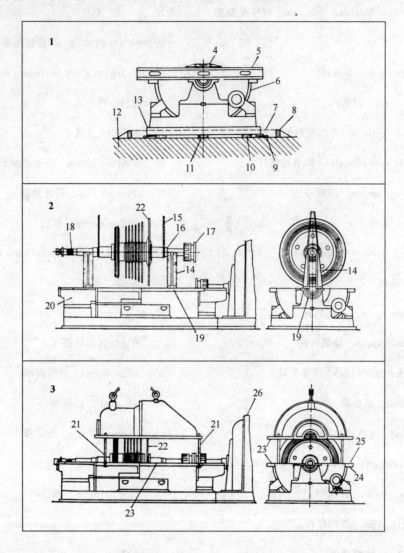

1	erection of turbine 汽轮机安装	14	guide bracket 导向架
2	guide brackets for rotor 转子导向架	15	slung 起吊绳
3	guide pillars for turbine cover 汽轮机汽缸盖导向柱	16	rotor 转子
4	spirit level 水平尺	17	coupling 联轴器
5	straight edge 直规,直尺	18	labyrinth packing 迷宫式轴封
6	turbine cylinder 汽缸	19	bearing 轴承
7	level of grouting 灌浆面	20	lower half of turbine 汽轮机下汽缸
8	dam of wood 模板	21	guide pillar 导向柱
9	wedge 楔形垫铁,斜垫板	22	blade 叶片
10	chock 垫板,垫铁	23	main horizontal joint 主水平接合面
11	packing pieces 平垫板,平垫铁	24	flange 法兰
12	foundation 基础	25	thin strips 密封垫片
13	bedplate 底板	26	driven machine 从动机

5.3 燃气轮机 Gas Turbine

5.3.1 燃气轮机结构 Constitution of Gas Turbine

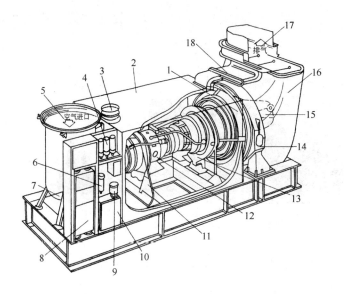

1 turbine 动力涡轮
2 package 箱装体
3 draft air inlet 通风空气进口
4 hand grenade 灭火瓶
5 air inlet 空气进口
6 lubricant filter 润滑油过滤器
7 splitter vane cascade 导流叶栅
8 control panel 控制盘
9 air separator of lubricant 滑油空气分离器
10 lubricator 润滑器
11 front holder of gas generator 燃气发生器前支架
12 gas generator 燃气发生器
13 back holder of gas generator 燃气发生器后支架
14 turbine holder 动力涡轮支架
15 output shaft 输出轴
16 exhaust volute 排气涡壳
17 exhaust 排气
18 draft air outlet 通风空气进口

5.3.2 燃气轮机发电机组 Gas Turbine Power Plant

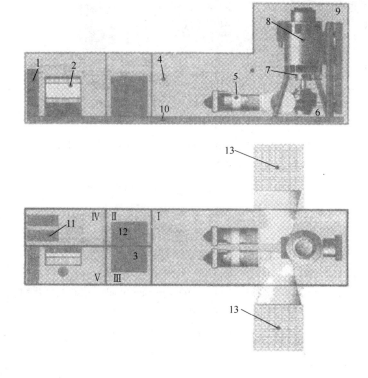

1 generator control cubicle 发电机控制室
2 control board 控制面板
3 power house transformer 电力变压器室
4 container 箱体
5 motors 电机
6 reductor 减压器
7 compensating coupling 补偿联轴器
8 generator 发电机
9 auxiliary ventilator 辅助通风装置
10 frame 外框架
11 pump units 泵单元
12 accumulating battery 蓄电池
13 module heat exchangers 换热器模块

219

5.3.3 燃气轮机简单循环发电厂 Gas Turbine Simple Cycle Power Plant

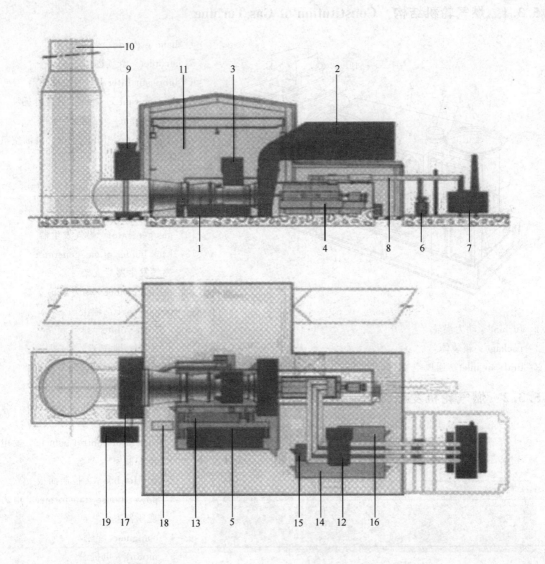

1 gas turbine 燃气轮机
2 air intake block 空气过滤器
3 blow-off silencer 消音器
4 generator 发电机
5 auxiliaries block 辅助设备
6 auxiliary transformer 辅助变压器
7 main transformer 主变压器
8 generator bus duct 发电机母线
9 coolers 冷却器
10 exhaust stack 烟囱
11 machine hall 厂房
12 generator breaker module 发电机断路模块
13 control valve block 控制系统柜
14 control module 控制模块
15 starting module 启动模块
16 battery module 蓄电池模块
17 fuel gas block 燃气系统
18 NO_x water injection block 控制氮氧化物储水系统
19 fuel oil block 燃油系统

5.3.4 燃气轮机-蒸汽轮机联合循环发电厂 Gas Turbine-Steam Turbine Combined Cycle Power Plant

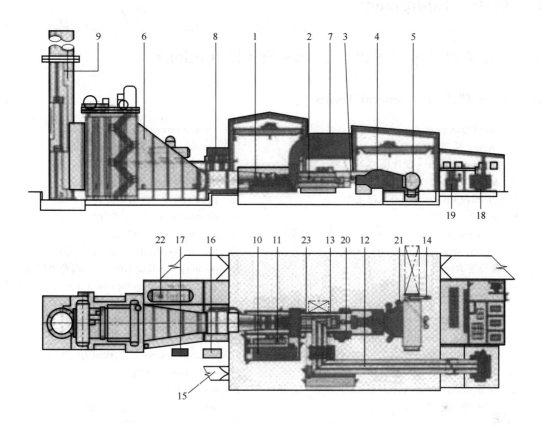

1	gas turbine 燃气轮机	14	cooling water pipe 冷凝水管
2	generator 发电机	15	NO_x water injection block 控制氮氧化物注水系统
3	clutch 离合器	16	fuel oil block 燃油系统
4	steam turbine 蒸汽轮机	17	fuel gas block 燃气系统
5	condenser 蒸汽凝结器	18	main transformer 主变压器
6	heat recovery boiler 余热锅炉	19	auxiliary transformer 辅助变压器
7	air intake block 空气过滤器	20	lube oil of steam turbine 蒸汽轮机润滑油箱
8	rotor air coolers 转子冷却器	21	condenser pump 冷凝水泵
9	stack 烟囱	22	feedwater tank 补充给水柜
10	auxiliary block of gas turbine 燃气轮机辅助设施	23	electrical and control module 电气系统控制模块
11	control valve block 控制系统柜		
12	generator bus duct 发电机母线		
13	generator breaker module 发电机断路模块		

6 设备　Equipment

6.1　设备设计常用术语　Terms and Definitions

6.1.1　一般术语　General Terms

1　calculation pressure　计算压力
2　creep　蠕变
3　design pressure　设计压力
4　design temperature　设计温度
5　effective thickness　有效厚度
6　fatigue　疲劳
7　finite element analysis（FEA）　有限元分析
8　general primary membrane stress　一次总体薄膜应力
9　limit analysis　极限分析
10　local structural discontinuity　局部结构不连续
11　maximum allowable working pressure（MAWP）　最高允许工作压力
12　membrane stress　薄膜应力
13　minimum design metal temperature　最低设计金属温度
14　minimum required fabrication thickness　最小成型厚度
15　normal stress　法向应力
16　nominal thickness　名义厚度
17　operating cycle　工作循环
18　operating pressure　操作压力
19　peak stress　峰值应力
20　plastic analysis　塑性分析
21　plastic hinge　塑性铰
22　plastic instability　塑性不稳定载荷
23　primary bending stress　一次弯曲应力
24　primary local membrane stress　一次局部薄膜应力
25　primary stress　一次应力
26　ratcheting　棘轮现象
27　secondary stress　二次应力
28　shakedown　安定性
29　stress cycle　应力循环
30　stress intensity　应力强度
31　test pressure　试验压力

6.1.2　设备类型术语　Types of Equipment

1　compressors　压缩机
2　heat exchangers　换热器
3　horizontal vessels　卧式容器
4　pumps　泵
5　reactors　反应器
6　rotating equipment　动设备
7　spherical tanks　球形储罐
8　static equipment　静设备
9　towers/columns　塔器
10　vertical，Flat-bottomed steel storage tanks　立式平底储罐
11　vertical vessels　立式容器
12　vessels　容器

6.1.3　压力容器分类术语　Classification for Categories of Pressure Vessels

1　category Ⅰ　第Ⅰ类压力容器
2　category Ⅱ　第Ⅱ类压力容器
3　category Ⅲ　第Ⅲ类压力容器
4　categories of pressure vessels　压力容器类别
5　heat-exchanger vessels（symbol E）　换热压力器（代号E）
6　reactor vessels（symbol R）　反应压力容器（代号R）
7　separation vessels（symbol S）　分离压力容器（代号S）
8　storage vessels（symbol C, for spherical tanks, symbol B）　储存压力容器（代号C，球罐代号B）

6.2 塔 Column

6.2.1 板式塔 Plate Column

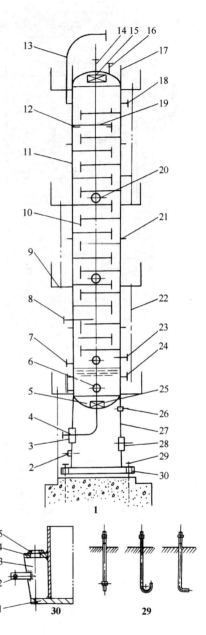

1 schematic drawing of plate column (tray column, trayed tower) assembly 板式塔装配简图
2 nameplate 铭牌
3 pipe opening and support plate 引出孔及支承板
4 liquid outlet 液体出口
5 ellipsoidal head 椭圆形封头
6 level gauge connection 液面计接口
7 pressure gauge connection 压力计接口
8 feed inlet 进料口
9 platform 平台
10 downcomer (downflow, downspout) 降液管，降液板
11 shell 壳体
12 tray support ring 塔盘支承圈
13 top davit 塔顶吊柱
14 vapor outlet 蒸气出口
15 demister 除沫器，破沫网
16 vent 放空口
17 lifting lug 吊耳
18 reflux inlet 回流口
19 tray 塔盘
20 manhole 人孔
21 insulation support ring 保温支承圈
22 ladder 直梯，梯子
23 reboiler return 重沸器（物料）返回口
24 thermometer connection 温度计接口
25 vortex breaker (anti-cavitation baffle) 防涡流挡板（破涡流器）
26 vent hole （裙座）排气管（口）
27 skirt 裙座
28 skirt access opening (access hole) （裙座）检查孔

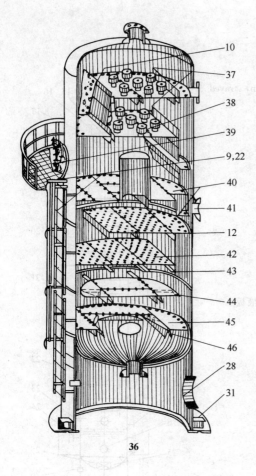

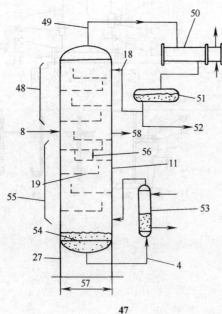

29　anchor bolt　地脚螺栓
30　anchor bolt chair　地脚螺栓座
31　base ring　基础环，底板
32　earth lug　接地板
33　gusset plate　筋板
34　compression ring　压环，环形盖板
35　washer　垫板
36　diagrammatic sketch of tower tray assembly　塔盘结构示意图
37　weir and seal plate　堰及密封板
38　typical bubble cap　典型泡罩
39　manhole with davit and cover plate　人孔（带吊柱及平盖）
40　accumulator tray with center stack and draw off box　集油塔盘（带中心管和抽液箱）
41　draw off nozzle　抽液管
42　perforated shower tray　多孔喷淋塔盘
43　channel truss　槽形构架
44　disc tray　圆形塔盘
45　donut tray　环形塔盘
46　trapezoidal truss　梯形构架
47　distillation tower arrangement　蒸馏塔的布置
48　rectifying section　精馏段
49　overhead vapor　塔顶蒸气
50　condenser　冷凝器
51　receiver drum　接收罐
52　distillate　馏出物
53　reboiler　再沸器，重沸器
54　bottom products　塔底产物
55　stripping section　提馏段，汽提段
56　trays spacing　塔盘间距
57　diameter　直径
58　side draw　侧线抽出馏分

6.2.1.1 液流型式 Liquid-Flow Patterns

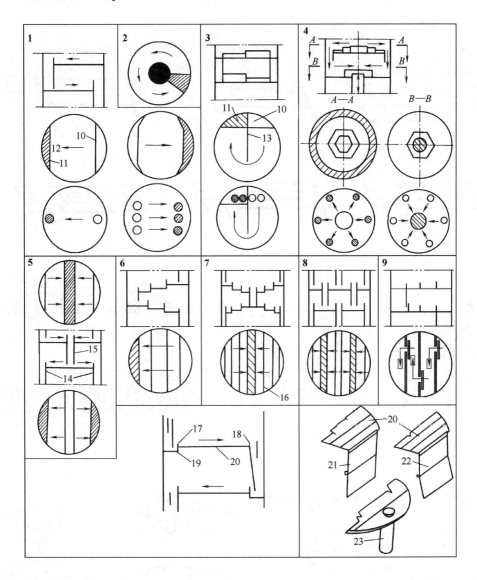

1 cross flow 横流
2 circumferential flow 环流
3 reverse flow 回转流,折流,U形流
4 radial flow 径流
5 split flow (double pass) 双流
6 cascade 阶梯流
7 double pass cascade 双阶梯流
8 four pass 四流
9 tortuous flow 曲折流
10 inlet downcomer 入口降液管
11 outlet downcomer 出口降液管
12 single flow 单流
13 baffle 折流挡板
14 side downcomer 侧面降液管
15 center downcomer 中间降液管
16 intermediate weir 中间堰
17 inlet weir 入口堰
18 outlet weir 出口堰
19 recessed seal pan (recessed pool) (凹)受液盘
20 tray 塔盘,塔板
21 straight downcomer (straight skirt) 直降液板
22 tapered (angle) downcomer (bent skirt) 斜降液板
23 circular downcomer 圆形降液管

6.2.1.2 塔盘型式　Types of Trays

(1) 浮阀塔盘　Valve Trays

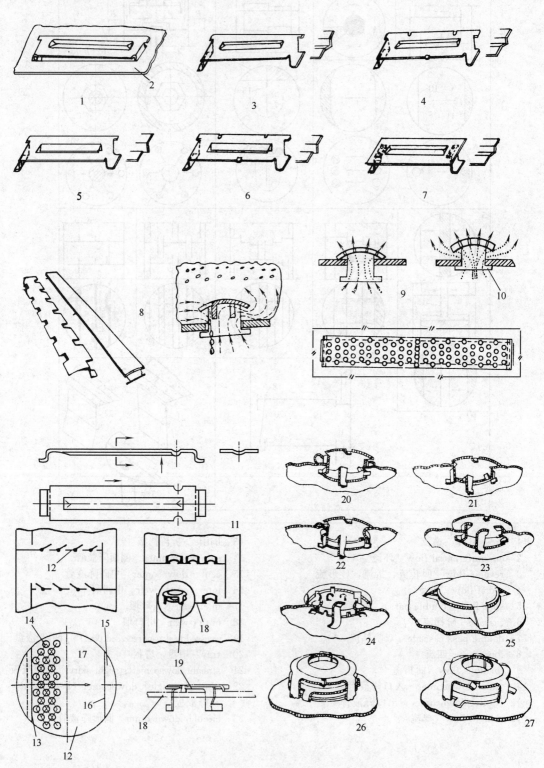

227

1	Nutter float valve tray 纳特条形浮阀塔盘	24	type V-6 V-6 型（浮阀）
2	deck（floor） 塔盘板	25	type V-0 V-0 型（浮阀）
3	P（plain）type Nutter float valve P（平面）型纳特条形浮阀	26	type A-1 A-1 型（浮阀）
		27	type A-2 A-2 型（浮阀）
4	D（dimpled）type Nutter float valve D（点接触）型纳特条形浮阀	28	T type flexitray（flexible tray） T 型盘式浮阀塔盘
5	L（louvered）type Nutter float valve L（直通）型纳特条形浮阀	29	A type flexitray（flexible tray） A 型盘式浮阀塔盘
6	DL（dimpled and louvered）type Nutter float valve DL（点接触和直通）型纳特条形浮阀	30	type T-9 T-9 型（浮阀）
		31	type T-Venturi 文丘里型（浮阀）
		32	cross flow valve tray 错流式浮阀塔盘
7	F（full cover）type Nutter float valve F（全覆盖）型纳特条形浮阀	33	pipe valve tray 管式浮阀塔盘
		34	pipe support 管式浮阀支架
8	Chepos valve tray 切普斯条形浮阀（穿流）塔盘	35	pipe valve 管式浮阀，管阀
		36	hy-contact valve tray 锥心浮阀塔盘
9	K_2a slat valve-sieve tray K_2a 长条形筛孔浮阀塔盘	37	float valve-sieve tray 浮阀-筛孔塔盘
		38	Glitsch combined valve and sieve tray 格里奇浮阀-筛孔混合塔盘
10	K_2a slat valve K_2a 长条形（筛孔）浮阀		
11	BDP valve tray BDP 条形浮阀	39	overflow weir 溢流堰
12	float valve tray 浮阀塔盘	40	hole （筛）孔
13	inlet weir 入口堰	41	turbo-float tray （TFT）波纹浮动塔盘，浮动角钢塔盘
14	float valve 浮阀		
15	downcomer 降液管	42	cross element 齿形横挡，横条
16	exit weir 出口堰	43	seal damper 闸式调节板，可调降液板
17	valve area 布阀区域	44	tray element 塔盘浮动元件，浮动角钢
18	ballast valve tray 重盘式浮阀塔盘	45	stopper 限制器，限位板
19	ballast valve 重盘式浮阀	46	support bar 支持杆，支持梁
20	type V-1 V-1 型（浮阀）	47	seal pan 受液盘
21	type V-2 V-2 型（浮阀）	48	square float valve 方形浮阀
22	type V-3 V-3 型（浮阀）	49	grid-valve tray 链网式浮阀塔盘
23	type V-4 V-4 型（浮阀）		

(2) 泡罩塔盘　Bubble Cap Trays

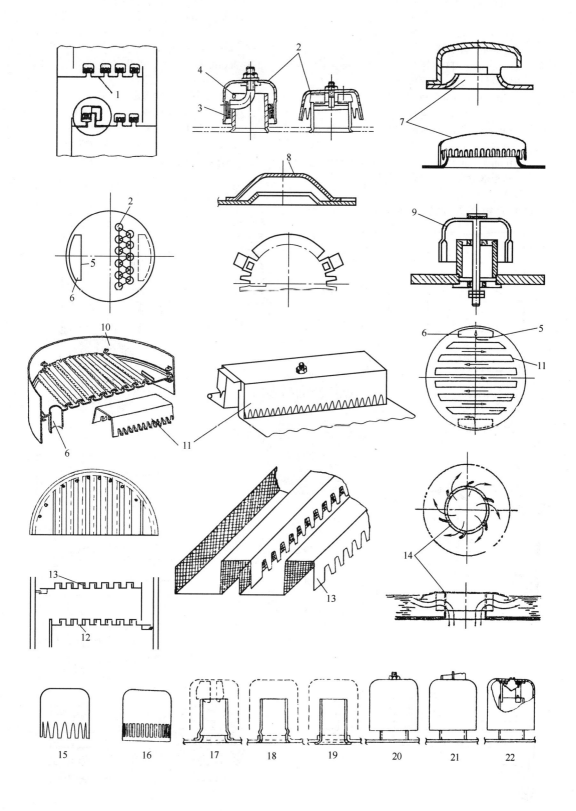

1　bubble cap tray（bell cap tray）　泡罩塔盘
2　bubble cap（bell cap）　泡罩
3　slot　齿缝
4　(vapor) riser　升气管
5　exit weir　出口堰
6　downcomer　降液管
7　flat bubble cap　扁平泡罩
8　squat bubble cap　伞形泡罩
9　rotating bell cap　旋转泡罩
10　tunnel（channel-cap）type tray　隧道式塔盘
11　tunnel cap（channel-cap，rectangular cap）槽形泡罩
12　uniflux tray　S形塔盘，单流式泡罩塔盘
13　"S" member　S形（塔盘）元件
14　vaned bubble cap　带导向叶片的圆泡罩
15　open slot　开口齿缝
16　closed slot　闭口齿缝
17　set on riser　外套式升气管
18　swaged riser　内插式升气管
19　pull up riser　穿插式升气管
20　bolt or stud hold down cap　螺栓（或双头螺柱）压紧式泡罩
21　wedge hold down cap　楔紧式泡罩
22　quarter turn cap　卡口式泡罩

（3）筛板塔盘　Sieve Trays

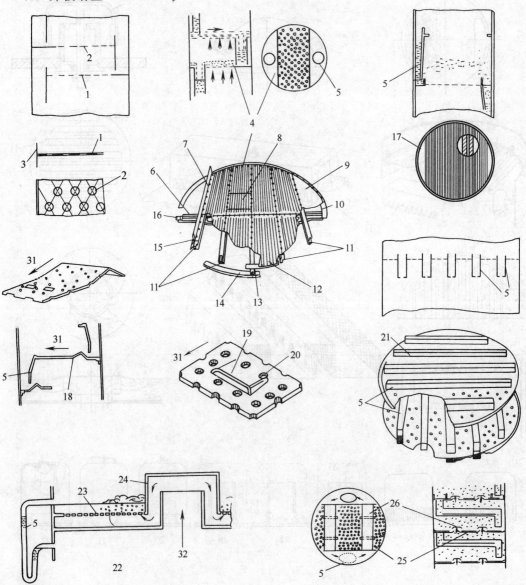

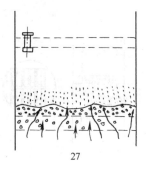
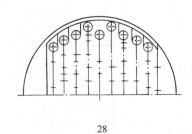
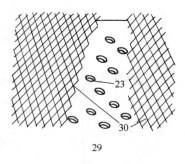

27　　　　　　　　28　　　　　　　　29

1　sieve tray（perforated tray）　筛板塔盘
2　hole　筛孔
3　exit weir　出口堰
4　sieve（perforated）tray with downcomer　有降液管的筛板塔盘
5　downcomer　降液管
6　major beam　主梁
7　weir　溢流堰
8　manway　人孔通道
9　stilling area　不鼓泡区，静止区
10　tray support ring　塔盘支持圈
11　minor beam support clamp　支梁卡子，副梁卡子
12　peripheral ring clamp　支承圈上的卡子，支承圈上的压紧件
13　subsupport tray ring　塔盘加强圈
14　subsupport angle ring　辅助角钢圈，加强角钢圈
15　minor beam　支梁，副梁
16　major beam clamp　主梁卡子
17　turbogrid with downcomer　有降液管的波纹式筛板（塔盘）
18　Linde sieve tray　林德筛板塔盘，导向筛板塔盘
19　directional slot　导向孔
20　drilling　钻孔
21　M-D（multi-downcomer）sieve tray　多降液管筛板塔盘，M-D筛板塔盘
22　Hagbarth tray　塔外降液管式泡罩-筛板塔盘
23　sieve plate　筛板
24　cap　泡罩
25　West tray　韦斯特筛板-槽形泡罩塔盘
26　tunnel cap　槽形泡罩，条形泡罩
27　float sieve tray　浮动筛板塔盘
28　new vertical sieve tray　新型立式筛板塔盘，新VST塔盘
29　modified sieve type tray　改型筛板塔盘
30　expanded metal　多孔拉制金属网板
31　liquid　液流
32　vapor　气流

（4）穿流型塔盘和喷射型塔盘　Dual-Flow Trays and Jet Trays

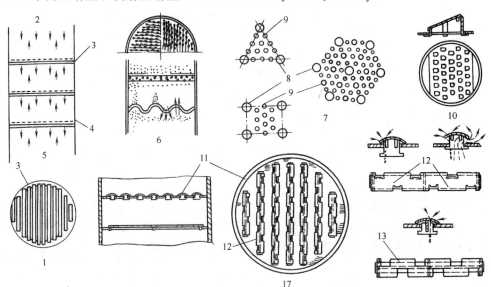

231

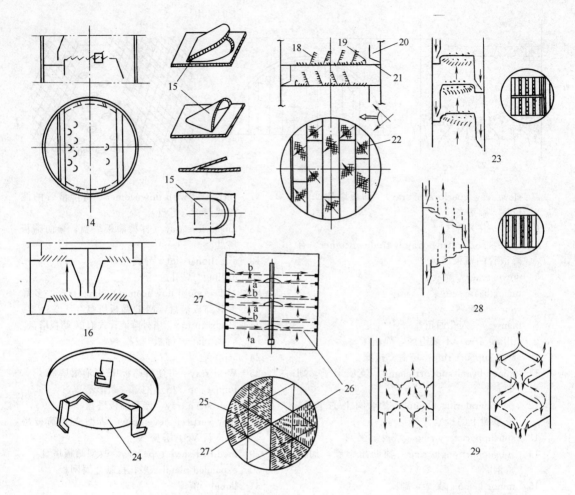

1 Turbogrid tray 波纹型穿流（淋降）栅板塔盘
2 liquid 液流
3 grid support ring 栅板支承圈
4 shell 塔壳
5 vapor 蒸气
6 ripple tray （穿流式）波纹筛板塔盘
7 arrangement of float valve and hole 浮阀和筛孔的排列形式
8 float valve 浮阀
9 hole 筛孔
10 directional float tray 浮舌塔盘，浮动舌形塔盘
11 slotted tray 条形浮阀穿流塔盘
12 K_1 type slat valve K_1 型条形浮阀
13 K_1a type slat valve K_1a 型条形浮阀
14 jet tray 喷射型塔盘，舌形塔盘
15 tab 舌片
16 double-pass jet tray 双流型舌形塔盘
17 perform tray 网孔塔盘，定向孔塔盘
18 flow breaker 挡沫板，碎流挡板
19 inlet weir 入口堰
20 downcomer (downspout) 降液管
21 seal pan 受液盘
22 expanded metal 多孔拉制金属网板
23 Venturi tray 文丘里塔盘
24 dual-flow float valve 穿流浮阀
25 standard Kittel plate 条孔网状塔盘，标准Kittel塔盘，基特尔标准式塔盘
26 lower plate 下板
27 upper plate 上板
28 cascade tray 阶梯式塔盘
29 directed cascading liquid tray 导向阶梯式液流塔盘

6.2.1.3 塔盘的支承　Supports of Tray

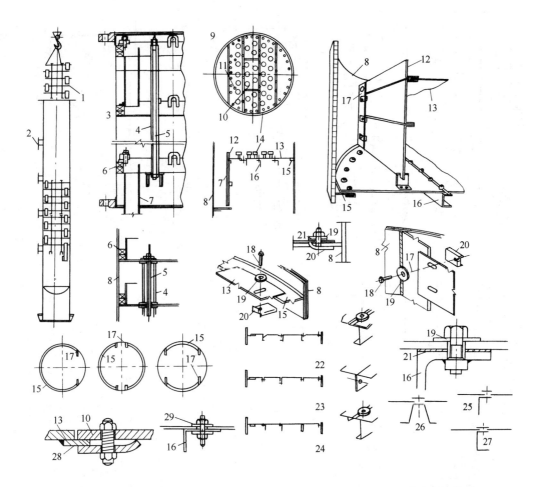

1　package tray　整装式塔盘
2　handhole　手孔
3　cartridge tray　整节塔盘
4　spacer　定距管
5　rod　拉杆
6　packing　填料
7　downcomer　降液管，降液板
8　tower wall　塔壁
9　separated tray　分块式塔盘
10　passageway cover plate　通道盖板
11　weep hole　泪孔，滴水孔
12　weir　溢流堰
13　tray sheet　塔板，塔盘板
14　bubble cap　泡罩
15　support ring　支承圈
16　support beam　支承梁
17　downcomer bar　降液板连接板
18　screw　螺钉
19　washer　垫圈
20　hold-down clamp　卡子
21　gasket　垫片
22　overlap type　搭接自身梁式（塔盘）
23　flange connection　折边自身梁式（塔盘）
24　beam support type（joist assembly）　梁支承式（塔盘）
25　angle truss（minor beam）　角钢支梁（副梁）
26　trapezoidal truss（minor beam）　梯形钢支梁（副梁）
27　channel truss（minor beam）　槽钢支梁（副梁）
28　backing strip　垫板
29　irregular gasket　异形垫板

6.2.1.4 塔底结构及重（再）沸器 Bottom Structures and Reboiler

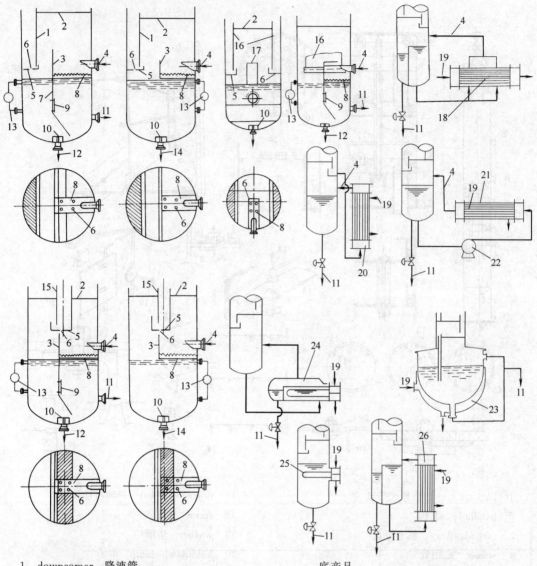

1　downcomer　降液管
2　bottom tray　底层塔盘
3　target　防冲板
4　reboiler return　重沸器（物料）返回口
5　seal pan　受液盘，液封盘
6　weep hole　泪孔，滴水孔
7　baffle　隔板，挡板
8　splash deck　防溅溢流板
9　manhole　人孔
10　vortex breaker (anti-cavitation baffle)　破涡流器，防涡流板
11　bottoms　塔底产品
12　reboiler feed　重沸器进料
13　level controller　液面调节器
14　reboiler feed and bottoms　重沸器进料和塔底产品
15　center downcomer　中间降液管
16　side downcomer　侧面降液管
17　splash plate　防溅板
18　horizontal thermo-syphon reboiler　卧式热虹吸重沸器
19　heat carrier　载热体
20　once pass reboiler　一次通过式重沸器
21　forced circulation reboiler　强制循环重沸器
22　pump　泵
23　jacketed kettle　夹套重沸釜
24　kettle-type reboiler　釜式重沸器，罐式重沸器
25　internal reboiler　内置重沸器
26　vertical thermo-syphon reboiler　立式热虹吸重沸器

6.2.1.5 进料和抽出 Feed and Draw-off

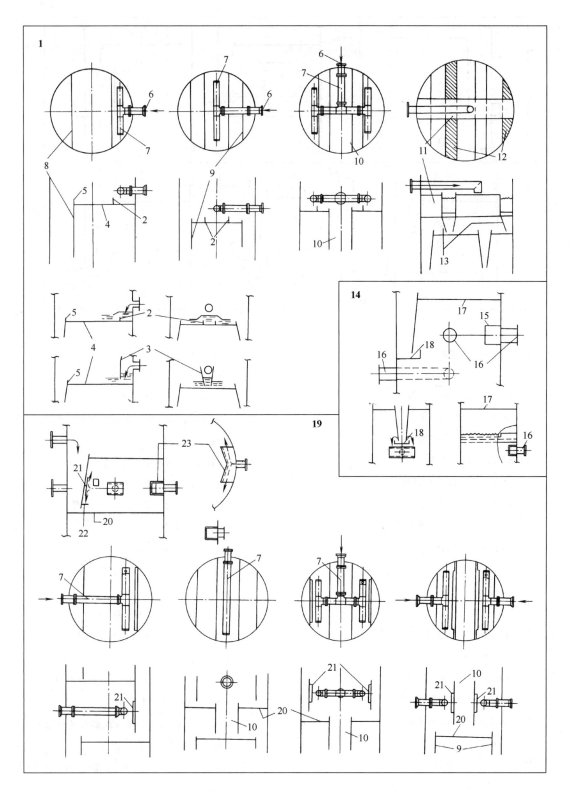

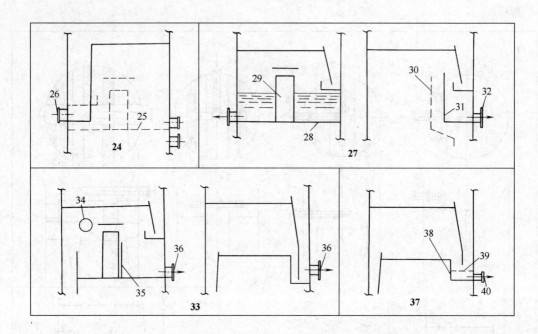

1	inlet feed 进料	22	guard 防护板
2	inlet weir 入口堰	23	channal baffle 槽形挡板
3	inlet baffle 入口挡板	24	total trap-out 全抽出
4	top tray 顶层塔盘	25	total draw chimney tray 有升气管的全抽出盘
5	outlet weir 出口堰	26	liquid draw nozzle 液体抽出口
6	reflux inlet 回流入口	27	total draw-off 全抽出
7	internal piping 内部接管	28	draw tray 抽出塔盘
8	downcomer 降液管	29	chimney 升气管
9	side downcomer 侧面降液管	30	spill baffle 溢流挡板
10	center downcomer 中间降液管	31	draw pan （弓形）抽出盘
11	feed trough 进料槽	32	draw nozzle 抽出口
12	feed area 进料区	33	partial draw-off 部分抽出
13	distribution baffle 分配挡板	34	return nozzle （重沸器）返回口
14	bottom vapor baffle 塔底蒸气入口	35	partition 隔板
15	vapor inlet baffle 蒸气入口挡板	36	partial draw-off nozzle 部分抽出口
16	vapor inlet 蒸气入口	37	water draw-off 水抽出
17	bottom tray 底层塔盘	38	weld-in draw pan 焊入式抽水斗（盘）
18	seal pan (recessed pan) 受液盘,液封盘	39	perforated plate 多孔板
19	intermediate feed 中间进料	40	water draw-off nozzle 水抽出口,水抽出接管口
20	feed tray 进料塔盘		
21	insulation plate 隔热板		

6.2.2 填料塔 Packed Column

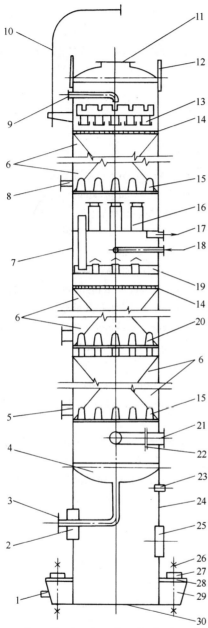

schematic drawing of packed column (tower) assembly 填料塔装配简图

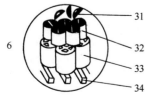

1 earth lug 接地板

2 pipe opening and support plate 引出孔及支承板
3 liquid outlet 液体出口
4 ellipsoidal head 椭圆形封头
5 manhole 人孔
6 packing 填料
7 shell 壳体
8 handhole 手孔
9 liquid inlet 液体入口
10 top davit 塔顶吊柱
11 gas outlet 气体出口
12 lifting lug 吊耳
13 liquid distributor 液体分布器
14 hole-down grid 格栅式填料压板
15 packing support plate 填料支承板
16 liquid collector 液体收集器
17 by-product 副产品
18 feed for distillation 蒸馏进料
19 liquid redistributor 液体再分布器
20 combined packing support plate and liquid redistributor 组合的填料支承板与液体再分布器
21 gas inlet 气体入口
22 internal flange and piping 内部法兰和接管
23 vent hole （裙座）排气管（口）
24 skirt 裙座
25 skirt access opening (access hole) （裙座）检查孔
26 anchor bolt 地脚螺栓
27 washer 垫板
28 compression ring 压环，环形盖板
29 gusset plate 筋板
30 base ring 基础环，底板
31 small packing 小型填料
32 medium size packing 中型填料
33 large packing 大型填料
34 support bar 支承杆

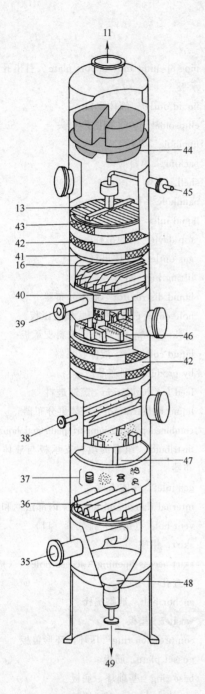

35	reboiler return	再沸器返回口
36	vapor injection packing support plate	气体喷射式填料支承板
37	random packings	散装填料
38	vapor feed	液体闪蒸进料
39	liquid feed	液体进料
40	ring channel with drainage	集液槽
41	support grid	填料支承栅板
42	structured packings	规整填料
43	locating grid	填料压圈
44	demister	除雾器，除沫器
45	reflux from condenser	回流入口
46	trough-type liquid distributor	槽式液体分布器
47	nozzle type liquid distributor	盘式液体分布器
48	vortex breaker	防涡流挡板
49	bottom product	塔釜出料口

diagrammatic sketch of packed column with various types of internals
填料塔示意图（带各种形式内件）

6.2.2.1 填料型式 Types of Packing

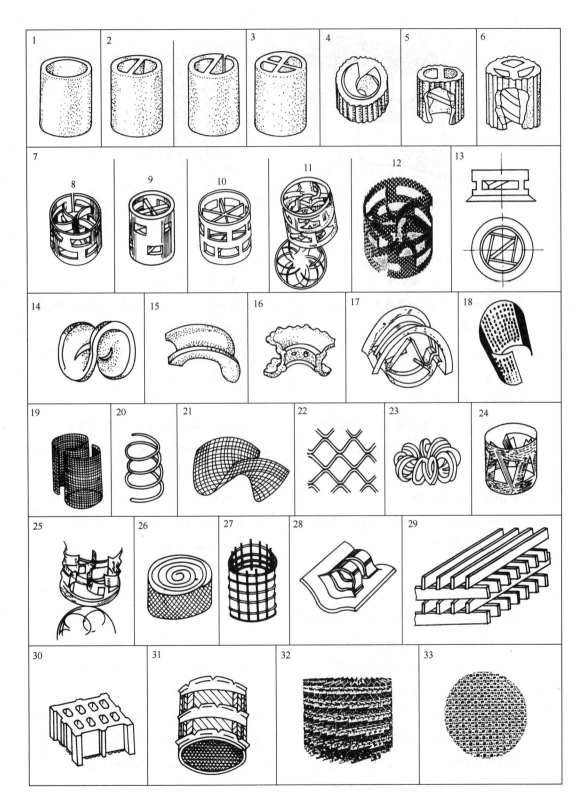

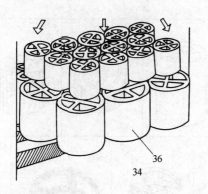
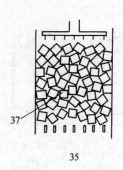

1　Raschig ring　拉西环
2　Lessing ring　θ环，勒辛环
3　cross-partition ring　十字格环
4　single spiral ring　单螺旋环
5　double spiral ring　双螺旋环
6　triple spiral ring　三螺旋环
7　Pall ring　鲍尔环
8　metallic Pall ring　金属鲍尔环，钢鲍尔环
9　Pall ring made of ceramic　瓷鲍尔环
10　Pall ring made of plastic　塑料鲍尔环
11　Hy-Pak　改进鲍尔环
12　Pall ring made of expanded metal　拉伸金属板网鲍尔环
13　cascade mini ring　阶梯环填料
14　Berl saddle　弧鞍型填料，伯尔鞍
15　Intalox saddle　矩鞍型填料，印泰劳格斯鞍
16　super Intalox saddle　改进矩鞍填料
17　Intalox metal packing　金属鞍环填料
18　Cannon packing　卡侬填料，坎农填料，金属凸刺填料
19　Dixon ring packing　θ网环填料，θ型填料，狄克松环填料
20　single turn helix packing　单螺旋形填料

21　McMahon saddle packing　金属网鞍形填料，麦克马洪填料
22　Spray packing　板条网孔填料，斯普雷帕克填料
23　Teller rosette packing（Tellerette）　特勒花环填料
24　ring packing　环填料
25　Leva packing　半环填料，莱瓦填料
26　Goodloe packing　缠绕金属网填料，古德洛填料
27　wire mesh packing　金属丝网环形填料
28　Interpack type　英特帕克填料
29　wood grid　木格
30　grid tile（drip-point grid）　格子砖
31　wire web packing　波纹网填料，丝网波纹填料
32　grid packing　格栅填料
33　square grid packing　方格栅填料
34　stacking　整齐排列
35　dumping　乱堆
36　structured packing　规整填料
37　random packing　散装填料

6.2.2.2 液体分布器和再分布器　Liquid Distributors and Redistributors

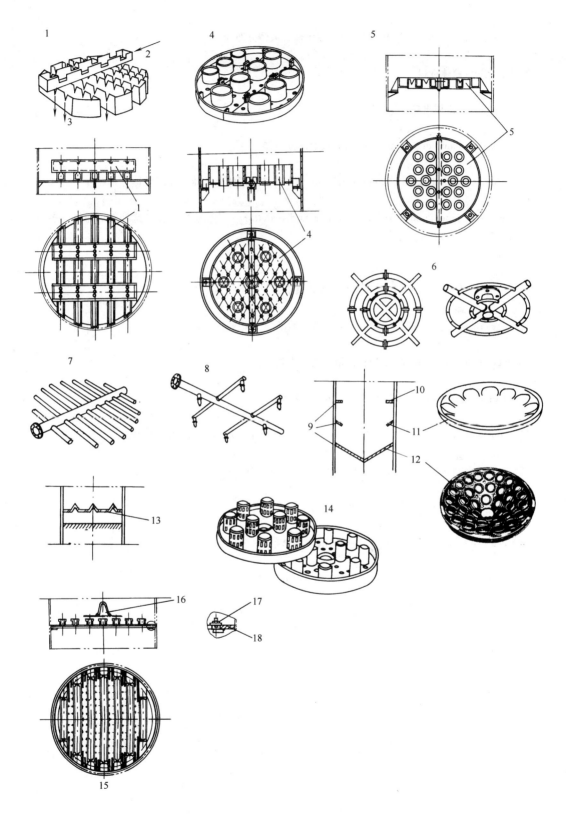

1　trough-type（box type）distributor　溢流槽式液体分布器

2　liquid in　液体进口

3　liquid out　液体出口

4　sieve type liquid distributor　筛孔盘式液体分布器

5　weir-riser（nozzle type）liquid distributor　溢流盘式液体分布器，盘式液体分布器

6　perforated pipe liquid distributor（ring sets distributor）　（环状）多孔管式液体分布器，多环管喷淋器

7　straight-tube sparger　水平引入管排管式喷淋器

8　showerhead type liquid distributor　莲蓬头喷洒器

9　wiper liquid redistributor　挡板式液体再分布器

10　straight side wiper　水平旁挡板（再分布器）

11　cone side wiper（wall wiper）　锥形旁挡板（再分布器），导液锥，改进分配锥

12　pyramid liquid redistributor　锥形液体再分布器，分配锥

13　weir-type liquid redistributor　堰式液体再分布器

14　bell-cap liquid redistributor　钟罩式液体再分布器

15　beam type liquid redistributor　梁型液体再分布器

16　beam type vapor injection support plate　梁型气体喷射式支承板

17　clamp　卡子

18　support ring　支承圈

6.2.2.3 填料压板和支承板 Packing Hold-Down Plates and Support Plates

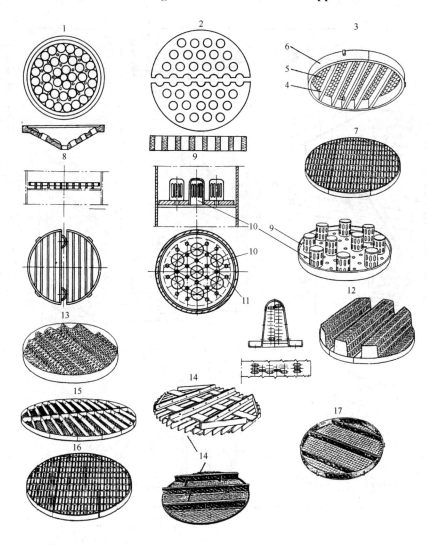

1　pyramid support plate　锥形支承板
2　perforated support plate　多孔式支承板
3　flat mesh support plate　金属网支承板
4　rib　筋板
5　wire mesh screen　金属丝网
6　flat ring　扁钢圈
7　support grid (grating)　栅板，格栅式支承板，蓖条支承板
8　improved support grid　改进的栅板支承板
9　perforated vapor riser type support plate (cap type support plate, separate flow type support plate, vapor injection type support plate)　钟罩型气体喷射式支承板
10　vapor riser　升气管
11　liquid downflow　降液孔
12　beam type vapor injection support plate　梁型气体喷射式支承板
13　expanded metal support plate　波纹板网支承板
14　hold-down plate　填料压板
15　wire mesh cover plate　丝网压板
16　hold-down grid (grating)　格栅式填料压板，栅条压板
17　bed limiter　床层限制板

6.2.3 楼梯（梯子）和平台 Stair and Platform

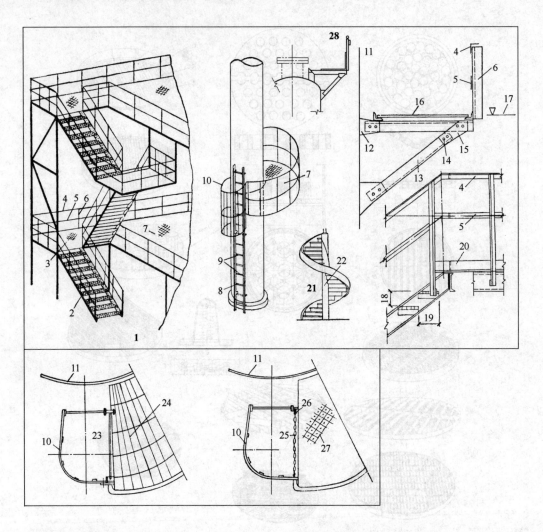

1	structure 框架	16	flooring （平台）铺板
2	stair 楼梯	17	platform elevation 平台标高
3	stair landing 中间平台，楼梯平台	18	riser 踏级高度
4	top rail (handrail) 扶手，栏杆	19	tread width 踏板宽度
5	mid rail 护腰，栏杆，横挡	20	toe plate 踢脚板
6	post (handrail post) 栏杆立柱	21	winding staircase 盘梯，旋梯
7	platform 平台	22	newel 旋梯中柱
8	ladder 直梯	23	safety gate 安全门
9	rung 直梯踏步，梯蹬	24	grating 笆子板
10	cage hoop 护罩	25	safety chain 安全链
11	vessel wall 器壁	26	snap 钩扣
12	lug 连接板	27	checkered plate (checkered sheet) 花纹钢板，网纹钢板
13	brace angle 斜撑		
14	rivet 铆钉	28	top platform 塔顶平台
15	gusset plate 连接板，结点板		

6.2.4 塔附件 Tower Attachments

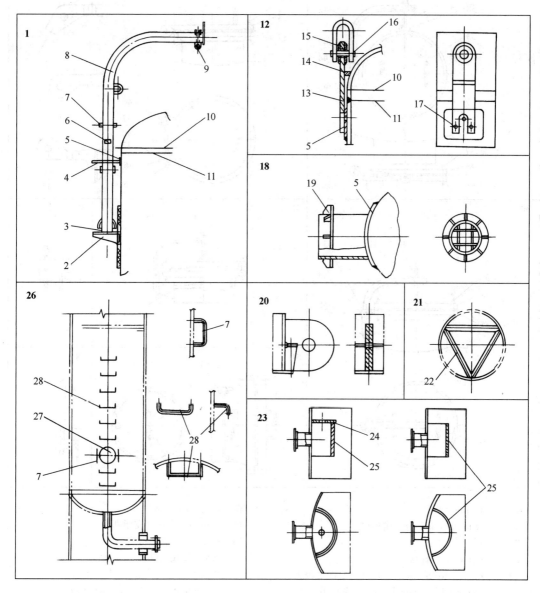

1	top davit 塔顶吊柱		13	lug plate 吊耳板
2	support bracket 支架		14	connecting plate 连接板
3	cap 防雨罩		15	shackle 吊钩，钩子，钩环
4	guide plate 导向板		16	shackle pin 钩环销
5	pad plate 垫板		17	vent hole 排气孔
6	name plate 铭牌		18	trunnion 轴式吊耳
7	handle 把手，手柄		19	rib 筋板
8	davit pipe 吊柱管，吊杆		20	tailing lug 尾部吊耳
9	hook 吊钩		21	skirt stiffener 裙座加强撑，刚性梁
10	tangent line 切线		22	pipe 钢管
11	welding line 焊缝线		23	deflector 进料口挡板
12	lifting lug 吊耳		24	cover plate 盖板

245

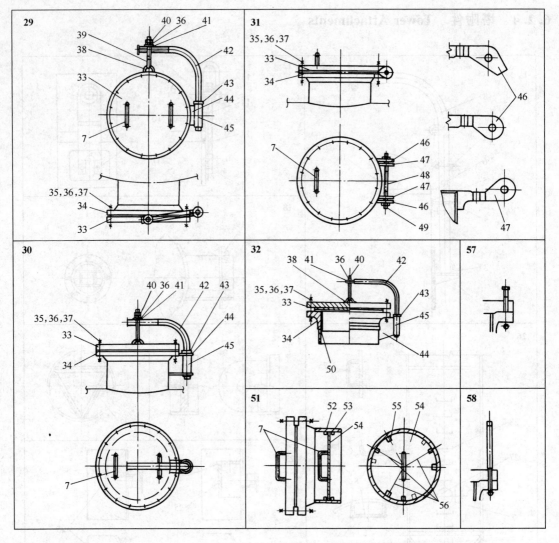

25	half pipe 半管		43	ring 环
26	internal ladder 内部爬梯		44	support plate 支承板
27	manhole 人孔		45	sleeve 套筒
28	ladder rung 梯蹬		46	cover hinge (lug) 盖轴耳
29	manhole with horizontal davit 垂直吊盖人孔		47	flange hinge (lug) 法兰轴耳
30	manhole with vertical davit 水平吊盖人孔		48	pin (hinge bolt) 轴销（回转螺栓）
31	manhole with hinge 回转盖人孔		49	cotter (split pin) 销（开口销）
32	manhole from stainless steel 不锈钢人孔		50	sleeve liner 衬筒
33	blind flange 法兰盖		51	discharge opening 卸料孔
34	flange 法兰		52	outside block 外挡块
35	stud bolt 双头螺柱		53	inside block 内挡块
36	nut 螺母		54	retaining plate 挡板
37	gasket 垫片		55	notch 凹槽
38	lifting ring 吊环		56	equally spaced 均布
39	eye bolt 吊环螺栓，吊钩		57	jack bolt 起重螺栓
40	lock nut 锁紧螺母		58	taper pin 锥销
41	washer 垫圈			
42	davit arm 转臂			

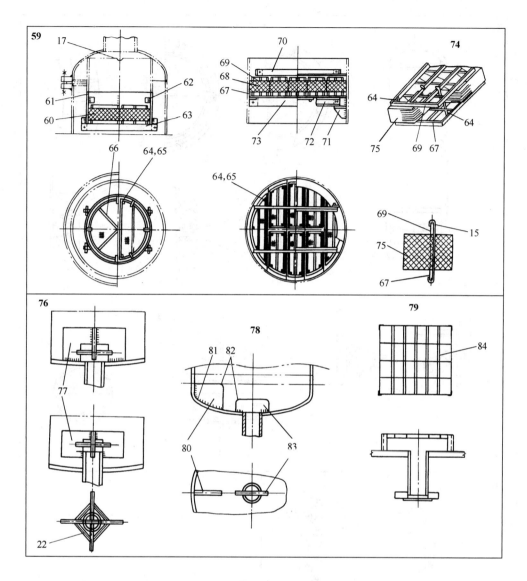

59	wire mesh demister（mist extractor） 丝网除沫器	72	channel beam 槽钢梁
60	wound mesh pad 缠绕破沫网垫	73	angle beam 角钢梁
61	gas riser 升气管	74	diagrammatic sketch of the mesh strip structure 网块结构示意图
62	stopper 挡块	75	wire mesh screen 丝网
63	flat bar（iron） 扁钢	76	vortex breaker 防涡流挡板
64	spacer 定距杆	77	cross baffle 十字形挡板
65	round rod（bar, iron） 圆钢	78	flat baffle 扁钢型挡板
66	cross bar 十字架	79	grating baffle 格栅式挡板
67	support grid 支承栅板	80	radial anti-swirl baffle 径向防涡流挡板
68	mesh strip 破沫网条块	81	tack weld 点焊
69	hold-down grid 压栅板	82	round corner 圆角
70	hold-down bar 压条	83	center vortex breaker 中心防涡流挡板
71	support rib 支承筋	84	grating 格栅

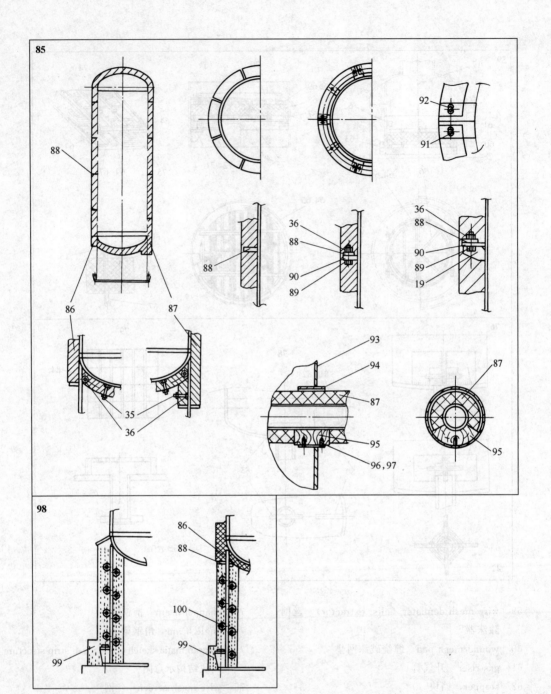

85　insulation support　保温支承
86　hot insulation　保温
87　cold insulation　保冷
88　support ring　支承圈
89　bolt　螺栓
90　lug　支耳板
91　hole　圆孔
92　slotted hole　长圆孔
93　skirt　裙座
94　pipe opening　引出孔
95　wooden pillow　木垫块
96　wood screw　木螺钉
97　stainless steel　不锈钢
98　fire protection support　防火支承
99　fire protection　防火层
100　hex nut　六角螺母

6.2.5 塔结构示例 Column Examples

6.2.5.1 CO_2 吸收塔 CO_2 Absorber

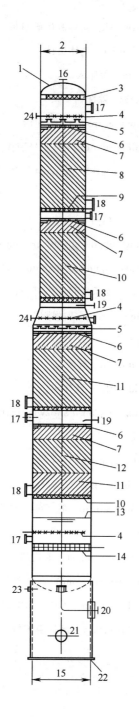

1　ellipsoidal head　椭圆形封头
2　inside diameter（I.D.）　内径
3　demister　除沫器
4　sparger　喷洒器
5　liquid distributor　液体分布器
6　hold-down grid（grating）　笼子压板，栅条压板
7　packing　填料
8　bed #1　第一层
9　packing support plate-vapor injection support plate and liquid redistributor　填料支承板-气体喷射式支承板和液体再分布器
10　bed #2　第二层
11　bed #3　第三层
12　bed #4　第四层
13　max. liquid level　最高液位
14　vortex breaker　防涡流挡板，破涡流器
15　anchor B.C.（bolt circle）　地脚螺栓中心圆
16　vapor outlet　气体出口
17　manhole　人孔
18　unloading connection　卸料孔
19　vapor sample　气体取样口
20　liquid outlet　液体出口
21　skirt access opening（access hole）（裙座）检查孔
22　anchor bolt　地脚螺栓
23　vent hole　排气管（口）
24　liquid in　液体进口

6.2.5.2 再生塔/CO₂ 汽提塔 Regenerator/CO₂ Stripper

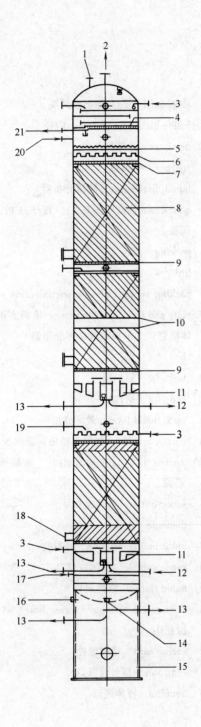

1　rupture disk　防爆膜爆破盘板
2　vapor outlet　气体出口
3　liquid in　液体入口
4　demister　除沫器
5　distribution trough　分配槽
6　liquid distributor #1　1号液体分布器
7　hold-down grid（grating）　笼子压板，栅条压板
8　packing　填料
9　packing support plate-vapor injection support plate and liquid redistributor　填料支承板-气体喷射式支承板和液体再分布器
10　stiffening ring　加强圈
11　drawoff pan　泄流盘
12　liquid drain　放液口
13　liquid outlet　液体出口
14　vortex breaker（antiswirl baffle）　防涡流挡板，破涡流器
15　skirt　裙座
16　vent hole　排气管（口）
17　reboiler return　重（再）沸器（物料）返回口
18　unloading catalyst　催化剂卸出口
19　vapor sample connection　气体取样口
20　liquid/vapor in　液体/气体入口
21　liquid overflow　液体溢流口

6.3 反应设备　Reactor

6.3.1 氨合成塔　Ammonia Converter

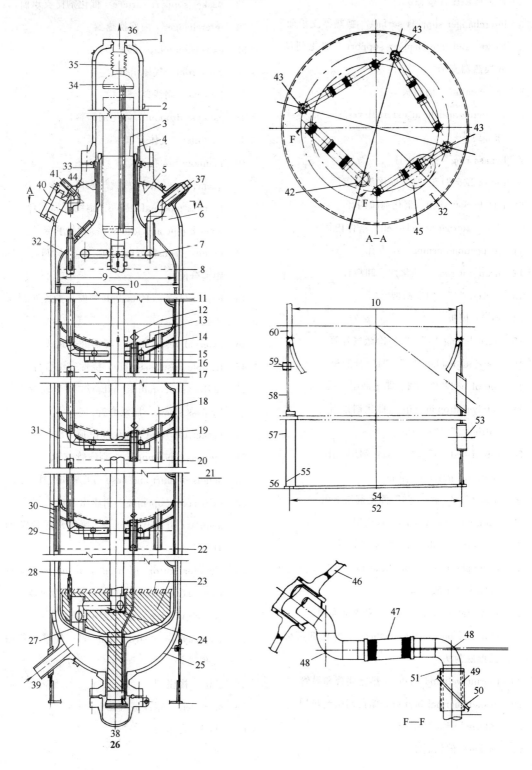

1	face of flange external vessel 塔体法兰面	32	toriconical top basket head 催化剂筐准锥形顶封头
2	shell metal temperature instrument block 壳体金属温度计接触块	33	basket support surface 催化剂筐支承面
3	interchanger support surface 换热器支承面	34	interchanger 内部换热器
4	basket and interchanger support 催化剂筐和换热器支承	35	expansion joint 膨胀节
5	anti rotation 防转板	36	gas outlet 气体出口
6	hemispherical head external vessel 塔体半球形封头	37	by pass 副线口
7	bypass ring 旁路环管	38	catalyst dropout 催化剂卸料口
8	top of bed "1" "1"催化剂层上限	39	gas inlet 气体入口
9	I. D. basket shell 催化剂筐内径	40	manhole 人孔
10	I. D. external vessel shell 塔体内径	41	quench gas inlet 冷激气入口
11	inspection opening 检查孔	42	by pass pipe and sleeve 副线管和套管
12	drop out rod （催化剂）卸料杆	43	quench pipe and leader 冷激管和导管
13	screen "1" "1"筛网	44	n-slots (equally spaced) n个长圆孔（等距离分配）
14	grid 栅板	45	basket top head manhole 催化剂筐顶封头人孔
15	quench ring "1" "1"冷激气环管	46	external vessel head 容器外壳封头
16	distributor plate "1" "1"分配板	47	flexible hose assembly 挠性管组合件
17	top of bed "2" "2"催化剂层上限	48	S. R. (short radius) elbow 短半径弯管
18	catalyst dropout pipe 催化剂排出管	49	by pass sleeve 副线套管
19	baffle "2" "2"挡板	50	basket top head 催化剂筐顶封头
20	top of bed "3" "3"催化剂层上限	51	by pass pipe support 副线管支承
21	center of gravity 重心	52	details of skirt and base 裙座和底座详图
22	top of bed "4" "4"催化剂层上限	53	skirt access opening 裙座出入口
23	alumina tabular balls 氧化铝球	54	diameter of bolt circle （地脚）螺栓节圆直径
24	gas return collector 气体回集器	55	chamfer bottom 6″ back to blend with base ring 底部向后倒角6″与底环圆滑过渡
25	hemispherical bottom head external vessel 塔体半球形底封头	56	bottom of base ring 底环底面
26	sectional elevation view 立视剖面图	57	manway for catalyst removal 卸催化剂用人孔
27	elliptical bottom basket head 催化剂筐椭圆形底封头	58	skirt 裙座
28	thermowell and leader 热电偶管和导管	59	skirt vent 裙座通气孔
29	basket thermal barrier 催化剂筐绝热层	60	shell 壳体
30	external vessel 塔体		
31	basket 催化剂筐		

6.3.2 尿素合成塔　Urea Reactor

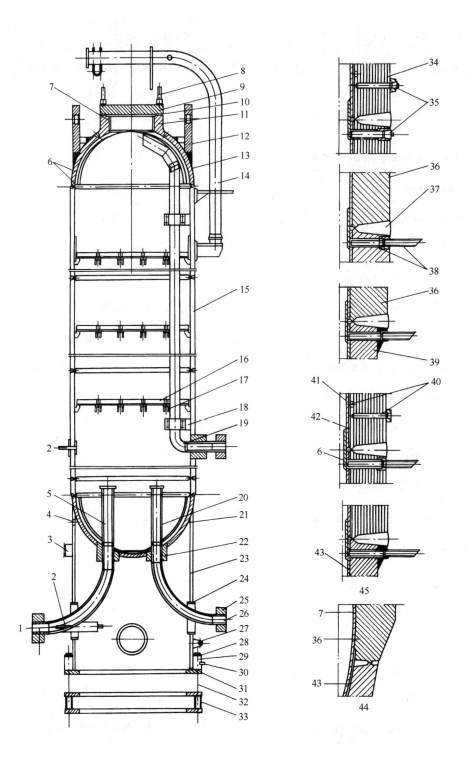

1　NH_3 inlet　氨气入口
2　thermowell　热电偶套管
3　nameplate　铭牌
4　skirt vent　裙座排气孔
5　NH_3 distributing pipe　氨气体分布管
6　cover plates　盖板
7　lining　衬里
8　stud bolts　双头螺栓
9　manhole cover　人孔盖
10　gasket　垫片
11　manhole　人孔
12　top lifting lugs　顶部吊耳
13　hemispherical top head　顶部半球形封头
14　top davit　顶部吊柱
15　shell　壳体
16　trays　塔盘
17　brackets for trays　塔盘托架
18　clips　卡子
19　elbow　弯头
20　CO_2 distributing pipe　二氧化碳气体分布管
21　hemispherical bottom head　底部半球形封头
22　reinforcement element　补强元件
23　skirt　裙座
24　skirt access opening　裙座出入口
25　flange　法兰
26　CO_2 inlet　二氧化碳气体入口
27　tailing lug　尾部吊耳
28　compression ring　压环
29　gusset plates　筋板
30　earth lug　接地板
31　base ring　基础环
32　anchor bolts/nuts/washers　地脚螺栓/螺母/垫圈
33　template　模板
34　layered shell　多层壳体
35　plugs for shipping　运输用塞子
36　solid shell　单层壳体
37　weld　焊缝
38　leak detector　检漏装置
39　solid head　单层封头
40　vent holes　排气孔
41　inner cylinder　内筒
42　dummy　盲层
43　overlay, weld deposited　堆焊层
44　type 1 solid shell attachment to solid head
　　型式 1：单层壳体和单层封头的连接
45　type 2 layered shell attachment to solid head
　　型式 2：多层壳体和单层封头的连接

6.3.3 环管反应器 Loop Reactor

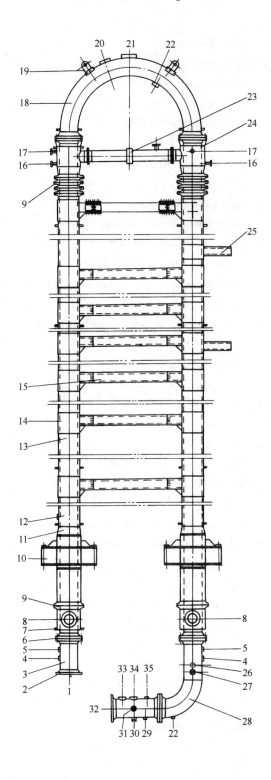

1　pump suction　泵吸入口
2　flange　法兰
3　inner　内管
4　killer injection　杀死剂注入口
5　temperature connection　温度计口
6　gasket　垫片
7　insulation supports　保温支承
8　jacket water inlet/outlet　夹套水进/出口
9　axial expansion　轴向膨胀节
10　reactor support　反应器支座
11　conical section　锥形短节
12　nameplate　铭牌
13　jacket　夹套
14　reinforcing pipe　补强管
15　connecting beam　连接梁
16　water surge drum connection　膨胀水槽连接口
17　jacket vent　夹套放空口
18　180° elbow　180°弯头
19　lifting lug　吊耳
20　vent to BDL　放空至界区
21　safety valve　安全阀
22　slurry inlet/outlet　淤浆入口/出口
23　expansion　膨胀节
24　jacket vent　夹套放空口
25　pipe support　管子支承
26　H_2 sample point　氢气取样口
27　nitrogen inlet　氮气入口
28　90° elbow　90°弯头
29　sample point　取样口
30　reactor emptying　反应器排空口
31　blow down bottom　底部排放口
32　pump discharge　泵排出口
33　feed　进料口
34　P.T　压力变送器口
35　spare　备用口

6.3.4 氧氯化反应器 Oxychlorination Reactor

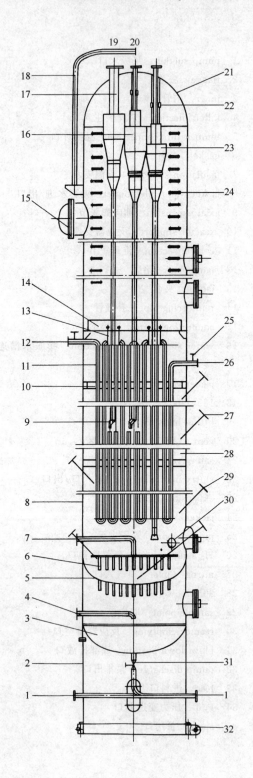

1　drain　排净口
2　skirt　裙座
3　elliptical head　椭圆封头
4　recycle gas inlet　循环气入口
5　distribution tray　分布盘
6　distributor　分布器
7　HCl, O_2 inlet　氯化氢、氧气入口
8　shell　壳体
9　trickle valve　滴流阀
10　fixed position frame　定位架
11　cooling coil bundle　冷却盘管管束
12　cooling coil outlet　冷却盘管出口
13　suspended support for cooling coil　冷却盘管悬挂式支承
14　cooling coil bundle support　冷却盘管管束支承
15　manhole　人孔
16　2nd stage cyclone　二级旋风分离器
17　3rd stage cyclone　三级旋风分离器
18　davit　吊柱
19　gas outlet　气体出口
20　vent　放空口
21　hemispherical head　半球形封头
22　suspended support for cyclone　旋风分离器悬挂式支承
23　1st stage cyclone　一级旋风分离器
24　internal ladder rung　内部爬梯
25　cooling coil vent　冷却盘管放空口
26　cooling coil inlet　冷却盘管入口
27　pressure gauge　压力计口
28　catalyst bed (fluidized bed)　催化剂床层（流化床）
29　thermometer　温度计口
30　catalyst transfer　催化剂输送口
31　ball valve　球阀
32　tailing lug　尾部吊耳

6.3.5 聚合釜 Polymerizer

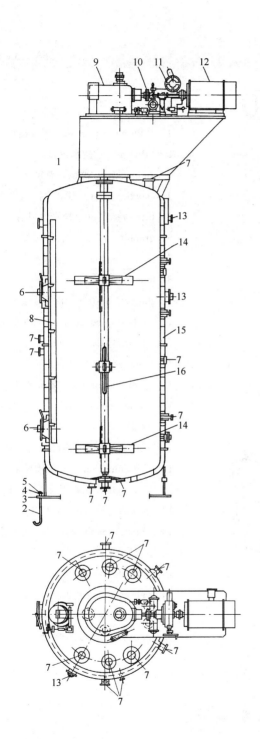
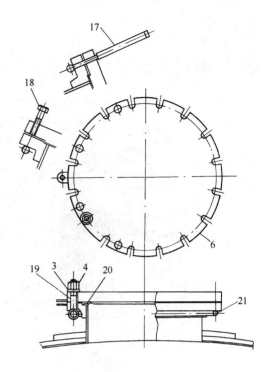

1 vessel 容器
2 anchor bolt 地脚螺栓
3 washer 垫圈
4 nut 螺母
5 lock nut 锁紧螺母
6 manhole cover 人孔盖
7 nozzle connection 接管口
8 baffle 挡板
9 reduction gear 变速齿轮
10 coupling 联轴器
11 torque convertor 扭矩变换器
12 motor 电动机
13 inspection 检查（孔）
14 agitator 搅拌器
15 jacket 夹套
16 key 键
17 taper pin 锥形销
18 jack bolt 起重螺栓
19 hinge bolt 回转螺栓
20 gasket 垫片
21 support 支承

6.3.6 隔膜电解槽 Diaphragm Cells

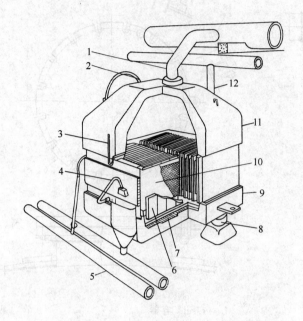

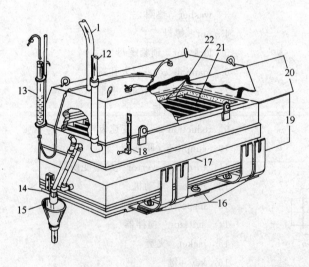

1 chlorine gas outlet 氯气出口
2 brine inlet 盐水进口
3 sight glass 视镜，看窗
4 caustic outlet 氢氧化钠出口
5 brine feed line 盐水给料管
6 graphite anodes 石墨阳板
7 anode conductor 阳极导体
8 insulator 绝缘体
9 concrete bottom 混凝土底
10 asbestos covered cathode 包石棉的阴极
11 concrete top 混凝土顶盖
12 hydrogen outlet 氢气出口
13 brine feed assembly 盐水给料装置
14 cell base 电解槽底板
15 cell liquor cup 电解槽碱液杯
16 anode grid lugs 阳极格栅挂耳
17 copper gird bar 铜箍条
18 anolyte manometer 阳极液压力计
19 cathode assembly 阴极组件
20 cell head 电解槽头部
21 carbon electrodes 碳极
22 cathode screens 阴极网

6.3.7 变换炉 Shift Converter

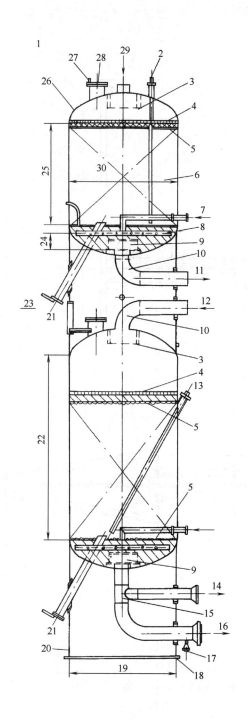

1　elevation　立视图
2　tubing for thermocouple　热电偶套管
3　distributor　分配器
4　grating　格子板
5　screen　筛网
6　catalyst (promoted iron oxide)　催化剂（活性氧化铁）
7　steam inlet　蒸汽入口
8　distributor header pipe　分配器集管
9　outlet　出口
10　90° long elbow　90°长半径弯头
11　vapor outlet　（高变）气体出口
12　intercooler vapor return　中间冷却器（高变）气体返回口
13　thermowell　热电偶套管
14　bottom outlet　底部出口
15　reduced tee　缩颈三通
16　vapor outlet　（低变）气体出口
17　drain　排液口
18　base ring　基础圈
19　bolt diameter circle　底脚螺栓节圆（直径）
20　skirt　裙座
21　catalyst dropout　催化剂卸出口
22　bottom compartment　下部变换室
23　center of gravity　重心（位置）
24　top of skirt　裙座顶部
25　top compartment　顶部变换室
26　elliptical head　椭圆形封头
27　lifting lug　吊耳
28　manhole　人孔
29　vapor inlet　（转换）气体入口
30　I.D. (inside diameter)　内径

6.4 造粒设备 Prilling Equipment

6.4.1 （造粒塔）总图及造粒喷头组装图 General Assembly and Prill-Spray Assembly

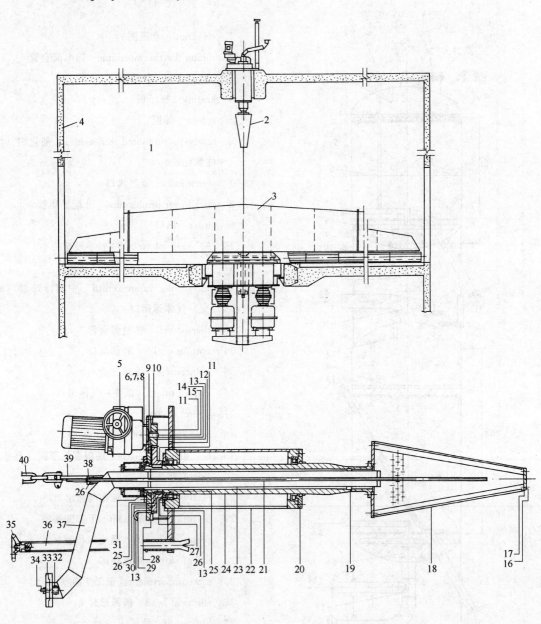

1	prill tower general assembly 造粒塔总图	7	washer 垫圈
2	prill spray assembly 造粒喷头组装	8	nut 螺母
3	reclaimer 扒料机	9,16,20	cover 盖
4	concrete tower body 混凝土塔体	10	screw 螺钉
5	motor reducer variator 电动机减速箱	11	key 键
6	stud end 双头螺柱	12	plug 丝堵

13	tooth 齿轮		27	oil sealed ring 油封环
14，31	ring 圈（环）		29	air breather 空气透气管
15	bolt 螺栓		30	distance box 间隔环
17，28	gasket 垫片		32	pin 销
18	prill bucket 造粒筒		33	flange 法兰
19	bearing 轴承		34	clamping bolt 夹紧螺栓
21	level indication pipe 液面指示管		35	handwheel 手轮
22	support prill-bucket 造粒筒支承套		36	column 立柱
23	charging funnel 进料筒		37	turn tube 弯管
24	external shaft 外轴		38	gland 压盖
25	outer safety ring 安全外挡圈		39	lifting yoke 吊钩
26	O-ring O形密封环		40	chain 链

6.4.2 造粒塔扒料机　Prill Tower Reclaimer

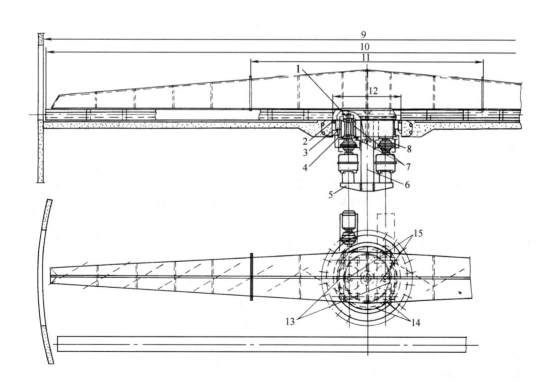

261

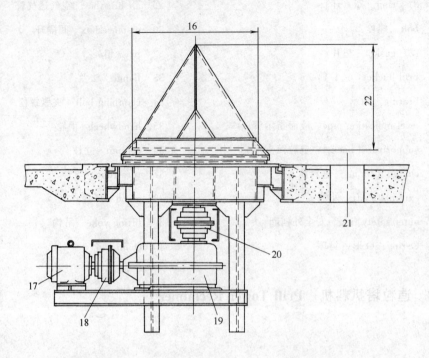

1	pinion 小齿轮	13	lubrication for 2 pinions 2个小齿轮润滑系统
2	ball race slewing rim single row 单列滚珠轴承外圈	14	lubrication system of ball race slewing rim 滚珠轴承外圈的润滑系统
3	foundation frame 基础构架	15	lubrication for 2 self-aligning roller bearings 2个自定心滚柱轴承用的润滑系统
4	compressed air connection 压缩空气连接	16	reclaimer arm 扒料机臂
5	drive unit support 传动装置支架	17	motor 电动机
6	support of speed controller 速度调节器支架	18	turbo coupling with flexible cam 带弹性凸轮的透平联轴器
7	lubrication for pinion and self-aligning roller bearing 小齿轮和自定心滚柱轴承用的润滑系统	19	gearing 齿轮传动装置
8	support of drive shaft 传动轴支架	20	coupling 联轴器
9	clearance 净空	21	axis of opening 开孔的轴线
10	max. length of reclaimer 扒料机最大长度	22	max. height of arm 臂的最大高度
11	center part 中央部分		
12	outside inner frame 内框架外径		

6.5 换热设备　Heat Exchanger

6.5.1 换热器的类型　Types of Heat Exchangers

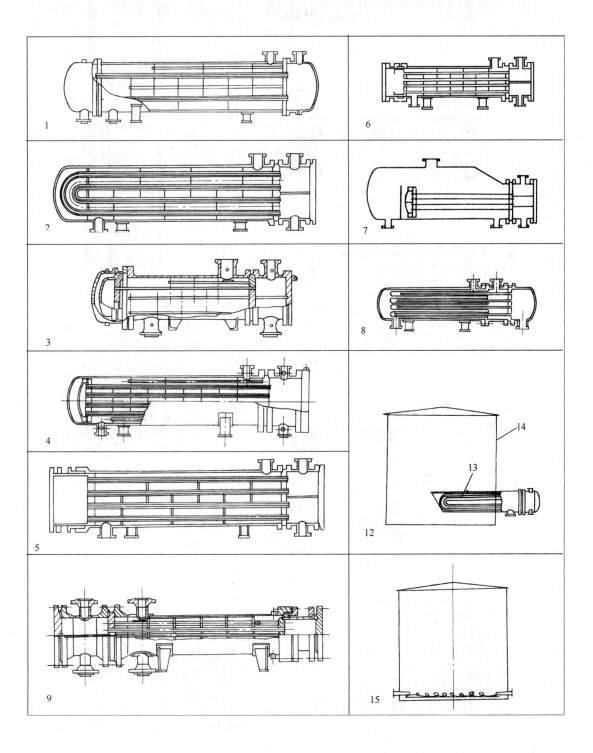

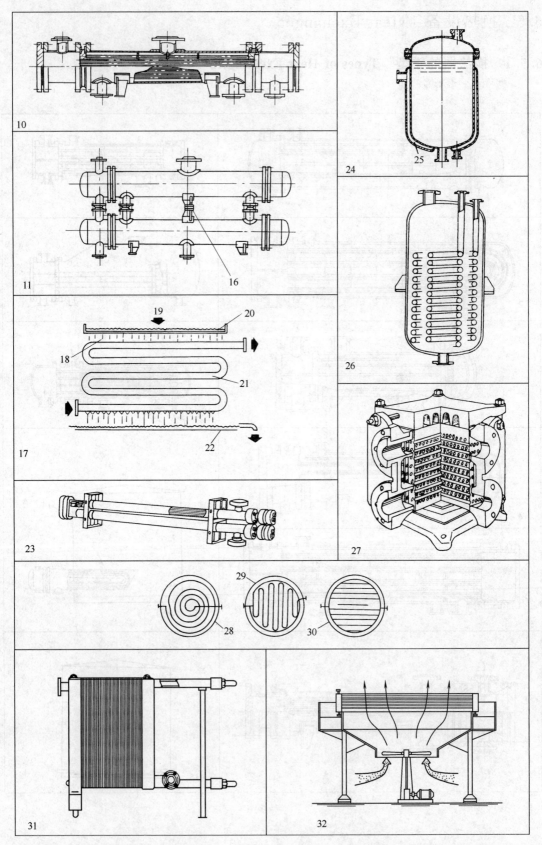

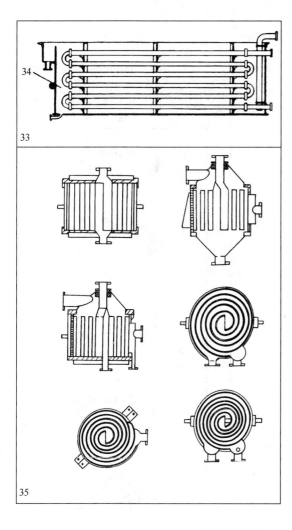

(1~11 tubular heat exchanger 管壳式换热器)

1　fixed tubesheet exchanger　固定管板式换热器

2　U-tube heat exchanger　U形管式换热器

3　internal-floating head exchanger　内浮头换热器，浮头式换热器

4　pull-through floating head exchanger　可抽式浮头换热器

5　outside packed floating head exchanger　（外）填料函式浮头换热器

6　externally sealed floating tubesheet exchanger (packed floating tubesheet with lantern ring exchanger)　外密封浮动管板式换热器（带灯笼环的填料函式浮动管板换热器）

7　kettle type reboiler　釜式重沸器

8　bayonet type exchanger　内插管式换热器

9　outside packed floating head exchanger with two shell pass　填料函双壳程换热器

10　packed floating tubesheet exchanger with split flow device　填料函分流式换热器

11　stacked exchanger　重叠式换热器

12　tank suction heater　储罐抽吸加热器

13　U-tube bundle　U形管束

14　tank　储罐

15　tank coil heater　盘管式储罐加热器

16　shim　调整垫片

17　horizontal film type cooler (cascade drip cooler)　卧式水淋冷却器

18　uniform film　均匀水膜

19　cooling water　冷却水

20　distributing trough　水分配槽

21　coil　盘管

22　water receiver pool　水收集槽，蓄水槽

23　double-pipe exchanger　钢制管式换热器，套管换热器

24　jacketed type exchanger　夹套式换热器

25　jacket　夹套

26　spiral tube exchanger　螺旋管式换热器

27　graphite block heat exchanger　块式石墨换热器

28　spiral coil　螺旋盘管

29　hairpin　U形盘管

30　ring header type　圆环集合管

31　plate exchanger　板式换热器

32　air cooled heat exchanger　空气冷却器，空冷器

33　submerged-pipe coil exchanger　浸没式盘管换热器

34　water box　水箱

35　spiral plate exchanger　螺旋板式换热器

6.5.2 管壳式换热器　Tubular Heat Exchanger

（1）前端管箱、壳体及后端管箱的型式　Types of Front End Channels, Shells and Rear End Channels

1		2		3	
A	4	E	9	L	18
B	5	Q	10	M	19
C	6	F	11	N	20
		G	12	P	21
		H	13	S	22 (钩圈式浮头)
N	7	I	14	T	23
D	8	J	15	U	24
		K	16	W	25
		O	17		

1　front end stationary head types　前端管箱型式

2　shell types　壳体型式

3　rear end head types　后端结构型式

4　channel and removable flat cover　平盖管箱

5　bonnet（integral cover）　封头管箱

6　channel integral with tubesheet and removable flat cover（for removable tube bundle）　用于可拆管束与管板制成一体的管箱

7　channel integral with tubesheet and removable flat cover　与管板制成一体的固定管板管箱

8　special high pressure closure　特殊高压管箱

9　one pass shell　单程壳体

10　condenser shell with one entry and one exit　单进单出冷凝器壳体

11　two pass shell with longitudinal baffle　具有纵向隔板的双程壳体

12　split flow　分流

13　double split flow　双分流

14　U-tube exchanger　U形管式换热器

15　divided flow（or condenser shell）　无隔板分流（或冷凝器壳体）

16　kettle type reboiler　釜式重沸器

17　outer flow diversion　外导流

18　fixed tubesheet like "A" stationary head　与"A"相类似固定管板结构

19　fixed tubesheet like "B" stationary head　与"B"相类似固定管板结构

20　fixed tubesheet like "N" stationary head　与"N"相类似固定管板结构

21　outside packed floating head　填料函式浮头

22　floating head with backing device　钩圈式浮头

23　pull-through floating head　可抽式浮头

24　U-tube bundle　U形管束

25　externally sealed floating tubesheet exchanger　外密封浮动管板换热器

(2) 换热器零部件名称 Components Nomenclature of Heat Exchanger

1	internal floating head exchanger	内浮头换热器,浮头式换热器
2	outside packed floating head exchanger	(外)填料函式浮头式换热器
3	externally sealed floating tubesheet exchanger	外密封浮动管板式换热器
4	kettle type floating head reboiler	釜式浮头重沸器
5	U-tube heat exchanger	U形管式换热器
6	fixed tubesheet exchanger	固定管板式换热器
7	lifting lug	吊耳
8	channel cover (flat cover)	管箱盖板,平盖
9	stationary head flange (channel flange)	管箱法兰
10	instrument connection	仪表接口
11	stationary head nozzle	管箱接管
12	pass partition	分程隔板
13	stationary tubesheet	固定端管板
14	shell nozzle	壳体接管
15	impingement plate	防冲板
16	removable tube bundle	可拆管束
17	shell	壳体
18	tierod and spacer	拉杆和定距管
19	transverse baffle	(横向)折流板
20	support plate	支持板
21	floating head backing device	浮头钩圈
22	vent connection	排气口
23	shell cover	外头盖
24	floating tubesheet	浮动管板
25	floating head cover	浮头盖
26	drain connection	排液口
27	floating head cover flange	浮头法兰
28	shell cover flange (rear head end flange)	外头盖法兰
29	shell flange-rear head end	外头盖侧法兰,后头盖端壳体法兰
30	saddle	鞍座
31	shell flange-stationary head end/shell flange	内浮头换热器,浮头式换热器 (固定端)壳体法兰
32	stationary head-channel (flat cover channel)	(固定端)平盖管箱
33	by-pass seal	旁路挡板
34	rear head end gasket	外头盖垫片
35	channel/bonnet gasket	管箱垫片
36	shell gasket next to channel side	管箱侧垫片
37	skid way	滑道
38	packing box	填料函
39	packing	填料
40	packing gland	填料压盖
41	floating tubesheet skirt	浮动管板裙
42	floating head cover-external	外浮头盖
43	slip-on backing flange	活套靠背法兰,活套法兰
44	split shear ring	剖分剪切环
45	lantern ring	套环,灯笼环
46	eccentrically conical shell	偏心锥壳
47	liquid level connection	液位计接口
48	weir	堰板
49	intermediate baffle	中间挡板
50	U-tube U-换热管	
51	inner flow-diversion sleeve	内导流筒
52	longitudinal baffle	纵向隔板(纵向折流板)
53	fixed support saddle	固定鞍式支座
54	sliding support saddle	活动鞍式支座
55	expansion joint	膨胀节
56	shell side	壳程
57	stationary head-bonnet/bonnet (internal cover)	封头管箱
58	tube side	管程
59	support bracket	耳式支座
60	channel/bonnet cylinder	管箱壳体(短节)

(3) 管板 Tubesheet

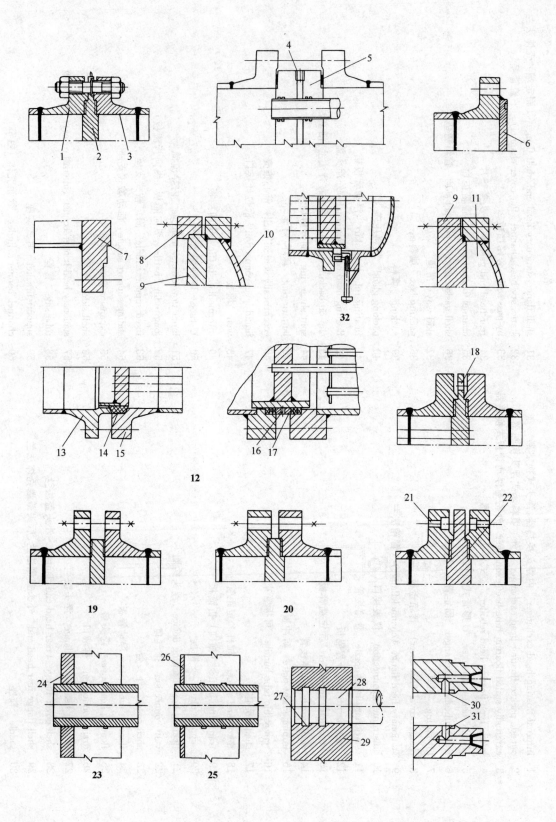

1	channel flange 管箱法兰	17	lantern ring 套环，灯笼环
2	stationary tubesheet 固定端管板	18	pulling eyes hole (eyebolt hole) 环首螺钉孔
3	shell flange 壳体法兰	19	unconfined joint 非限制式连接
4	vent or drain 放气口或排液口	20	confined joint 限制式连接
5	double tubesheet 双管板	21	jackscrew hole 顶丝孔
6	thinner tubesheet 薄管板	22	gasket 垫片
7	fixed tubesheet 固定管板	23	clad tubesheet 复合管板
8	floating head backing device 浮头钩圈	24	cladding 复合层
9	floating tubesheet 浮动管板	25	faced tubesheet 衬层管板
10	spherically dished head 球冠形封头	26	weld deposited overlay 堆焊层
11	floating head cover 浮头盖	27	tube hole 管孔
12	single box packed floating tubesheet 单填料函式滑动管板	28	tubesheet 管板
		29	tube hole groove 管孔槽
13	packing gland 填料压盖	30	vent 排气口
14	packing 填料	31	gain 排液口
15	packing box 填料函	32	double box packed floating tubesheet 双填料函式滑动管板
16	weep hole 泄漏孔，检查孔		

(4) 管箱 Channel

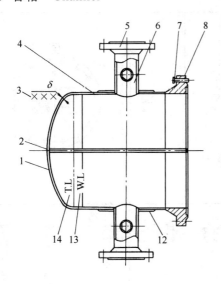 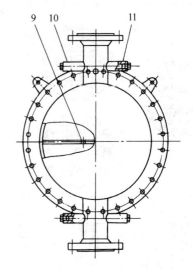

1 ellipsoidal head/convex head 椭圆形封头/凸形封头
2 pass partition 分程隔板
3 minimum thickness after forming 成形后最小厚度
4 channel/bonnet cylinder 管箱圆筒（短节）
5 nozzle flange 接管法兰
6 nozzle 管子
7 jack screw 顶丝
8 channel flange 管箱法兰
9 weep hole 泪孔
10 pressure gage connection 压力计连接口
11 thermometer connection 温度计连接口
12 pad reinforcement (reinforcing ring) 补强圈
13 welding line (W.L) 焊缝线
14 tangent line (T.L) 切线

（5）管子-管板连接、膨胀节及其他零件　Tube-to-Tubesheet Joints, Expansion Joints and Other Parts

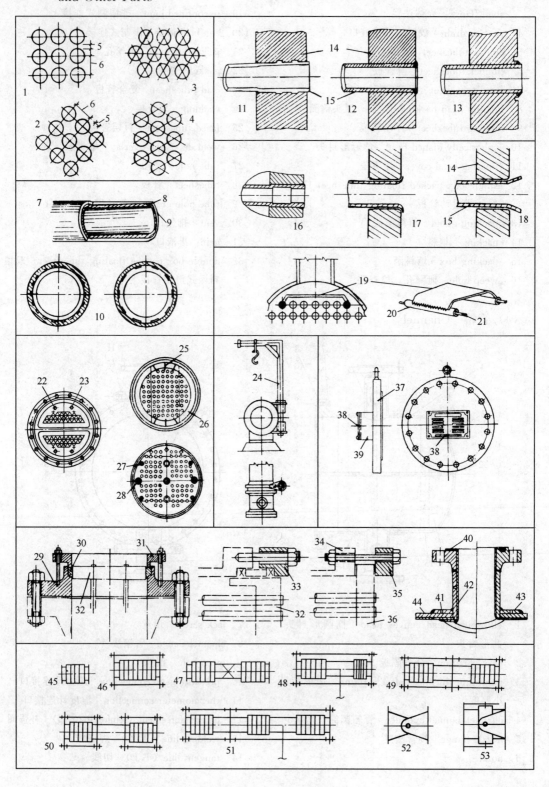

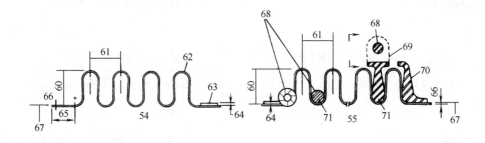

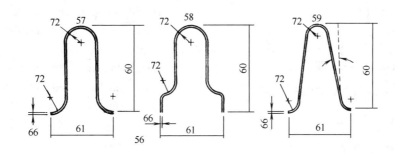

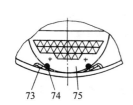 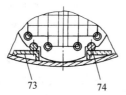 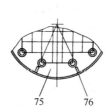 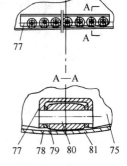

（1～4 tube pattern 管子排列形式）

1　square pattern　正方形排列

2　rotated square pattern　转角正方形排列

3　triangular pattern　三角形排列

4　rotated triangular pattern　转角三角形排列

5　ligament width　孔桥宽度

6　tube pitch　管间距

7　duplex tube　双层管

8　inner tube　内管

9　outer tube　外管

10　duplex leak-detector tube　可检漏双层管

11　strength expanded joint　强度胀接

12　strength welded joint　强度焊

13　seal welded joint　密封焊

14　tubesheet　管板

15　tube　管子

16　strength welded with light expanded joint
　　强度焊加贴胀

17　beaded joint（belled joint）　翻边接头

18　flared joint　扩口接头，喇叭口接头

19　impingement plate　防冲板

20　spacer　定距管

21　tie rod　拉杆

22　eyebolt　环首螺钉

23 plug 螺塞
24 davit 塔顶吊柱，吊柱
25 seal strip（by pass seal） 旁路挡板
26 segmental baffle 弓形折流板
27 tie rod with spacer 带定距管的拉杆
28 dummy tube 挡管
29 test ring 试压环
30 packing 填料
31 packing gland 填料压盖
32 floating tubesheet 浮动管板
33 test gland 试压压盖
34 shell flange 壳体法兰
35 test flange 试压法兰
36 stationary tubesheet 固定端管板
37 channel cover（flat cover） 管箱盖板，平盖
38 name plate 铭牌
39 bracket 铭牌座
40 weld deposited facing（overlay） 堆焊面（层）
41 weep hole 检查孔
42 lining 衬里
43 pad reinforcement 补强圈
44 shell 壳体
45 expansion joint 膨胀节
46 single expansion joint with tie rod 带有拉杆的单膨胀节
47 double expansion joint with intermediate anchor 带有中间固定管架的双膨胀节
48 pressure balanced expansion joint 压力平衡式膨胀节
49 universal expansion joint with overall tie rod 带有全长拉杆的万能膨胀节
50 universal expansion joint with short tie rod 带有短拉杆的万能膨胀节
51 universal pressure balanced expansion joint 万能压力平衡式膨胀节
52 hinged expansion joint 铰链式膨胀节
53 gimbal expansion joint 万向接头式膨胀节
54 unreinforced bellows 无加强的波形管（膨胀节）
55 reinforced bellows 加强的波形管（膨胀节）
56 typical convolution profile 典型的波形断面
57 "U" profile U形（波形）断面
58 "U-span" profile U形开跨（波形）断面
59 "V" profile V形（波形）断面
60 convolution depth 波的深度
61 bellows pitch 波纹管（膨胀节）节距
62 bellows 波形管
63 collar 套箍
64 collar thickness 套箍厚度
65 tangent section 切线段
66 nominal material thickness of one ply 一层材料的公称厚度
67 outside of cylinder tangent 圆筒形切线段的外径
68 reinforcing ring 加强环
69 equalizing ring 稳定环
70 end equalizing ring 端部稳定环
71 cross sectional metal area of one reinforcing member 一个加强元件的金属横截面积
72 convolution form radius 波的成形半径
73 guide way 支承导轨
74 round steel skid bar 圆钢滑条
75 baffle/support plate 折流板/支承板
76 (flat steel) skid strip （扁钢）滑道
77 channel 槽钢
78 end plate 挡板
79 roller 滚子
80 shaft (stainless steel) 轴（不锈钢）
81 shell 壳体

（6）折流板、纵向隔板和拉杆-定距管结构　Transverse Baffles, Longitudinal Baffles and Tie Rod-Spacer Construction

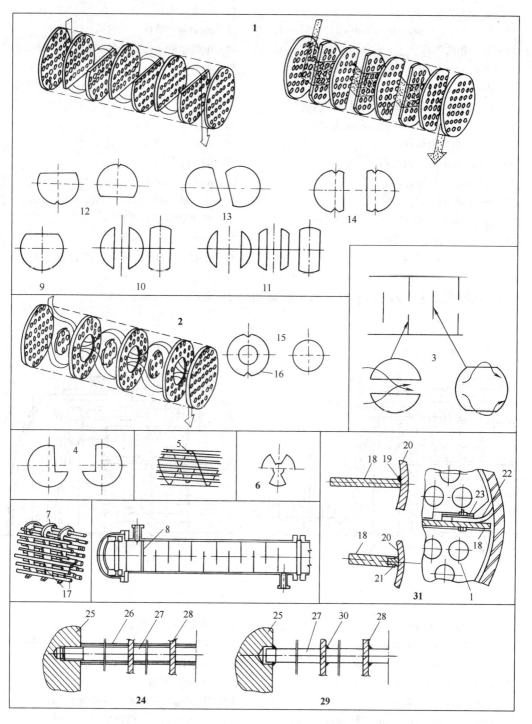

(1～7　transverse baffle　折流板)
1　segmental baffle　弓形折流板
2　disk-ring baffle　圆盘-圆环形折流板
3　window-cut baffle　窗口式折流板
4　dam baffle　堰形折流板
5　helical baffle　螺旋形折流板

6　vane type baffle　叶片式折流板
7　orifice baffle　孔式折流板
8　support plate　支持板
(9～11　types of segmental baffles　弓形折流板的型式)
9　single segmental baffle　单弓形折流板
10　double segmental baffle　双弓形折流板
11　triple segmental baffle　三弓形折流板
(12～14　baffle cut for segmental baffle　弓形折流板切口)
12　horizontal cut　水平切口
13　rotated cut　转角切口
14　vertical cut　竖直切口
15　disc（disk）　圆盘
16　ring（doughnut）　圆环
17　tube bundle　管束
18　longitudinal baffle　纵向隔板
19　weld　焊缝
20　shell　壳体
21　packing　填料
22　sealing strip　密封带
23　seating plate　压紧板
24　tie rod-spacer construction　拉杆-定距管结构
25　tubesheet　管板
26　spacer　定距管
27　tie rod　拉杆
28　baffle　折流板
29　tie rod-baffle tack welding construction　拉杆-折流板点焊结构
30　tack welding　点焊
31　longitudinal baffle seal　纵向隔板的密封

(7) 钩圈式封头　Floating Head with Backing Device

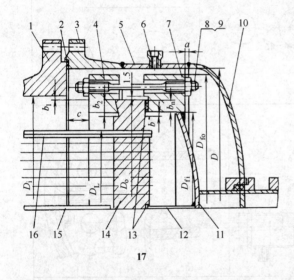 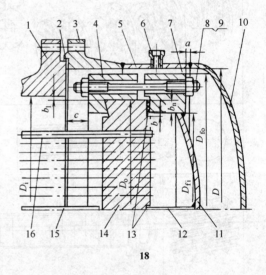

17　　　　　　　　　　　　　18

1　shell flange（rear head end）　外头盖侧法兰
2　cover gasket（rear head end）　外头盖垫片
3　shell cover flange　外头盖法兰
4　backing device　钩圈
5　skirt segment　短节
6　vent or drain connection　排气口或放液口
7　floating head cover flange　浮头法兰
8　stud　双头螺柱
9　nut　螺母
10　shell cover　封头
11　spherically dished head　球冠形封头
12　pass partition　分程隔板
13　gasket　垫片
14　floating tubesheet　浮动管板
15　dummy tube　挡管
16　tube　换热管
17　backing device type A　A形钩圈
18　backing device type B　B形钩圈

6.5.3 套管式换热器和刮面式换热器　Double-Pipe Heat Exchanger and Scraped-Surface Exchanger

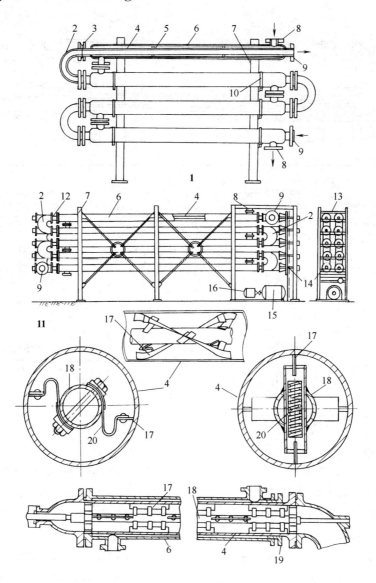

1	fixed type double pipe heat exchanger　固定式套管换热器	11	scraped surface exchanger　刮面式换热器（套管结晶器）
2	return bend　回弯头	12	stuffing box　填料函，填料箱
3	flange　法兰	13	drive chain　传动链
4	inner pipe　内管	14	sprocket　链轮
5	support piece　支承块，支持块	15	reducer　减速箱，减速机
6	outer pipe (shell pipe)　外管	16	motor　电动机
7	support　支架，支柱	17	scraper blade　刮板，刮刀
8	shell nozzle　壳程管口	18	shaft　轴
9	inner pipe nozzle　管程管口	19	gland　填料压盖
10	U bolt　U形螺栓	20	spring　弹簧

6.5.4 套管式纵向翅片换热器 Double Pipe Longitudinal Finned Exchanger

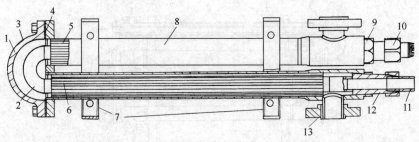

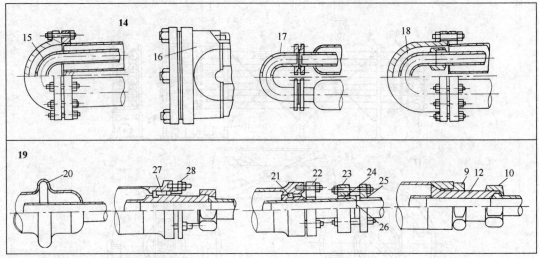

1	shell cover 壳体端盖		16	return bend housing 回弯头箱
2	return bend 回弯头，U形弯管		17	flanged return bend 法兰连接式回弯头
3	vent or drain 放气口或排液口		18	unionfitted return bend 活接头连接式回弯头
4	shell cover gasket 壳体端盖垫片			
5	longitudinal fin (axial fin) 纵向翅片		19	fixed end closure 固定端密封结构
6	tube (inner pipe) 内管		20	bellows expansion joint 波形膨胀节
7	movable shell support 可移动的壳体支座		21	sealing ring 密封圈
8	shell 壳体		22	compression flange 压紧法兰
9	cone plug nut 锥面密封螺母		23	fintube fitting flange 翅片管连接法兰
10	union nut 活接螺母		24	split ring 对开环
11	straight adapter 直管管接头		25	stub end flange 翻边短管活套法兰
12	cone plug 锥面密封短节		26	gasket 垫片
13	shell nozzle 壳体管口		27	packing 填料
14	return end closure 回弯端封闭结构		28	gland 填料压盖
15	welded return bend 焊接回管头			

6.5.5 板式换热器 Plate Exchangers

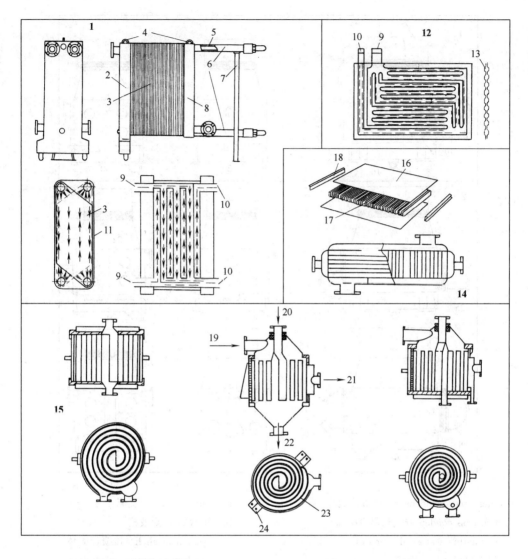

1 plate exchanger 板式换热器
2 frame plate 固定板
3 heat transfer plate 传热板
4 hanging hook 吊钩
5 guide bar 导向杆
6 distance collar 定距套管
7 guide support 导向杆支架
8 pressure plate 压紧板，压板
9 fluid inlet 液体入口
10 fluid outlet 液体出口
11 gasket 垫片
12 platecoil 压焊板换热器
13 fluid passage 液体通道
14 plate-fin heat exchanger 板翅式换热器
15 spiral-plate exchanger 螺旋板式换热器
16 flat metal sheet 金属平板
17 corrugated fin 波形翅片
18 channel 槽钢
19 hot fluid inlet 热物料入口
20 cold fluid inlet 冷物料入口
21 cold fluid outlet 冷物料出口
22 hot fluid outlet 热物料出口
23 metal strips 金属带
24 support 支承

6.5.6 蒸汽表面冷凝器，凝汽器 Steam Surface Condensers

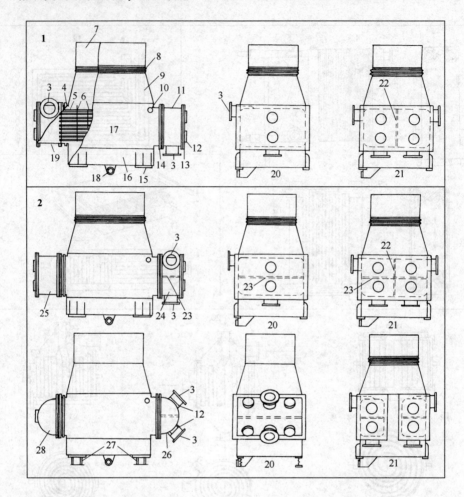

1	single pass condenser 单程冷凝器
2	two pass condenser 双程冷凝器
3	circulating water inlet or outlet 循环水入口或出口
4	shell expansion joint 壳体膨胀节
5	tube 管子
6	tube support plate 管子支持板
7	exhaust connection 排气接口
8	exhaust neck expansion joint 排气过渡段膨胀节
9	exhaust neck 排气过渡段
10	venting outlet 放气口
11	inlet waterbox 入口水箱
12	handhole 手孔
13	waterbox cover 水箱盖板
14	tube sheet 管板
15	support feet 支座
16	hotwell 凝液槽
17	condenser shell 冷凝器壳体
18	condensate outlet 冷凝液出口
19	outlet waterbox 出口水箱
20	non-divided waterbox 非分隔式水箱
21	divided waterbox 分隔式水箱
22	water dividing partition 水箱隔板
23	water pass partition 水箱分程隔板
24	inlet-outlet waterbox 带进出口的水箱
25	return waterbox 回水水箱
26	bonnet type inlet-outlet waterbox 带进出口的头盖式水箱
27	spring support 弹簧支座
28	bonnet type return waterbox 头盖式回水水箱

6.5.7 蒸发器 Evaporators

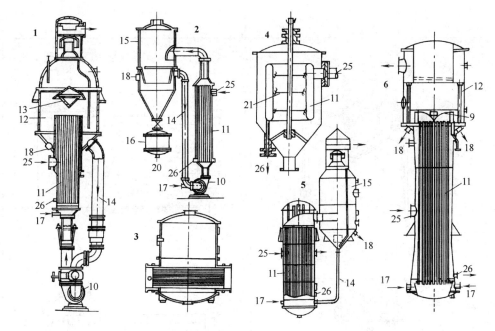

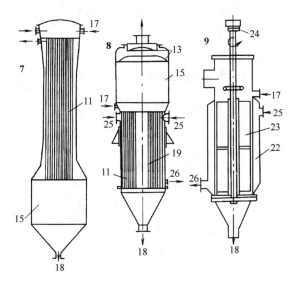

6　long tube evaporator　长管蒸发器
7　falling-film evaporator　降膜蒸发器
8　short tube evaporator　短管蒸发器
9　agitated film evaporator　搅拌膜式蒸发器
10　circulation pump　循环泵
13　demixter (demister)　除沫器
14　circulation tube　循环管
15　separator (vapor space)　分离室
16　filter　过滤器
17　feed　进料
18　concentrate　浓缩液
19　central downcomer (central well down take)　中心降液管
20　filtrate outlet　滤液出口
21　rotating brush　转动刷子
22　steam jacket　蒸汽夹套
23　agitator　搅拌器
24　belt pulley　带轮
25　steam inlet　蒸汽入口
26　condensate outlet　冷凝液出口

[1, 2　forced-circulation (F.C.) evaporator　强制循环蒸发器]
1　submerged tube F.C. evaporator　浸没管束强制循环蒸发器
2　Oslo-type crystallizer　奥斯陆型结晶器
3　horizontal tube evaporator　卧管蒸发器，横管蒸发器
4　evaporator with rotating brush　带转动刷的蒸发器
5　upward-flow evaporator (climbing-film evaporator)　上流（升膜）式蒸发器
11　heating element　加热器
12　flash chamber　蒸发器，闪蒸器

281

6.5.8 空冷器、空气冷却器 Air-Cooled Heat Exchangers

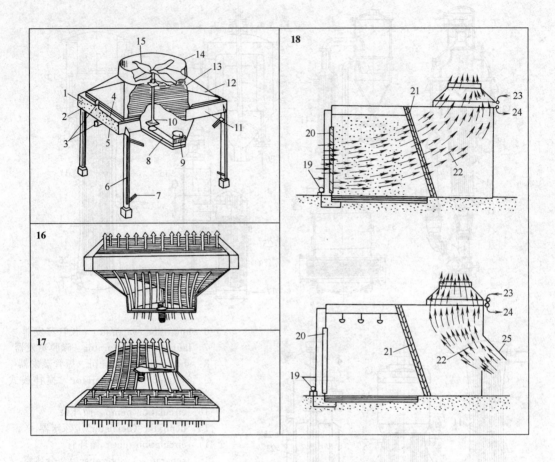

1　header　管箱
2　tube plug　管箱丝堵
3　nozzle　接管口
4　vent　放气口
5　drain　排液口
6　supporting column　支柱
7　brace　斜撑
8　V-belt drive　V形三角带传动
9　electric motor　电动机
10　bearing　轴承
11　channel frame　（管束）槽钢侧梁，槽钢框架
12　plenum　风筒，风斗
13　finned tube bundle　翅片管束
14　fan ring　风扇圈
15　fan　风机
16　forced draft　鼓风式，送风式
17　induced draft　引风式
18　combine-air humidified air cooler　增湿空冷器，联合空冷器
19　water pump　水泵
20　louver　（固定）百叶窗
21　mist eliminator　除雾器
22　air　空气
23　hot fluid in　热流体入口
24　cold fluid out　冷流体出口
25　shutter　活动百叶窗

（1）空冷器的组合形式　Bay Arrangements of Air-Cooled Heat Exchanger

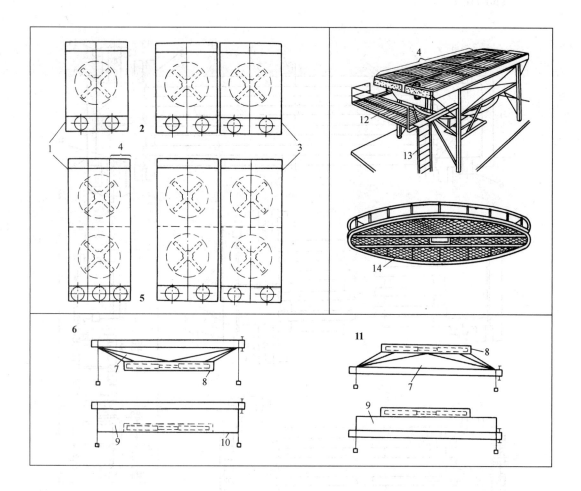

1	tube bundle　管束①	8	fan ring　风扇圈
2	one-bay② unit③　单跨空冷器组	9	panel plenum　方形风筒
3	one-fan bay　单风机跨空冷器	10	fan deck　风筒底板
4	two-bay unit　双联空冷器组	11	induced plenum　引风式，诱导式
5	two-fan bay　双风机跨空冷器	12	walkway　通道，走道
6	forced draft　鼓风式	13	ladder　直梯，梯子
7	transition plenum　斗形风筒	14	fan blade guard　风机护罩

---

① tube bundle：管束；管箱、管子和侧梁（框架）的组合件。
② bay：跨置于同一跨构架上，且配有相应风筒和其他附件的一片或几片管束，共用一台风机或几台风机。
③ unit：组置于一跨构架或几跨构架上作为单独用途的一片管束或几片管束。

283

(2) 管束和头盖（管箱）的典型结构　Typical Construction of Tube Bundles and Headers

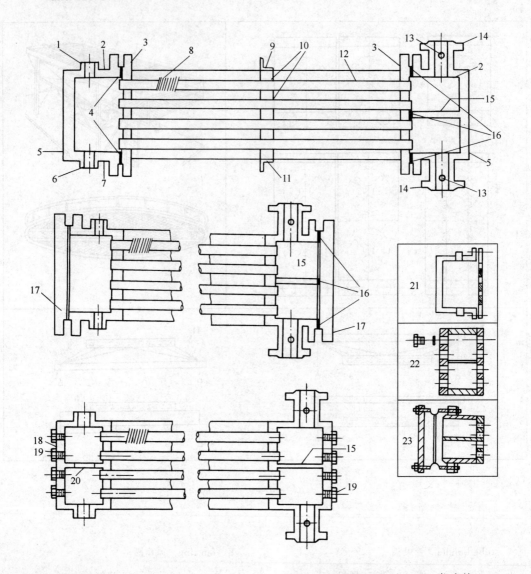

1	vent 放气口	13	instrument connection 仪表接口
2	top plate 顶板	14	nozzle 管口
3	tube sheet 管板	15	pass partition 管程隔板，分程隔板
4	unconfined gasket 非限位垫片	16	confined gasket 限位垫片
5	removable bonnet 可拆式头盖	17	removable cover plate 可拆卸盖板
6	drain 排液口，放尽口	18	plug sheet 丝堵板
7	bottom plate 底板	19	plug 丝堵
8	finned tube 翅片管	20	stiffener 筋板，加强筋板
9	tube keeper 管支承	21	bonnet header 封闭式头盖
10	tube spacer （翅片）管定距支承件	22	plug header 丝堵式头盖
11	tube support cross-member 管支承横梁	23	cover-plate header 盖板式头盖
12	side frame 侧梁		

(3) 空冷器的驱动装置　Drive Arrangements for Air Cooler

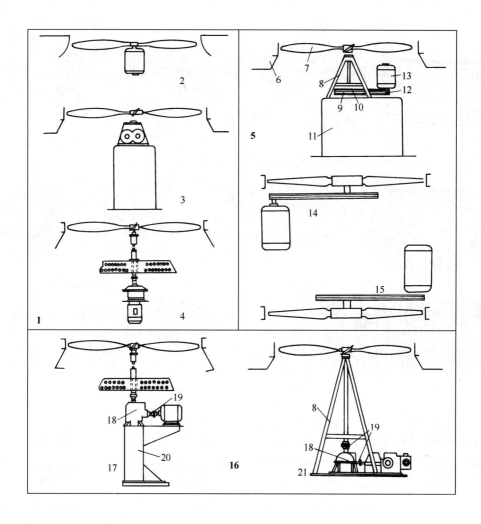

1	direct drive　直接驱动	12	V-belt　V形带
2	electric motor direct drive　电动机直接驱动	13	motor　电动机
3	hydraulic motor drive　液压马达驱动	14	bottom-suspended V-belt drive　下悬式 V 形带传动
4	electric gear head motor drive　齿轮减速电动机驱动	15	top suspended V-belt drive　上悬式 V 形带传动
5	V-belt drive　V 形带传动	16	gear drive　减速器传动
6	fan ring　风扇圈	17	right-angle gear drive　锥齿轮传动
7	fan　风扇、风机	18	gear (reduction gear)　齿轮箱，减速齿轮箱
8	tripod support (fan support)　三脚架（风机支架）	19	coupling　联轴器
9	sheave　带轮	20	pedestal　底座，支座
10	bearing　轴承	21	steam turbine drive or gasoline engine drive　蒸汽透平驱动或汽油（发动）机驱动
11	concrete pedestal　混凝土底座		

(4) 翅片 Fins

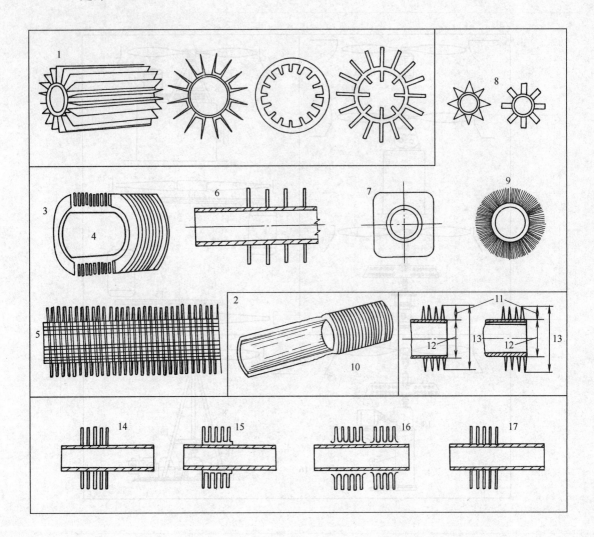

1　longitudinal fin　纵向翅片
2　transverse fin　横向翅片
3　high fin　高翅片
4　bimetallic finned tube　双金属翅片管
5　helical fin　螺旋翅片
6　disc type fin　圆盘式翅片
7　stamped fin　冲压方形翅片
8　star type fin　星形翅片
9　spine type fin　针形翅片
10　low fin　低翅片
11　fin hight　翅片高度
12　root diameter　（翅片）根径
13　fin outside diameter　翅片外径
14　tension warpped fin　张力紧固翅片
15　tension warpped foot fin　张力紧固 L 形翅片
16　duplex continuous integral fin　双金属整体轧制翅片
17　imbedded fin　镶嵌式翅片

（5）空冷器的温度控制　Temperature Control of Air Cooler

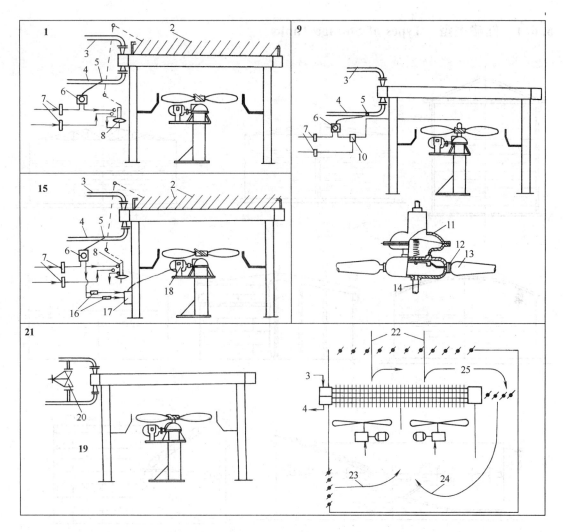

1	top louvers control　顶部百叶窗式调节		13	fan blade　风机叶片
2	adjustable shutter　可调百叶窗		14	drive shaft　驱动轴
3	process stream in　工艺流体流入，工艺介质流入		15	combination control　（百叶窗及变速马达）联合式调节
4	process stream out　工艺流体流出，工艺介质流出		16	pneumatic-electric relay　气动继电器
5	thermocouple　热电偶		17	motor controller　电动机转速调节器
6	pneumatic temperature controller　气动温度调节器		18	two-speed electric motor　双速电动机
			19	by-pass control　副线调节，旁路调节
7	pressure regulator and filter　压力调节器和过滤器		20	by-pass valve　旁通阀，副线阀
			21	air-cooler recirculation system　空冷器热风再循环系统
8	pneumatic shutter operator　百叶窗气动机构		22	hot air out　热空气流出
9	automatic pitch control　自动调角风机调节		23	cold ambient air　冷空气
10	pneumatic booster relay　气动升压继动器		24	warm air　暖空气
11	air motor（diaphragm）　风动马达，膜头		25	recirculated hot air　循环热空气
12	hub　毂			

6.6 储罐 Storage Tanks

6.6.1 储罐类型 Types of Storage Tanks

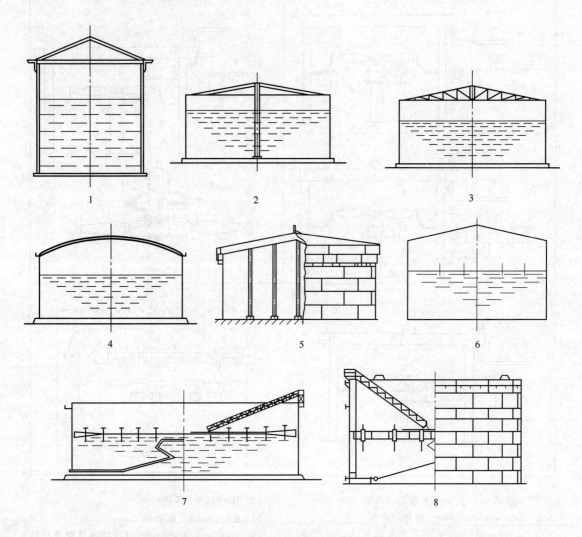

1　self-supporting cone roof tank　自支承锥顶罐
2　(columns) supported cone roof tank　(柱) 支承式锥顶罐
3　(trusses) supported cone roof tank　(桁架) 支承式锥顶罐
4　self-supporting dome roof tank　自支承拱顶罐
5　litter roof tank (expansion roof tank)　升顶罐
6　internal floating roof tank (covered floating roof tank)　内浮顶罐 (带盖浮顶罐)
7　external floating roof tank　外浮顶罐
8　double-deck-type external floating roof tank　双盘式外浮顶罐

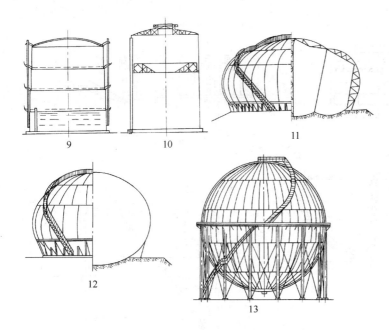

9 telescopic gas holder (fluid seal gas holder) 湿式气柜

10 stationary gas holder (dry seal gas holder) 干式气柜

11 noded Hortonspheroid (sph-eroid) 多弧滴形罐

12 plain spheroid tank 椭球罐

13 spherical tank 球罐

6.6.2 固定顶储罐 Fixed-Roof Tanks

6.6.2.1 储罐零部件名称 Nomenclature of Tank Parts

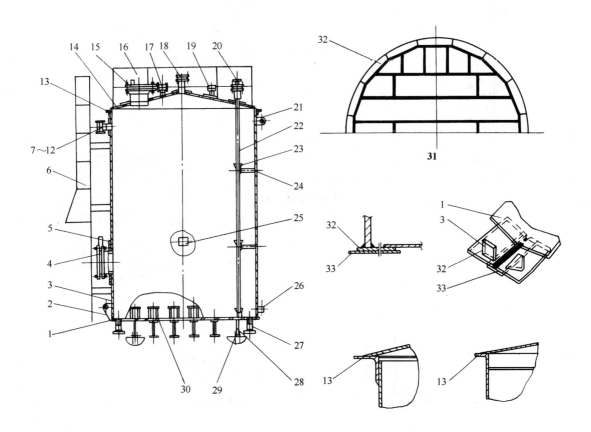

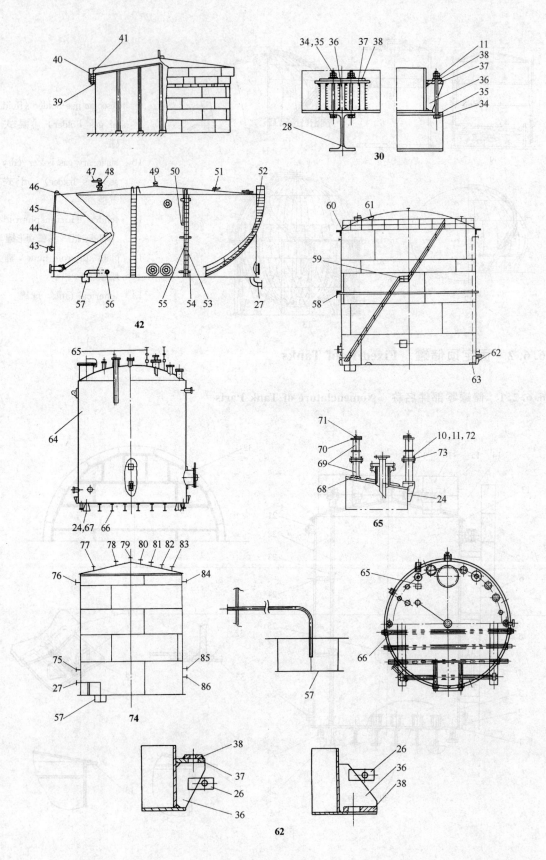

1	bottom plate	底板
2	tailing lug	尾部吊耳
3	shell plate	壁板，罐壁
4	shell manhole	罐壁人孔
5	reinforcing pad	补强圈
6	ladder	直梯，梯子
7	shell nozzle	罐壁管口
8	nozzle flange	接管法兰
9	bland flange	法兰盖
10	bolt	螺栓
11	nut	螺母
12	gasket	垫片
13	top angle	包边角钢
14	cone roof	锥顶
15	roof manhole	罐顶人孔
16	roof walkway and handrail	罐顶走道和护栏
17	spare(W/B.F)	备用口（带盲板）
18	test branch	测试口
19	roof nozzle	罐顶管口
20	level measurement	液位测量口
21	lifting lug	吊耳
22	dip tube	插入管
23	pipe guide	导向管
24	rib	筋板，拉筋
25	nameplate	铭牌
26	earth lug	接地板
27	drain	排净口，排液口
28	I beam	工字梁
29	anchor bolt	地脚螺栓
30	setting bolt case	安装螺栓座
31	bottom plate plan	底板排板图
32	annular plate	边缘板
33	backing strip	背板
34	setting bolt	安装螺栓
35	washer	垫圈
36	gusset	筋板，连接板
37	cover plate	盖板
38	pad plate	垫板
39	liquid seal trough	液封槽
40	weather skirt	防雨裙
41	dip skirt	液封裙
42	nomenclature of tank accessories	储罐附件名称
43	winch	绞盘
44	swing line	摆动式抽油管，升降式抽油管
45	cable	钢丝绳
46	sheave	滑轮
47	conservation vent	呼吸阀
48	flame arrester	阻火器
49	free vent	放空口，通气口
50	float	浮子
51	gauge hatch	量油口
52	spiral stairway	盘梯
53	tank gauge	液位标尺
54，65	bracket	支架
55	level indicator	液位指示器
56	antifreeze valve	防冻阀
57	sump	排水槽，集水槽
58	intermediate wind girder	抗风圈，中间加强圈
59	platform and walkway	平台和走道
60	compression ring	抗压环
61	dome roof	拱顶
62	anchor bolt setting	地脚螺栓座
63	impingement plate	防冲板
64	lining	衬层
66	grillage beam	格排梁
67	base plate	底板
68	pad	垫块
69	channel	槽钢
70	support plate	支承板
71	liner plate	调整垫板
72	spring washer	弹簧垫圈
73	connection plate	连接板
74	nozzle nomenclature	管口名称
75	sample connection	取样口
76	foam chamber (connection)	消防口,泡沫接口
77	return	回流口
78	high level alarm	高液位报警口
79	safety valve (relief valve)	安全阀口
80	level transmitter (LT)	液位变送器口
81	pressure transmitter (PT)	压力变送器口
82	vapor balancing line	蒸气平衡口
83	back balance	返回平衡口
84	overflow	溢流口
85	TG	温度计口
86	utility connection	公用工程口

6.6.2.2 罐顶结构 Roof Structure

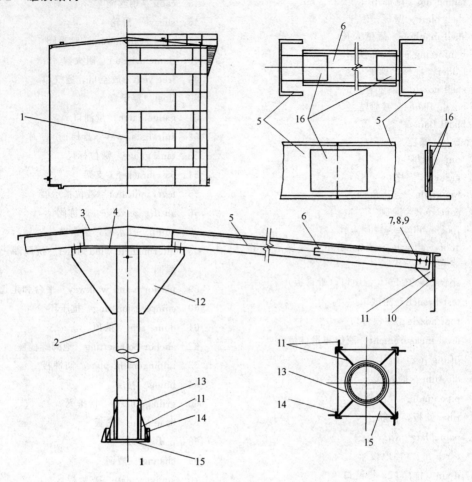

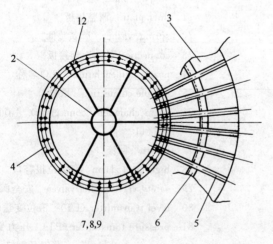

1 column support structure 柱支承结构
2 center column 中心支柱
3 roof plate 顶板
4 center ring 中心环
5 rafter 橡子,檩条
6 girder 桁架
7 stud bolt 双头螺柱
8 nut 螺母
9 washer 垫圈
10 pad 垫板
11 gusset 连接板
12 rib 筋板
13 column guide 支柱套筒
14 guide 定位板
15 base plate 底板
16 reinforcing plate 加强板

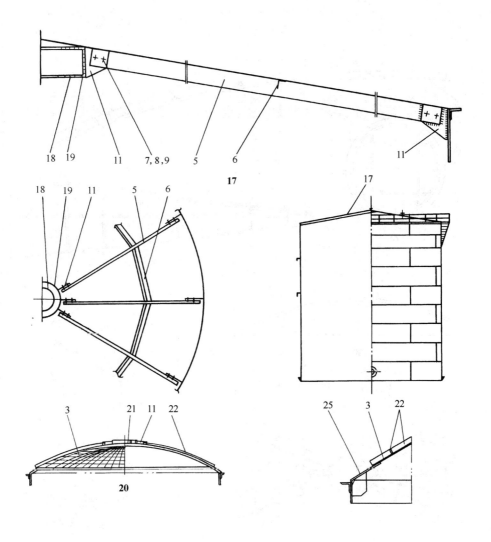

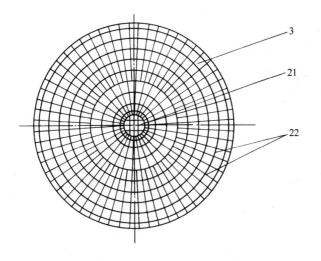

17　roof structure　罐顶钢结构
18　stiffener　加强筋，加强支肋，加筋肋
19　ring　圆环
20　dome roof　拱顶
21　center plate　中心顶板
22　rib　罐顶肋条
23　center plate detail　中心顶板详图
24　development of roof plate　顶板展开图
25　compression ring　抗压环
26　purlin(e)　檩条，桁条
27　center ring　中心环
28　clip　支架
29　wind bracing　抗风拉杆
30　plan of roof plate　顶板平面图

293

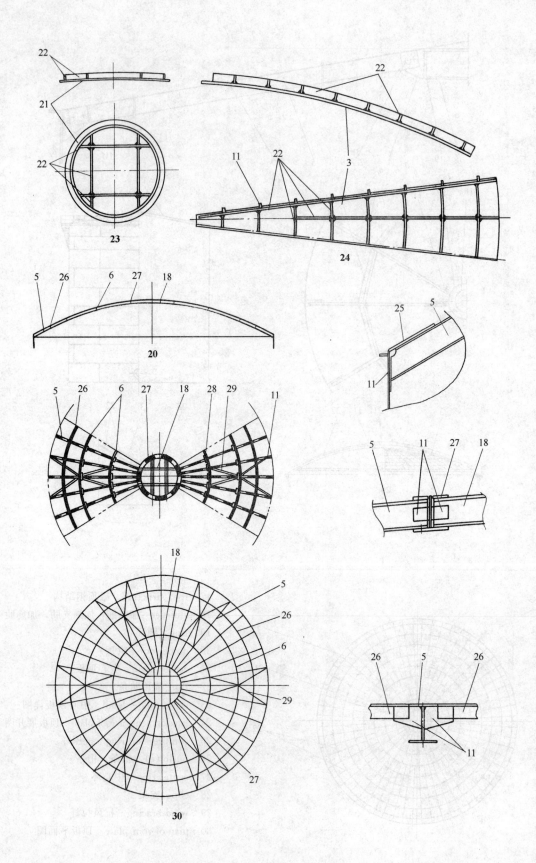

6.6.3 内浮顶罐 Internal Floating Roof Tank

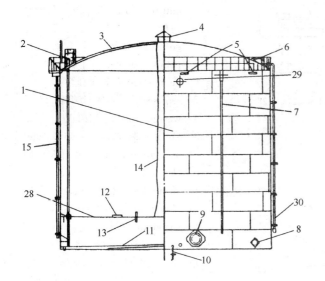

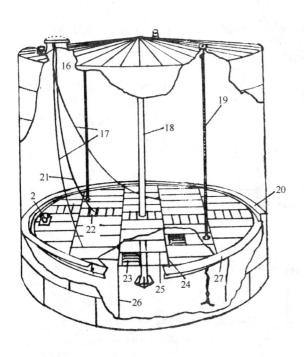

1 shell plate 罐壁
2 gauge hatch 量油口
3 fixed roof 固定罐顶
4 vent 通气孔，放空口
5 shell vent 罐壁通气孔
6 roof manhole 罐顶人孔
7 semi-fixed foam injection system 半固定泡沫喷射装置
8 shell nozzle 罐壁管口
9 shell manhole 罐壁人孔
10 tank earth 罐体接地线
11 heater 加热器
12 deck manhole 浮盘人孔
13 roof support 浮顶支柱
14 deck earth 浮盘接地线
15 overflow 溢流管
16 aludeck internal floating cover tank 浮盘内浮顶罐
17 bonding wire 静电导线
18 central column 中央支柱
19 guide wire 导向钢丝绳
20 Z-seal Z形密封
21 nylon rope 尼龙绳
22 degasing hole 放气孔
23 float （浮顶）浮动元件，浮子
24 beam 横梁
25 spacer 隔条
26 deck support 浮盘立柱
27 peripheral angle 周边角钢
28 roof deck 内浮盘
29 high level alarm 高液位警报器
30 level gauge 液面计

295

6.6.4 外浮顶罐 External Floating Roof Tank

6.6.4.1 外浮顶罐零部件名称 Nomenclature of External Floating Roof Tank Parts

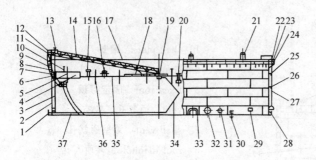
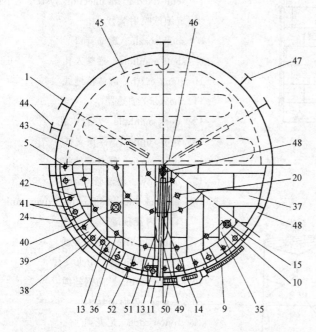
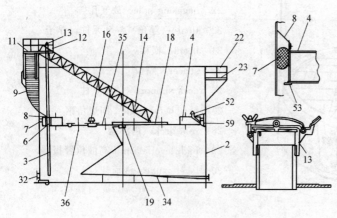

1　roof drain nozzle　浮顶排水口
2　shell plate　罐壁
3　antirotation guide pole　防转导向杆
4　pontoon　浮船
5　pontoon support　浮船支柱
6　guide sleeve　导向套管
7　roof seal　浮顶密封装置
8　weather shield　挡雨板
9　spiral stairway and intermediate/medium platform　盘梯及中间平台
10　foam dam　泡沫挡板
11　top platform　顶部平台
12　handrail　栏杆
13　gauge hatch　量油孔
14　rolling ladder　转动扶梯
15　automatic bleeder vent　自动通气口
16　manhole with pressure valve　装有压力阀的人孔
17　top angle　包边角钢
18　rolling ladder runway　转动扶梯导轨
19　roof drain sump　浮顶中央集水槽
20　emergency drain　紧急排水口
21　foam maker（pourer）　泡沫发生器
22　railing　扶手，栏杆
23　walkway　人行通道，走道
24　wind girder　抗风加强圈
25　intermediate wind girder　中间抗风加强圈
26　stiffen ring　加强圈
27　level indicator（float type）　液位指示器（浮子型）
28　drain nozzle　排水口
29　name plate　铭牌
30　tank earth　罐体接地线

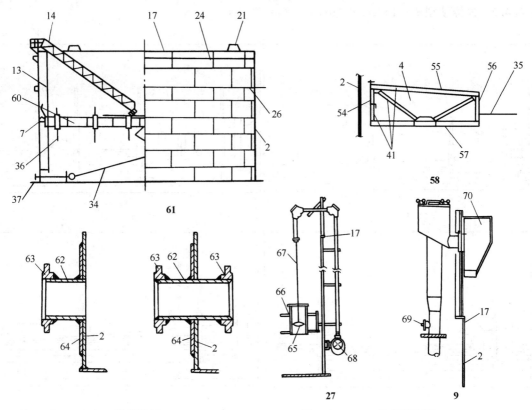

31	mixer nozzle 搅拌器口	52	rim vent 边缘通气孔
32	shell manhole 罐壁人孔	53	roof stopper 浮板定位板
33	flush type cleanout door 齐平型清扫孔	54	outer rim 外边缘板
34	roof drain pipe system 浮顶排水管系统	55	upper plate 上板
35	roof deck (deck plate) 浮顶板	56	inner rim 内边缘板
36	deck support 浮顶立柱	57	lower plate 下板
37	bottom plate 底板	58	pontoon section 浮船剖面图
38	pontoon manhole 浮船人孔，船舱人孔	59	rim seal 边缘密封
39	roof compartment 船舱	60	double-deck roof 双盘顶
40	deck manhole 浮顶人孔	61	double-deck-type floating roof tank 双盘式浮顶罐
41	pontoon frame 浮船支架		
42	pontoon bulkhead plate 船舱隔板	62	nozzle 管口
43	deck support leg 浮顶立柱支腿	63	flange 法兰
44	oil outlet nozzle 出油口	64	reinforcing pad (reinforcing plate) 补强圈
45	heater coil 加热器（加热盘管）	65	float 浮子
46	flexible hose 软管	66	float well 浮子导向套
47	oil inlet nozzle 进油口	67	wire 钢丝绳
48	annular plate 边缘板	68	level transmitter 液位变送器
49	wheel for rolling ladder 转动扶梯轮子	69	air inlet 空气入口
50	top stairway 顶部走道	70	box deflector 泡沫导流室
51	deck earth 浮盘接地线		

297

6.6.4.2 外浮顶型式　External Floating Roof Types

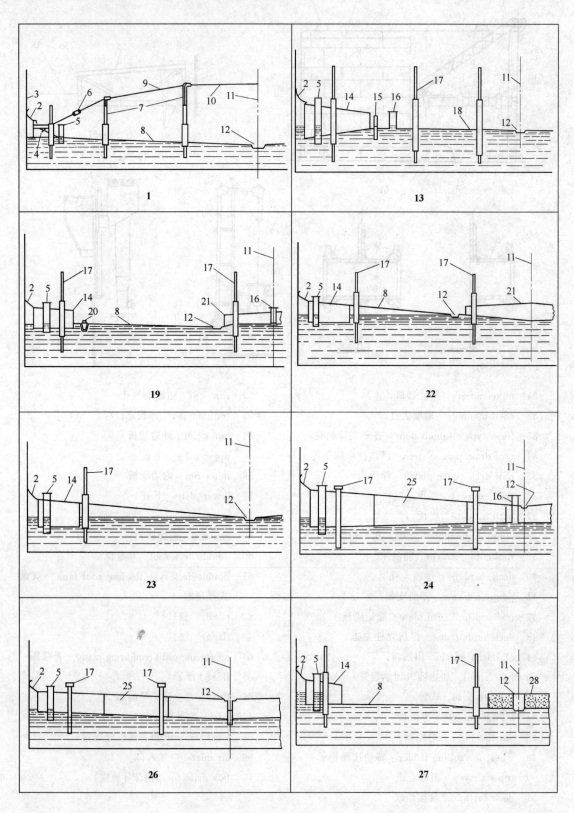

1 trussed pan-type floating roof   构架连接的盘式浮顶

2 seal   密封

3 tank shell   罐壁

4 gusset plate   角撑板

5 gage hatch   量油孔，检尺口

6 turnbuckle   花篮螺钉，松紧扣

7 roof support and truss post   浮顶支柱和桁架柱

8 deck   浮盘

9 truss rod   构架拉杆

10 center truss ring   构架中心环

11 center of roof   浮顶中心线

12 drain   排水坑

13 annular pontoon floating roof with single center deck   （周边浮船）单盘式浮顶

14 pontoon   浮船

15 emergency drain   紧急排水口

16 vent   自动通气阀

17 roof support   浮顶支柱

18 center deck   中央浮盘，单盘板

19 low deck type floating roof with center pontoon   带中央浮船的低盘式浮顶

20 emergency siphon   紧急虹吸排水装置

21 center pontoon   中央浮船

22 high deck type floating roof with center pontoon   带中央浮船的高盘式浮顶

23 high deck type floating roof without center pontoon   （无中央浮船）高盘式浮顶

24 floating roof with converging double deck   收敛双盘式浮顶

25 double deck pontoon   双盘浮船

26 floating roof with parallel double deck   平行双盘式浮顶

27 center weighted pontoon type floating roof   中央加重浮船式浮顶

28 center weight   中央（浮船）加重物

6.6.4.3 外浮顶罐的密封型式 Seal Types of External Floating Roof Tank

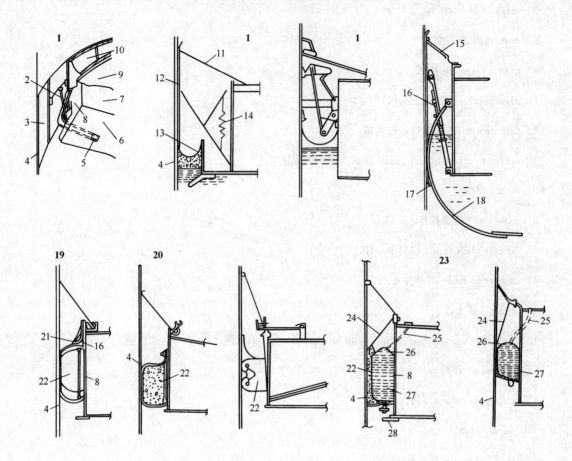

1	mechanical seal (metallic seal) 机械密封，金属密封	14	spring 弹簧
2	pantagraph hanger 动臂吊杆，剪刀叉	15	fabric-reinforced rubber 夹布橡胶（防护板）
3	sealing ring 密封圈	16	hanger 吊杆
4	tank shell 罐壁	17	sliding steel shoe （密封板）滑动钢靴
5	weight 重锤	18	pusher 压紧杆
6	bottom deck （船舱）底板	19	soft seal 软密封
7	bulkhead 船舱隔板	20	foam seal 泡沫塑料密封
8	rim 边缘板	21	envelope 密封袋
9	top deck （船舱）顶板	22	polyurethane foam 聚氨酯泡沫塑料
10	continuous seal 环形密封带，连续密封板	23	tube seal 管式密封
11	weather shield 挡雨板	24	scarf band 吊带
12	sealing plate 密封板	25	flexible hose （充液）软管
13	rubber slab 橡胶（密封）板	26	seal pipe 密封管
		27	liquid 液体
		28	bumper 限位板

6.6.5 球形储罐及其他低压大型储罐 Spherical Tanks and Other Low-Pressure Storage Tanks

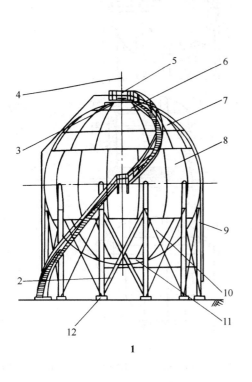

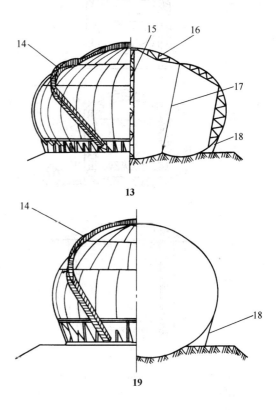

1 spherical tank 球形储罐
2 tie rod 拉杆
3 platform 平台
4 lightning rod 避雷针
5 relief valve 安全泄放阀
6 top crown 上极板
7 north (upper) temperate zone plate 北温带板
8 equator zone plate 赤道带板
9 support 支柱
10 south (lower) temperate zone plate 南温带板
11 bottom crown 下极板
12 base plate 底板
13 noded Hortonspheroid (spheroid) 多弧滴形罐
14 stairway 盘梯
15 center support 中央支柱
16 truss 桁架
17 tie 拉杆，拉筋
18 bracket (girder) 托架，撑架
19 plain spheroid 椭球罐

6.6.6 低温储罐 Refrigerated Storage Tanks

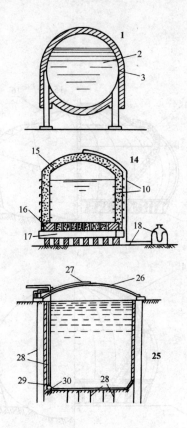
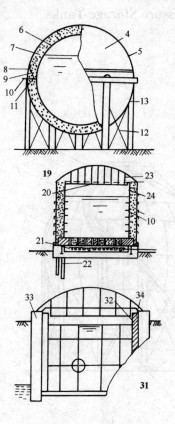

1	spherical single-wall tank 单壳球罐	19	suspension deck LNG tank 吊顶式液化天然气储罐
2	LPG（liquefied petroleum gas） 液化石油气	20	suspension deck 拉顶
3	insulation 保冷层	21	heater （电）加热器
4	spherical double-wall tank 双壳球罐	22	pilling 桩
5	outer tank 外罐，外壳	23	suspension bar 拉杆
6	inner tank 内罐，内壳	24	glass wool insulation 玻璃棉保冷层
7	pearlite 珍珠岩	25	underground storage tank (inground storage tank, frozen earth reservoir) 地下储罐
8	compression ring （承）压圈	26	roof insulation 顶盖保冷层
9	hanger rod 吊杆	27	spray line 喷淋管
10	stiffener ring 加强圈，补强圈	28	freeze tube 冻结管
11	sway rod 抗震水平拉杆	29	pump 泵
12	tie rod 拉杆，拉筋	30	foot valve 底阀
13	support column 支柱	31	membrane（LNG）tank 膜式储罐
14	flat bottomed cylindrical LNG tank 平底圆筒形液化天然气储罐	32	membrane 膜板
15	outer tank roof 外罐顶盖	33	concrete ring 混凝土圈
16	anchor bolt 地脚螺栓，锚固螺栓	34	composite shell segment 组合式罐壁预制件
17	floating foundation slab 浮床式基础板		
18	ground level (G.L) 地平面		

6.7 衬里设备 Lining Equipment

6.7.1 橡胶衬里设备 Rubber Lining Equipment

(1) 橡胶衬里 Rubber Lining

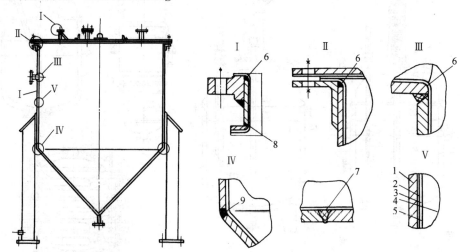

1 steel sheet 钢壳体
2 primer 底涂料
3 adhesive 胶黏剂
4 seam 搭接缝
5 rubber sheet 橡胶板
   soft rubber 软质胶
   hard rubber 硬质胶
6 grind to 3 mm min. radius 打磨圆角（最小半径为 3mm）
7 grind weld flush with vessel walls 打磨焊缝与容器壁齐平
8 grind weld flush with nozzle I.D. 打磨焊缝与接管内径齐平
9 radius fillet weld by grinding 角焊缝打磨圆角

(2) 硫化 Vulcanization

① 硫化罐硫化 Vulcanization by Autoclave

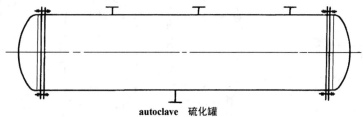

autoclave 硫化罐

② 自然硫化 Self Vulcanization

(3) 橡胶衬里有关术语 Term of Rubber Lining

1 solution 溶剂
2 cleaner 清洁剂
3 hardener 固化剂
4 diluent 稀释剂
5 elongation at tear 扯断伸长率
6 rubber cement 胶泥
7 workshop rubber lining 车间衬胶
8 on site rubber lining 现场衬胶
9 application 衬胶施工

6.7.2 搪玻璃设备 Glass-Lined Equipment

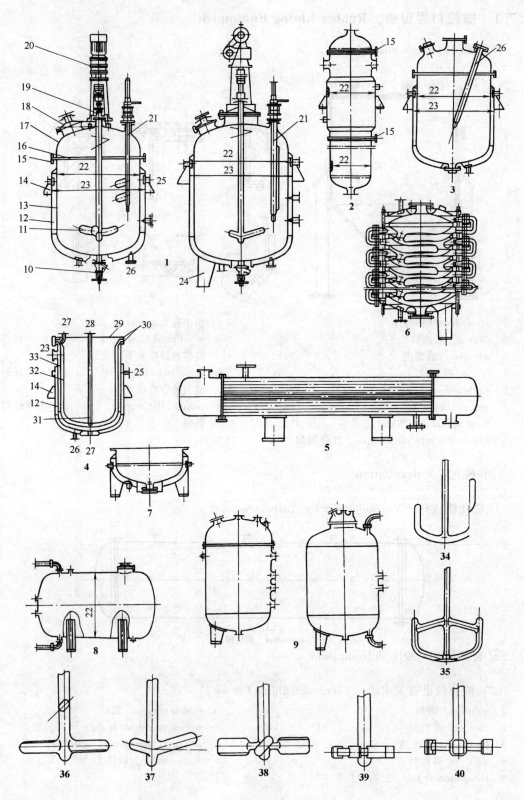

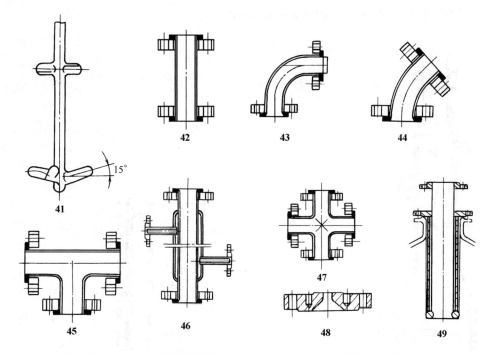

(1～9 glass-lined equipment 搪玻璃设备)
1 glass-lined reactor 搪玻璃反应器
2 glass-lined column 搪玻璃塔
3 glass-lined distillator 搪玻璃蒸馏罐
4 glass-lined sleeve type heat exchanger 搪玻璃套筒式换热器
5 glass-lined tubular heat exchanger 搪玻璃管壳式换热器
6 glass-lined plate condenser 搪玻璃碟片式冷凝器
7 glass-lined evaporator 搪玻璃蒸发器
8 horizontal glass-lined storage tank 卧式搪玻璃储罐
9 vertical glass-lined storage tank 立式搪玻璃储罐
10 glass-lined discharge valve 搪玻璃放料阀
11 glass-lined agitator 搪玻璃搅拌器
12 jacket 夹套
13 glass-lined shell 搪玻璃壳体
14 lugs 耳式支座
15 bolts/nuts 螺栓/螺母
16 glass-lined flange 搪玻璃法兰
17 glass-lined elliptical heat 搪玻璃椭圆封头
18 glass-lined manhole with sight glass 带视镜搪玻璃人孔
19 shaft seal 轴密封
20 reducer 减速机
21 glass-lined thermowell 搪玻璃温度计套管
22 glass-lined shell inside diameter 搪玻璃壳体内径
23 jacket inside diameter 夹套内径
24 legs 支腿
25 steam inlet 蒸汽入口
26 steam outlet 蒸汽出口
27 medium inlet 介质入口
28 medium outlet 介质出口
29 glass-lined cover 搪玻璃盖
30 gasket 垫片
31 glass-lined outlet pipe 搪玻璃出料管
32 glass-lined inside cylinder 搪玻璃内筒
33 glass-lined outside cylinder 搪玻璃外筒
(34～41 type of glass-lined agitator 搪玻璃搅拌器的型式)
34 anchor agitator 锚式搅拌器
35 gate agitator 框式搅拌器
36 paddle agitator 桨式搅拌器
37 impeller agitator 叶轮式搅拌器
38 propellor agitator 推进式搅拌器
39 disk-turbine agitator 圆盘搅拌器
40 floating head agitator 浮头式搅拌器
41 combination agitator 综合式搅拌器
(42～49 type of glass-lined agitator 搪玻璃管件的型式)
42 pipe 直管
43 90° elbow 90°弯头
44 45° elbow 45°弯头
45 tee 三通
46 jacketed pipe 带夹套直管
47 cross 四通
48 dip pipe 插入管
49 reducing flange 异径法兰

6.7.3 衬砖设备 Brick Lining Equipment

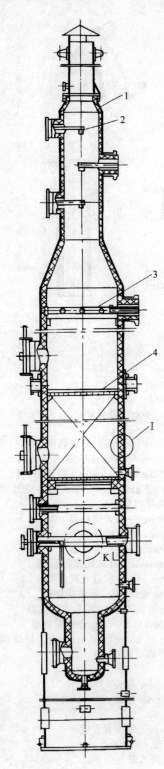
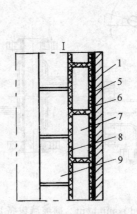

1 steel sheet　钢壳体

2 spray nozzle (material: PTFE)　喷嘴（材料：聚四氟乙烯）

3 liquid distributor (material: PPH)　液体分布器（材料：硬聚丙烯）

4 grid (material: PVDF)　格栅（材料：聚偏氟乙烯）

5 epoxy resin　环氧树脂

6 modified furan resin reinforced with glass fibre　改性呋喃树脂玻璃钢

7 acidproof ceramic brick　耐酸瓷砖

8 acidproof mortar　耐酸胶泥

9 carbon brick　碳砖

6.8 非金属设备 Non-metal Equipment

6.8.1 石墨设备 Graphite Equipment

6.8.1.1 石墨设备的类型 Type of Graphite Equipment

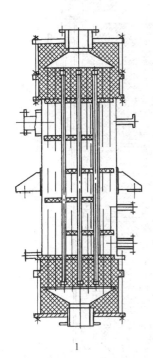

1

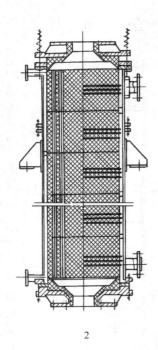

2

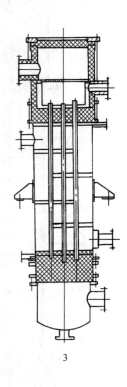

3

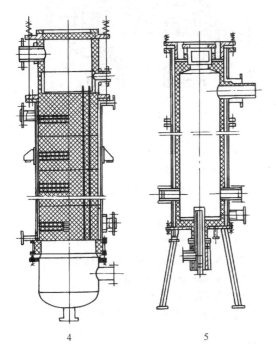

4　　5

1　shell and tube graphite heat exchanger　列管式石墨换热器

2　cylindrical block graphite heat exchanger 圆块式石墨换热器

3　graphite shell and tube falling-film absorber 列管式石墨降膜吸收器

4　graphite cylindrical block falling-film absorber 圆块式石墨降膜吸收器

5　water jacketed graphite hydrochloride synthesize furnace　水套式石墨氯化氢合成炉

307

6.8.1.2 石墨设备零件名称　Components Nomenclature of Graphite Equipment

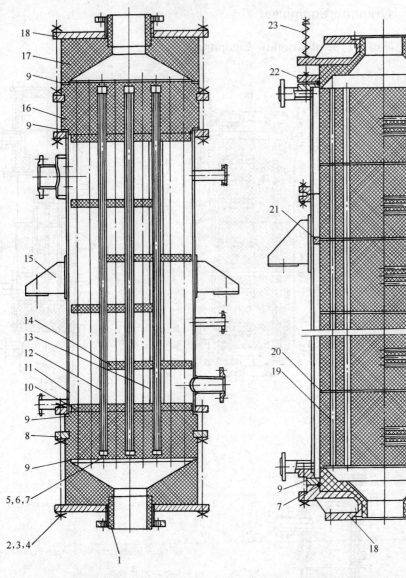

1	joint pipe　连接管	13	fixed position tube　定距管
2	screw pole　双头螺栓	14	baffle　折流板
3	nut　螺母	15	lug support　支座
4,6	washer　垫圈	16	tubesheet　管板
5	packing nut　压紧螺母	17	bonnet　封头
7	"O" sealing ring　O形密封环	18	plate　盖板
8	auxiliary flange　卡环	19	heat exchanger block　换热块
9	gasket　垫片	20	sealing ring　密封圈
10	auxiliary tubesheet　辅助管板	21	baffle　折流环
11	steel shell　钢壳体	22	sealing flange　密封法兰
12	heat exchanger tube　换热管	23	spring　弹簧

6.8.2 玻璃钢设备　FRP (Fibre Reinforce Plastics) Equipment

6.8.2.1 玻璃钢设备的类型　Type of FRP Equipment

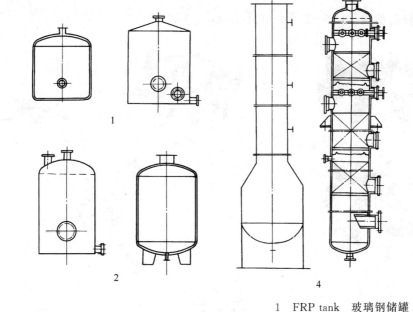

1　FRP tank　玻璃钢储罐
2　FRP vertical vessel　立式玻璃钢容器
3　FRP horizontal vessel　卧式玻璃钢容器
4　FRP tower　玻璃钢塔

6.8.2.2 玻璃钢管连接型式　Type of FRP Pipe Joint

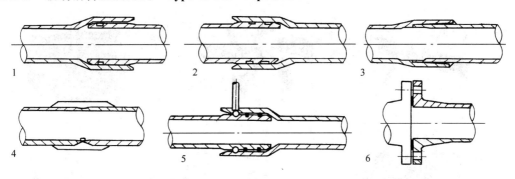

1　spigot-and socket joint with single sealing ring　单密封圈承插连接
2　spigot-and socket joint with double sealing ring　双密封圈承插连接
3　spigot-and socket joint with glueing　承插粘接
4　butt joint　对接
5　spigot-and socket joint with double "O" ring and "O" locking key made of nylon rod　承插 O 密封连接尼龙棒 O 键锁口
6　flange joint　法兰连接

6.9 混合设备　Mixing Equipment

6.9.1 搅拌器型式　Types of Agitator

(1) 搅拌器型式（1）　Types of Agitator（Ⅰ）

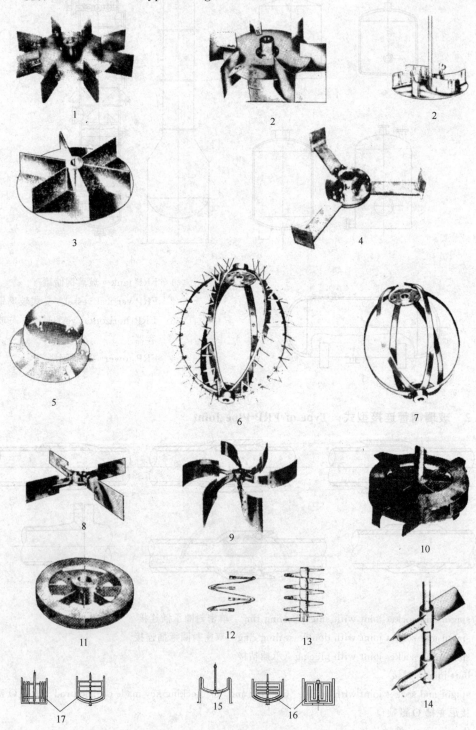

17

1　flat blade turbine　平叶片涡轮
2　curved blade turbine　弯曲叶片涡轮
3　lifter turbine　提升式涡轮

4　radial propeller　径向螺旋桨
5　flat blade turbine with stabilizing ring　带稳定环的平叶片涡轮
6　studded-cage beater　钉头笼式打浆器
7　plain cage beater　平边笼式打浆器
8　vertical flat-blade turbine impeller　垂直平叶片涡轮桨
9　vertical curved blade turbine impeller　垂直弯叶片涡轮桨
10　shrouded turbine　闭式涡轮
11　turbine with 3, 4 or 6 radial baffles　有三、四或六块径向挡板的涡轮
12　ribbon impeller　螺带式搅拌桨
13　vertical screw　立式螺杆(搅拌桨)
14　flat blade pitched paddle　倾斜的平叶桨
15　anchor impeller　锚式搅拌桨
16　double-acting paddle impeller　双动搅拌桨
17　gate paddle impeller　门框式搅拌桨

(2) 搅拌器型式（2）　Types of Agitator (Ⅱ)

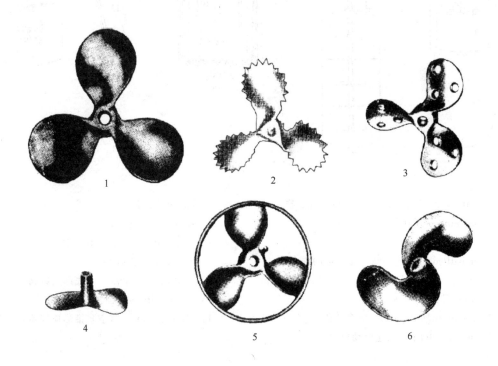

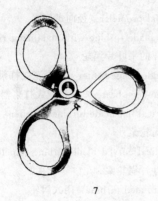

1 marine propeller 船用螺旋桨
2 saw-toothed propeller 锯齿形螺旋桨
3 perforated propeller 多孔式螺旋桨
4 folding propeller 折叶螺旋桨
5 propeller with ring guard 带护环的螺旋桨
6 weedless propeller 防缠式螺旋桨
7 cut-out propeller 镂空式螺旋桨

6.9.2 混合（搅拌）槽 Mixing Tanks

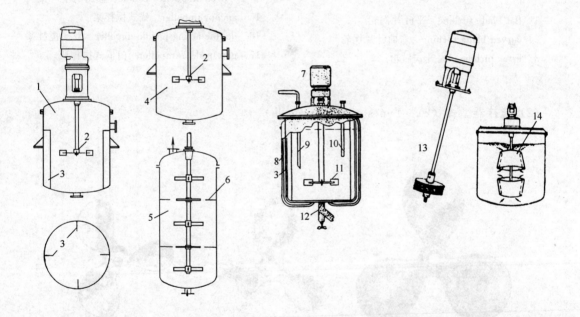

1 baffled mixing tanks 带挡板的混合槽
2 agitators 搅拌器
3 baffle 挡板
4 unbaffled mixing tank 无挡板的混合槽
5 compartmented agitator 隔室式混合器，多室混合器
6 compartment spacer 混合式隔板
7 motor reducer 减速电动机
8 jacket 夹套
9 feed pipe 进料管
10 thermo-well 温度计套管
11 impeller 搅拌桨
12 bottom flush valve 底部放料阀
13 disk agitator 圆盘搅拌器
14 cone agitator （空）锥形搅拌器

6.9.3 管道混合器 Line Mixers (Flow Mixers)

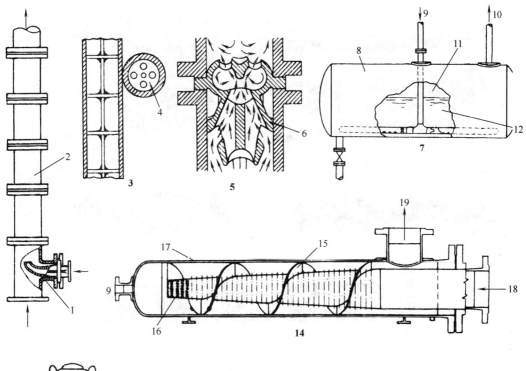

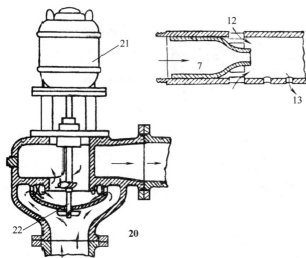

1 elbow jet mixer 弯头喷嘴混合器
2 orifice column 孔板混合柱
3 orifice mixer 孔板混合器，孔流混合器
4 orifice plate （锐）孔板
5 nozzle mixer 喷嘴混合器
6 nozzle 喷嘴
7 injector mixer 引射混合器
8 settling drum 澄清罐，沉降罐
9 oil in 油入口
10 oil out 油出口
11 washed oil 清净油
12 wash liquid 洗涤液
13 dispersion 分散孔
14 duo-sol crude mixer 双溶剂润滑油精制用原料混合器
15 spiral plate 螺旋（导流）板
16 perforation 开孔
17 shell 壳体
18 solvent 溶剂
19 mixture 混合物
20 agitated line mixer 搅拌式管道混合器
21 motor 电动机
22 agitator 搅拌器

6.9.4 静止混合器 Static Mixers

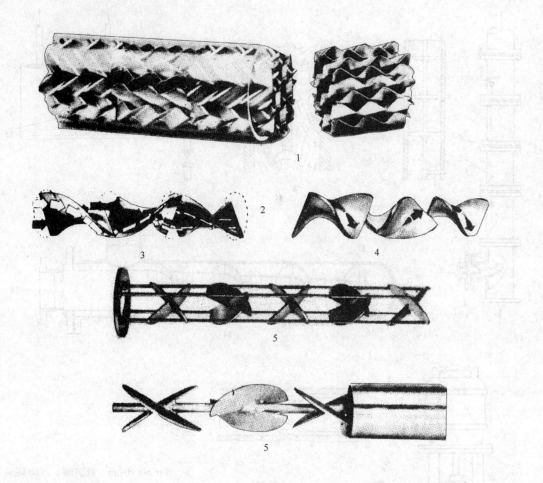

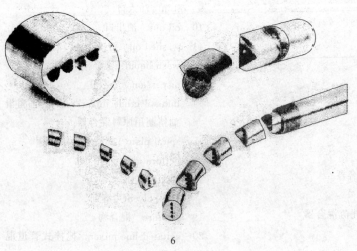

1 Sulzer static mixer 苏尔兹型静止混合器
2 Kenics static mixer 凯尼克型静止混合器
3 flow division 分割流动
4 radial mixing 径向混合
5 Dow motionless mixer 道氏静止混合器
6 Dow interfacial surfacegenerator 道氏界面表面发生器

6.9.5 双螺杆连续混合机 Double Screw Continuous Mixer

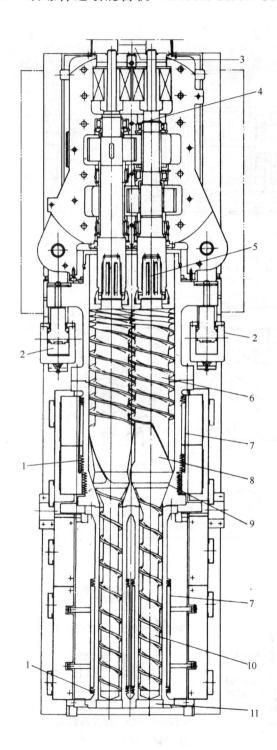

1　cylinder cooling fin　缸体冷却翅片
2　slot clearance adjustment hydraulic cylinder　调整缝隙的液压缸
3　main shaft　主轴
4　driving unit　传动装置
5　spline　花键，轴
6　feed section　进料部分
7　band heater　带式加热器
8　mixing section　混合部分
9　conical section　锥体部分
10　discharge section　出料部分
11　strainer or die　筛板或模板

6.9.6 膏状物料及黏性物料混(拌)合设备　Paste and Viscous-Material Mixing Equipments

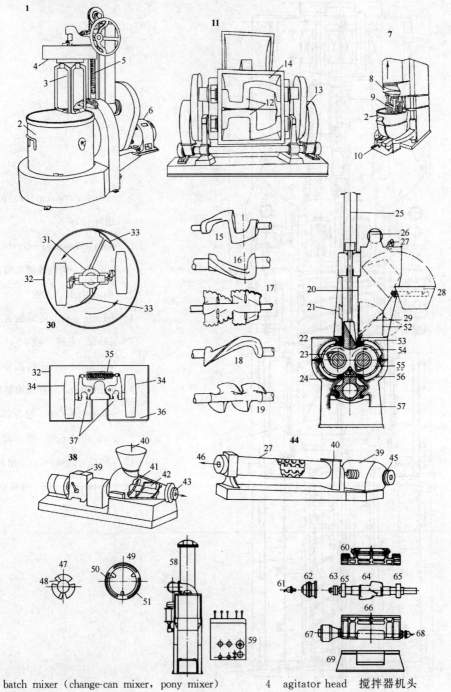

1　batch mixer (change-can mixer, pony mixer)
　　间歇式混料器，分批混合器
2　rotating change can　更换式转动混料罐，
　　间歇式转动混料槽
3　agitator blade　搅拌桨，搅拌器叶片
4　agitator head　搅拌器机头
5　lift　提升机构
6　motor　电动机
7　beater mixer　打浆拌和器
8　planetary gear　行星齿轮

9	beater 打浆器,搅拌(打)器	43	extrusion die (die head) 模头
10	cart 拌和罐拖车	44	ko-kneader (continuous kneader) 连续揉搓式混料机,单螺杆挤压式捏合机
11	double-arm kneader mixer 双桨捏和机		
12	counter rotating blades 逆向回转桨叶	45	drive pulley 传动带轮
13	gearing 传动装置	46	discharge 卸料口
14	rectangular trough 矩形料槽	47	screw 螺杆
15	sigma blade S形桨叶	48	interrupted flight 间断螺旋片
16	dispersion blade L形桨叶,散料式桨叶	49	barrel 筒体
17	multiwiping overlap blade 多齿形桨叶	50	stationary teeth 固定齿
18	single-curve blade 扭曲式桨叶	51	jacket 夹套
19	double nabe blade 麻花形桨叶	52	cleaning flap 清理门
20	feed hopper 加料仓	53	mixer body (top part) 混合机体(顶部)
21	feed opening for powders 粉末加料口	54	mixing chamber (top part) 混合室(顶部)
22	floating weight 浮动重块	55	mixing chamber (bottom part) 混合室(底部)
23,64	rotor 转子		
24	discharge device (sliding door) 出料装置(滑动门)	56	mixer body (bottom part) 混合机体(底部)
		57	machine base 机座
25	ram cylinder 压锤缸	58	feed hopper with ram 带压锤的加料仓
26	exhaust fan 排气扇	59	control station 控制台
27	light 灯	60	mixer body top part with mixing chamber top part 混合机顶(包括混合室顶部)
28	feed bucket 加料斗		
29	hopper door 加料仓门	61	cooling and heating connection 冷却和加热接口
30	muller mixer (pan muller) 碾轮混料机		
31	turret 转动架	62	thrust bearing case 推力轴承箱体
32	crib (pan) 碾压槽	63	thrust bearing 推力轴承
33	plow 刮板	65	self-aligning roller bearing 自定位滚柱轴承
34	muller wheel 碾轮		
35	spring 弹簧	66	mixer body bottom part with mixing chamber bottom part 混合机底(包括混合室底部)
36	bedplate 底板		
37	adjusting screw 调节螺钉		
38	roto-feed mixer (mixer-extruder) 混合-挤压机	67	rotary activator for drop door 出料门旋转机构
		68	cooling and heating connection for the drop door 出料门冷却和加热接口
39	gear box 齿轮箱,变速箱		
40	feed 进料		
41	conical chamber 锥形混料室	69	machine base 机座
42	helical agitator 螺旋搅拌器		

6.9.7 固体混合机械 Solids Mixing Machines

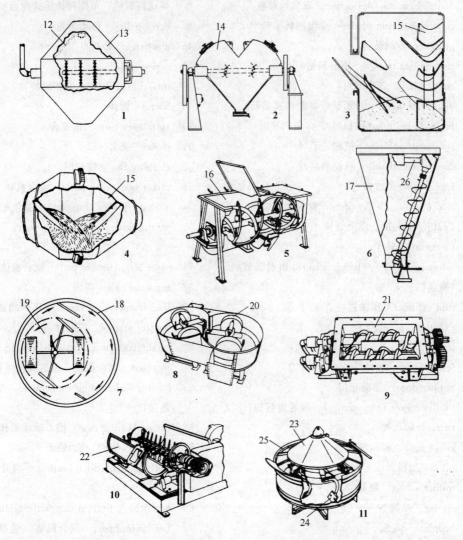

(1, 2 agglomerate-breaking devices 结块破碎机)
1 double cone mixer 双锥鼓混合器
2 twin shell mixer 双筒混合器
3 horizontal drum mixer 卧式转鼓混合机
4 double-cone baffled tumbler 双锥挡板转鼓
5 ribbon mixer 螺带式混合器
6 vertical screw mixer 立式螺旋混合器
7 batch muller 间歇式碾磨机
8 continuous muller 连续式碾磨机
9 twin rotor mixer 双转子混合器
10 single rotor mixer 单转子混合器
11 turbine mixer 透平式混合机，涡轮式混合机
12 spray nozzle 喷洒嘴
13 tumbler 转鼓，转筒
14 unbaffled tumbler 无挡板转鼓
15 baffle 挡板
16 stationary shell or trough 固定外壳或料槽
17 conical tank 锥形罐
18 pan (rotates clockwise) 混合槽（顺时针转动）
19 muller turret (rotates counter clockwise) 碾轮转动架（逆时针转动）
20 muller 滚轮，碾轮
21 cylindrical shell 筒形壳
22 split casing 对开式外壳
23 housing 罩子
24 plow shares 犁铧
25 circular trough 圆槽
26 orbiting arm （做圆周运动的）旋转臂

6.10 萃取器 Extractors

6.10.1 浸提设备 Leaching Equipments

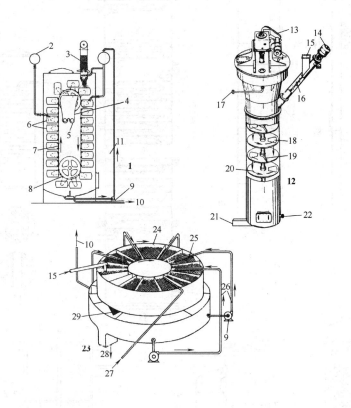

1 Bollman extractor 博尔曼浸提装置
2 pure solvent tank 纯溶剂罐
3 dry flake 干料片
4 hopper for leached solid 渣料出料斗
5 paddle conveyor (screw conveyor) 螺旋输送器
6 bucket with perforated bottom (basket) 漏底吊斗
7 chain 链条
8 sprocket 链轮
9 pump 泵
10 full miscella 富液
11 half miscella 半贫液
12 rotating blade extractor 转盘式浸提塔
13 contactor drive 转盘驱动装置
14 feed drive 加料驱动装置
15 solid feed 固体加料口
16 screw feeder 螺旋给料机
17 liquid overflow （抽提）液溢流
18 plate opening （转）盘开口
19 scraper 刮板
20 path of solids 固体路径
21 extracted solids discharge conveyor 渣料排送器
22 liquid feed 液体进料
23 Rotocel extractor 罗托西尔浸提装置，转动隔仓式浸提器
24 rotating cell 旋转料室，旋转料仓
25 spray 喷雾器
26 interstage liquid 中间循环液
27 fresh solvent 新鲜溶剂
28 leached solid (marc) 浸提后的渣料
29 hinged screen bottom 铰链式筛网底

6.10.2 连续萃取设备，连续抽提设备 Continuous Contact (Differential Contact) Equipments

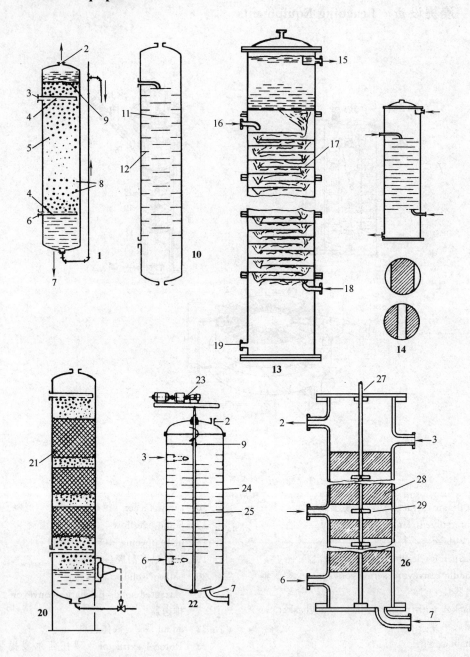

1 spray tower 喷淋塔
2 light liquid out 轻液出口
3 heavy liquid in 重液入口
4 distributor 分布器，分配器
5 continuous phase 连续相
6 light liquid in 轻液入口
7 heavy liquid out 重液出口
8 dispersed phase 分散相
9 interface 界面
10 disk-and-doughnut baffle tower 盘-环式挡板塔
11 disk 圆盘，盘形挡板

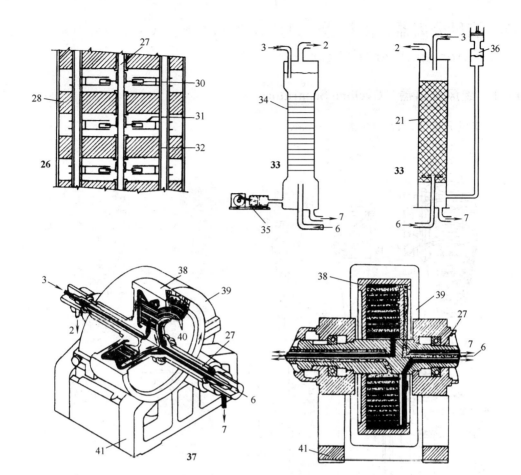

12	doughnut 圆环，环形挡板		夏伯尔萃取塔
13	side-to-side baffle tower 边到边（单流式）挡板塔	27	rotating shaft 转动轴
		28	wire mesh packing 金属丝网填料
14	center-to-side baffle tower 中到边（双流式）挡板塔	29	turbine agitator 涡轮搅拌器
		30	turbine impeller 涡轮搅拌桨
15	extract 萃取液，抽出液	31	horizontal stationary baffle 水平固定挡板
16	feed 进料	32	tie rod 拉杆
17	segmental baffle 圆缺挡板	33	pulsed column 脉冲（萃取）塔
18	solvent 溶剂	34	perforated plate 多孔板，筛板
19	raffinate 抽余液，萃余液	35	pump pulse generator 往复脉冲发生器
20	packed tower 填料塔	36	air pulser 空气脉冲（发生）器
21	packing 填料	37	centrifugal extractor 离心式萃取机
22	rotating disk contactor 转盘萃取塔	38	rotor 转子
23	variable speed drive 变速驱动装置	39	rotor guard 转子壳体
24	stator ring 定环	40	perforated shell 多孔（同心）壳体
25	rotor disk 转盘	41	base 底座
26	Scheibel extractor 有搅拌器的萃取塔，		

321

6.11 旋风分离器、澄清器、过滤器和离心机 Cyclone, Decanter, Filter and Centrifuger

6.11.1 旋风分离器 Cyclone Separators

(1) 旋风分离器（1） Cyclone Separators（Ⅰ）

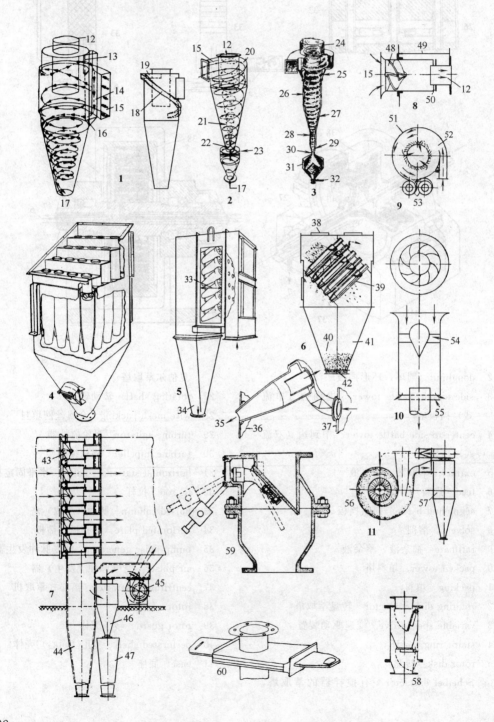

1	Van Tongeren cyclone	范·汤格恩旋风分离器
	cyclone with dust shave-off	带粉体旁路进口的旋风分离器
2	helical entry cyclone (Duclon cyclone)	螺旋进入式旋风分离器，杜康旋风分离器
3	involute entry cyclone	渐开线进入式旋风分离器
4	multiclone collector	（立式）多管旋风分离器
5	Dustex miniature collector assembly	德斯泰格斯式多斜管旋风分离器
6	multiple cyclone	（有预沉降室的）多管式旋风分离器
7	multicellular straight through cyclone	多室直通式旋风分离器
8	fixed impeller straight through cyclone	固定叶轮直通式旋风分离器
9	Buell-Van Tongeren scroll collector	布埃尔·范·汤格恩涡卷形除尘器
10	uniflow cyclone	单流旋风分离器
11	collector combining induced draught fan	联合式除尘器（连有引风机）
12	clean gas outlet	净化气体出口
13	dust shave-off	粉体旁路进口
14	by-pass dust channel	粉尘旁路
15	dust laden gas inlet	含尘气体入口
16	by-pass re-entry opening	旁路返回口
17	dust outlet	粉尘出口，粉体出口
18	helical by-pass dust channel	螺旋式粉尘旁路流道
19	volute type shave-off	涡旋式粉体旁路进口
20	helical top (helical inlet)	螺旋顶板，螺旋式入口
21	vortex	旋流
22	vortex shield	破涡流板
23	dust trap	集尘斗，灰斗
24	outlet head	出口管帽
25	upper cone	上锥体
26	upper cylinder	上筒体
27	middle cone	中间锥体
28	middle cylinder	中间筒体
29	lower cone	下锥体
30	tail piece	尾管
31	receptacle	收集器，灰斗
32	blast gate	风闸，吹风口
33	gas outlet duct	气体排出通道
34	dust hopper	粉尘斗，集尘斗
35	dust	粉尘，粉体，灰尘
36	hopper tube sheet	粉斗固定板
37	inlet tube sheet	入口管固定板
38	inspection door	检查门，检查口
39	tubeblock	管式分离器组
40	centrifugation hopper	离心分离储尘斗，细粉尘斗
41	decantation hopper	沉降储尘斗，粗粉尘斗
42	dust slide	卸尘滑板
43	cell guide vane	通道导流叶片
44	coarse dust hopper	粗粉尘斗
45	booster fan	辅助吸风机
46	secondary cyclones	二级旋风分离器（组）
47	fine dust hopper	细粉尘斗
48	fixed vane	固定式导流叶片
49	dust header	粉尘集箱
50	annular dust slot	粉尘环槽，粉尘环隙流道
51	primary dust separator	一级粉尘分离器
52	variable inlet control damper	可调入口挡板
53	secondary dust collectors	二级粉尘收集器
54	swirl vane	旋流叶片
55	deflector ring	导流环，导流板
56	induced draught fan	引风机
57	cyclone	旋风分离器
(58~60)	cyclone hopper valves	旋风分离器底阀，料斗阀
58	double flap valve	双翻板阀，双翼阀
59	counter weighted flap valve	重锤式翻板阀，重锤式翼阀
60	push-pull valve	插板阀

(2) 旋风分离器(2)　Cyclone Separators（Ⅱ）

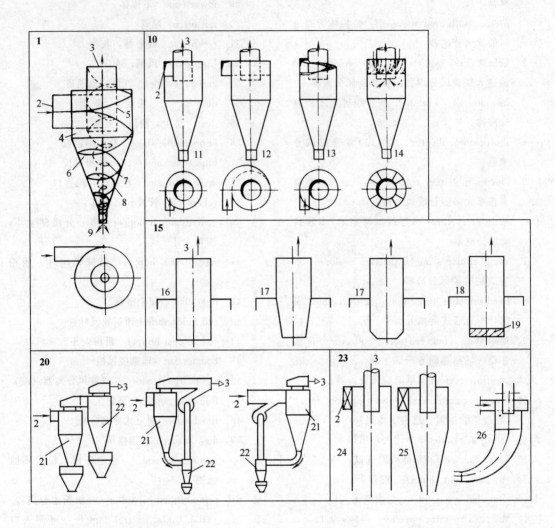

1　cyclone　旋风分离器
2　dust-laden gas inlet　含尘气体入口，含粉体气体入口
3　clean-gas outlet　净化气体出口
4　body　筒体
5　inner cylinder　气体出口管，内筒
6　outer vortex　外旋流
7　inner vortex　内旋流
8　cone　锥体
9　dust outlet　粉尘出口，粉体出口
10　cyclone entries　旋风分离器进入型式
11　tangential entry　切线进入
12　curved entry　螺线进入
13　wrap around entry　螺旋线进入
14　axial entry　轴线进入
15　cyclone exit patterns　旋风分离器出口型式
16　straight exit pipe　直出口管
17　sloping exit pipe　带锥度的出口管
18　straight exit pipe with vanes　装有叶片的直出口管
19　vane　叶片
20　cyclone arrangements in series　旋风分离器的串联布置
21　primary cyclone　（第）一级旋风分离器
22　secondary cyclone　（第）二级旋风分离器
23　cyclone trunk patterns　旋风分离器筒体型式
24　straight cylinder　直筒体
25　cylinder and cone　筒锥体
26　bent cone　弯锥体

6.11.2 沉降罐、澄清器 Gravity Settlers (Decanters)

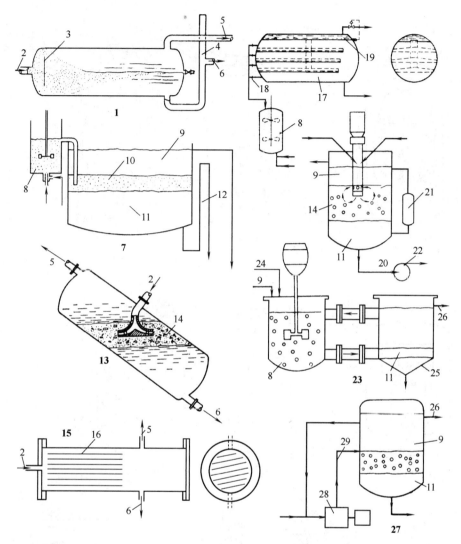

1　gravity settler　沉降罐，澄清器
2　dispersion in　不混溶液体入口
3　slotted impingement baffle　开槽防冲挡板
4　siphon break　破虹吸（管）
5　light liquid out　轻液出口
6　heavy liquid out　重液出口
7　vertical decanter　立式澄清器，立式沉降罐
8　mixer　混合器
9　solvent　溶剂
10　dispersion band　分散带，分散混合层
11　aqueous　水溶液
12　gravity leg　液封腿
13　settler of Edeleanu　爱德利努式澄清器
14　emulsion zone　乳浊区，乳浊带
15　baffled settler　挡板式澄清器，挡板式沉降罐
16　baffle　挡板
17　baffled settler　挡板式澄清器，挡板式沉降罐
18　orifice　孔板
19　interface level　界面
20　internal mixer-settler　内部混合-沉降罐
21　level controller　液位调节器
22　interstage pump　中间泵
23　mixer-settler　混合-沉降罐
24　slurry feed　悬浊浆料
25　cone bottom settling tank　锥底沉降罐
26　overflow　溢流
27　pump mixer-settler　泵送混合-沉降罐
28　mixing pump　混合泵
29　mixed phases　混合相

6.11.3 气体洗涤器 Gas Scrubbers

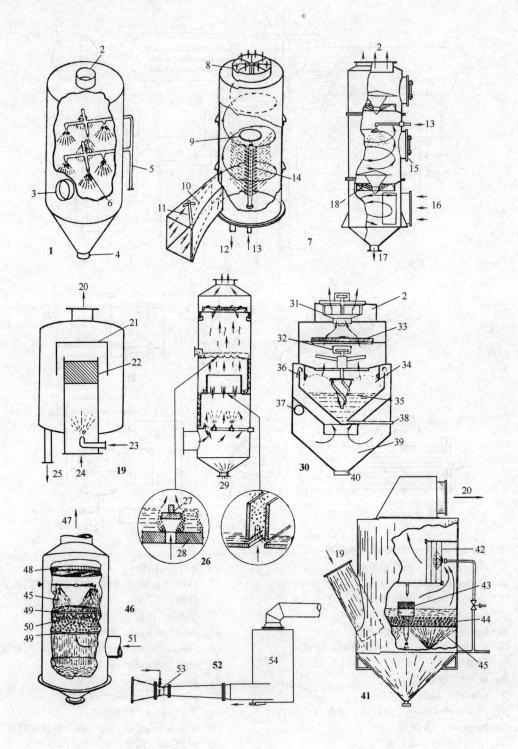

1 spray tower 喷淋塔
2 clean air outlet 净化空气出口
3 dirty air inlet 脏空气入口
4 water and sludge drain 水和泥渣排放口
5 supply water piping 供水管
6 spray jet 喷头
7 cyclone spray scrubber（centrifugal spray scrubber） 湿式旋风除尘器
8 antispin vane 消旋叶片
9 core buster disk 中心挡盘
10 swinging inlet damper 摆动式入口挡板
11 handle 手柄
12 water outlet 水出口
13 water inlet 水入口
14 spray manifold 多喷嘴管
15 inspection door 检查门，检查孔
16 fume inlet 烟气入口
17 sludge outlet 泥渣出口
18 fixed vane 固定叶片，固定式（旋流）叶片
19 fibrous-bed scrubber 纤维填充床洗涤器
20 clean gas out 净化气体出口
21 cap 罩，帽
22 irrigated pad 湿滤（填充）床
23 liquid in 液体入口
24 dirty gas 脏气体
25 slurry out 淤浆出口
26 impingement plate scrubber 撞击挡板式洗涤塔
27 impingement baffle 撞击挡板
28 sieve plate 筛孔板
29 dirty water discharge 脏水排出口
30 scrubber equipped with vertical rotor 装有立式转子的洗涤器
31 exhaust fan 排气风扇，排风机
32 enclosed belt tunnel 封闭式传动皮带通道
33 entrainment separator 雾沫分离器
34 spray zone 溅雾区
35 vertical rotor 立式转子
36 baffle 挡板
37 contaminated air inlet 污染空气入口
38 drain 排液口，放净口
39 dry precleaner 粗粉尘预沉降器
40 coarse particles out 粗粒粉尘出口
41 packed-bed scrubber 填充床洗涤器
42 zig-zag entrainment separator 曲径式雾沫捕集器，曲径式雾沫分离器
43 overflow pipe 溢流管
44 marble packed bed 石球填充床
45 spray 雾沫，水雾
46 fluidized-bed scrubber（Aerotec floating-bed scrubber） 流化床洗涤器
47 cleaned gas 净化气体
48 mist eliminator 除沫器
49 retaining grid 限位格栅
50 floating bed 浮动床
51 dust-laden gas 含尘气体
52 Venturi scrubber 文丘里洗涤器，文丘里除尘器
53 Venturi 文丘里管
54 cyclone separator 旋风分离器

6.11.4 过滤机 Filters

6.11.4.1 压滤机 Pressure Filters

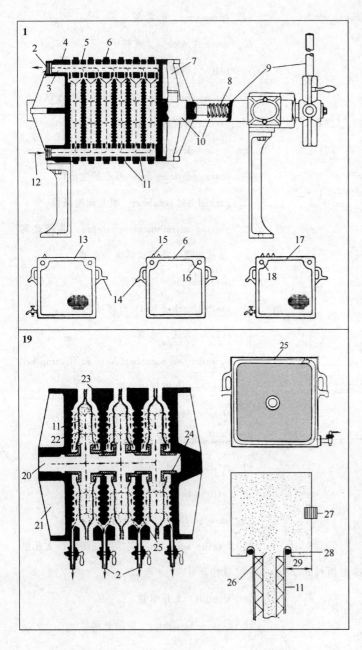

1　plate and frame filter press　板框（式）压滤机
2　filtrate outlet　滤液出口
3　channel (path)　通道
4　fixed head　固定端板
5　plate　滤板
6　frame　滤框
7　movable head　活动端板
8　hand screw　手动压紧丝杠
9　closing device　（自动）压紧装置
10　side rail　侧轨，横杆
11　filter cloth　滤布
12　slurry　滤浆
13　non-wash plate　非洗板
14　side lug　支耳
15　button　（标志）块
16　feed inlet　滤浆入口，原料入口
17　wash plate　洗液板
18　wash inlet　洗涤水入口，洗液入口
19　recessed plate filter (chamber press)　箱式压滤机，槽板式压滤器
20　mixture inlet　滤浆入口，混合物入口
21　head　机头
22　cake　滤饼
23　rim　凸缘
24　screwed union　（滤布）压紧螺套
25　recessed plate　带槽滤板
26　calked gasket-recessed filter plate　嵌条-垫片密封式槽板压滤机的滤板
27　sealing gasket　密封垫
28　calking strip　滤布嵌条，滤布张紧条
29　cake recess　滤饼槽

6.11.4.2 叶滤机 Pressure Leaf Filters

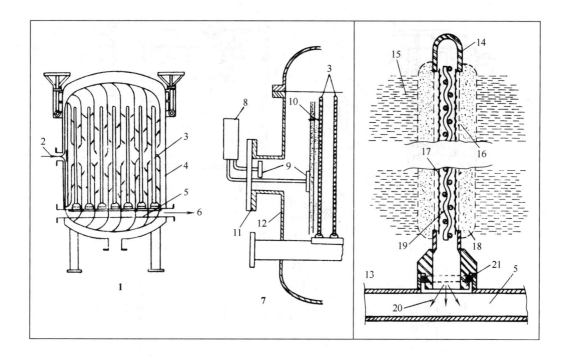

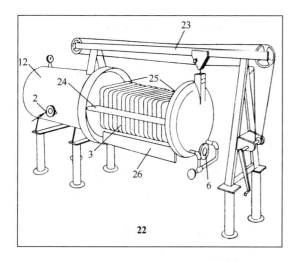

1　vertical pressure leaf filter　立式叶滤机
2　slurry inlet　滤浆入口
3　leaf　滤叶
4　upright cylindrical pressure tank　立式圆筒形压力罐
5　filtrate manifold　滤液集合管
6　filtrate outlet　滤液出口
7　cake-thickness detector　滤饼厚度测试装置
8　differential pressure switch　差压开关
9　pressure-sensing element　测压元件
10　cake thickness　滤饼厚度
11　flange opening　法兰连接式开口
12　filter tank　滤筒
13　precoated wire filter leaf　预涂层式滤叶
14　binding　卡框
15　unfiltered liquid　未过滤液
16　precoat　预涂层
17　filter cloth　滤布
18　filter cake　滤饼
19　chamber screen　铁丝网
20　filter liquid　滤液
21　O-ring　O形环
22　horizontal-tank pressure leaf filter　卧式叶滤机
23　overhead rail (track)　高架导轨
24　side bar　横梁
25　leaf assemble (leaf pack)　滤叶组件
26　scavenger leaf　残液滤片

6.11.4.3 袋式过滤器 Bag Filers

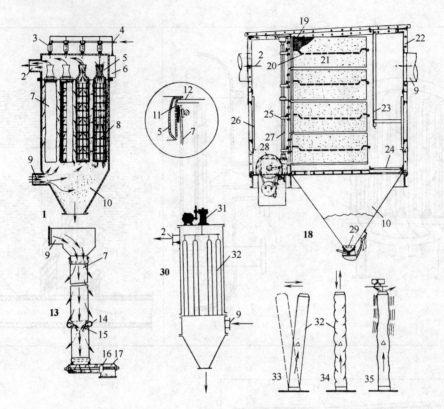

(1～13 felt-fabric filter 毛毡过滤器)
1 reverse-jet filter （轮换）反吹式毡袋过滤器
2 clean air outlet 净化空气出口
3 solenoid valve 电磁阀
4 compressed-air manifold 压缩空气集合管
5 Venturi nozzle 文丘里喷嘴
6 timer 定时器
7 felt filter tube 毡过滤管，过滤毡袋
8 wire retainer 金属网架
9 dusty air inlet 含尘空气入口
10 hopper 漏斗，集尘斗
11 collar 翻边套管
12 tube sheet 毡管固定板，文丘里管固定板
13 Hersey reverse jet filter （滑环）反吹式毡袋过滤器，赫西反吹式毡袋过滤器
14 reverse-jet blow ring (cleaning ring) 反吹环
15 high pressure air 高压空气
16 choke cone 阻流锥
17 tension spring 推簧，张紧弹簧

(18～30 woven-fabric filter 布过滤器)
18 envelope type cloth filter 布套式过滤器
19 screen 筛网
20 screen beater 筛网振打器，筛网敲打器
21 cloth 滤布
22 casing 箱体，外壳
23 removable baffle 可拆挡板
24 wire mesh walkway 金属丝网走道
25 rocker arm 摇臂
26 man door 检修门
27 rocker shaft 摇臂轴
28 rapping mechanism 敲击机构，敲击装置
29 hopper valve 料斗底阀，漏斗底阀
30 bag type cloth filter 布袋式过滤器
31 shaking mechanism 振动装置，振动机构
32 bag （布）袋
33 gentle sideways movement 侧向摆动
34 vertical movement 垂向抖动
35 sideways vibration 侧向振动，颤动

6.11.4.4 转鼓真空过滤机 Rotary-Drum Vacuum Filter

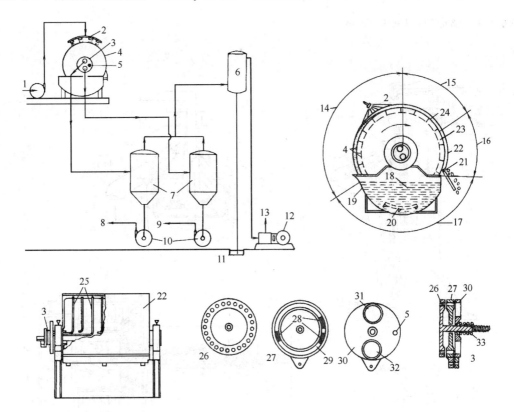

1	wash water pump 洗涤水泵	19	slurry trough 滤浆槽
2	wash spray 水喷头	20	agitator 搅拌器
3	rotary valve 转盘阀，分配头	21	knife (doctor blade) 刮刀
4	rotary drum 转鼓，转筒	22	cake 滤饼
5	air connection 空气接管	23	compartment 隔室
6	moisture trap 水分离器	24	filter medium 过滤介质，过滤材料
7	vacuum receiver 真空罐，真空受液罐	25	internal pipe 内部连接管
8	wash 洗液	26	rotating plate 旋转板
9	filtrate 滤液	27	stationary intermediate plate 中间固定板
10	pump 泵	28	division bridge 分配块
11	barometric seal 水封	29	annulus 环形槽沟
12	dry vacuum pump 干式真空泵	30	stationary plate 固定板
13	air out 空气出口	31	wash vacuum connection 洗液与真空系统的接口
14	washing zone 洗涤区		
15	sucked dry zone 吸干区	32	filtrate vacuum connection 滤液与真空系统的接口
16	dislodged zone 去饼区		
17	filter zone 过滤区，吸滤区	33	spring 弹簧
18	slurry 滤浆		

6.11.5 离心式分离机 Centrifugal Separator
6.11.5.1 离心机 Centrifuges

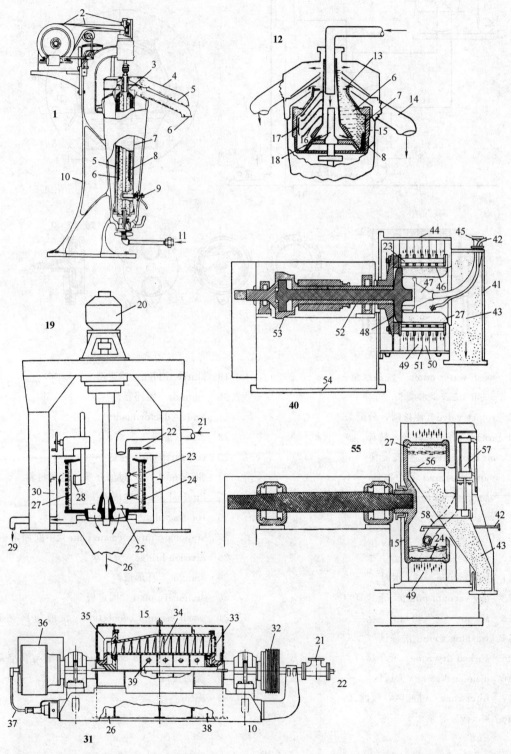

1 tubular-bowl centrifuge 管式高速离心机，超速离心机
2 driving mechanism 驱动装置
3 ring dam 环形堰板
4 discharge cover 排液管
5 light liquid 轻液
6 heavy liquid 重液
7 rotating bowl 转鼓
8 solid 固体
9 brake 制动器
10 frame 机架
11 liquid inlet 液体入口
12 disk centrifuge 转盘式离心机
13 adjustable dam 可调堰
14 discharge spout 排液管
15 bowl （转）鼓
16 disk 金属盘
17 disk stack 叠装盘组
18 spindle 轴
19 top-suspended basket centrifugal，top-suspended centrifugal filter 上悬式转鼓离心过滤机
20 motor 电动机
21 feed slurry 滤浆进料
22 wash inlet 洗液入口
23 rotating basket 转鼓
24 filter medium 过滤介质，过滤材料
25 removable valve plate 提升式（卸料）阀板
26 solids discharge 滤渣卸出口
27 cake 滤饼
28 adjustable unloader knife 可调式刮刀
29 liquid drawoff 液体排出
30 curb 外壳
31 cylindrical-conical helical-conveyor centrifuge (solid-bowl centrifuge) （卧式）螺旋卸料离心机
32 driven sheave 传动带轮
33 adjustable filtrate port 可调滤液口
34 scroll-conveyor 螺旋输送器
35 solids-discharge port 固体卸料口
36 conveyor drive 输送器驱动装置
37 overload release 过载断路器
38 filtrate discharge 滤液出口
39 feed port 进料口
40 continuous centrifugal filter 连续式离心过滤机
41 access door 检修门，检修孔
42 wash feed pipe 洗液导管
43 cake discharge chute 滤饼卸出槽
44 filtrate housing 滤液室
45 slurry feed pipe 滤浆进料管
46 screen 滤网
47 feed funnel 进料斗，受料斗
48 pusher 活塞推送（卸料）器
49 filtrate 滤液
50 wash liquid 洗液
51 wet housing separator 液室隔板
52 piston rod （卸料）推杆
53 hydraulic servo motor 液压伺服马达
54 base 机座
55 batch centrifugal filter 间歇式离心过滤器
56 peeler knife 刮刀
57 hydraulic cylinder 液压缸
58 feed duct （滤浆）进料导管

6.11.5.2 双鼓真空离心过滤机 Double-Bowl Vacuum Centrifuge

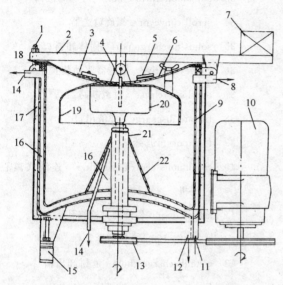

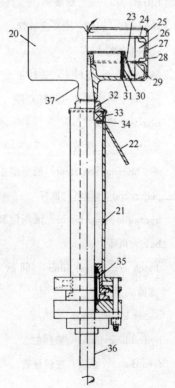

1 swing bolt 活节螺栓
2 cover yoke 顶盖轭架
3 electric heaters 电热器
4 gel feed 胶料进口
5 sight glass 视镜，看窗
6 dished cover 碟形顶盖
7 counter weight 平衡锤
8 vacuum supply 接真空
9 vacuum tank 真空罐
10 electric motor 电动机
11 heating water inlet 加热水入口
12 gel outlet 胶料出口
13 V-belt drive 三角带传动
14 heating water outlet 加热水出口
15 vibration isolator 减振器，隔振体
16 water jacket 水夹套
17 insulation 保温层
18 gasket 垫片
19 curved baffle 曲面挡板
20 centrifuge basket 离心过滤转鼓
21 shaft column 轴承支柱
22 conical stiffener 加强锥，锥形加固架
23 conical liquid diversion baffle 锥形导液板
24 basket lip 转鼓唇缘
25 basket skirt 转鼓裙
26 liquid surge baffles 液体缓冲挡板
27 solid collecting chamber 固体存积腔
28 outer solid bowl 整体外转鼓
29 ring baffle 环形挡板
30 inner perforated bowl 带孔内转鼓
31 filter paper 滤纸
32 slinger 轴承罩
33 top ball bearing 顶部滚珠轴承
34 bearing retainer 轴承支圈，轴承挡圈
35 mechanical seal 机械密封
36 shaft 轴
37 basket hub 转鼓毂

6.11.5.3 静止叶片型离心式分离器 Stationary

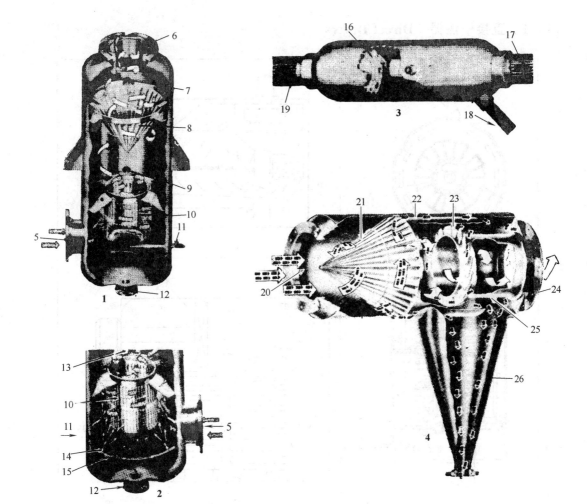

1 scrubber with internal feed 中心喷淋的洗涤器
2 scrubber with spray ring 喷淋型洗涤器
3 line type centrifugal separator 管道型离心式分离器
4 stationary vane centrifugal separator 静止叶片离心式分离器
5 contaminated gas or vapor inlet 玷污的气体或蒸气进口
6 clean gas or vapor outlet 清洁的气体或蒸气出口
7 drying stage drain 干燥段排液口
8 drying tuyere 干燥段风嘴
9 scrubbing stage drain 洗涤段排液口
10 scrubbing tuyere 洗涤段风嘴
11 scrubbing liquid inlet 洗涤液进口
12 bottom drain 底部排液口
13 scrubbing stage drain 洗涤段排液口
14 spray nozzle 喷嘴
15 spray ring 喷淋环
16 stationary centrifugal separating element 静止的离心分离元件
17 clean air, steam or gas 清洁的空气、蒸汽或气体（出口）
18 drain liquids and solids 液体及固体排出口
19 contaminated air, steam or gas 玷污的空气、蒸汽或气体（出口）
20 inlet 进口
21 helicoid tuyere 螺旋面风嘴
22 by-pass area 旁路区域
23 vane blade 叶片
24 outlet 出口
25 tailpipe 尾管
26 hopper 集尘斗

6.12 干燥器 Dryers

6.12.1 直接干燥器 Direct Dryers

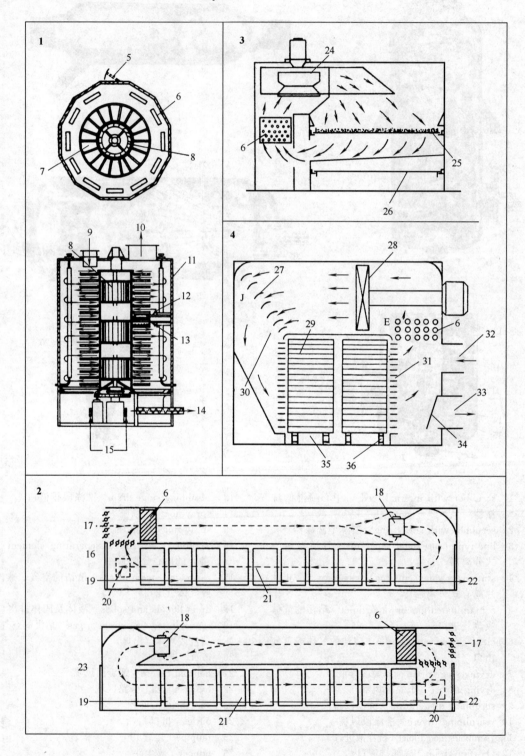

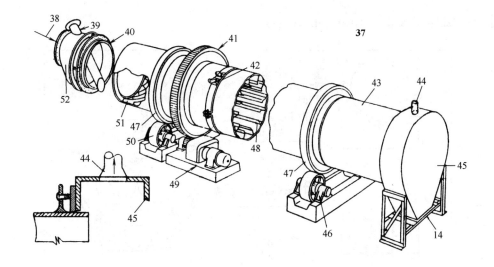

1 turbo-tray dryer 转盘式干燥器
2 tunnel dryer 隧道式干燥器，洞道式干燥器
3 through-circulation dryer 带式干燥器
4 tray dryer 箱式干燥器
5 access door 检修门
6 heater 加热器
7 rotating annular shelf 转盘，环状转盘
8 turbo-type fan 涡轮式送风机
9 wet feed 湿进料，湿料
10 exhaust air 排气
11 insulated housing 保温外套
12 wiper 刮料片，刮刀
13 leveler 刮平器
14 dry discharge 干料排出
15 variable-speed drive 变速驱动装置
16 countercurrent tunnel dryer 逆流式隧（洞）道干燥器
17 fresh air inlet 新鲜空气入口
18 blower 风机
19 wet material in 湿料进口
20 exhaust-air stack 排气烟道，排气烟囱
21 truck 小车
22 dry material out 干料出口
23 parallel current tunnel dryer 并流式隧（洞）道干燥器
24 air circulating fan 空气循环风机
25 conveyor 传送机，传送带
26 conveyor return 返回传送带
27 turning vane 转向叶片，导向叶片
28 adjustable-pitch fan 调角风机
29 shallow tray 浅盘
30 plenum chamber 干燥室干燥箱
31 adjustable baffles 调节挡板
32 air-inlet duct 空气入口通道，进气通道
33 air-exhaust duct 空气排出通道，排气通道
34 damper 挡板
35 rack （料盘）架
36 truck wheel 车轮
37 rotary dryer 转筒干燥器
38 gas inlet 气体入口
39 feed chute 进料斜管，送料滑槽
40 friction-seal 摩擦密封
41 girt gear 齿圈，齿轮
42 knocker （防黏结）振击器，敲击器
43 dryer shell 干燥器筒体，转筒
44 gas outlet 气体出口
45 breeching 尾端（防干燥器筒体退滑）
46 trunnion 耳轴
47 riding ring 轮圈，轮箍
48 lifting flight 升举式抄板
49 drive assembly 驱动装置总成（组件）
50 thrust roll 支承滚子，托轮
51 spiral flight 螺旋式导料板
52 inlet head 入口端帽

6.12.2 间接干燥器 Indirect Dryers

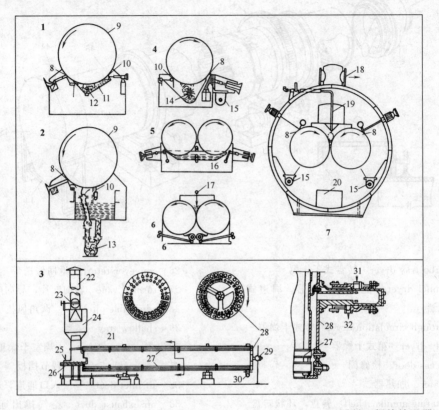

1	drum dryer 滚筒干燥器	15	conveyor （干料）螺旋输送器
2	single drum dryer with dip-feed 浸没布料式单滚筒干燥器	16	shallow pan 浅料盘，浅料槽
3	single drum dryer with pan-feed 喷洒布料式单滚筒干燥器	17	feed pipe 进料管
		18	vapour outlet 蒸汽出口
4	single drum dryer with splash-feed 飞溅布料式单滚筒干燥器	19	pendulum feed 摆动式进料管
		20	manhole 人孔
5	double drum dryer with dip-feed 浸没布料式双滚筒干燥器	21	steam-tube (rotary) dryer 蒸汽（加热）管回转式干燥器
6	double drum dryer with top-feed 顶部加料式双滚筒干燥器	22	natural-draft stack 自然抽风筒
		23	damper 挡板
7	vacuum double drum dryer 双滚筒真空干燥器	24	dust drum 粉尘筒，粉尘捕集器
		25	wet material feed 湿物料进料
8	knife 刮刀	26	screw feeder 螺旋给料器
9	revolving drum 滚筒，转筒	27	steam-heated tube 蒸汽加热管
10	spreader 布料器	28	steam manifold 蒸汽分配盘，蒸汽分配头
11	feed pan 料槽	29	steam neck (steam joint) 进汽接头
12	agitator 搅拌器	30	dried material discharge conveyor 干物料卸料输送器
13	pump 泵	31	steam in 蒸汽入口
14	splash roll 溅料轮	32	condensate out 冷凝水出口

6.12.3 喷雾干燥器　Spray Dryers

6.12.3.1 喷雾干燥装置　Spray Dryer Installation

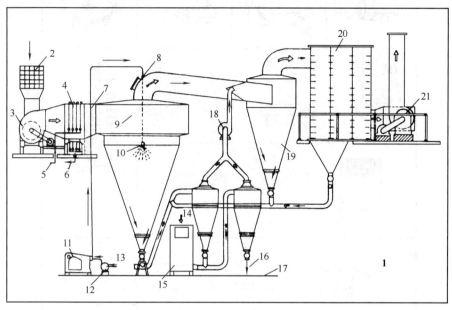

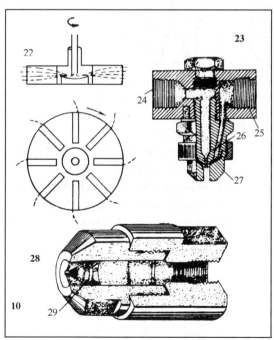

1　spray dryer installation　喷雾干燥装置
2　air filter　空气过滤器
3　primary fan　主风机
4　air heater　空气加热器
5　steam　水蒸气
6　condensate　冷凝水
7　hot air duct　热空气导管，热风道
8　spray pipe guide　喷雾管导向套
9　drying chamber　干燥室
10　atomizer　雾化器，雾化喷头
11　high pressure pump　高压泵
12　booster pump　增压泵，接力泵
13　feed　进料
14　ambient air　环境空气，大气
15　air conditioner　空气调节器
16　product　产品
17　floor level　地平面
18　conveying fan　输送风机
19　cyclone dry collector　干粉旋风分离器
20　bag filter　袋式过滤器
21　exhaust fan　排风机
22　centrifugal disk atomizer　（离心）导流转盘式雾化器
23　two-fluid nozzle　气动雾化喷头
24　liquid feed　液体进口
25　gas feed　气体进口
26　fluid nozzle　流体喷嘴
27　air nozzle　空气喷嘴
28　pressure nozzle　压力（雾化）喷头，机械（雾化）喷头
29　core　（雾化）槽芯

6.12.3.2 雾化喷头、喷雾嘴、雾化器 Spray Nozzles (Atomizers)

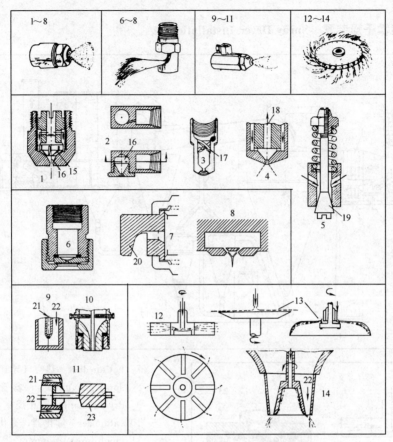

[1~8 pressure nozzle 压力（雾化）喷头，机械（雾化）喷头]

1 grooved core (swirl-spray nozzle) 槽芯旋涡式（喷头）
2 whirl-chamber hollow cone 空锥（形）旋涡室式（喷头）
3 solid cone 整体锥式（喷头）
4 bypass (spill) 回流式（喷头）
5 poppet 动头式（喷头）

6~8 fan-spray nozzle 雾化膜喷头
6 oval orifice 椭圆孔式（喷头）
7 deflector jet 偏流喷射式（喷头），导流喷射式（喷头）
8 impinging jet 冲击喷射式（喷头）

[9~11 two-fluid nozzles (pneumatic nozzle) 气动雾化喷头]
9 two fluid nozzle （内混式）气动雾化喷头
10 converging pneumatic nozzle 外混式气动雾化喷头
11 pneumatic impingement nozzle 冲击式气动雾化喷头

(12~14 rotary atomizer 旋转式雾化器，离心式雾化器)
12 vaned rotating disk 导流转盘式（雾化器）
13 spinning disk 转盘式（雾化器）
14 rotary cup (air-blast bowl atomizer) （鼓风）转杯式（雾化器）

15 grooved core （雾化）槽芯
16 swirl chamber 旋涡室
17 insert 内插（雾化）件
18 return flow line 回流通道
19 poppet 动头
20 curved deflector 弧形导流器，弧形偏流板
21 liquid 液体
22 air 空气
23 impinger 冲击（雾化）器

6.12.4 气流（气动）输送干燥器 Pneumatic Conveyor Dryers

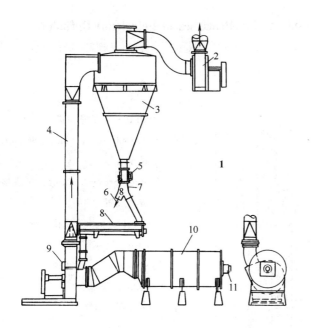

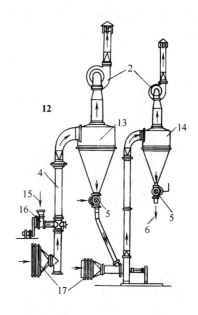

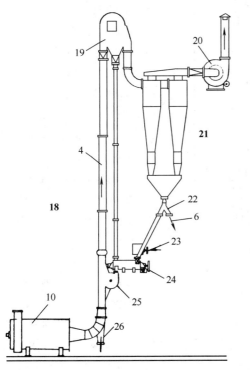

1　single-stage pneumatic conveyor dryer　单段气流输送干燥器
2　vent fan　排风机
3　cyclone　旋风分离器
4　duct　风道，风筒
5　airlock　气闸，旋转阀
6　finished product　成品
7　dry divider　干料分流器
8　wet feed mixer　湿（进）料混合器
9　cage mill（disintegrator）　笼式磨碎机
10　air heater　空气加热器
11　burner　烧嘴，燃烧器
12　two-stage pneumatic-conveyor dryer　两段气流输送干燥器
13　wet-stage cyclone　湿段旋风分离器
14　dry-stage cyclone　干段旋风分离器
15　wet feed　湿（进）料
16　wet feeder　湿料加料器
17　steam air heater　蒸汽加热式空气加热器
18　Strong-Scott flash dryer　斯特朗-斯科特气流闪蒸干燥器
19　classifier　料粒分选器
20　system fan　（干燥）装置排风机
21　multiple cyclone　并联旋风分离器
22　splitter valve　分配阀，分流阀
23　feed-all　进料
24　turbulizer-backmixer　涡流-回混器
25　dispersion sling　抛掷分散器，扬送器
26　tramp discharge　沉底料排出口

6.13 其他静设备及部件 Other Equipments and Components

6.13.1 石油炼制中的流化过程 Fluidization Processes in Petroleum Refinery

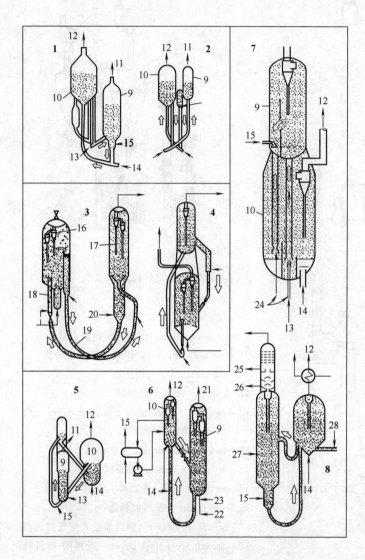

5 Shell unit 壳牌装置
6 SOD hydroformer unit SOD加氢重整装置
7 Kellogg orthoflow unit 凯洛格正流式装置
8 fluid coking unit 流化焦化装置
9 reactor 反应器
10 regenerator 再生器
11 product 产物
12 flue gas 烟气
13 feed 进料
14 air 空气
15 steam 蒸汽
16 cyclone 旋风分离器
17 dip leg 料腿
18 standpipe 立管
19 U bend transfer line U形管
20 stripping steam 汽提蒸汽
21 reformed gas 重整气
22 hydrogen-rich recycle gas 富氢循环气
23 naphtha feed 石脑油进料
24 plug valve 塞阀
25 gas oil 瓦斯油
26 slurry recycle 焦浆循环,油浆循环
27 pitch feed 沥青进料
28 quench water 骤冷水,急冷水

1 SOD (Standard Oil Development) model Ⅱ catalytic cracking unit SOD Ⅱ型催化裂化装置
2 SOD model Ⅲ catalytic cracking unit SOD Ⅲ型催化裂化装置
3 SOD model Ⅳ catalytic cracking unit SOD Ⅳ型催化裂化装置
4 UOP (Universal Oil Products) stacked unit UOP烟囱式装置

6.13.1.1 流态化 Fluidization

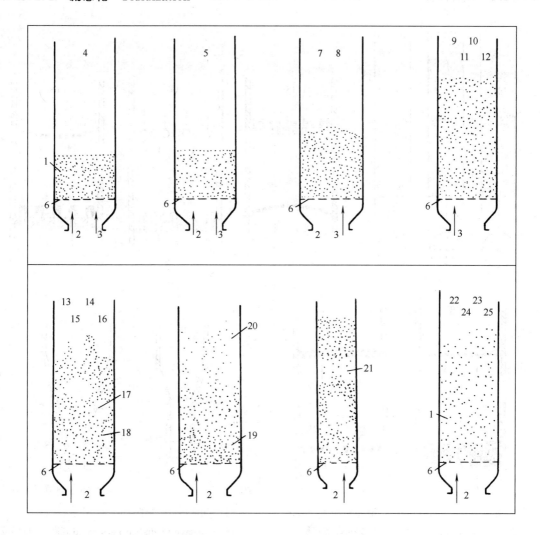

1　fine particle　微细颗粒，细粉颗粒
2　gas　气体
3　liquid　液体
4　fixed bed　固定床
5　expanded bed　膨胀床
6　distributor　分布器
7　incipiently fluidized bed　初始流化床
8　minimum fluidization　临界流态化
9　particulately fluidized bed　散式流化床
10　homogeneously fluidized bed　均一流化床
11　smoothly fluidized bed　平稳流化床
12　liquid fluidized bed　液体流化床
13　aggregative fluidized bed　聚式流化床
14　heterogeneously fluidized bed　非均一流化床
15　bubbling fluidized bed　鼓泡流化床
16　gas fluidized bed　气态流化床
17　bubbling　气泡
18　channeling　沟流
19　dense-phase fluidized bed　密相（流化）床
20　dilute-phase fluidized bed　稀相（流化）床
21　slugging　腾涌
22　disperse-phase fluidized bed　分散相流化床
23　lean-phase fluidized bed　稀相流化床
24　broad fluidization　广义流态化
25　moving bed　移动床

6.13.1.2 流化床分布器 Distributors for Fluidized Bed

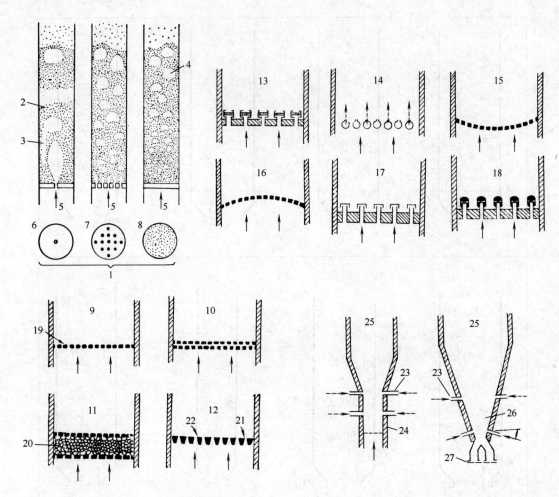

1 quality of fluidization as influenced by type of gas distributor 受气体分布器型式影响的流化质量
2 slugging 腾涌
3 channelling 沟流
4 bubble 气泡
5 gas 气体
6 single orifice plate 单孔板
7 multiorifice plate 多孔板
8 sintered plate 烧结板
(9～18 types of distributors 分布器型式)
9 single perforated plate 单层多孔板
10 staggered perforated plate 错列式多孔板
11 sandwiched packed bed 填充夹层
12 grate bars plate 炉栅板
13 multiple filter plate 复式过滤器板
14 pipe grid 管栅
15 concave or dished perforated plate 凹形或碟状多孔板
16 convex perforated plate 凸形多孔板
17 bubblecaps 泡罩
18 nozzles 喷嘴
19 wire mesh 丝网
20 granular material 粒状材料
21 grate bars 炉栅
22 slits 缝
23 side mixing nozzles 侧向混合嘴
24 side wall 侧壁
25 Winkler generator 温克勒发生炉
26 teetered bed 搅拌床层,浅流化床层
27 ash removal 除灰

6.13.2 壳体和封头　Shells and Heads

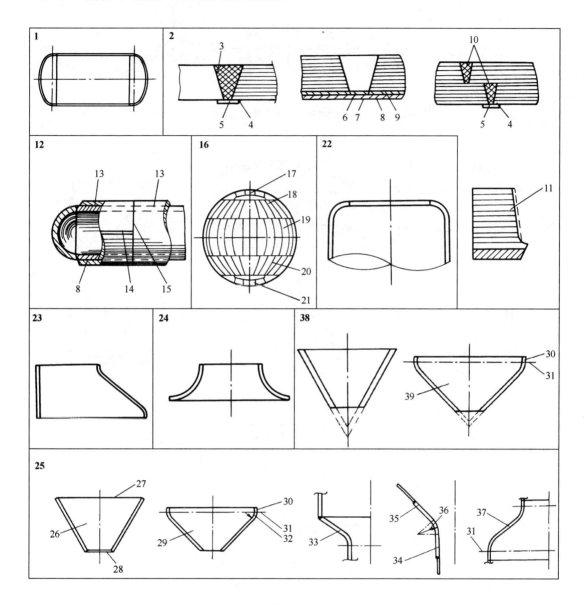

1　single wall cylindrical shell　单层圆筒
2　wrapped layered cylindrical shell　多层包扎圆筒
3　buttered weld　打底焊层
4　tack weld　定位焊
5　backing strip　垫板
6　dummy insert　垫板
7　longitudinal inner shell seams　内筒纵焊缝
8　inner shell　内筒
9　dummy layer　盲层
10　longitudinal layer seams　层板纵焊缝
11　welded cladding　堆焊层
12　shrink fit cylindrical shell　热套圆筒
13　outer shell　外筒
14　longitudinal outer shell seams　外筒纵焊缝
15　circumferential outer shell seams　外筒环焊缝
16　spherical shell　球壳
17　top crown　上极板

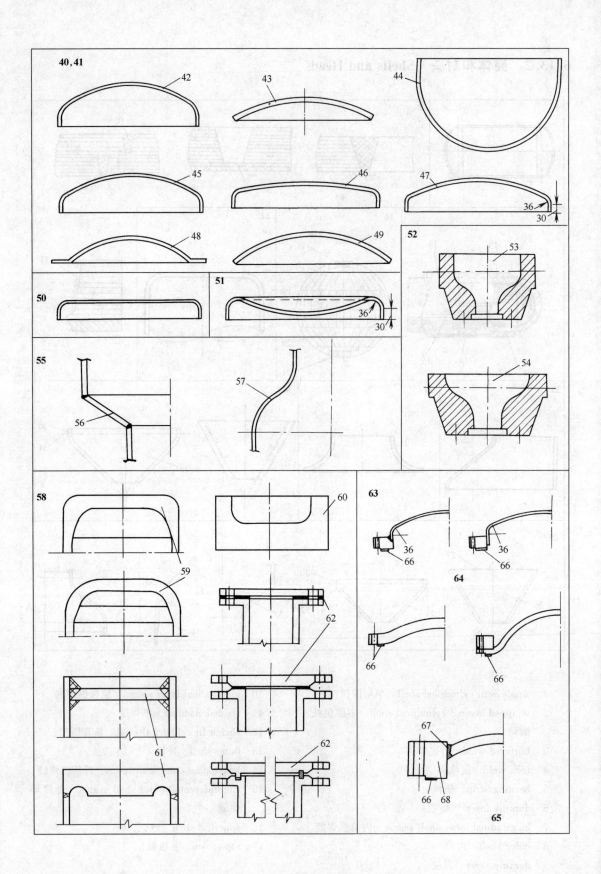

18	north (upper) temperate zone plate 北温带板	44	hemispherical head 半球形封头
19	equator zone plate 赤道带板	45	torispherical head 碟形封头
20	south (lower) temperate zone plate 南温带板	46	flanged shallow dished head 带折边浅碟形封头
21	bottom crown 下极板	47	flanged standard dished head 带折边标准碟形封头
22	jacket shell 夹套壳体	48	flared and dished head 平边碟形封头
23	eccentric cone 偏心圆锥体	49	dished only head 无边碟形封头
24	flare 喇叭口	50	flanged-only head 折边平封头
25	conical shell 锥壳	51	flanged and reverse dished head 带折边倒碟形封头
26	conical section without knuckle 无折边锥壳	52	forged reduced head 锻制紧缩口封头
27	large end of conical section 锥壳大端	53	forged reduced head with skirt 带直边的锻制紧缩口封头
28	small end of conical section 锥壳小端	54	forged reduced head without skirt 不带直边的锻制紧缩口封头
29	toriconical section with knuckle at large end 大端折边锥壳	55	reducer 变径段
30	straight flange 直边	56	conical section 锥形壳体
31	tangent line 切线	57	reducer of reverse-curve type 反向曲线变径段
32	knuckle 折边，过渡段	58	circular flat cover and noncircular flat cover 圆形平盖与非圆形平盖
33	toriconical section with knuckle at small end 小端折边锥壳	59	as an integral part with cylinder 与圆筒成一体（平盖）
34	reinforcing segment at cylinder section 圆筒加强段	60	butt-welded with cylinder 与圆筒对接（平盖）
35	reinforcing segment at conical section 锥壳加强段	61	fillet welded or other welded types with cylinder 与圆筒角焊或其他焊接（平盖）
36	knuckle radius 转角半径	62	bolted connection 螺栓连接（平盖）
37	toriconical section with knuckle 折边锥壳	63	dished head with bolting flange 带法兰的凸形封头
38	conical head 锥形封头	64	slip-on flange type 平焊法兰型
39	flanged and conical dished head 带折边锥形封头	65	integral-flange type 整体式法兰型
40	formed head and spinning head 成型封头与旋压封头	66	gasket 垫片
41	convex head 凸形封头	67	full penetration construction 全焊透结构
42	ellipsoidal head 椭圆形封头	68	centroid 形心
43	spherically dished head 球冠形封头		

6.13.3 破沫器 Demister

6.13.3.1 破沫器及其应用 Demister and Its Applications

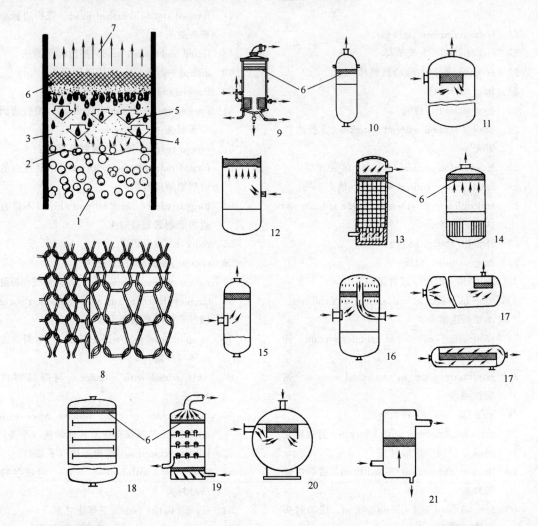

1 bubble 气泡
2 liquid surface 液体表面
3 rising gas stream 上升气流
4 fine spray 雾沫
5 liquid drop 液滴
6 wire mesh demister 破沫网，丝网除沫器
7 gas free from liquid entrainment 无液体雾沫夹带的气体
8 knitted wire mesh 编织（金属）丝网
9 evaporator 蒸发器
10 batch kettle 间歇操作釜
11 oil-gas separator 油气分离器
12 open separator 敞开式气液分离器
13 packed tower 填料塔
14 crystallizer 结晶器
15 vertical separator 立式气液分离器
16 in-line gas scrubber 管道气液分离器
17 horizontal separator 卧式气液分离器
18 absorber 吸收塔
19 fractionating tower 分馏塔，精馏塔
20 spherical separator 球形分离器
21 knock-out drum 气液分离罐，缓冲罐

6.13.3.2 破沫网的安装和纤维除雾气 Installation of Mesh and Fiber Mist Eliminator

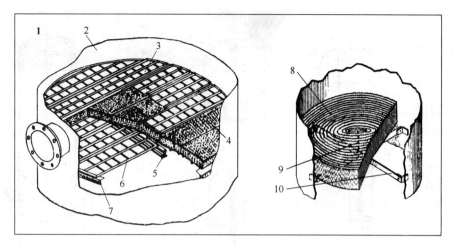

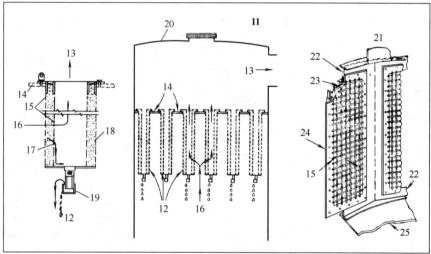

1 installation of mesh 破沫网的安装
2 vessel 容器
3 hold-down bar 压条
4 mesh strip 破沫网条块
5 intermediate support 中央支架，中间横梁
6 bottom support bar 底部支架
7 angle support 角钢支承（圈），支持圈
8 wound mesh pad 缠绕破沫网垫
9 skewer pin 插杆
10 cross bar 十字架
11 fiber mist eliminator 纤维除雾器
12 filter candle (filter element) 滤筒
13 clean gas out 净化气体出口
14 support plate （滤筒）固定板，支持板
15 screen 筛网
16 misty gas 带雾沫气体
17 liquid drainage 排液
18 fiber packing (fiber bed) 纤维填充物，纤维填充床
19 liquid seal pot 液封筒
20 absorption tower 吸收塔
21 H-V Brink mist eliminator H-V 布林克除雾器
22 element supporting structure （过滤）元件支承架
23 glass-fiber packing 玻璃棉填充物
24 gas flow 气流
25 cone bottom 锥底

6.13.4 设备支座 Supports of Equipments

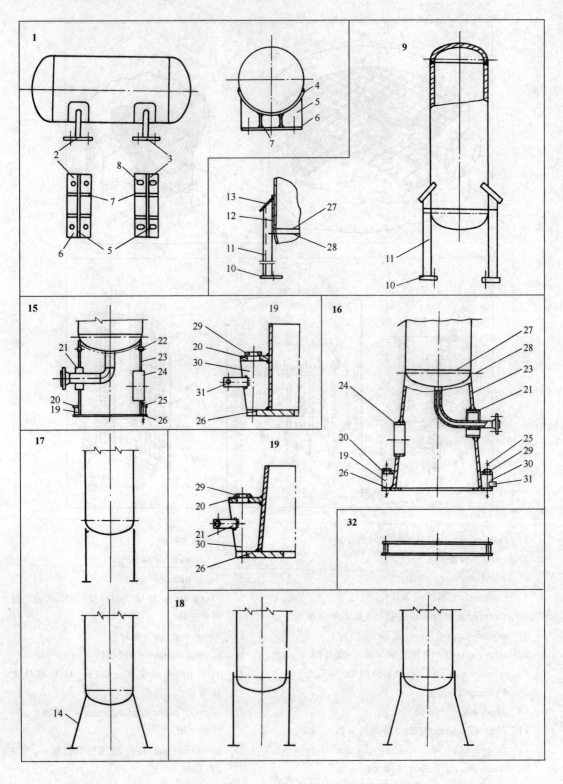

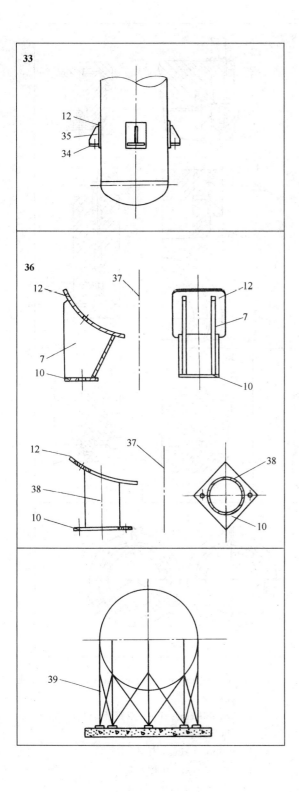

1　saddle support　鞍式支座
2　fixed saddle　固定鞍座
3　sliding saddle　滑动鞍座
4　wear plate　磨耗增强板
5　web plate　腹板
6　bearing plate　底板
7　rib　筋板
8　slotted hole　槽孔，长圆孔
9　leg support　腿式支座
10　base plate　底板
11　leg　支腿，支柱
12　pad plate　垫板
13　cover plate　盖板
14　skirt support　裙式支座
15　cylindrical skirt　圆筒形裙式支座
16　tapered conical skirt support　圆锥形裙式支座
17　right circular skirt　对接式裙座
18　external lapping skirt　搭接式裙座
19　anchor bolt chair　地脚螺栓座
20　compression ring　压环，环形盖板
21　pipe opening　引出孔
22　vent hole　排气管
23　skirt　裙座，塔裙
24　skirt access(access hole)　（裙座）检查孔
25　anchor bolt　地脚螺栓
26　base ring　基础环，底板
27　weld line　焊缝线
28　tangent line　切线
29　washer　压板
30　gusset plate　筋板
31　earth lug　接地板
32　template　模板
33　lug support　耳式支座
34　lug plate　支耳底板
35　lug　支耳
36　bracket support　支承式支座
37　vessel center line　容器中心线
38　steel pipe　钢管
39　support column　支柱

351

6.13.5 密封型式 Types of Seal

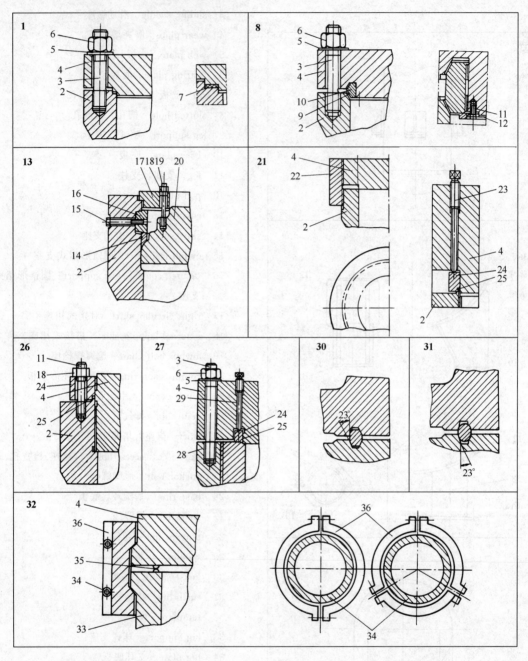

1	metallic flat gasket seal 金属平垫密封	9	soft gasket or soft metallic wire 软垫片或金属丝
2	bolted cylinder end 筒体端部	10	double-cone ring 双锥环
3	main bolt 主螺栓	11	bolt 螺栓
4	flat cover 平盖	12	support ring 托环
5	washer 垫圈	13	Wool seal 伍德密封
6	main nut 主螺母	14	sealing gasket 压垫
7	flat gasket 平垫片	15	stretching bolt 拉紧螺栓
8	double-cone ring seal 双锥密封		

16　combined ring　四合环
17　constraining ring　牵制环
18　nut　螺母
19　constraining bolt　牵制螺栓
20　top cover　顶盖
21　cassale seal　卡扎里密封
22　threaded socket　螺纹套筒
23　check bolt　顶紧螺栓
24　seating ring　压环
25　sealing gasket　密封垫
26　inner threaded cassale seal　内螺纹卡扎里密封
27　improved cassale seal　改良卡扎里密封
28　flanged of bolted cylinder end　筒体端部法兰
29　seating bolt　预紧螺栓
30　octagonal ring seal　八角垫密封
31　oval ring seal　椭圆垫密封
32　clip fastened structure　卡箍紧固结构
33　cylinder end　筒体端部
34　clamp bolt and nut　紧固螺栓和螺母
35　sealing gasket　密封环
36　clip　卡箍

6.13.6　非圆形截面容器　Vessels of Noncircular Cross Section

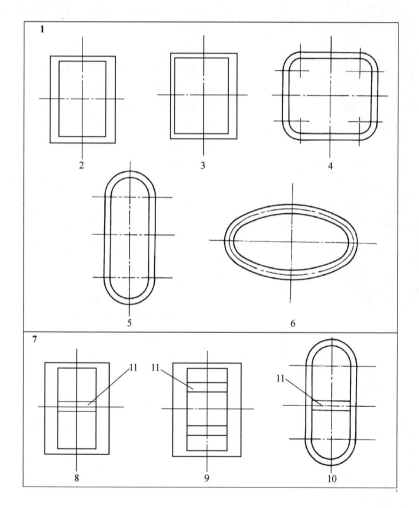

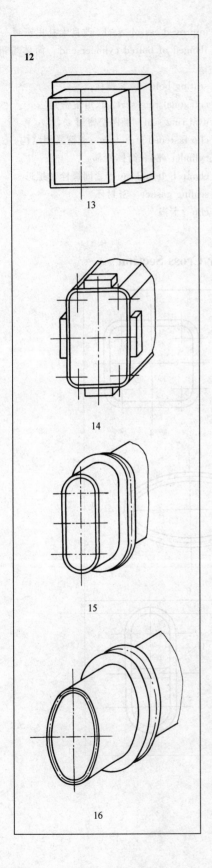

1 non-reinforced vessels of noncircular cross section　无加强的非圆形截面容器
2 vessel of symmetric rectangular cross section　对称矩形截面容器
3 vessel of asymmetric rectangular cross section　非对称矩形截面容器
4 vessel of rectangular cross section with rounded-corners　带圆角的矩形截面容器
5 vessel of obround cross section　长圆形截面容器
6 vessel of elliptical cross section　椭圆形截面容器
7 stayed vessels of noncircular cross section　拉撑加强的非圆形截面容器
8 vessel of symmetric rectangular cross section single-stayed at mid-length　单拉撑加强的对称矩形容器
9 vessel of symmetric rectangular cross section double-stayed at one third points　双拉撑加强的对称矩形容器
10 vessel of obround cross section single-stayed at mid-length　单拉撑加强的长圆形容器
11 stay　拉撑
12 externally reinforced vessels of noncircular cross section　外加强的非圆形截面容器
13 externally reinforced vessels of symmetric rectangular cross section　外加强的对称矩形容器
14 externally reinforced vessel of rectangular cross section with rounded-corners　外加强带圆角的矩形容器
15 externally reinforced vessel of obround cross section　外加强的长圆形截面容器
16 externally reinforced vessel of elliptical cross section　外加强的椭圆形截面容器

6.13.7 其他容器及零部件名称　Nomenclature of Other Vessels and Components

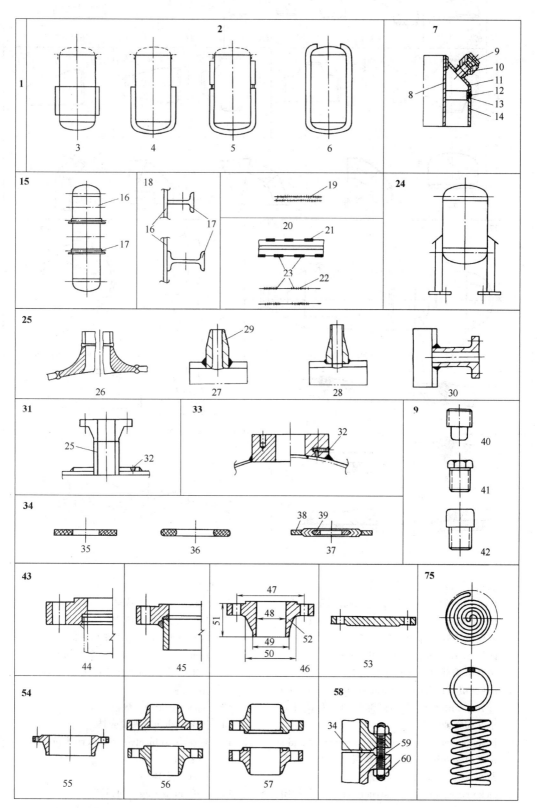

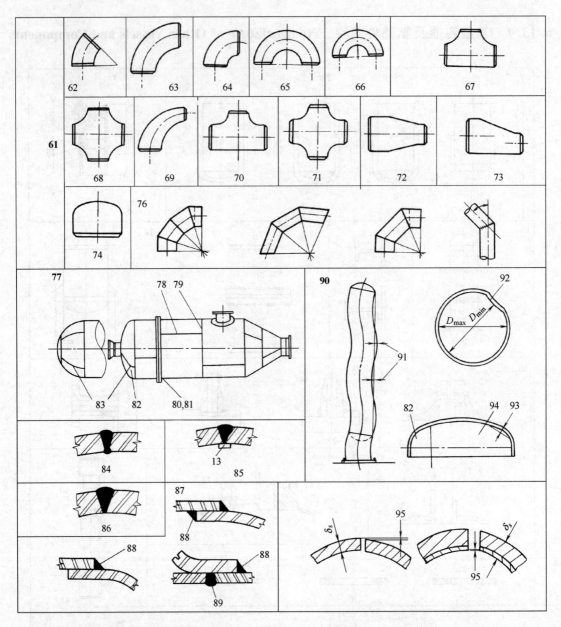

1 jacketed vessel 夹套容器
2 types of jacketed vessel 夹套容器的型式
3 jacket is located at cylinder portion only 夹套仅位于圆筒部分
4 jacket covers a part of cylinder and one head 夹套覆盖部分圆筒和一个封头
5 jacket of cylinder portion is equipped with strut or ring stiffener 在圆筒部分的夹套设有支承或加强环
6 jacket covers whole cylinder, one head and a part of another head 夹套覆盖圆筒，一个封头及另一封头的一部分
7 jacketed enclosing constructions 夹套封闭件结构
8 inner vessel 内壳体
9 plug 螺塞，丝堵
10 screwed coupling 螺纹管接头
11 jacketed enclosing 夹套封闭件
12 full penetration welded construction 全焊透结构
13 backing strip 垫板
14 jacket shell 夹套壳体
15 vessels subjected to external pressure 外压容器
16 shell 壳体
17 stiffening ring 加强圈

18	types of attaching stiffening rings to shell 加强圈与壳体的连接形式	58	bolting 螺栓连接
19	continuous welds 连续焊	59	stud bolt 双头螺柱
20	intermittent welds 间断焊	60	nut 螺母
21	stagger 交错	61	steel butt-welding seamless pipe fittings 钢制对焊无缝管件
22	in-line 并排	62	45° elbow 45°弯头
23	adjacent welds 相邻焊缝	63	90° long radius elbow 90°长半径弯头
24	vessels subjected to internal pressure 内压容器	64	90° short radius elbow 90°短半径弯头
25	nozzle 接管	65	180° long radius return 180°长半径弯头
26,27	inserted nozzle 插入式接管	66	180° short radius return 180°短半径弯头
28	abutted nozzle 安放式接管	67	straight tee 等径三通
29	forged nozzle 锻制接管	68	straight cross 等径四通
30	long welding neck 整体式锻制接管，LWN 法兰	69	90° long radius reducing elbow 90°长半径异径弯头
31	reinforcing pad 补强圈	70	reducing tee 异径三通
32	tell-tale hole 检漏孔	71	reducing cross 异径四通
33	stud-pad (studding outlet, pad flange) 凸缘	72	concentric reducer 同心异径接头，同心大小头
34	gasket 垫片	73	eccentric reducer 偏心异径接头，偏心大小头
35	non-metallic gasket 非金属软垫片	74	cap 管帽
36	double-jacketed gasket (metal-jacket gasket) 金属包垫片	75	coil 盘管
37	spiral wound gasket 缠绕垫片	76	mitered elbow 虾米腰，焊接弯头
38	outer ring 外加强环	77	types of welded joints 焊接接头类型
39	inner ring 内加强环	78	longitudinal joint 纵向接头
40	square head plug 方头丝堵	79	circumferential joint 环向接头
41	hex head plug 六角头丝堵	80	juncture joint of flange-to-shell 法兰与壳体的连接接头
42	round head plug 圆头丝堵	81	non-butt welded joint 非对接接头
43	flange for pressure vessel 压力容器法兰	82	convex head 凸形封头
44	A-type socket-weld flange 甲型平焊法兰	83	patched joints 拼焊接头
45	B-type socket-weld flange 乙型平焊法兰	84	double-welded butt joint 双面焊对接接头
46	welding neck flange 长颈对焊法兰	85	single-welded butt joint with backing strip 带垫板的单面焊对接接头
47	bolt circle diameter 螺栓中心圆直径	86	single-welded butt joint without use of backing strip 不带垫板的单面焊对接接头
48	inside diameter (diameter of bore) 内径	87	lap joint 搭接接头
49	diameter of hub at point of welding (diameter at small end of hub) （法兰）颈部小端直径	88	fillet weld 角焊缝
		89	plug weld 塞焊
50	diameter of hub at base (diameter at large end of hub) （法兰）颈部大端直径	90	tolerance and deviation 公差与偏差
51	overall length (length through hub) 总高	91	straightness tolerance of shell 壳体直线度允差
52	hub 法兰颈部	92	out-of-roundness 圆度
53	blind flange 法兰盖	93	shape deviations of inner surface 内表面形状偏差
54	flange facing 法兰密封面		
55	raised face (R.F.) 平密封面，突面	94	template 样板
56	male and female face 凹凸密封面	95	alignment offset 对口错边量
57	tongue and groove face 榫槽密封面		

6.14 耐火衬里结构 Refractory Lining Constructions

6.14.1 陶瓷纤维衬里结构 Ceramic Fiber Lining Construction

6.14.1.1 陶瓷纤维毯结构 Ceramic Fiber Blanket Layered Construction

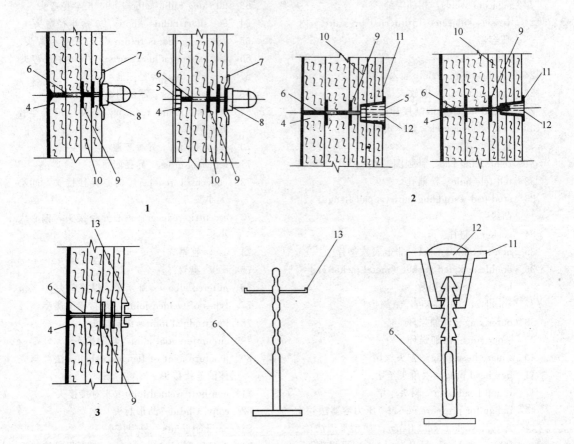

1　stud anchor protected with ceramic cap nut　压盖螺母固定结构
2　stud anchor protected with ceramic cup-lock filled with moldable ceramic fiber　陶瓷杯转卡固定结构
3　stud anchor with turn clip　转卡压盖固定结构
4　back-up layer　背衬
5　special nut　专用螺母
6　twist stud anchor　锚固钉
7　special washer　专用垫圈
8　ceramic cup nut　陶瓷杯螺母
9　speed clip　快速夹子
10　ceramic fiber blanket　陶瓷纤维毯
11　ceramic cup-lock　陶瓷杯
12　ceramic cup-lock cover　陶瓷杯盖
13　turn clip　旋转卡子

6.14.1.2 陶瓷纤维模块结构 Ceramic Fiber Modular Construction

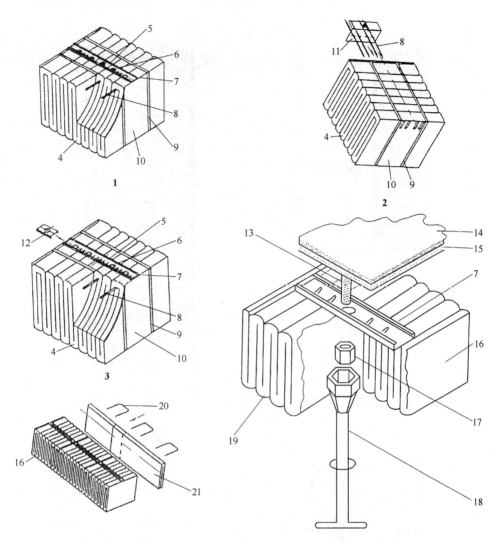

1 centric hole suspended module construction 中心孔吊装式模块结构
2 interpose module construction 剌分式模块结构
3 runner module construction 滑槽式模块结构
4 ceramic fiber blanket 陶瓷纤维毯
5 steel clip 钢夹
6 compressor 压紧件
7 channel frame 槽形架
8 support beam 支承棒
9 bandaging 捆扎带
10 plywood board 胶合（保护）板
11 angle steel 角钢
12 slip block 滑块
13 stud anchor 锚固钉
14 casing plate 炉墙板
15 back-up layer 背衬
16 ceramic fiber module 陶瓷纤维模块
17 special nut 专用螺母
18 box wrench 套筒扳手
19 hot face 迎火面
20 "U" pin U形钉
21 folded batten ("U" type blanket) U形折叠（压缩）条

6.14.2 砖炉衬结构 Fire Brick Lining Construction

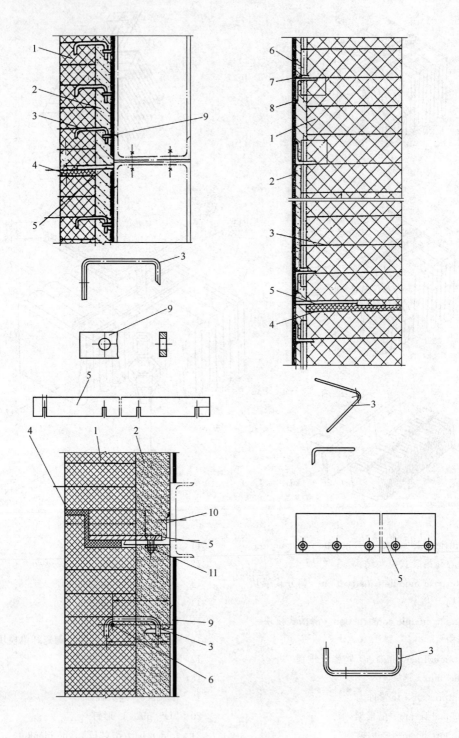

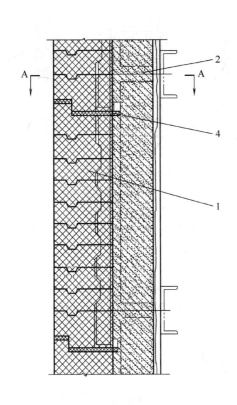
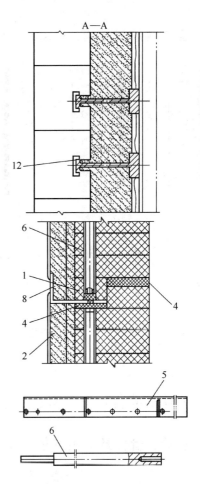
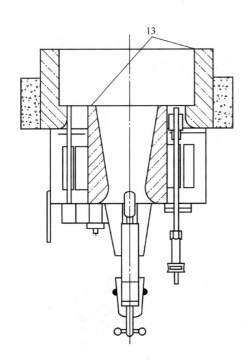

1 firebrick 耐火砖
2 back-up layer 背衬
3 tieback brick anchor 拉砖用钩
4 ceramic fiber（blanket） 陶瓷纤维（毯）
5 support lintel（shelve for support） 托砖板
6 tie rod 拉杆
7 washer 垫圈
8 beam support for tie rod 拉杆座
9 tieback plate 锚板
10 lintel base（angle or plate） 支承板
11 screw and nut 螺钉和螺母
12 tieback brick hanger 拉砖架
13 burner tile 烧嘴砖

6.14.3 浇注料炉衬结构 Castable Lining Construction

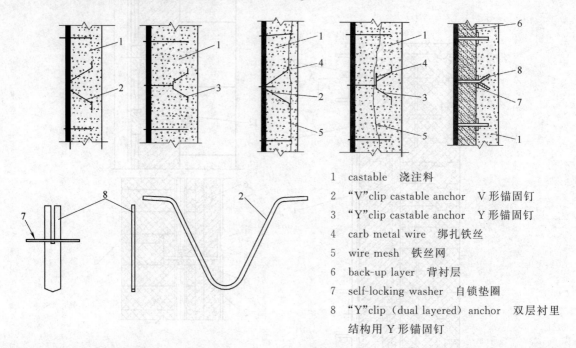

1 castable 浇注料
2 "V"clip castable anchor V形锚固钉
3 "Y"clip castable anchor Y形锚固钉
4 carb metal wire 绑扎铁丝
5 wire mesh 铁丝网
6 back-up layer 背衬层
7 self-locking washer 自锁垫圈
8 "Y"clip (dual layered) anchor 双层衬里结构用Y形锚固钉

6.15 炉用零部件 Furnace Components

6.15.1 炉管、管件及联箱（集合管） Tubes, Fitting and Header (Manifold)

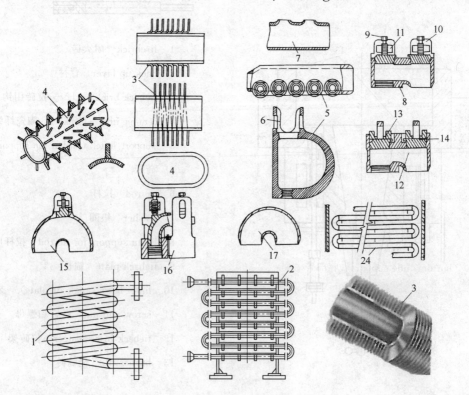

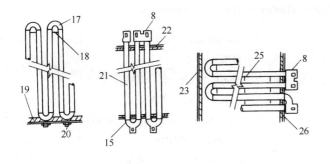

1　helical coil　螺旋盘管
2　serpentine coil　蛇形盘管
3　transverse round fin tube　横向环形翅片管
4　stud tube　钉头管
5　plug-type header　堵头式联管箱
6　plug seat　堵头座圈
7　extrude header　拔头式集合管
8　H return bend（mule-ear closure）box　H形回弯头（骡耳压紧式）
9　set screw　压紧螺钉

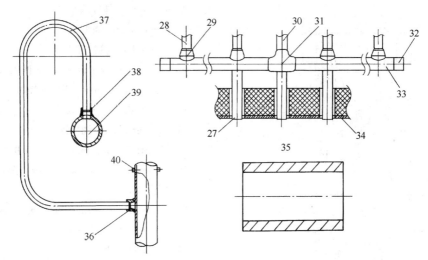

10　mule ear　骡耳
11　dog　回弯头支架
12　H return bend(bull plug closure)　H形回弯头(强力螺母压紧式)
13　bull nut　强力螺母
14　thrust ring　止推环，止推垫圈
(15～17　U return bend　U形回弯头)
15　180°plug header　带堵头的180°回弯头
16　furnace fitting with removable U bend　可拆卸式U形弯头
17　blind U-bend　盲式U形弯头
18　tube support　炉管支架
19，34　floor　炉底
20　bottom tube support　底部管支架
21　vertical tube　立管
22　furnace arch　炉顶
23　side wall　侧墙
24　horizontal tube　卧管
25　intermediate tube sheet　中间管板
26　end tube sheet　端部管板
27　pipe sleeve（guide pipe）　套筒(导向管)
28　catalyst tube　催化剂管
29　branch connection　支管管接头
30　riser　上升管
31　riser tee　上升管三通
32　manifold end cap　集气管端管帽
33　outlet manifold　出口集气管
35　radiant coil tube（harp）　辐射段盘管(竖琴管排)
36　pigtail connection（weldolet or sockolet）　猪尾管管接头
37　pigtail　猪尾管
38　weldolet（sockolet）　管接头
39　inlet manifold　入口集气管
40　support trunnion　支承耳轴

6.15.2 燃烧器（烧嘴）和附属设备 Burners and Accesseries

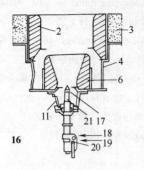
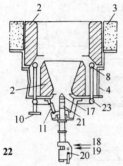
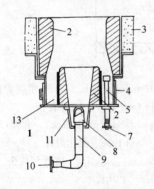
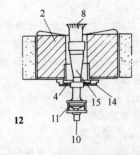
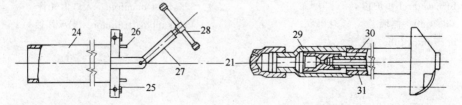

1　gas burner　气体烧嘴，瓦斯燃烧器，瓦斯烧嘴
2　muffle block/tile　烧嘴砖
3　furnace floor　炉底
4　secondary air register　二次风门
5　gas pilot　长明灯
6　plenum/windbox　风箱
7　pilot air shutter　长明灯调火门
8　gas tip　气体喷头
9　burner gun　烧嘴枪
10　gas inlet　燃料气入口
11　primary air register　一次调风门
12　side wall burner　侧壁燃烧器
13　sight glass port　视镜
14　burner plate　烧嘴固定板
15　Venturi tube（mixer）　文丘里管（混合器）
16　steam atomizing oil burner　蒸汽雾化油燃烧器，蒸汽雾化油烧嘴
17　oil nozzle（oil tip）　油喷头，油喷嘴
18　steam inlet　蒸汽入口
19　oil inlet　油入口
20　detaching gear　可卸（油枪）接头
21　oil gun　油枪
22　combination gas and oil burner　油气联合燃烧器，油气联合烧嘴
23　gas manifold　气体集合管，瓦斯集合管
24　guide tube　导向套管
25　oil gun receeiver　油枪固定板
26　guide pin　导向销
27　clevis　插杆
28　clevis handle　插杆手柄
29　atomizer　雾化器
30　oil tube　油管
31　steam tube　蒸汽管

6.15.3 管板及管架 Tube Sheet and Tube Supports

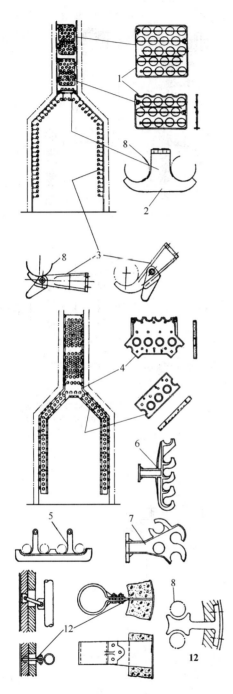

4　end tube sheet　端部管板

5　single-row roof intermediate tube support　单排顶管中间管架

6　intermediate tube support（side wall, single row）　中间管架（侧壁，单排）

7　intermediate tube support（side wall, double row）　中间管架（侧壁，双排）

8　shield tube（shock tube）　遮蔽管

（9～13　tube support for vertical tube　立管管架）

9　bottom tube support　管底支架，管下部支架

10　top tube support　管顶吊架

11　bottom guide（bottom pin and socket type guide）　下部导向架

12　intermediate guide　中间导向架

13　top guide　上部导向架

14　weld　焊缝

15　tube　炉管

16　hanging bolt　吊杆螺栓

17　return bend　回弯头

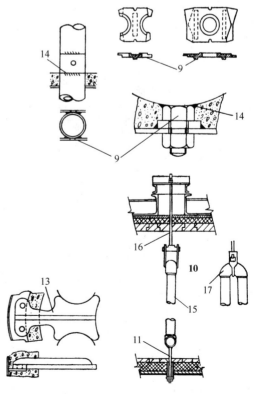

（1～7　tube support for horizontal tube　卧管管架，水平管管架）

1　tube sheet　管板

2　shield shock tube support　遮蔽管管板

3　intermediate tube support　中间管架

6.15.4 其他炉用附件　Furnace Accessories

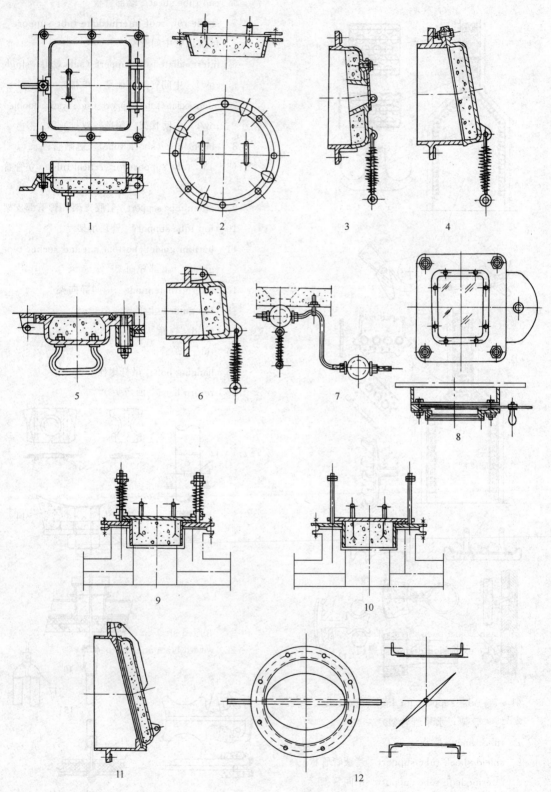

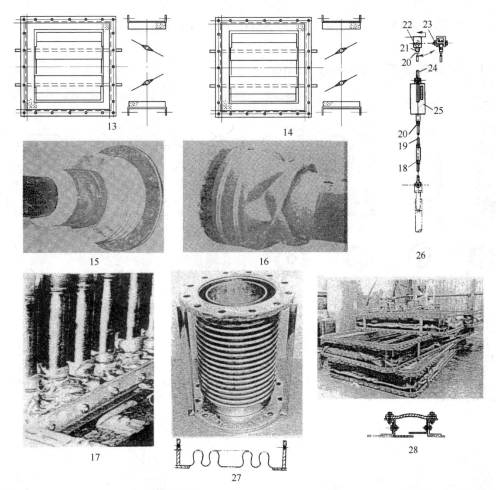

[1，2 manhole (access hole) 炉门]
1　square-type manhole　方形炉门
2　round-type manhole　圆形炉门
(3～8　cast peep door (observation hole)　看火门，视孔)
3　side turn peep door　侧翻看火门
4　rotation-open peep door　回转看火门
5　round-type peep door with packing seal　压紧圆形看火门
6　round-type peep door　圆形看火门
7　floor sphericity peep door　炉底球形看火门
8　flash board peep door　抽板看火门
(9～11　explosion door　防爆门)
9　arch spring explosion door　炉顶弹簧防爆门
10　arch explosion door　炉顶防爆门
11　side-wall explosion door with packing seal　侧壁带填防爆门
(12～14　damper　挡板，风门)
12　single blade damper　单片式挡板
13　multiouver opposed blade damper　多片对开式挡板
14　multiouver parallel blade damper　多片平行式挡板
(15～17　seal boots　密封套)
15　bellows type seal boots　波形密封靴
16　cone type seal boots　锥形密封靴
17　multi-tube bellows　多管波型密封装置
18　turn buckle　松紧螺母
19　hanger rod　拉杆
20　lock-nut　锁紧螺母
21　forged steel clevis　马蹄形挂钩
22　welded beam attachment　焊接在梁上的附件
23　bolt，nut and washer　螺栓、螺母和垫片
24　welded eye rod　焊接环杆
25　spring hanger　弹簧吊架
26　spring hanger assembly　弹簧吊架组合件
(27，28　expansion joint　膨胀节)
27　metal expansion joint　金属膨胀节
28　elastomeric expansion joint　橡胶膨胀节

6.16 炉型及结构　Types and Structure of Furnace

6.16.1 管式加热炉　Tubular Heater

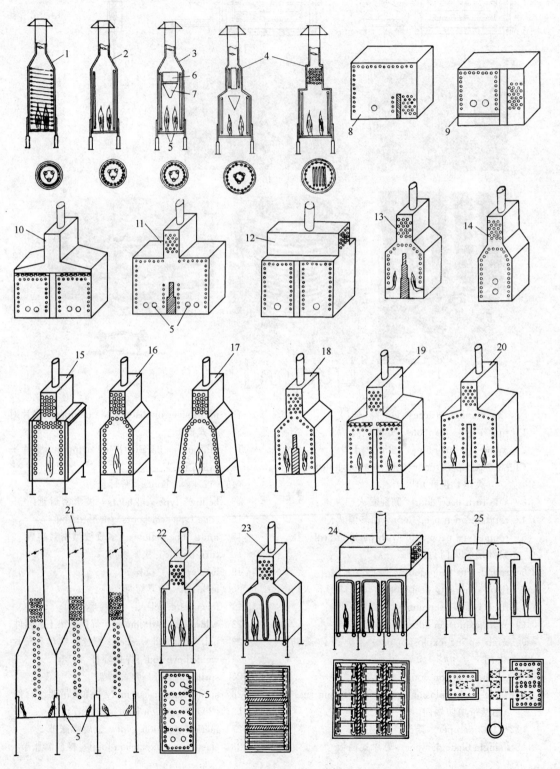

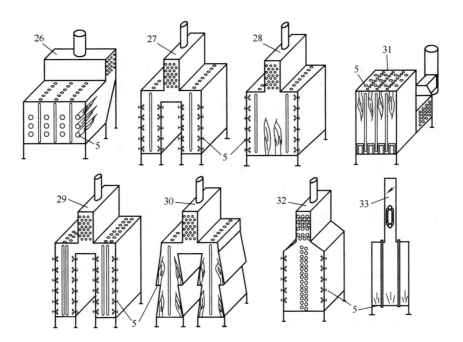

[1～4 vertical cylindrical heater (circular updraft heater) 立式圆筒炉]

1 all-radiant type (helical coil) heater 全辐射型（螺旋管）炉

2 all-radiant type (vertical tube) heater 全辐射型（立管）炉

3 combined radiant and convection type isoflow furnace (Petro-Chem) 辐射-对流式等流圆筒炉（石油化工型）

4 separated radiant and convection type isoflow furnace (Petro-Chem) 辐射-对流分段式等流圆筒炉（石油化工型）

5 burner 燃烧器，烧嘴

6 baffle sleeve 折流套筒

7 reradiating cone (radiant cone) 反射锥，辐射锥

(8～14 box type heater 箱式炉)

8 down flow-convection heater 对流室烟气下行箱式炉

9 up flow-convection heater 对流室烟气上行箱式炉

10 large box heater (Lummus) 大型箱式炉（Lummus 型）

11 multi-combustion chamber heater (Lummus) 多室箱式炉（Lummus 型）

12 multi-combustion chamber heater 多室箱式炉

13 double up-fired heater (UOP) 双室底烧式炉

14 up draft heater 上抽式炉

(15～21 cell type heater 单元组合式炉)

15 straight-up heater (Born) 立式炉（Born 型）

16 up-draft heater (Esso) 上抽式炉（Esso 型）

17 A-frame heater (Kellogg) A 型炉，A 型立式炉（Kellogg 型）

18 double up-fired heater (UOP) (updraft heater) 双室上抽式炉（UOP 型）

19 multi-combustion chamber heater (Lummus) 多室立式炉（Lummus 型）

20 multi-combustion chamber heater (Esso) 多室立式炉（Esso 型）

21 equilux furnace 等热通量炉，均热炉

22 vertical box type heater (Esso) 立式箱型炉（Esso 型）

23 hoop type heater 环形管立式炉，环形炉管重整炉

24 wicket type heater 门式炉

25 BIPM type heater 巴达维亚国际石油公司式炉（带独立对流室）BIMP 式炉

26 Alcom heater Alcom 型多组箱式炉

27 radiant wall type heater (Lummus) 辐射墙式炉（Lummus 型乙烯裂解炉）

28 radiant wall with up fired heater (Esso) 带底烧的辐射墙式炉（Esso 型制氢炉）

29 radiant wall type heater (Topsφe) 辐射墙

369

式炉（Topsφe 型制氢炉）
30　(Foster Wheeler) terrace heater (Foster Wheeler)
　　　梯台炉
31　ICI top fired heater　ICI 型顶烧炉

32　radiant wall type heater (Selas)　辐射墙式炉
33　bottom fired heater (Chemico)　底烧炉(Chemico 型制氢炉)

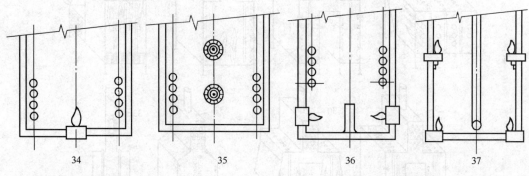

34　up-fired　底烧式
35　endwall fired　端烧式

36　sidewall fired　侧烧式
37　sidewall fired multi-level　多层侧烧式

6.16.2　典型加热炉结构　Typical Structure of Heater

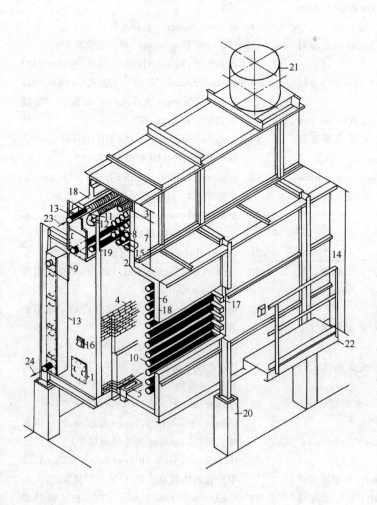

1　access door　人孔门
2　arch　炉顶
3　breeching　尾部烟道（过渡段）
4　bridgewall　桥墙（挡火墙）
5　burner　燃烧器
6　casing　炉壳
7　convection section　对流室
8　corbel　折流体（防烟气短路）
9　crossover　转油线（跨越管）
10　tubes　管子
11　extended surface　扩大表面积管（翅片或钉头管）
12　return bend　180°弯头（回弯头）
13　header box　弯头箱
14　radiant section　辐射室
15　shield section　遮蔽段
16　observation door　观察门（视孔）
17　tube support　管架
18　refractory lining　耐火衬里
19　tube sheet　管板
20　pier　基础柱墩
21　stack　烟囱
22　platform　平台
23　process in　工艺介质入口
24　process out　工艺介质出口

6.16.3 转化炉 Reformers

6.16.3.1 一段转化炉/转化炉 Primary Reformer

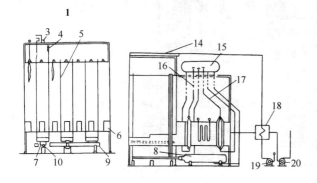

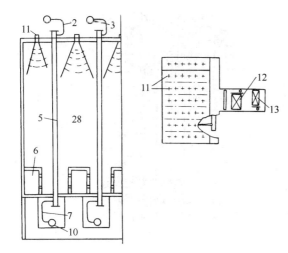

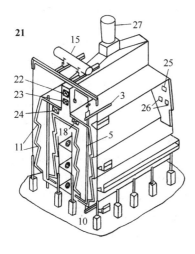

1 power gas type reformer 气体原料转化炉
2 inlet pigtail 入口猪尾管
3 inlet manifold 入口集合管
4 spring hanger 弹簧吊架
5 catalyst tube 催化剂管
6 flue gas duct 烟道
7 outlet pigtail 出口猪尾管
8 reforming gas boiler 转化气废热锅炉
9 outlet header transfer line 出口总管
10 hot water 热集气管
11 burner 燃烧器，烧嘴
12 steam superheater 蒸汽过热器
13 flue gas waste boiler 烟道气废热锅炉
14 burner air header 燃烧空气总管
15 steam drum 汽包
16 riser 上升管
17 downcomer 下降管，降水管
18 air preheater 空气预热器
19 air blower (force draft fan) 空气鼓风机
20 induce draft fan （烟道气）引风机
21 terrace type reforming furnace 梯台式转化炉
22 boiler feedwater heater 锅炉给水加热器
23 steam generator 蒸汽发生器
24 feed preheater 进料预热器
25 explosion door 防爆门
26 peep hole 视孔，看火孔，观察孔
27 stack 烟囱
28 ICI reforming furnace ICI型转化炉

371

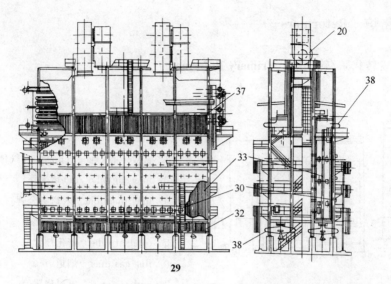
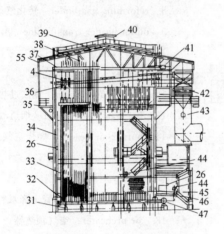
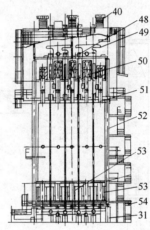

29　Topsφe reforming furnace　Topsφe 型转化炉
30　flat-flame burner　平焰燃烧器
31　pier　基础柱墩
32　reforming gas manifold　转化气集气管
33　catalyst tube（reforming tube）催化剂管（转化管）
34　end wall　端墙
35　plateform　平台
36　air duct　风道
37　inlet（process gas）manifold　入口集气管
38　pigtail pipe　猪尾管
39　rain protection　防雨棚
40　louvers for ventilation　通风百叶窗
41　inlet（process gas）header　入口集合管
42　hot air duct damper　热空气风道蝶阀
43　hot air duct　热空气风道
44　access door（manhole）人孔
45　breeching　过渡段
46　balance burner　平衡烧嘴
47　reforming gas header　转化气集合管
48　fuel gas manifold　燃料气集气管
49　insulation for pigtail（pipe）猪尾管外保温
50　combustion air duct manifold　燃烧空气集气管
51　arch（main）burner　拱顶烧嘴（主烧嘴）
52　side wall　侧墙
53　flue gas tunnel　炉底烟道
54　stairway　梯子
55　eaves gutter　檐沟

6.16.3.2 二段转化炉 Secondary Reformer

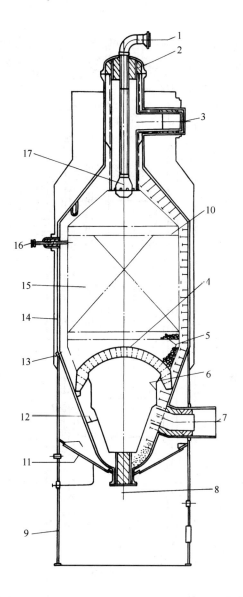

1　air steam in　空气蒸汽入口
2　bubbled alumina　泡沫氧化铝
　　light weight, high purity fused aluminum oxide bubbled type castable refractory　轻质、高纯度熔融氧化铝泡沫型浇注料
3　process gas inlet　工艺气入口
4　bed support dome　床层支承拱
5　high alumina spheres　高含量氧化铝（刚玉）球
6　internal shroud　内罩
7　vapor outlet　气体出口
8　manhole　人孔
9　skirt　裙座
10　hexagonal tile　六角砖
11　bottom jacket closure　夹套底封盖
12　castable thickness　浇注料厚度
13　steam vent pipe　蒸汽排气管
14　water jacket　水夹套
15　catalyst　催化剂（触媒）
16　thermowell　温度计套管
17　air mixer burner　空气混合器

6.16.3.3 换热式转化炉 Reforming Exchanger

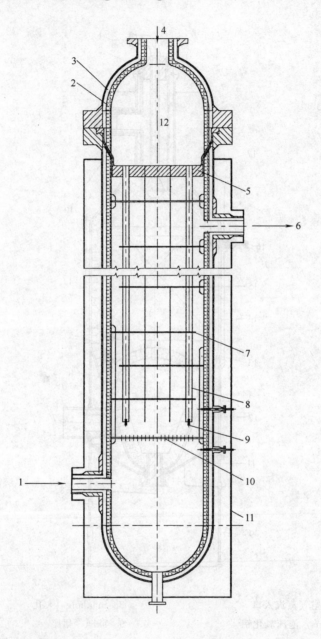

1	autothermal reformer effluent 自热式转化炉出口介质	7	baffle plate 折流板
2	dense layer 重质层	8	catalyst tube 催化剂管
3	back-up layer 隔热层	9	tube cap 管帽
4	feed+steam 原料+水蒸气	10	prorated distributor plate 多孔的分布盘
5	skirt tubesheet 裙式管板	11	water jacket 水夹套
6	to heat recovery 去热回收	12	KBR type reforming exchanger KBR型换热式转化炉

6.16.4 裂解炉 Cracking Furnace

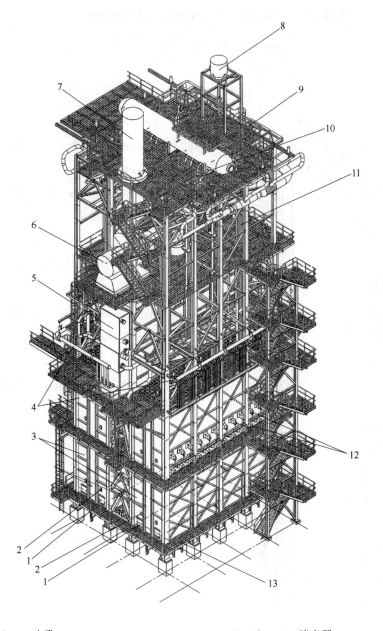

1	access door 人孔		8	silencer 消音器
2	peep hole 视孔		9	steam drum 汽包
3	radiant section 辐射段		10	riser 上升管
4	cross over 转油线		11	linear cooler 线性急冷换热器
5	convection section 对流段		12	sidewall burner 侧墙烧嘴
6	flue gas fan 引风机		13	floor burner 炉底烧嘴
7	stack 烟囱			

6.16.5 气化炉 Gasifier

6.16.5.1 粉煤加压气化炉 Pulverized-coal Pressurized Gasifier

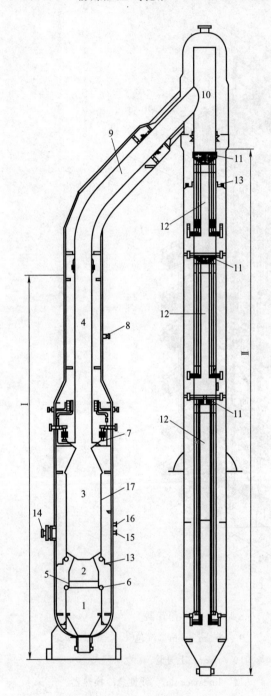

Ⅰ gasifier and quench pipe 气化室和激冷管
Ⅱ syngas cooler 合成气冷却器
1 slag bath 渣池
2 conical slag screen 锥形渣屏
3 reactor 反应室
4 quench pipe membrane wall 激冷管膜式壁
5 heat skirt 圆筒形渣屏
6 spay ring 喷淋环
7 high-velocity quench 高速激冷箱
8 rapping device 敲击器
9 transfer duct 输气管
10 gas reversing chamber 气体返回室
11 support cross 十字支承架
12 syngas cooler wall 合成气冷却器水冷壁
13 sealing 密封隔离罩
14 coal burner 煤烧嘴
15 ignitor 点火烧嘴
16 start-up burner 开工烧嘴
17 reactor membrane wall 反应室膜式壁

6.16.5.2 水煤浆气化炉 Coal-water Slurry Gasifier

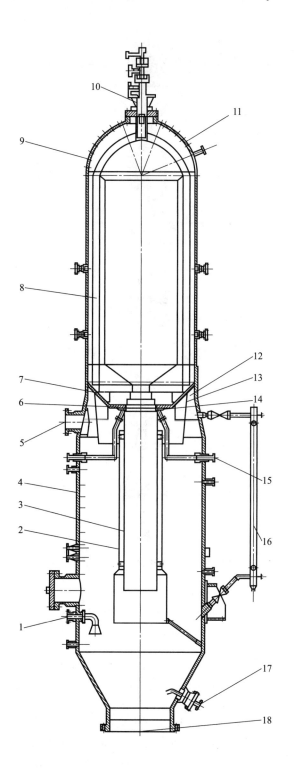

1 blowdown water outlet 黑水出口
2 draft tube 抽气管
3 dip tube 下降管
4 vessel shell 承压壳体
5 quench gas outlet 合成气出口
6 quench ring flange 激冷环法兰
7 refractory support assembly 托砖盘
8 refractory lining 耐火衬里
9 temperature element holding apparatus 表面热电偶固定件
10 burner 烧嘴
11 spherical head 球形封头
12 gusset 托板
13 gas shield 气体遮蔽板
14 quench ring 激冷环
15 quench water inlet 激冷水入口
16 level transmitter channel 液位指示器联箱
17 lock hopper recircling water inlet 锁斗循环水进口
18 slag & water outlet 渣水出口

6.16.5.3 加压固定床煤气化炉 Pressurized Fixed-bed Coal Gasifier

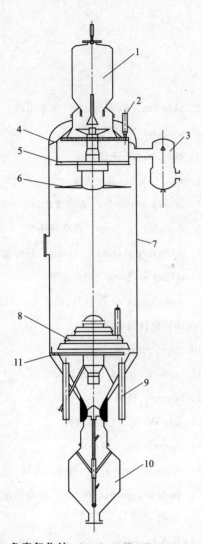

鲁奇气化炉 Lurgi gasification reactor

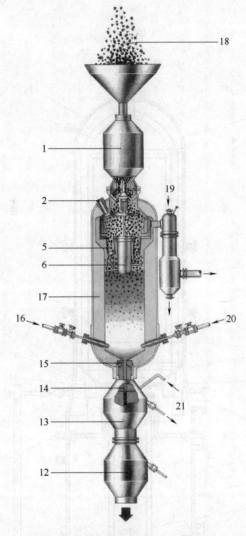

BGL 气化炉 British Gas-Lurgi gasifier

1 coal lock 煤锁斗
2 driver 上部传动装置
3 raw gas offtake & wash cooler 原料气出口 & 洗涤冷却器
4 water jacket 水夹套
5 coal distributor 布煤器
6 stirrer 搅拌器
7 vessel shell 炉体
8 grate 炉箅
9 rotating grate driver 炉箅传动装置
10 ash lock 灰锁
11 scrapper 刮刀
12 slag lock hopper 渣锁斗
13 slag quench chamber 渣激冷室
14 molten slag 熔渣
15 slag tag 液态排渣口
16 steam and oxygen inlet 蒸汽和氧气入口
17 refractory lining 耐火衬里
18 feed/coal 原料/煤
19 aqueous liquor 洗涤液
20 coal fines/liquors/oils, tars 煤粉/水/焦油
21 circulating quench water 循环激冷水

6.16.5.4 流化床煤气化炉　Fluidized-bed Coal Gasifier

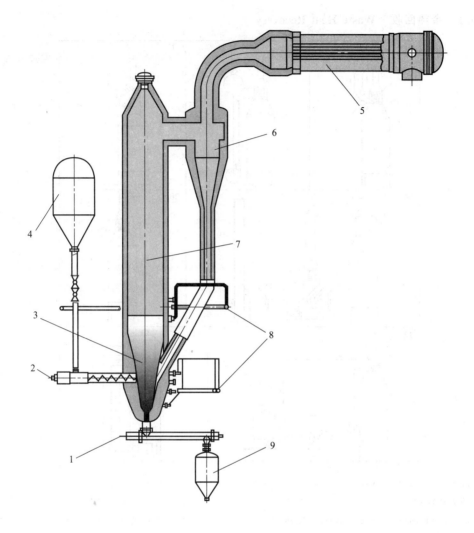

1　cooling screw　冷却螺杆

2　feed screw　进料螺杆

3　fluidized bed　流化床

4　charge bin　进料锁斗

5　waste heating boiler　废热锅炉

6　cyclone　旋风分离器

7　gasification chamber　气化室

8　gasification agents　气化介质（氧气/蒸汽或空气）入口

9　slag lock hopper　渣锁斗（收集器）

6.16.6 余热回收和废热锅炉　Waste Heat Recovery and Waste Heat Boiler

6.16.6.1 余热回收　Waste Heat Recovery

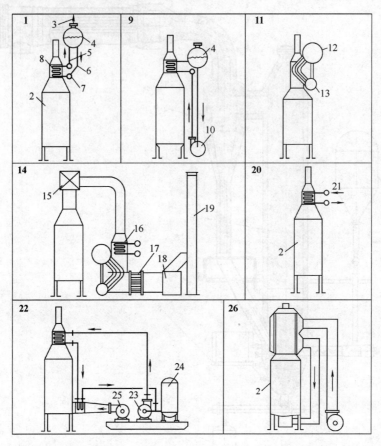

1　natural circulation waste-heat boiler　自然循环废热锅炉
2　heater (furnace, fired heater)　加热炉，火焰加热炉
3　saturated steam　饱和蒸汽
4　steam drum　汽包
5　riser　上升管
6　downcomer　下降管，降水管
7　header　联管箱，集合管
8　boiler tube　锅炉管
9　forced-circulation waste-heat boiler　强制循环废热锅炉
10　circulation pump　循环泵
11　natural-circulation waste-heat boiler with inclined tubes　倾斜炉管自然循环废热锅炉
12　upper drum　上汽包
13　lower drum　下汽包
14　waste-heat boiler serving heaters　几台加热炉共用的废热锅炉
15　common duct　公用烟道
16　superheater　过热器
17　economizer　废气预热器
18　induced draft fan　引风机
19　stack　烟囱
20　economizer oil heater　节油器油加热炉
21　heat carrier　载热体
22　secondary heat recovery with closed-oil circulating system　采用密闭油循环系统的二次热回收
23　oil circulating pump　循环油泵
24　expansion tank　（油）缓冲罐
25　forced draft fan　鼓风机
26　combustion-air preheater　燃烧空气预热器

6.16.6.2 CO 燃烧废热锅炉 CO Firing Waste Heat Boiler

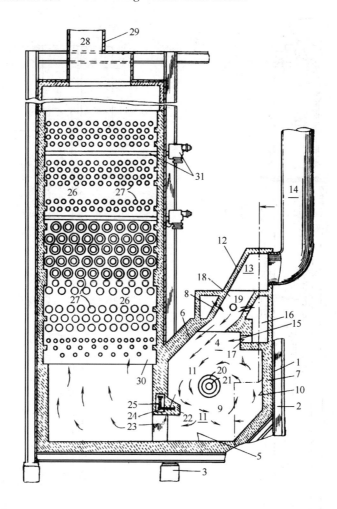

1	furnace 炉膛	17	air inlet port 空气进口
2	steel frame 钢框架	18	end gas conduit 尾气烟道
3	foundation pier 基础墩	19	premix air port 预混空气喷入口
4	side wall 侧墙	20	auxiliary burner 辅助燃烧器
5	floor 底面	21	burner air passage 燃烧器空气通道
6	roof 顶	22	lip 凸缘
7	outward wall 外墙	23	baffle plate 挡板
8	bridge wall 火墙	24	wide flange member 宽凸缘构件
9	combustion chamber 燃烧室	25	heat-sink passage 散热通道
10	high quality refractory brick 优质耐火砖	26	heat-recovery section 热回收段
11	gas stream 烟气流	27	coil 蛇形管，盘管
12	gas shroud 烟气罩	28	stack 烟囱
13	gas plenum 烟气室	29	hood 烟囱罩
14	gas duct 烟气通道	30	screen tube 屏蔽管
15	air jet 空气射流	31	soot blower 吹灰器
16	air plenum 空气室		

6.16.6.3 第一废热锅炉 Primary Waste Heat Boiler

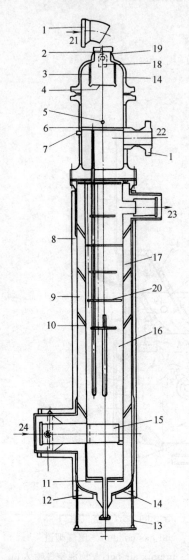

1 R. FW. N. FLG. (raised face welding neck flange) 凸面对焊法兰
2 safety type anchor shackle 安全型锚钩环
3 tie rod 拉杆
4 perforated baffle 多孔挡板
5 lifting eye for inner bundle 内管束吊环
6 face of tube sheet 管板平面
7 vent (plugged) 放气孔（装丝堵）
8 water jacket 水夹套
9 bubble alumina castable 泡沫铝矾土浇注料
10 expansion joint 膨胀节
11 void space 空间
12 drain hole 排污孔
13 skirt 裙座
14 elliptical head 椭圆封头
15 inlet gas distributor 进气分配器
16 liner 衬筒
17 cardboard sleeve 厚纸板套筒
18 lifting lugs 吊耳
19 stub end 短管接头
20 baffle 折流板，挡板
21 boiler water inlet 锅炉水入口
22 boiler water outlet 锅炉水出口
23 process gas outlet 工艺气体出口
24 process gas inlet 工艺气体入口
25 tubesheet 管板
26 outer tube 外管
27 inner tube (bayonet tube) 内管（刺刀管）
28 "nail" spacer 定距钉

6.16.6.4 第二废热锅炉 Secondary Waste Heat Boiler

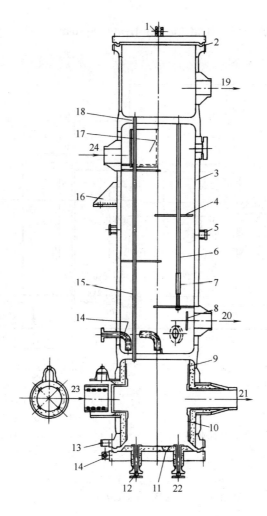

1	lug for lifting cover 管箱盖吊耳	13	guide support 导向支架
2	tongue and groove type flange 榫槽式法兰	14	blowdown pipe 泄放管
3	shell 壳体	15	pipe 管子
4	baffle 折流板，挡板	16	support 支耳
5	trunnion 轴式吊耳	17	outlet weir 出口堰
6	tie rods 拉杆	18	tube sheet 管板
7	spacers 定距管	19	P.G. (process gas) outlet 工艺气体出口
8	impingement baffle 防冲挡板	20	B.W. (boiler water) outlet 锅炉水出口
9	bubble alumina castable 泡沫铝矾土浇注料	21	bypass 旁通管（副线）
10	cardboard 厚纸板	22	blowdown nozzle 泄放管口
11	castable retaining studs 浇注料衬里用锚固钉	23	P.G. (process gas) inlet 工艺气体入口
12	ball joint nozzle and blind flange 球形密封面连接，带盲板	24	B.W. (boiler water) inlet 锅炉水入口

6.16.7 热载体加热炉 Heat Carrier Heater

6.16.7.1 热载体加热炉系统 Heat Carrier Heater System

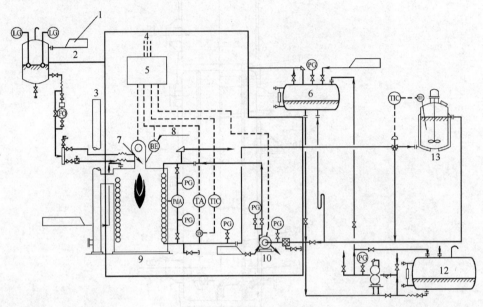

1 oil inlet 油入口
2 spare oil tank 备用油槽
3 stack 烟囱
4 power 电源
5 burner control panel 燃烧器控制盘
6 expansion tank 膨胀槽
7 burner 燃烧器
8 flame detector 火焰检测器
9 heat carrier heater 热载体加热炉
10 circulation pump 循环泵
11 oil ejection pump 注油泵
12 drain tank 排放槽
13 heat consumer 热量用户

6.16.7.2 水（油）浴炉 Water (Oil) Bath Heater

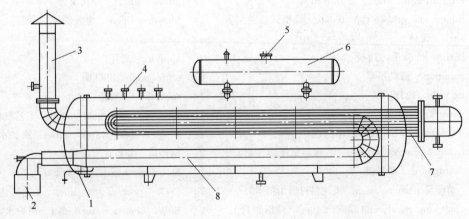

1 removable fire-tube 可拆式火管部件
2 burner 燃烧器
3 stack 烟囱
4 thermometer 温度计口
5 fill hatch 装填口
6 expansion tank 膨胀槽
7 removable coil 可拆式盘管（预热盘管）
8 fire-tube 火管

6.16.7.3 热载体加热炉 Heat Carrier Heater

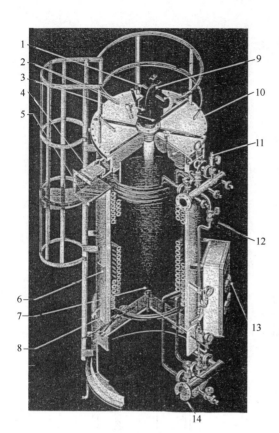

1 burner 燃烧器
2 sight glass 视镜
3 flame detector 火焰检测器
4 explosion door 防爆门
5 flue gas outlet 烟气出口
6 name plate 铭牌
7 manhole 人孔
8 refractory lining 耐火衬里
9 blower 鼓风机
10 purge steam connection 灭火蒸汽口
11 safety valve 安全阀
12 heat carrier inlet 热载体入口
13 burner control panel 燃烧器控制盘
14 heat carrier outlet 热载体出口

7 转动机器 Rotary Machine

7.1 泵 Pump

7.1.1 各种形式的泵 Various Types of Pumps

(1) 各种形式的泵（1） Various Types of Pumps（Ⅰ）

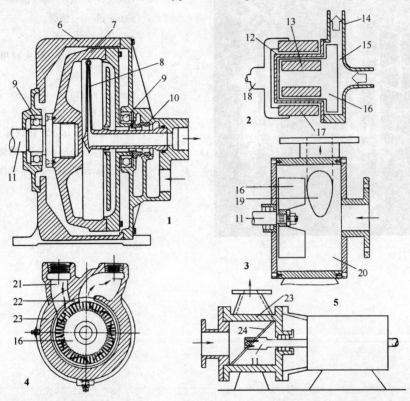

1　rotor jet rotating casing pump　回转壳体泵
2　hermetically sealed magnetic drive pump　封闭式电磁泵
3　vortex pump　旋涡泵，涡流泵
4　turbine pump　涡轮泵
5　inclined rotor pump　斜转子泵
6　rotor housing　转子壳体
7　rotor　转子
8　pitot tube　皮托管
9　bearing　轴承
10　mechanical seal　机械密封
11　shaft　泵轴
12　rear casing　后泵壳
13　impeller magnet　带叶轮的磁铁
14　front casing　前泵壳
15　fluid pumped　泵送流体
16　impeller　叶轮
17　driving magnet　驱动磁铁
18　motor　电动机
19　throat　喉部
20　casing chamber　泵壳腔
21　fluid particles　流体质点
22　stripper　限流板
23　casing　泵壳
24　plate　刮板

(2) 各种形式的泵 (2)　Various Types of Pumps (Ⅱ)

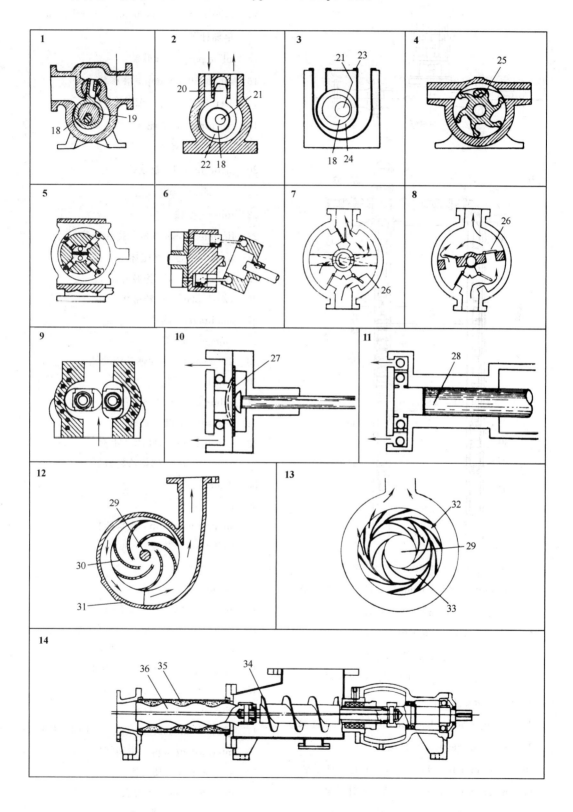

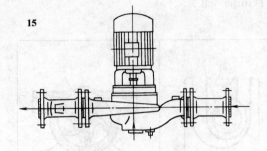
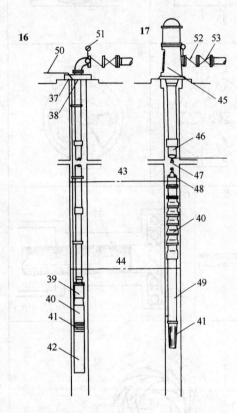

1 cam-and-piston pump 凸轮-活塞泵
2 cam pump (roller pump) 凸轮泵，滚柱泵
3 squeegee pump 挤压泵，胶管泵
4 neoprene pump 氯丁橡胶转子泵
5 radial-plunger pump 径向柱塞泵
6 swash-plate pump 甩板泵，轴向柱塞泵
7 hand pump (wobble pump) 手摇泵
8 quadruple-action hand pump 四作用手摇泵
9 root rotary pump 鲁特旋转泵
10 metering pump diaphragm 膜片式计量泵
11 metering pump piston 活塞式计量泵
12 volute centrifugal pump 蜗壳式离心泵
13 diffuser centrifugal pump 导叶式离心泵
14 progressive cavity pump (spiral pump) 单螺杆泵
15 inline pump 管道泵
16 submersible pump 潜水泵
17 deep well pump 深井泵
18 eccentric 偏心轮
19 piston 活塞
20 pivot journal 枢轴颈
21, 48 shaft 轴
22 roller 滚柱
23 flexible rubber tube 挠性胶管
24 squeeze ring 挤压环
25 neoprene vane 橡胶叶轮
26 swash-plate 旋转甩板
27 diaphragm 膜片
28 plunger 柱塞
29 drive shaft 驱动轴，传动轴
30 impeller 叶轮
31 volute casing 蜗壳
32 diffuser 导叶，扩散器
33 impeller vane 叶片，叶轮
34 screw feeder 螺旋给料器
35 stator 定子
36 rotor 转子
37 supporting clamp 支承压板
38 well cover 井盖
39, 52 check valve 止回阀
40 pump 泵
41 strainer 粗滤器，滤水网
42 submersible motor 潜水（防水）电机
43 natural water level 天然水位
44 pumping water level 泵抽水位
45 motor frame 电机座
46 discharge pipe 排水管
47 guard of intermediate pipe 中间管保护套
49 intermediate pipe 中间管
50 suction pipe 吸入管
51 pressure gage 压力表
53 gate valve 闸阀

（3）各种形式的泵（3） Various Types of Pumps（Ⅲ）

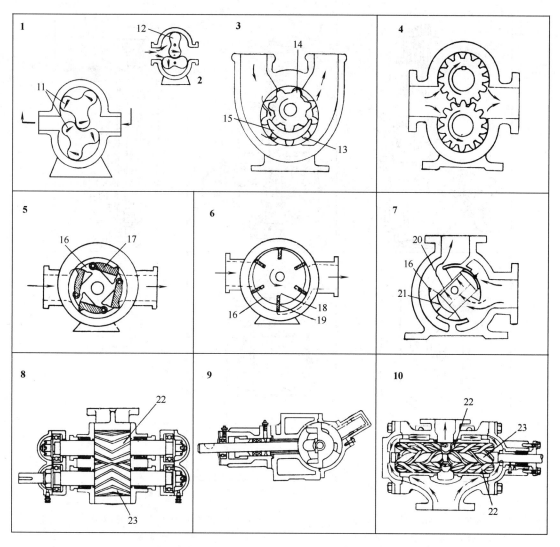

1	three lobe pump 凸叶转子泵	13	internal gear 内齿轮
2	two lobe pump 双凸叶转子泵	14	gear 传动齿轮
3	internal gear pump 内齿轮泵	15	crescent 新月型腔
4	external gear pump 外齿轮泵	16	rotor 转子
5	swinging vane pump 转叶泵	17	swinging vane 转叶
6	sliding vane pump 滑板泵	18	guide slot 导向槽
7	shuttle block pump 滑块泵	19	sliding vane 滑片
8	two-screw pump 双螺杆泵	20	shuttle block 滑块
9	universal-joint pump 万向接泵	21	piston 活塞
10	three-screw pump 三螺杆泵	22	idle rotor 从动转子
11	three-lobe rotor 三凸叶转子	23	power rotor 主动转子
12	two-lobe rotor 二凸叶转子		

(4) 各种型式的泵（4）　Various Types of Pumps（Ⅳ）

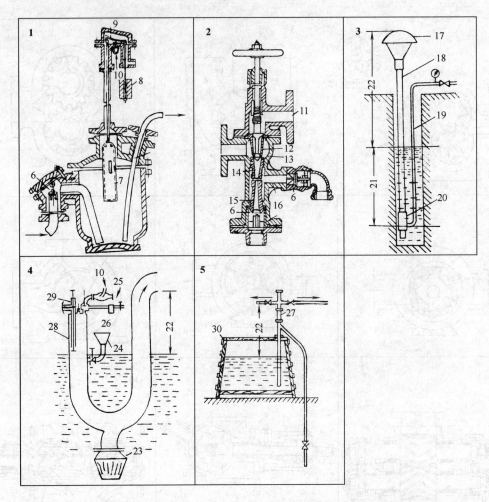

1	automatic acid egg　自动酸蛋	16	condenate outlet　冷凝液出口
2	steam jet ejector　蒸汽喷射器	17	separator　分离器
3	air-lift-pump　气升泵，空气升液器	18	air and liquid mixer　气液混合器
4	Humphrey gas pump　汉弗莱气爆水泵，内燃水泵	19	air pipe　空气管
		20	mixer　混合器
5	siphon　虹吸管	21	submerged length　沉浸深度
6	check valve　止回阀	22	delivery lift　扬水高度，提升高度
7	float　浮桶	23	bottom valve　底阀
8	balance weight　平衡锤	24	funnel　加水斗
9	lever　杠杆	25	gas in　煤气入口
10	air in　进气口	26	igniter　点火器
11	steam inlet　蒸汽入口	27	sight indicator　观察罩
12	steam nozzle　蒸汽喷嘴	28	cylinder　圆筒
13	suction chamber　吸入室	29	discharge valve　排气阀
14	mixing nozzle　混合喷嘴	30	vessel　储槽
15	diffuser　扩散管		

7.1.2 离心泵 Centrifugal Pump

(1) 离心泵（1） Centrifugal Pump（Ⅰ）

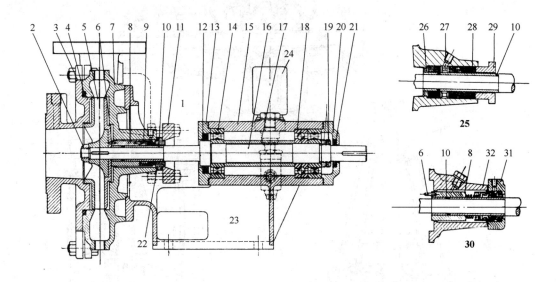

1	centrifugal pump 离心泵		16	shaft 轴
2	impeller nut 叶轮螺母		17	spacer sleeve 定距套
3	suction cover 吸入端泵盖		18	thrust bearing 止推轴承
4	casing O-ring 壳体O形环		19	bearing lock nuts 轴承锁紧螺母
5	impeller 叶轮		21	outer oil seal 外侧（轴承）油封
6	seal driving 动环传动销，密封套拨杆		22	seal joint 密封压环
7	casing 泵壳		23	support 泵支架
8	seal drive sleeve 动环传动套筒，密封传动套		24	constant level oiler 恒液位油杯
9	mechanical seal 机械密封		25	soft packing 软填料
10	shaft sleeve 轴套		26	neckbush 颈衬套
11	seal end plate 密封端压盖		27	lantern ring 灯笼环
12	inner oil seal 内侧（轴承）油封		28	gland packing 填料
13，20	bearing cap 轴承盖		29	gland 压盖
14	journal bearing 径向轴承		30	mechanical seal 机械密封
15	bearing housing 轴承箱		31	seal cover 密封盖
			32	seal 密封

（2）离心泵（2） Centrifugal Pump（Ⅱ）

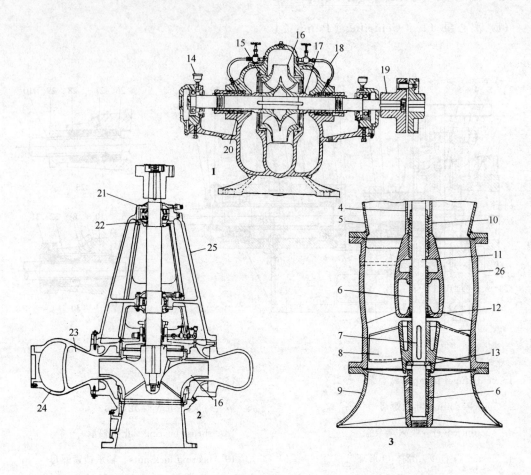

1	horizontal single-stage double-suction volute pump 卧式单级双吸蜗壳泵	13	thrust collar 止推环
2	vertical-shaft end-suction double volute casing pump 立式端吸双蜗壳泵	14	oil cup 油杯
		15	casing ring 泵体密封环
		16	impeller 叶轮
3	vertical wet-pit diffuser pump bowl 带扩压器的立式深井泵	17	shaft sleeve 轴套
		18	seal piping 密封液管
4	shaft-enclosing tube 轴套管	19	coupling 联轴器
5	column pipe 柱状套管	20	seal cage 液封腔
6	bearing bushing 轴承套（衬）	21	shaft collar 轴环
7	propeller key 叶轮键	22	bearing spacer 轴承隔离圈，轴承定距环
8	propeller 叶轮，桨式叶轮	23	double volute 双蜗壳
9	suction bowl 吸入筒	24	casing 泵壳
10	connector bearing 接头轴承	25	frame 悬架
11	pump shaft 泵轴	26	discharge bowl 排出筒
12	seal 密封		

(3) 离心泵 (3) Centrifugal Pump (Ⅲ)

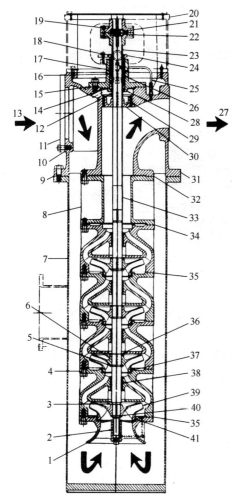

vertical multistage centrifugal pump with barrel casing
具有筒型泵壳的立式多级离心泵

1　sectional bell　钟形吸入口
2　lower sleeve bearing　下套筒轴承
3　first stage snap ring　第一级卡圈
4　diffuser　扩压器
5　first stage impeller retaining collar　第一级叶轮挡圈
6　second stage impeller and above　第二级叶轮并往上类推
7　tank　筒体
8　spacer column　定距接管
9　intermediate gasket　中垫片
10　blind flange gasket　法兰盲板垫片
11　blind flange　法兰盲板
12　lower shaft sleeve　下轴套
13　inlet　进口
14　stuffing box bushing　填料函衬套
15　upper gasket　上垫片
16　intermediate shaft sleeve　中轴套
17　seal cage　密封腔
18　gland bolt　压盖螺栓
19　complete coupling　整套联轴器
20　motor support column　电动机支柱
21　motor half coupling lock nut　电动机半联轴器的锁紧螺母
22　pump half coupling lock nut　泵联轴器的锁紧螺母
23　upper shaft sleeve　上轴套
24　gland　压盖
25　complete piping　连通管
26　packing　填料
27　discharge　出口
28　stuffing box　填料函
29　nozzle head bushing　接管衬套
30　balance disc　平衡盘
31　nozzle head　接管
32　lower gasket　下垫片
33　shaft with key　带键的轴
34　second stage snap ring and above　第二级卡圈并依次类推
35　diffuser ring　扩压环
36　upper sleeve ring　上部轴套密封环
37　first stage impeller　第一级叶轮
38　first stage impeller retaining ring　第一级叶轮挡圈
39　suction bell ring　钟形吸入罩

(4) 离心泵（4）——屏蔽泵 Centrifugal Pump (Ⅳ)—Canned Pump

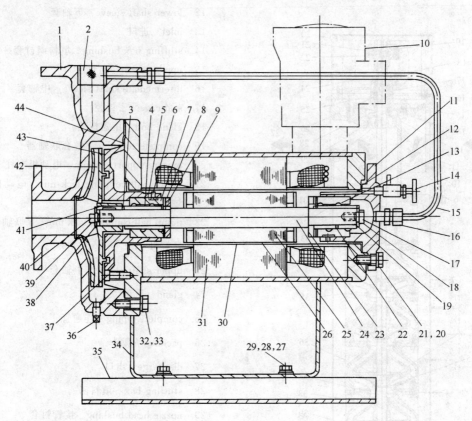

1	casing 泵体	16	union 活接头
2	filter 过滤器	17, 21, 29, 32, 37, 40 screw 螺钉	
3	adjustable gasket 调整垫片	18, 39 thrust nut 止推螺母	
4, 22	shaft sleeve 轴套	24	shaft 轴
5, 19, 20, 27, 28, 33, 38 gasket 垫片		25	rotor 转子
6	cock screw 紧定螺钉	26	rotor can 转子屏蔽套
7, 13	bearing 轴承	30	stator can 定子屏蔽套
8, 23	thrust collar 推力环	31	stator 定子
9	pin 销	34	pump support 机架
10	terminal box 接线盒	35	baseplate 底座
11	RB cover RB端盖	36	plug 塞子
12, 44	seal washer 密封垫圈	41	key 键
14	vent valve 排气阀	42	impeller 叶轮
15	circulation pipe 循环管	43	FB cover FB端盖

（5）离心泵（5）——磁力泵　Centrifugal Pump（Ⅴ）—Magnetic Pump

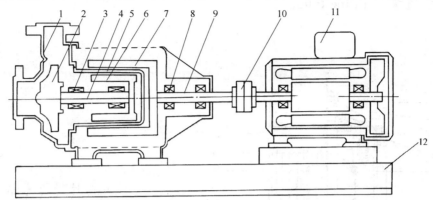

1　casing　泵体
2　impeller　叶轮
3　journal bearing　滑动轴承
4　inner pump shaft　泵内轴
5　shroud　隔离套
6　inner magnetic body　内磁钢
7　outer magnetic body　外磁钢
8　thrust bearing　止推轴承
9　driving shaft　驱动轴
10　coupling　联轴器
11　motor　电机
12　baseplate　底座

7.1.3　轴流泵　Axial Flow Pump

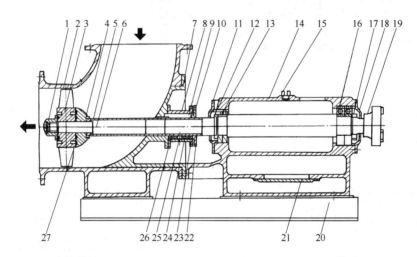

1　impeller nut　叶轮螺母
2　elbow　泵体
3　impeller　叶轮
4　cover　封头
5　shaft sleeve　前轴套
6　shaft　轴
7　flange support　法兰支架
8　mechanical seal housing　机械密封箱
9　mechanical seal　机械密封
10　mechanical seal gland　机械密封压盖
11，18　bearing gland　轴承压盖
12，17　nut　螺母
13，16　bearing　轴承
14　bearing housing　轴承箱
15　screw plunger　螺塞
19　coupling　联轴器
20　baseplate　底座
21　bearing housing gland　轴承箱盖
22　packing gland　填料压盖
23　packing housing　填料箱
24　packing　填料
25　lantern ring　灯笼环
26　shaft sleeve　轴套
27　O-ring　O形环

7.1.4 管道泵 Inline Pump

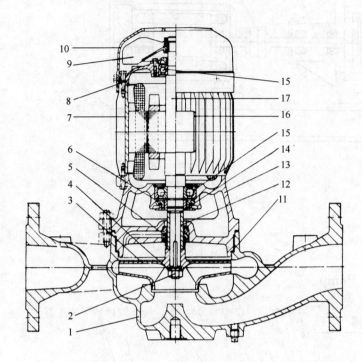
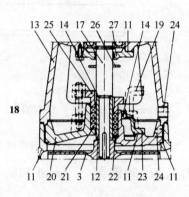

1	volute casing	蜗壳
2	impeller nut	叶轮螺母
3	impeller	叶轮
4	lock washer	锁紧垫圈
5	mechanical seal	机械密封
6	labyrinth seal	迷宫式密封
7	motor casing	电动机壳体
8	casing stud	壳体螺柱
9	fan impeller	风扇叶轮
10	cowl	风扇罩
11	flat gasket	平垫片
12	shaft protecting sleeve	轴套
13	bearing bracket	轴承架
14	O-ring	O形环
15	deep groove ball bearing	深槽球轴承
16	terminal box	接线盒
17	rotor	转子
18	soft-packed stuffing box	软填料填料函
19	cooling water outlet and inlet	冷却水进出口
20	discharge cover	排出端泵盖
21	stuffing box packing	填料
22	throat ring	喉部垫环, 填料衬环
23	cooling compartment cover	冷却室盖
24	leakage fluid drain	排液口
25	stuffing box gland	填料压盖
26	thrower	甩水圈, 抛油环
27	bearing cover	轴承盖

7.1.5 混流泵　Mixed Flow Pump

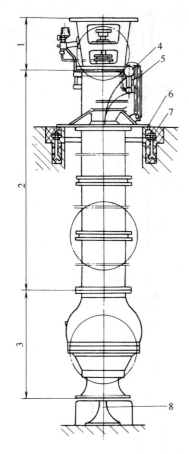
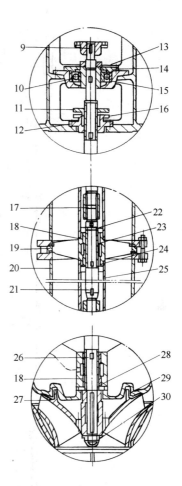

1　drive stool　驱动座，（电动机）支座
2　rising main　扬水主管
3　pump part　泵部分
4　pressure-vacuum gauge　压力-真空表
5　stop cock　断流旋塞
6　foundation ring　基础环，底座
7　foundation bolt　地脚螺栓
8　entry cone　入口锥体
9　driving shaft　驱动轴
10　anti-friction bearing　减磨轴承，滑动轴承
11　stuffing box gland　填料函压盖
12　stuffing box　填料函
13　loose collar　松配合套
14　bearing cover　轴承盖
15　bearing housing　轴承座
16　stuffing box housing　填料函
17　threaded coupling　螺纹联轴器
18　bearing bush　轴承瓦
19　bearing bracket　轴承架
20　riser pipe　扬水管，提升管
21　intermediate shaft　中间轴
22　retaining ring　卡环
23　shaft protecting sleeve　轴护套
24　O-ring　O形环
25　shaft protecting pipe　轴护管
26　pump shaft　泵轴
27　impeller　叶轮
28　radial shaft seal ring　径向轴封环
29　casing wear ring　泵壳耐磨环
30　impeller nut　叶轮螺母

7.1.6 斜流泵 Diagonal Flow Pump

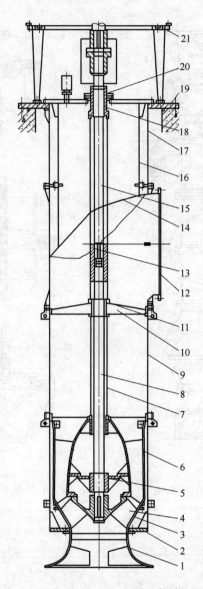

1	suction bell 吸入喇叭口	12	discharge elbow 吐出弯管
2	external column (lower) 外接管（下）	13	internal column (upper) 内接管上
3	impeller house 叶轮室	14	guide vane 导流片
4	impeller 叶轮	15	shaft (upper) 主轴上
5	guide bearing (lower) 导轴承（下）	16	guide vane column 导流片接管
6	diffuser casing 导叶体	17	external column (upper) 外接管上
7	internal column (lower) 内接管（下）	18	guide bearing (upper) 导轴承上
8	shaft (lower) 主轴（下）	19	supporting plate 支承板
9	external column (middle) 外接管（中）	20	stuffing box 填料函
10	bearing bracket 轴承支架	21	motor support 电机支座
11	guide bearing (middle) 导轴承（中）		

7.1.7 螺杆泵 Screw Pump

7.1.7.1 单螺杆泵 Progressive Cavity Pump

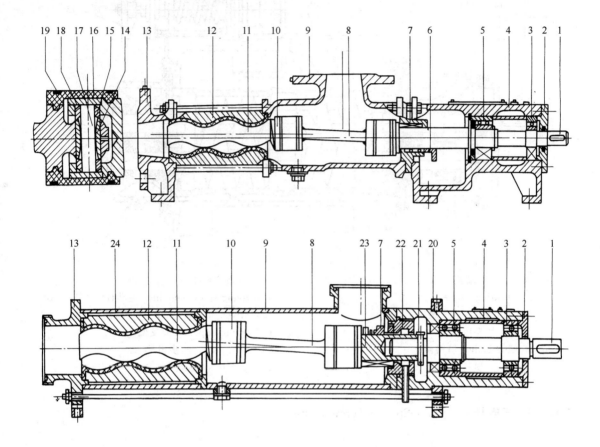

1	drive shaft 传动轴		13	discharge 出口
2	bearing cover 轴承压盖		14	pin bush 销套
3	single-row bearing 单列轴承		15	pin ring 销环
4	bearing housing 轴承箱		16	cover sleeve 护套
5	double-row bearing 双列轴承		17	pin 销
6	stuffing seal 填料密封		18	retaining sleeve 护圈
7	mechanical seal 机械密封		19	clamping band 带卡
8	coupling rod 连接杆		20	tray 托盘
9	suction casing 入口		21	connection pin 销轴
10	cardan join 万向节		22	mechanical seal cover 密封压盖
11	rotor 螺杆		23	hollow shaft 空心轴
12, 24	bush 衬套			

7.1.7.2 双螺杆泵 Twin Screw Pump

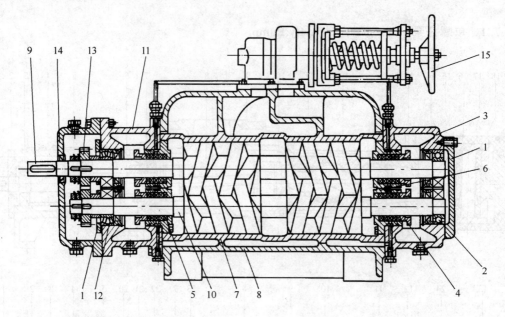

1 end cover   端盖
2, 12 bearing   轴承
3 rear support   后支架
4 gland   填料压盖
5 stuffing body   填料函
6 stuffing   填料
7 insert   衬套
8 casing   泵体
9 driving screw-spindle   主动杆
10 driven screw-spindle   从动杆
11 front support   前支架
13 time gear   同步齿轮
14 gear box   齿轮箱
15 safety-valve   安全阀

7.1.7.3 三螺杆泵 Three-Spindle Screw Pump

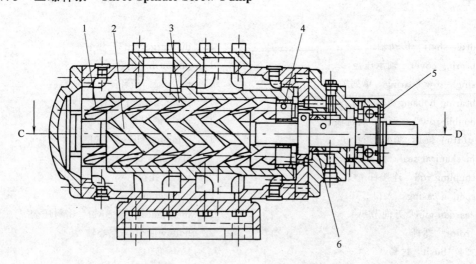

1 driven screw-spindle   从动杆
2 driving screw-spindle   主动杆
3 insert   衬套
4 balance insert   平衡套
5 bearing   轴承
6 mechanical seal   机械密封

7.1.8 计量泵　Metering Pump

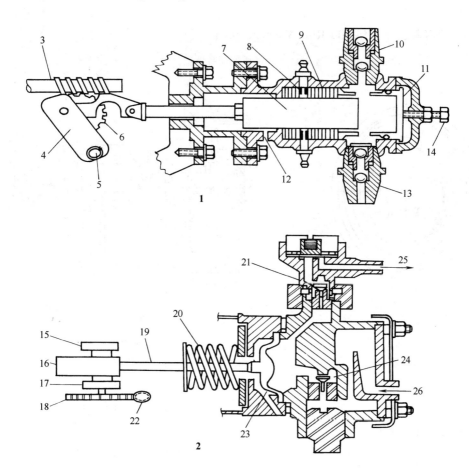

1　plunger (piston) type metering pump　柱塞（活塞）式计量泵
2　diaphragm type metering pump　隔膜式计量泵
3　motor shaft　电动机轴
4　yoke　摆动臂
5　sliding pivot　滑动枢轴
6　gear　齿轮
7　plunger　柱塞
8　lantern ring　灯笼环
9　packing　填料
10　discharge valve stack　出口阀座
11　packing follower　填料随动件，填料压紧盖
12　packing drain　填料漏液排出口
13　suction valve stack　吸入阀座
14　packing take-up blot　填料压紧螺栓
15，17　bearing　轴承
16　eccentric　偏心轮
18　drive gear　驱动齿轮
19　push rod　推杆
20　return spring　复位弹簧
21　discharge valve　排出阀
22　motor shaft with worm　带蜗杆的电动机轴
23　diaphragm　隔膜
24　suction valve　吸入阀
25　outlet　出口
26　inlet　入口

401

7.1.9 真空泵 Vacuum Pump

7.1.9.1 水环真空泵 Liquid Ring Vacuum Pump

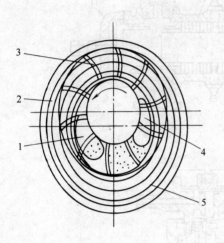
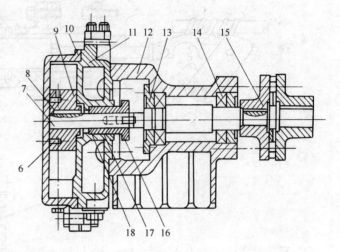

1　suction　吸入口
2　cylinder　气缸
3　impeller　叶轮
4　discharge　排出口
5　liquid ring　液环
6　balance hole　平衡孔
7　shaft　轴
8　key　键
9　working impeller　工作轮
10　cover　泵盖
11　casing　泵体
12　support　机座
13,16　gland　压盖
14　bearing　轴承
15　flexible coupling　弹性联轴器
17　packing　填料环
18　bush　衬套

7.1.9.2 旋片式真空泵 Vane Vacuum Pump

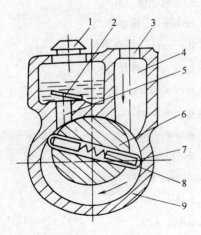
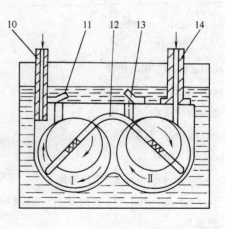

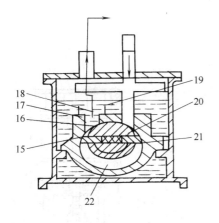
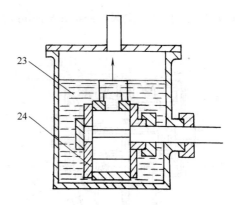
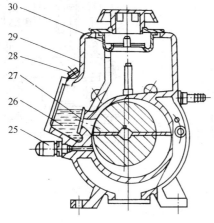
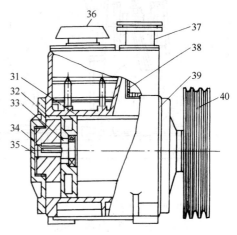

1，36	outlet	排气口	18	valve	活门
2，27	outlet valve	排气阀	20	upper chamber	上面空腔
3	inlet valve	吸气阀	21	column chamber	圆柱形腔
4	inlet channel	吸气管	23	oil reservoir	储油槽
5	outlet channel	排气管	24	surface sealed by oil	用油密封表面
6，22	rotor	转子	25	gas ballast valve	气镇阀
7	vane	旋片	26	oil	油
8	spring	弹簧	28	hexagon plug	六角塞
9，29	casing	泵体	30	oil and vapour separator	油气分离器
10	reference gas valve	掺气阀	31	big rotor	大转子
11	outlet valve for low vacuum stage	低真空级排气阀	32	small rotor	小转子
12	balance pipe	连通管	33	shaft seal	轴封
13	outlet valve for high vacuum stage	高真空级排气阀	34	shaft	轴
			35	gland	端盖
14，19	inlet pipe	进气管	37	inlet	吸气口
15	chamber	空腔	38	strainer valve	过滤阀
16	surface sealed by oil	用油密封表面	39	pump cover	泵侧盖
17	slot way	槽路	40	troughed belt wheel	槽带轮

7.1.9.3 往复式真空泵 Piston Vacuum Pump

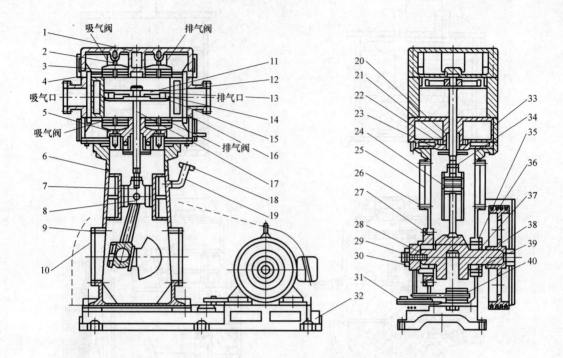

1	upper casing 上泵盖		21	oil seal gasket 油封垫圈
2	valve upper push rod 气阀上顶杆		22	oil seal liner 油封衬
3	valve cover 气阀压盖		23	oil seal liner cover 油封衬压盖
4	valve 气阀		24	crosshead pin 十字头销
5	valve lower push rod 气阀下顶杆		25	small oil window sight glass 小油窗视镜
6	frame 机身		26	small oil window cover 小油窗压盖
7	crosshead 十字头		27	crankshaft blank cap 曲轴闷盖
8	connection rod 连杆		28	crankshaft connection key 曲轴连接键
9	oil window cover 油窗压盖		29	connection link 连接杆
10	oil window sight glass 油窗视镜		30	oil pump 油泵
11	piston washer 活塞垫圈		31	oil drain pipe 放油管
12	piston 活塞		32	baseplate 底座
13	small piston ring 小活塞环		33	lower casing 下泵盖
14	big piston ring 大活塞环		34	piston rod locknut 活塞杆缩紧螺母
15	elastic ring 弹性圈		35	crankshaft through cover 曲轴通盖
16	pump body 泵壳		36	belt roller 带轮
17	piston rod 活塞杆		37	reinforced oil seal 骨架油封
18	oil-fill in pipe cover 加油管盖		38	crankshaft 曲轴
19	oil-fill in pipe 加油管		39	bearing 轴承
20	oil seal 油封		40	oil filter 滤油器

7.1.10 气动隔膜泵 Air Operated Pump

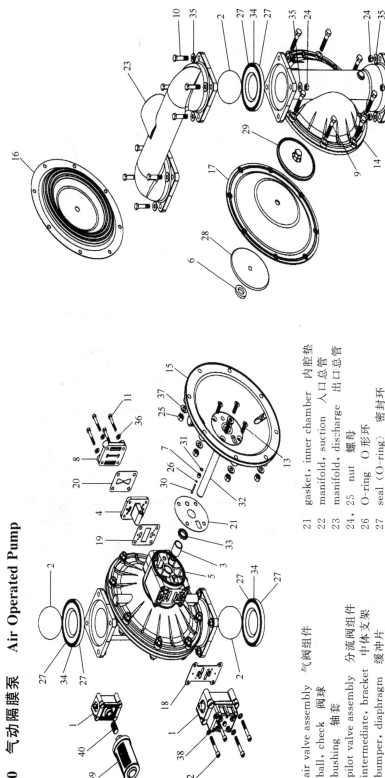

1	air valve assembly	气阀组件	
2	ball, check	阀球	
3	bushing	轴套	
4	pilot valve assembly	分流阀组件	
5	intermediate, bracket	中体支架	
6	bumper, diaphragm	缓冲片	
7	bushing, plunger	柱塞衬套	
8	cap, air inlet	进气罩	
9～13	capscrew	螺钉	
14	chamber, outer	外腔	
15	chamber, inner	内腔	
16	diaphragm, overlay	叠层膜片	
17	diaphragm	隔膜	
18	gasket, air valve	气阀垫	
19	gasket, pilot valve	分流阀垫	
20	gasket, air inlet	空气入口垫片	
21	gasket, inner chamber	内腔垫	
22	manifold, suction	入口总管	
23	manifold, discharge	出口总管	
24、25	nut	螺母	
26	O-ring	O形环	
27	seal (O-ring)	密封环	
28	plate, inner diaphragm	内隔膜	
29	plate, outer diaphragm	外隔膜	
30	plunger, actuator	柱塞执行机构	
31	ring, retaining	支承环	
32	rod, diaphragm	柱塞杆	
33	seal, diaphragm rod	柱塞杆密封圈	
34	seat, check ball	阀座	
35～38	flat washer	平垫	
39	muffler	消声器	
40	nipple	管接头	

7.1.11 往复泵 Reciprocating Pump

7.1.11.1 活塞式往复泵 Reciprocating Piston Pump

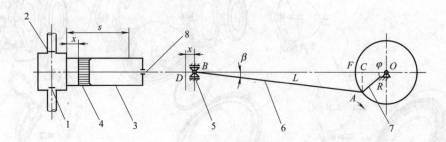

1 suction valve 入口阀
2 discharge valve 出口阀
3 frame body 机身
4 piston 活塞
5 crosshead 十字头
6 connection rod 连杆
7 crankshaft 曲轴
8 stuffing box 填料函

7.1.11.2 双作用蒸汽往复泵 Duplex Acting Steam-Driven Reciprocating Pump

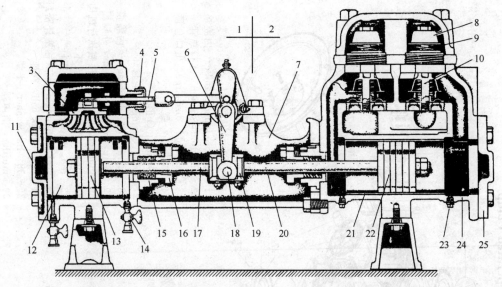

1 steam end 蒸汽端
2 hydraulic end (fluid end) 水力端（流体输送端）
3 slide valve 滑阀
4 gland 压盖
5 valve rod 阀杆
6 connecting rod 连杆
7 internal piece 内支座
8 discharge valve 出口阀
9 valve spring （阀）弹簧
10 suction valve 入口阀
11 end cover 端盖
12 cylinder 汽缸
13 steam piston 蒸汽端活塞
14 cock 旋塞
15 packing 填料
16 gland 压盖
17, 20 piston rod 活塞杆
18 crosshead pin 十字头销
19 crosshead 十字头
21 cylinder liner 汽缸衬
22 liquid piston 液端活塞
23 plug 丝堵，旋塞
24 pump body 泵体
25 pump cover 泵盖

7.1.11.3 双作用活塞式往复泵（皮碗式） Double Action Reciprocating Pump (Bucket Type)

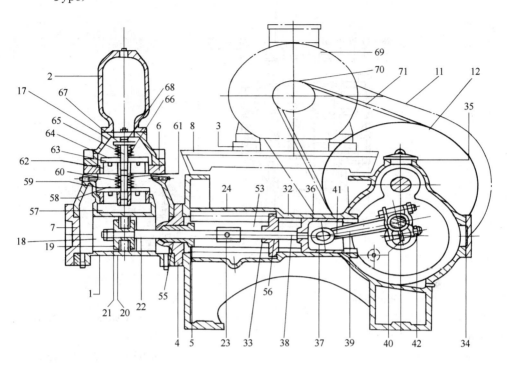

1	pump body 泵体		21	bucket cup 皮碗
2	delivery air vessel 出口储气缸		22	bucket rod and nut 活塞杆和螺母
3	motor support rail 电动机支承导轨		23	rod coupling with set screw 带紧定螺钉的活塞杆联轴器
4	bucket rod stuffing box 活塞杆填料箱		24	trunk 机身，筒体
5	stuffing box gland 填料函盖，压盖		25	cover (pulley side) 泵盖板（带轮侧）
6	delivery valve deck plate 出口阀定位板		26	countershaft bearing cover (pulley side) 传动轴轴承压盖（带轮侧）
7	pump back cover 泵后盖		27	trunk side cover (blank side) 泵盖板（非驱动侧）
8	slide rail 滑轨			
9	suction wing valve 翼形吸入阀		28	countershaft bearing cover (gear side) 传动轴轴承压盖（齿轮侧）
10	wing valve seat 翼形阀座			
11	vee rope guard V形带护罩		29	countershaft bearding (pulley side) 传动轴轴承（带轮侧）
12	vee rope pulley for pump 泵用V形带轮			
13	delivery wing valve 翼形出口阀		30	countershaft bearing (gear side) 传动轴轴承（齿轮侧）
14	delivery wing valve seat 翼形出口阀座			
15	suction wing valve spring 翼形吸入阀弹簧		31	crankshaft bearing cover 曲轴轴承压盖
16	delivery wing valve spring 翼形出口阀弹簧		32	trunk stuffing cover 泵填料压盖
17	pump top cover 泵顶盖		33	trunk stuffing cover nut 泵填料压盖螺母
18	pump liner 泵衬里		34	trunk end cover 泵端盖
19	bucket plate 活塞压板，皮碗压板			
20	bucket center 活塞中心体，皮碗中心体			

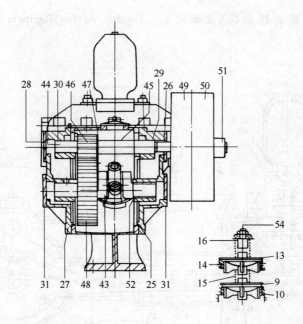

35	trunk oil cap 泵润滑油杯盖		手孔盖
36	crosshead 十字头	54	stop for delivery wing valve 翼形出口阀限位器（挡板）
37	crosshead pin 十字头销	55	packing for bucket rod 活塞杆填料
38	crosshead rod 十字头杆	56	packing for crosshead rod 十字头杆填料
39	connecting rod 连杆	57	suction disc valve seat 盘形吸入阀座
40	connecting rod cap 连杆箍，连杆大头瓦盖	58	suction disc valve 盘形吸入阀
41	connecting rod small end bush 连杆小端衬套	59	valve top plate 阀顶板
42	connecting rod nut and bolt 连杆螺母和螺栓	60	valve spindle 阀杆
		61	valve spring 阀弹簧
43	crankshaft 曲轴	62	valve seat 阀座
44	countershaft 传动轴	63	delivery disc valve 盘形出口阀
45	countershaft sleeve 传动轴套	64	delivery disc valve top plate 盘形出口阀顶板
46	pinion 小齿轮		
47	pinion shroud 小齿轮侧板	65	delivery disc valve spindle 盘形出口阀杆
48	spur wheel 正齿轮	66	delivery disc valve spring 盘形出口阀弹簧
49	fast pulley 固定轮	67	spindle cap 阀杆帽
50	loose pulley 惰轮	68	valve stop 阀升程限制器
51	collar 轴环	69	electric motor 电动机
52	crankshaft bearing 曲轴轴承	70	vee rope pulley for motor 电动机V形带轮
53	trunk handhole cover 筒体手孔盖，机身	71	vee rope V形带

7.1.12 喷射泵 Jet Pumps

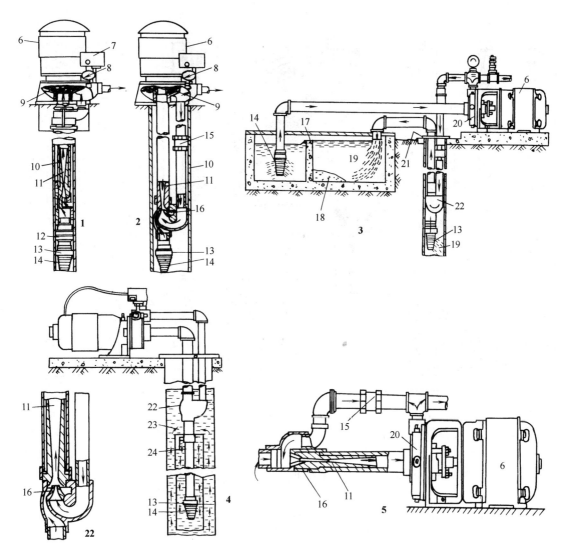

1 one-pipe jet pump 单管喷射泵
2 two-pipe jet pump 双管喷射泵
3 settling chamber (removes sand from water) 沉降池（去掉水中的沙）
4 enclosing foot valve (helps prevent entrance of air bubbles) 套筒式底阀（避免吸入气泡）
5 connecting jet near the pump (to give a steep HQ curve) 泵附近装的喷射器（以获得陡峭的扬程-流量特性曲线）
6 motor 电动机
7 pressure switch 压力开关
8 pressure gauge 压力表
9 pump impeller 泵叶轮
10 wall casing 套管、井管
11 jet diffuser 喷射扩散器
12 well packing 井筒填料
13 foot valve 底阀
14 strainer 粗过滤器
15 slip coupling 滑套接头
16 jet nozzle 喷嘴
17 dam 隔墙，挡水墙
18 sand 沙
19 sandy water 含沙的水
20 single-stage centrifugal pump 单级离心泵
21 pipe clamp 管卡
22 jet assembly 喷射器
23 perforated casing cap 多孔套筒帽
24 tail pipe 尾管

7.1.13 喷射装置　Ejector Units

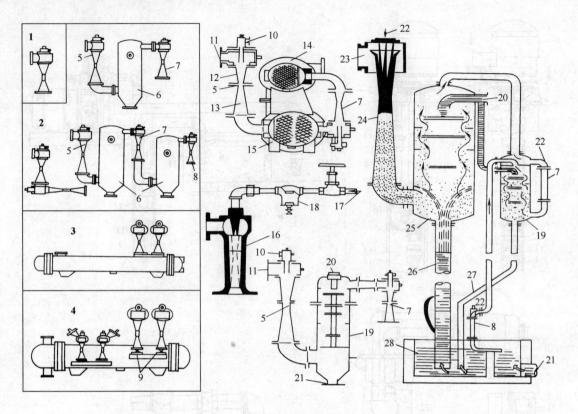

1　single stage, single element ejector　单级单联喷射器
2　multi-stage, single element ejector　多级单联喷射器
3　single stage, twin element ejector　单级双联喷射器
4　multi-stage, multiple element ejector　多级多联喷射器
5　first stage ejector　第一级喷射器
6　intercondenser　中间冷凝器
7　second stage ejector　第二级喷射器
8　third stage ejector　第三级喷射器
9　isolating interstage valve　级间隔断阀
10　steam inlet　蒸汽入口
11　vapour inlet　蒸气入口
12　diffuser inlet　扩散管混合段,扩散管收缩段
13　diffuser discharge　扩散管扩压段
14　surface aftercondenser　表面式后冷凝器
15　surface intercondenser　表面式中间冷凝器
16　single stage non-condensing ejector　非冷凝式单级喷射器
17　motive steam inlet　工作蒸汽入口,动力蒸汽入口
18　strainer　粗过滤器
19　barometric intercondenser　大气式中间冷凝器
20　water inlet　水入口
21　water outlet　水出口
22　operating steam　工作蒸汽
23　air-vapour inlet　空气-蒸气(混合气体)入口
24　booster　主喷射器
25　booster condenser　主冷凝器
26　barometric leg　大气腿
27　tail pipe　尾管
28　hot well　热水(水封)槽,水封槽

喷射器的结构 Ejector Structures

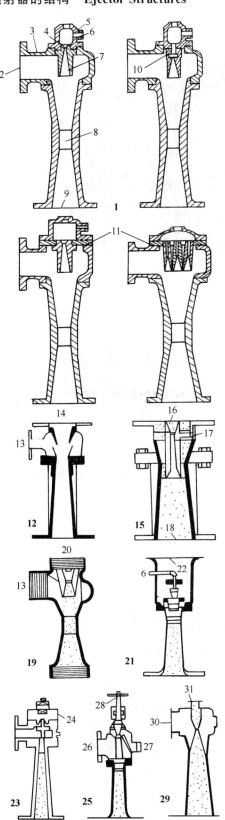

1 steam jet ejector 蒸汽喷射器
2 suction 吸入口
3 suction chamber 吸入室
4 nozzle throat 喷嘴喉部
5 steam chest 蒸汽箱，蒸汽室
6 steam inlet 蒸汽入口
7 steam nozzle 蒸汽喷嘴
8 diffuser throat 扩散管喉部
9 discharge 排出口
10 nozzle extension 喷嘴连接管
11 nozzle plate 喷嘴固定板
12 liquid mixing eductor 液体喷射混合器
13 liquid suction 液体吸入口
14 pressure liquid 压力液体，高压液流
15 desuperheater 过热蒸汽冷却器
16 superheated steam inlet 过热蒸汽入口
17 water suction 水吸入口
18 desuperheated steam outlet 降低过热度的蒸汽出口
19 jet siphon 蒸汽喷液泵，喷射虹吸器
20 condensable vapour 可凝蒸气
21 steam jet blower 蒸汽喷射送风器
22 air inlet 空气入口
23 steam jet vacuum pump 蒸汽抽空器，蒸汽喷射真空泵
24 live steam inlet 新鲜蒸汽入口
25 jet compressor 喷射压气器
26 gas suction 气体吸入口
27 gas inlet 气体入口
28 regulating spindle 调节杆
29 fume scrubber 烟雾洗涤器
30 entrained vapour suction 有夹带物的气体吸入口
31 pressure water inlet 压力水入口

7.1.14 皮托管泵 Pitot Tube Pump

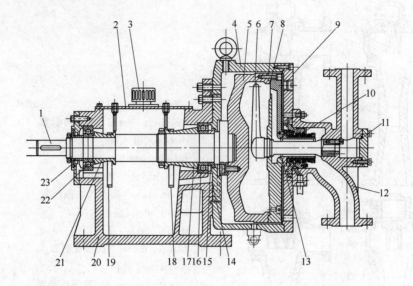

1　pump shaft　泵轴
2　oil tank plate　油箱盖
3　air breather　空气滤清器
4　casing　泵壳
5　rotor cover　转子腔
6　pitot tube　集流管
7　rotor　叶轮
8，11，13，14　O ring　O形圈
9　pump end plate　泵端盖
10　mechanical seal　机械密封
12　inlet and outlet　进出水体
15，23　oil seal ring　封油环
16，22　bearing　轴承
17，21　shaft sleeve　轴套
18，19　oil ring　油环
20　bearing housing　轴承箱

7.2　压缩机、鼓风机和风机　Compressors, Blowers and Fans

7.2.1　螺杆压缩机　Screw Compressor

7.2.1.1　无油螺杆压缩机　Dry Screw Compressor

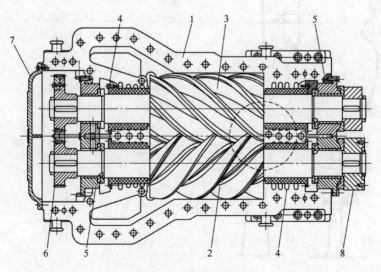

1　casing　壳体
2　male rotor　阳转子
3　female rotor　阴转子
4　shaft seal　轴封
5　radial/thrust bearing　径向/推力轴承
6　timing gear　同步齿轮
7　end cover　端盖
8　drive shaft　驱动轴

7.2.1.2　喷油螺杆压缩机　Oil-injected Screw Compressor

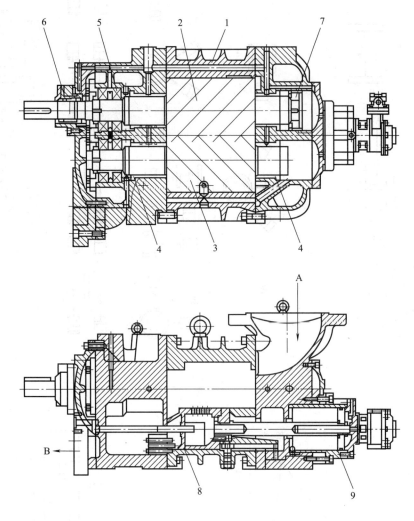

1　casing　壳体
2　male rotor　阳转子
3　female rotor　阴转子
4　radial bearing　径向轴承
5　thrust bearing　止推轴承
6　shaft seal　轴封
7　hydraulic thrust compensating piston　液压推力补偿活塞
8　capacity-control slide valve　负荷控制滑阀
9　double-acting hydraulic piston　双作用液压活塞
A　inlet　进气口
B　outlet　排气口

7.2.2 往复式压缩机　Reciprocating Compressor

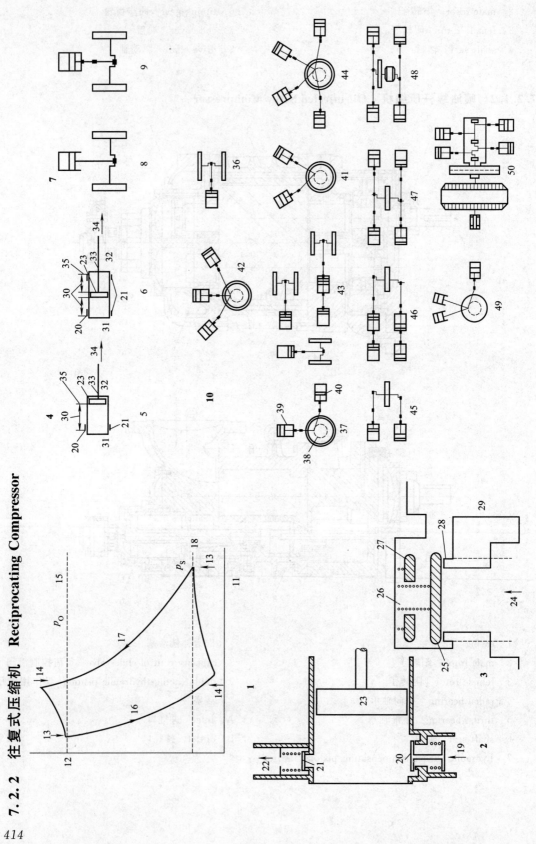

1 compressor indicator diagram 压缩机示功图
2 schematic drawing of a compressor cylinder 压缩机气缸示意图
3 sectional view of suction valve 吸入阀剖面图
4 cylinder action 气缸动作
5 single acting cylinder 单作用气缸
6 double acting cylinder 双作用气缸
7 reciprocating compressor types 往复式压缩机的类型
8 single-acting positive displacement compressor 单作用容积式压缩机
9 double-acting positive displacement compressor 双作用容积式压缩机
10 various types of cylinder arrangement (压缩机) 气缸布置型式
11 volume 容积
12 pressure 压力
13 valve closed 阀关闭
14 valve opening 阀打开
15 discharge pressure 排出压力
16 expansion 膨胀（过程）
17 compression 压缩（过程）
18 suction pressure 吸入压力
19 flow low pressure 低压气流
20 suction valve 吸入阀
21 discharge valve 排出阀
22 flow high pressure 高压气流
23 piston 活塞
24 gas to valve 流向阀的气体
25 valve plate 阀盘
26 return spring 复位弹簧
27 damper plate 升程限制器
28 seat 阀座
29 compressor cylinder 气缸
30 stroke 行程
31 cylinder head end 气缸盖侧
32 crank end 曲轴侧
33 piston rod 活塞杆
34 to driver 接驱动机
35 end of stroke 下止点（行程末端）
36 horizontal reciprocating compressor 卧式往复压缩机
37 angle reciprocating compressor 角式往复压缩机
38 crank shaft 曲轴
39 vertical cylinder 立式气缸
40 horizontal cylinder 卧式气缸
41 V- or Y-type reciprocating compressor V形或Y形往复压缩机
42 W-type reciprocating compressor W形往复压缩机
43 single-frame (straight-line) reciprocating compressor 单列（直线型）往复压缩机
44 semiradial reciprocating compressor 半圆辐射型往复压缩机
45 duplex reciprocating compressor 双列往复压缩机
46 duplex tandem steam-driven reciprocating compressor 双列串联蒸汽驱动往复压缩机
47 duplex four-cornered steam-driven reciprocating compressor 双列四缸对置式蒸汽驱动往复压缩机
48 four-cornered motor-driven reciprocating compressor 四缸对置式电机驱动往复压缩机
49 angle or L-type integral gas or oil engine driven compressor 角型或L型燃气发动机或柴油机驱动的压缩机
50 horizontal opposed reciprocating compressor 卧式对动往复压缩机

7.2.2.1 活塞式压缩机   Piston Compressor

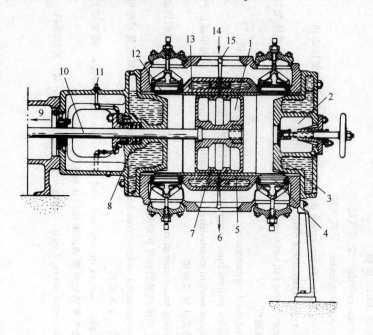

1 piston 活塞
2 head end clearance pocket 气缸端盖全隙腔
3 cylinder head 气缸盖
4 discharge valve 排气阀
5 cylinder 气缸（体）
6 gas outlet 气体出口
7 piston ring 活塞环
8 packing 填料（函）
9 to crank 接至曲拐
10 piston rod 活塞杆
11 packing vent or purge 填料函放空孔或吹扫孔
12 suction valve 吸气阀
13 cylinder liner 气缸衬套
14 inlet 气体进口
15 piston lubrication 活塞注油口

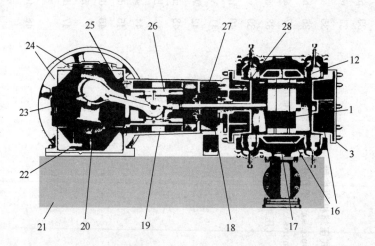

16 cylinder barrel, head and air passage water jacketed for cooling 冷却用气缸缸体、气缸盖和空气通道夹套
17 air passage 空气通道
18 distance piece (allows access to packing and oil-wiper rings) 定距隔板（通过它维修填料和刮油圈）
19 crosshead guide 十字头滑道
20 counterweights 平衡重
21 foundation 基础
22 screened oil suction 过滤油吸入口
23 crankpin and main bearing 曲柄销及主轴承
24 frame 机座
25 die-forged steel connecting rod 模锻钢连杆
26 crosshead 十字头
27 wiper rings (keep crankcase oil out of cylinder) 刮油圈（防止曲柄箱内的油进入气缸）
28 full-floating metallic packing (self-adjusting) 全浮动式金属填料（自动调整）

7.2.2.2 迷宫式压缩机 Labyrinth Compressor

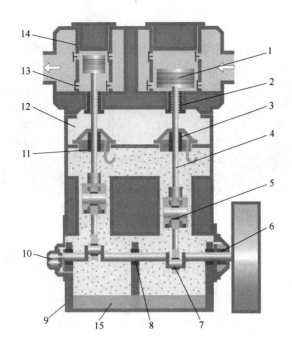

■ cooling water 冷却水

■ gas compressed 气体压缩
  flow region 流动区域

▨ lube oil area 润滑区域

■ process gas 工艺气体

1 piston with labyrinth seal 带迷宫密封的活塞
2 piston rod stuffing with labyrinth seal 带迷宫密封的活塞杆填料
3 oil scraper 刮油环
4 piston rod 活塞杆
5 crosshead 十字头
6 mechanical seal 机械轴封
7 crank shaft 曲轴
8 main bearing 主轴承
9 lube oil tank 润滑油箱
10 gear-type lube oil pump 齿轮式润滑油泵
11 guide bearing 导轴承
12 isolation chamber 中间隔体
13 air valve 气阀
14 cylinder 气缸
15 lube oil 润滑油

7.2.2.3 隔膜式压缩机 Diaphragm Compressor

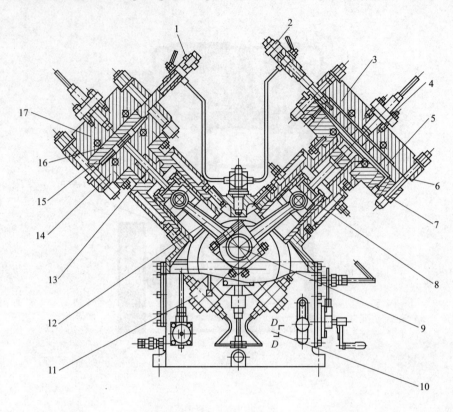

1 second stage pressure regulating valve 二级调压阀
2 first stage pressure regulating valve 一级调压阀
3 first stage cylinder cover 一级缸盖
4 first stage piston 一级活塞
5 first stage diaphragm 一级膜片
6 first stage oil distributor 一级配油盘
7 first stage cylinder 一级缸体
8 cross head 十字头
9 crank shaft 曲轴
10 crank box 曲轴箱
11 oil pump 补油泵
12 rod 连杆
13 second stage plunger 二级柱塞
14 second stage cylinder 二级缸体
15 second stage oil distributor 二级配油盘
16 second stage diaphragm 二级膜片
17 second stage cylinder cover 二级缸盖

7.2.3 低密度聚乙烯（超）高压压缩机　High Pressure Compressor for Low Density Polythylene Process

1	large mechanically driven compressor for very high pressures	大型机械驱动超高压压缩机
2	different types of driving mechanism	驱动机构的各种形式
3	applied to smaller	适用于小型装置上
4	applied to smaller opposed compressor	适用于小型对置压缩机
5	applied to large opposed compressor with front and hind crosshead	大型对置式带主、副十字头的压缩机
6	applied to large frame-crosshead compressor with intermediate guide	适用于大型框架式十字头带中间导向装置的压缩机
7	applied to large frame-crosshead compressor without intermediate guide	适用于大型框架式十字头无中间导向装置的压缩机
8	hind crosshead	副十字头
9	front crosshead	主十字头
10	connecting rod	连杆
11	intermediate guide	中间导向装置
12	frame-crosshead	框形十字头
13	intermediate rod	中间导杆
14	delivery port	排气口
15	suction port	吸入口
16	distance piece	隔离室
17	frame	机身
18	frame top cover	机身顶盖
19	complete connecting rod with bearings and bolts	连杆组件（带轴承和螺栓）
20	blind cover	盲盖
21	complete crosshead frame with slipper	十字头组件
22	crosshead bolt with nut	十字头螺栓及螺母
23	frame tie bolts	机身紧固螺栓
24	upper crosshead frame	十字头上体
25	tungsten carbide plunger	碳化钨柱塞
26	cylinder with shrunk liner	带热压缸套的气缸
27	central valve	中心型组合阀
28	cylinder head	气缸头
29	cylinder assembling bolt	气缸安装螺栓
30	intermediate crosshead frame	十字头中间体
31	crosshead slipper with lower frame	十字头滑板系块
32	hind counter guide	副十字头导轨
33	hind guiding plate	副十字头导板
34	big end bearing	大端轴承盖
35	hind crosshead metal	副十字头（导轨）镶条
36	crankpin	曲柄销
37	fork connecting rod	叉式连杆
38	crosshead pin	十字头销
39	small end bearing	小端轴承盖
40	connecting rod bolt with nut	连杆螺栓及螺母
41	foundation bolt hole	基础螺栓孔
42	front counter guide	主十字头导轨
43	front crosshead metal	主十字头（导轨）镶条
44	front guiding plate	主十字头导板
45	base	底座
46	cylinder bottom	气缸座
47	complete metallic packing	金属填料组件
48	cylinder support	缸体支承
49	crane	吊车
50	jig for plug in or out	装卸用工具

420

7.2.3.1 高压气缸和中心型组合阀 High-Pressure Cylinder and Central Valve

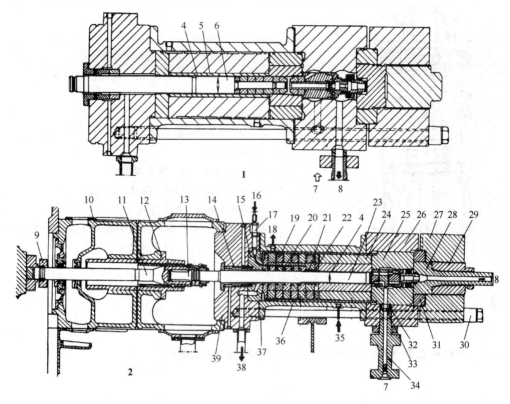

1	piston type high-pressure cylinder 活塞型高压气缸	20	packing rings 填料环
2	plunger type high-pressure cylinder 柱塞型高压气缸	21	guide ring 导向环
		22	pressure breaker ring 减压环
3	suction and delivery valves (central valve) 吸排气组合阀（中心型组合阀）	23	tungsten carbide plunger 碳化钨柱塞
		24	shrunk liner 热压缸套
4	cylinder 气缸	25	core 阀芯
5	tungsten carbide liner 碳化钨衬套	26	cylinder valve head 气缸阀盖
6	piston 活塞	27	cylinder head cover 气缸盖
7	suction port 进气口	28	spacer with pipe socket 带管座的管接头
8	delivery port 排气口	29	central screw 中心螺纹顶杆
9	coupling ring 连接环	30	cylinder bolt with nut 气缸螺栓（带螺母）
10	distance piece 隔离室	31	hydraulic piston 液压活塞
11	intermediate guide rod 中间导杆	32	spring 弹簧
12	guide liner 导向套	33	lens ring 透镜垫片
13	compression bar 压缩杆	34	suction pipe socket 吸入管管接头
14	complete stuffing box 填料函组件	35	cylinder cooling oil inlet 气缸冷却油入口
15	bottom plate to packings 填料底板	36	packing bush 填料衬套
16	inner oil inlet 内部油入口	37	cooling cylinder 冷却外缸套
17	injection non-return valve 注油止回阀	38	leakage gas 漏泄气体
18	cylinder cooling oil outlet 气缸冷却油出口	39	inner cylinder cover 内气缸盖
19	complete metallic packing 金属填料组件	40	suction valve body 吸入阀座
		41	seal ring set 密封环组

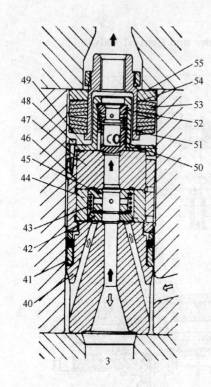

42 suction valve plate 吸入阀板
43 suction valve spring 吸入阀弹簧
44 tungsten carbide guide ring 碳化钨导向环
45 distance ring 隔环
46 guide sleeve 导套
47 delivery valve body 排出阀座
48 screw 螺栓
49 valve catcher case 排出阀压罩
50 delivery valve poppet 菌状排出阀芯
51 delivery spring 排出弹簧
52 valve catcher with washer 排出阀挡（带垫圈座）
53 Belleville washer 贝式弹簧垫圈
54 spring seat 弹簧座
55 pressing device 阀门压紧装置

7.2.3.2 卸荷阀及其他阀 Unloading Valve and Other Valves

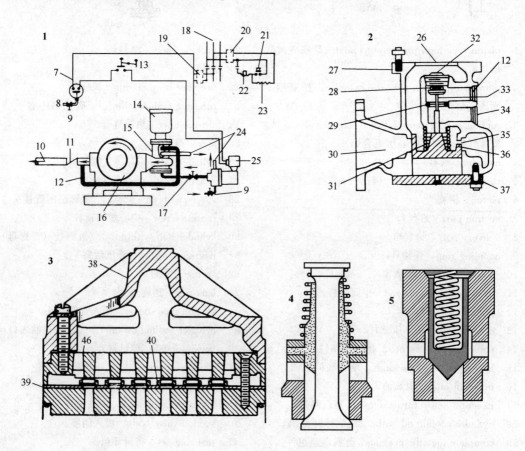

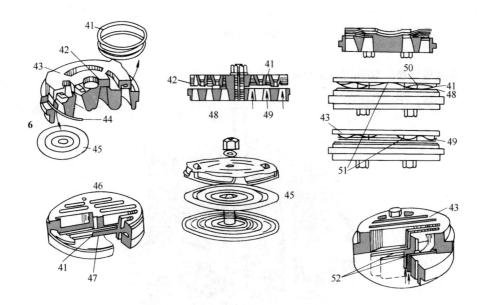

1	wiring schematic for unloading valve 卸荷阀配线原理图	26	air inlet 空气入口
2, 15	unloading valve 卸荷阀	27	valve body 阀体
3	channel valve 槽形阀	28	unloading valve spring 卸荷阀弹簧
4	poppet valve 菌状阀，提升阀	29	unloading valve bushing 卸荷阀衬套
5	thimble valve 针形阀	30	compressor inlet 压缩机入口
6	valve system for positive displacement single-acting compressors 单作用容积式压缩机的阀门组件	31	piston valve spring 活塞式阀弹簧
7	pressure switch circuit 压力开关回路	32	retaining bushing 制动衬套
8	air from receiver 来自储气罐的空气	33	pressure release valve 卸压阀
9	hand valve 手控阀	34	vent connection 放空管接头
10	discharge piping 排出管	35	bleed port 漏气孔
11	check valve 止回阀	36	piston inlet valve 活塞式进口阀
12	relief by-pass line 安全阀旁通线	37	valve cover 阀盖板
13	manual switch 手控开关	38	crab 蟹爪式阀罩
14	intake filter silencer 进口过滤消音器	39	seat plate 阀座板
16	compressor 压缩机	40	channel 槽形阀片
17	hand throttle valve 手控节流阀	41	valve spring 阀弹簧
18	three-phase AC power 三相交流电源	42	cushion pocket 缓冲弹簧槽
19	transformer 变压器	43	valve plate 阀护套
20	starter transformer 启动变压器	44	retainer ring 扣环
21	start button 启动按钮	45	valve plate 阀板
22	stop button 停止按钮	46	stop plate 升程限制器
23	starter holding coil 启动自保持线圈	47	valve channel 阀片槽
24	vent line 放空管线	48	valve closed 阀关闭
25	three-way solenoid pilot valve 电磁三通导流阀	49	valve open 阀打开
		50	reaction plate 波形板弹簧（反作用板）
		51	support point 支点
		52	flexible valve strip 弹性阀带

7.2.4 离心压缩机 Centrifugal Compressor

7.2.4.1 高速压缩机 High Speed Compressor

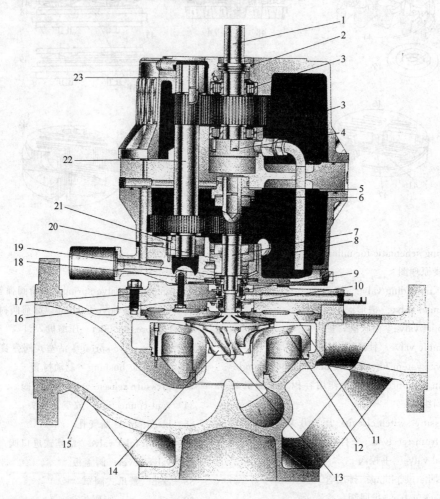

1	input shaft 输入轴		13	lock nut 锁紧螺帽
2	input shaft sealing 输入轴密封		14	impeller 叶轮
3	input shaft bearing 输入轴轴承		15	compressor casing 压缩机壳体
4	lube oil pump 润滑油泵		16	seal rotary ring, double or tandem 密封动环，双端面或串联式
5	bearing 轴承		17	sealing box 密封腔
6	thrust bearing 推力轴承		18	mechanical seal 机械密封
7	high speed shaft 高速轴		19	seal rotary ring 密封动环
8	high speed shaft bearing 高速轴轴承		20	thrust bearing 推力轴承
9	sleeve, gear box side 轴套，齿轮箱侧		21	bearing, intermediate shaft 轴承，中间轴
10	sleeve, process side 轴套，工艺侧		22	middle shaft 中间轴
11	diffuser cover 扩散器盖板		23	bearing, intermediate shaft 轴承，中间轴
12	diffuser 扩散器			

7.2.4.2 水平剖分式离心压缩机 Horizontally Split Centrifugal Compressor

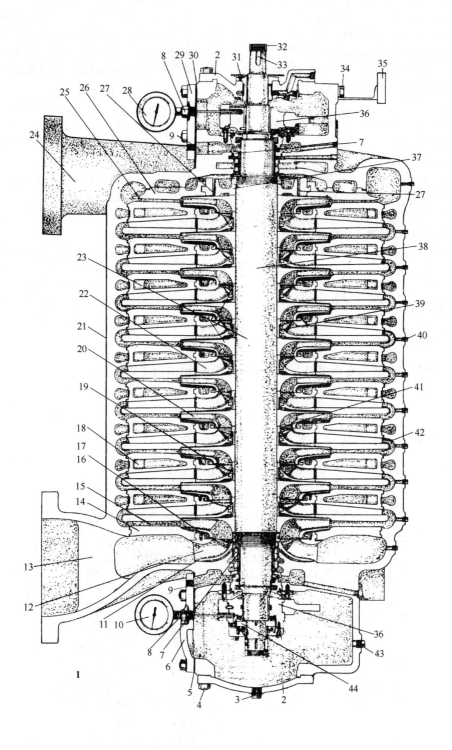

1 horizontally split centrifugal compressor (uncooled casing) 非冷却型水平剖分式离心压缩机
2 end cover 端盖
3 plug 堵头
4 end cover bolt 端盖螺栓

5	intake end gasket 进气端检查孔盖垫片		25	diffuser spacer 扩压器定距块
6	intake inspection cover 进气端检查孔盖		26	discharge wall 排出口挡板
7	inner seal assembly 内密封组件		27	balance piston labyrinth 平衡活塞迷宫
8	connector hub 接管颈		28	discharge end thermometer 出口端温度计
9	bearing isolation chamber 油封室		29	discharge inspection cover 出口端检查孔盖
10	intake end thermometer 进气端温度计		30	discharge end gasket 出口端检查孔盖垫片
11	inlet guide face & split O-ring 可分进口导流气O形环		31	shaft seal 轴封
12	inlet guide vane 进口导流叶片		32	coupling to driver 与驱动装置连接
13	suction 吸入口		33	keyway 键槽
14	inlet wall 进口挡板		34	flexible support bolt 挠性支座螺栓
15	diffusion passage 扩压器		35	flexible support 挠性支座
16	holding screw 紧固螺钉		36	integrally-supported shaft bearing 整体支托的径向轴承
17	labyrinth 迷宫		37	balance piston 平衡活塞
18	diaphragm 隔板		38	shaft 轴
19	impeller spacer 叶轮定距环		39	interstage labyrinth seal 级间迷宫式密封
20	impeller 叶轮		40	inter-stage drain 级间排液孔
21	casing assembly 壳体组件		41	impeller eye 叶轮上的（气体）入口
22	interstage guide vane 级间导流叶片		42	return bend 回流弯道
23	rotor assembly 转子组件		43	drain plug 放油丝堵
24	discharge 排出口		44	thrust bearing 推力轴承

7.2.4.3 径向剖分式离心压缩机 Radially Split Centrifugal Compressor

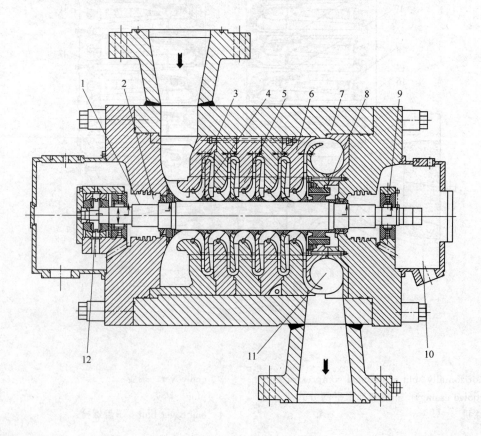

1	shaft 轴		7	outer casing 外筒体
2	shaft seal 轴封		8	balance drum 平衡毂
3	impeller wear ring 叶轮密封环		9	radial bearing 径向轴承
4	impeller 叶轮		10	bearing housing 轴承箱
5	interstage sealing 级间密封		11	volute 蜗室
6	diaphragm 隔板		12	thrust bearing 推力轴承

7.2.4.4 整体齿轮式压缩机 Integrally Geared Centrifugal Compressor

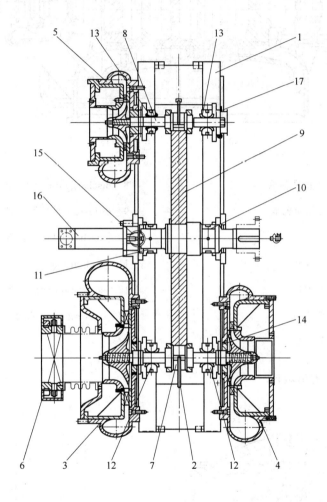

1	gear box 齿轮箱		10	bearing for gear (motor side) 大齿轮轴承（电机侧）
2	internal hole 内管路		11	bearing for gear (compressor side) 大齿轮轴承（压缩机侧）
3	first stage shell 一级定子		12, 13	tilting pad bearing 可倾瓦轴承
4	second stage shell 二级定子		14	sealing component 密封组
5	third stage shell 三级定子		15	coupling 接头部
6	inlet guide vane 进口调节器		16	centrifugal pump 离心泵
7	first and second stage rotor 一、二级转子		17	end cover 端盖
8	third stage rotor 三级转子			
9	gear 大齿轮			

427

7.2.4.5 轴流压缩机　Axial-flow Compressor

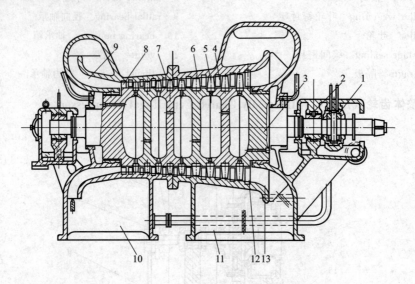

1　thrust bearing　止推轴承
2　radial bearing　径向轴承
3　rotor　转子
4　stator cascade　静叶
5　rotor cascade　动叶
6　front casing　前气缸
7　rear casing　后气缸
8　discharge director　出口导流器
9　diffuser　扩压器
10　discharge line　出气管
11　suction line　进气管
12　suction director　进气导流器
13　convergence　收敛器

7.2.5 鼓风机　Blowers

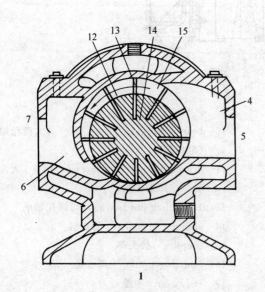
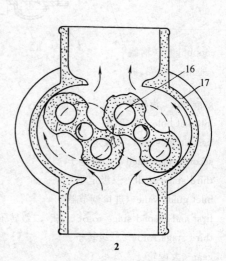

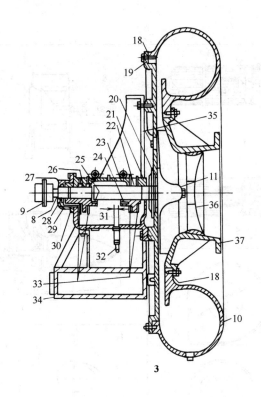

3

1	sliding-vane rotary blower 滑片式鼓风机	19	backplate 后盖板
2	Roots blower (Lobe-type blower) 罗茨式鼓风机	20	packing box 填料函
3	centrifugal blower 离心式鼓风机	21	impeller end oil guard 叶轮端挡油圈
4	inlet port 吸入孔	22	impeller end bearing sleeve 叶轮端轴承
5	inlet 进气口	23	bearing housing 轴承箱
6	discharge port 排出口	24	impeller end thermometer 叶轮端温度计
7	discharge 排气口	25	coupling end thermometer 轴承端温度计
8	shaft 轴	26	coupling end bearing sleeve 联轴器端轴套
9	coupling 联轴器	27	thrust bearing housing 推力轴承箱
10	casing 机壳	28	coupling end oil guard 联轴器端挡油圈
11	impeller 叶轮	29	thrust bearing 推力轴承
12	counter clockwise (CCW) 逆时针	30	thrust bearing orifice 推力轴承进油孔
13	sliding vane 滑片	31	oil level 油面
14	rotor 转子	32	to oil reservoir 去油箱
15	annular cell 环形气室	33	oil ring 甩油环
16	mating lobed impeller 配对"8"行转子	34	baseplate 底座
17	cylinder or casing 气缸或机壳	35	inspection hole plug 检查孔丝堵
18	gasket 垫片	36	inlet guide vane 进口导叶
		37	inlet connection 入口接管口

429

7.2.6 风机 Fans

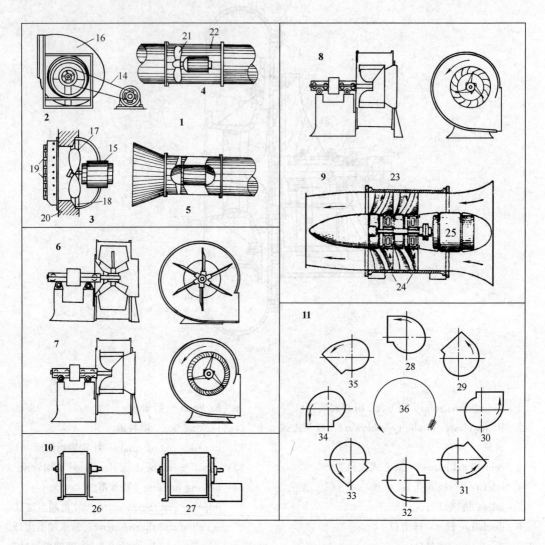

1 ventilating and industrial fans 通风机和工业用风机
2 centrifugal fan 离心式风机
3 propeller fan 旋浆式风机
4 tubeaxial fan 管装式轴流风机
5 vaneaxial fan 翼式轴流风机
6 straight-blade (steel-plate) fan 直叶式风机
7 forward-curved (sirocco-type) fan 前曲叶式风机
8 backward-curved-blade fan 后向曲叶式风机
9 two-stage axial-flow fan 两级轴流式风机
10 standard centrifugal fan arrangement designation 离心式风机标准配置标志
11 designation of direction of rotation and discharge （风机）旋转方向与排风口方向的标志
12 wheel 叶轮
13 centrifugal fan inlet side view 离心式风机进风端侧视图
14 belt drive 带传动
15 direct connection drive 直联传动
16 scroll type of housing 蜗壳
17 propeller (or disc) wheel 浆式叶轮，圆盘叶轮
18 mounting ring or plate 安装环，安装板
19 shutter 活动百叶窗
20 wall 墙

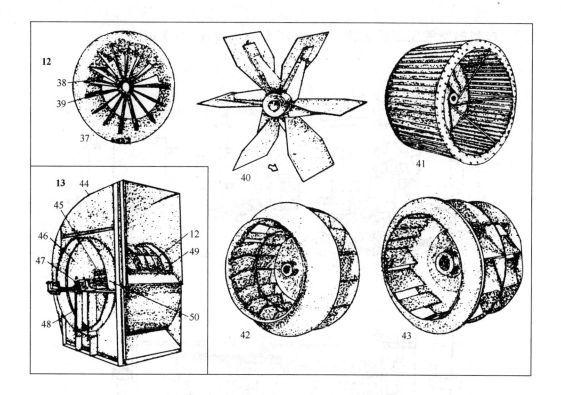

21	axial flow wheel 轴流叶轮	37	stationary inlet guide vane 固定式进口导流叶片
22	cylinder 圆筒机壳		
23	guide vane 导向冀片	38	stationary inlet vane 固定式进口叶片
24	rotating blade 转动叶片	39	inlet bell 钟形进风口
25	motor 电动机	40	long shaving wheel 长刮板式叶轮
26	SWSI (single width, single inlet) 单幅（叶轮）单进风口	41	single width wheel with forward curved blades 单幅前曲叶式叶轮
27	DWDI (double width, double inlet) 双幅（叶轮）双进风口	42	backward curved blades 后向曲叶
		43	double inlet, double width wheel, backward blades 双进风、双幅后向叶式叶轮
28	top horizontal 上水平送风		
29	top angular up 上仰角送风	44	housing 机壳
30	up blast 由下往上送风	45	shaft 轴
31	bottom angular up 下仰角送风	46	inlet 进风口
32	bottom horizontal 下水平送风	47	sleeve bearing 套筒轴承
33	bottom angular down 下倾角送风		anti-friction bearing 滚动轴承
34	down blast 由上往下送风	48	bearing support 轴承支架
35	top angular down 上倾角送风	49	cut-off 阻流板
36	counter-clockwise (CCW) 逆时针	50	inlet collar 挡圈

7.2.7 典型的空气压缩机装置 Typical Air Compressor Unit

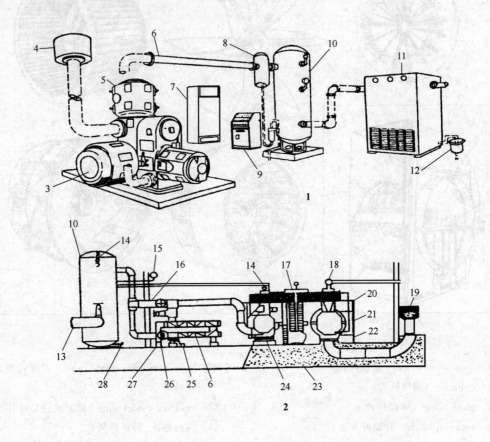

1 installation drawing   安装（示意）图
2 sectional view   剖视图
3 direct connected motor   直联电动机
4 inlet air filter   进口空气过滤器，滤气器
5 two-stage air compressor   两段空气压缩机
6 aftercooler   后冷（却）器
7 motor starter   电动机启动器
8 separator   油水分离器
9 local control panel   就地控制盘
10 receiver (air receiver)   储气槽
11 air dryer   空气干燥器
12 automatic condensate drain   自动排凝器
13 pipe   管道
14 safety valve   安全阀
15 pressure gage   压力表
16 aftercooler by-pass   后冷（却）器旁通
17 intercooler   中间冷却器
18 unloader valve   卸荷阀
19 air intake filter   空气进口过滤器
20 water inlet   上水口
21 open funnel   漏斗
22 drain   排水口
23 foundation   基础
24 air compressor   空气压缩机
25 automatic condensate trap   自动排凝阀
26 cooling water   冷却水
27 drain   排水口
28 receiver blow-off   储气罐吹扫口

7.3 搅拌器 Mixer

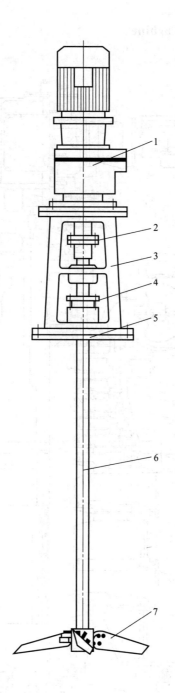

1 gear box 减速机
2 coupling 联轴器
3 frame 机架
4 stuffing box 填料
5 baseplate 底座
6 shaft 轴
7 mixer 搅拌器

7.4 驱动机 Driver

7.4.1 蒸汽轮机 Steam Turbine

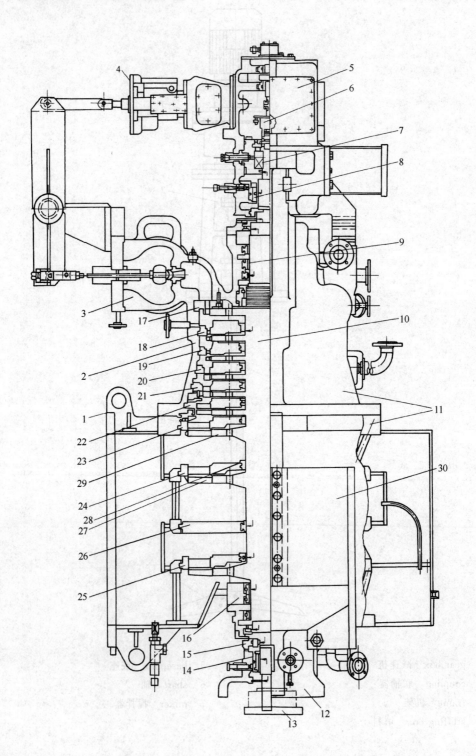

1 exhaust cylinder 排汽缸
2 upper cylinder 上汽缸
3 adjustable valve 调节阀
4 speed-adjusting mechanics 调速机构
5 front bearing housing 前轴承箱
6 protector 保护器
7 thrust bearing 推力轴承
8 front bearing 前轴承
9 high pressure end steam lock 高压端汽封
10 rotor 转子
11 below cylinder 下汽缸
12 back bearing housing 后轴承箱
13 coupling 联轴器
14 back bearing 后轴承
15 oil sealing 油封
16 low pressure end steam lock 低压端汽封
17 nozzle 喷嘴
18～25 baffle board 隔板
26 divide board 分流隔板
27 adjustable impeller 动叶片
28 baffle board steam lock 隔板汽封
29 impeller 叶轮
30 flexible support board 挠性支承板

7.4.2 蒸汽轮机的分类与结构 Classification and Structure of Steam Turbines

（1）蒸汽轮机的分类（1） Classification of Steam Turbines（Ⅰ）

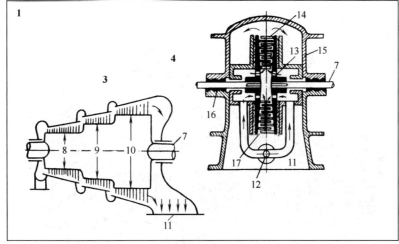

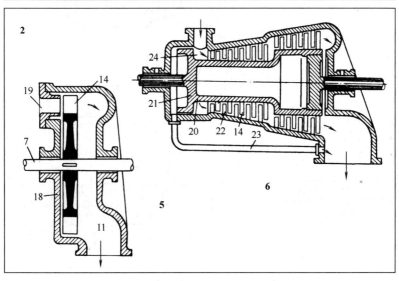

1　classification according to the direction of steam flow　按蒸汽流动方向分类
2　classification according to the principle of action of steam　按蒸汽工作原理分类
3　axial turbine　轴流式汽轮机
4　radial turbine　径流式汽轮机
5　impulse turbine　冲动式汽轮机
6　reaction turbine　反作用式汽轮机
7　shaft　轴
8　high-pressure rotor　高压转子
9　medium-pressure rotor　中压转子
10　low-pressure rotor　低压转子
11　exhaust pipe　排气管
12　fresh steam supply　新蒸汽供汽
13　disc　叶轮
14　moving blades　动叶片
15　casing　汽缸，机壳
16　bearing　轴承
17　rows of moving blades　动叶片排
18　stator　机体
19　nozzle　喷嘴
20　rotor drum　转鼓
21　balance piston　平衡活塞
22　guide blades　导向叶片
23　pressure equaliser steam pipe　均压蒸汽室
24　annular steam chamber　环形蒸汽室

（2）蒸汽轮机的分类（2）　Classification of Steam Turbines（Ⅱ）

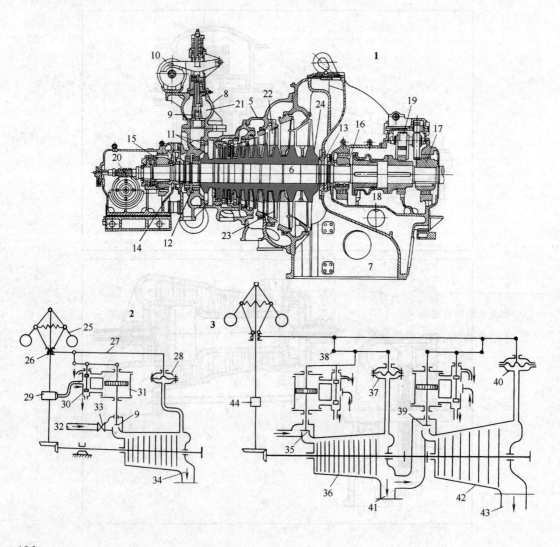

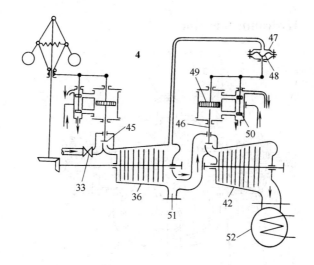

1	condensing turbine 凝汽式汽轮机		24	disc 叶轮
2	back pressure turbine 背压式汽轮机		25	centrifugal speed governor 离心调速器
3	back pressure turbine with a pass out 抽汽背压式汽轮机		26	sleeve 套筒
			27	lever 杠杆
4	condensing turbine with extraction 抽汽凝汽式汽轮机		28，47	pressure regulator 调压器
			29，44	oil pump 油泵
5	casing 汽缸		30	pilot valve 错油阀，导阀，控制阀
6	shaft 主轴		31	servomotor 伺服马达
7	exhaust pipe 排气管		32	live steam 新蒸汽
8	valve chest 阀腔		33	main stop valve 主汽阀
9，45	regulating valve 调节阀		34	back pressure 背压
10	valve actuator 阀门执行机构		35	fresh steam regulating valve 新蒸汽调节阀
11	regulating stage 调速级		36	H. P. cylinder 高压汽缸
12	H. P. labyrinth packing 高压迷宫式轴封		37，40	back pressure regulator 背压调节器
13	L. P. labyrinth packing 低压迷宫式轴封		38	fulcrum point 支点
14	front bearing casing 前轴承箱		39	pass-out regulating valve 排汽（背压）调节阀
15	front combined journal and thrust bearing 前轴颈及止推联合轴承		41	to the consumer 至用户
16	rear journal bearing 后轴颈轴承		42	L. P. cylinder 低压汽缸
17	generator bearing 发电机轴承		43	exhaust pipe 排汽管
18	coupling 联轴器		46	second regulating valve 第二抽汽调节阀
19	shaft turning gear 盘车齿轮		48	membrane 薄膜
20	worm reduction gearing for speed governor 调速器蜗轮蜗杆减速齿轮		49	servomotor piston 伺服马达活塞
			50	pilot valve piston 错油阀活塞
21	distribution box 配气室		51	intermediate extraction 中间抽汽
22	nozzle 喷嘴		52	condenser 凝汽器
23	blade 叶片			

7.4.3 液力透平 Hydraulic Turbine

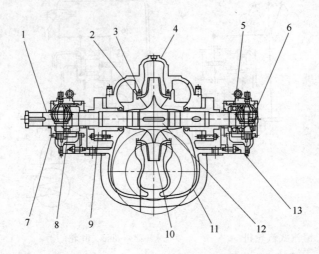

1　shaft　主轴
2　impeller wear ring　叶轮口环
3　pump wear ring　泵体口环
4　top casing　上泵体
5　radial/thrust bearing　径向/支承轴承
6　constant oil indicator　恒位油杯
7　bearing cover　轴承压盖
8　radial bearing　径向轴承
9　mechanical seal　机械密封
10　impeller　叶轮
11　bottom casing　下泵体
12　key　键
13　bearing housing　轴承箱

7.4.4 燃气轮机 Gas Turbine

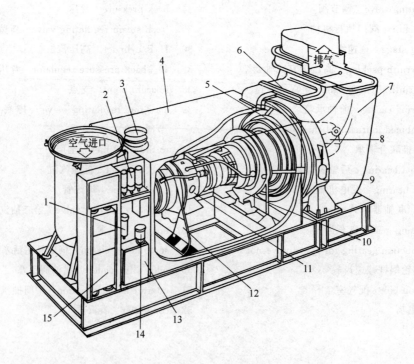

438

1 oil filter 油过滤器
2 hand grenade 手动灭火瓶
3 air inlet 空气进口
4 box body 外壳
5 power turbine 动力涡轮
6 air outlet 空气出口
7 exhaust turbine 排气蜗壳
8 output shaft 输出轴
9 power turbine support 动力涡轮支架
10 back gas generator support 燃气发生器后支架
11 gas generator 燃气发生器
12 front gas generator support 燃气发生器前支架
13 oil reservoir 油箱
14 oil-gas separator 油气分离器
15 control panel 控制盘

7.4.5 烟气轮机　Flue Gas Turbine

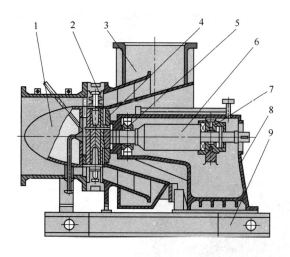

1 suction 进气
2 shroud 围带
3 casing 壳体
4 air seal assembly 气封组装
5 radial supporting bearing 径向轴承
6 rotor 转子
7 radial bearing 径向止推轴承
8 bearing housing 轴承箱
9 baseplate 底座

7.4.6 膨胀机/再压缩机　Expander/Compressor

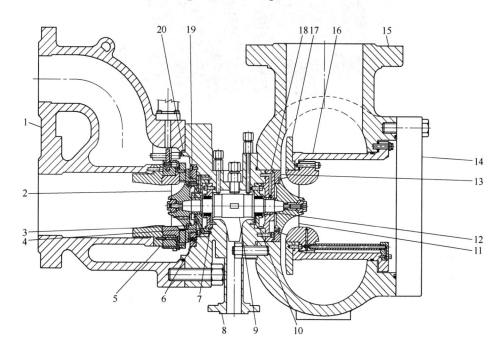

439

1 expander housing 膨胀机外壳	11 compressor wheel 压缩机叶轮
2 expander wheel 膨胀机叶轮	12 compressor shaft seal ring 压缩机轴密封环
3 expander wheel seal 膨胀机叶轮端密封	13 compressor wheel seal 压缩机叶轮端密封
4 O-ring O形圈	14 compressor inlet spool 压缩机进口短节
5 flow guide, expander discharge 膨胀机出口导叶	15 compressor housing 压缩机外壳
6 pressure ring guide 压力环导向件	16 diffuser bracket 扩压器外壳
7 expander bearing 膨胀机轴承	17 diffuser wings 扩压器叶片
8 bearing housing 轴承箱	18 compressor shaft seal 压缩机轴密封
9 shaft 轴	19,20 seal ring 密封圈
10 compressor bearing 压缩机轴承	

7.4.7 柴油机 Diesel Engine

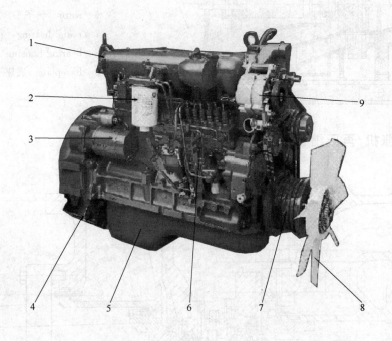

1 inlet piping 进气管
2 diesel filter 柴油过滤器
3 starter 启动机
4 flywheel cover 飞轮壳
5 bottom cover 底壳
6 injection pump 喷油泵
7 bumper 减震器
8 fan 风扇
9 charging generator 充电发电机

7.5 转动设备辅机 Rotating Equipment Auxiliaries

7.5.1 液力耦合器 Fluid Coupling

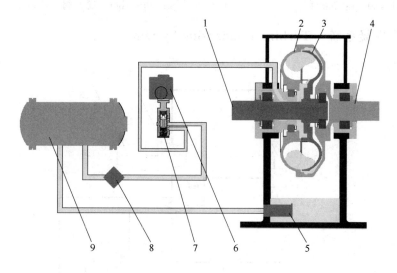

1　input shaft　输入轴
2　turbo　涡轮
3　pump wheel　泵轮
4　output shaft　输出轴
5　rough filter　粗滤
6　electrical actuator　电动执行器
7　control valve　控制阀
8　fine filter　细滤
9　oil cooler　油冷器

7.5.2 齿轮箱 Gear

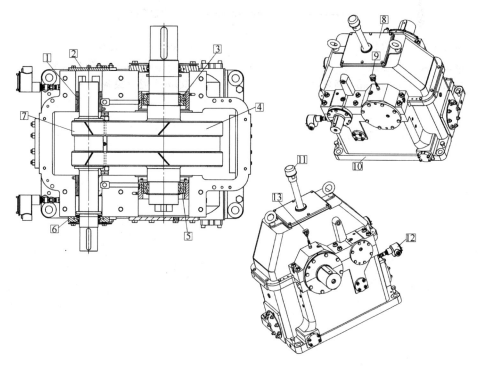

1　journal bearing　径向轴承
2　end cover　端盖
3，5　rolling bearing　滚动轴承
4　assembly，gear/shaft　组装齿轮轴
6　labyrinth seal　迷宫密封
7　pinion　小齿轮
8　inspection cover　检修盖
9，12，13　Rtd Pt100　铂热电阻
10　housing　齿轮箱外壳
11　breather　通气阀

7.5.3　透平冷凝系统　Turbine Condensing System

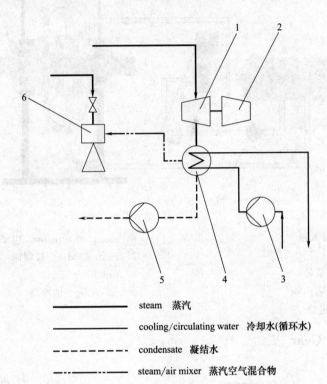

steam　蒸汽
cooling/circulating water　冷却水(循环水)
condensate　凝结水
steam/air mixer　蒸汽空气混合物

1　steam turbine　汽轮机
2　compressor　压缩机
3　circulating water pump　循环水泵
4　condenser　冷凝器
5　condensate pump　凝结水泵
6　ejector　抽气设备

8 管道工程 Piping Engineering

8.1 设备及管道布置 Equipment and Piping Layout

8.1.1 设备布置 Equipment Layout

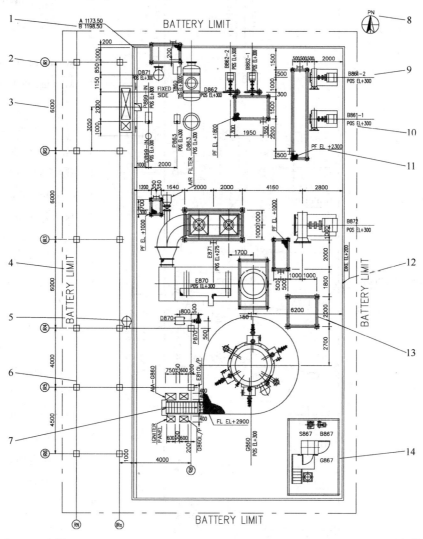

1 coordinate 坐标
2 column number 柱号
3 position dimension 定位尺寸
4 battery limit 界区
5 ladder 直爬梯
6 pipe rack 管廊
7 stair 斜爬梯
8 plant north 装置北向
9 equipment item number 设备位号
10 POS EL(point of support elevation) 支承点标高
11 PF EL (platform elevation) 平台标高
12 dike 围堰
13 steel structure 钢结构
14 compressor shelter 压缩机厂房

8.1.2 管道布置图,配管图 Piping Layout Drawing

8.1.2.1 管道平面布置 Piping Plan Layout

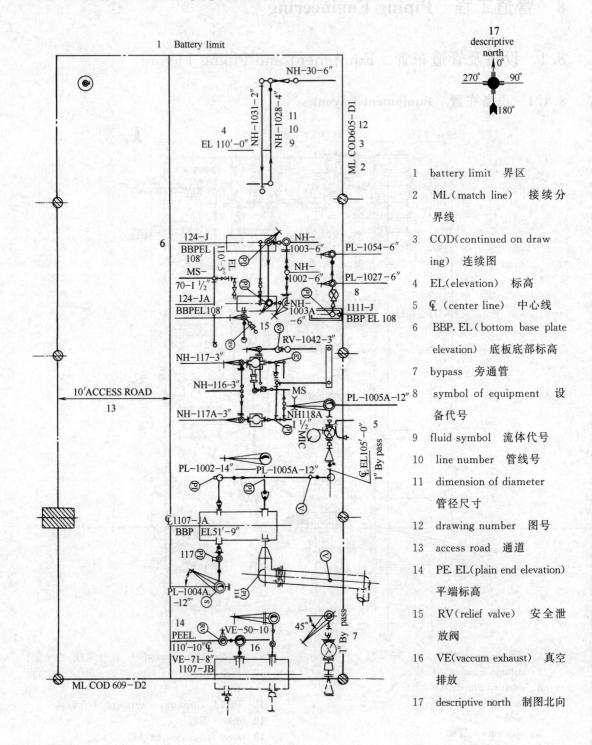

1 battery limit 界区
2 ML(match line) 接续分界线
3 COD(continued on drawing) 连续图
4 EL(elevation) 标高
5 ℄(center line) 中心线
6 BBP.EL(bottom base plate elevation) 底板底部标高
7 bypass 旁通管
8 symbol of equipment 设备代号
9 fluid symbol 流体代号
10 line number 管线号
11 dimension of diameter 管径尺寸
12 drawing number 图号
13 access road 通道
14 PE.EL(plain end elevation) 平端标高
15 RV(relief valve) 安全泄放阀
16 VE(vaccum exhaust) 真空排放
17 descriptive north 制图北向

8.1.2.2 立面图 Section Drawing

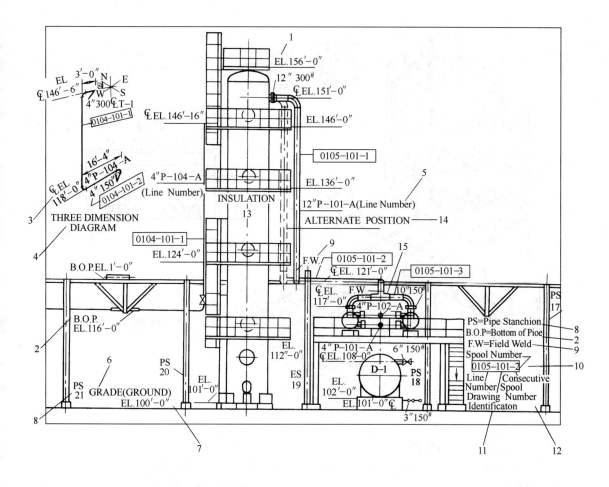

1 EL（elevation）　标高
2 B.O.P.（bottom of pipe）　管底
3 ℄（center line）　中心线
4 three dimension diagram, isometric diagram　（管段）空视图，轴测图，管段图
5 line number　管号
6 grade　地坪
7 ground EL.　地面标高
8 pipe stanchion（PS）　管架
9 F.W.（field weld）　现场焊接
10 spool number　管段号
11 drawing identification　管段所在图号
12 consecutive spool number　管段顺序号
13 insulation　保温层
14 alternate position　另一种布置位置
15 Ecc. Red.（eccentric reducer）　偏心异径管，偏心大小头

445

8.1.2.3 管道布置——装置配管 Piping Layout—Installation Piping

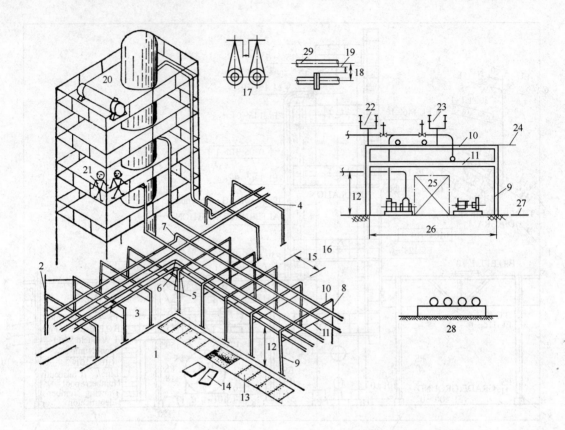

1　yard piping　装置内配管
2　yard piping rack, yard piping support　装置内管架
3　yard support bent, yard steel bent　管廊，一排管架
4　one level yard　单层管架
5　no elevation difference　标高相等
6　flat turn　等高度改变走向
7　changing direction, change elevation　改变走向，改变标高
8　yard bank　单个管架
9　yard column　管架立柱
10　top yard bank　管架上梁
11　bottom yard bank　管架下梁
12　headroom　（高度方向的）净空
13　trench　（管）沟，（电缆）沟
14　cover plate, trench cover　沟盖板
15　spacing between yard support bents　管架间距
16　lines span　管道跨距
17　valve spacing　阀门净空
18　line spacing　管道净空
19　clearance (gap)　净空，净距
20　structure　框架，构筑物
21　operator　操作工
22　platform　操作平台
23　catwalk　检修走道，（油罐管架顶上的）人行栈桥
24　cantilever　悬臂支架
25　access road　检修通道
　　walkway　走道，通行道
26　yard span (yard steel span)　管架跨距
27　grade (grade elevation, ground level, ground elevation)　地面标高
28　pipe sleeper　管墩
29　bare line　裸管，不保温管

8.1.3 管道组装图（轴测图，管段图） Erection Diagram（Pipe Line Isometric Diagram）

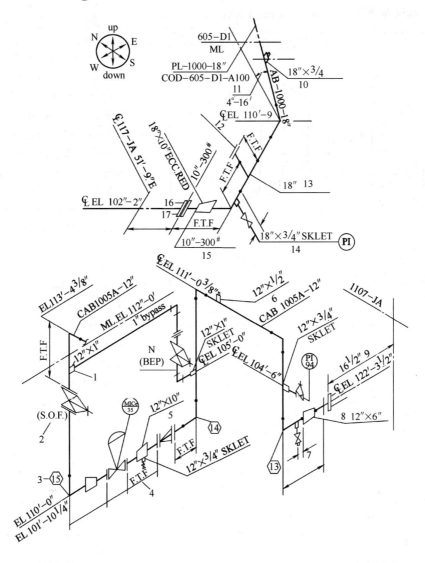

1 sockolet 承插支管台
2 S. O. F.（slip on flange） 滑套法兰
3 number of pipe support 管架号
4 F. T. F.（fitting to fitting） 管件至管件
5 ECC. RED.（eccentric reducer） 偏心异径管，偏心大小头
   F. O. B. TYP.（flat on bottom type） 底平
6 THLET.（thredolet） 螺纹支管台
7 MIN.（minimum） 最小
8 CONC. RED.（concentric reducer） 同心异径管，同心大小头
9 REF.（reference） 参考尺寸
10 THLET. W/P（thredolet with plug） 螺纹丝堵
11 HOR.（horizontal angle） 水平角
12 BLD. FLG.（blind flange） 盲板
13 "T" Type STR.（strainer） T形过滤器
14 N(BEP)T/S GA. VA.（nipple both ends plain thread/socket gate valve） 平螺纹两端短管
15 W. N. F.（weld neck flange） 对焊法兰
16 inlet 入口管
17 R. F.（raised face） 凸面(密封面)

8.1.4 伴热及夹套 Tracing Lines and Jacket

8.1.4.1 蒸汽伴热管 Steam Tracing Lines

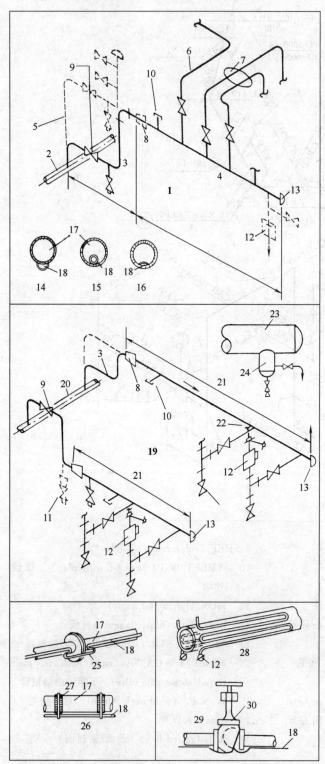

1　steam tracing supply line　蒸汽伴热供汽管
2　steam supply header　供汽总管
3　tracing supply header, subheader　伴热蒸汽主管，分主管
4　manifold　集合管，分配管
5　alternate arrangement　另一种管线布置方式
6　tubing to traced lines　去伴热的管子
7　tie in bundles　捆扎成束
8　reducer or swage nipple　异径管或异径短接
9　header isolation valve（shut-off valve）主管切断阀
10　unused spares　备用接头
11　drain valve with plug　带丝堵的排放阀
12　trap（if manifold is pocketed）疏水器（当集合管积水时安装）
13　cap　管帽
14　half moon jacketing　半月形伴热套管
15　unitrace　内套管伴热
16　internal tracing　（管）内伴热
17　process line　（被伴热）工艺管线
18　tracer pipe　伴热管
19　steam tracing condensate lines　蒸汽伴热冷凝液管
20　condensate return header　冷凝液回水总管
21　condensate manifold　冷凝液集合管
22　piping from traced line　接自伴热管
23　steam main　主蒸汽管，蒸汽主管
24　condensate pot　冷凝液包
25　expansion loop　热补偿弯管
26　spacer tracing（to avoid hot spots）隔离伴热（防止局部过热）
27　asbestos tape　石棉带
28　multiple tracers　多程伴热管
29　wrap round tracing　缠绕伴热
30　traced valve　伴热阀门

8.1.4.2 热水伴热管 Hot Water Tracing Lines

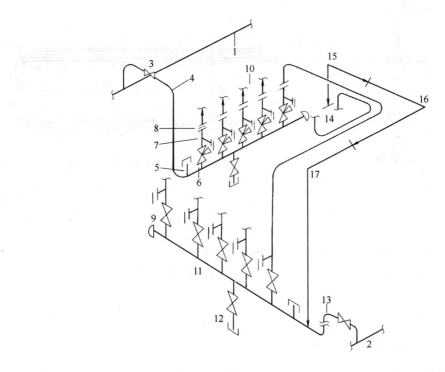

1	hot water main（supply）　热水主管（供）	10	tracer supply lines　伴管供给管线
2	hot water main（return）　热水主管（回）	11	tracer return manifold　伴管回水站
3	block valve　根部阀	12	drain valve　排液阀
4	tracer manifold supply　伴管分配站供水	13	tracer manifold return　伴热站回水
5	spare nozzle　备用口	14	traced item　被伴热项
6	tracer manifold　伴管分配站	15	tracer supply　伴管供
7	purge nozzle　吹扫口	16	tracer　伴热管
8	restriction orifice　限流孔板	17	tracer return　伴管回
9	valve　阀门		

8.1.4.3 夹套管 Jacket Lines

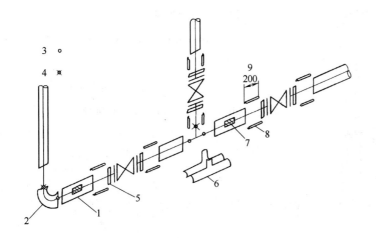

1　jacket pipe　夹套管
2　jacket elbow　夹套弯头
3　field shop weld　现场车间焊
4　field weld（FW）　现场焊
5　flat ring plate　环形端板
6　split tee　分开成两半的三通
7　guide plate　导向板
8　split pipe　分开成两半的管子

449

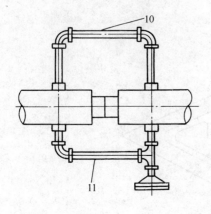 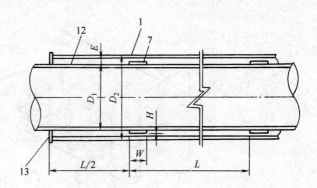

9　min.　最小
10　jump over piping for steam or hot water　蒸汽或热水跨接管
11　jump over piping for condensate or hot water　冷凝液或热水跨接管
12　internal pipe　内管
13　end plate　端板

8.1.4.4　电伴热　Electric Heat Tracing

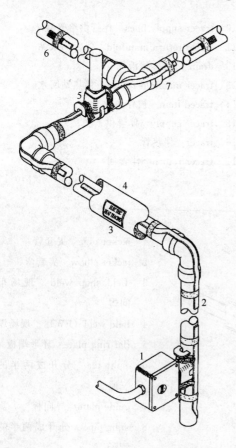

1　electric power wiring box　电源接线盒
2　self-regulate and control tracing tape　自调控伴热带
3　electric tracing label　电伴热标签
4　hot insulation layer and other protect layer　保温层及其他外保护层
5　T-type tracing tape connect box　T形伴热带联接盒
6　the end of tracing tape　伴热带的尾端
7　polyester fiber tape　聚酯纤维带

8.1.5 急救冲洗和洗眼站 Safety Shower and Eyewash Station

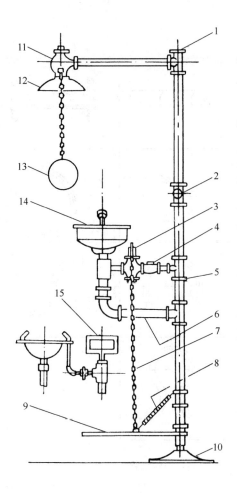
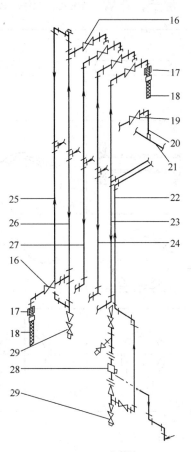

hose station 软管站

1 supply connection 供水接头
2 alternate supply 备用水源
3 brass whistle valve 铜笛阀
4 screwdriver stop 改锥旋动阀
5 plugged 堵住
6 galvanized pipe 镀锌管
7 pull chain 拉链
8 steel spring 钢制弹簧
9 aluminium foot pedal 铝制踏板
10 floor flange stand 底盘、地轴架
11 self-closing valve 自闭阀
12 shower head, deluge head 莲蓬头
13 pull chain and ring 拉链和拉环
14 stainless steel bow with twin fountain heads 带两个喷水头的不锈钢盆
15 panic bar 紧急手把、事故手把

16 block valve 切断阀
17 temporary breakaway type hose coupling 快速接头、软管临时接头
18 hose 软管
19 root valve 根部阀
20 branch 支管
21 header 总管
22 condensate to header (closed system) 凝液去总管（闭式系统）
23 steam 蒸汽
24 plant air 工厂空气
25 anti-freeze by-pass 防冻旁通管
26 water 水
27 inert gas 惰性气体
28 steam trap 蒸汽疏水器
29 drain and test connection 排水和检查接头

8.2 管道材料 Piping Material

8.2.1 管子,弯管 Pipe, Pipe Bends

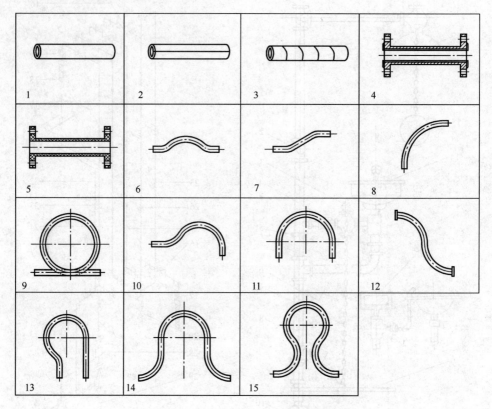

(1~5 pipe 管子
  steel pipe 钢管
  cast iron pipe 铸铁管
  clad pipe 复合管
  carbon steel pipe 碳钢管
  alloy steel pipe 合金钢管
  stainless steel pipe 不锈钢管
  austenitic stainless steel pipe 奥氏体不锈钢管
  ferritic alloy steel pipe 铁合金钢管
  wrought-steel pipe 轧制钢管
  wrought-iron pipe 锻铁管
  water-gas steel pipe 水煤气钢管)
1 seamless [SMLS] steel pipe 无缝钢管
  hot-rolling seamless pipe 热轧无缝钢管
  cold-drawing seamless pipe 冷拔无缝钢管
2 welded steel pipe 焊接钢管
  electric-resistance welded steel pipe (ERW) 电阻焊钢管
  electric-fusion (arc)-welded steel-plate pipe 电熔(弧)焊钢板卷管
3 spiral welded steel pipe 螺旋焊接钢管
4 lined pipe 衬里管
5 plastic pipe 塑料管
(6~15 fabricated pipe bends 预制弯管)
6 crossover bend 交叉弯曲管
7 offset bend 偏置弯曲管
8 quarter bend 90°弯管
9 circle bend 环形弯曲管
10 single offset quarter bend 单侧偏置90°弯管
11 half bend 回弯管,回折管
12 "S" bend S形弯管
13 single offset "U" bend 单侧偏置回弯管
14 expansion "U" bend U形补偿器
15 double offset expansion "U" bend U形回弯补偿器

8.2.2 管件 Pipe Fitting

8.2.2.1 钢焊接管件 Steel-Welding Fittings

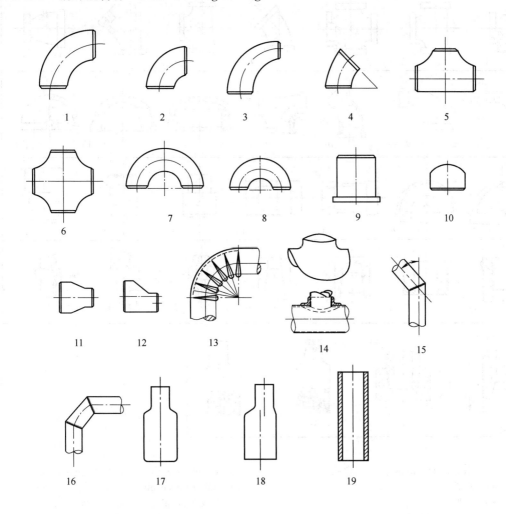

1　long radius 90° elbow　90°长半径弯头
2　short radius 90° elbow　90°短半径弯头
3　long radius reducing 90° elbow　90°长半径异径弯头
4　45° elbow　45°弯头
5　straight tee　等径三通
　　reducing tee　异径三通
6　straight cross　等径四通
　　reducing cross　异径四通
7　long radius 180° return　180°长半径弯头
8　short radius 180° return　180°短半径弯头
9　stub end (lap joint)　管端突缘（松套法兰用）
10　cap　管帽
11　concentric reducer　同心异径管
12　eccentric reducer　偏心异径管
13　corrugated bend　折皱弯管
14　reinforcing saddles　马鞍形补强
(15，16　mitre bend　斜接弯管，虾米（腰）弯头)
15　45° one mitre bend　45°一次斜接弯头
16　90° two mitre bend　90°二次斜接弯头
(17，18　reducing nipple, swage nipple　短节)
17　concentric reducer swage nipple　同心异径短节
18　eccentric reducer swage nipple　偏心异径短节
19　nipple　短节

8.2.2.2 螺纹管件 Threaded Fittings

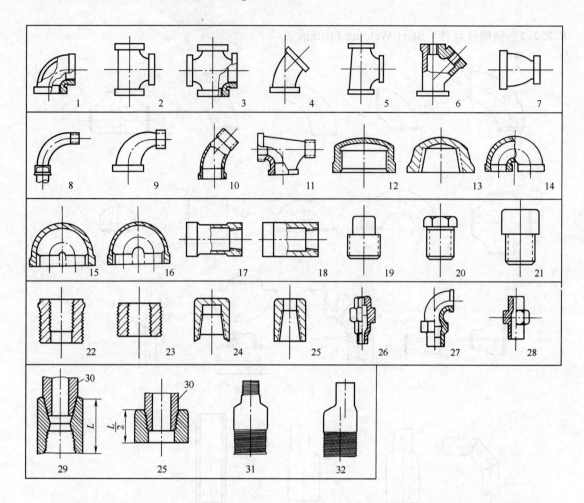

1 90°(threaded) elbow 90°(螺纹)弯头
2 tee 三通
3 cross 四通
4 45° elbow 45°弯头
5 reducing tee 异径三通
6 45° Y-branches (straight size) 45° Y形三通（等径）
7 reducing coupling 异径管接头
8 street elbow 带内外螺纹弯头
9 reducing elbow 异径弯头
10 45° reducing elbow 45°异径弯头
11 reducing tee (reducing on outlet) 异径三通（分支口为异径）
12 cap with recess 带退刀槽的管帽
13 cap without recess 不带退刀槽的管帽
14 open pattern return bend 敞开式180°弯头
15 close pattern return bend 封闭式180°弯头
16 medium pattern return bend 普通型180°弯头
17 coupling with band 带箍管接头
18 coupling without band 无箍管接头
19 square head plug 方头丝堵
20 hex head plug 六角头丝堵
21 round head plug 圆头丝堵
22 hex head bushing 六角头内外螺纹缩接
23 flush bushing 内外螺纹缩接
24 cap 管帽
25 threaded half coupling 螺纹半管接头
26 male and female nipple, screw end male and female union 螺纹端内外螺纹活接头
27 elbow union 弯头活接头
28 union 活接头
29 threaded full coupling 螺纹管接头
30 threaded nipple 螺纹短节
31 threaded concentric reducer swage nipple 螺纹同心异径短节
32 threaded eccentric reducer swage nipple 螺纹偏心异径短节

8.2.2.3 支管台与承插管件 Outlet and Socket Welding Fittings

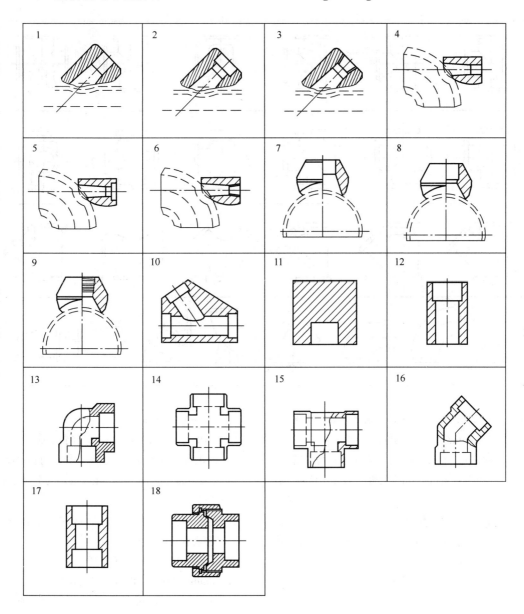

1 butt welding 45° latrolet 对焊 45°斜接支管台
2 socket welding 45° latrolet 承插焊 45°斜接支管台
3 thread 45° latrolet 螺纹 45°斜接支管台
4 weldolet elbow outlet 对焊弯头支管台
5 sockolet elbow outlet 承插焊弯头支管台
6 thredolet elbow outlet 螺纹弯头支管台
7 weldolet 焊接支管台
8 sockolet 承插焊支管台
9 thredolet 螺纹支管台
10 socket welding 45° lateral tee 承插焊 45°斜三通
11 socket welding cap 承插焊管帽
12 socket welding half coupling 承插焊半管接头
13 socket welding 90° elbow 承插焊 90°弯头
14 socket welding cross 承插焊四通
15 socket welding tee 承插焊三通
16 socket welding 45° elbow 承插焊 45°弯头
17 socket welding coupling 承插管接头
18 socket welding union 承插焊活接头

8.2.2.4 法兰管件 Flanged Fittings

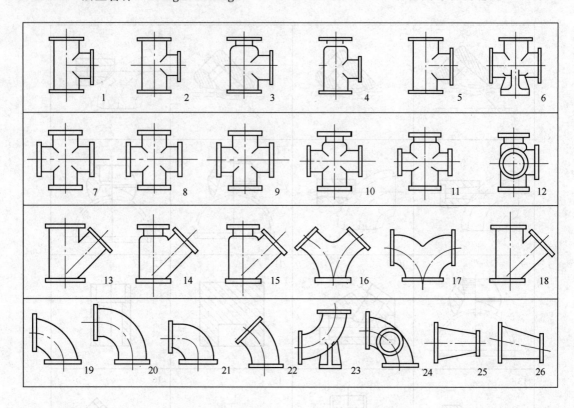

1 (flanged) straight tee （法兰）等径三通
(2~5 reducing tees 异径三通)
2 reducing on outlet 分支口为异径的三通
3 reducing on one run 主管一端为异径的三通
4 reducing on one run and outlet 主管一端和出口一端为异径的三通
5 reducing on both runs (bullhead) 主管两端为异径的三通（双头式）
6 base tee 带支座三通
(7~12 cross 四通)
7 straight cross 等径四通
8 reducing on one outlet 一个分支口为异径的四通
9 reducing on both outlet 分支口两端为异径的四通
10 reducing on one run and outlet 一个主管端和一个分支端为异径的四通
11 reducing on one run and both outlet 主管一端和分支口两端为异径的四通
12 side outlet cross straight size 分支带侧出口的等径四通
(13~18 reducing laterals 异径斜三通)
13 reducing on branch 支管为异径的斜三通
14 reducing on one run 主管一端为异径的斜三通
15 reducing on one run and branch 主管一端和支管为异径的斜三通
16 true "Y" 等径 Y 形三通
17 double branch elbow 双支管弯头
18 45° lateral straight size 45°等径斜三通
19 90° radius elbow 90°弯头
20 90° long radius elbow 90°长半径弯头
21 90° reducing elbow 90°异径弯头
22 45° elbow 45°弯头
23 90° base elbow 带支座90°弯头
24 side outlet 90° elbow straight size 带侧出口的90°等径弯头
(25, 26 reducer 异径管, 大小头)
25 concentric reducer 同心异径管
26 eccentric reducer 偏心异径管

8.2.2.5 塑料压接管接头 Plastics Compression Joints

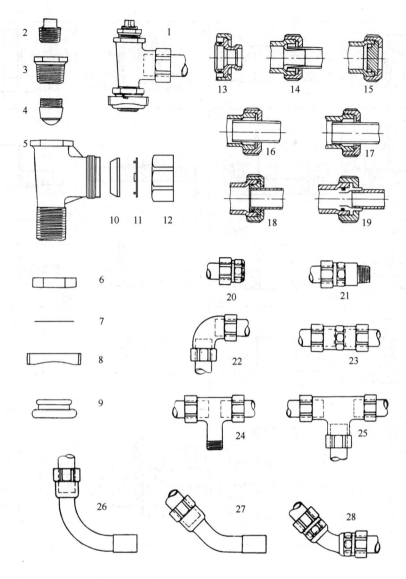

1	swivel tee 旋转三通	15	stop end 封头
2	plug 丝堵，堵头	16	flex-lock assembly 挠性-锁紧连接件
3	reducing bush 内外丝大小头	17	flexible end assembly 挠性端头连接件
4	nylon stopper 尼龙堵头	18	thread adaptor 螺纹管接头
5	body （三通）本体	19	stepped thread adaptor 异径螺纹管接头
6	backnut 锁紧螺母	20	coupling cap 管帽
7	tin disc 白铁片盘	21	adaptor 管接头
8	saddle washer 鞍形垫圈	22	90° elbow 90°弯头
9	rubber gasket 橡皮垫片	23	coupling 管接头，管箍
10	rubber gasket 橡皮密封垫	24	syphon pot tee 虹吸三通
11	disc 盘	25	all flex tee 软管三通
12	nut 螺母	26	90° coupling bend 90°弯管接头
13	reducer assembly 异径连接体	27	135° coupling bend 135°弯管接头
14	positive end assembly 刚性端头连接件	28	135° double coupling bend 135°弯管双端接头

8.2.3 法兰、法兰密封面 Flanges, Flange Facings

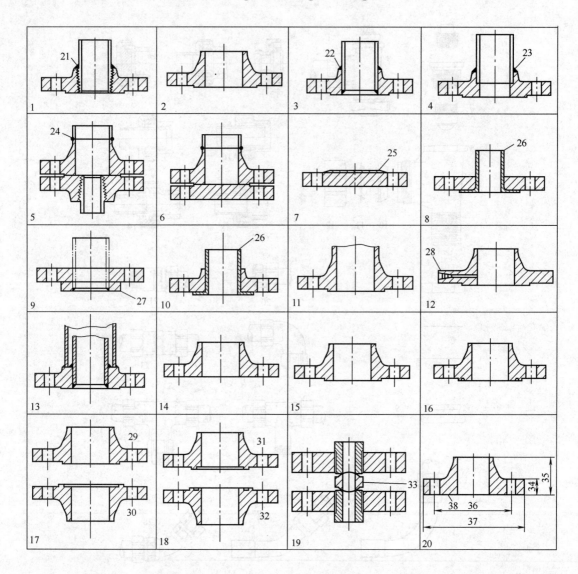

(1~11 flange type 法兰型式)
1 threaded flange 螺纹法兰
2 welding neck flange 对焊法兰
3 slip-on flange (SO) 滑套法兰（平焊法兰）
4 socket welding flange 承插焊法兰
5 reducing threaded flange 异径螺纹法兰
6 blind flange; blind 法兰盖
7 lining blind flange 衬里贴面法兰盖，贴面法兰盖

[8~10 lap joint flange 松套法兰（活套法兰）]
8 loose plate flange 松套板式法兰

9 loose plate flange (welding-on collar) 平焊环松套板式法兰
10 lapped flange 带颈松套法兰
11 integral pipe flange 整体管法兰
12 orifice flange 孔板法兰
13 jacketed flange 夹套法兰
(14~19 flange facing 法兰密封面)
14 flat face; full face (FF) 全平面
15 raised face (RF) 突面
16 ring joint face (RJ) 环连接面
17 large female and male face 大凹凸面

		small female and male face 小凹凸面	28	tap hole 取压孔
18		large tongue and groove face 大榫槽面	29	male face（MF） 凸面
		small tongue and groove face 小榫槽面	30	female face（FMF） 凹面
19		lens face 透镜式密封面	31	tongue face 榫面
20		dimension of flange 法兰尺寸	32	groove face 槽面
21		seal-welded 密封焊	33	lens gasket 透镜垫
22		welded front and back 前后焊	34	thickness of flange 法兰厚度
23		fillet-welded 角焊	35	length through hub 法兰高度
24		butt-welded (to pipe) 对焊（与管子）	36	diameter of bolt circle 螺栓孔中心圆直径
25		protective disc 法兰盖贴面	37	outside diameter of flange 法兰外径
26		stub end 管端突缘，翻边管接头	38	facing finish 法兰面加工
27		welding-on collar 平焊环		roughness 粗糙度

8.2.4 管螺纹 Pipe Threaded

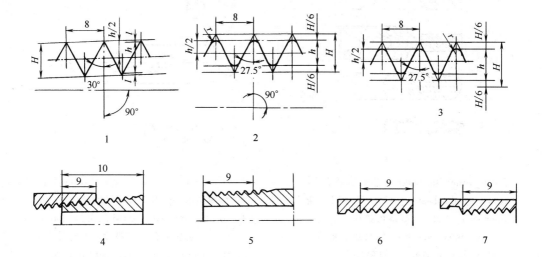

1　60° American standard taper pipe thread（NPT）　美国标准60°锥管螺纹牙型

2　55° taper pipe thread　55°圆锥管螺纹牙型

3　55° cylindrical pipe thread　55°圆柱管螺纹牙型

4　60° American standard taper pipe thread（NPT）　美国标准60°锥管螺纹

5　55° taper pipe external thread　55°圆锥管外螺纹

6　55° cylindrical pipe internal thread　55°圆柱管内螺纹

7　55° taper pipe internal thread　55°圆锥管内螺纹

8　pitch of thread　螺距

9　effect threads length　有效螺纹长度

10　outside threads length　外螺纹长度

8.2.5 垫片 Gasket

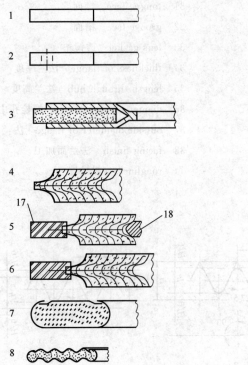
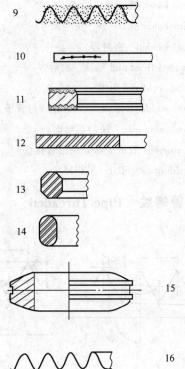

1 flat non-metallic gasket 非金属平垫片
 compressed asbestos gasket 压缩石棉垫片
 fiber sheet gasket 纤维板垫片
 rubber coated asbestos 涂橡胶石棉板
 rubber sheet gasket 橡胶垫片
 cloth-inserted rubber sheet gasket 编织物充填橡胶垫片
 PTFE impregnated asbestos gasket 浸聚四氟乙烯的石棉垫片
2 full face non-metallic gasket (for flat face flanges) 非金属全平面垫片（全平面法兰用）
3 plastic jacket gasket 塑料包覆垫
 PTFE envelope gasket 聚四氟乙烯包覆垫
(4～6 spiral-wound metal gasket, non-metallic filler 金属缠绕式垫片，非金属填充)
4 spiral-wound metal gasket 金属缠绕垫片
5 spiral-wound metal gasket, with inner and outer ring 金属缠绕垫片，带内外定位环
6 spiral-wound metal gasket, with outer ring 金属缠绕垫片，带外定位环
7 flat metal jacketed gasket, non-metallic filler 金属包覆平垫片，非金属填充
8 corrugated metal jacketed gasket, non-metallic filler 金属包覆波纹垫片，非金属填充
9 flexible graphite corrugated metal gasket 柔性石墨金属波齿垫片
10 flexible graphite metal gasket 柔性石墨复合垫片，金属增强石墨垫片
11 serrated metal gasket with PTFE (or with flexible graphite) 金属齿形聚四氟乙烯复合垫（或石墨复合）
12 flat metal gasket 金属平垫片
(13～15 ring joint gasket 环槽式垫圈片)
13 octagonal ring gasket 八角形垫圈片
14 oval ring gasket 椭圆垫，椭圆垫片
15 lens gasket 透镜垫，透镜垫片
16 solid-metal serrated gasket 整体金属齿形垫片
17 outer ring (external guide ring) 外定位环
18 inner ring (internal guide ring) 内定位环

8.2.6 管道用紧固件　Piping Fastener

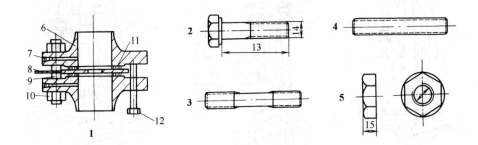

1　bolting orifice flange　孔板法兰的螺栓连接
2　machine bolt, hexagonal head bolt　机制螺栓，六角头螺栓
　　coarse thread　粗牙螺纹
　　fine thread　细牙螺纹
3　stud bolt threaded at both end　双头螺柱
4　stud bolt threaded full length　全螺纹螺柱
5, 10　nut　螺母
6　orifice flange　孔板法兰
7　tap hole　取压孔
8　orifice plate　孔板
9　stud bolt　双头螺柱
11　gasket　垫片
12　jack screw　顶丝螺栓
13　length of bolt　螺栓长度
14　diameter of bolt　螺栓直径
15　high of nut　螺母高度

8.2.7 阀门　Valve

8.2.7.1 阀杆与阀盖结构　Valve Stem and Bonnet Structure

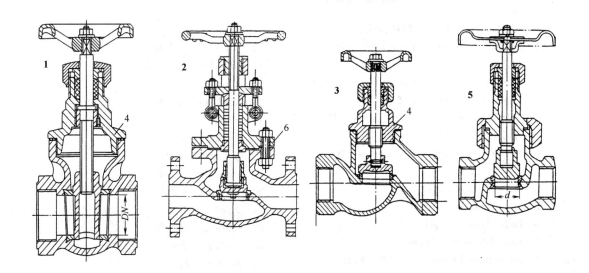

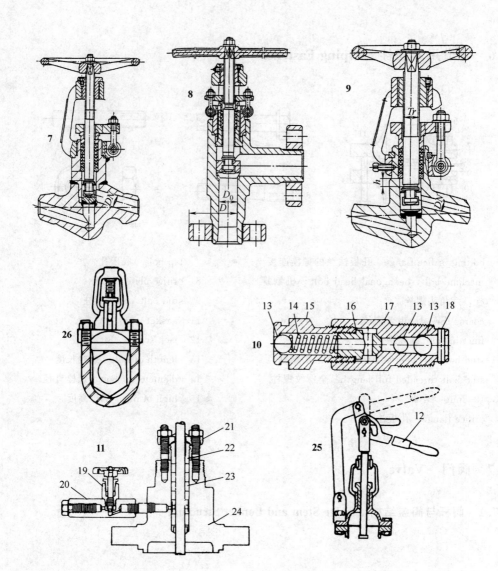

(1~3, 25　stem types　阀杆型式)
1　IS NRS (inside screw non-rising stem)　暗杆内螺纹
2　OS & Y (rising stem outside screw and yoke) 支架式明杆外螺纹
3　ISRS (inside screw rising stem)　明杆内螺纹
(4~9, 26　standard bonnet designs　标准阀盖结构)
4　screwed bonnet　螺纹阀盖
5　union bonnet　活接阀盖
6　bolted bonnet　闩接阀盖
7　welded bonnet　焊接阀盖
8　pressure seal bonnet　压力密封阀盖
9　integral bonnet　整体阀盖
10　double-ball grease injector　双球注油器
11　grease injector at lantern ring　套环上的注油器
12　hand lever　手柄
13　ball check　止回球
14　spring　弹簧
15　head housing　阀头
16　needle　针
17　valve housing　阀体
18　pin　销钉
19　stop valve　切断阀
20　injector　注油器
21　gland　压盖
22　follower　随动件
23　packing　填料
24　bonnet　阀盖
25　sliding stem quick opening　快开滑动杆
26　U-bolt bonnet　U形螺栓拴接阀盖

8.2.7.2 阀门 **Valves**

(1) 阀门（1） Valves（Ⅰ）

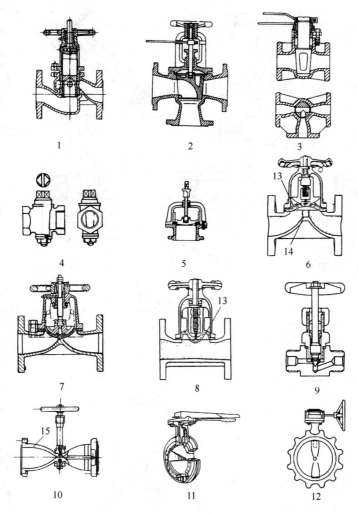

1 piston type regulating valve 柱塞式调节阀
2 change valve 换向阀
3 three way plug valve (T-port valve) 三通旋塞阀
4 cock 旋塞
5 flat valve 盖阀
6 weir type diaphragm valve 堰式隔膜阀
7 rubber lined diaphragm valve 橡胶衬里隔膜阀
8 through conduit type diaphragm valve 直通式隔膜阀
9 needle valve (pintle valve) 针形阀
10 pinch valve 夹紧式胶管阀
11 wafer style butterfly valve 对夹式蝶阀
12 lug style butterfly valve 凸耳式蝶阀
13 flexible reinforced diaphragm 挠性加强膜片
14 weir 堰
15 flexible elastomer tube 挠性合成橡胶管

(2) 阀门 (2) Valves (Ⅱ)

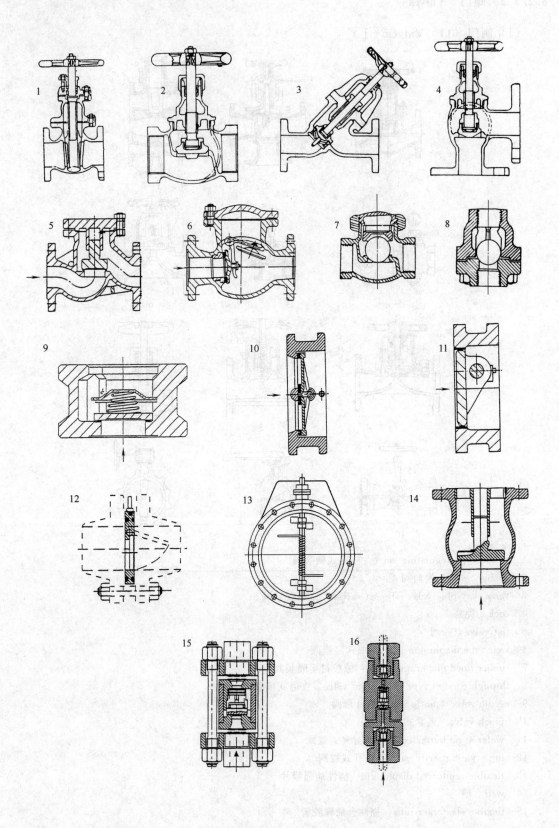

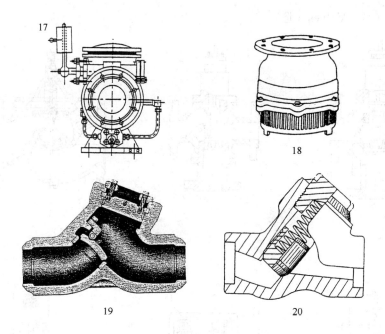

1　gate valve　闸阀

2　globe valve　截止阀

3　Y-globe valve　Y形截止阀

4　angle valve　角阀

[5～17　check valve (retaining valve, non-return one-way valve)　止回阀，单向阀]

5　lift check valve　升降式止回阀，摇板式止回阀

6　swing check valve　旋启式止回阀

7　ball check valve (for horizontal piping)　球式止回阀（水平管用）

8　ball check valve (for vertical piping)　球式止回阀（垂直管用）

9　wafer lift check valve　对夹升降式止回阀

10　wafer dual plate check valve　对夹双板止回阀

11　wafer butterfly check valve　对夹蝶式止回阀

12　wafer swing check valve (thin type)　对夹旋启式止回阀（薄型）

13　slow-closing butterfly check valve　缓闭蝶形止回阀

14　vertical lift check valve　立式升降止回阀

15　high pressure wafer lift check valve　高压对夹升降式止回阀

16　super high pressure check valve　超高压止回阀

17　slow-close swing check valve　缓闭旋启式止回阀

18　foot valve　底阀

19　tilting disc check valve　倾斜板止回阀

20　Y-pattern lift check valve　Y形升降式止回阀

(3) 阀门 (3) Valves (Ⅲ)

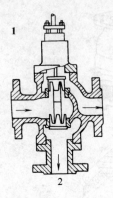 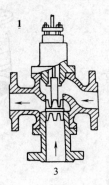 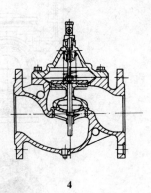

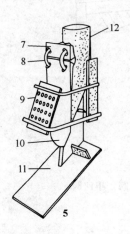 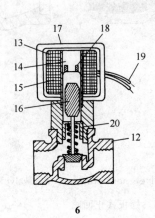 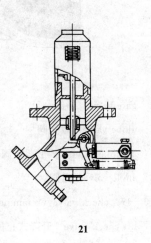

1	three way valve 三通阀	11	diffuser plate 分配板
2	divergent 分流（式）	12	body 阀体
3	convergent 合流（式）	13	solenoid coil 电磁线圈
4	pressure reducing valve, pilot controlled 先导式减压阀	14	stationary core 固定铁芯
		15	core tube 芯管
5	trickle valve (for cracking unit) 滴流阀（催化裂化装置用）	16	movable core 活动铁芯
		17	housing 外壳
6	solenoid valve 电磁阀	18	shading coil 校正线圈
7	hinge support bracket 铰链环支座	19	coil connection 线圈接头
8	hinge ring 吊环	20	spring retainer 弹簧支座
9	valve plate stop 阀板挡板	21	tank emergency shut-off valve 车用紧急切断阀
10	valve plate 阀板		

(4) 阀门 (4)　Valves (Ⅳ)

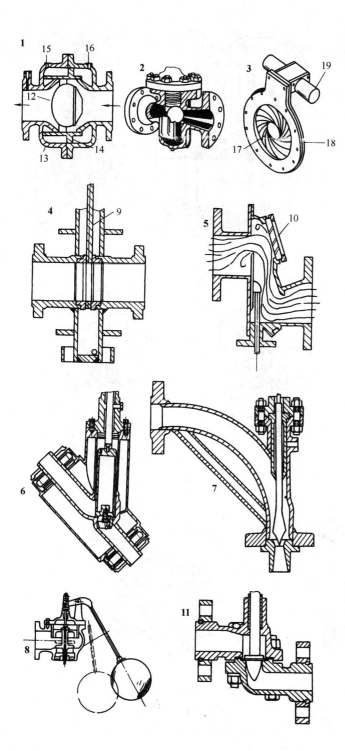

1　positioned plug valve　定位阀塞调节阀
2　cage positioned ball valve　笼式定位球调节阀
3　variable-orifice damper valve　变孔调节阀
4　slab type sliding gate valve　平板式滑动闸阀
5　guillotine sliding gate valve　闸刀式滑动闸阀
6　vacuum-jacketed Y-valve　真空夹套 Y 形阀
7　sweep angle valve　清刮式角阀
8　float-ball self-closing valve　浮球式自动关闭阀，自力式浮球阀
9　sliding plate　滑板
10　inspection window　窥视窗
11　self-draining valve　自动排液阀
12　plug　阀塞，阀芯
13　cylinder　阀筒体
14　piston　活塞
15　closing port　关阀（信号）接口
16　opening port　开阀（信号）接口
17　adjustable orifice　可调孔口
18　wafer valve body　薄片型阀体
19　actuator　驱动器

467

（5）阀门（5） Valves（Ⅴ）

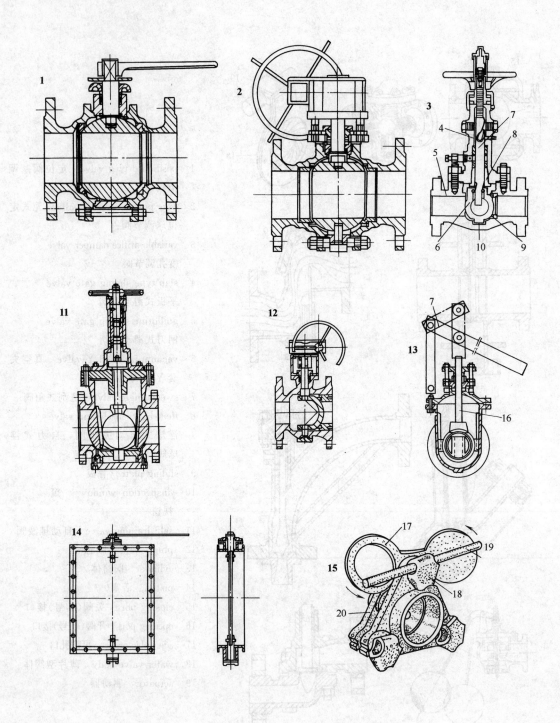

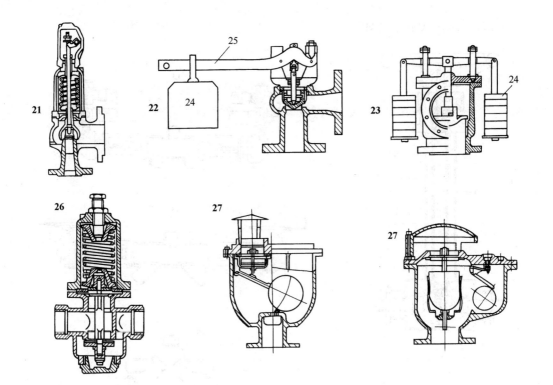

1 floating ball valve (Three piece) 浮动球阀（三片式）
2 trunnion ball valve (trunnion mounted) 固定球阀（装有底轴的）
3 orbit ball valve 轨道球阀
4 bonnet 阀盖
5 body 阀体
6 ball 球体
7，16 stem 阀杆
8 bonnet sleeve 阀盖衬套
9 seat ring 阀座密封圈
10 trunnion bushing 耳轴衬套
11 whole slant metal seal ball valve 撑开式金属密封球阀
12 V-eccentric ball valve V形偏心球阀
13 quick action valve 速动阀
　 quick opening valve 快开阀
　 quick closing valve 快关阀
　 quick closing emergency valve 事故快关阀
14 handle operated rectangle butterfly valve 手动矩形蝶阀
15 line blind valve 管道盲板阀
17 spectacle plate 8字盲板
18 machine bolt 机制螺栓
19 bar (handle) 把手
20 flange 法兰
[21～23 safety valve (escape valve protection valve) 安全阀]
21 spring safety valve 弹簧安全阀
22 lever and weight safety valve 杠杆重锤式安全阀
23 dead-weight safety valve 重锤安全阀
24 weight 重锤
25 lever 手杆
26 (pressure) reducing valve 减压阀
27 composite air release valve 复合式排气阀

(6) 阀门 (6) Valves (Ⅵ)

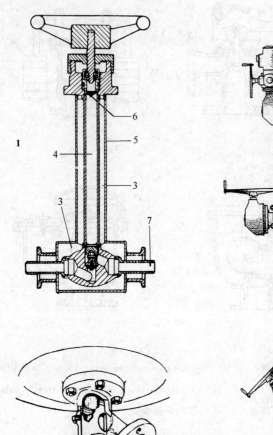
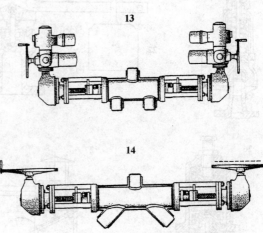
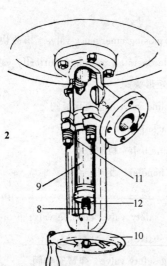
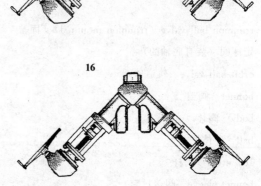

1 cryogenic stop valve 低温用切断阀
2 flush bottom outlet valve 罐底排污阀，放净阀
3 vacuum chamber 真空室
4 hollow operating stem 管状阀杆
5 vacuum jacket 真空夹套
6 thermal insulation 绝热层
7 cryogenic pipeline 低温管道
8 open yoke 开式支架
9 stainless steel plunger 不锈钢柱塞
10 hand-wheel 手轮
11 compression spring 压缩弹簧
12 non-rising stem 暗杆
(13～16 diverter valve 换向阀)
13 90° diverter valve 90°换向阀
14 45° diverter valve 45°换向阀
15 T-type diverter valve T形换向阀
16 M-type diverter valve M形换向阀

8.2.7.3 闸阀 Gate Valve

(1) 闸阀 (1) Gate Valve (Ⅰ)

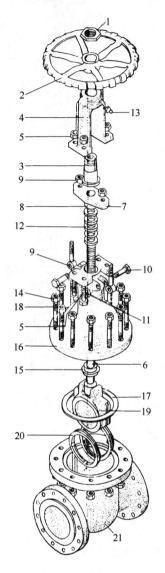

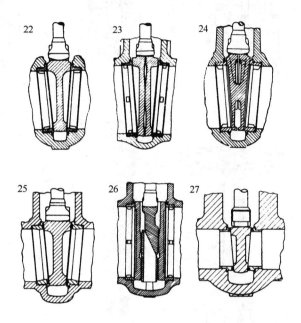

1 hand wheel nut 手轮螺母
2 hand wheel 手轮
3 stem nut 阀杆螺母
4 yoke 支架
5 yoke bolting 支架螺栓
6 stem 阀杆
7 gland flange 压盖法兰,填料压盖
8 gland 填料压盖
9 gland eyebolt and nut 压盖活节螺栓和螺母
10 gland lug bolts (eyebolt pin) 压盖凸耳螺栓 (活节螺栓销钉)
11,14 nut 螺母
12 stem packing 阀杆填料
13 drain plug 放油丝堵
15 backseat bushing (bonnet stem bushing) 上密封座套筒 (阀杆套筒)
16 bonnet 阀盖
17 bonnet gasket 阀盖垫片
18 bonnet bolts and nuts 阀盖螺栓,螺母
19 gate (disc wedge) 闸板,楔形闸板
20 renewable seat ring 可更换的阀座密封圈
21 body 阀体
22 solid wedge 楔形单闸板
23 split wedge 楔形双闸板
24 flexible wedge 挠性楔形闸板
25 teflon seat ring 聚四氟乙烯密封圈
26 double disc, parallel seat 平行双闸板
27 teflon insert in seat ring 聚四氟乙烯镶嵌密封圈

471

(2) 闸阀（2）　Gate Valve（Ⅱ）

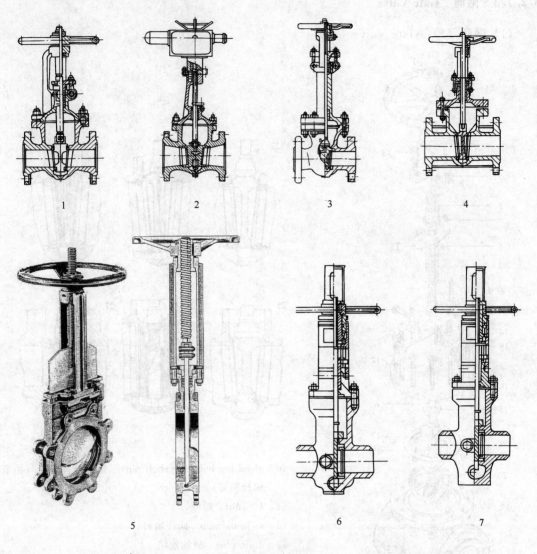
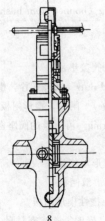

1　solid wedge gate valve　楔形单闸板闸阀
2　split wedge gate valve　楔形双闸板闸阀
3　cryogenic service gate valve　低温闸阀
4　jacket gate valve　保温夹套闸阀
5　flat gate valve；knife gate valve　刀型闸阀
（6，7　gate of non-flow-guiding hole　不带导流孔闸板阀）
6　general slab type gate　普通型平板闸阀
7　slab type gate of adjust type　调节型平板闸阀
8　slab type gate with flow-guiding hole　带导流孔平板闸阀

8.2.7.4 波纹管密封闸阀 Bellows Sealed Gate Valve

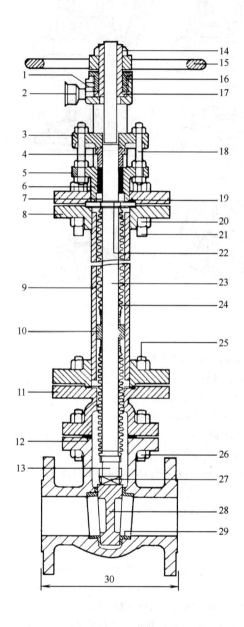

1　set screw　定位螺钉
2　lubricator　油杯
3　gland flange　压盖法兰
4　gland　填料压盖
5　upper bonnet　上阀盖
6　gland packing　填料
7　insert　垫圈
8　flange　法兰
9　flange tube　法兰管
10　center joining sleeve　中间连接套
11　lower bonnet　下阀盖
12，22　gasket　垫片
13　coupling　连接器
14　hand-wheel nut　手轮螺母
15　hand-wheel　手轮
16　yoke nut　支架螺母
17　yoke sleeve　支架衬套
18　gland studs and nuts　压盖双头螺柱和螺母
19　bellows plate　波纹管端板
20　flange tube nuts　法兰管螺母
21　flange tube stud bolt　法兰管双头螺柱
23　spindle　阀杆
24　bellows　波纹管
25　bonnet flange studs　法兰阀盖双头螺柱
26　bonnet studs and nuts　阀盖双头螺柱和螺母
27　tee head　T形接头
28　wedge　楔形闸板
29　body rings　密封座圈，阀体密封圈
30　face to face (flanged end)　法兰面至面长度（法兰端），结构长度

473

8.2.7.5 截止阀 Globe Valve

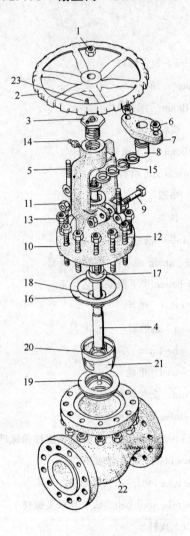

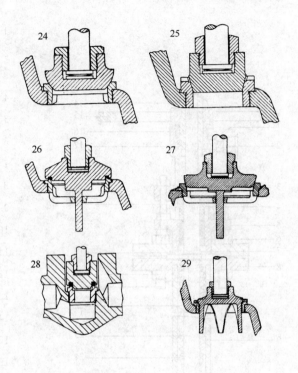

14　grease injector　注油器
15　packing　填料
16　bonnet gasket　阀盖垫片
17　bonnet stem bushing（backseat bushing）　阀杆套筒（上密封座套筒）
18　disc nut　阀瓣螺母
19　seat ring（facing）　密封圈，阀瓣密封面
20　thrust plate　止推板
21　disc　阀瓣，阀盘
22　body　阀体
23　locking pin　锁紧销
24　ball type seat disc（metal disc）　球形座阀瓣
25　plug type seat and disc without pilot（composition disc）　无导向旋塞型阀座和阀瓣（拼合阀盘）
26　globe type disc　球心型阀瓣（大阀用）
27　plug type seat and disc with pilot　带导向塞型阀座和阀瓣
28　globe type disc　球心型阀瓣（小阀用）
29　throttling type seat and disc（V-port）　节流型阀座和阀瓣（V形开口）

1　wheel nut　手轮螺母
2　hand-wheel　手轮
3　yoke sleeve　支架螺纹套筒，阀杆螺母，阀轭套
4　stem　阀杆
5　eyebolt　压盖活节螺栓
6　eyebolt nut　压盖活节螺母
7　gland flange　压盖法兰
8　gland　填料压盖
9　eyebolt pin　活节螺栓销钉
10　bonnet bolted　阀盖
11，13　nut　螺母
12　bonnet bolted　阀盖双头螺柱

8.2.7.6 止回阀 Check Valve

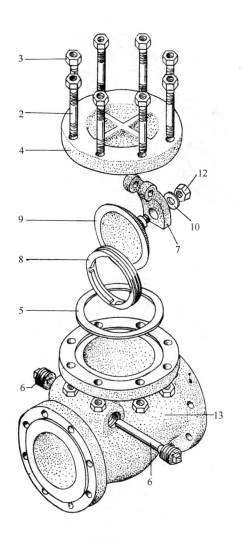
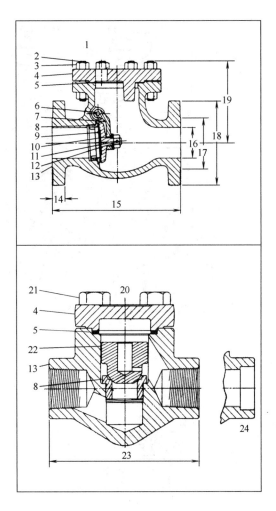

1	swing check valve 旋启式止回阀		13	body 阀体
2	bonnet stud bolt (cover stud bolt) 阀盖双头螺柱		14	flange thickness 法兰厚度
			15	face to face 法兰面到面长度（指法兰阀）
3	bonnet stud nut 阀盖螺母		16	boby end port 阀体通径
4	bonnet cap (cover) 阀盖		17	diameter of raised face 突面直径
5	bonnet gasket 阀盖垫片		18	outside diameter of pipe flange 管法兰外径
6	hinge pin (disc arm pin) 摇板销轴		19	center to top 阀体中心线至顶部距离
7	hinge (disc arm) 摇臂		20	lift check valve 升降式止回阀
8	seal ring 密封圈		21	bonnet cap screw 阀盖螺钉
9	disc (clack) 阀盘		22	ball type disc 球面阀盘
10	disc washer 阀盘垫圈		23	end to end 端面间长度（指螺纹端阀、承插焊接阀）
11	disc nut pin 阀盘螺母销钉			
12	disc nut 阀盘螺母		24	socket weld end 承插焊接端

475

8.2.7.7 球阀 Ball Valve

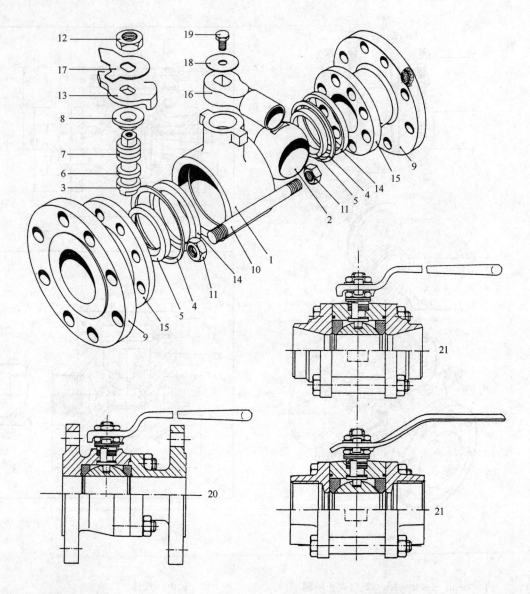

1　body　阀体
2　ball　球体
3　stem　阀杆
4　body gasket　阀体密封垫片
5　ball seat　阀座
6　stem gasket　阀杆垫圈
7　packing gland　填料
8　stem seal follower compression ring　阀杆密封件压环
9　body end　阀体端部
10　body bolt　阀体螺柱
11　body nut　阀体螺母
12　stem nut　阀杆螺母
13　stop plate　止动件
14　centre ring　定心环
15　body flange　阀体法兰
16　handle　手柄
17　lock ring　锁定环
18　handle washer　手柄垫圈
19　handle screw　手柄螺钉
20　full port type　通径型
21　reduced port type　缩径型

8.2.7.8 旋塞阀 Plug Valve

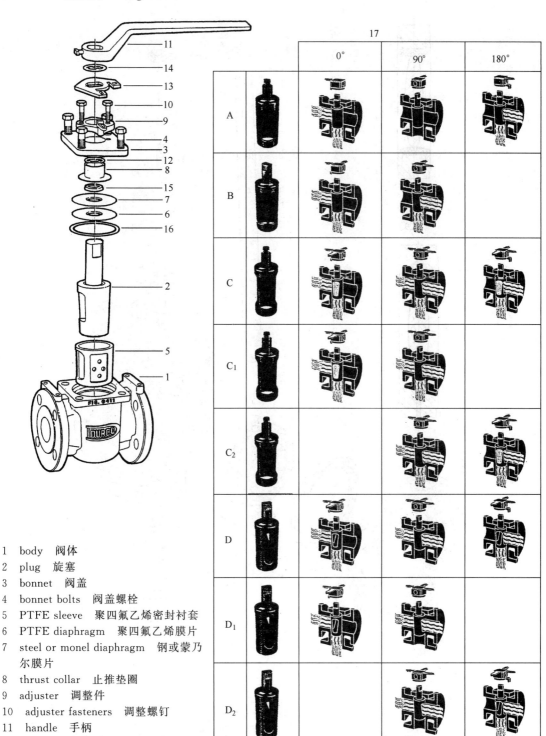

1 body 阀体
2 plug 旋塞
3 bonnet 阀盖
4 bonnet bolts 阀盖螺栓
5 PTFE sleeve 聚四氟乙烯密封衬套
6 PTFE diaphragm 聚四氟乙烯膜片
7 steel or monel diaphragm 钢或蒙乃尔膜片
8 thrust collar 止推垫圈
9 adjuster 调整件
10 adjuster fasteners 调整螺钉
11 handle 手柄
12 grounding spring 止推弹簧
13 stop collar 限位环
14 stop collar retainer 限位环固定垫圈
15 packing 填料
16 gasket 垫片
17 plug port arrangement for bottom three way plug valve 三通旋塞阀的旋塞通道形式

8.2.7.9 蝶阀 Butterfly Valve

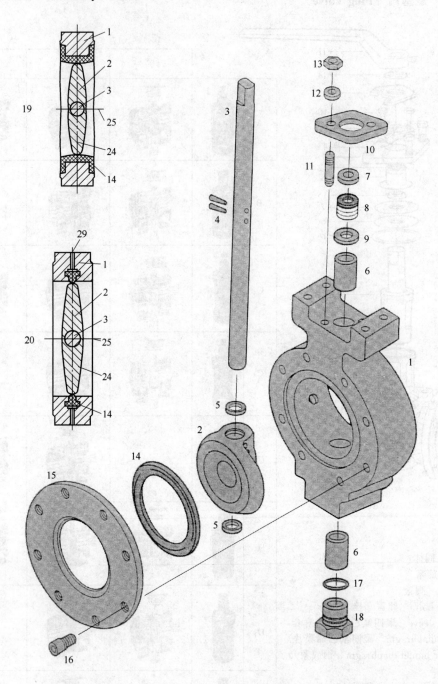

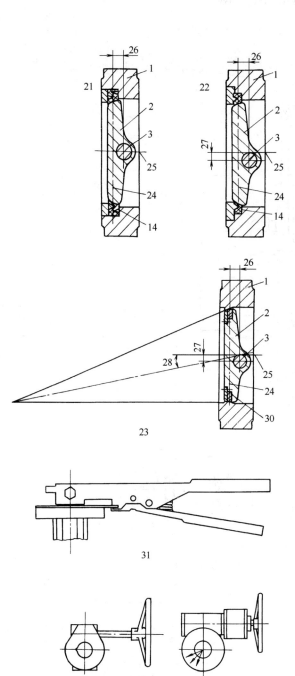

1　body　阀体
2　disc　阀板
3　stem, shaft　阀杆
4　taper pin　圆锥销
5　washer　阀板垫圈
6　bearing　轴承组件
7　packing ring　填料环
8　stem seal follower　阀杆密封件
9　thrust washer　止推垫圈
10　packing gland　填料压盖
11　gland bolting　填料压盖螺柱
12　lock washer　锁紧垫圈
13　nut　螺母
14　seat　阀座
15　seat retainer ring　阀座压环
16　seat retainer ring bolting　阀座压环螺栓
17　"O" ring　O形圈
18　positioned plug　定位旋塞
19　center line butterfly valve　中线蝶阀
20　seal butterfly valve charged with air　充压密封蝶阀
21　single eccentric butterfly valve　单偏心蝶阀
22　double eccentric butterfly valve　双偏心蝶阀
23　third eccentric butterfly valve　三偏心蝶阀
24　disc sealing　阀板密封截面
25　body center line　阀体中心线
26　first eccentricity　一次偏心
27　second eccentricity　二次偏心
28　third eccentricity　三次偏心
29　pressure charged input　充压口
30　stainless steel　不锈钢垫圈
31　handle actuator　手柄传动装置
32　single level worm gear actuator　一级蜗轮驱动装置
33　two level worm gear actuator　二级蜗轮驱动装置

8.2.7.10 弹簧安全泄压阀 Spring Safety-Relief Valve

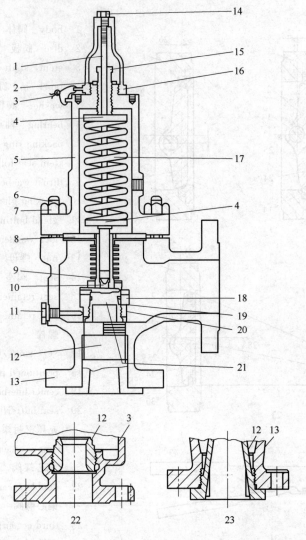

1	stem 阀杆		13	body 阀体
2	lock-nut 锁紧螺母，防松螺母		14	gag screw (adjusting screw) 调节螺钉
3	car seal (lead seal, wire seal) 铅封		15	compression screw 压紧螺钉
4	spring step (spring button) 弹簧挡片		16	cap 护罩
5	bonnet 阀盖		17	spring 弹簧
6	body studs 阀体双头螺柱		18	disc holder 阀盘座
7	stud nuts 螺母		19	disc retainer 阀盘护圈
8	guide 导向盘		20	adjusting ring 调整环
9	bellows 波纹管		21	disc 阀盘
10	holden insert 阀芯垫块		22	semi-nozzle construction 半嘴结构
11	ring pin 定位销		23	full nozzle construction 全嘴结构
12	nozzle 喷嘴			

8.2.7.11 液面控制浮球阀 Pilot Operated Ball Float Valve

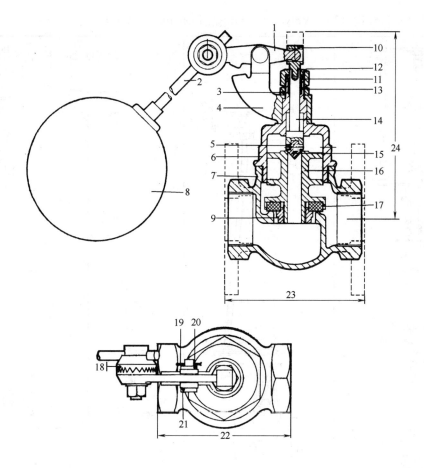

1	lever arm 杠杆臂		14	stem 阀杆
2	float arm 浮球连杆		15	disc 阀盘
3	lock nut 锁紧螺母		16	piston 活塞
4	fulcrum arm 支点臂		17	rubber piston insert 橡胶活塞垫圈
5	disc pin 阀盘销		18	lever connector 杠杆连接节
6	cover (bonnet) 阀盖		19	cotter pin 开口销
7	gunmetal body 炮铜阀体		20	fulcrum pin washer 支点销垫圈
8	copper float 铜浮球		21	fulcrum pin 支点销钉
9	retaining cap 支持环，固定环		22	end to end 端面间长度（指螺纹阀，承插焊接阀）
10	stem eye 阀杆孔		23	face to face 法兰面至面长度（指法兰阀）
11	gland 压盖		24	center to top 阀中心至顶部距离
12	gland nut 压盖螺母			
13	gland packing 填料			

8.2.7.12 阀门操纵机构 Valve Operating Mechanisms

(1) 阀门操纵机构（1） Valve Operating Mechanisms（Ⅰ）

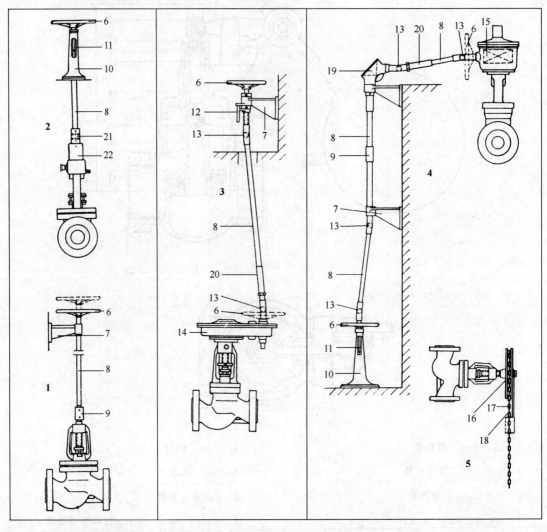

(1~3 operating position above the valve 操作位置在阀门上方)
1 direct mounted 直接延长阀杆
2 extension spindle taken through a floor 穿过楼板的延伸杆
3 the valve is offset to the floor column in one or both direction 阀门在一个或两个方向偏离地轴架
4 operating position below the valve 操作位置在阀门下方
5 chain valve 链轮阀
6 hand-wheel 手轮
7 support bearing 支承架
8 extension spindle 延伸杆
9,21 coupling 连接器
10 floor stand (floor column) 地轴架，落地式支座
11 indicator 阀开度指示器
12 padlock 扣锁
13 universal joint 万向接头
14 spur gearing 正齿轮传动
15 bevel gearing 伞齿轮传动
16 chain wheel 链轮
17 chain 链条
18 adjustable sprocket rim 可调节的链轮圈
19 skew bevel gearing 斜伞传动齿轮
20 flexible joint 挠性接头
22 yoke 支架

（2）阀门操纵机构（2）　　Valve Operating Mechanisms（Ⅱ）

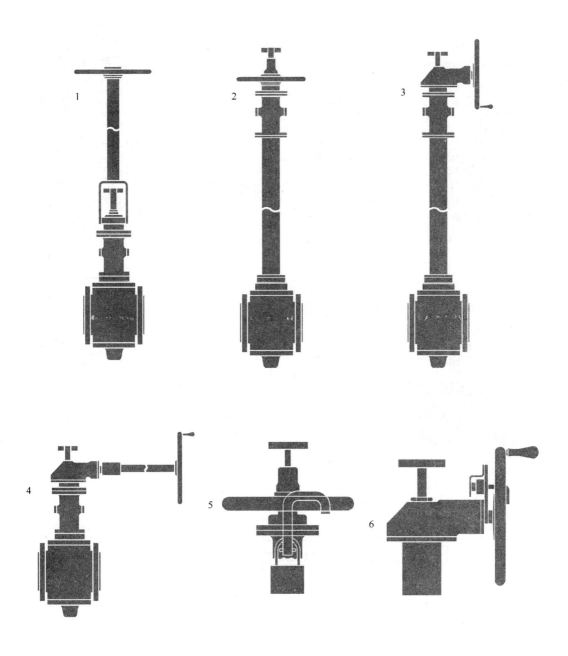

（1～4　extensions　伸长杆）
1　hand extension valve　手轮伸长杆阀门
2　hand extension valve for underground burial　埋地式手轮伸长杆阀门
3　gear extension valve for underground burial　埋地式齿轮传动伸长杆阀门
4　gear extension valve　齿轮传动伸长杆阀门
（5，6　locking devices　锁住装置）
5　locking device for hand operators　手轮操作锁住装置
6　locking device for gear operators　齿轮传动锁住装置

8.2.7.13 蒸汽疏水阀（器）和空气疏水阀（器） Steam Traps and Air Traps

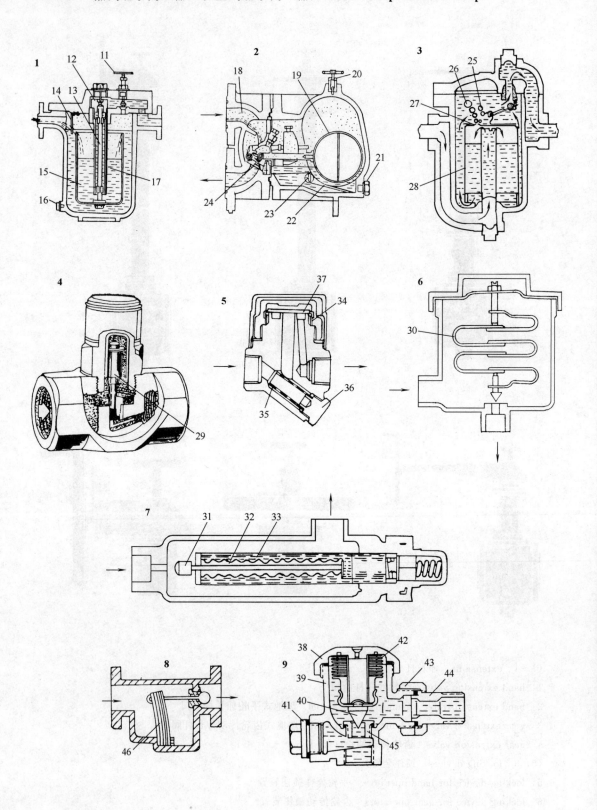

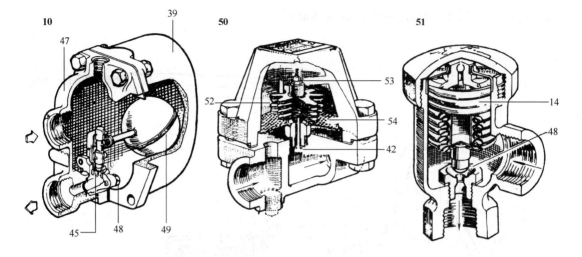

(1～3　mechanical trap　机械型疏水阀)
1　open bucket steam trap　浮桶式蒸汽疏水阀
2　float steam trap　浮球式蒸汽疏水阀
3　inverted bucket steam trap　倒吊桶式（钟形浮子式）蒸汽疏水阀
(4，5　thermodynamic trap　热动力式疏水阀)
4　impulse steam trap　脉冲式蒸汽疏水阀
5　disc steam trap with automatic blow-off device　带自动排放装置的盘式蒸汽疏水阀
(6～9　thermostatic trap　恒温型疏水阀)
6　metal expansion steam trap　金属膨胀式蒸汽疏水阀
7　liquid expansion steam trap　液体膨胀式蒸汽疏水阀
8　bimetallic expansion steam trap　双金属膨胀式蒸汽疏水阀
9　thermostatic trap　热静力式疏水阀
10　compressed air trap　压缩空气疏水阀
11　air vent valve　放气阀
12，45　valve seat　阀座
13　needle　阀针
14　baffle　挡板
15　bucket　浮桶
16，21，41　plug　丝堵
17　center pipe　中间管
18　condensate inlet　凝液入口
19　float　浮球
20　air vent　排气阀

22　lever　杠杆
23　crank shaft　曲轴
24，29，31，40　disc　阀盘
25　air bubble　气泡
26　steam bubble　蒸汽泡
27　air vent　排气孔
28　invert bucket　倒吊桶，钟形浮子
30，34，46　bimetal　双金属片
32，42　bellows　波纹管
33　liquid　液体
35　screen　纱网
36　screen holder　滤网座
37　disc plate　阀板
38　bonnet　压盖
39　body　壳体
43　coupling　管接头
44　union nipple　螺纹短管
47　cover　阀盖
48　valve plug　阀芯
49　ball float　浮球
50　bimetallic pressure balanced steam trap　双金属压力平衡式蒸汽疏水阀
51　balanced pressure type steam trap　压力平衡式蒸汽疏水阀
52　bimetal crosses　十字形双金属片
53　adjustment　调节（螺钉）
54　strainer　滤网

8.2.7.14 热膨胀阀 Thermo Expansion Valve

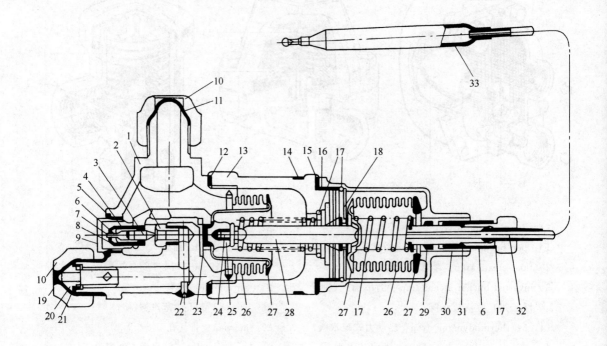

1	body 阀体		18	bellows keeper 波纹管座
2	seat 阀座		19	cap 帽
3	needle 阀针		20	strainer 过滤器
4	spring for needle 针簧		21	strainer washer 过滤器垫圈
5	needle case 阀针壳体		22	set screw 固定螺钉
6	adjusting bolt 调节螺栓		23	connector 连接头
7	pin 销		24	spring stopper 弹簧挡块
8	adjusting nut 调节螺母		25	lock nut 锁紧螺母
9	seal nut 密封螺母		26	bellows 波纹管
10	flare nut 扩口螺母		27	bellows fittings 波纹管接头
11	seal cap 密封帽		28	connector 连接杆
12	body gasket 阀体垫片		29	O-ring O形环
13	bakelite cover 绝缘木盖		30	bushing 套筒
14	name plate 铭牌		31	washer 垫圈
15	cover gasket 罩子垫片		32	capillary tube 毛细管
16	cover 罩子		33	remote bulb 遥控温包
17	spring 弹簧			

8.2.8 管道特殊件 Pipe Specialty

8.2.8.1 过滤器，阻火器 Strainer, Flame Arrester

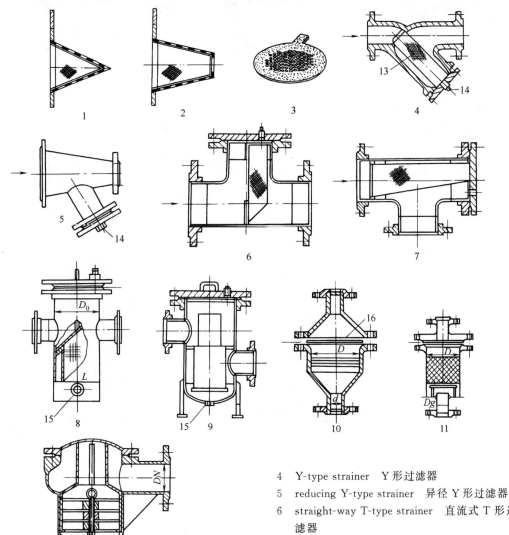

(1～3 temporary strainer 临时过滤器)
1 cone type strainer (fastigium) 锥形过滤器（尖顶）
2 cone type strainer (flat roof) 锥形过滤器（平顶）
3 metal edge-type filter 算式过滤器，多孔金属片过滤器
4 Y-type strainer Y形过滤器
5 reducing Y-type strainer 异径Y形过滤器
6 straight-way T-type strainer 直流式T形过滤器
7 side-way T-type strainer 折流式T形过滤器
8 basket Type strainer 篮式过滤器
9 supporting basket strainer of filter buckets 多滤桶支承型篮式过滤器
10 metal screen flame arrester 金属丝网阻火器
11 gravel flame arrester 砾石阻火器
12 explosion-proof flame arrester 防爆阻火器
13 screen 滤网
14 vent plug 排放丝堵
15 drain hole 排液口
16 element of metal foil 金属滤网

8.2.8.2 分离器，消声器，取样冷却器 Separator, Silencer, Sample Cooler

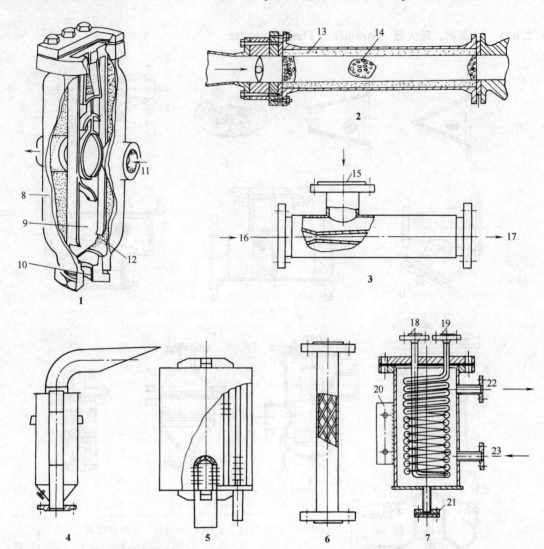

1	separator 汽水分离器	13	sound absorber 吸声材料
2	silencer 消声器，消音器	14	perforated acoustic construction 多孔吸声结构
3	steam-water mixer 汽水混合器	15	steam inlet 蒸汽入口
4	gas silencer 气体消声器，气体消音器	16	water inlet 进水口
5	steam silencer 蒸汽消声器，蒸汽消音器	17	water outlet 出水口
6	static mixer 静态混合器	18	stream inlet 流体入口
7	sample cooler 取样冷却器	19	stream outlet 流体出口
8	body 壳体	20	lug 支耳
9	baffle 挡板	21	drain 排液口
10	drain 放液口	22	cooling water outlet 冷却水出口
11	liquid mist and gas 气体与水雾（进入）	23	cooling water inlet 冷却水入口
12	liquid droplet 液滴		

8.2.8.3 视镜 Sight Glass

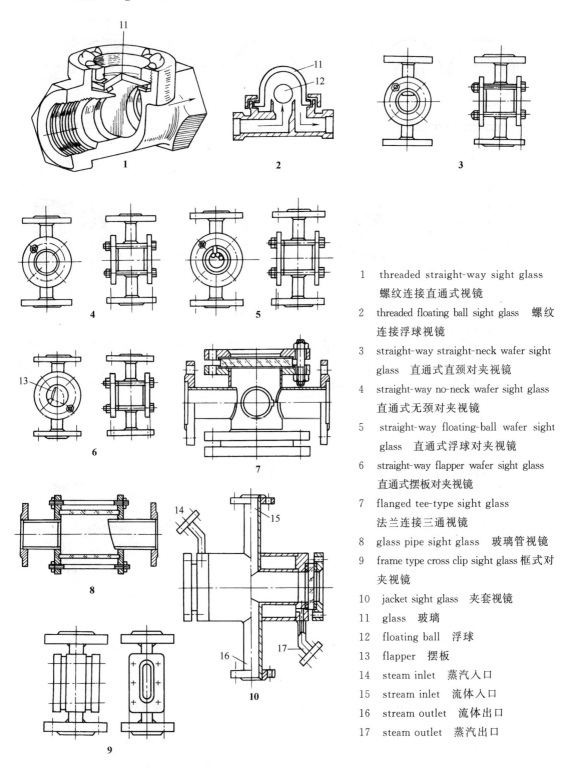

1 threaded straight-way sight glass 螺纹连接直通式视镜
2 threaded floating ball sight glass 螺纹连接浮球视镜
3 straight-way straight-neck wafer sight glass 直通式直颈对夹视镜
4 straight-way no-neck wafer sight glass 直通式无颈对夹视镜
5 straight-way floating-ball wafer sight glass 直通式浮球对夹视镜
6 straight-way flapper wafer sight glass 直通式摆板对夹视镜
7 flanged tee-type sight glass 法兰连接三通视镜
8 glass pipe sight glass 玻璃管视镜
9 frame type cross clip sight glass 框式对夹视镜
10 jacket sight glass 夹套视镜
11 glass 玻璃
12 floating ball 浮球
13 flapper 摆板
14 steam inlet 蒸汽入口
15 stream inlet 流体入口
16 stream outlet 流体出口
17 steam outlet 蒸汽出口

8.2.9 管道特殊元件 Pipe Special Element

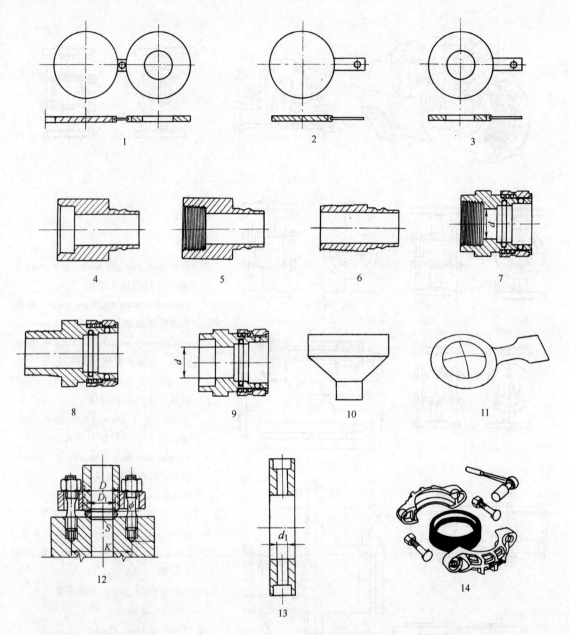

1　spectacle blind，figure 8 blind　8字盲板
2　blank　插板，盲板
3　spacer　垫环
(4~6　hose connection　软管接头)
4　socket welding end　承插焊端
5　threaded end　螺纹端
6　male threaded end　外螺纹端
(7~9　quick coupling　快速接头)
7　threaded end　螺纹端
8　male threaded end　外螺纹端
9　socket welding end　承插焊端
10　drain funnel　排液漏斗
11　rupture disc　爆破膜
12　pad type flange　盘座式法兰，拧入式法兰
　　special flange　特殊法兰
13　drip ring　排液环
14　Victaulic coupling　唯特利接头，卡箍接头

8.2.10 填料 Packings

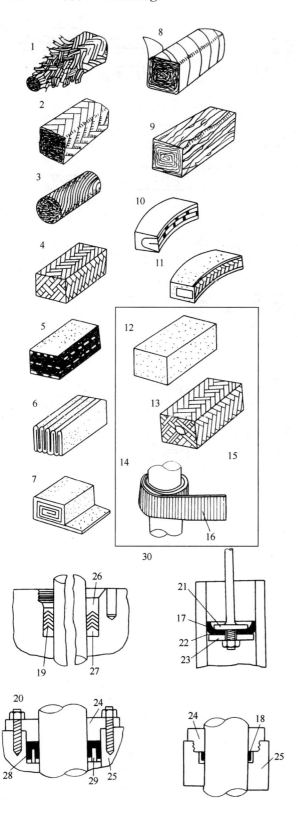

(1~7 fibrous compression packing 压紧纤维填料)

1. square braided packing 方形编织填料
2. plaited packing 褶叠编织填料
3. twisted packing 麻花填料
4. interbraid packing 交叉编织填料
5. laminated cloth packing 叠层织物填料
6. folded cloth packing 折叠织物填料
7. rolled cloth packing 卷压织物填料

(8~11 metallic packing 金属填料)

8. metal foil spiral wrapped packing 金属箔螺旋卷制填料
9. metal foil crinkled and twisted packing 绉状金属箔卷制填料
10. metal core packing 金属芯填料
11. braided core packing 编织芯填料
12. lubricated plastic packing 润滑塑料填料
13. dry bonded plastic packing 塑料黏结填料
14. flexible graphite packing 挠性石墨填料
15. plastic core 塑料芯
16. graphite foil 石墨薄片
17. cup packing 杯形填料
18. flange packing 凸缘式填料
19. V-ring packing V形环填料
20. U-seal U形环密封填料
21. inside follower 内随动件，内填料压紧件
22. heel 根部
23. back plate 底板，填料压板
24. backing ring 上密封环
25. gland 压盖
26. female support 凹形垫环
27. male support 凸形垫环
28. U-ring U形环（填料）
29. pedestal ring 支托环
30. lip packing 唇形密封件

8.2.11 管子端部连接 End Connection in Tubing and Pipe

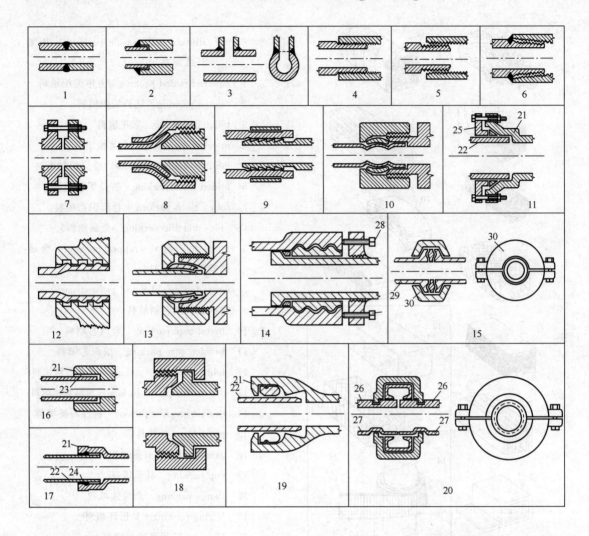

1. butt weld　对焊
2. socket weld　承插焊
3. branch weld　支管焊接
4. taper pipe to straight coupling thread　圆锥管螺纹与圆柱管螺纹管接头连接
5. taper pipe thread　锥管螺纹连接
6. seal-welded taper pipe thread joint　锥管螺纹密封焊连接
7. flanged joint　法兰连接
8. flared-fitting　扩（大管）口连接
9. clamped-insert joint　插入夹紧连接
10. compression-fitting joint　压紧连接
11. packed-gland joint　填料函式连接
12. expanded joint　胀接
13. bite-type fitting joint　咬紧连接
14. pressure-seal joint　压紧封接
15. V-clamp joint　V形管卡连接，夹紧环连接
16. soldered joint　软钎焊连接
17. poured joint　承插浇注连接
18. union　活接头
19. push-on joint　推紧连接
20. grooved joint　沟槽连接
21. bell　承口
22. spigot　插口
23. solder or cement　钎料或水泥
24. poured compound　浇注材料
25. follower　随动件
26. cut-groove type　切削槽式
27. rolled-groove type　滚压槽式
28. jack bolt　顶丝
29. stub end　翻边短管
30. clamp ring　夹紧环

8.3 管道机械 Piping Mechanics

8.3.1 金属软管 Metal Flexible Hose

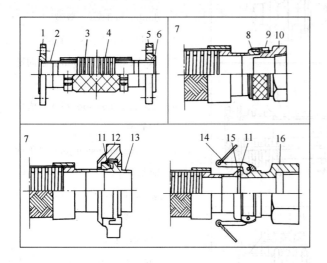

1 SO-flange 滑套法兰
2 collar pipe 接管
3 corrugated hose of stainless steel 不锈钢波纹管
4 stainless steel braiding 不锈钢网套
5 LJ-flange 松套法兰
6 welding rim 密封座
7 joint 接头
8 plug bush 插套
9 cutting sleeve 卡套
10 plug 插轴
11 gasket 密封垫
12 outside joint 外接头
13 inside joint 内接头
14 lever 控制杆
15 female joint 阴接头
16 male joint 阳接头

8.3.2 金属补偿器（膨胀节） Metal Expansion Joint

8.3.2.1 填函式伸缩节 Packed Slip Expansion Joint

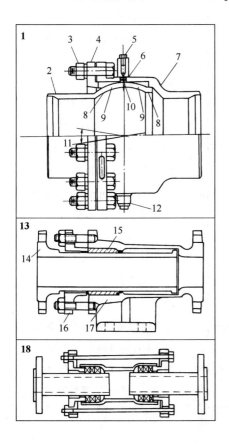

1 ball joint 球形补偿器
2 ball 球
3 flange bolt 法兰螺栓
4 retainer flange 固定法兰
5 plunger 柱塞
6 packing cylinder 柱塞填料缸
7 socket 承插节
8 compression seal 密封垫
9 containment seal 密封圈
10 injectable packing 密封填料
11 angular flex 转动角度
12 1/2" coupling and plug 1/2"半管接头，堵头
13 single action packed slip joint 单向滑动填料函式接头
14 slip pipe 滑动管
15 packing 填料
16 gland 填料压盖
17 sleeve 套管
18 double action packed slip joint 双向滑动填料函式接头

493

8.3.2.2 波纹补偿器（膨胀节） Bellows Expansion Joint

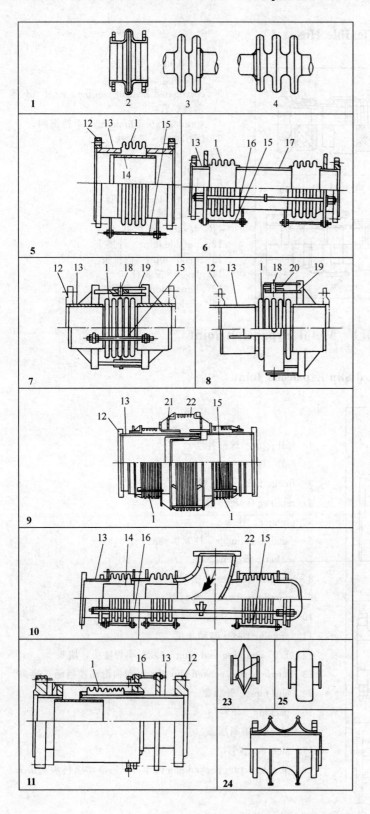

1　bellow　波纹管
2　single bellow　单波
3　double bellows　双波
4　multiple bellows　多波
5　axial expansion joint　轴向膨胀节
6　lateral expansion joint　横向膨胀节
7　hinged expansion joint　角向膨胀节
8　gimbal expansion joint　万向铰膨胀节
9　straight pressure balanced expansion joint　直管压力平衡膨胀节
10　bend pressure balanced expansion joint　曲管压力平衡膨胀节
11　externally pressurized expansion joint　外压膨胀节
12　flange　法兰
13　end pipe　端接管
14　internal sleeve　内衬筒
15　fixed bolt　固定拉杆
16　tie-rod　大拉杆
17　connecting pipe　中间接管
18　hinge pin　铰链销
19　hinge pad　铰链板
20　floating gimbal ring　浮动万向环
21　end plate　端板
22　balanced bellows　平衡波纹管
23　formed heads　锥盘型
24　convex type　内凸型
25　flat plates　平板型

8.3.3 管道支吊架　Pipe Hangers and Pipe Supports

8.3.3.1 管道支架　Pipe Supports

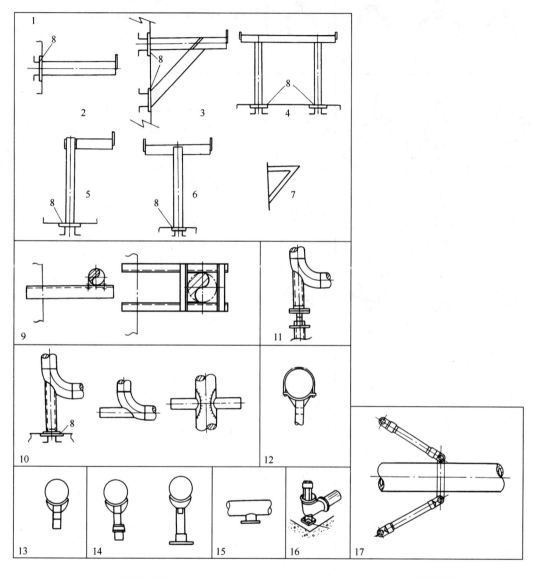

1	resting support　承重支架	10	trunnion support　耳轴支架
2	cantilever support　悬臂支架	11	adjustable (adj.) support　可调支架
3	triangular support　三角支架	12	pipe stanchion　鞍式管支柱
4	Π-type support　门形支架	13	pipe saddle support　鞍式支架
5	L-type support　L形支架	14	adj. pipe saddle support　可调鞍座管支架
6	T-type support　T形支架	15	stool　平管支架
7	welded bracket　焊接墙架	16	base elbow　弯管支架，弯头支架
8	inserted element　预埋件	17	sway brace　斜拉架
9	guide support　导向支架		

8.3.3.2 管道吊架　Pipe Hangers

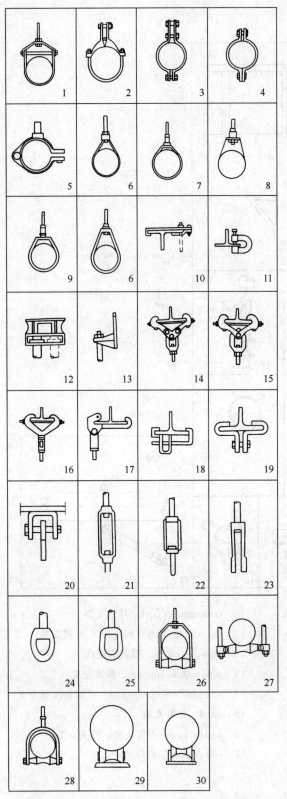

1　adj. steel clevis hanger　钢制可调 U 形管吊
2　yoke type pipe clamp　轭式管夹
3　steel double bolt pipe clamp　钢制双螺栓管卡
4　steel pipe clamp　钢制管卡
5　split pipe clamp　对开式管卡
6　adj. band hanger　可调带式管吊
7　adj. swivel pipe ring (solid ring type)　可调旋转整环式管吊
8　adj. swivel pipe ring (band ring type)　可调旋转带环式管吊
9　adj. swivel pipe ring (split ring type)　可调旋转对开式管吊
10　top I-beam clamp　工字钢顶卡
11　C-clamp　C 形卡
12　malleable concrete insert　（可锻）铸铁混凝土预埋件
13　side beam bracket　梁侧吊架
14　steel wide face clamp with eye nut　带环形螺母宽缘工字钢吊卡
15　steel I-beam clamp with eye-bolt　带环形螺母工字钢吊卡
16　malleable beam clamp with extension piece　带延伸件的钢梁吊卡
17　side I-beam or channel clamp　工字钢侧卡，槽钢卡
18　side I-beam clamp　工字钢侧卡
19　center I-beam clamp　工字钢中间卡
20　clevis bracket　插式吊耳
21　steel turnbuckle　钢花篮螺栓
22　swivel turnbuckle　可调花篮螺栓
23　steel clevis　钢 U 形吊卡
24　socket only for split ring　拼合环的套节
25　steel weldless eye nut　钢制整体式环形螺母
26　single rod roll type hanger　单杆滚轮管吊
27　two rod roll type hanger　双杆滚轮管吊
28　adj. swivel pipe roll　可调旋转滚轮管吊
29　roller　滚轮
30　rolling bearing　滚动支座

8.3.3.3 弹簧架 Spring Supports

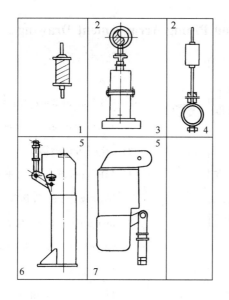

1　spring cushion　弹簧减震器
2　variable spring support　可变弹簧支架
3　resting type variable spring support　可变弹簧托架
4　variable spring hanger　可变弹簧吊架
5　constant spring support　恒力弹簧支架
6　resting type constant spring support　恒力弹簧托架
7　constant spring hanger　恒力弹簧吊架

8.3.3.4 管架零部件 Attachments of Piping Supports

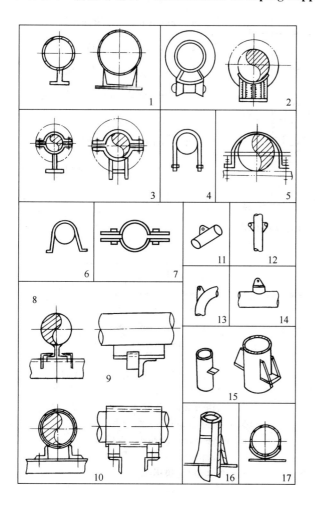

1　shoe　管托
2　saddle shoe　鞍形管托
3　clamp shoe　管夹式管托
4　U-bolt　U形螺栓
5　U-band　U形管卡
6　long clamp　长管卡
7　riser clamp　立管管卡
8　typical standard guide support　常用导向管架
9　angle guide for tee support　T形管托的角钢导向架
10　tube type guide　套管式导向管架
11　ears　吊耳
12　raiser ear　立管吊耳
13　elbow lug　弯头吊耳
14　cylindrical lug　圆筒式吊耳
15　raiser lug　立管支耳
16　skirt　裙式管座
17　shield　保护板

8.4 地下管道 Underground Piping

8.4.1 地下管平面布置图 Underground Piping Arrangement Drawing

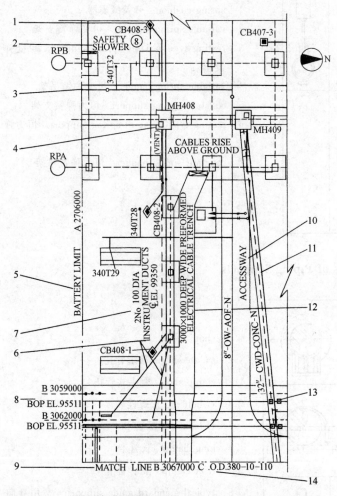

1. CB (catchbasin) 雨水口
2. safety shower 安全淋浴
3. drip funnel 漏斗
4. MH (manhole) 检查井
5. battery limit 界区线
6. instrument duct 仪表电缆套管
7. CL EL. (center line elevation) 中心标高
8. BOP EL. (bottom elevation of pipe) 管底标高
9. match line 接续分界线
10. accessway 通道
11. underground pipe 地下管道
12. electrical cable trench 电气电缆沟
13. guard post 防撞柱
14. C.O.D. (continued on drawing) 接续

8.4.1.1 地下管道图例 Underground Piping Legend

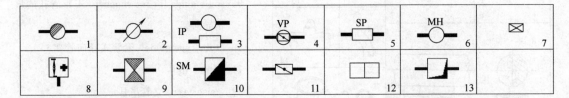

1. fire hydrant 消火栓
2. fire monitor 消防炮
3. instrument pit 仪表井
4. valve pit 阀门井
5. septic tank 化粪池
6. manhole 检查井
7. fire box 消火栓箱
8. eye washer & safety shower 洗眼器及安全淋浴
9. overflow pit 溢流井
10. sealed manhole 水封井
11. valve direct buried 直埋阀门
12. oil separation tank 隔油池
13. sump pit 集水坑

8.4.1.2 地下管道管线代号 Underground Pipe Line Code

1	CW cooling water 循环水	7	FW fire water 消防水
2	CWD contaminated water drainage 污染水排水	8	IW industrial water 生产给水，工厂用水
3	CWR cooling water return 循环（冷却水）回水	9	IWW industrial wastewater 生产废水，工业废水
4	CWS cooling water supply 循环（冷却水）给水	10	OW oily water 含油水
		11	PW potable (drinking) water 饮用水
5	CWT cooling water tempered 循环（冷却水）中间回水	12	RW raw water 原水
		13	RWS rainwater sewer 雨水排水
6	CWW clean wastewater 清净废水	14	SAS sanitary sewer 生活污水
		15	DW demineralized water 脱盐水

8.4.2 地下管常见附属设施及构筑物 Common Facility and Structure of U/G Piping

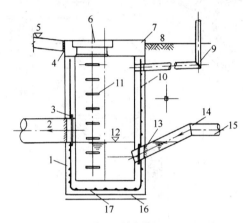

manhole 检查井

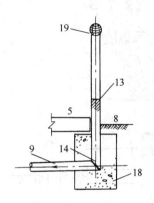

vent pipe 通气管

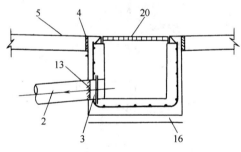

catchbasin 雨水口

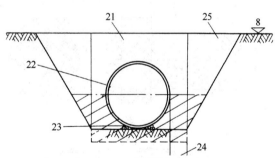

pipe bedding 管道基础

1	wall 井壁	14	mitre 斜接弯管
2	outlet 出口	15	sealed pipe 水封管
3	puddle flange 翼环	16	blinding 垫层
4	expansion joint 伸缩缝	17	base 井底板
5	paving level 铺砌地坪	18	mass concrete support 素混凝土支墩
6	heavy duty/light duty cover 重型/轻型井盖	19	galvanized wire balloon 镀锌钢丝球网
7	cover slab 盖板	20	grating 箅子板
8	grade 地坪	21	backfill soil 回填土
9	vent pipe 放空管	22	sand/fine aggregate 沙子/细料
10	bar 钢筋	23	granular bed 管基粒料
11	ladder 爬梯	24	working space 作业空间
12	liquid level 液位	25	pipe trench 管沟
13	coating and wrapping 涂漆及缠带		

8.5 绝热 Insulation

8.5.1 管道绝热 Piping Insulation

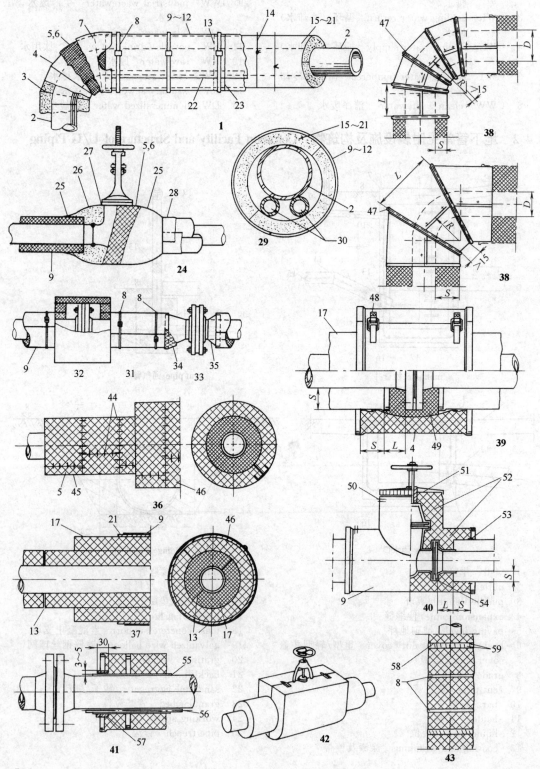

1　piping insulation　管道绝热（保温，保冷）
　　hot insulation　保温
　　cold insulation　保冷
2　carbon steel piping　碳钢管道
3　mastic gum　玛琉脂胶黏剂
4　pipe insulation segmental　管子保温预制块
5　wire mesh　金属丝网，铁丝网
6　glass fabric　玻璃布
7　mastic weatherproof coating　玛琉脂涂层，玛琉脂防潮层
8　bands　箍带
9　metal jacketing；metal weatherproofing　金属保护层（防风雨）
10　cellular-glass　泡沫玻璃
11　polykraft moisture (vapor) barries　多层牛皮纸防潮层
12　aluminum alloy jacketing　铝合金板保护层
　　galvanized (sheet) iron; galvanized plain sheet jacketing　镀锌铁皮保护层
13　lap sealer　搭接封口，搭接密封
14　wire ties　铁丝箍紧
15　centrifugal glass wool pipe shell　离心玻璃棉管壳
　　rock wool pipe shell　岩棉管壳
　　aluminosilicate pipe shell　硅酸铝管壳
16　calcium silicate pipe shell　硅酸钙管壳
17　cellular glass pipe shell; foam glass pipe shell　泡沫玻璃管壳
　　polyurethane foam pipe shell　泡沫聚氨酯管壳
　　polystyrene foam pipe shell; cellular polystyrene pipe shell　泡沫聚苯乙烯管壳
18　diatomaceous earth block　硅藻土保温块
19　expanded perlite block　膨胀珍珠岩块
20　glass wool blanket (board)　玻璃棉毡（板）
21　mineral wool blanket　矿渣棉毡
22　longitudinal lap　纵向搭接
23　circumferential lap　环向搭接
24　valves without flange　非法兰连接式阀门（的保温）
25　finishing cement　抹面水泥
26　insulating cement　保温水泥，隔热水泥
27　seal around bonnet with mastic sealing compound　用玛琉脂胶黏剂封固阀盖四周
28　lap metal jacketing over finishing cement　在抹面水泥层外的金属保护罩
29　insulation for double tracer　双伴热管保温
30　tracers　伴热管
31　insulation at flanges　法兰的绝热
32　insulated flanges　绝热的法兰
33　uninsulated flanges　不绝热的法兰
34　mastic and glass fabric　玛琉脂和玻璃布
35　metal jacketing end covers　金属保护端罩
36　single or multi-layer hot insulation　单层或多层保温
37　single or multi-layer cold insulation　单层或多层保冷
38　cladding for elbows　弯头的外保护层
39　hot and cold insulation-removable flange cap for preventing formation of condensate　防冷凝可拆卸法兰的保温与保冷
40　cold insulation-removable valve cap for insitu foaming　现场发泡可移动阀盖的保冷结构
41　insulation structure for pipes with electric tracing　管道电伴热的保温结构
42　hot insulation-removable valve cap　可拆卸阀盖的保温结构
43　heat insulation-fixing of insulation materials on vertical pipes　立管保温材料的固定
44　wire hooks for joining the insulation materials　保温钩钉
45　blanket (1st layer)　棉毡（第一层）
46　blanket (2nd layer)　棉毡（第二层）
47　crimping　咬接接缝
48　snap catch　搭扣
49　filling with loose mineral fibres　填充松散矿棉
50　self-tapping screw　自攻螺钉
51　rubber hose　橡胶软管
52　foamed insulation material　泡沫保温材料
53　sealing　密封
54　plastic foil　塑料隔层
55　thermal foil　隔热层
56　electric tracing　电伴热带
57　hot insulation interleaf fixed with tie wire　用铁丝捆扎的固定保温的插入层
58　spacer, support ring　隔环，支承环
59　insulation fixed with tie wire　用铁丝捆扎固定的保温层

8.5.2 立式容器的外部隔热 External Thermal Insulation for Vertical Vessel

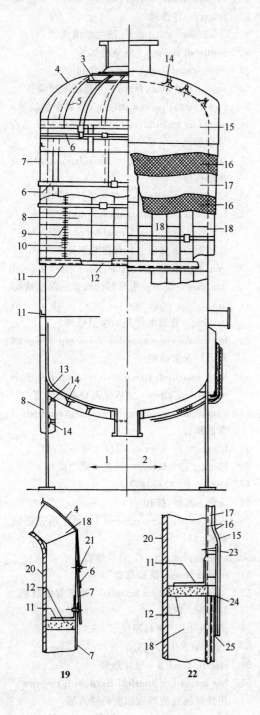

1 blanket insulation with metal jacketing 保温毡保温（带金属保护层）
2 block insulation with mastic weatherproofing 保温块保温（玛琉脂保护层）
3 steel ring 钢环
4 metal jacketing 金属保护层
5 lap sealer 搭接封口
6 circumferential band 环箍
7 corrugated metal jacketing 波纹金属保护层
8 blanket insulation 保温毡
9 wire lacing 捆扎铁丝
10 lock rings 骑马环（锁环）
11 angle ring support 角钢支承环
12 loose mineral fiber 矿渣棉
13 air space 气隙
14 clip 保温卡
15 mastic weatherproof coating 玛琉脂保护层
16 wire mesh or glass fabric 铁丝网或玻璃布
17 insulating or finishing cement 保温水泥层或水泥抹面
18 insulation block 保温块
19 expansion joint 膨胀节，伸缩接头
20 vessel wall 容器壁
21 self tapping screws 自攻螺钉
22 expansion joint-block insulation with mastic weatherproofing 保温块保温-玛琉脂防护层的伸缩接头
23 resin sized paper 涂树脂的纸
24 hardware cloth 金属丝网
25 clearance for expansion 伸缩缝隙

8.6 工程图常用缩写词（按字母顺序排列） Abbreviations for Use on Drawing (in Alphabetical Order)

缩 写	意 义 英 文	意 义 中 文
AB	anchor bolt	地脚螺栓
ABR SW	air break switch	空气切断开关
ABS	absolute	绝对
AC	alternating current	交流电
ACB	air circuit breaker	空气自动断路器
ACI	American Concrete Institute	美国混凝土学会
ACS	American Chemical Society	美国化学学会
ADPT	adapter	连接头
AGA	American Gas Association	美国气体协会
AGMA	American Gear Manufacturers Association	美国齿轮制造厂协会
AIA	American Institute of Architects	美国建筑师协会
AIChE	American Institute of Chemical Engineers	美国化学工程师学会
AISC	American Institute of Steel Construction	美国钢结构学会
AISI	American Iron and Steel Institute	美国钢铁学会
AL	aluminum	铝
ALM	alarm	报警器
ALT.	alternate	备用
ALT	altitude	高度
AM	ammeter	安培计
AMB	ambient	环境
AMP	ampere	安（培）
AMT	amount	总数，数量
ANSI	American National Standards Institute	美国国家标准学会
APHA	American Public Health Association	美国公共健康协会
API	American Petroleum Institute	美国石油学会
APPROX	approximate	近似
AREA	American Railway Engineering Association	美国铁道工程协会
ARRGT	arrangement	布置，排列
ASB	asbestos	石棉
ASCE	American Society of Civil Engineers	美国土木工程师学会
ASHVE	American Society of Heating and Ventilating Engineers	美国采暖通风工程师学会
ASM	American Society for Metals	美国金属学会
ASME	American Society of Mechanical Engineers	美国机械工程师学会
ASPH	asphalt	沥青
ASRE	American Society of Refrigerating Engineers	美国制冷工程师学会
ASSY	assembly	组件，总装
ASTM	American Society for Testing and Materials	美国材料与试验协会
ATM	atmosphere	大气压
AUTO	automatic	自动
AUX	auxiliary	辅助
AVC	asbestos and varnished cambric insulated wire	石棉和浸渍黄蜡布绝缘电缆
A/G	above ground	地上

续表

缩　写	意　义 英　文	意　义 中　文
AVG	average	平均
AWG	American wire gage	美国线规
AWS	American Welding Society	美国焊接协会
AWWA	American Water Works Association	美国水工协会
BAR	barometer	气压计
BB	bolted bonnet	螺栓连接阀帽
B-B	back to back	背到背
BBL	barrel	桶
BC	bolted cover	螺栓连接阀盖
B. C.	between centers	中心间距
BC	bolt circle	螺栓中心圆
BE	baume	波美度
BE	beveled end	坡口端
BEP	both ends plain	两端平端
BET	both ends threaded	两端螺纹端
BH	boiler house	锅炉房
BHN	brinell hardness number	布氏硬度值
BHP	brake horsepower	制动马力
BKR	breaker	断电器，断路器
BL	battery limit	界区
BLD	blind	盲板
BLDG	building	建筑物
BLK	blank	空白
BLK.	block	区
BLNKT	blanket	覆盖层
B. M.	bending moment	弯矩
BM	bench mark	基准点
BOM	bill of material	材料表
BOT	bottom	底
BP	back pressure	背压
B. P.	base plate	底板
B. PT.	boiling point	沸点
BPD	barrels per day	每天桶数，桶/天
BPH	barrels per hour	每小时桶数，桶/时
BPSD	barrels per stream day	每连续工作日桶数
BRKT	bracket	托架，牛腿，墙架
BRS	brass	黄铜
BRZ	bronze	青铜
BS	British standard	英国标准
B. S.	both sides	两边
BSWG	brown and sharp wire gage	BS线规
BW	butt weld	对焊
BWG	Birmingham wire gage	伯明翰线规
C. S. C	car seal close	铅封关，未经允许不得开启
C. S. O	car seal open	铅封开，未经允许不得关闭

续表

缩写	意义 英文	意义 中文
℃	degree Centigrade	度（摄氏）
C. A.	chromel-alumel (thermocouples)	铬铝（热电偶）
CA	corrosion allowance	腐蚀裕量
CALC	calculate	计算
CAT	catalog	目录，样本
CAT.	catalyst	催化剂，触媒
CB	catch basin	集水池，雨水井
C-C	center to center	中心到中心
C. C.	copper-constantan (thermocouples)	康铜（热电偶）
CC	cubic centimeter	立方厘米
C-E	center to end	中心到端部
CEM	cement lined	水泥衬里
CENT.	centrifugal	离心
C-F	center to face	中心到面
CFH	cubic feet per hour	立方英尺/时
CFM	cubic feet per minute	立方英尺/分
CFS	cubic feet per second	立方英尺/秒
CFW	continuous fillet weld	连续填角焊
CG	center of gravity	重心
CG.	centigram	厘克
CHAM	chamfer	倒角
CHAN	channel	通道，槽钢
CHKD PL	checkered plate	网纹板
CI	cast iron	铸铁
CIRC.	circulate	循环
CIRC	circumference	周边
C. K.	chromel-KA2 (thermocouples)	铬镍合金-KA2（热电偶）
CKT	circuit	回路
C. L.	center line	中心线
CL	class	等级
CLG	ceiling	天花板
CLNC	clearance	间隙，净空
cm	centimeter	厘米
C. M.	circular mil	密耳圆[①]
CND	conduit	导管
CNDS	condensate	冷凝液
C. O.	change order	变动顺序
CO	clean out	清扫
CO.	company	公司
COD.	continued on drawing	接续图
COEF	coefficient	系数
COL	column	柱，列，塔
COMB.	combination	组合，联合
COMBU	combustion	燃烧
COML	commercial	商业的
COMPL	complete	完全

续表

缩 写	意 义 英 文	中 文
CWP	cool work pressure	冷态工作压力
COMPR	compressor	压缩机
COMPT	compartment	室，舱
CON	concentric	同心的
CONC	concrete	混凝土
CONDEN.	condenser	冷凝器
COD.	condition	情况，条件
COD	conductor	导体
CONN	connect, connecting	连接
CONST.	constant	常数
CONST	construction	结构
CONT	control	控制，调节
CONTD	continued	接续
CONTV	control valve	调节阀，控制阀
CORP.	corporation	公司，企业
CORR	corrugate	波纹的
C. P.	candle power	烛光
cP	centipoise	厘泊
CPLG	coupling	联轴器，管接头，管箍
CPS	cycles per second (Hertz)	每秒周数（赫兹）
C. S.	carbon steel	碳钢
CS.	centistokes	厘沲
CSA	Canadian Standards Association	加拿大标准协会
CST	cast steel	铸钢
CSTG	casting	铸造，铸件，浇注
CT	center tap	中心抽头
C. T.	current transformer	变流器
CTE	coal Tar Enamel Lined	柏油层衬里
CTR	center	中心
CTSK	countersunk	埋头孔
CTWT	counterweight	砝码，平衡锤
CUFT	cubic feet	立方英尺
CUIN.	cubic inch	立方英寸
CUYD	cubic yard	立方码
CYL	cylinder	气缸，圆柱体
D	density	密度
dB	decibel	分贝
DBL	double	双的，复式的
DC	direct current	直流
DEG	degree	度
DEPT	department	部门，工段
DET	detail	详图
DF	drinking fountain	喷嘴式饮水龙头
DIA	diameter	直径
DIAG.	diagram	图表
DIM.	dimension	尺寸
DN	dominal diameter	公称直径
DP	design Pressure	设计压力
DT	design Temperature	设计温度

续表

缩写	意义英文	意义中文
DISCH	discharge	出料,卸料
DISTR	distribution	分配,配电
DIV	division	刻度,分部
DN	down	下
DO.	ditto	同上
DO	draw out	抽出
DP	dew point	露点
DP DT	double pole, double throw	双刀双掷
DP ST	double pole, single throw	双刀单掷
DRN	drain	排液
DR.	drill	钻孔
DR	drive	驱动
DSGN	design	设计
DWG	drawing	图纸
E	east	东
ECC	eccentric	偏心的
ECON	economizer	省煤器,废气预热器
EF	electric furnace (steel)	电炉(钢)
EFF	efficiency	效率
EFW	electric fusion welding	电熔焊
EL	elevation	标高
ELEC	electric	电
ELL	elbow	弯头
ELLIP	ellipsoidal	椭圆体
EMER	emergency	事故
ENCL	enclosure	外壳,盒子
ENGR	engineering	工程
ENGR STD	engineering standard	工程标准
EP	explosion proof	防爆
EPA	Environmental Protection Agency	美国环境保护署
EPDM	ethylene propylene diene monomer	三元乙丙橡胶
EQ	equation	公式
EQUIP.	equipment	设备
ERW	electric resistance welding	电阻焊
EST	estimate	估计
ETL	effective tube length	有效管长
EXH	exhaust	排气
EXIST	existing	现有的
EXP	expansion	膨胀
EXP JT	expansion Joint	膨胀节,补偿器
EXT	external	外部
EXTEN	extension	伸长,延长
°F	degree Fahrenheit	度(华氏)
FABR	fabricate	制造
FB	full bore	通径
FBQ	firebox quality	燃烧室质量

续表

缩 写	意 义	
	英 文	中 文
FEED	front end engineering design	前端工程设计
F. D.	floor drain	地面排水口
FDN	foundation	基础
FDW	feed water	给水
FE	flanged end	法兰端
F-F	face to face	面至面
FF	flat face	平面
F. H.	fire hose	消防带，水龙带
FH	flat head	平头
FIG.	figure	图
FIN.	finish	加工
FL	floor	楼板
FLA	full load amperes	全负荷安培
FLG	flange	法兰
FOB	free on board	船上交货
FP	freezing point	冰点
FPM	feet per minute	英尺/分
FPRF	fireproof	防火
FPS	feet per second	英尺/秒
FQ	flange quality	法兰质量
FREQ	frequency	频率
FRT	freight	货运
FS	far side	远侧
FST	forged steel	锻钢
ft	feet	英尺
FTG	fitting	管件
FTG.	footing	底座
FTLB	foot pounds	磅-英尺
F. W.	field weld	现场焊接
FW	fresh water, fire water	新鲜水，消防水
GA	gage	量规
gal	gallon	加仑
GALV	galvanize	镀锌，电镀
GASO	gasoline	汽油
GE	groove end	槽端
GEN	general	普通的，总的
GENR	generator	发电机，发生器
GL	glass	玻璃
GOV	governor	调速器
GPH	gallons per hour	加仑/时
GPM	gallons per minute	加仑/分
GPS	gallons per second	加仑/秒
GR	grade	等级
GRD	ground	地面
GO	gear operation	齿轮操作
GRP	glass Reinforced Plastic	玻璃钢
GRPH	graphite	石墨
GRTG	grating	格，栅

续表

缩写	意义 英文	意义 中文
GSKT	gasket	垫片
HAZ	heat affected zone	热影响区
H. C.	hand control	手动控制
HC.	hose connection	软管接头
HC	hydrocarbon	烃，碳氢化合物
HCAP	high capacity	大容量
HDR	header	集合管，主管
HEX	hexagon	六角形，六角体
HF	hardfaced	表面硬化
HH	hand hole	手孔
HHV	higher heating value	高热值
HI	hydraulic Institute	水力学会
HIC	hydrogen induced cracking	氢致开裂
HOR	horizontal	水平的
HP	high performance	高性能
H. P.	high pressure	高压
HP	horsepower	马力
HPT	high point	高点
HR.	handrail	栏杆
H. R.	hot rolled	热轧
HR	hour	小时
HRP	handrail post	栏杆立柱
HT	high temperature	高温
HTR	heater	加热炉（器）
HVY	heavy	重
HYD	hydraulic	水力的，液压的
HYDRO	hydrostatic	流体静力学的
IC	iron-constantan (thermocouples)	铁-康铜（热电偶）
ID	inside diameter	内径
IEEE	Institute of Electrical & Electronics Engineers	电力和电子工程师学会
IES	Illuminating Engineering Society	照明工程学会
IF	inside frosted	内部磨砂
IG	imperial gallons (British)	法定标准加仑（英）
IHP	indicated horsepower	指示马力
IN.	inch	英寸
IN	inlet	入口
INC.	incorporated	联合的，包括
INCL	include	包括
IND	indicate	指示，指出
IND.	induction	感应
INS.	insulation	绝热，绝缘
IR	inner ring	内环
ISO	International Organization for Standardization	国际标准化组织

续表

缩 写	意 义	
	英 文	中 文
INS	integral seat	整体座
INST	instrument	仪表
INSTL	installation	装置
INSTR	instructions	规程,说明书
INSTV	instrument valve	仪表阀
INT.	internal	内部
INT	intersect	相交
INTER	intermediate	中间的
INTMT	intermittent	间歇的,断续的
INV	invert	反置
IPCEA	Insulated Power Cable Engineers Associated	绝缘电缆工程师联合会
IPS	iron pipe size	铁管尺寸
ISA	Instrument Society of America	美国仪表学会
ISS	inserted stainless steel seat	镶嵌不锈钢座
JCT	junction	接合
JT	joint	接头
°K	degree Kelvin	开尔文度
K	kip (kilopound,1000 lb)	千磅
KB	knee brace	斜撑
kg	kilogram	公斤
KO	knockout	敲落孔
KR	knuckle radius	转向半径
kV	kilovolt	千伏
kVA	kilovolt-ampere	千伏·安
kVAh	kilovolt-ampere hour	千伏·安·小时
kW	kilowatt	千瓦
kWh	kilowatt hour	千瓦时
LAV	lavatory	厕所,卫生间
LB	pound	磅
LC	locked closed	关闭状态锁定
LEP	large end plain	大头平端
LG	length	长度
LG.	long	长
LH	left hand	左手
LHV	lower heating value	低热值
LIN	linear	线性,直线的
LINFT	lineal feet	英尺长
LIQ	liquid	液体
LJ	lap joint	搭接
LN	logarithm (to base e)	对数(e为底)
LNG.	lining	衬里
LNG	liquified natural gas	液化天然气
LO	locked open	开启状态锁定
LOC	locate, location	位置,部位
LTCS	low temperature carbon steel	低温碳钢

续表

缩写	意义英文	意义中文
LOG	logarithm (to base 10)	对数（10 为底）
LONG.	longitude, longitudinal	经度，纵向，轴向
LP	low pressure	低压
LPG	liquified petroleum gas	液化石油气
LPT	low point	低点
L. R.	load ratio	负荷比
LR	long radius	长半径
LRA	locked rotor amperes	转子制动电流，堵转电流
LT	light	灯，轻的
LTG	large-tongue & groove	大榫槽
LTG.	lighting	照明
LTR	letter	字母
LUB	lubricate	润滑剂
m	meter	米
MACH	machine	机器
MAINT	maintenance	维修
MAN	manual	手动的
MATL	material	材料
MAWP	maximum allowable working pressure	最大容许工作压力
MAX	maximum	最大
M. C.	moment connection	瞬时连接
MC	multiple contact	多点接触
MCFH	thousand cubic feet hour	千立方英尺小时
MCM	thousand circular mils	千密耳圆
MECH	mechanical	机械的
MEP	mean effective pressure	平均有效压力
MF	male and female	凸面和凹面
MFD	manufactured	制造
MFG	manufacturing	制造（业）的
MFR	manufacture, manufacturer	制造厂，制造者
mg	milligram	毫克
M-G	motor generator	电动机，发电机
MGD	million gallons per day	百万加仑/天
MH	manhole	人孔，检查井
MI	malleable iron	可锻铸铁
MIG	metal arc welding (inert gas)	金属极电弧焊（惰性气体保护）
MIN	minimum	最小
MIN.	minute	分钟
MISC	miscellaneous	其他，杂项
MK	mark	记号，标记
mm	millimeter	毫米
MMA	manual metal arc welding	手动金属电弧焊
MO	month	月
MOL WT	molecular weight	分子量
MTO	material take-off	材料统计
MR	material requisition	材料清购文件

续表

缩 写	意 义 英 文	中 文
MPH.	miles per hour	英里/时
MPH	moles per hour	摩尔/时
MSCFH	thousand standard cubic feet hour	千标准立方英尺小时
MSS	Manufacturers Standardization Society of the Valve and Fittings Industry	阀门和管件工作制造者标准化学会
MT	magnetic particle examination	磁粉检验
MTD	mean temperature difference	平均温差
MTD.	mounted	安装好的
MTG	mounting	安装，固定
MTR	material test report	材料实验报告
mW	milliwatt	毫瓦
M. W.	mineral wool	矿渣棉
Min. W	minimum wall	最小壁（厚）
N	north	北
NACE	National Association of Corrosion Engineers	美国腐蚀工程师协会
NAM	National Association of Manufacturers (US)	国家制造业协会（美）
NATL	national	国家的，全国的
NB	nominal bore	公称孔
NBFU	National Board of Fire Underwriters (US)	国家火灾保险局（美）
NBS	National Bureau of Standards (US)	国家标准局（美）
NC	American national coarse thread	美国国标粗牙
NC.	no comment	无说明
N. C.	normally closed	正常时关闭
NEC	National Electrical Code (US)	国家电气规范（美）
NEG	negative	负
NEMA	National Electrical Manufacturers Association (US)	国家电气制造业协会（美）
NESC	National Electrical Safety Code (US)	国家电气安全规范（美）
NEUT	neutral	中性
NDT	nondestructive test	无损试验
NF	American national fine thread	美国国标细牙
N. F.	near face	近的一面
NF.	not furnished	不提供
NFPA	National Fire Protection Association (US)	美国消防协会（美）
NIP	nipple	短管
NL	neoprene lined	氯丁橡胶衬里
NO	normally open	正常时开启
NOM	nominal	公称的
NOR	normal	正常的
NOZ	nozzle	喷嘴，接管
NPS	national pipe size	公称直径
NPSH	net positive suction head	净正吸入压头
NPT	American national taper pipe thread	美国国家锥形管螺纹
NS	near side	近侧
NUM	number	号码，数字
NV	needle valve	针阀
OR	outer ring	外环
OSBL	outside battery limit	界区环

续表

缩写	意义 英文	中文
OA	overall	全部的，总的
OCT	octagon	八角形
OD	outside diameter	外径
OET	one end threaded	一端螺纹
OH	open hearth (steel)	平炉（钢），开式炉
O-O	out to out	外廓尺寸
OPER	operating	操作的，控制的
OPP	opposite	相反
OPR	operate	操作，控制
ORF	orifice	孔板
OSHA	Occupational Safety and Health Act	职业安全和健康法规
OSL	outstanding leg	伸出肢，挑梁
OS&Y	outside screw and yoke	外螺纹阀杆及阀轭
OVHD	overhead	架空的，上面的
OZ	ounce	盎司
P	plug	塞，栓，丝堵
Par.	parallel	平行，并联
PARA	paragraph	段，节，款
PATT	pattern	型式
PB	push button	按钮
PBSTA	push button station	按钮站
PC.	piece	零件，件
P.C.	point of curvature	弯曲点
PC	pulsating current	脉动电流
PCF	pounds per cubic foot	磅/立方英尺
PF MK	piece mark	零件标记
PD	pitch diameter	节径
PE	plain end	平端
PI	point of intersection	相交点
PILCDSTA	paper insulated, lead covered double steel tape armored cable	纸绝缘铅包双钢带铠装电缆
PILCSWA	paper insulated, lead covered single wire armored cable	纸绝缘铅包单股铠装电缆
PL	plate	板
PLATF	platform	平台
PNEU	pneumatic	气动的
POS	positive	正（数），正（极）
POSN	position	位置
POSUP	point of support	支承点
PPH	pounds per hour	磅/时
PPM	parts per million	百万分之一
PPS	pounds per second	磅/秒
PRESS.	pressure	压力
PROJ	project	设计，方案，项目
PS	pipe support	管架
PID	piping and instrument diagram	管道及仪表流程图

续表

缩 写	意 义	
	英 文	中 文
PTFE	polytetrafluoroethylene	聚四氟乙烯
PSF	pounds per square foot	磅/平方英尺
PSI	pounds per square inch	磅/平方英寸
PSIA	pounds per square inch absolute	磅/平方英寸（绝）
PSIG	pounds per square inch gage	磅/平方英寸（表）
PT.	point	点
PT	liquid penetrant examination	液体渗透检验
PVC	polyvinyl chloride	聚氯乙烯
PVCL	polyvinyl chloride lined	聚氯乙烯衬里
PWHT	post weld heat treatment	焊后热处理
QT	quart	夸脱
QTY	quantity	数量
QUAL	quality	质量
R	radius	半径
RAD.	radial	径向的，放射式的
RAD	radiator	辐射器，散热器
RDGEN	regenerator	蓄热器，再生器
REC.	receiver	受槽
REC	record	验收人，接收器
RECIP	reciprocate	往复的
RECIRC	recirculate	再循环
RECOM	recommended	推荐的
RECP	receptacle	容器，仓库
RED.	reducer	异径管，大小头
REF	reference	参考，基准，引用规范，文献
REF.	refinery	炼厂
REFR	refractory	耐火材料
REG	regulator, regular	调节器，正常的
REINF	reinforce	补强，加强
REQD	required	需要，要求
RET WT	retiring wall thickness	壁厚的报废厚度
REV	revise	修订
RF	raised face	突面
RG	ring gasket	环形垫
R/H	relative humidity	相对湿度
RH	right hand	右手
RHL	heat resisting, rubber insulated and lead covered cable	耐热橡胶绝缘铅包电缆

续表

缩写	意义 英文	意义 中文
RJ	ring joint	环接
RL	rubber lined	衬橡胶
RO	restriction orifice	限流孔板
RPM	revolutions per minute	转/分
RPS	revolutions per second	转/秒
RR	railroad	铁路
RS	requisition sheet	申请单，请购单
RT	radiographic examination	射线照相检验
RTFE	reinforced polytetrafluoroethylene	增强聚四氟乙烯
RV	relief valve	安全泄液阀
S	south	南
SAW	submerged arc welding	埋弧焊
SAE	Society of Automotive Engineers	汽车工程师学会
SAT.	saturate	饱和
SB	stud bolt	双头螺柱
SCF	standard cubic feet	标准立方英尺
SCFM	standard cubic feet per minute	标准立方英尺/分
SCH	schedule	管表号
SE	stub end	管端突缘
SECT	section	节，段
SEP.	separate	分开的
SEP	small end plain	小端平端
SET	small end threaded	小端螺纹端
SG	safety gate	安全门
SH ABS	shock absorber	减振器，振动吸收器
SK	sketch	草图
SLV	slide valve	滑阀
SMLS	seamless	无缝
SN.	snubber	缓冲器
SN	swaged nipple	异径短节
SNG	synthetic natural gas	合成天然气
SO.	steam out	蒸汽出口
SOF	slip on flange	滑套法兰，平焊法兰
SP	static pressure	静压
SP DT	single pole, double throw	单刀双掷开关
SPEC	specification	说明，规范
SPGR	specific gravity	相对密度
SPHT	specific heat	比热容
SPST	single pole, single throw	单刀单掷开关
SQ	square	方形，平方
SQFT	square foot	平方英尺
SQIN.	square inch	平方英寸
S. R.	short radius	短半径
SR	stress relief	应力消除
SSF	saybolt seconds furol (viscosity)	塞波特量油黏度计秒数
SS	stainless steel	不锈钢
SSU	saybolt seconds universal (viscosity)	通用塞波特黏度计秒数

续表

缩写	意义 英文	意义 中文
ST	structural tee	结构三通
STA	station	站
STD	standard	标准
STD WT	standard weight	标准重量
STG	small tongue and groove	小榫槽
STIFF.	stiffener	加强肋
STL	stellite	司太立公金，钴基合金
STM	steam	蒸汽
ST WY	stairway	梯子
SUCT	suction	吸入
CUPHTR	superheater	过热器
SV	safety valve	安全泄气阀
S.W.	salt water	盐水
SW	socket weld	承插焊
SW.	switch	开关
SYM	symmetrical	对称的
SYMB	symbol	符号
SYS	system	系统
T	tee	三通
T	ton	吨
TB	top and bottom	顶和底
TB.	trolley beam	吊车梁
TC	thermocouple	热电偶
TE	threaded end	螺纹端
TEMA	tubular Exchanger Manufacturers Association	管式换热器制造者协会
TEMP	temperature	温度
TERM.	terminal	末端的，终端，接线端子
TG	tongue and groove	榫槽
THD	threaded	螺纹的
THK	thickness	厚度
TIG	tungsten inert gas welding (inert gas)	钨极电弧焊（惰性气体保护）
TL	tangent line	切线
TOP	top of pipe	管顶
TOS	top of support	支架顶部
TPO	toe plate only	只有踢脚板
TRANS.	transfer	传送，变换
TRANS	transformer	变压器
TRUN	trunnion	轴颈，耳轴
TT	tell tale	信号器
T-T	tangent to tangent	切线到切线
TURB	turbine	汽轮机，透平机
TYP	typical	标准的，典型的

续表

缩 写	意 义	
	英 文	中 文
U/G	underground	地下
UH	unit heater	供热机组，单元加热器
UL	Underwriter's Laboratories，Inc.	保险商联合研究所
UN	union	活接头
UNC	unified national coarse thread	国家统一标准粗牙螺纹
UNF	unified national fine thread	国家统一标准细牙螺纹
UR	urinal	小便池
USA	United States of America	美利坚合众国，美国
USG	United States standard gage	美国标准量规
UT	unitrace	单伴管
UTRC	ultrasonic testing	超声波检验
UTS	ultimate tensile stress	极限拉应力
VAC	vacuum	真空
VAR	variable	可变的，变量
VARN	varnish	清漆
VCL	varnished cambric，insulated wire lead covered cable	清漆黄蜡布绝缘线铅包电缆
VEL	velocity	速度
VERT	vertical	垂直的，立式
VIR	vulcanized rubber insulated wire	硫化橡胶绝缘电线
VIRLCSWA	vulcanized rubber insulated lead covered single wire armored cable	硫化橡胶绝缘铅包单股铠装电缆
VISC	viscosity	黏度
VIT	vitreous	玻璃的，上釉的
VOL	volume	体积，容积
W.	Watt	瓦特
W	west	西
W/	with	和
WB	welded bonnet	焊接阀帽
WC.	water closet	卫生间
WC	weld cap	焊帽
WD.	width	宽
WF	wide flange	宽缘
WHSE	warehouse	仓库
WN	weld neck	焊颈，高颈
WNF	weld neck flange	高颈法兰，对焊法兰
WP.	weatherproof	全天候的，防风雨的
W. P.	working point	工作点
WP	working pressure	工作压力，操作压力
WT	wall thickness	壁厚
WT.	weight	重量
XH	extra heavy	加重
XMTR	transmitter	变送器，发射器
XR	X-ray	X射线
XS	extra strong	加厚管
XXS	double extra strong	双加厚管，特厚管
YD	yard	码，场地，管廊
YP	yield point	屈服点
YR	year	年

① 直径为1密耳的金属丝面积单位；1密耳＝0.001英寸。

8.7 防腐 Anticorrosion

8.7.1 内防护 Internal Anticorrosive Coating

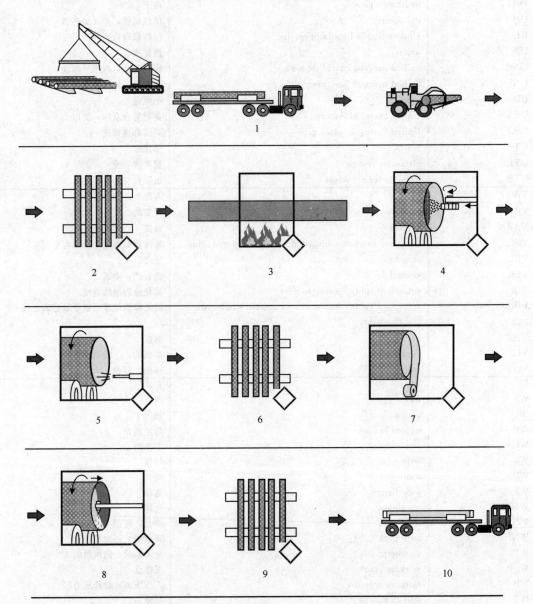

1　transportation of bare pipe from stockpile to coating plant　裸管运至加工厂
2　incoming racks and foam plug　装架，堵头
3　preheat oven　预热
4　blast cleaning　喷砂除锈
5　dust and abrasive removal　清除表面污垢
6　inspection racks　备检
7　cutback masking　端部保护
8　paint application　内防腐
9　coating inspection　涂层检验
10　transportation to stockpile　运至仓库

8.7.2 三层聚乙烯涂层 Three Layer Polyethylene Coating

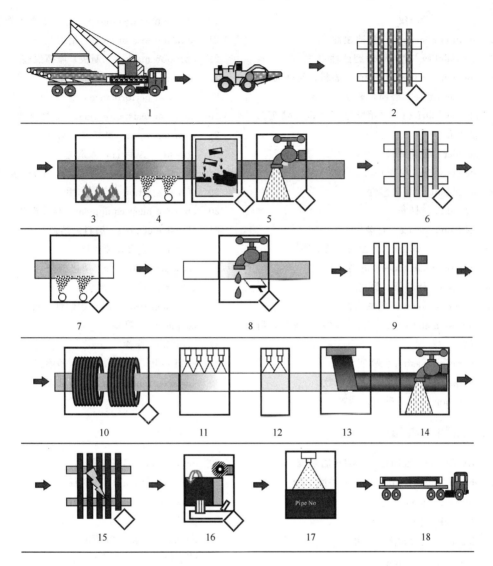

1. transportation of bare pipe from stockpile to coating plant 裸管运至加工厂
2. incoming racks, foam plug insertion and water wash 装架，堵头，水洗
3. preheat oven 预热
4. 1st blast cleaning 一次喷砂除锈
5. acid wash and water rinse (if necessary) 酸洗，水洗（如需要）
6. grinding racks 备磨
7. 2nd blast cleaning 二次喷砂除锈
8. chromate application 铬酸洗
9. inspection racks and internal blow-out station 备检，管内吹扫
10. induction heating 感应加热
11. FBE application 喷涂熔结环氧粉末
12. adhesive application 涂胶
13. PE application PE 涂层
14. water quench 水冷
15. holiday detection and coating inspection 漏涂检测和涂层检验
16. cutback cleaning 除垢
17. stencil application 涂蜡
18. transportation to stockpile 运至仓库

8.7.3 表面处理 Surface Preparation Vocabulary

1. abbatre 凹凸纹
2. ablation resistance 耐冲蚀性
3. abradability 磨耗性,磨蚀性
4. abradant, abrading agent 磨料,研磨剂
5. abrade 磨耗,磨损,磨蚀
6. abraded surface （用喷砂或喷丸）清理过的表面
7. abrader 磨耗试验机
8. abrade wear 磨损
9. abradibility 可研磨性
10. abrading 打磨
11. abrading device 打磨机
12. abrading powder 磨蚀粉,研磨粉
13. abrading substance 磨料
14. abraser 研磨器,磨料
15. abrasiometer 磨耗试验机
16. abrasion and wear 磨耗,磨损,擦伤,研磨
17. abrasion blasting 喷砂
18. abrasion coefficient 磨耗系数
19. abrasion cloth 砂布,砂带
20. abrasion concealing coating 耐磨罩面涂层
21. abrasion loss of gloss 磨失光泽
22. abrasion mark （制板）摩擦痕迹
23. abrasion pattern 磨纹
24. abrasion performance 耐磨（耗）性
25. abrasion resistance index 耐磨指数
26. abrasion value 磨耗量,磨耗值
27. abrasite 刚铝石,人造金刚砂
28. abrasive 磨料,研磨剂,磨耗的,磨蚀的
29. abrasive action 磨损作用
30. abrasive blast equipment 喷砂设备
31. abrasive cement 磨料黏结剂
32. coating 涂层,覆盖层
33. cathodic protection 阴极保护
34. colour (color) 颜色
35. corrosion resistance 耐蚀性,抗蚀性
36. lacquering 喷漆
37. painting 涂漆
38. sand mill, sand grinder 砂磨机
39. surface preparation 表面处理

8.7.4 涂料 Paint

1. AA system 酸固化丙烯酸树脂系统
2. abalyn, methyl abietate 松香酸甲酯
3. adhesive 胶黏剂
4. abies oil 冷杉油
5. abietate 松香酸酯（或盐）
6. abietic acid 松香酸
7. abietic anhydride 松香酸酐
8. abietinol 松香醇
9. abietyl 松香酸基
10. abietylamine 松香胺
11. Abraham consistometer 亚伯拉罕稠度计
12. abasin oil 桐油
13. acetylacetone 乙酰丙酮
14. acid-proof paint 耐酸漆
15. acrylic baking enamel 丙烯酸烘干磁漆
16. alcohol 乙醇
17. aldehyde 乙醛
18. aldehyde resin 聚醛树脂
19. aldol resin 醇醛树脂
20. alkail-proof paint 耐碱漆
21. alkamine, amino-alcohol, amino-alkyd 氨基醇
22. alkyd coatings 醇酸树脂涂料
23. alkyd enamel 醇酸瓷漆
24. alkyd resin 醇酸树脂
25. alkyd resin varnish 醇酸（树脂）清漆
26. allyl alcohol 烯丙醇
27. aluminium rosinate 松香酸铝
28. amide 氨化物
29. amine salt 胺盐
30. amyl carbamate 氨基甲酸戊酯
31. anti-corrosive paint, rust-proof paint 防腐漆,防锈剂
32. antirust paint 防锈漆
33. bassanite 石膏
34. benzoin 安息香
35. bituminous paint 沥青漆
36. calcium resinate 松香酸钙

37	cellulose 纤维素		55	mother-of-pearl pigment 云母珠光颜料
38	colour former 成色剂		56	organic silicon paint 有机硅漆
39	colouring agent 染料，颜料，着色剂		57	oxalic acid 乙二酸
40	colouring matter 色料		58	phenolic paint 酚醛涂料
41	emulgator，emulgent，emulsifier 乳化剂		59	pinoline，abietene 松香烯
42	emulsion paint 乳胶漆		60	plate oil 凹版墨料
43	enamel 瓷漆		61	polyacrylic resin 聚丙烯酸树脂
44	epoxy resin paint 环氧树脂漆		62	polyurethane paint 聚氨酯漆
45	ethinoic acid 乙二磺酸		63	primary coat，primer 底漆
46	ethyl abietate 松香酸乙酯		64	thermo-paint 测温漆
47	ethylene perchloride paint 过氯乙烯漆		65	universal primer 万能底漆
48	ferrous oxide 氧化亚铁		66	urushiol resin paint 漆酚树脂漆
49	finishing coat 面漆		67	varnish 清漆
50	glycol ester 乙二醇酯		68	white metal 白涂料
51	glyoxal 乙二醛		69	zinc iron oxide 氧化铁锌
52	heat-proof paint 耐热漆		70	zinc oxide pigment 氧化锌颜料
53	hydroxy ethyl cellulose 羟乙基纤维素		71	zinc resinate 松香酸锌
54	inorganic zinc-rich paint 无机富锌漆			

8.7.5 腐蚀 Corrosion

1	ablation 沸蚀，消蚀		7	gaseous corrosion 气相腐蚀
2	cavitation corrosion 气蚀		8	hydrogen embrittlement 氢脆
3	contact corrosion 接触腐蚀		9	intergranular corrosion 晶间腐蚀
4	corrosion 腐蚀		10	pitting 点蚀
5	couple corrosion 电化腐蚀		11	stress corrosion 应力腐蚀
6	crevice corrosion 缝隙腐蚀		12	electro-chemical corrosion 电化学腐蚀

8.7.6 其他常用词汇 Others

1	abaiser 象牙黑			作用
2	abatement pollution 污染防治		20	API 消蚀性能指数
3	Abbe refractometer 阿贝折射计		21	basso 电解凹印版
4	abbertite 黑沥青		22	blue 蓝色的
5	Abbe value 阿贝值		23	brown 棕色的，褐色的
6	abhesion 脱黏		24	cathodic protection 阴极保护
7	ablated weight 烧蚀（物质的）重量		25	coating 涂层，覆盖层
8	ablating rate 烧蚀率，消蚀率		26	colour（color）颜色
9	ablating surface 烧蚀（表）面		27	dark 深色的
10	ablation performance index 烧蚀性能指数		28	green 绿色的
11	ablative coatings 烧蚀涂料，消蚀涂料，烧蚀隔热涂料		29	grey（gray）灰色的
			30	light 淡（浅）色的；轻的
12	ablative composite material 烧蚀复合材料		31	malachite green 孔雀绿
13	ablative layer 烧蚀层		32	orange colour 橙色的
14	ablative thermal protection 烧蚀热防护		33	painting 涂漆
15	ablator 烧蚀材料，烧蚀剂		34	pigment 颜料
16	Abney effect 阿布尼效应		35	purple 紫色的
17	abnormal density 反常密度		36	red 红色的
18	abnormal test condition 特殊试验条件		37	spanish red 氧化铁红色
19	above-the-melt polymerization 熔融聚合		38	yellow 黄色的

8.8 管道应力分析 Piping Stress Analysis

8.8.1 应变 Strain

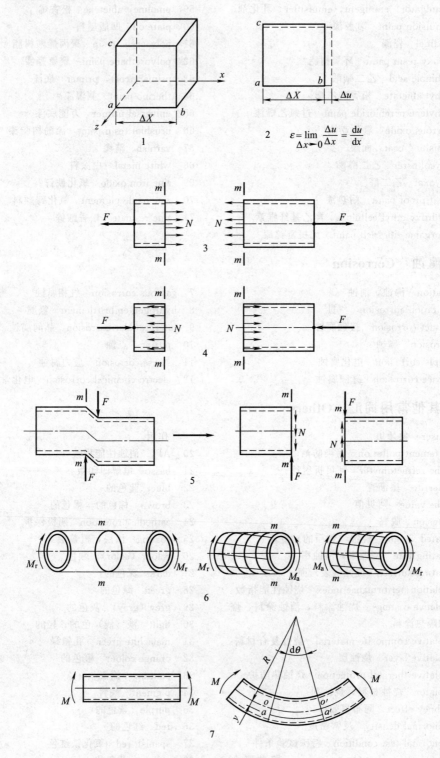

$$2 \quad \varepsilon = \lim_{\Delta x \to 0} \frac{\Delta u}{\Delta x} = \frac{du}{dx}$$

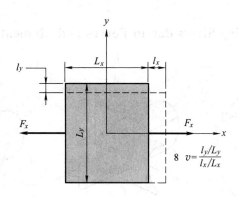

1 strain 应变
2 longitudinal strain 线应变
3 tension 拉伸
4 compression 压缩
5 shear 剪切
6 torsion 扭转
7 bending 弯曲
8 Poisson's Ratio 泊松比

8.8.2 应力 Stress

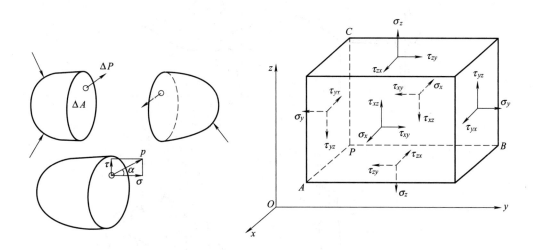

1 $P = \lim\limits_{\Delta A \to 0} \dfrac{\Delta P}{\Delta A}$

2 $\sigma = p \cos \alpha$

3 $\tau = p \sin \alpha$

4 $\sigma_{ij} = \begin{bmatrix} \sigma_x & \tau_{xy} & \tau_{xz} \\ \tau_{yx} & \sigma_y & \tau_{yz} \\ \tau_{zx} & \tau_{zy} & \sigma_z \end{bmatrix}$

1 stress 应力
2 normal stresses (perpendicular to the cross section) 正应力（垂直于截面的应力）
3 shear stress (parallel to the cross section) 剪应力（平行于截面的应力）
4 stress tensor 应力张量

8.8.3 管道应力 Piping Stress due to Forces and Moments

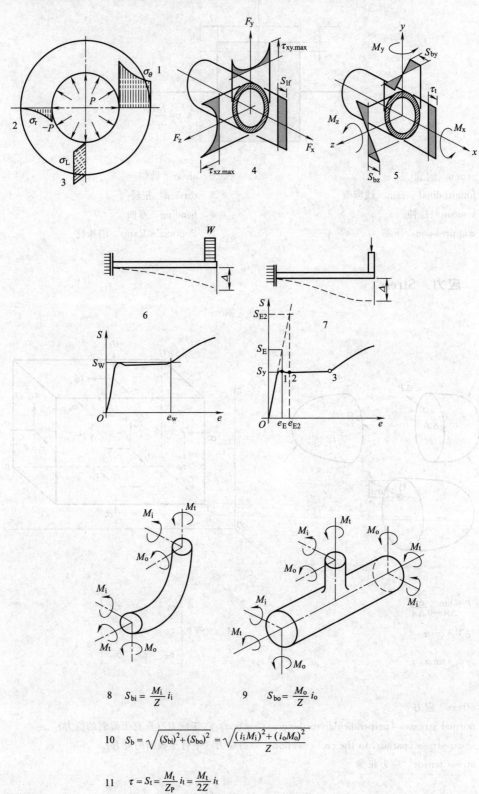

8   $S_{bi} = \dfrac{M_i}{Z} i_i$

9   $S_{bo} = \dfrac{M_o}{Z} i_o$

10   $S_b = \sqrt{(S_{bi})^2 + (S_{bo})^2} = \dfrac{\sqrt{(i_i M_i)^2 + (i_o M_o)^2}}{Z}$

11   $\tau = S_t = \dfrac{M_t}{Z_P} i_t = \dfrac{M_t}{2Z} i_t$

1	hoop stress due to internal pressure 内压引起的周向应力 S_θ	6	sustained stress 持续应力
2	radial stress due to internal pressure 内压引起的径向应力 S_r	7	displacement stress (self-limiting) 位移应力（有自限性）
3	axial stress due to internal pressure 内压引起的轴向应力 S_L	8	in-plane bending stress 平面内弯曲应力
		9	out-plane bending stress 平面外弯曲应力
4	stress due to forces 力引起的应力	10	combined bending stress 组合弯曲应力
5	stress due to moments 力矩引起的应力	11	torsion stress 扭转应力

8.8.4 管道应力分析 Piping Stress Analysis

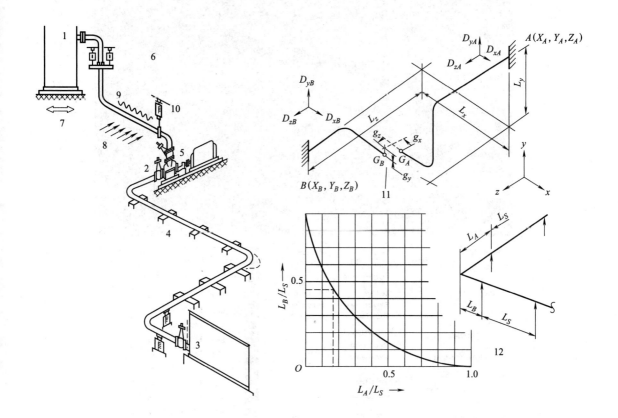

1 vessel flexibility and local stresses 设备管口柔性和局部应力
2 flange load and leakage 法兰荷载和泄漏
3 shell deflection and rotation 罐壁位移与转角
4 support and restraint 支承和约束
5 machinery load 动设备荷载
6 equipment arrangement and piping routing 设备和管道布置
7 seismic effect 地震效应
8 wind effect 风载效应
9 effecting of vibration and shock load 振动和冲击荷载效应
10 spring selection 弹簧设计
11 cold spring 冷紧
12 piping spans 管道跨距

8.8.5 管道应力分析输入条件 Input of Piping Stress Analysis

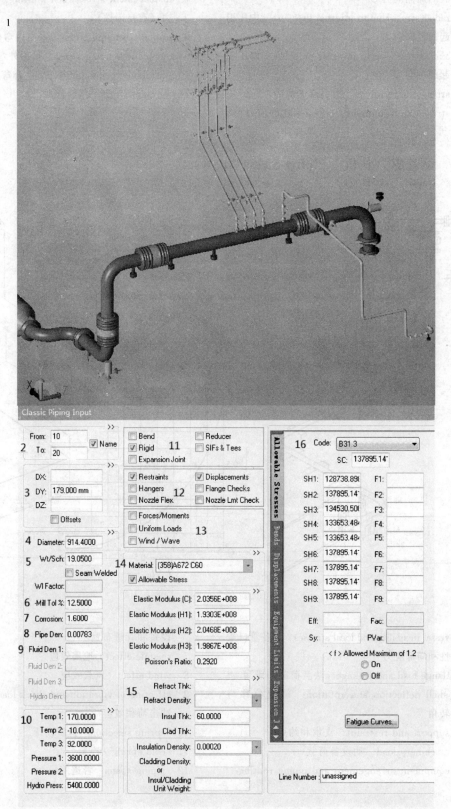

1	model of pipe Stress analysis 管道应力模型	9	services density 物料密度
2	element nodes 单元节点	10	process cases 工艺条件
3	dimension of the element 单元尺寸	11	fitting 管道元件
4	pipe diameter 管径	12	type of restraint 约束形式
5	pipe thickness 管壁厚	13	additional load 附加载荷
6	mill tolerance 机械偏差	14	material of pipe and fitting 管道和管件材料
7	corrosion allowance 腐蚀余量	15	insulating layer 保温层
8	pipe density 管密度	16	piping code 管道材料规范

8.8.6 管道柔性 Piping Flexibility

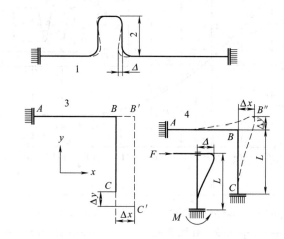

1 pipe expansion loop 管道膨胀弯
2 loop leg length 膨胀弯长度
3 free expansion 自由膨胀
4 constrained expansion 受限膨胀

8.8.7 管道动态效应 Dynamic Effects on Piping

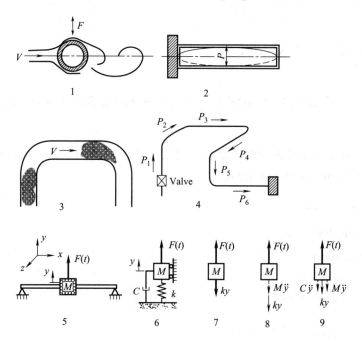

1　vortex shedding　旋涡脱落
2　acoustic pulsation　声学脉动
3　slug flow　柱塞流
4　pressure wave　压力波
5　the elastic simply-supported beam with lumped mass　集中质量的弹性简支梁
6　spring-mass model　刚度-质点模型
7　static forces　静态荷载
8　dynamic without damping　无阻尼振动
9　dynamic with damping　阻尼振动

8.8.8　疲劳曲线　Fatigue Curves

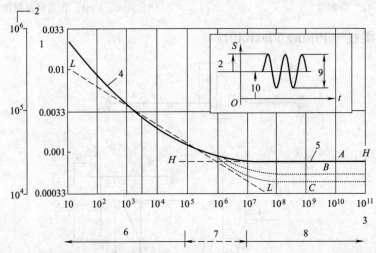

1　allowable strain amplitude　容许应变振幅
2　elastic equivalent stress amplitude　弹性等效应力幅
3　number of operating cycles　循环次数
4　design fatigue curve　设计疲劳曲线
5　endurance limit　疲劳极限
6　low cycle fatigue　低周疲劳
7　transition stage　过渡段
8　high cycle fatigue　高周疲劳
9　fatigue stress range　疲劳应力范围
10　mean stress　平均应力

8.9　三维模型　3D Model

8.9.1　三维管道设计软件专用词汇　Vocabulary for 3D Piping Design Software

8.9.1.1　Smart Plant 3D 管道模型分层管理词汇　Hierarchical Management Vocabulary for Smart Plant 3D Piping Model

—Piping System
　—Pipeline System
　　—Pipe Run
　　　—Features
　　　—Parts/Components
　　　—Ports
　　　—Connections

Piping System	管道系统	Parts	部件
Pipeline System	管线系统	Ports	端口
Pipe Run	管段	Components	元件
Features	特征	Connections	连接点

8.9.1.2 PDMS 模型分层管理词汇　Hierarchical Management Vocabulary for PDMS Model

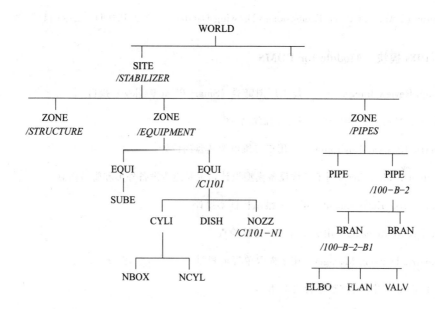

1　World　世界
2　Site　工厂
3　Zone　分区
4　Equipment　设备
5　Pipe　管子
6　Subequipment　子设备
7　Cylinder　圆柱体
8　Dish　椭圆封头，碟形封头
9　Nozzle　管嘴，管口
10　Branch　分支
11　Box　盒子，箱体
12　Elbow　弯头
13　Flange　法兰
14　Valve　阀门

8.9.1.3 PDMS 数据库类型词汇　Database Type Vocabulary for PDMS

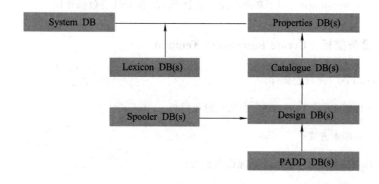

System Database 系统数据库

Properties Database（s）—PROP 特性数据库

Lexicon Database（s）/Dictionary Database（s）—DICT 用户定义属性数据库

Catalogue Database（s）—CATA 元件数据库

Spooler Database（s）/Isodraft Database（s）—ISOD 管段数据库

Design Database（s）—DESI 设计数据库

Produciton of Annotated and Dimensioned Drawing Database（s）—PADD 二维图数据库

8.9.1.4 PDS 模块 Module for PDMS

1 Design Review Integrator 智能工厂浏览器（Smart Plant Review）接口

2 Drawing Manager 用于生成平、剖面配置图

3 Electric Raceway Environment 用于三维电缆托盘的设计

4 Equipment Modeling 用于三维设备模型设计，主要建造设备外形和管口信息

5 Frameworks Environment 用于三维结构框架设计

6 Interference Manager 用于三维工厂碰撞检查

7 Isometric Drawing Manager 用于提取单管轴测图

8 PE-HVAC 用于三维暖通管道设计

9 Piping Design Data Manager 主要用于三维设计数据检查及二维与三维间的数据校验

10 Piping Designer 用于三维配管设计

11 Pipe Stress Analysis 管道应力分析接口

12 Pipe Support Designer 用于管道支吊架设计

13 Project Administrator 项目控制和管理

14 Reference Data Manager 参考数据库管理

15 Report Manager 用于提取各种材料报告

16 Schematic Environment 用于高阶段设计绘制 P&ID 图和仪表数据管理

8.9.1.5 建立设备模板 Create Equipment Template

1 Input Macro File 读取宏文件

2 Parametric Equipment Template 参数化设备模板

3 Primitive 搭积木方式

4 Standard Equipment Template 标准设备模板

8.9.1.6 建立基本体 Create Primitives

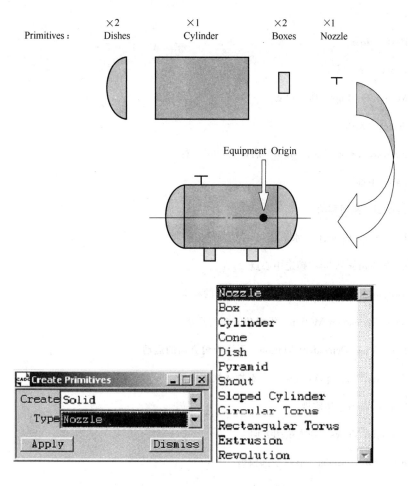

Primitives 基本体
Nozzle 管口
Box 箱体
Cylinder 筒体
Cone 圆锥体
Dish 碟形封头
Pyramid 金字塔锥体
Snout 管
Sloped Cylinder 斜锥体
Circular Torus 圆环面
Rectangular Torus 矩形环面
Extrusion 拉伸体
Revolution 旋转体
Equipment Origin 设备原点

8.9.2 Smart Plant Review 软件专用词汇 Specific Vocabulary for Smart Plant Review

8.9.2.1 按钮 Button

Open 打开文件

Save 保存

Print 打印

	Cut　剪切
	Copy　复制
	Paste　粘贴
	Properties　属性
	Motion Settings　移动设置
	Select　选择
	Forward/Back，Left/Right　前，后，左，右
	Pan　平移
	Lateral　侧面移动
	Horizontal Encircle　水平环绕
	Lock Center Point　锁定中心点
	Motion Lock Elevation　锁定高度方向的移动
	View Dependent Motion　沿视角方向的运动
	Level View Dependent Motion　不沿视角方向的运动
	Activate Clipping Volume　动态物体视深选择
	Clip Volume　切割视深
	Common Views　基本视图方向
	Fit View to Model　显示全部模型
	Fit View to Object　显示所选物体
	Fit View to Volume　显示所选范围
	Center View　居中显示
	Place Eye　放置视线原点
	Place Center　放置视线中心
	Zoom In/Out　放大、缩小
	View Setting　视窗设定
	Save and Recall Setting　保存、重启设置
	Snapshot View　视窗截图
	Measurement　测量
	One-Click Measurement Mode　单击测量模式
	Snaplock Measurement　两点锁定的测量

Icon	English	Chinese
	Surface Measurement	表面测量
	Shortest Measurement	最短距离
	Move Measurement	移动测量
	Restore Position	恢复测量位置
	New Measurement Collection	新的测量收集功能
	Delete Last Measurement	删除最近一次测量结果
	Delete Active Collection Measurement	删除活动集的测量
	Delete All Measurements	删除所有测量结果
	Delete all Measurements Collection	删除所有测量集

8.9.2.2 菜单 Menu

File Edit View Tools Display Sets Animation Motion Tags Accessories Window Help

File 文件

命令	快捷键	中文
Open...	Ctrl+O	打开
Close		关闭
Save	Ctrl+S	保存
Save As...		另存为
Add Vue Files...		添加Vue文件
Load Vue Files...		加载Vue文件
Remove Vue Files...		移除Vue文件
Import	▶	输入
Export	▶	输出
Page Setup...		页面设置
Print Preview		打印预览
Print...	Ctrl+P	打印

Edit 编辑

命令	快捷键	中文
Undo	Ctrl+Z	撤销
Redo	Ctrl+Y	恢复
Select		选择
Select Filter	▶	选择过滤器
Select Settings...		选择设置
Cut	Ctrl+X	剪切
Copy	Ctrl+C	复制
Paste	Ctrl+V	粘贴
Find Object	Ctrl+F	查找目标
Next Object	Shift+N	下一个目标
Previous Object	Shift+P	前一个目标
Match	Shift+M	匹配
File Position...		文件位置

View 视图

Fit ▶	视图显示
Place ▶	位置
Common ▶	通用
Zoom ▶	放大缩小
Settings... F6	设置
Display ▶	显示
Advanced ▶	高级
Photo-Realism ▶	图片写实
Save and Recall... Ctrl+F6	保存和恢复
Snapshot...	截图
Full Screen F11	全屏
✓ Menu Bar	菜单栏
✓ Status Bar	状态栏
Toolbars ▶	工具栏
Customize...	自定义
✓ Project Manager Alt+0	模型管理器
✓ Position Control Alt+1	位置控制
✓ Perspective Angle Alt+2	视角
✓ Encircle Radius Alt+3	水平环绕角度
✓ Common Views Alt+4	基本视角
Clip Volume Alt+5	切视深
Performance Control Alt+6	显示性能控制
Properties Alt+7	属性

Tools 工具

Materials ▶	材料
Measure ▶	测量
Hide Objects	隐藏目标
Hide Level... F4	隐藏层
Level Settings... Ctrl+F4	层设置
Collision Detection ▶	碰撞检查
Data Annotation ▶	数据注释
Volume Annotation ▶	图形注释
Format Readouts...	格式读取器
Refresh Data	刷新数据
Get External Data ▶	获取外部数据

534

Display Sets 显示设置

Show		显示
Shade		隐藏
Dim		减弱
Material		材料
Move		移动
Show Only		只显示
Hide Only		只隐藏
Edit Definition...	F3	编辑自定义
Assign Material...	Ctrl+F2	赋材料属性
Edit Position...	Ctrl+F3	编辑位置
Reset Position		重置位置
Auto-Define...		自动定义
New	▶	新建
Delete		删除
Rename		重命名
Reverse Dim		颜色反置

Animation 动画

Display Sets	▶	显示设置
ScheduleReview	▶	进度回放
Playback/Capture Settings	Alt+9 ▶	回看、捕捉设置
Display Sets Type		显示设置类型
ScheduleReview Type		进度回放类型
Key Frame Type		关键帧类型
Toolbar		工具栏
Player	Alt+8	播放器
Sequencer	Alt+Q	定序器
Play	R	播放
Play Backward	B	回放
Pause	Space	暂停
Stop	Shift+S	停止
Step Forward	Shift+R	快进
Step Backward	Shift+B	快退
Go to Start	Ctrl+R	回到开始
Go to End	Ctrl+B	跳至结尾
Go to Date...	Ctrl+Space	去到某一位置
Loop		循环
Pong		
Record	Ctrl+Shift+R	录制

Motion 动作

Eye Point		视点
Display Set Motion Mode		显示设置运动模式
Move	▶	移动
Settings...	Ctrl+F7	设置
Lock Center Point		锁中心点
Lock Elevation		锁高度
Mouse Drag Modes	▶	鼠标拖动模式
Positioning Modes	▶	定位模式
Directional Modes	▶	方向模式

Tags 标签

Place	▶	放置
Edit...		编辑
Delete...		删除
Find...		查找
"Find Tags" Results		查找标签的结果
Next	F10	下一个
Previous	Ctrl+F10	前一个
Go To...		前往指定位置
Display	▶	显示

Accessories 附件

Load Accessories		加载附件
Text Annotations	▶	文字注释
Set Window Size		设置窗口尺寸
Paint		涂色

Windows 窗口

Main	Ctrl+1	主视图
Plan	Ctrl+2	俯视图
Elevation	Ctrl+3	立面图
Text	Ctrl+4	文字信息
View Layout	▶	视窗个数
Cascade		串联
Arrange Icons		图标布置
Tile Horizontally		横向放置图标
Tile Vertically		竖向放置图标
Arrange All	Ctrl+F11	全部重排
Restore All		全部重置
Scroll Bars		滚动条
Refresh All	Ctrl+F5	全部刷新
Refresh	F5	刷新

9 工业自动化仪表 Process Measurement and Control Instrument

9.1 温度计 Thermometer

9.1.1 测温元件 Thermometer Element

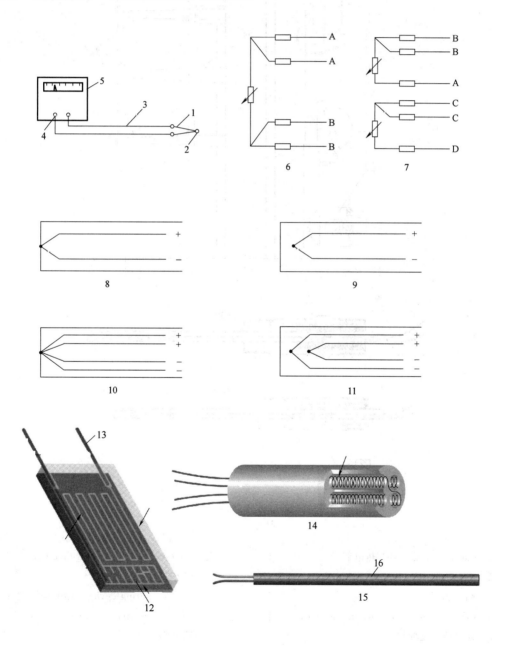

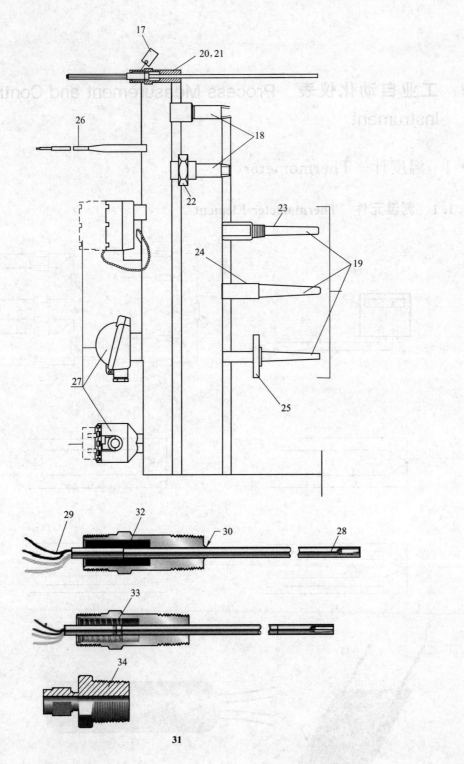

1 thermocouple wire 热电偶丝
2 measuring junction (hot junction) 测量端（热端）
3 extension wire (extension lead wire) 补偿导线（延伸线）
4 reference junction (cold junction) 基准点（冷端）
5 emf measuring device 热电势测量表
6 single element，4-wire RTD 单支四线制热电阻

7　double element，3-wire RTD　双支三线制热电阻
8　single element，grounded thermocouple　单支接地热电偶
9　single element，ungrounded thermocouple　单支非接地热电偶
10　double element，grounded，unisolated thermocouple　双支接地非绝缘热电偶
11　double element，ungrounded，isolated thermocouple　双支非接地绝缘热电偶
12　thin film element　薄膜式热电阻
13　element leads　引线
14　wire-wound element　绕丝式热电阻
15　armored rtd/armored thermocouple　铠装热电阻/铠装热电偶
16　sensor sheath　传感器铠装层
17　identification tag　设备标牌
18　extension　传感器延伸段/上保护套管
19　thermowell　温度计保护套管
20　sensor adapter assembly　传感器安装连接件
21　coupling-nipple　联合接头
22　union-nipple　联合活动接头
23　threaded thermowell　螺纹连接的温度计保护套管
24　socket weld thermowell　承插焊温度计保护套管
25　flanged thermowell　法兰连接的温度计保护套管
26　lead wire extensions and seals　引线延伸件和密封套
27　connection head　接线盒
28　sensor immersion length　传感器插入长度
29　conduit entry　电缆接口
30　process connection　工艺接口
31　sensor mounting style　传感器安装方式
32　compact welded adapter　焊接式安装接头
33　spring loaded adapter　压簧式安装接头
34　compression fitting　卡套式安装接头

9.1.2　温度变送器　Temperature Transmitter

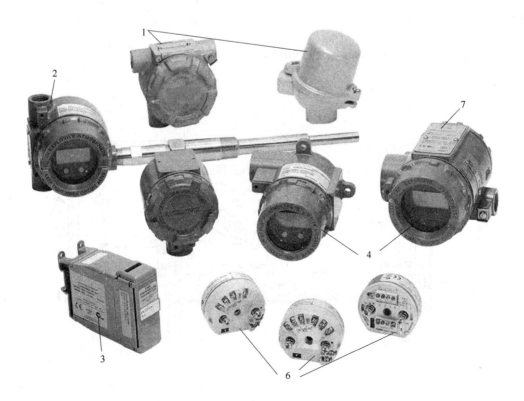

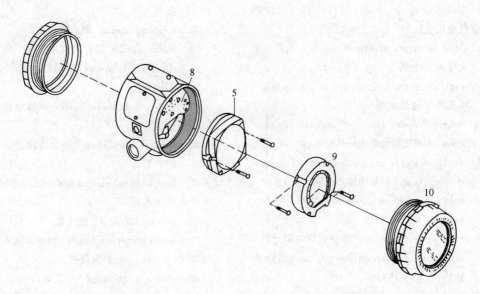

1	head mount (integral) temperature transmitter 顶装式（一体化）温度变送器	5	electronics module 电子模块（温度变送单元）
2	field mount temperature transmitter 现场安装式温度变送器	6	cover with wiring diagram 带接线简图的变送器盖
3	rail mount (control room) temperature transmitter 轨道安装式（控制室）温度变送器	7	name plate 铭牌
		8	housing with terminal block 带端子块的变送器外壳
4	head mount (integral) temperature transmitter with LCD meter 顶装式（一体化）带液晶显示表的温度变送器	9	LCD meter 液晶显示表
		10	cover with glass 带玻璃面板的变送器盖

9.1.3　辐射高温计　Radiation Pyrometer

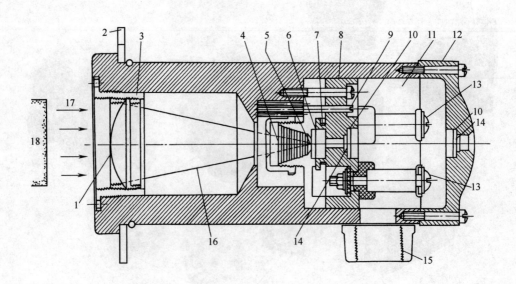

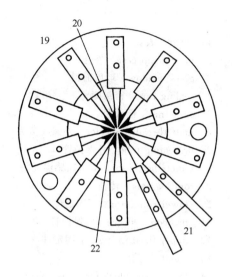
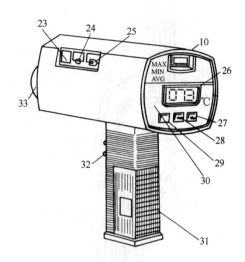

1　lens　透镜，物镜
2　mounting ring　安装环
3　interchangeable lens assemble　通用透镜组件
4　aperture　小孔
5　fixed aperture　固定小孔
6　thermopile　热电堆
7　compensating coil　补偿线圈
8　thermopile housing　热电堆外壳
9　calibration adjustment　调校机构
10　sighting window　视窗
11　terminal compartment　端子盒
12　terminal compartment cover　端子盒盖
13　terminal　端子
14　sealed sighting lens　密封式目镜
15　conduit fitting　导线管管件
16　cone of rays　光锥
17　radiation energy　辐射能
18　target　被测物体
19　radiant thermopile　辐射热电堆
20　small thermocouple　小热电偶
21　tiny thermocouple junction　微型热电偶接头
22　group of thermocouples　热电偶堆
23　power switch　电源开关
24　analog signal output　模拟信号输出
25　out power input　外接电源输入
26　LCD screen　液晶显示屏
27　ε increase key　ε 递增键
28　ε decrease key　ε 递减键
29　function select key　功能选择键
30　function mark　功能标志
31　battery box cover　电池盒盖
32　measure switch　测量开关
33　dust-proof cover　防尘盖

9.1.4　双金属温度计　Bimetal Thermometer

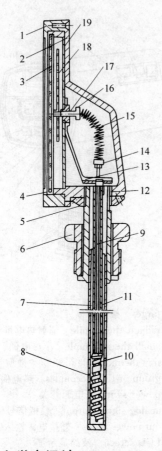

1 cover 表盖
2 instrument case 外壳
3 glass window 玻璃面板
4 sealing ring 密封圈
5 circle nut 圆螺母
6 male thread connector 外螺纹接头
7 protecting tube 外保护管
8 bimetallic sensor 双金属感温元件
9 rotating axis 转轴
10 lower connector 下固定件
11 protective sleeve 内保护管
12 upper fixed fitting 上固定件
13 stand 支架
14 lower fixed block for adjustable-angle spring 转角弹簧可动件
15 adjustable-angle spring 转角弹簧
16 upper fixed block for adjustable-angle spring 转角弹簧上固定件
17 spindle 心轴
18 pointer 指针
19 dial 刻度盘

9.1.5 光学高温计 Radiation Pyrometer

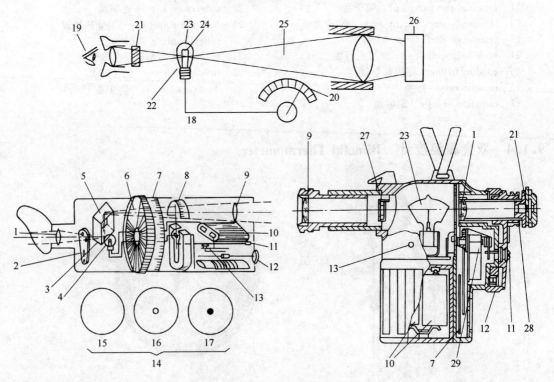

1	ocular lens 目镜	16	too high 太高
2	red grass slide 红色玻璃滑片	17	too low 太低
3	test mark 试验标志	18	schematic of an optical pyrometer 光学高温计原理图
4	lamp 灯泡		
5	prisms 棱镜	19	observer's eye 观测者的眼睛
6	optical wedge 光楔	20	scale spans 标度
7	temperature scale 温度标尺	21	red filter 红色滤光玻璃
8	ammeter scale 电流表标尺	22	fixed source 固定光源
9	objective lens 物镜	23	comparison lamp 对比灯
10	battery 电池	24	filament 灯丝
11	ammeter zero adjuster 电流表调零螺钉	25	unknown brightness 待测亮度
12	rheostat 变阻器	26	glowing object 炽热物体
13	switch 开关	27	absorb glass 吸收玻璃
14	photometry match 光度匹配	28	ocularlens position adjuster 目镜定位螺母
15	correct 正确	29	ammeter 电流表

9.2 压力测量仪表 Pressure Instruments

9.2.1 电接点压力表和压力开关 Pressure Gauge with Electric Contact and Pressure Switch

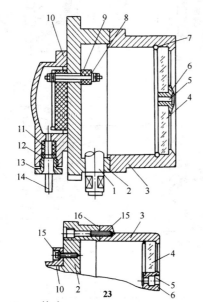
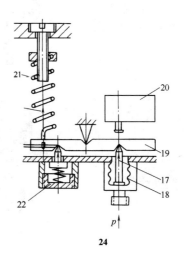

1	connection 接头		六角螺钉
2	housing 壳体	14	leading cable (four core) 引入电缆（四芯）
3	cover 盖	15	hexagon socket cap head screw 圆柱头内六角螺钉
4	glass 表玻璃		
5	adjustment 调节柱	16	thin gasket 薄垫片
6	shaft protecting sleeve 轴套	17	adjustment rod 调节杆
7	spring ring 弹簧圈	18	bellows 波纹管
8	conduct screw 导电螺钉	19	lever 杠杆
9	insulate sheath 绝缘套管	20	micro-switch 微动开关
10	terminal block 端子盒	21	balance spring 平衡弹簧
11	seal gasket 密封垫圈	22	differential spring 差动弹簧
12	gasket 垫圈	23	pressure gauge with electric contact 电接点压力表
13	hexagon socket cap head screw 圆柱头内	24	pressure switch 压力开关

9.2.2 压力表 Pressure Gauge

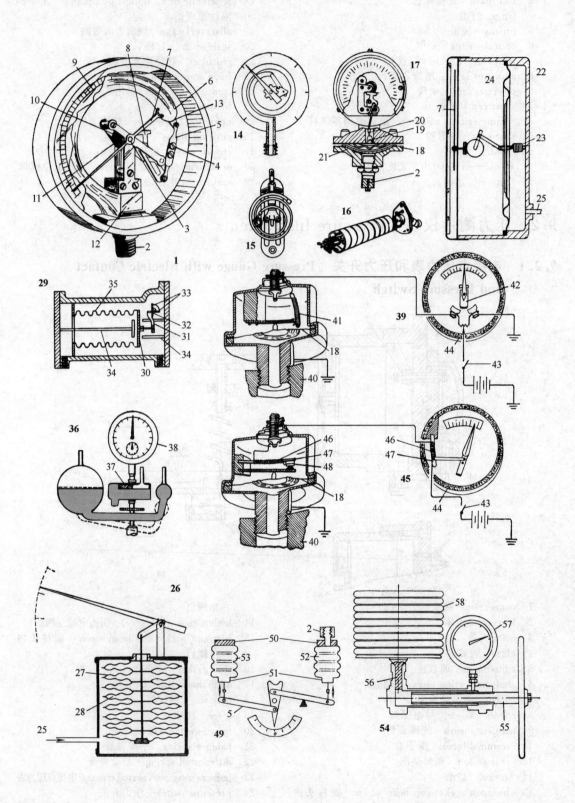

1	Bourdon tube pressure gage	弹簧管（波登管）压力表
2	process connector	工艺过程测量接头
3	zero adjustment	调零螺钉
4	range adjustment	量程调节螺钉
5	linkage	连杆
6	Bourdon tube	波登管，弹簧管
7	pointer	指针
8	quadrant (sector gear)	扇形齿轮
9	dial	标尺，度盘
10	pinion	小齿轮
11	hairspring	游丝
12	socket	支座
13	tip	自由端
14	spiral tube pressure gage	螺旋管压力表
15	flat spiral Bourdon tube	扁螺旋弹簧管
16	helical type Bourdon tube	管状螺旋弹簧管
17	stiff diaphragm pressure gage	硬质膜片压力表
18	diaphragm	膜片
19	spheric hinge	球铰链
20	coupling link	顶杆
21	protecting diaphragm	保护膜片
22	slack diaphragm pressure gage	低程膜片压力表
23	spring	弹簧
24	annular diaphragm	环形膜片
25	pressure inlet	压力入口
26	capsule pressure gage	膜盒压力表
27	corrugated diaphragm	波纹膜片
28	diaphragm stack	膜片组
29	high pressure bellows manometer	高压波纹管式压力计
30	free end	自由端
31	torque tube lever	扭力杆
32	torque tube assembly	扭力管组件
33	flexible metal strip	弹性金属片
34	stop	限位止挡
35	bellows unit	波纹管
36	Stalhane micromanometer	斯达尔汗微压计
37	cistern	储液槽
38	micrometer gage	微压表
39	remote pressure gage	遥测压力表
40	pressure tap	取压口
41	slide wire resistor	滑线电阻
42	permanent magnet	永久磁铁
43	switch	开关
44	indicating unit	指示装置
45	remote pressure switch	遥控压力开关
46	heater coil	加热线圈
47	bimetal strip	双金属条
48	breaker point	断开接点
49	absolute pressure bellows indicator	波纹管式绝对压力计
50	fixed end	固定端
51	movable end	活动端
52	process bellows	测量波纹管
53	evacuated and sealed reference bellows	抽空并密封的参比波纹管
54	static weight pressure calibrator	静重压力校验器（活塞式压力校验台）
55	screwed ram	丝杠
56	piston	活塞
57	calibrated pressure gage	受验压力表
58	weight	砝码

9.2.3 压力变送器 Pressure Transmitter

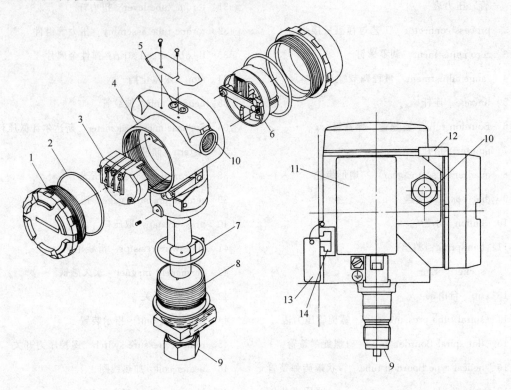

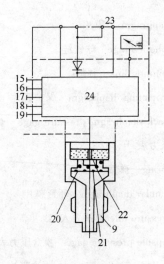

1　cover　盖
2　cover O-ring　盖上 O 形密封圈
3　terminal block　端子块
4　housing　表头外壳
5　span and zero adjustment　量程和零位调节
6　circuit board　电路板
7　module O-ring　模块上 O 形密封圈
8　sensor module　传感器模块
9　process connection　工艺连接
10　electrical connection　电气连接
11　name plate　铭牌
12　lock plate　锁紧片
13　label　标签
14　channel for mounted bolt　安装螺栓槽
15　lower range value　下限值
16　increase damping　增加阻尼
17　decrease damping　减少阻尼
18　span　量程
19　write over interlock　写入过量联锁
20　isolate diaphragm　隔离膜片
21　filled liquid　灌充液
22　sensing diaphragm　传感膜片
23　test instrument　测试仪表
24　microprocessor-based electronics　基于微处理器的电子模块

9.2.4 基地式压力指示调节器 Field Installed Type of Pressure Indicating and Controller

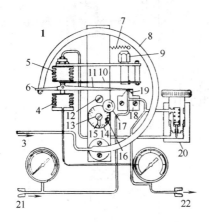
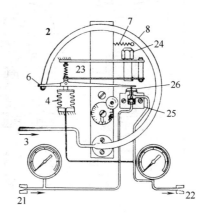

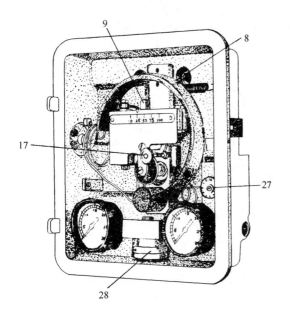

1 proportional integral mechanism 比例积分机构
2 differential gap mechanism 差动间隙机构
3 controlled variable 被测变量
4 proportional bellows 比例波纹管
5 integral bellows 重定波纹管，积分波纹管
6 overrange lever 超程杠杆
7 burden plate tension spring 弹簧板拉簧
8 burden tube 弹簧管
9 proportional band adjustment knob 比例度（带）调节旋钮
10 proportional leaf spring 比例簧片
11 flapper 挡板
12 burden plate post 弹簧板支架
13 setting screw 整定螺钉
14 set point scale 设定值刻度盘
15 pointer 给定指针
16 set point scale alignment screw 设定值标尺调整螺钉
17 set point adjustment knob 设定值点调节旋钮
18 nozzle block 喷嘴组件
19 nozzle 喷嘴
20 damping unit 阻尼部件
21 air supply 供气
22 output 输出
23 alignment spring 调整弹簧
24 differential-gap adjustment screw 差动间隙调节旋钮
25 pilot block 放大器组件
26 exhaust block 排气组件
27 retegral time adjustment 积分时间调节旋钮
28 booster relay 功率放大器

9.2.5 压力表的安装 Installation of Pressure Gage

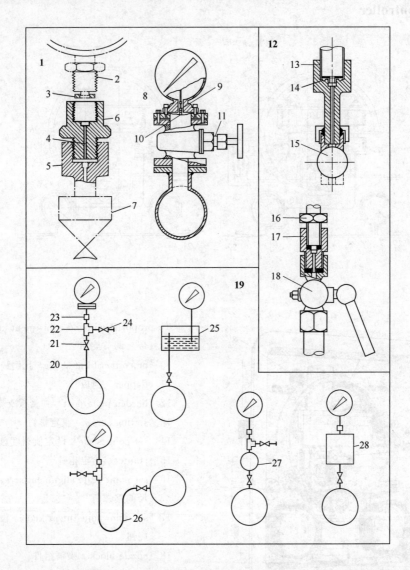

1	damper protecting the gauge against fluctuating 压力表阻尼保护	15, 21	globe valve 接管式截止阀
2	pressure gauge 压力表	16	P. G. (pressure gauge) shank 压力表六角头
3	gasket 密封垫	17	adaptor 转换接头
4	restriction hole 节流孔	18	globe valve 压力表截止阀
5	pressure gauge connector 压力表接头	19	local connections （压力表）现场连接形式
6	snubber 减振接头	20	welded nipple 焊接短管
7	isolating valve 隔断阀	22	tee 三通
8	flanged diaphragm gauge 法兰式膜片压力表	23	gauge coupling 压力表接头
9	spheric hinge 球铰链	24	veat discharge valve 泄压阀
10	diaphragm 膜片	25	sealing pot 隔离罐
11	gauge valve 压力表阀门	26	siphon 冷凝弯管
12	pressure gauge connections 压力表接头	27	pig tail type siphon 回弯冷凝管
13	hexagon shank 六角头	28	pulsation damper 缓冲罐
14	gramophone groove 密封线水线		

9.3 流量仪表 Flow Meter

9.3.1 科氏力质量流量计 Coriolis Mass Flow Meters

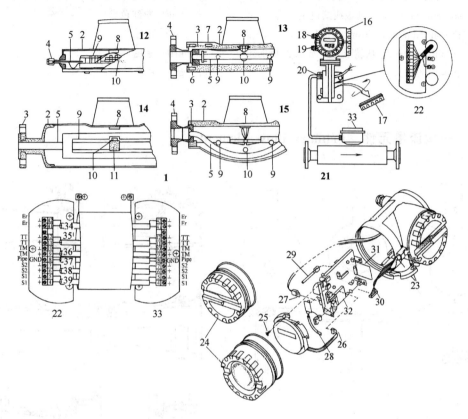

1 Coriolis mass flow meters sensor 科氏力质量流量计传感器
2 housing, vessel 外壳，容器
3 supporter 支架
4 process connect 工艺接头
5 measure tube 测量管
6 gasket 垫片
7 plug 插头
8 cable connect 电缆接头
9 electromagnetic phase shift detector 电磁式相位检测仪
10 exciter 励磁系统
11 torsion mode balanced 自平衡系统
12 single curved tube 单弯管
13 single straight tube 单直管
14 two straight tube 双直管
15 two slightly curved tube 双微弯管
16 terminal area 端子室
17 cover of the transmitter connection housing 变送器接线室盖
18 power supply cable gland 电源电缆入口密封部件
19 signal cable gland 信号电缆入口密封部件
20 sensor to transmitter special cable gland 传感器至变送器的专用电缆入口密封部件
21 remote version transmitter Coriolis mass flow meter 分体变送器型科氏力质量流量计
22 transmitter connection housing 变送器接线室
23 Allen screws 爱仑螺钉（锁紧螺钉）
24 cover of the electronics area of the transmitter housing 变送器电子部件室盒盖
25 mounting screws of display module 显示模块装配螺钉

549

26	plug of the ribbon cable of display module 显示模块带状电缆的插头	31	transmitter electronics board 变送器电子板
		32	replaceable data module 可替换数据的模块
27	2-pole plug of the power supply cable 电源电缆的两极插头	33	sensor connection housing 传感器接线室
		34	brn 棕色
28	electrode signal cable 电极信号电缆	35	wht 白色
29	coil current cable 励磁线圈电流电缆	36	pnk 粉红色
30	ribbon cable plug from terminal area to the amplifier board 从端子室至放大器板的带状电缆的插头	37	yel 黄色
		38	grn 绿色
		39	gry 灰色

9.3.2　一次流量元件　Primary Flow Element

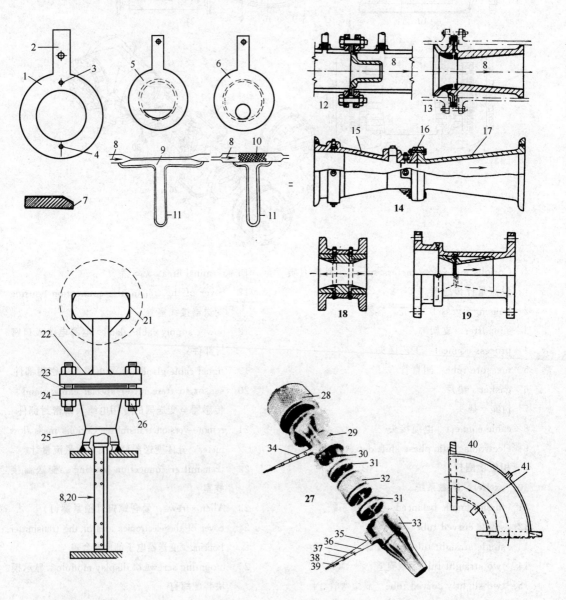

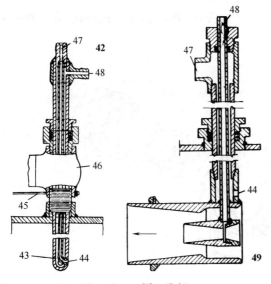
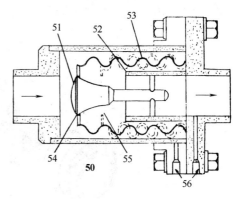

1 concentric orifice plate   同心孔板
2 tab   尾柄
3 vent   排气孔
4 drain   放液孔
5 segmental orifice plate   圆缺孔板
6 eccentric orifice plate   偏心孔板
7 sharp square edge   （孔板）锐边
8 flow direction   流向
9 capillary tube   毛细管
10 porous plug   多孔塞
11 manometer   压力计
12 flow nozzle   流量喷嘴
13 Lo-loss tube   芦罗斯管（低压损管之一种）
14 Venturi tube   文丘里管
15 inlet core   入口段
16 throat   文氏管喉口
17 outlet core   出口段
18 Gentile tube   坚梯耳管
19 Dall tube   道尔管
20 Annubar Sensor   阿牛巴检测元件
21 instrument head   仪表表头
22 sensor flange   安装检测元件的法兰
23 gasket   法兰密封垫片
24 weld neck flange   对焊法兰
25 weld coupling   焊接管箍
26 threaded fitting   螺纹接头
27 laminar flow element   层流元件
28 filter   过滤器
29 housing   壳体
30 run rod   通杆
31 straightening vane   导直管
32 main matrix metering section   主要阵列计量段
33 downstream housing   下游壳体
34 upstream tap   上游取压口
35 downstream tap   下游取压口
36 pulsation absorber   脉动吸取器
37 "O" ring and gland   O形环和压盖
38 gland spring   压盖弹簧
39 hose nipple   软管接头
40 elbow taps   肘管取压
41 pressure tap   取压口取压接头
42 Pitot tube   皮托管
43 static pressure hole   静压口
44 high-pressure impact hole   动压口
45 flow direction indicator   流向指示器
46 gate valve   闸阀
47 high-pressure connection   高压接头
48 low-pressure connection   低压接头
49 Pitot-Venturi tube (double-venturi pitot tube)   文丘里皮托管
50 Gilflo meter   吉尔福罗流量计
51 static control member   静压调节件
52 overrange stop   超程止挡
53 bellows   波纹管
54 measuring orifice at no flow   无流量时的检测孔板位置
55 measuring orifice at a flow   有流量时的检测孔板位置
56 differential pressure taps   差压取压点

9.3.3 流量测量用差压变送器 Difference Pressure Transmitter for Flow Measurement

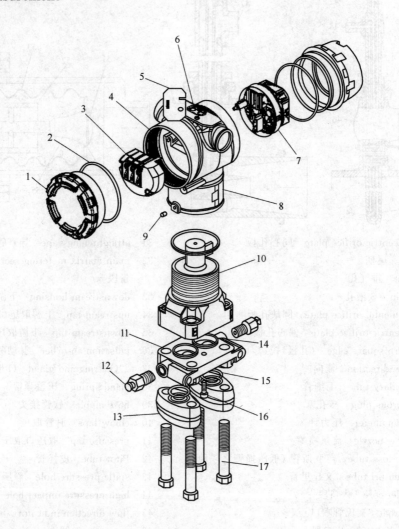

1 cover 接线盒盖
2 cover O-ring 接线盒O形密封圈
3 terminal block 接线端子
4 electronics housing 电子装置外壳
5 cover for local configuration buttons cover 就地组态按钮盖
6 local configuration buttons 就地组态按钮
7 electronics board 电子线路板
8 nameplate 铭牌
9 housing rotation set screw 外壳旋转止动螺钉
10 sensor module 传感器模块
11 coplanar flange 共平面法兰
12 drain/vent valve 排净/排气阀门
13 flange adapters 法兰适配器
14 process O-ring 工艺连接O形密封圈
15 flange adapter O-ring 法兰适配器O形密封圈
16 flange alignment screw 法兰调整螺钉
17 flange bolts 法兰螺栓

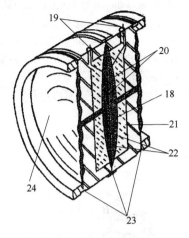

18　sensing diaphragm　传感膜片
19　leading wire　引线
20　capacitance plate　电容极板
21　rigid insulator　刚性绝缘体
22　filled liquid　灌充液
23　welded seal　焊接密封
24　isolate diaphragm　隔离膜片

9.3.4　流速式流量计　Fluid Velocity Meter

9.3.4.1　旋涡流量计　Vortex Flow Meter

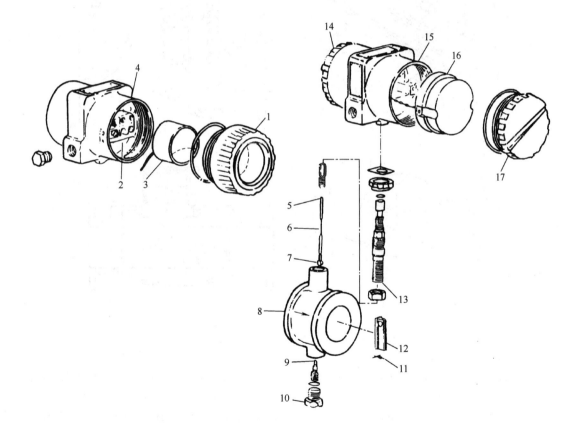

1　end cover (with indicator)　端盖（带指示器）
2　terminal block　接线板
3　linear optional plug-in flow rate indicator (analog output only)　线性可选择的插入式流量指示器（仅为模拟输出）
4　electronics housing (flow rate indicator end)　电子部分外壳（流量指示器端）
5　coax connector　同轴接头
6　sensor assembly　敏感元件组件
7　detector　探测器
8　flow meter body　流量计本体
9　pin　销子

553

10	cap screw 有帽螺钉		15	electronics housing (output module end) 电子部分外壳（输出组件端）
11	flow dam 流量堰			
12	vortex shedding element 漩涡元件		16	output module assembly (three versions available) 输出组件装配单元（有三种型式）
13	mechanical connector 机械接头			
14	end cover (without indicator) 端盖（无指示器）		17	end cover 端盖

9.3.4.2　涡轮流量计　Turbine Meter

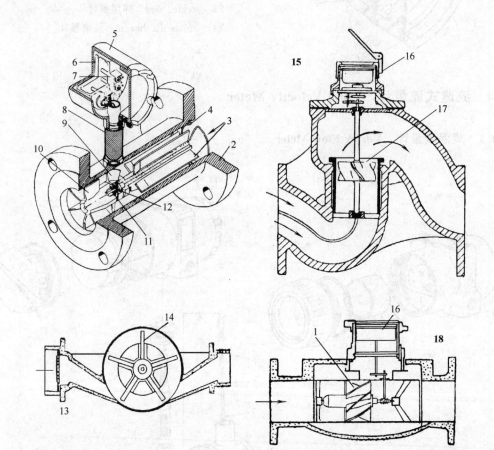

1	horizontal helix 水平螺翼		11	shaft 主轴
2	meter body 表体		12	thrust bearing 止推轴承
3	upstream stator 上游支架，上游固定架		13	fan type inferential meter 叶轮式流量表
4	retaining ring 挡圈		14	multi-blade fan 叶轮组，外叶叶轮
5	preamp (preamplifier) 前置放大器		15	vertical helix type inferential meter 垂直螺翼式流量表
6	amp unit 放大单元			
7	terminal block 端子板		16	counter 计数器
8	pick-off coil 感应线圈		17	vertical helix 垂直翼轮
9	rotor blade (impeller) 叶片，叶轮		18	horizontal helix inferential meter 水平螺翼式流量计
10	downstream stator 下游固定架			

9.3.5 容积式流量计 Positive Displacement Type Flow Meters

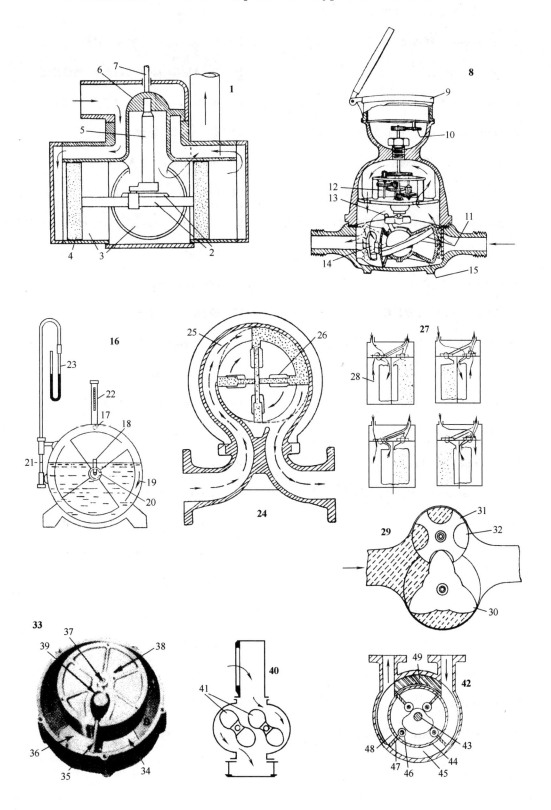

1 reciprocating piston meter (metering pump) 往复式活塞流量计，计量泵
2 scotch yoke 制动架
3 cylinder 活塞缸
4 piston 活塞
5 central crankshaft 中心曲轴
6 rotary valve 旋转阀
7 drive to register 指示表驱动轴
8 nutating disk meter 转盘式流量计，盘形流量计
9 register 指示表
10 counter gear 计数齿轮
11 nutating disk 转盘
12 gear train 齿轮组
13 disc spindle 转盘心轴
14 radial partition 径向分隔件
15 working chamber 工作室
16 wet gas meter 湿式气体流量计
17 outlet 出口
18 inlet 入口
19 direction of rotation 旋转方向
20 inlet slot 入口缝
21 level gage 液面计
22 thermometer 温度计
23 manometer 压力计
24 rotary vane meter 旋翼式流量计

25 cylindrical working chamber 圆筒形工作室
26 vane 翼片
27 bellows meter 皮囊流量计
28 flexible leathers chamber 挠性皮囊
29 duplex rotor meter 双转子流量计
30 gear teeth 轮齿
31 fluted rotor 带槽转子
32 rotor lobe 转子耳槽
33 rotary (oscillating) piston meter 旋摆式活塞流量计
34 inlet port 进气口
35 diaphragm 隔板
36 outlet port 出气口
37 spindle 心轴
38 oscillating piston 摆动活塞
39 roller 滚柱
40 lobed rotor meter 腰轮流量计
41 impeller 叶轮
42 sliding-vane meter 滑片式流量计
43 main shaft 主轴
44 rotor 转子
45 measuring chamber 测量室
46 cam roller 凸轮滚子
47 sliding vane 滑片
48 housing 外壳
49 block 隔离块

9.3.6 可变面积式流量计 Variable Area Type Flow Meters

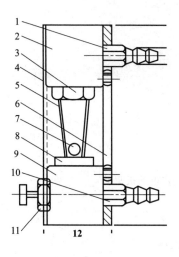
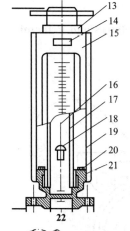
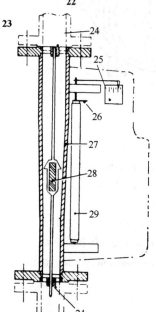
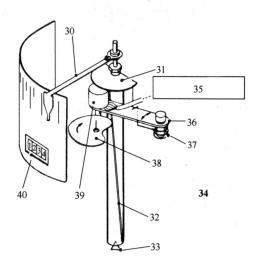

1 outlet 流出嘴
2 base (up) 上基座
3 pole 推压杆
4 case 有机罩壳
5 taper glass tube 锥形玻璃管
6,19 fixed plate 支承板
7 float 浮子
8 baseplate 下底座
9 base (down) 下基座
10 inlet 流入嘴
11 needle valve 针形阀
12 glass tube rotameter 玻璃管转子流量计
13 base 基座
14 label 标牌
15 case 罩壳
16 guide 导杆
17 taper glass tube 锥形玻璃管
18 float 浮子
20 bolt 螺栓
21 cover 压盖
22 glass tube rotameter (flange type) 法兰连接玻璃管转子流量计
23 metallic rotameter 金属转子流量计
24 float stop 浮子止挡
25 flow indicator scale 流量指示器刻度盘

26　flow sensing cam　流量传感凸轮
27　metal metering tube　金属流量计管
28　magnet embedded in metering float　嵌在流量计浮子中的磁铁
29　magnetic position converter　磁浮子位置转换器
30　pointer for direct indication　直接指示用指针
31　characterized cam　特性化凸轮
32　magnetic helix　磁性螺旋柱
33　bearing　轴承
34　electronic transmitter　电动变送器
35　amplifier and relay　放大器和继电器
36　pivot arm　支枢臂
37　spring return　复位弹簧
38　timing disc (synchronous motor driven)　定时圆盘（同步电动机驱动）
39　sensing coil　感应线圈
40　counter　计数器

9.3.7　电磁流量计　Electro-magnetic Flow Meters

9.3.7.1　电磁流量计及测量原理　Electro-magnetic Flow Meters and Principle of Measurement

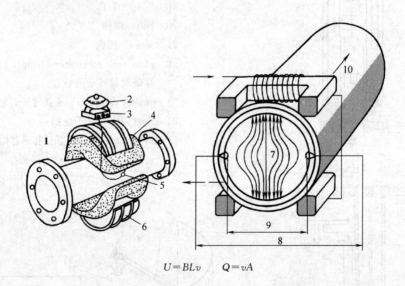

$$U = BLv \qquad Q = vA$$

1　electro-magnetic flow meters　电磁流量计
2　current transformer　电流信号转换器
3　terminate box　接线盒
4　copper coil　铜线圈
5　measuring electrode　测量电极
6　iron circuit　磁轭
7　magnetic induction B　磁场 B
8　induced voltage U　电势 U
9　distance between electrodes L　电极间距离 L
10　flow velocity v　流速 v
volume flow Q　体积流量 Q
pipe cross-section A　管截面积 A

9.3.7.2 电磁流量计电极 Electro-magnetic Flow Meters Electrode

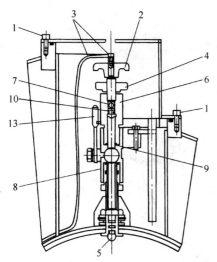
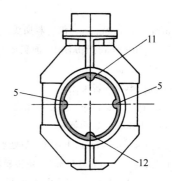

1 Allen screw 爱仑螺钉（锁紧螺钉）
2 rotary arm 旋臂
3 electrode cable 电极电缆
4 knurled nut (counter nut) 凸边螺母（反螺纹螺母）
5 measuring electrode 测量电极
6 holding cylinder 支承套筒
7 locking pins (rotary arm) 锁紧销（旋臂）
8 ball valve housing 球阀外壳
9 gasket (holding cylinder) 垫片（支承套筒）
10 coil spring 线圈弹簧
11 empty pipe detection electrode 空管检测电极
12 reference electrode 参考电极
13 shut-off valve (ball valve) 切断阀（球阀）

9.3.7.3 电磁流量计变送器 Electro-magnetic Flow Meters Transmitter

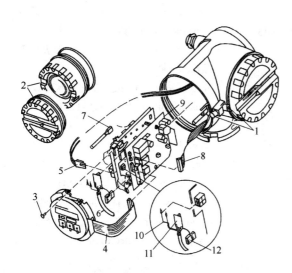

3 mounting screws of display module 显示模块装配螺钉
4 plug of the ribbon cable of display module 显示模块带状电缆的插头
5 2-pole plug of the power supply cable 电源电缆的两极插头
6 electrode signal cable 电极信号电缆
7 coil current cable 励磁线圈电流电缆
8 ribbon cable plug from terminal area to the amplifier board 从端子室至放大器板的带状电缆的插头
9 transmitter electronics board 变送器电子板
10 EPD (empty pipe detection) electrode cable 空管检测电极电缆
11 cable board 电缆板
12 replaceable data module cable 可替换数据的模块电缆

1 Allen screws 爱仑螺钉（锁紧螺钉）
2 cover of the electronics area of the transmitter housing 变送器电子部件室盒盖

9.3.8 热式质量流量计 Thermal Mass Flow Meters

1　transducer connector　传送器插头
2　pre-amplifier circuit board　前置放大器线路板
3　shroud　盖板
4　housing cover　表头盖
5　liquid crystal display　液晶显示器
6　display holder　显示器座
7　microprocessor circuit board　微处理器线路板
8　main sensor terminal board　主传感器端子板
9　housing cable loom　表头电缆插头
10　flowcell screws and washers　与流体腔室固定的螺栓和垫圈
11　housing seal　表头密封圈
12　pre-amplifier connector　前置放大器插头
13　transducer seal　传送器密封圈
14　housing screws and washers　与表头固定的螺栓和垫圈
15　process screw fitting　工艺连接螺纹接头
16　housing collar　与表头连接的法兰
17　insertion rod　插入棒
18　flowcell　流体腔室
19　terminal compartment cover　接线端子盒盖
20　insertion sensor　插入式传感器
21　wafer flowcell　夹持连接式流体腔室
22　flanged flowcell　法兰连接式流体腔室
23　compact sensor electronic housing　一体式电子表头
24　remote sensor electronic housing　分体式电子表头
25　calibrated transducer assembly　校准后的传送器组件

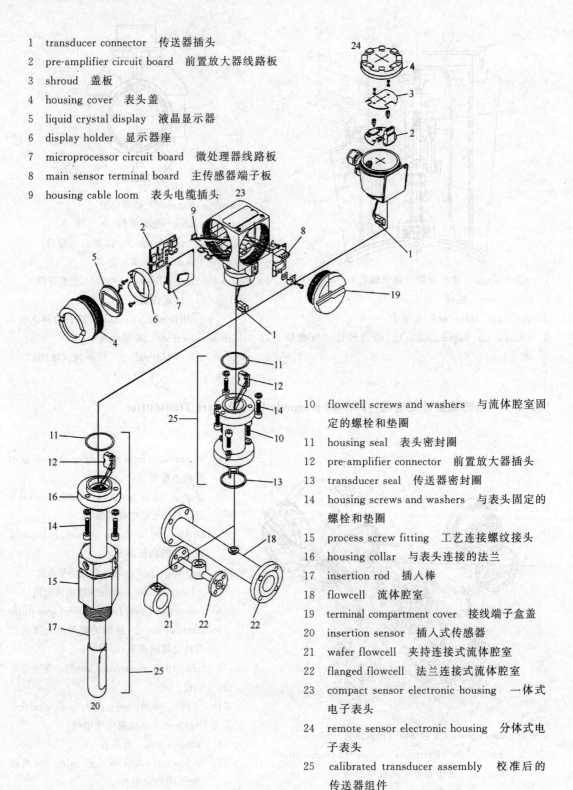

9.3.9 差压流量计的安装 Installation of Head Meters

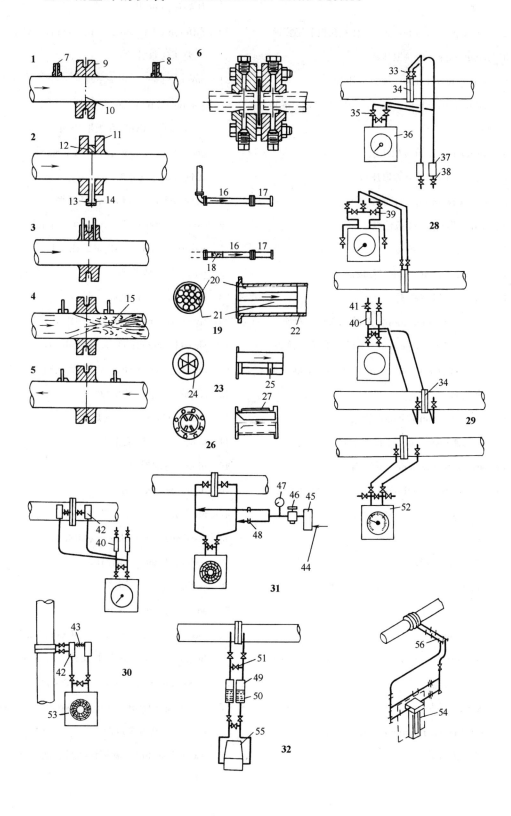

1	pipe taps (full flow taps, line taps) 管线取压法	29	differential flowmeter in liquid service 液体差压流量计
2	carrier pattern orifice plate 环室孔板节流装置	30	differential flowmeter in steam service 蒸汽差压流量计
3	flange taps 法兰取压法	31	purge system 吹洗系统
4	vena contract taps 缩脉取压法	32	sealing system 隔离系统
5	radius taps (throat taps) 径距取压法，喉道取压法	33	orifice tap valve 孔板取压阀
6	corner taps 角接取压法	34	detecting element (sensing element) 检测元件（敏感元件）
7	upstream tap 上游取压口	35	three valve manifolds 三阀组
8	downstream tap 下游取压口	36	flow registering instrument 流量记录仪
9	orifice flange 孔板法兰	37	moisture sump (sediment chamber, knockout pot) 凝液收集器
10	orifice plate 孔板		
11	adjacent flange 环室孔板法兰	38	drain valve 排凝阀，排液阀
12	carrier 环室孔板	39	five valve manifolds 五阀组
13	upstream connection 上游（取压）接头	40	air collecting vessel (air chamber) 集气器
14	downstream connection 下游（取压）接头	41	release valve (vent valve) 泄气阀（放空阀）
15	vena contracta (minimum section) 缩脉（最小截面）	42	condensate pot 冷凝器
16	upstream run 上游直管段	43	lagging (insulate) 绝热层
17	downstream run 下游直管段	44	purge fluid 吹洗流体
18	straightening vane 导直器	45	filter 过滤器
19	multitube straightening vane 多管导直器	46	regulator 减压阀
20	sheet metal tube 薄金属管	47	pressure gage 压力表
21	metal pipe 金属管	48	restricted orifice plate 限流孔板
22	pipeline 管线	49	sealing chamber 隔离罐
23	sector type vane 扇形导流叶片	50	sealing liquid 隔离液
24	gusset 角撑板	51	equalizing valve 平衡阀
25	ring support 支持圈	52	mechanical mercury meter 机械式水银压差计
26	straightening vane for sewage service 下水道用导直器	53	bellows meter 波纹管式差压流量计
27	cover 盖板	54	glass tube manometer 玻璃管压差计
28	differential flowmeter in gas service 气体差压流量计	55	diaphragm transmitter (differential pressure cell) 膜式差压变送器
		56	meter lead line 流量计测量引线（管）

9.4 液位测量仪表 Level Meter

9.4.1 直读式就地液位指示仪 Locally Mounted Direct Reading Level Gages

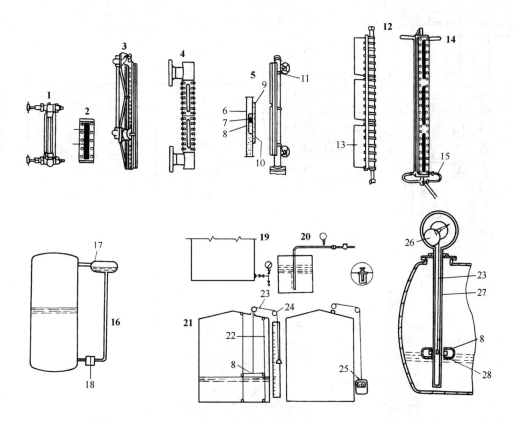

1 tubular gage glass 玻璃管液位计
2 reflex level gage 反射式液位计
3 transparent level gage with illuminator 带照明的透光液位计
4 transparent level gage 透光式液位计
5 armored magnetic level gage 金属管磁浮子式液位计
6 magnetic steering device 磁导向机构
7 magnet 磁铁
8 float 浮标，浮子
9 indicator 指示标尺
10 indicator wafers 指示板
11 indicator clamped to float chamber 固定在浮子室上的指示板
12 non-frosting type level gage 防霜式液位计
13 frost extension 防霜装置
14 heating type level gage 加热式液位计
15 jacket steam pipe 夹套蒸汽管
16 hydrostatic level-measuring system 静压式液位测量系统
17 condensate pot 凝液罐
18 differential pressure measuring element 差压测量元件
19 direct hydrostatic head type level gage 直接静压头式液位计
20 air-bubble system hydrostatic head level gage 吹气静压式液位计
21 tape float level gage 钢带浮标液位计
22 guide wire 导向绳
23 tape 钢带
24 sheave 滚轮
25 gage head 表头
26 magnetic-bond float gage 磁力耦合式浮子液位计
27 non-magnetic guide tube 非磁性导向管
28 follower magnet 随动磁铁

9.4.2 液位测量用差压变送器 Difference Pressure Transmitter for Liquid Level Measurement

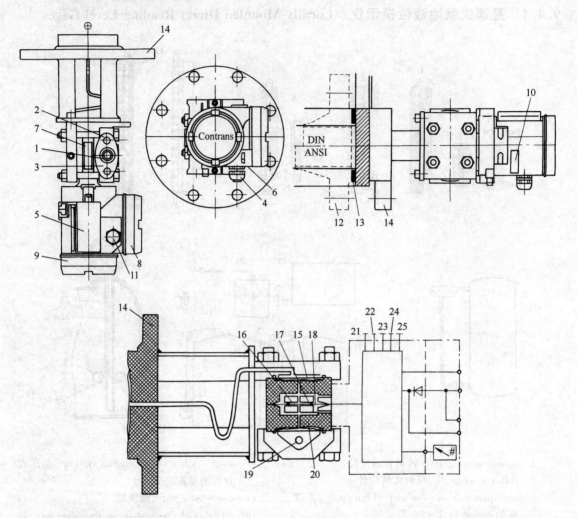

1	female screw for process connector 工艺接口用内螺纹	13	user installation gasket 用户安装用垫片
2	female screw for fitting bolt 固定螺栓用内螺纹	14	process flange 工艺连接法兰
3	elliptic flange 椭圆法兰	15	higher pressure side 高压侧
4	electric connector 电气连接	16	measure chamber 测量室
5	name plate 铭牌	17	sensing diaphragm 传感膜片
6	lock 锁扣	18	over range diaphragm 过载膜片
7	measure unit trademark 测量机构标牌	19	lower pressure side 低压侧
8	cover for terminal chamber 接线端子室表盖	20	measure body 测量体
9	cover for digital indicator 数字指示器表盖	21	lower range value 下限值
10	label 标签	22	increase damping 增加阻尼
11	plug 丝堵	23	decrease damping 减少阻尼
12	user installation flange 用户安装用法兰	24	span 量程
		25	write over interlock 写入过量联锁

9.4.3 浮筒式液位计　Displacement Type Level Meter

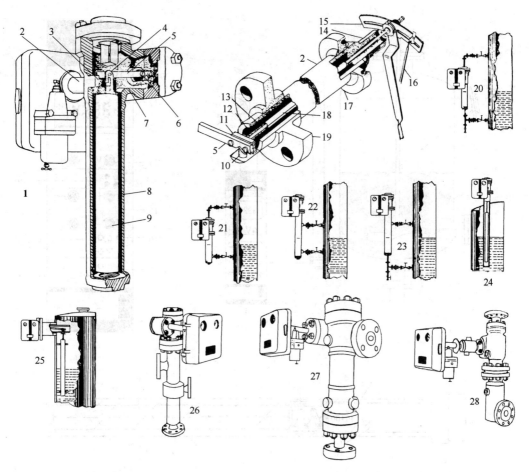

1　transmitter　变送器
2　torque tube housing　扭力管外壳
3　limit stop　限位止挡
4　hanger　挂钩
5　torque arm　扭力臂
6　knife-edge bearing　刀刃支架
7　integral mechanism chamber　整体机构室
8　displacer chamber (displacer cage)　浮筒室
9　displacer　浮筒
10　knife edge　刀刃
11　torque arm block　扭力臂卡头
12　torque tube bushing　扭力管内外螺纹接头
13　torque tube rod　扭力管心轴
14　torque tube housing extension　扭力管外壳延长段
15　reversing arc　弧形换向板
16　control link　调节杆，控制杆
17　case mounting flange　表体安装法兰
18　torque tube　扭力管
19　torque tube housing flange　扭力管外壳法兰
20　top and bottom screwed connections　顶底螺纹连接方式
21　top and side screwed connections　顶侧螺纹连接方式
22　side and side screwed connections　侧螺纹连接方式
23　side and bottom screwed connections　侧底螺纹连接方式
24　top vessel flanged connection　容器顶部法兰连接方式
25　side vessel flanged connection　容器侧壁法兰连接方式
26　displacer chamber with distributing ring　带分配环的浮筒室
27　high pressure level controller　高压液面调节器
28　complete steam jacketing　蒸汽夹套式（液位计）

9.4.4 伽马射线液位计 Gamma Radiometric Level Meter

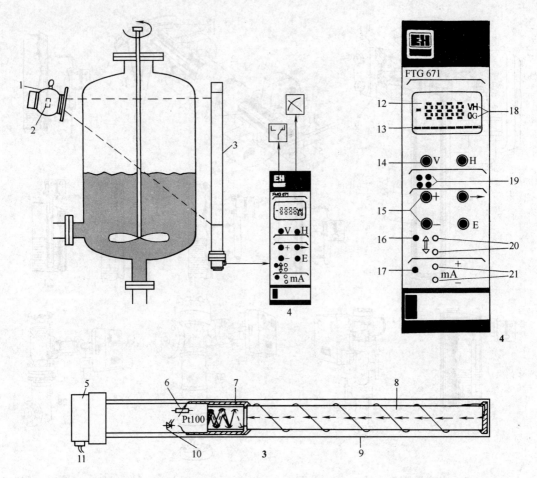

1 source container 伽马射线源密封盒
2 source 伽马射线源
3 detector 检测器
4 transmitter 变送器
5 detector head 检测器头
6 temperature sensor 温度测量元件
7 photomultiplier 光电倍增器
8 scintillation rod produces flashes when gamma radiation impinges on it 当伽马射线照射它时,火花棒就闪烁
9 outer stainless steel encapsulation 不锈钢外套
10 reference diode 参照用的发光二极管
11 connection to transmitter 连接至变送器
12 measured value display showing level in % at V0H0 在V0H0矩阵点处显示的以百分数表示的液位测量值
13 bar chart 棒图
14 matrix selection keys 矩阵点选择键
15 parameter entry keys 参数输入键
16 green communications LED—lights when communication 绿色通信发光二极管——通信时灯亮
17 alarm relay LED—lights on fault condition, relay de-energies 报警继电器发光二极管——事故时继电器失电,灯亮
18 matrix position Indicator 矩阵点位置显示
19 limit relay LED—red lit, relay de-energized green lit, relay energisued 极限继电器发光二极管——继电器失电红灯亮,继电器通电绿灯亮
20 commulog sokets 通信插口
21 test sockets for 0/4...20mA, analogue output 0/4～20mA 模拟量输出测试插口

9.4.5 雷达液位计　Radar Level Meter

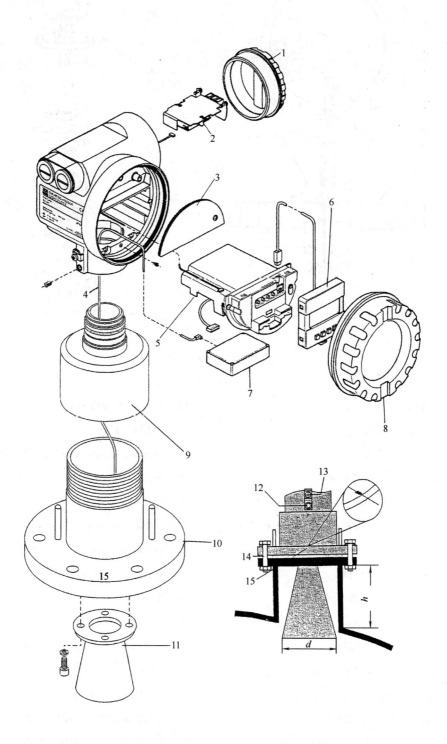

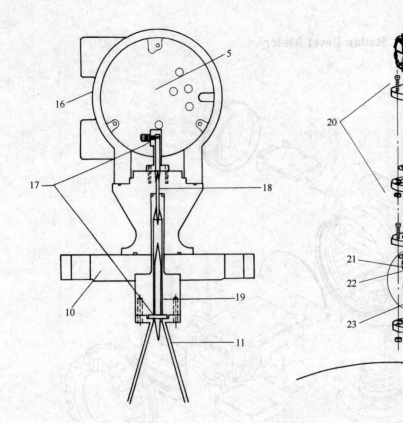

1 lid 盖子
2 terminal 端子块
3 cover plate 盖板
4 antenna cable 天线电缆
5 radar electronics module 雷达电子模块
6 operating and display module 操作和显示模块
7 HF module 高频模块
8 lid with window 带玻璃的表盖
9 adapter for housing complete with O-rings 带O形圈的表头适配器
10 flange assembly 法兰组件
11 antenna horn 喇叭形天线
12 locking screw 锁紧螺钉
13 ground terminal 接地端子
14 datum point for measurement 测量基准点
15 alignment mark points to the tank wall 针对罐壁的标记
16 aluminum gauge housing 铝制表头
17 waveguide 波导管
18 alumina 氧化铝陶瓷
19 Teflon vapor seal 四氟塑料气封
20 mounting bolts/nuts 安全螺栓/螺母
21 SST window ring 不锈钢窗环
22 O-ring O形环
23 process O-ring 工艺连接O形环
24 spiral-wound gasket (non-wetted) 缠绕垫 (不接触介质)
25 stand off pipe or spool piece (non-wetted) 支承管或短管 (不接触介质)
26 Teflon window 四氟塑料窗
27 process flange 工艺连接法兰

9.4.6 伺服液位计 Servo Level Gauge

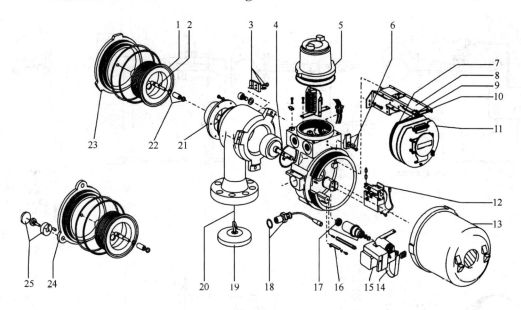

1 drum with stainless steel wire＋shaft and circlip   磁鼓带不锈钢钢丝＋轴和挡圈
2 drum shaft   磁鼓轴
3 seal assembly   密封组件
4 key   键
5 kit containing higher terminal set   高级终端成套设备
6 lock bracket with screws for medium pressure   中压型螺纹锁紧支架
7 option board for foundation fieldbus communication   现场总线板卡
8 support bracket for backplane assembly   底板组装支架
9 GPS printed circuit board   电源板
10 back-plane assembly   底板组件
11 printed circuit board with communication channel   通信电路板
12 force transducer   力传感器
13 electronic compartment cover   电气接线室盖
14 motor board   马达板卡
15 motor assembly   马达组建
16 transport bracket   传送支架
17 ball bearing   滚珠轴承
18 IR connector (chassis part)   红外线接头（底盘部分）
19 displacer   浮子
20 measuring wire stainless steel   不锈钢测量钢丝
21 magnet cap   电磁帽
22 set drum bearings   固定磁鼓密封圈
23 drum compartment cover medium pressure   中压型磁鼓室盖
24 drum compartment cover high pressure and chemical   高压型和化学型磁鼓室盖
25 lock bracket for high pressure and chemical   高压型和化学型锁紧支架

9.4.7 液位调节器及液位开关 Level Controllers and Switches

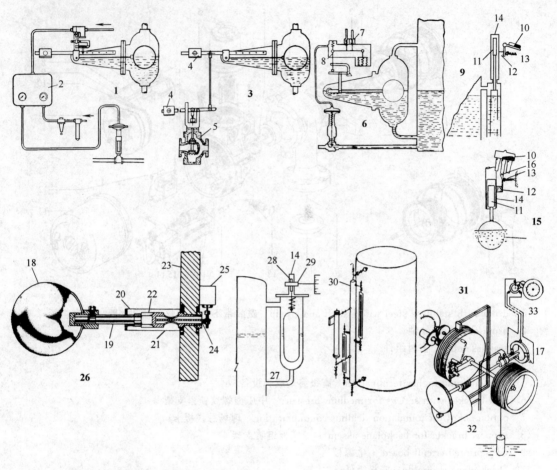

1 external cage-mounted ball float controller 外浮球式液位调节器
2 controller 调节器
3 float actuated level valve 浮球驱动液位控制阀
4 counterweight 平衡重锤
5 balanced control valve 平衡型控制阀
6 ball float with pilot valve 带先导阀的浮球式液位调节器
7 pilot valve 先导阀
8 reduction mechanism 变换机构
9 magnetically actuated displacer switch 磁动式浮筒液位开关
10 mercury switch 水银开关
11 magnetic armature 衔铁
12 permanent magnet 永久磁铁
13 tension (return) spring 拉力（复位）弹簧
14 non-magnetic tube 非磁性管
15 magnetically actuated ball-float switch 磁动式浮球液位开关
16 pivot 支枢
17 transmitter 变送器
18 float 浮球
19 float arm 浮球杆
20 limit-stop bracket 止挡托架
21 flattened section 扁平部分，扁平段
22 packless flexible shaft 无填料挠性轴
23 standard flange 标准法兰
24 float-arm extension tongue 浮球杆延伸舌片
25 air pilot 气动放大器
26 float level transmitter 浮球液位变送器
27 magnetically coupled displacer unit 磁力耦合式浮筒装置
28 drive magnet 驱动磁铁
29 magnet follow 随动磁铁
30 standpipe (pipe column) 立管
31 indicator for precise indication of level of oil in tanks 精密油罐液面计
32 two phase induction motor (reversing) 两相感应电动机（可逆电动机）
33 distant indicator 远距离指示器

9.4.8 气动液位调节器 Pneumatic Liquid Level Controller

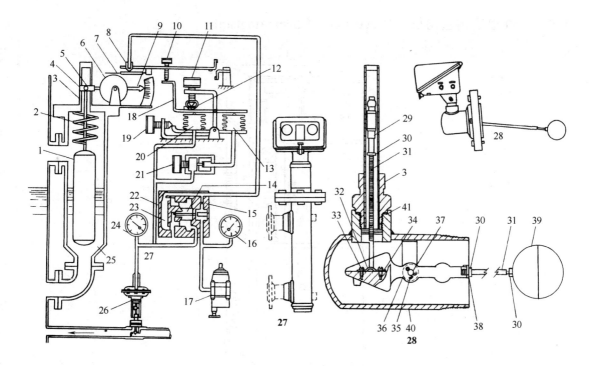

1	displacer 浮筒		22	relay 继动器
2	range spring 量程弹簧		23	chamber 腔室
3	enclosing tube 密封管		24	output gauge 输出压力表
4	magnet 磁极		25	cage 浮筒室
5	attraction ball 吸引球		26	control valve 控制阀
6	cam 凸轮		27	pneumatic displacer liquid level controller 气动浮筒液位调节器
7	flapper 挡板			
8	pilot nozzle 先导放大器喷嘴		28	pneumatic float ball liquid level controller 气动浮球液位调节器
9	indicator 指针			
10	zero adjustment 零位调节螺钉		29	attraction sleeve 吸引套筒
11	proportional band adjustment 比例带调节螺钉		30	hex nut 六角螺母
			31	stem 支杆
12	proportional band spring 比例带弹簧		32	stem retaining bracket 支杆座
13	reset bellows 重定波纹管		33	screw 螺钉
14	exhaust valve 排气阀		34	bracket 支架
15	orifice 孔板		35	cotter pin 固定销针
16	supply air gauge 供气压力表		36	washer 垫圈
17	filter regulator 过滤器减压阀		37	pivot pin 芯轴
18	actuating lever 作用力杠		38	lock washer 锁紧垫圈
19	set point adjustment 设定点调节螺钉		39	float 浮球
20	feed back bellows 反馈波纹管		40	body 壳体
21	reset valve 重定阀		41	E-tube gasket 管用E形垫片

9.5 过程分析仪表 Process Analyzer

9.5.1 过程气相色谱仪 Process Gas Chromatograph

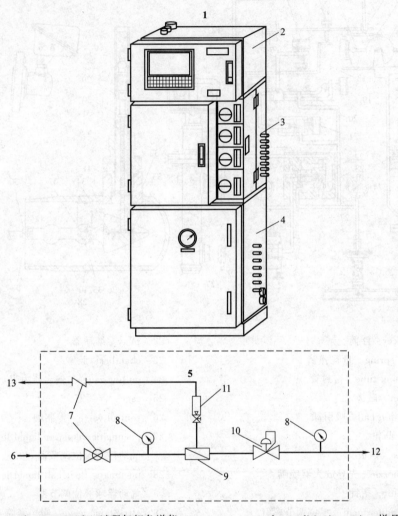

1 process gas chromatograph 过程气相色谱仪
2 control unit 控制单元
　pressurized explosion-proof endosure 正压防爆箱体
　LCD panel with multi-functional menu 带多功能菜单的液晶显示面板
　LED indication and membrane keyboard LED 显示及触摸键盘
3 analysis process unit 分析处理单元
　thermostatic oven 恒温加热炉
　TCD or FID detectors TCD 或 FID 检测器
　column flow programming and valve with actuator 色谱柱流量编程和切换阀
4 sample conditioning unit 样品处理单元
5 typical sample conditioning system 典型的样品处理系统
6 sample gas in 采样气体进入
7 valves 阀
8 pressure gauges 压力表
9 sampling filter 进样过滤器
10 regulator 减压阀
11 rotameter 转子流量计
12 sample gas to analyzer 采样气体至分析器
13 fast loop or fast by-pass return 快速回路或快速旁路返回点

9.5.2 非色散红外线气体分析仪　Non-dispersive Infra-red Gas Analyzer（NDIR）

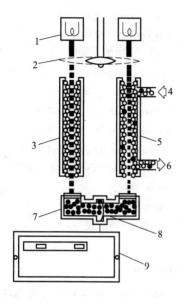

1　infrared light source　红外光源
2　chopper　切光片
3　reference cell　参比池
4　sample in　采样气体入口
5　sample cell　测量池
6　sample out　采样气体出口
7　detector　检测器
8　film capacitor detector　薄膜电容检测器
9　front panel　分析仪盘面

9.5.3 氧分析器　Oxygen Analyzer

9.5.3.1 氧化锆分析仪　Zirconium Oxide Oxygen Analyzer

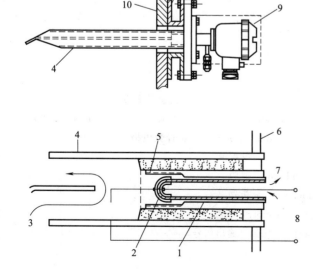

1　zirconium oxide element　氧化锆元件
2　electrode　电极
3　sample gas　采样气体
4　pipe for sample gas　采样气体导流管
5　heater　加热器
6　flange　法兰
7　reference gas　参比气
8　output　输出
9　detector　检测器
10　burner wall　炉壁

573

9.5.3.2 顺磁式氧分析仪 Paramagnetic Oxygen Analyzer

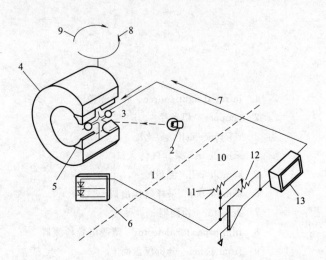

1 detector/magnet assembly 检测器/磁力组件
2 source lamp 光源
3 test body 试验体
4 magnet 磁铁
5 shaded pole pieces (4) 4块磁极
6 dual photocell 双光电管
7 restoring current 复位电流
8 restoring torque 复位力矩
9 displacement torque 位移力矩
10 control assembly 控制组件
11 zero 零点
12 span 量程
13 readout 指示仪

9.5.3.3 电化学式氧分析仪 Electrochemical Oxygen Analyzer

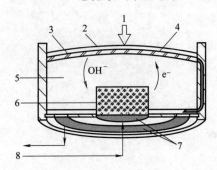

1 oxygen in the sample gas 采样气体中氧气
2 gas permeable membrane 气体渗透薄膜
3 thin electrolyte layer 薄电离层
4 cathode 阴极
5 electrolyte 电离
6 anode 阳极
7 contact plate 接触板
8 current signal 电流信号

9.5.3.4 溶解氧分析仪 Dissolved Oxygen Analyzer

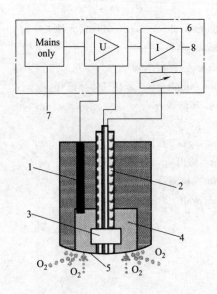

1 silver reference electrode 银参考电极
2 gold working electrode (cathode) 金工作电极（阴极）
3 silver counter-electrode 银计数电极
4 electrolyte 电解液
5 membrane 薄膜
6 analyzer 分析器
7 power supply 电源
8 output signal 输出信号

9.5.4 热导式分析仪 Thermal Conductivity Analyzer

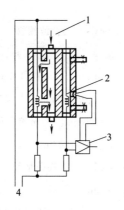

1 sample gas 采样气体
2 measuring resistance (Pt) 测量电阻（铂电阻丝）
3 amplifier 放大器
4 constant current 恒电流

9.5.5 pH 分析仪 pH Analyzer

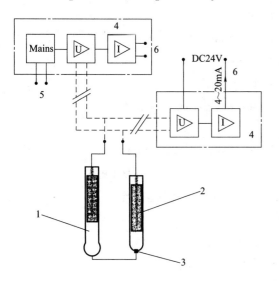

1 measuring electrode 测量电极
2 reference electrode 参考电极
3 diaphragm 薄膜
4 analyzer 分析仪
5 power supply 电源
6 output signal 输出信号

9.5.6 电导仪 Conductivity Analyzer

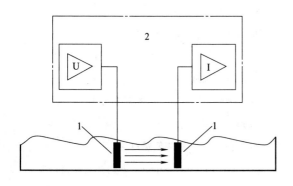

1 electrodes 电极
2 analyzer 分析仪

9.6 控制阀 Control Valve

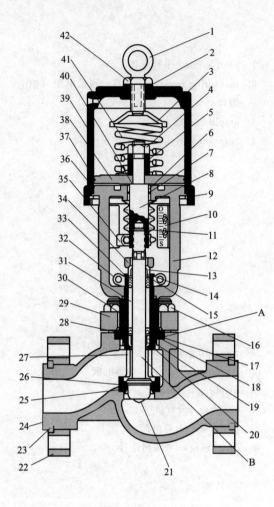

1　lifting ring　起吊环
2　adjusting screw gasket　调整螺钉垫片
3　spring button　弹簧压片
4　piston retaining nut　活塞夹持螺帽
5　actuator stem spacer　执行机构输出杆衬套
6　actuator stem bushing　执行机构输出杆套筒
7　actuator stem O-ring　执行机构输出杆O-形环
8　actuator stem　执行机构输出杆
9　cylinder retaining ring　气缸夹持环
10　stroke plate　行程标尺
11　stem bellows　阀杆波纹管
12　yoke　支架
13　upper stem guide　上阀杆导向
14　upper stem guide liner　上阀杆导向衬套
15　half clamp spacer　防挤出压环
16　bonnet flange stud/nut　上阀盖法兰螺栓/螺母
17　bonnet gasket　上阀盖垫片
18　lower packing　下填料
19　lower stem guide　下阀杆导向
20　lower stem guide liner　下阀杆导向衬套
21　plug　阀芯
22　end flange　法兰连接端
23　half ring　半环
24　body　阀体
25　seat ring gasket　阀座垫片
26　seat ring　阀座
27　seat retainer　阀座夹持器
28　anti-extrusion spacer　防挤出压环
29　bonnet flange　上阀盖法兰
30　bonnet　上阀盖
31　packing spacer（s）　填料套
32　yoke clamp　支架夹具
33　upper packing　上填料
34　gland flange　压盖法兰
35　stem clamp　阀杆夹具
36　yoke O-ring　支架O形环
37　piston O-ring　活塞O形环
38　piston　活塞
39　cylinder　气缸
40　piston stem O-ring　活塞杆O形环
41　spring　弹簧
42　adjusting screw　调整螺钉
A　lower stem guide retainer　下阀杆导向夹持器
B　stem　阀杆

9.6.1 气动执行机构 Pneumatic Actuator

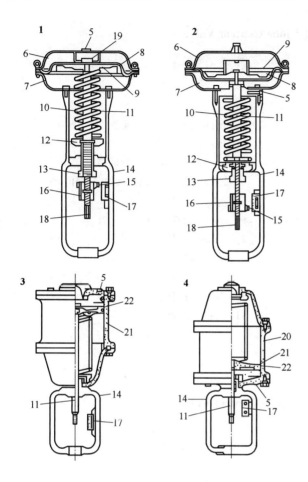

1. direct acting diaphragm actuator 正作用薄膜执行机构
2. reverse acting diaphragm actuator 反作用薄膜执行机构
3. direct acting piston actuator 正作用活塞执行机构
4. reverse acting piston actuator 反作用活塞执行机构
5. air connection 气信号接口
6. upper casing 上盖
7. lower casing 下盖
8. diaphragm 膜片
9. diaphragm plate 膜盘（硬芯）
10. actuator spring 执行机构弹簧
11. actuator stem 执行机构输出杆
12. spring seat 弹簧座
13. spring adjustor 弹簧调整器
14. yoke 支架
15. travel indicator 行程指针
16. stem connector 阀杆连接件
17. indicator scale 行程标尺
18. valve stem 阀杆
19. lock nut 锁紧螺母
20. cylinder 气缸
21. piston 活塞
22. piston ring 活塞环

9.6.2 电动执行机构 Electric Actuator

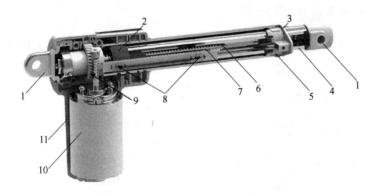

1. front / rear clevis 前/后连接叉
2. drive nut 驱动螺母
3. wiper 刮片
4. inner tube 内套管
5. outer tube 外套管
6. safety stop 安全止推机构
7. spindle 轴
8. limit switches 限位开关
9. gear 齿轮
10. servo motor 伺服电机
11. motor housing 电机外壳

577

9.6.3 控制阀型式 Control Valve Type

9.6.3.1 直通单座控制阀 Single-Ported Globe Control Valve

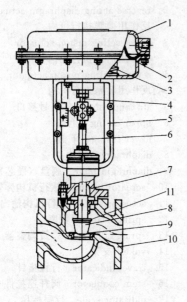

1 diaphragm casing 膜盖
2 diaphragm 膜片
3 actuator spring 执行机构弹簧
4 actuator stem 执行机构输出杆
5 yoke 支架
6 stem 阀杆
7 bonnet 上阀盖
8 plug 阀芯
9 seat ring 阀座
10 body 阀体
11 cage 阀笼

9.6.3.2 直通双座控制阀 Double-Ported Globe Control Valve

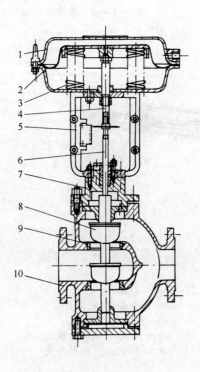

1 diaphragm casing 膜盖
2 diaphragm 膜片
3 actuator spring 执行机构弹簧
4 actuator stem 执行机构输出杆
5 yoke 支架
6 stem 阀杆
7 bonnet 上阀盖
8 plug 阀芯
9 seat ring 阀座
10 body 阀体

9.6.3.3 角形控制阀 Angle Control Valve

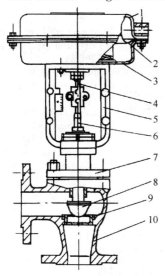

1 diaphragm casing 膜盖
2 diaphragm 膜片
3 actuator spring 执行机构弹簧
4 actuator stem 执行机构输出杆
5 yoke 支架
6 stem 阀杆
7 bonnet 上阀盖
8 plug 阀芯
9 seat ring 阀座
10 body 阀体

9.6.3.4 三通控制阀 Three-Way Control Valve

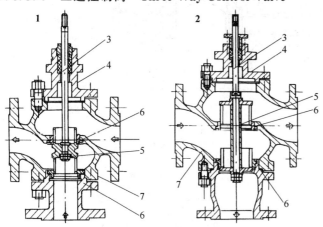

1 three-way valve for mixing operation 三通合流阀
2 three-way valve for diverting operation 三通分流阀
3 stem 阀杆
4 bonnet 上阀盖
5 plug 阀芯
6 seatring 阀座
7 body 阀体

9.6.3.5 控制蝶阀 Butterfly Control Valve

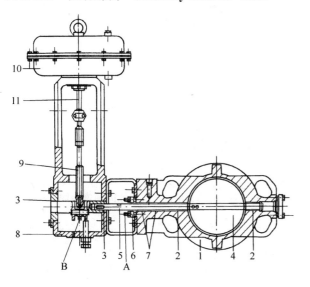

1 body 阀体
2 internal bearing 内部轴承
3 external bearing 外部轴承
4 disc 阀板
5 shaft 轴
6 stuffing box 填料函
7 packing 填料
8 york 支架
9 link rod 连杆
10 pneumatic actuator 气动执行机构
11 actuator stem 执行机构推杆
A packing gland 填料压盖
B splined lever 花键连杆

9.6.3.6 偏心旋转控制阀 Rotary Control Valve

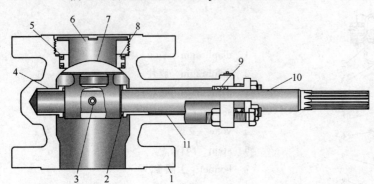

1 valve body 阀体
2 thrust washer 止推垫圈
3 taper and expansion pins 锥形膨胀销钉
4 bearing 轴承
5 face seals 表面密封
6 retainer 保持架
7 valve plug 阀芯
8 seat ring 阀座
9 packing 填料
10 valve shaft 阀轴
11 bearing stop 轴承止挡

9.6.3.7 V形控制球阀 V-Ball Control Valve

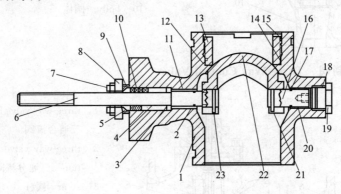

1 body 阀体
2 shaft O-ring 轴O形环
3 thrust bearing 止推轴承
4 packing spacer 填料套
5 packing follower 填料压盖
6 shaft 轴
7 packing nut 填料螺母
8 gland flange 密封法兰
9 packing stud 填料螺栓
10 packing 填料
11 shaft bearing 传动轴承
12 seals 密封
13 retainer 夹具
14，15 retainer O-ring 夹具O形环
16 post 支柱
17 post O-ring 支承O形环
18 O-ring O形环
19 plug 丝堵
20 post bearing 支承轴承
21 post pin 支承销钉
22 ball 球
23 shaft pin 轴销钉

9.6.3.8 控制球阀 Ball Control Valve

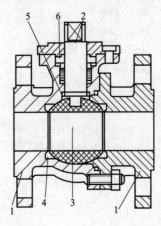

1 body segments 阀体
2 ball shaft 球阀轴
3 ball 球阀芯
4 seat ring 阀座
5 V-ring packing V形环填料
6 Belleville spring washers Belleville弹簧垫圈

9.6.4 控制阀上阀盖 Control Valve Bonnet

9.6.4.1 波纹管密封型上阀盖 Bonnet with Guardian Bellows Seal

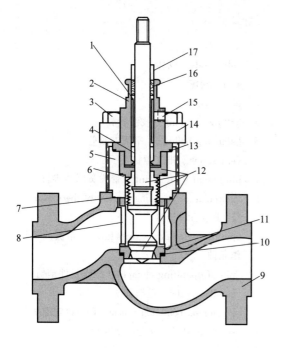

1　packing spacers　填料隔套
2　bonnet　上阀盖
3　bonnet flange bolt　上阀盖法兰螺栓
4　lower guide　下阀杆导向
5　sleeve　套筒
6　seal ring gasket　密封环垫片
7　sleeve gasket　套筒垫片
8　retainer　保持架
9　body　阀体
10　seat ring gasket　阀座环垫片
11　seat ring　阀座环
12　metal bellows plug assembly　金属波纹管阀芯组件
13　bonnet gasket　上阀盖垫片
14　bonnet flange　上阀盖法兰
15　tell-tale tap（purge plug）　监测口（吹气丝堵）
16　packing　填料
17　upper stem guide　上阀杆导向

9.6.4.2 低温型上阀盖 Cryogenic Bonnet

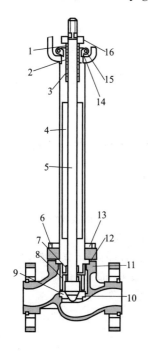

1　upper stem guide　上阀杆导向
2　packing　填料
3　packing spacer　填料隔套
4　extended bonnet　伸长型上阀盖
5　plug　阀芯
6　bonnet flange　上阀盖法兰
7　lower stem guide　下阀杆导向
8　seat retainer　阀座保持架
9，11　seat ring　阀座环
10　seat ring gasket　阀座环垫片
12　bonnet gasket　上阀盖垫片
13　bonnet flange bolting　上阀盖法兰螺栓
14　yoke clamp　支架夹具
15　yoke clamp bolting　支架夹具螺栓
16　gland flange　压盖法兰

581

9.6.5 自力式调节阀 Self-Operated Regulator

9.6.5.1 自力式温度调节阀 Temperature Regulator

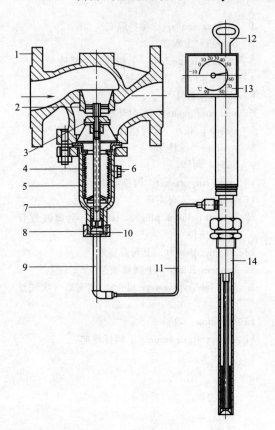

1 valve body 阀体
2 seat 阀座
3 plug 阀芯
4 bellows housing 波纹管腔室
5 balancing bellows 平衡波纹管
6 vent screw 排气螺钉
7 plug stem with spring 带弹簧阀芯推杆
8 connection for operating element of thermostat 与温度操作元件连接
9 operating element with bellows 带波纹管操作元件
10 pin of operating element 操作元件销钉
11 capillary tube 毛细管
12 key for set point adjustment 调整设定点钥匙
13 set point dial 设定点刻度盘
14 temperature sensor 温度传感器

9.6.5.2 自力式流量调节阀 Flow Regulator

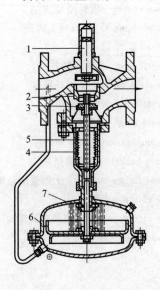

1 restriction to adjust flow set point 流量限流设定
2 seat 阀座
3 plug 阀芯
4 metal bellows 金属波纹管
5 plug stem 阀芯推杆
6 operating diaphragm 操作膜片
7 positioning spring 定位弹簧

9.6.5.3 自力式差压调节阀 Differential Pressure Regulator

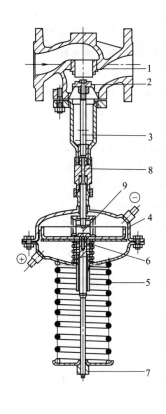

1　seat　阀座
2　plug　阀芯
3　plug stem　阀芯推杆
4　operating diaphragm　操作膜片
5　positioning spring　定位弹簧
6　force limiting device　限力装置
7　set point adjustment　设定点调整
8　distance piece　间隔件
9　overload protection　过载保护

9.6.5.4 自力式压力调节阀 Pressure Regulator

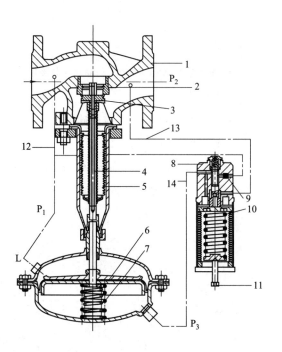

1　body　阀体
2　seat　阀座
3　plug　阀芯
4　plug stem　阀芯推杆
5　balancing bellows　平衡波纹管
6　operating diaphragm　操作膜片
7　positioning spring　定位弹簧
8　pilot valve　先导阀
9　sieve　滤网
10　set point springs　设定点弹簧
11　set point adjustmenter　设定点调整器
12　control line (upstream pressure)　上游压力控制管线
13　control line (downstream pressure)　下游压力控制管线
14　control line (control pressure)　压力控制管线

9.6.6 控制阀附件　Control Valve Accessory

9.6.6.1 电/气转换器 I/P Converter (E/P Converter)

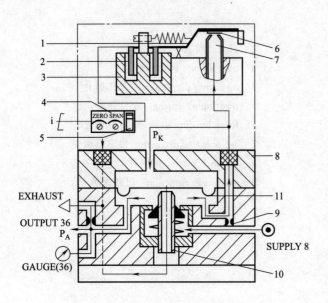

1　balance beam　平衡杆
2　plunger coil　柱塞式线圈
3　permanent magnet　永磁铁
4　zero point and span adjusters　零点和量程调整
5　switch-off electronics；slide switch　电子部分开关；滑动开关
6　flapper plate　挡板
7　nozzle　喷嘴
8　volume booster　气动放大器
9　fixed restrictor　固定节流器
10　sleeve　套管
11　diaphragm　套管

i　Current input　电信号输入（如 4~20mA）
SUPPLY 8　气源
OUTPUT 36　气信号输出
GAUGE（36）　气信号压力表

9.6.6.2 电/气阀门定位器 I/P Positioner

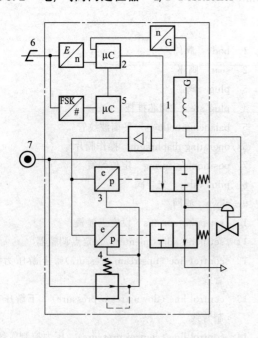

1　inductive displacement sensor　电感位移传感器

2，5　micro-controller　微控制器

3，4　3/2 valve　二位三通阀

6　frequency shift keying signal for communication　用于通信的频率变换键控信号

7　air supply　气源

9.6.6.3 数字式阀门定位器 Digital Positioner

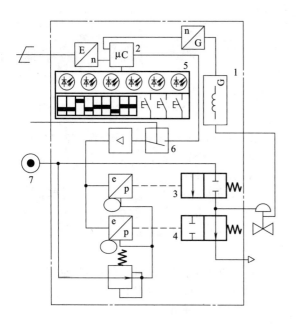

1 inductive displacement sensor 电感位移传感器
2 microcontroller 微型控制器
3 on-off valve for supply air 供气用开关阀
4 on-off valve for venting air 排气用开关阀
5 operating level 操作状态
6 forced fail-safe venting action 强制故障安全排放装置
7 air supply 气源

9.6.6.4 现场总线式阀门定位器 Fieldbus Positioner

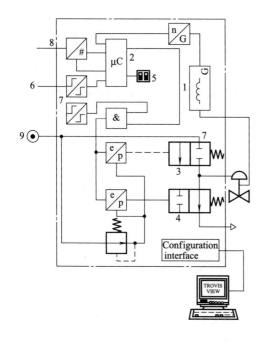

1 inductive displacement sensor 电感位移传感器
2 microcontroller 微型控制器
3 on-off valve for supply air 供气用开关阀
4 on-off valve for venting air 排气用开关阀
5 microswitch for write protection and simulation mode 写保护和模拟样式的微动开关
6 binary input 二进制输入
7 forced venting 强制排放
8 fieldbus interface module 现场总线接口模块
9 air supply 气源

9.6.6.5 限位开关 Limit Switch

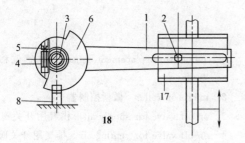

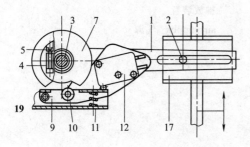

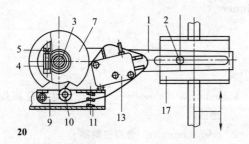

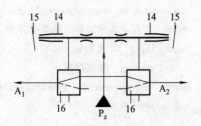

1　lever for valve travel　阀行程杠杆
2　pin　销钉
3　shaft　轴
4　switch case　开关盒
5　adjustment screw　调整螺钉
6　metal tag　金属凸轮
7　cam disc　凸轮盘
8　inductive pick-up　电感信号检测器
9　switch lever　开关杠杆
10　roller　卷轴
11　spring　弹簧
12　electric switching element　电动开关元件
13　pneumatic switch element　气动开关元件
14　nozzle with switch　带开关喷嘴
15　flapper　挡板
16　pneumatic microswitch　气动微动开关
17　plate for attachment to either actuator stem or plug stem　在执行机构推杆或阀芯推杆上的固定板
18　inductive limit switch　电感式限位开关
19　electric limit switch　电动限位开关
20　pneumatic limit switch　气动限位开关

9.6.6.6 电磁阀 Solenoid Valve

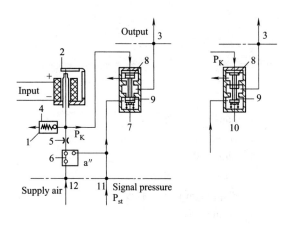
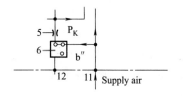

1 relay with coil 带线圈的继电器
2 flapper plate 挡板
3 air connection for PA PA 空气连接
4 nozzle 喷嘴
5 pressure limiter 限压器
6 restriction 节流
7 3/2-way valve 二位三通阀
8 switching diaphragm 开关膜片
9 control piston 控制活塞
10 3/2-way valve（reverse acting） 二位三通阀（反作用）
11 air connection for pressure to be switched in the line 开-关气动信号管线空气连接
12 air connection for supply 气源连接

9.6.6.7 阀位变送器 Position Transmitter

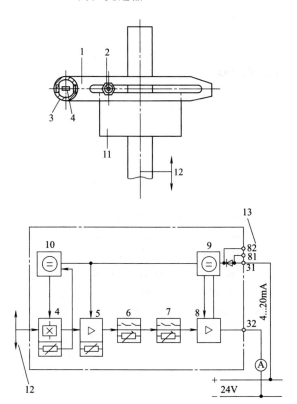

1 lever for valve travel 阀行程杠杆
2 coupling pin 连接销钉
3 solenoid system 电磁系统
4 sensor with temperature sensitive resistor 带热敏电阻的传感器
5 measuring amplifier 测量放大器
6 switch with potentiometer for coarse and fine ZERO adjustment 带电位器的开关，用于零点粗调和微调
7 switch with potentiometer for coarse and fine SPAN adjustment 带电位器的开关，用于量程粗调和微调
8 output stage 输出级
9 constant-voltage source 恒压源
10 constant-current source 恒流源
11 plate for attachment to the actuator or plug stem of the control valve 在控制阀执行机构或阀芯推杆上的固定板
12 travel 阀开度
13 test terminals 试验端子

587

9.6.6.8 气锁阀 Pneumatic Lock-up Valve

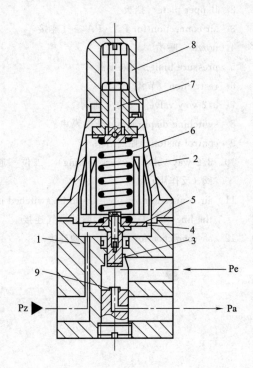

1 housing 外壳
2 cover 盖
3 plug 阀芯
4 diaphragm 膜片
5 diaphragm plate 膜板
6 spring 弹簧
7 spindle 轴
8 cap 帽
9 seat 阀座
Pa signal pressure output 信号压力输出
Pe signal pressure input 信号压力输入
Pz supply air 气源

9.6.6.9 空气过滤减压器 Air Pressure Reducing Station

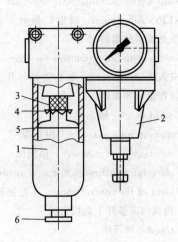

1 pressure filter 压力过滤器
2 supply pressure regulator 供气压力调节器
3 filter element 过滤器元件
4 serrated lock washer 带齿压紧垫圈
5 filter housing 过滤器外壳
6 drain plug 排放丝堵

9.6.6.10 一体化过滤减压阀 Integrated Filter Regulator

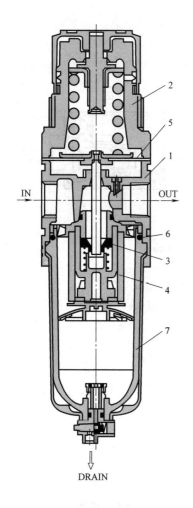

1 body 阀体
2 bonnet 阀盖
3 valve assembly 阀芯组件
4 filter element 滤芯
5 diaphragm assembly 膜片组件
6 bowl seal O-ring 杯密封O形圈
7 bowl assembly 杯组件
IN 入口
OUT 出口
DRAIN 排水口

9.6.6.11 气动放大器（增速继动器） Pneumatic Volume Booster（Booster Relay）

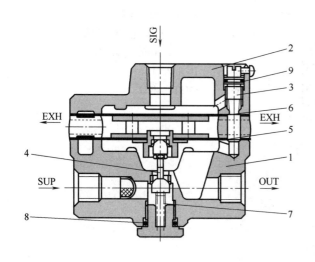

1 valve 阀体
2 cover 阀盖
3 throttle valve 节流阀
4 inner valve 内阀
5 diaphragm assembly 膜片组件
6 diaphragm 膜片
7 valve spring 内阀弹簧
8 O-ring O形圈
9 O-ring O形圈
SUP 气源入口
OUT 气源出口
SIG 信号
EXH 排放口

9.6.6.12　快速排放阀　Quick Exhaust Valve

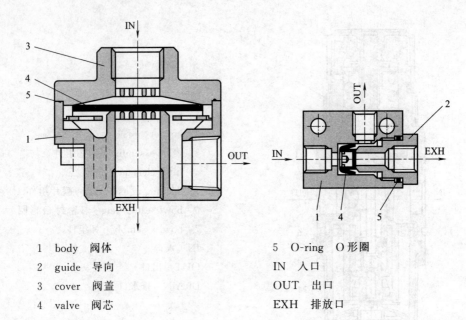

1　body　阀体
2　guide　导向
3　cover　阀盖
4　valve　阀芯
5　O-ring　O形圈
IN　入口
OUT　出口
EXH　排放口

9.6.6.13　止回阀　Check Valve

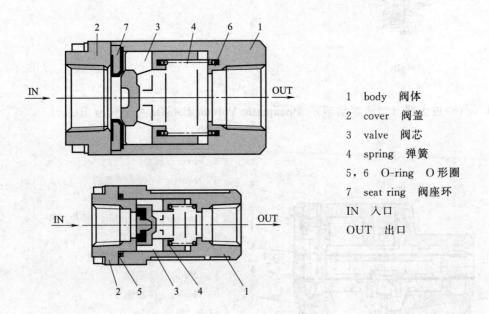

1　body　阀体
2　cover　阀盖
3　valve　阀芯
4　spring　弹簧
5，6　O-ring　O形圈
7　seat ring　阀座环
IN　入口
OUT　出口

9.6.7 安全泄压阀 Safety Relief Valve

9.6.7.1 弹簧式安全泄压阀 Spring Loaded Safety Relief Valve

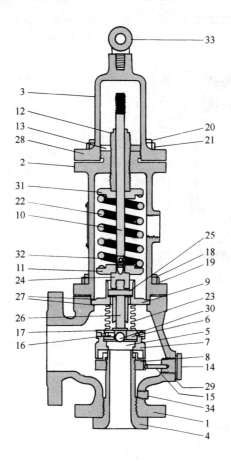

1	boby 阀体	18,19	body stud/nut 阀体双头螺栓/螺母
2	casing 弹簧罩	20,21	casing stud/nut 弹簧罩双头螺栓/螺母
3	cap 护罩	22	spring 弹簧
4	nozzle 喷嘴	23	bellows 波纹管
5	disc 阀芯	24	spindle head 阀杆端头
6	disc holder 阀芯座	25	piston 活塞
7	reaction hood 回座压力调整环	26	guide spindle 导向杆
8	blowdown ring 增压环	27	body gasket 阀体垫片
9	guide plate 导向盘	28	cap gasket 护盖垫片
10	spindle 阀杆	29	setting screw gasket 设定螺钉垫片
11	low spring cap 下弹簧挡片	30	ball 球
12	adjusting screw 调节螺钉	31	upper spring cap 上弹簧挡片
13	locking nut 锁紧螺母	32	grooved pin 销钉
14	setting screw 设定螺钉	33	eye bolt 带眼螺栓
15	setting screw rod 设定螺杆	34	drain plug 排放丝堵
16	tabwasher 垫圈		
17	pinning screw 压紧螺钉		

9.6.7.2 带导阀的安全泄压阀 Pilot Operated Safety Relief Valve

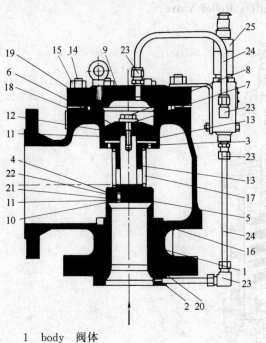

1 body 阀体
2 nozzle 喷嘴
3 guide 导向盘
4 disc holder 阀芯座
5 disc insert 阀芯垫块
6 piston 活塞
7 locknut 锁紧螺母
8 lift stop 上升止挡
9 cover 阀盖
10 retaining plate 固定板
11 disc holder seal 阀芯座密封
12 piston seal 活塞密封
13 guide rings 导向环
14,15 body stud/nut 阀体双头螺栓/螺母
16 lock ring 锁紧环
17 spring 弹簧
18 body gasket 阀体垫片
19 guide seal 导向密封
20 nozzle seal 喷嘴密封
21 retaining plate screw 固定板螺钉
22 counter sunk screw 反向沉头螺钉
23 fittings 管接头
24 tubes 管子
25 pilot 导阀

9.6.8 爆破片 Burst Disc

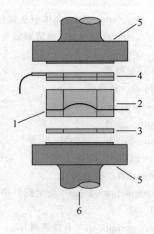

1 burst disc 爆破片
2 holder 夹持部件
3 gasket 垫片
4 remote indicator 远程指示
5 flange 法兰
6 flow direction 流向

9.7 盘装仪表 Panel Mounting Instrument

9.7.1 安全栅 Safety Barriers

9.7.1.1 用于变送器的安全栅 Safety Barriers for Transmitter

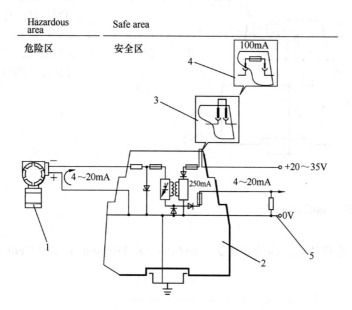

1 conventional or 'smart' 2-wire 4~20 mA transmitter 常规型或智能型二线制 4~20 mA 变送器
2 safety barrier 安全栅
3 link for loop disconnecting 回路断路链（回路跨线）
4 fuse for disconnecting 熔断器
5 load resistor 负载电阻

9.7.1.2 用于数字（开关）输入的安全栅 Safety Barriers for Digital (Switch) Inputs

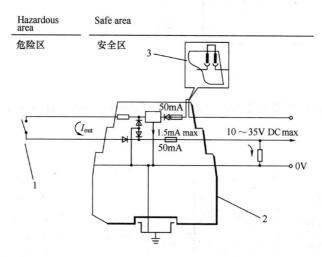

1 digital (switch) inputs 数字（开关）输入
2 safety barrier 安全栅
3 link for loop disconnecting 回路断路链（回路跨线）

9.7.1.3 用于数字（开关）输出的安全栅 Safety Barriers for Digital (Switch) Outputs

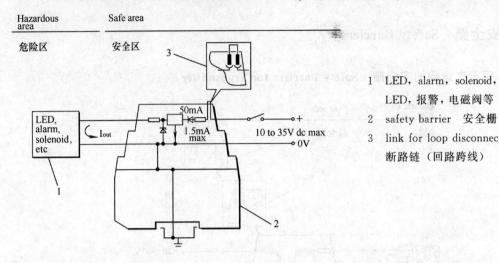

1 LED, alarm, solenoid, etc.　LED，报警，电磁阀等
2 safety barrier　安全栅
3 link for loop disconnecting　回路断路链（回路跨线）

9.7.2 隔离器 Isolator

9.7.2.1 用于变送器和控制器的隔离器 Isolator for Transmitter and Controller

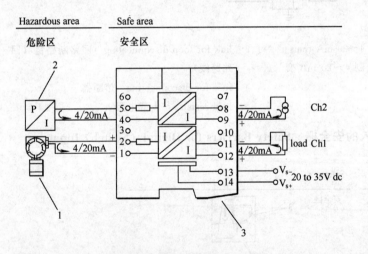

1 2-wire 4/20 mA transmitter
　二线制 4～20mA 变送器
2 4/20mA controller　4～20mA 控制器
3 isolator　隔离器

9.7.2.2 用于电磁阀/报警驱动器的隔离器 Isolator for Solenoid/Alarm Driver

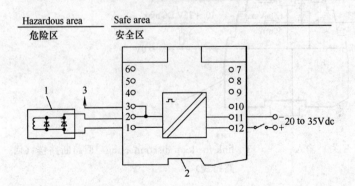

1 solenoid, alarm or other IS device
　电磁阀，报警或其他本安装置
2 isolator　隔离器
3 to earth leakage detector　接至漏电检测器

9.7.2.3 用于开关/趋近检测器的隔离器　Isolator for Switch/Proximity Detector

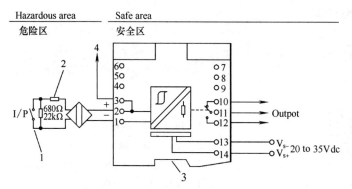

1　switch or proximity detector
　　开关或趋近检测器
2　resistors(only required for line fault detector)　电阻（仅用于线路故障检测）
3　isolator　隔离器
4　to earth leakage detector　接至漏电检测器

9.8　特殊仪表　Special Instrument

9.8.1　速度传感器　Speed Sensor

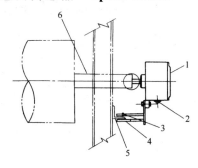 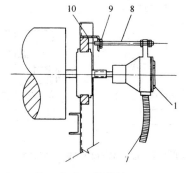

1　speed sensor　速度传感器
2　cable connector　电缆接头
3　spring　弹簧
4　arresting bracket　固定托架
5　angle iron　角铁
6　pulley shaft　带轮轴
7　flexible conduit　电缆保护软管
8　arrestor rod　固定杆
9　tension spring　拉紧弹簧
10　angle bracket　角形托架

9.8.2　轴位移变送器　Transmitter for Thrust Position

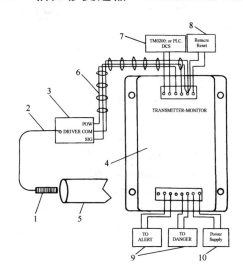

1　probe　测位移探头
2　extension cable　延伸电缆
3　driver　驱动器
4　transmitter for thrust position　轴位移变送器
5　shaft　轴
6　signal cable　信号电缆
7　digital display unit,PLC or DCS　数字显示单元，可编程控制器（PLC）或分散控制系统（DCS）
8　remote reset　远程设定
9　annunciator　信号报警器

10 power supply 电源

9.8.3 称重系统 Weighing System

9.8.3.1 负荷传感器 Load Cells

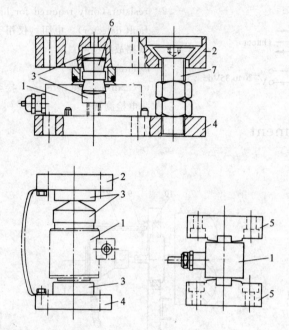

1. load cell 负荷传感器
2. top plate 顶板
3. force guiding piece 力导向部件
4. base plate 底板
5. pressure plate 压力板
6. self-aligning bolt 自调准螺栓
7. oscillation limitation and protection against raising up 限振和防止侧倾

9.8.3.2 称重系统 Weighing System

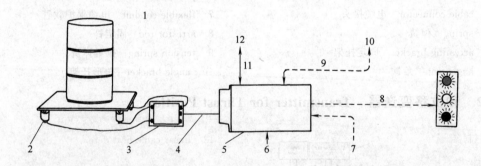

1. typical weighing system architecture 典型的称重系统结构
2. load cell 负荷传感器
3. junction box 接线箱
4. signal cable 信号电缆
5. weighing terminal 称重终端
6. power supply 电源
7. from push-button 按钮输入
8. control output 控制输出
9. RS-232 interface RS-232 接口
10. computer/printer/remote display unit 计算机/打印机/远程显示器
11. fieldbus interface 现场总线接口
12. PLC 可编程控制器

9.9 火灾自动报警系统 Automatic Fire Alarm System

9.9.1 火灾自动报警系统 Automatic Fire Alarm System

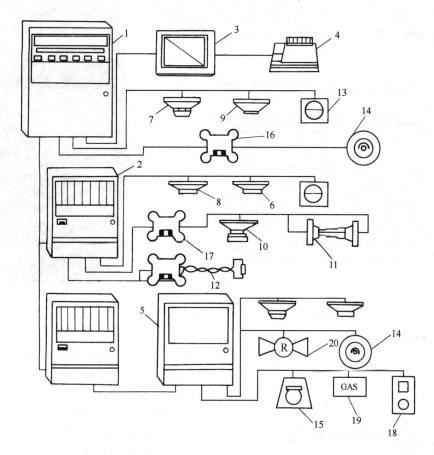

1　remote fire alarm controller　集中火灾报警控制器
2　local fire alarm controller　区域火灾报警控制器
3　graphic fire alarm display panel　图形式火灾显示盘
4　printer　打印机
5　fire fighting linkage controller　消防联动控制盘
6　combination photoelectric smoke heat detector　光电感烟探测器
7　rate-of-rise and fixed temperature heat detector　（电子）差定温感温探测器
8　optical/heat detector　烟温复合探测器
9　ionisation smoke detector　离子感烟探测器
10　ultraviolet flame detector　紫外火焰探测器
11　linear infrared beam smoke detector　线型红外光束感烟探测器
12　linear cable heat detector　线型缆式感温探测器
13　manual call point　手动报警按钮
14　alarm bell　警铃
15　audible and visual alarm　声光报警器
16　interface module　接口模块
17　control module　控制模块
18　instancy abort/release　紧急中止/释放
19　discharge indicator　放气指示
20　release control valve/monitor input　释放控制阀/监控输入

9.9.2 火灾报警控制盘 Fire Alarm Control Panel

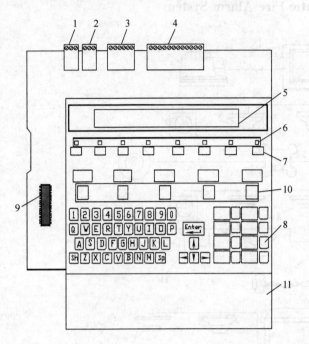

panel fabric drawing 盘面结构图

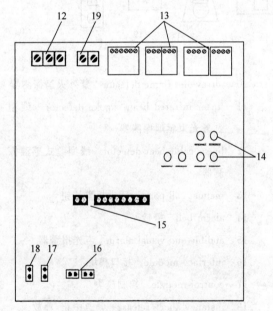

panel rearward node drawing 盘后端子图

1. printer interface 打印机接口
2. terminal interface 终端接口
3. mode interface 模式接口
4. SLC loop SLC 回路
5. 80 character LCD display 80 字符 LCD 显示
6. estate indicator Status lamp 状态指示灯
7. cut in label 插入式标签
8. program keyboard 编程键盘
9. erasable memory 可擦除存储器
10. control key button 控制按键
11. folding cover board 折叠盖板
12. aternating current power supply terminals 交流电源 AC
13. direct current power supply terminals 直流电源 DC
14. estate/failure status lamp 状态/故障指示灯
15. transformation module 转换模块
16. option interface 可选接口
17. electromagnetism switch 电磁开关
18. code input 代码输入
19. batteries terminals 电池

598

9.9.3 激光感烟探测器 Laser Smoke Detector

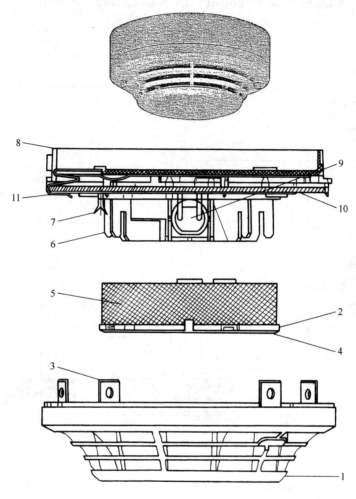

laser smoke detector fabric drawing　　激光感烟探测器结构图

1　protection cover　保护盖
2　detection chamber cover　探测室盖
3　lock up parcel　锁定片
4　RF screen meshwork　RF屏蔽网
5　insect barrier　防虫网
6　measure reference plate　测量室基片
7　insect barrier spring card　防虫网弹簧卡片
8　detector base　探头底座
9　laser diode　激光二极管
10　circuit board　电路板
11　installation base contact plate　安装底座接触板

9.10 工业电视监视系统 Industry Television Surveillance System

9.10.1 工业电视监视系统 Industry TV Surveillance System

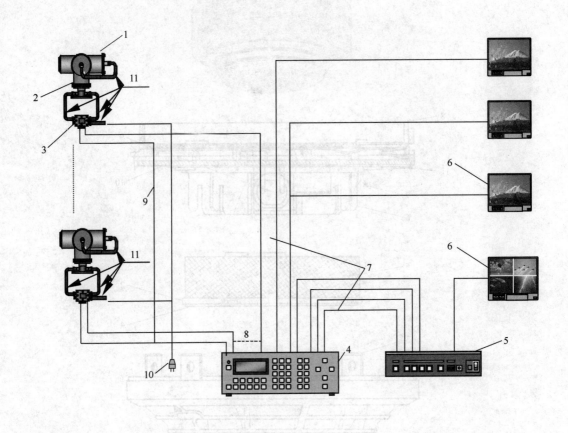

1 pickup camera of explosion proof type 防爆型摄像机
2 flameproof electricity pan/tilt 隔爆型电动云台
3 flameproof decoding controller 隔爆型解码控制器
4 matrix switcher 矩阵控制器
5 frame division 画面分割器
6 color monitor 彩色监视器
7 video output 视频输出
8 video input 视频输入
9 control line 控制线
10 alternating current 交流电源 AC
11 flexibility pipe (Ex) 防爆管

9.10.2 矩阵控制器 Matrix Switcher

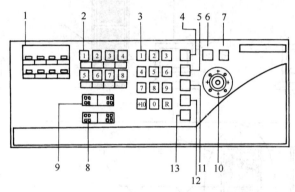

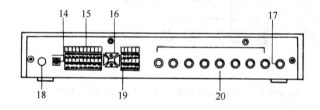

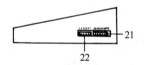

1 alarm lamp 报警灯
2 pickup camera push-button 摄像机按钮
3 keyboard 键盘
4 predefined button 预置按钮
5 swing push-button 摆动按钮
6 actuate push-button 开动按钮
7 stop push-button 停止按钮
8 change distance switch 变距开关
9 foci switch 焦距开关
10 control stick 控制棒
11 series push-button 连续按钮
12 switch push-button 切换器按钮
13 in support push-button 后备按钮
14 power supply input node 电源输入接口
15 alarm input/output node 报警输入/输出接口
16 buckle nip 扣夹
17 video output node 视频输出接口
18 power supply switch 电源开关
19 pickup camera control I/F 摄像机控制 I/F
20 video input node 视频输入接口
21 function switch 1 功能开关 1
22 function switch 2 功能开关 2

9.11 数字控制系统 Digital Control System

9.11.1 分散控制系统 Distributed Control System

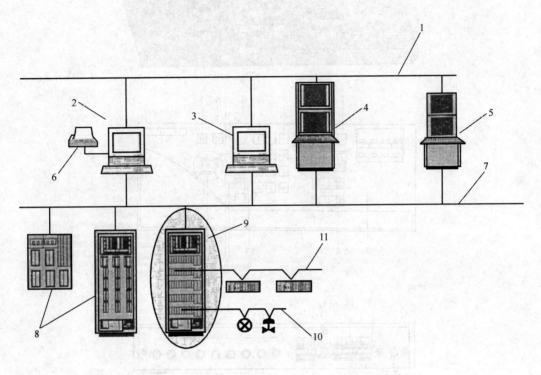

typical DCS system architecture 典型的 DCS 系统结构

1　ethernet　以太网

2　operator station　操作站

3　operator station/engineering station　操作站/工程师站

4　closing and on standing type operator station　封闭型落地式操作站

5　opening and on standing type operator station　开放型落地式操作站

6　printer　打印机

7　control bus　控制总线

8　process control station (for legacy field devices)　过程控制站（用于传统的现场设备）

9　process control station (for fieldbus devices)　过程控制站（用于现场总线设备）

10　fieldbus (Profibus or FF)　现场总线（Profibus or FF）

11　ER bus　扩展远程总线

9.11.2 现场总线控制系统 Fieldbus Control System（FCS）

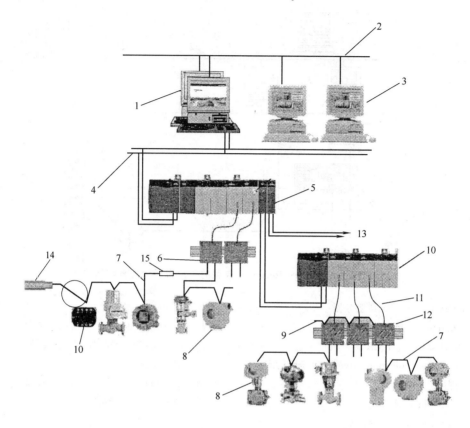

typical FCS system architecture 典型的现场总线控制系统结构

1 FCS server　FCS 服务器
2 Ethernet　以太网
3 FCS station　FCS 客户机
4 controlnet or Ethernet　控制网或以太网
5 controller, fieldbus interface modules and IOMs　控制器,现场总线接口和 I/O 模块
6 fieldbus RTPs unpowered　非供电型现场总线远程端子板
7 fieldbus H1(1), H2(2)　现场总线 H1(1), H2(2)
8 fieldbus devices　现场总线设备
9 24V DC power　24V DC 电源
10 fieldbus interface modules and IOMs　现场总线接口和 I/O 模块
11 RTP cables　远程端子板电缆
12 fieldbus RTPs（W/I.S. barrier）　现场总线远程端子板（带安全栅）
13 controlnet to more I/O racks　控制网接至更多的 I/O 机架
14 terminator　终端器
15 intrinsic safety barrier/arrester　本安信号安全栅/避雷器

9.11.3 工厂资源管理系统 Plant Resource Management System（PMR）

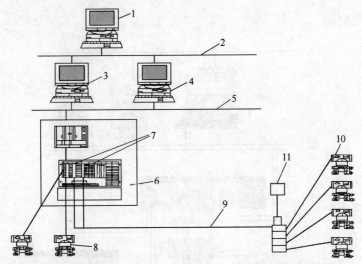

typical PMR system architecture 典型的工厂资源管理系统结构

1 PMR server　PMR 服务器
2 ethernet　以太网
3 PMR station　PMR 客户机
4 field communication server　现场通信服务器
5 control bus　控制总线
6 control station　控制站
7 redundant　冗余
8 legacy field devices　传统的现场设备
9 fieldbus(Profibus or FF)　现场总线（Profibus or FF）
10 fieldbus devices　现场总线设备
11 terminator　终端器

9.11.4 可编程控制器 Programmable Logic Controller（PLC）

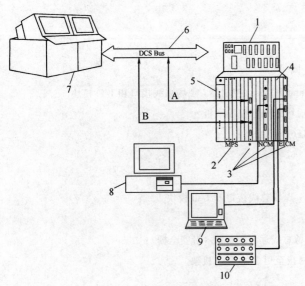

1 PLC chassis　PLC 机架
2 main processors　主处理器
3 communication modules　通信模块
4 I/O modules　I/O 模块
5 power modules　电源模块
6 DCS bus　DCS 通信总线
7 DCS system　DCS 系统
8 programming workstation　编程站
9 host computer　主计算机
10 annunciator　信号报警器

typical PLC system architecture 典型的可编程控制器系统结构

9.11.5 紧急停车系统 Emergency Shutdown System (ESD)

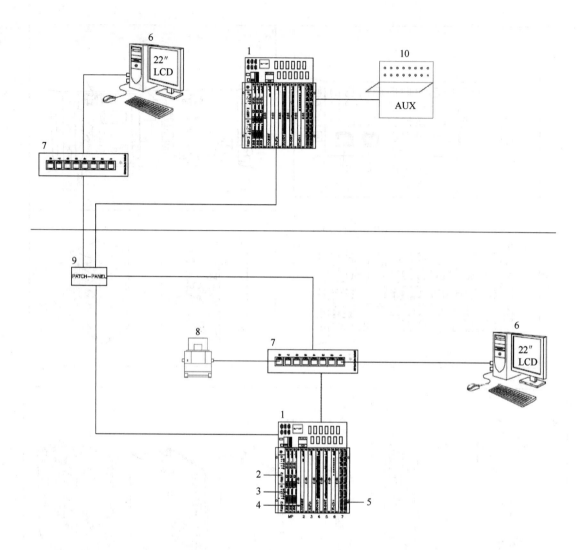

1　chassis　系统机架
2　chassis power　系统机架电源
3　CPU　系统处理器
4　I/O module　系统I/O模块
5　communication module　系统通信模块
6　operator station　操作站
7　photoelectric converter　光电转换器
8　printer　打印机
9　optical fiber　光纤
10　auxiliary console　辅操台

9.12 仪表盘，柜和操作台　Instrument Panels, Cabinets and Consoles

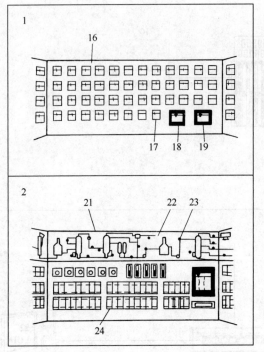
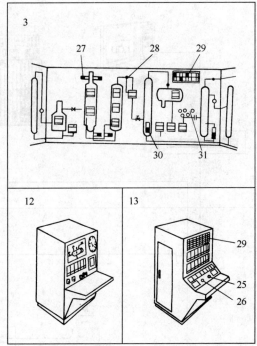
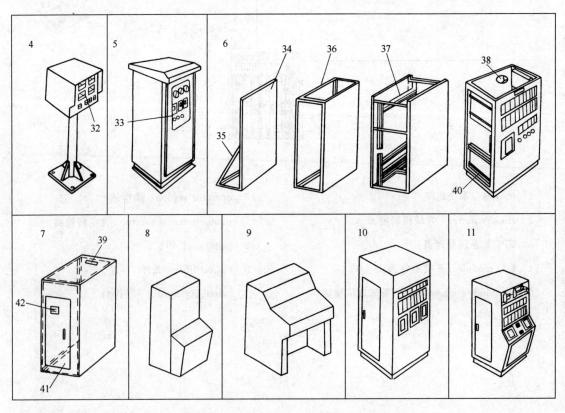

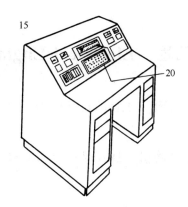

1	conventional panel	普通仪表盘	22	flow diagram	流程图
2	semi-graphic panel	半模拟仪表盘	23	alarm light	报警灯
3	full-graphic panel	全模拟仪表盘	24	narrow case instrument	条形仪表
4	outdoor stanchion box type	室外立柱箱式	25	test push button	试验按钮
5	outdoor cabinet type	室外柜式	26	acknowledge push button	消除按钮
6	self-standing open type	自立敞开式	27	valve position indicator	阀位指示仪
7	enclosed cabinet type	封闭柜式	28	control point	控制点
8	bench type	附接操纵台式	29	annunciator	报警器
9	operator console type	操作台式	30	level indicator	液位指示仪
10	cabinet type	柜式	31	valve position light	阀位指示灯
11	control console type	控制台式	32	field instrument	就地仪表
12	cabinet with desk type	柜式（带台面）	33	local control station	就地操作器
13	console with desk type	控制台式（带台面）	34	front panel	仪表盘正面
14	look-over type	观看式仪表盘	35	support	支承
15	control desk	控制台	36	angle frame	角钢框架
16	control panel	控制盘	37	universal channel	通用线槽
17	instrument	仪表	38	panel cooling fan	机柜冷却风扇
18	recorder	记录仪	39	universal panel light	通用柜内照明灯
19	temperature recorder	温度记录仪	40	enclosure eath bar	外壳接地排
20	pushbutton matrix selector	按钮式矩阵选择开关	41	cable gland plate	电缆格兰安装板
21	graphic display section	显示部分	42	filter	过滤器

607

10 电气工程 Electrical Engineering

10.1 旋转电机 Electrical Rotating Machine

10.1.1 直流电动机 Direct-Current Motor

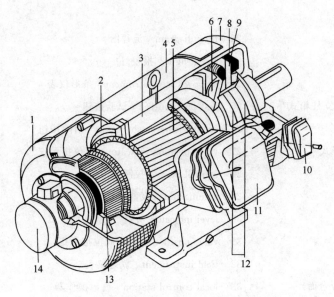

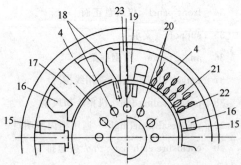

1 bearing bracket of non-drive end 非传动侧轴承架
2 stator winding 定子绕组
3 laminated core of stator 定叠片铁芯
4 frame 机座
5 rotor 转子
6 fan 风扇
7 bearing bracket of drive end 传动侧轴承架
8 rotor of fan motor 风扇电动机的转子
9 stator of fan motor 风扇电动机的定子
10 terminal box of fan motor 风扇电动机的端子盒
11 terminal box of main motor 主电动机端子盒
12 bearing of fan motor 风扇电动机轴承
13 commutator 整流子，换向器
14 tachogenerator 测速发电机
15 compole winding 整流极绕组
16 ventilating ducts 通风孔
17 main pole winding 主极绕组
18 main pole 主极
19 ventilating path 通风道
20 stator core（compole） 定子铁芯（整流极）
21 stator core（main pole）定子铁芯（主极）
22 compensating winding 补偿绕组
23 compole 整流极

10.1.2 直流发电机 Direct-Current Generator

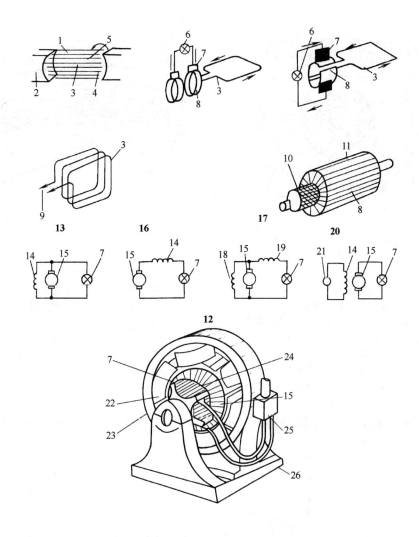

1   magnetic field   磁场
2   pole   磁极
3   coil   线圈
4   direction of induced voltage   感应电压方向
5   direction of rotation   旋转方向
6   load   负载
7   brushes   电刷
8   collecting rings   集电环
9   to commutator   至换向器，至整流子
10 commutator segments   换向器片，整流片
11 armature core   电枢铁芯
12 methods of field excitation   磁场激励方法
13 shunt   并联励磁
14 field   场
15 armature   电枢
16，19 series field   串联磁场
17 compound   复合励磁
18 shunt field   并联磁场
20 separate exciter   单独励磁机
21 exciter   励磁机
22 field poles and windings   磁极及绕组
23 yoke   磁轭
24 commutator   换向器，整流子
25 terminal box   端子箱
26 base   底座

10.1.3 AC 换向电机 AC Commutator Machine

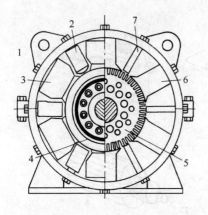
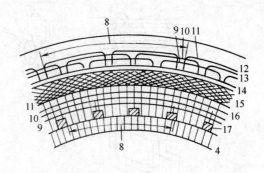

1　AC commutator machine　交流换向电机
2　compole spool　补偿磁极线圈，整流磁极线圈
3　main spool　主线圈
4　commutator　换向器，整流子
5　armature punching　电枢冲模片，电枢冲压硅钢片
6　main pole　主磁极
7　compole　补偿磁极，整流磁极
8　(360°) electrical degree　(360°) 电气度
9　phase 1　第 1 相
10　phase 2　第 2 相
11　phase 3　第 3 相
12　stator iron　定子铁芯
13　stator winding　定子绕组
14　air gap　气隙
15　rotor winding　转子绕组
16　brush connectors　电刷接头
17　brush　电刷

10.1.4 同步电动机的无刷励磁 Brushless Excitation of Synchronous Motor

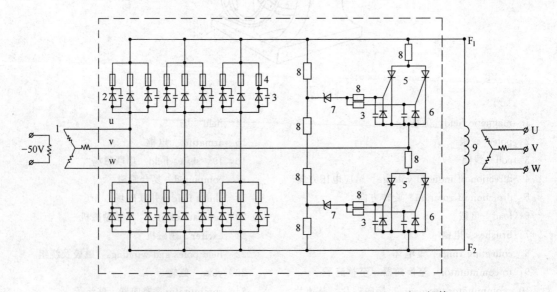

1　AC exciter　交流励磁机
2　silicon rectifier　硅整流元件
3　condenser　电容器
4　quick-break fuse　速熔断器
5　silicon controllable rectifier　晶闸管整流器
6　silicon diode　硅二极管
7　Zener diode　齐纳二极管
8　resistor　电阻器
9　field of synchronous motor　同步电动机的磁场

10.1.5 交流电动机 Alternating-Current Motor

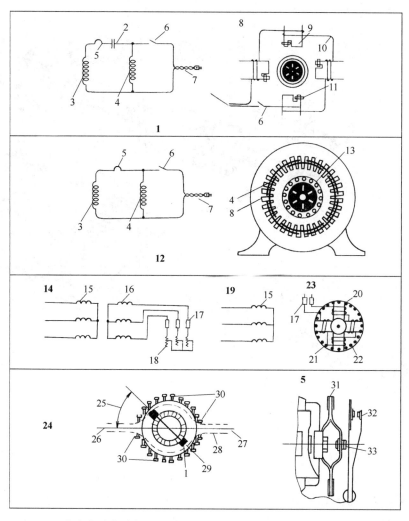

1	capacitor motor 电容启动电动机	18	external resistance 外电阻
2	capacitor 电容器	19	synchronous motor 同步电动机
3	starting winding 启动绕组	20	squirrel-cage winding 笼式绕组
4	running winding 运行绕组	21	DC winding 直流绕组
5	centrifugal switch 离心式开关	22	rotor 电动机转子
6	on-off 闭合-断开	23	DC supply 直流电源
7	AC voltage source 交流电源	24	repulsion motors 推斥电动机
8	shaded-pole motor 屏蔽电动机	25	axis of brushes is displaced from axis of stator poles 电刷轴线与定子轴线偏离一定角度
9	laminated poles 叠片极	26	north pole 北极
10	main winding 主绕组	27	south pole 南极
11	copper rings 铜环	28	main flux 主磁通量
12	split-phase motor 分相电动机	29	short-circuited brush 短路电刷
13	squirrel-cage armature 笼式电枢	30	flux 磁通量
14	polyphase induction motor 三相感应式电动机	31	fly ball 飞球
15	field (stator) coils 电枢（定子）线圈	32	governor contacts 控制触点
16	armature (rotor) coils 电枢（转子）线圈	33	riding contact 引导触点
17	slip rings 滑环		

10.1.6 滑环式感应电动机 Slip-Ring Type Induction Motor

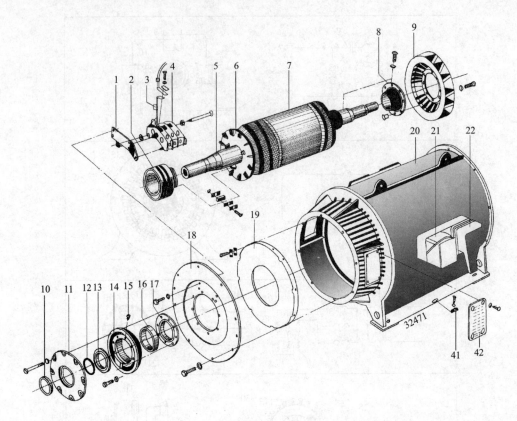

1 brushgear 电刷机构
2 slip ring assembly 滑环组件
3 carbon brush 电刷（碳刷）
4 brush-holder unit 刷握件
5 shaft 轴
6 balancing ring 平衡环
7 rotor core with winding 带绕组的转子铁芯
8 hub of internal fan with bushes 内风扇的叶毂及衬套
9 internal fan wheel 内风扇叶轮
10 V-shaped axial shaft sealing ring, drive end 传动端的V形轴向密封环
11 outer bearing cap, drive end 传动端的外轴承帽
12 snap-ring, drive end 传动端开口环
13 grease slinger, drive end 传动端的抛油圈
14 bearing housing, drive end 传动端的轴承箱
15 greasing nipple 加滑脂的螺纹接套
16 cylindrical roller (floating) bearing, drive end 传动端的滚柱（自动定心）轴承
17 inner bearing cap with felt rings, drive end 传动端的内轴承帽连同毡环
18 end shield, drive end 传动端的端罩
19 partition 隔板
20 stator frame with core and winding 定子机座连同铁芯和绕组
21 secondary terminal box 二次端子盒
22 primary terminal box 一次端子盒
23 end shield, non-drive end 非传动端的端罩

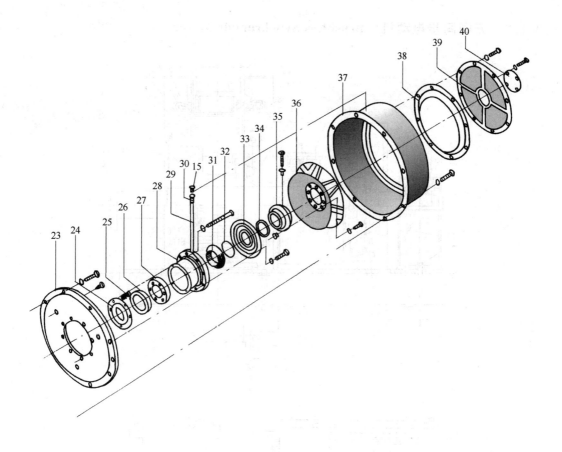

24	inner bearing cap with felt rings, non-drive end 非传动端的内轴承帽连同毡环		口环
25	compression spring 压紧弹簧	33	outer bearing cap, non-drive end 非传动端的外轴承帽
26	deep groove (locating) bearing, non-drive end 非传动端的深槽（定位）滚珠轴承	34	V-shaped axial shaft sealing ring, non-drive end 非传动端的 V 形轴向密封环
27	cylindrical roller (floating) bearing, non-drive end 非传动的滚柱（自动定心）轴承	35	hub of external fan with bushes 外风扇的叶毂及衬套
28	bearing housing, non-drive end 非传动端的轴承箱	36	external fan wheel 外风扇叶轮
29	grease supply tube 滑脂注入管	37	cowl 外壳
30	hexagon threaded gland 六角形螺纹压管	38	guide ring 导环
31	grease slinger, non-drive end 非传动端的抛油圈	39	screen 网
		40	cover 盖
		41	earth terminal strap 接地端片
32	snap-ring, non-drive end 非传动端的开	42	cover on servicing opening 检修孔盖板

613

10.1.7 无刷同步电动机 Brushless Synchronous Motor

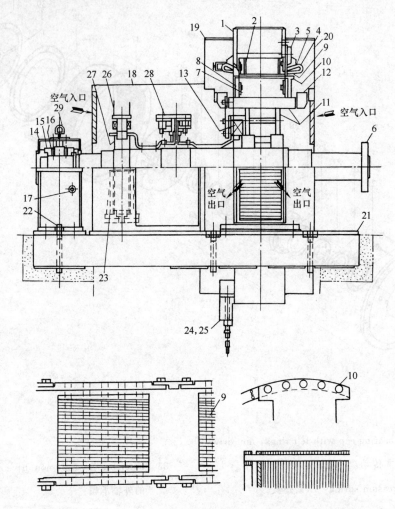

1	stator frame 定子机座	17	oil gauge 油表
2	stator core 定子铁芯	18	exciter shield 励磁设备外罩
3	stator collar 定子压圈	19,20	protecting shield 保护罩
4	stator winding 定子绕组	21	base-plate 底板
5	tie ring 缚线环	22	anchor bolt 地脚螺栓
6	shaft 轴	23	AC exciter terminal box 交流励磁机接线盒
7	laminated pole 叠片磁极	24	space heater terminal box 防潮加热器接线盒
8	rotor winding 转子绕组	25	terminal box for measuring coil temperature 线圈测温接线盒
9	damping winding bar 阻尼绕组条	26	stator for AC exciter 交流励磁机定子
10	damping wheel 阻尼环	27	rotor for AC exciter 交流励磁机转子
11	rotor yoke 转子轭	28	rotary rectifier 旋转整流器
12	fan blade 风叶	29	dial thermometer 圆盘指示温度计
13	discharge resistance 放电电阻		
14	pedestal bearing 轴承座		
15	babbit metal 轴承合金(巴氏合金)		
16	oil ring 油环		

10.1.8 线槽和绕组 Slots and Windings

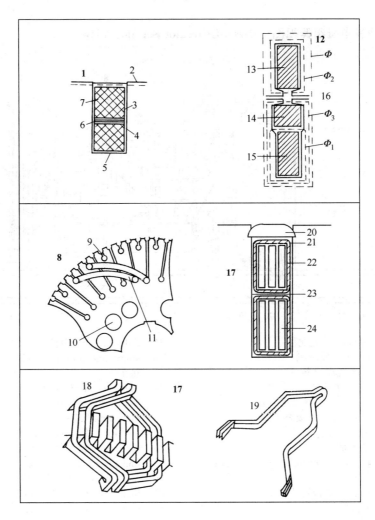

1	slot section 线槽断面	14	regulating winding 调节绕组
2	steel binding wire 绑扎钢丝	15	primary winding 一次绕组
3	slot liner 槽绝缘衬	16	(Φ、Φ_1、Φ_2、Φ_3) magnetic flux 磁通
4	cambric tap 黄蜡布带	17	commutator windings 换向器绕组,整流变速器绕组
5	slot bottom 槽底		
6	spacer between layer 层间绝缘间隔片	18	two-layer winding 双层绕组
7	conductor 导体	19	pre-formed single-turn coils 前行平波线圈,线模定型单绕线圈
8	induction motor 感应电动机		
9	slots for outer cage winding 笼式绕组外层线槽	20	wedge 槽楔
		21	slot liner 槽绝缘衬
10	ventilation holes 通风孔	22	main insulation 主绝缘
11	running winding 运行绕组	23	separator, spacer 层间绝缘板,绝缘间隔片
12	slot of polyphase commutator machines 三相换向器电动机线槽		
		24	cotton taped conductor 缠布带导体
13	secondary winding 二次绕组		

10.1.9 柴油发电机组 Diesel Generator Set

10.1.9.1 柴油发电机组侧视图 Diesel Generator Set Side View

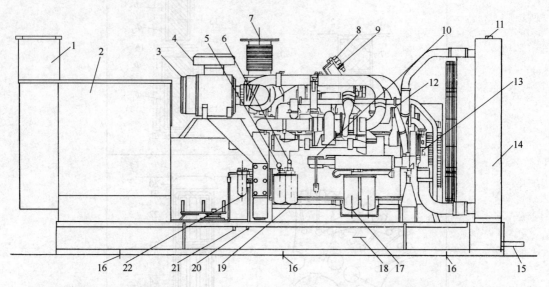

1 local control panel 控制屏
2 power generator 发电机
3 air cleaner 空气净化器
4 air filter 空气过滤器
5 fuel pressure gauge 燃油压力表
6 fuel priming pump 初级燃油泵
7 exhaust 排气管
8 oil filler 油料加入器
9 electronic governor control box 电子调速器控制箱
10 oil level gauge 油液位表
11 lifting eye 吊环
12 radiator cap 散热器盖
13 shunt line connection pipe tap 旁路管线分接头
14 radiator 散热器
15 fumes disposal tube 排烟管
16 mounting holes 安装孔
17 oil filter 机油过滤器
18 oil drain 机油排放口
19 fuel filter 燃油过滤器
20 fuel return 燃油回流管
21 fuel supply 燃油供给口
22 primary fuel filter 初次燃油过滤器

10.1.9.2 就地控制屏 Local Control Panel

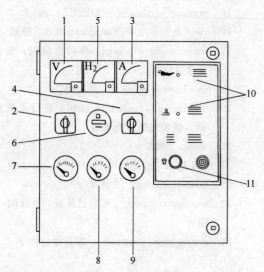

1 AC voltmeter 交流电压表
2 AC voltmeter selector switch 交流电压表选择开关
3 AC ammeter 交流电流表
4 AC ammeter selector switch 交流电流表选择开关
5 frequency meter 频率表
6 hours run meter 小时表
7 engine water temperature gauge 柴油机水温表
8 DC battery voltmeter 直流电池电压表
9 engine oil pressure gauge 机油压力表
10 fault indicator lamps 故障指示灯
11 lamp test pushbutton 灯测试按钮

10.1.9.3 柴油机左前视图 Diesel Engine Front and Left Side View

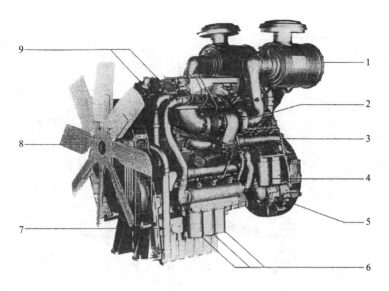

1	air cleaner 空气净化器	6	canister of the lubricating oil filter 润滑油过滤器罐
2	turbocharger 涡轮增压器		
3	exhaust manifold 排气阀组	7	crankcase breather 曲轴箱通气管
4	canisters of the fuel filter 燃油过滤器罐	8	fan 风扇
5	coolant pump 冷却剂泵	9	thermostat housings 自动调油温器外壳

10.1.9.4 柴油机右后视图 Diesel Engine Rear and Right Side View

1	turbocharger 涡轮增压器	6	flywheel housing 飞轮外壳
2	alternator 交流发电机	7	flywheel 飞轮
3	dipstick 量油计	8	starter motor 启动马达
4	filter cap of lubricating oil 润滑油过滤器盖	9	fuel lift pump 燃油泵
5	lubricating oil sump 润滑油油箱	10	air cleaner 空气清洁器

10.1.9.5 发电机 Power Generator

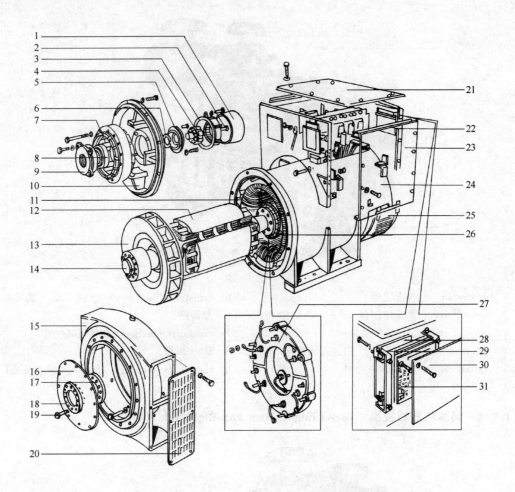

1　PMG (permanent magnet generator) cover　永磁罩
2　PMG rotor stud　永磁转子护套
3　PMG stator assembly　永磁定子组件
4　PMG rotor assembly　永磁转子组件
5　N. D. E. bearing cap　后轴承挡板
6　circlip N. D. E.　弹簧锁片
7　exciter stator　励磁定子
8　N. D. E. bearing　后轴承
9　N. D. E. bearing cartridge　后轴承座
10　N. D. E. (non drive end) bracket　后盖
11　main stator/bar assembly　主定子线圈组件
12　main rotor assembly　主转子线箱组件
13　fan　风扇
14　shaft　主轴
15　drive end bracket　前壳体
16　coupling disc　连接盘
17　coupling spacer　连接盘底座
18　coupling pressure plate　连接盘固定板
19　coupling bolt　连接螺栓
20　D. E. cover　风扇座挡板
21　terminal box lid　出线箱盖板
22　main terminal board　出线排固定支架
23　terminal box side panel　出线箱侧板
24　main frame　机座
25　N. D. E. cover　后防护挡板
26　exciter rotor　励磁转子
27　rotating rectifier assembly　旋转整流子
28　auxiliary terminal board　传感组件接线板
29　AVR mounting bracket　AVR安装底板
30　AVR cover plate　AVR盖板
31　AVR (automatic voltage regulator)　自动电压调整器

10.2 变压器 Transformer

10.2.1 油浸式变压器 Oil Immersed Transformer

10.2.1.1 油浸式变压器正视图 Oil Immersed Transformer Front View

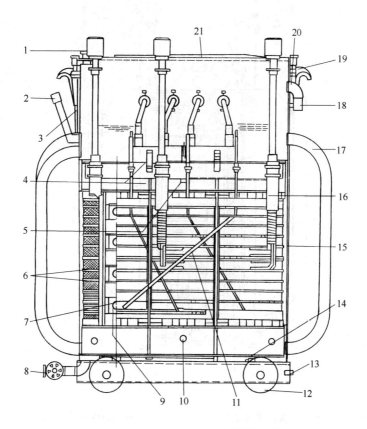

front view of transformer 变压器正视图

1　oil filling hole　充油孔，灌油孔
2　thermometer pocket　温度计插套
3　protected oil gauge　金属套管保护油位计
4　tie rod　拉杆
5　adjustable coil clamp　活动线圈卡
6　radial coil spacers　径向线圈间隔片
7　core leg clamp plates　铁芯夹板
8　oil drain valve　放油阀
9　insulated core　绝缘铁芯
10　bolts　固定螺栓
11　insulating tube　绝缘管
12　rollers　滚轮
13　earthing socket on L.V. (low voltage) side 低压侧接地插座
14　feet for locating core and coils　铁芯及线圈底座
15　steel plate tank　钢制油箱
16　tapping switches　分接开关，抽头开关
17　cooling tubes　散热管
18　free air ventilator　通风器
19　lifting lugs　吊耳
20　captive cover fixing bolts　吊挂紧回螺栓
21　domed weatherproof cover　穹形密封盖

10.2.1.2 油浸式变压器侧视图 Oil Immersed Transformer Side View

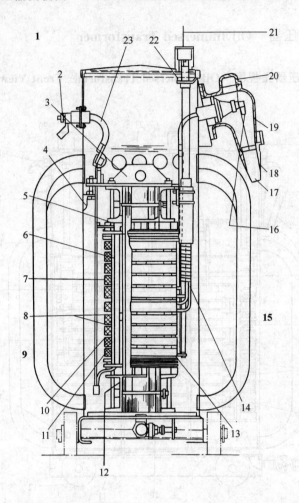

1	side view of transformer 变压器侧视图	13	coil stack end insulation 绕组端绝缘层
2	porcelain insulated studs with cable socket 带电缆卡的陶瓷绝缘双头螺栓	14	tapping leads to switches 至开关的抽头引线
3	core and coils lifting lug 铁芯和线圈吊耳	15	HV side 高压侧
4	anchorage lugs for core and coils 铁芯和线圈支耳角钢	16	porcelain insulated terminal 套管绝缘端子
5	top core clamp 顶部铁芯固定卡	17	drilled to suit customer's cable 配合用户电缆的钻孔
6	LV winding 低压绕组	18	armor clamp 钢铠夹子
7	HV winding 高压绕组	19	cable box 电缆箱
8	oil ducts 油槽	20	expansion chamber 膨胀室
9	LV side 低压侧	21	tapping switches 分接开关,抽头开关
10	LV insulating cylinder 低压绝缘筒	22	tap indicator plate 抽头指示牌
11	HV insulating cylinder 高压绝缘筒	23	flexible leads 挠性引线
12	bottom core clamps 底部铁芯夹紧夹		

10.2.1.3 油浸式变压器外视图 Oil Immersed Transformer Exterior

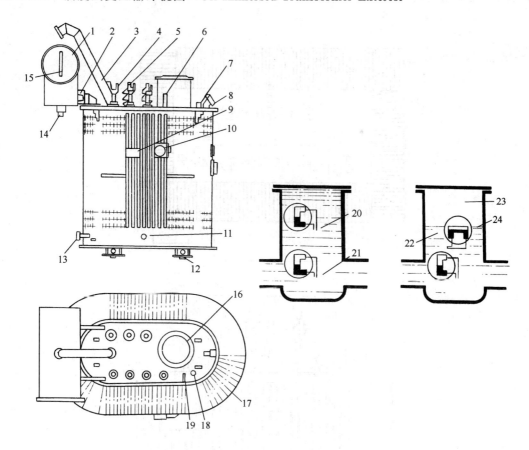

1 oil conservator 储油柜,油枕
2 gas relay (Buchholz relay) 气体继电器
3 safety vent 安全气道
4 low-voltage bushing insulators 低压套管
5 high-voltage bushing insulators 高压套管
6 low-voltage bushing insulators (neutral) 低压套管（零相）
7 lifting eye-bolts 吊环
8 lifting lugs 吊耳
9 name plate 铭牌
10 dial-type thermometer with alarm contact 度盘式信号温度计
11 earthing screw 接地螺栓
12 rollers 滚轮
13 oil drain valve 放油阀
14 dehydrator (dehumidizer) 吸湿器
15 oil level gauge 油面指示器
16 on-load tap changer 有载分接开关
17 radiator (cooling tubes) 散热器
18 mercury thermometer well (adapter) 水银温度计座
19 sensor 测温元件
20 alarm 报警（位置）
21 trip 自动跳闸机构
22 hinge 铰链
23 gas 气体
24 oil level 油面

10.2.2 油浸式密闭变压器 Sealed Oil Immersed Transformer

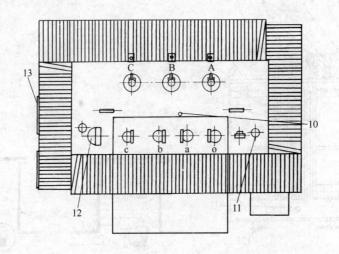

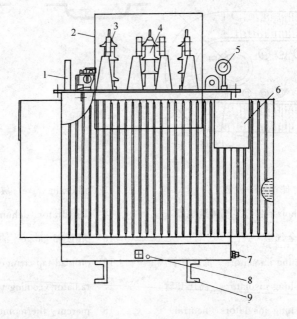

1	filling pipe 注油管	8	earthing screw 接地螺栓
2	bushing house 封闭罩	9	base framework 底座
3	LV bushing 低压套管	10	tap-changer 开关
4	HV bushing 高压套管	11	pressure-relief device 压力释放阀
5	signaling thermometer 测温装置	12	gas relay 气体继电器
6	terminal box 端子箱	13	name plate 铭牌
7	oil brain valve 活门		

10.2.3 树脂绝缘干式变压器 Cast Resin Transformer

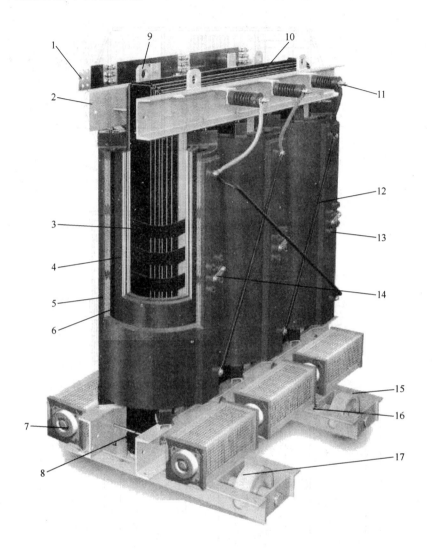

1	LV outlet busbar 低压出线铜排		10	upper yoke 上铁轭
2	clamping flame 夹件		11	HV terminal 高压端子
3	core limb 铁芯		12	HV connection bar 高压连接杆
4	LV winding 低压线圈		13	HV tapping 高压分接头
5	HV winding 高压线圈		14	HV tapping link 高压连接片
6	air duct 风道		15	base 底座
7	cooling fan 冷却风扇		16	earthing bolt 接地螺钉
8	supporting pad 垫块		17	bi-direction rollers 双向轮
9	lifting eye 吊环			

10.2.4 箱式变电站 Transformer Substation

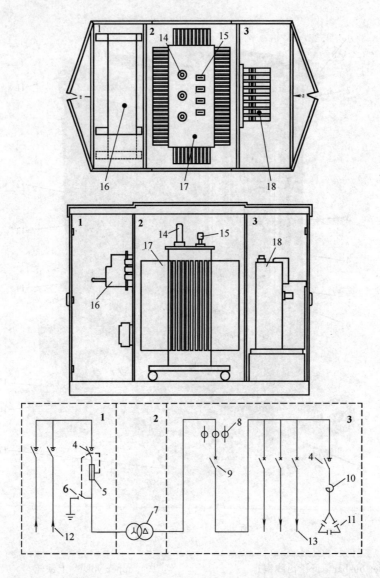

1	high voltage section 高压室	10	reactance 电抗器
2	transformer section 变压器室	11	compensation capacitor 补偿电容器
3	low voltage section 低压室	12	HV cable 高压电缆
4	load switch 负荷开关	13	LV cable 低压电缆
5	fuse 熔断器	14	HV bushing isolator 高压绝缘套管
6	earthing switch 接地开关	15	LV bushing isolator 低压绝缘套管
7,17	transformer 变压器	16	MV switchboard 中压开关柜
8	current transformer 电流互感器	18	LV switchboard 低压开关柜
9	circuit breaker 断路器		

10.3 整流器和电池 Rectifier and Battery

10.3.1 整流器 Rectifier

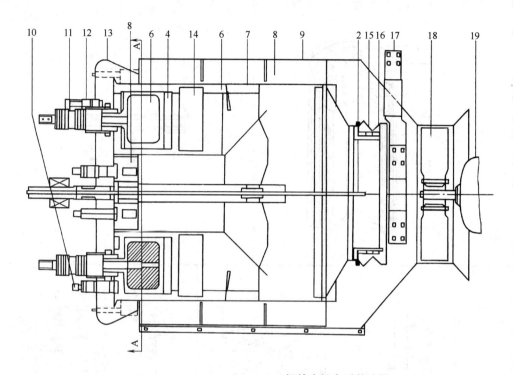

steel tank mercury arc rectifier 钢筒水银电弧整流器

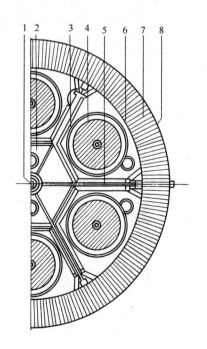

1 ignition anode 点火阳极
2 excitation anode 起弧阳极
3 control grid 控制栅极
4 main anode 主阳极
5 sector shaped anode shield 扇形整流阳极屏蔽
6 cylindrical vacuum tank 圆柱真空筒
7 copper fins 铜散热片
8 cylindrical air guide 圆筒形空气导向装置
9 grid seal 栅极密封
10 anode seal 阳极密封
11 vacuum valve 真空阀
12 air deflector 空气偏转板
13 deionizing baffle 去电离挡板
14 vitreous enamelled cathode seal 珐琅阴极密封
15 cathode protective cylinder 阴极保护筒
16 cathode lead 阴极引线
17 fan 风扇
18 fan motor 风扇马达

10.3.2 原电池　Primary Batteries

(1) 原电池(1)　Primary Batteries（Ⅰ）

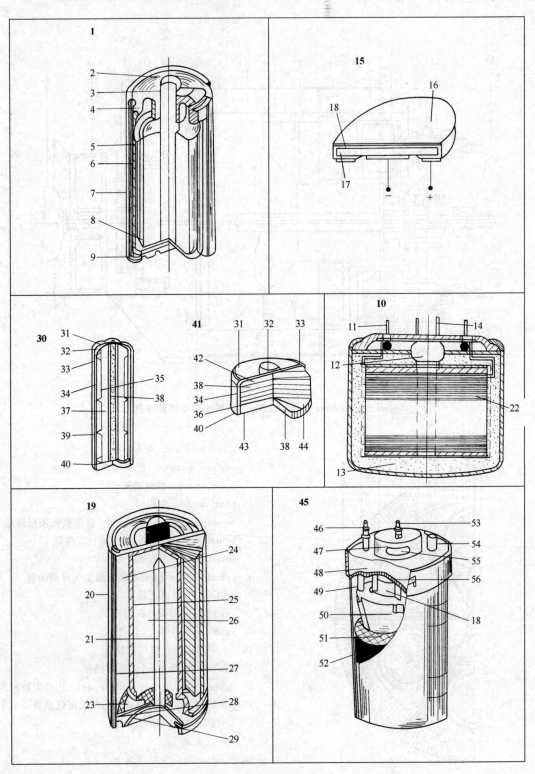

1	carbon-zinc battery	碳锌电池
2	top plate	顶部绝缘层
3	red ring	红环
4	polyethylene top sealing	聚乙烯顶部密封
5	vinyl inner tube	内部乙烯树脂管
6	zinc can	锌筒
7	outer metal jacket	金属外壳
8	guide paper	导纸
9	bottom plate (negative terminal)	底部极板（负极端子）
10	thermal battery	热电池
11	output terminal	引出端子
12	igniter	点火器
13	thermal insulator	保温层
14	ignition terminal	点火端子
15	solar battery	太阳能电池
16	P-type layer	P 型层
17	N-type bulk	N 型层
18	anti-reflection film	防反射层
19	alkaline manganese battery	碱性锰电池
20	plastic tube	塑料管
21	current collector	电流集电器
22	unit cell	单元电池
23	plastic	塑料膜
24	positive electrode mix	正极混合物
25	separator	隔板
26	gel negative electrode	凝胶体负极
27	positive electrode case	正极外壳
28	sealing washer	密封垫圈
29	metal spring	金属弹簧
30	cylinder type mercury battery	圆柱形水银电池
31	outer top	外顶部
32	inner top	内顶部
33	gasket	衬垫
34	inner can	内筒
35	depolarizer	去极化剂
36	absorbent sleeve	吸收筒
37	insulator spacer	绝缘垫片
38	electrolyte	电解液
39	zinc anode cylinders	锌阳极筒
40	outer can	外筒体
41	flat type mercury battery	扁形水银电池
42	zinc anode pellet	锌阳极子
43	gas vent	排气孔
44	barrier	阻挡层
45	air battery	空气电池
46	negative terminal	负极端子
47	cover	盖
48	compound	绝缘膏
49	insulation cover	绝缘盖
50	negative	负极
51	partition for regenerator	再生器隔板
52	regenerator	再生器
53	positive terminal	正极端子
54	electrolyte injection hole	电解液注入孔
55	air intake hole	空气吸入孔
56	battery cover	蓄电池盖

(2) 原电池（2）　Primary Batteries（Ⅱ）

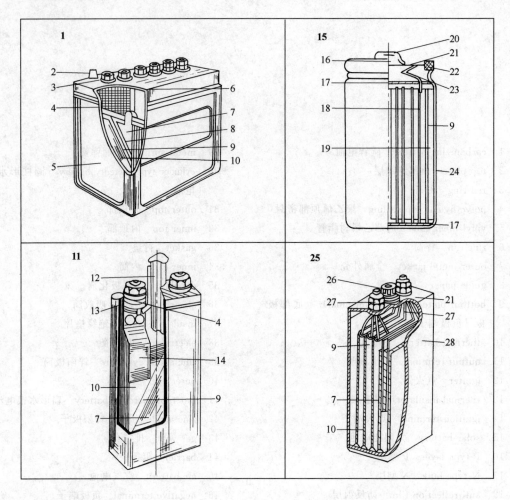

1	lead-acid storage battery　铅酸蓄电池		电池
2	terminal　端子	16	resealing relief valve　绝缘垫圈
3	vent plug　放气螺塞	17	insulating washer　绝缘垫圈
4	pole　电极	18	positive electrode　正极
5	container (ebonite)　蓄电池外壳（硬质橡胶）	19	negative electrode　负极
6	one piece cover　上盖	20	positive cap and terminal　正极引出头及接线端
7	positive plate　正极板	21	cover　外盖
8	glass mat　玻璃栅网	22	gasket　垫片
9	separator　隔板	23	welded positive tap　正极焊头
10	negative plate　负极板	24	separator　隔板
11	nickel-cadmium battery　镍镉蓄电池	25	silver oxide-zinc battery　银氧化锌蓄电池
12	filler and vent plug　注液螺塞	26	cap　引出头
13	cell cover　电池盖	27	negative current collector　负极集电器
14	current collector　集电器	28	case　外壳
15	sealed nickel-cadmium battery　密封镍镉蓄		

10.3.3 阀控式密封铅酸蓄电池 Stationary Lead Acid Valve Regulated Sealed Batteries

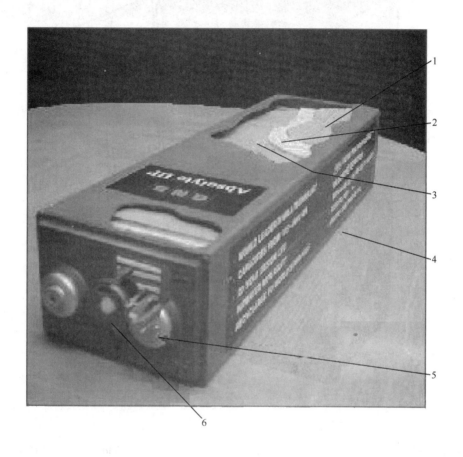

1 negative plate 负极板

2 glass fiber wool 玻璃纤维棉

3 positive plate 正极板

4 polypropylene battery enclosure and steel cover shell 聚丙烯电池壳、外套钢制壳体

5 low resistance pure copper terminal 低电阻纯铜芯端子

6 explosion proof safety valve 防爆安全阀

10.3.4 交流不间断电源 AC Uninterrupted Power Supply

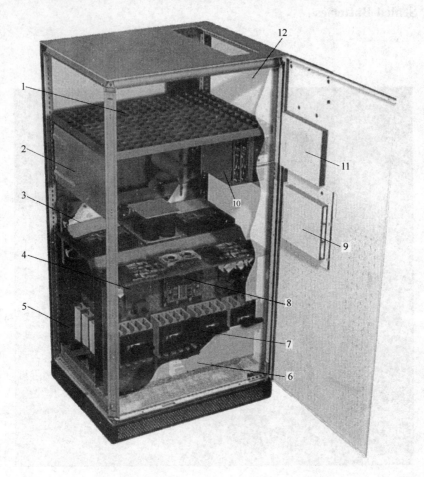

1 space for filter, wiring terminal, isolation transformer, power distribution switch 空间用于无源滤波器、出线端子、隔离变压器、输出配电开关
2 rectifier / charger 整流器/充电器
3 inverter 逆变器
4 protection unit 保护装置
5 power transformer and three phase reactor 电源变压器和三相电抗器
6 protection shield plate 防护挡板
7 input and output circuit breaker and wiring terminal 输入输出断路器和接线端子
8 static bypass switch 静态旁路开关
9 communication card slot 通信板插槽
10 electronic control card 电子控制板
11 monitor 显示器
12 AC power socket and modem AC 插座和调制解调器

10.3.5 直流不间断电源 DC Uninterrupted Power Supply

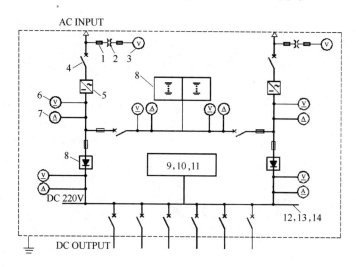

1	fuse 熔断器	8	sealed lead-acid battery 铅酸免维护电池
2	voltage transmitter 电压互感器	9	insulation monitor 绝缘监察
3	AC voltage meter 交流电压表	10	voltage monitor 电压监察
4	MCCB 断路器	11	flicker device 闪光装置
5	charger 充电器	12	flicker bus 闪光母线
6	DC voltmeter 直流电压表	13	control bus 控制母线
7	DC amperemeter 直流电流表	14	close bus 合闸母线

10.4 高中压开关装置 High and Middle Voltage Switchgear

10.4.1 110kV 六氟化硫气体封闭式组合电器 110kV SF$_6$ Gas-Insulated Switchgear

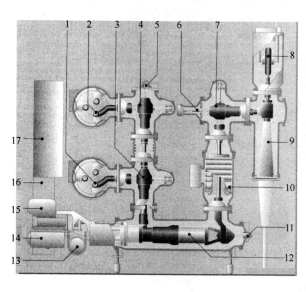

1 busbar Ⅰ 母线Ⅰ
2 busbar Ⅱ 母线Ⅱ
3 busbar disconnector Ⅰ 母线隔离开关Ⅰ
4 busbar disconnector Ⅱ 母线隔离开关Ⅱ
5,6,11 grounding switch 接地开关
7 cable isolator 电缆隔离开关
8 voltage transformer 电压互感器
9 cable sealing end 电缆密封接头
10 current transformer 电流互感器
12 circuit breaker 断路器
13 hydraulic storage cylinder 液压存储瓶
14 electrohydraulic operating unit 电动液压操作单元
15 oil tank 油罐
16 circuit-breaker control unit 断路器控制单元
17 local control cabinet 就地控制柜

10.4.2 110kV 空气绝缘高压组合电器 110kV Prefabricated Air Insulated Switchgear

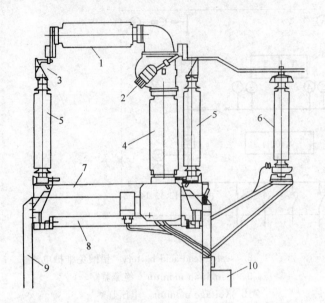

1 circuit breaker 断路器
2 current transformer 电流互感器
3 disconnecting switch 隔离开关
4 porcelain through insulator 支承瓷套
5 support insulator 支承绝缘子
6 lightening arrester 避雷器
7 fixing frame 固定支架
8 movable module chassis 移动模块底盘
9 support pole 支承柱
10 control cabinet 控制箱

10.4.3 中压六氟化硫气体绝缘开关装置 Medium Voltage SF_6 Gas-Insulated Switchgear

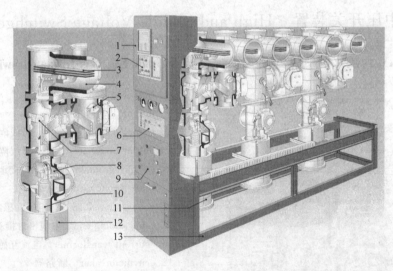

1 low voltage compartment 低压室
2 control unit 二次设备
3 bus bar 母线
4 cast aluminum box 铸铝箱
5 disconnector 隔离开关
6 operation mechanism and interlock of three position switch 三位置开关的操作机构及联锁
7 three position switch 三位置开关
8 circuit breaker poles with upper and lower bushing 带上、下衬套的断路器极
9 circuit breaker operation mechanism 断路器操作机构
10 vacuum circuit breaker 真空断路器
11 connector 连接器
12 current transformer 电流互感器
13 frame 柜架

10.4.4 金属封闭式开关装置 Metal-Enclosed Switchgear

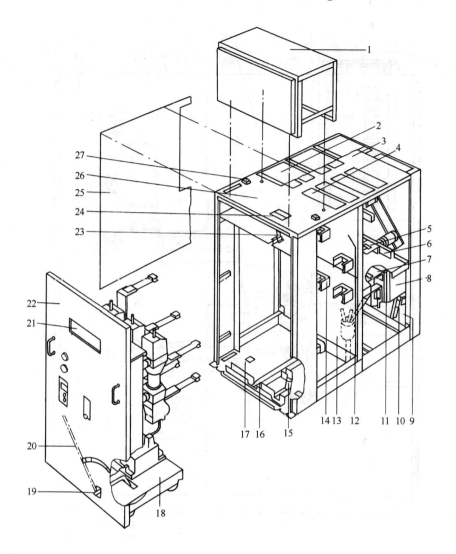

1	low-voltage compartment 低压室		15	interlocking channel 联锁槽
2	pressure relief plate 泄压板		16	center guide 中心导轨
3,26	top sheet 顶板		17	bottom sheet 底板
4	pressure relief cutouts 泄压孔		18	breaker truck 断路器架
5	top isolating contact 上隔离触头		19	locking 闭锁装置
6	busbar 母线,汇流排		20	lever 杠杆
7	bottom isolating contact 下隔离触头		21	inspection window 窥视窗
8	current transformer 电流互感器		22	front plate 面板
9	rear sheet 背板		23	latch 闩
10	earthing switch 接地开关		24	low-voltage socket 低压插座
11~13	partitions 隔板		25	side sheet 侧板
14	barrier box 隔弧盒		27	earth terminal 接地端子

10.4.5 中压金属铠装开关装置　Middle Voltage Metal-Clad Switchgear

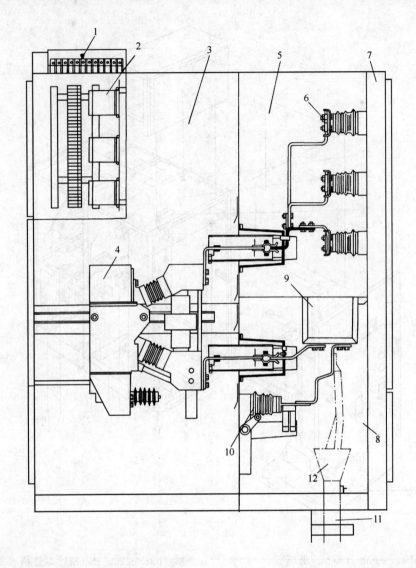

1	control bus bar compartment　小母线室	7	vent of cable chamber　电缆室出气道
2	relay compartment　继电器室	8	cable chamber　电缆室
3	wheeler chamber　手车室	9	current transformer　电流互感器
4	circuit breaker wheeler　断路器小车	10	earthing switch　接地开关
5	main bus bar compartment　主母线室	11	cable　电缆
6	main bus bar　主母线	12	zero sequence transformer　零序互感器

10.4.6 环网柜 Ring Switchgear

10.4.6.1 六氟化硫气体绝缘环网柜 SF₆ Gas Insulated Ring Switchgear

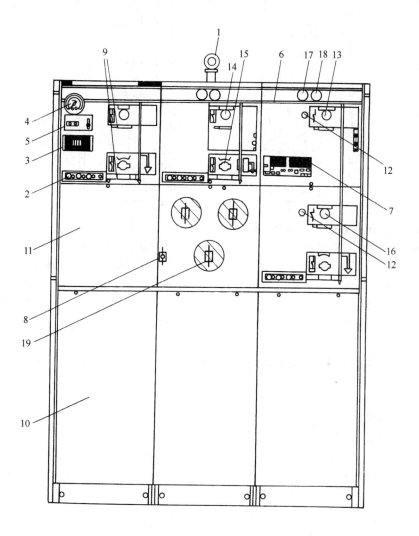

1 lifting eye 吊环
2 voltage indicator 电压指示器
3 earthing fault indicator 接地故障指示器
4 pressure indicator 压力指示器
5 name plate 铭牌
6 mimic diagram 模拟线路图
7 relay 继电器
8 fuse indicator 熔断器指示器
9 lock device on the panel 面板上的挂锁装置
10 cable chamber 电缆室
11 RTU installation compartment RTU安装室
12 lock 锁
13 circuit breaker handle hole 断路器操作孔
14 load switch handle hole 负荷开关操作孔
15 earthing switch handle hole 接地开关操作孔
16 disconnector handle hole 隔离开关操作孔
17 opening push-button 分闸按钮
18 closing push-button 合闸按钮
19 fuse chamber 熔断器室

10.4.6.2 金属封闭环网柜 Metal Clad Ring Switchgear

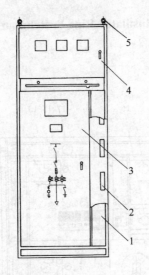

1 door of operation section 操作室小门
2 operation panel 操作圆盘
3 front lower door 前下门
4 door of relay section 继电器室门
5 lifting ring 吊环
6 busbar room 母线室
7 vacuum load switch 真空负荷开关
8 current transformer 电流互感器
9 earthing switch 接地开关
10 relay compartment 继电器室
11 operation mechanism 操作机构
12 fuse 熔断器

10.4.7 高压户外六氟化硫断路器 High Voltage Outdoor SF$_6$ Circuit Breakers

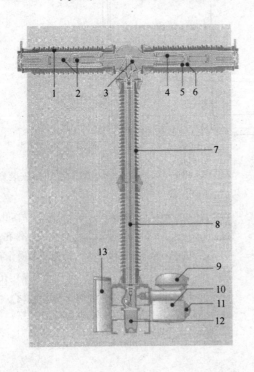

1 interrupter unit 断流器单元
2 arc-quenching nozzles 息弧嘴
3 bell-crank mechanism 钟形曲柄机构
4 moving contact 动触头
5 blast piston 吹弧活塞
6 blast cylinder 吹弧腔
7 insulator column 绝缘柱
8 operating rod 操作杆
9 oil tank 油罐
10 hydraulic operating mechanism 液压操动机构
11 on/off indicator 开/闭指示
12 filter 滤波器
13 control unit 控制单元

10.4.8 中压六氟化硫断路器 Medium Voltage SF₆ Circuit Breakers

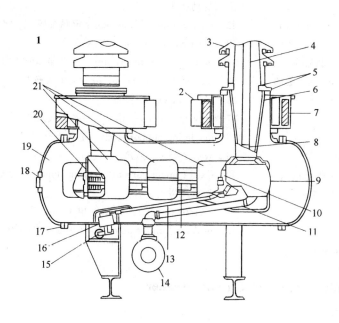

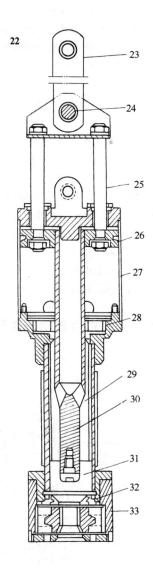

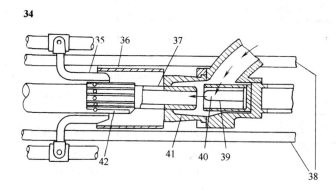

1 sectional view of a SF₆ circuit breaker SF₆ 断路器剖视图
2 current transformer terminal box 电流互感器端接箱
3 hollow upper porcelain containing SF₆ 内含 SF₆ 的上部瓷套管
4 hollow aluminum conductor 空心铝导体
5 gas tight joints 气密紧合面
6 lower insulator 下部绝缘套管
7 current transformer 电流互感器
8 gas filter 气体过滤器
9 internal high pressure reservoir 内部高压储气箱
10 blast valve 吹弧阀
11 insulating gas feed tube 绝缘气的进气管
12 moving contact coupling rods 动触头连杆
13 insulating operating rod 绝缘操作杆
14 high pressure storage receiver 高压储气容器
15 mechanism drive rod 传动机构的传动杆
16 shaft seal assembly 轴密封件
17 end cover with double "O" ring seal 双 "O" 形密封环
18 over pressure safety device 过压安全装置

19　low pressure tank containing SF_6　内含 SF_6 的低压罐

20　accelerating spring　加速弹簧

21　stress shields covering interrupter assembly　内装断流组件的受压外罩

22　fixed and moving contact system and arc-control chamber of SF_6 circuit breaker　SF_6 断路器的定触头和动触头系统及弧腔室

23　insulated moving-contact actuator　绝缘的动触头操动机构

24　steel connecting pin　钢连接销

25　guide　导杆

26　SF_6 compression piston to assist rapid arc extinction　促使快速消弧的 SF_6 压缩活塞

27　SF_6 compression cylinder　SF_6 压缩缸

28　moving contact current collector ring　动触头集电环

29　glass-fiber reinforced epoxy-resin arcing chamber　玻璃纤维增强的环氧树脂灭弧室

30　moving-contact attached to SF_6 compression piston　附着在 SF_6 压缩活塞上的动触头

31　tungsten-tipped moving contact　尖端镶钨的动触头

32　fixed-contact arcing ring　定触头环

33　fixed-contact holder　定触头座

34　part detail of interrupter　断流器部分详图

35　moving contact cross bar　动触头横臂

36　blast shield　吹弧罩

37　moving contact-orifice　动触头气流孔

38　insulating support rods　绝缘支杆

39　arcing horn　弧角

40　fixed contact current carrying fingers　定触头载流指

41　current transfer fingers　电流传导指

42　insulating contact tie rods　连接触头的绝缘杆

10.4.9　户内真空断路器　Indoor Vacuum Circuit Breaker

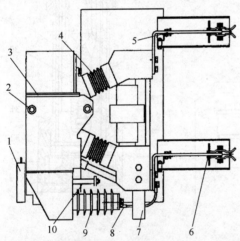

1　fixed part　固定部分
2　roller wheel　滚轮
3　earthing ear　接地体
4　vacuum circuit breaker　真空断路器
5　upper moving contact　上动触头
6　lower moving contact　下动触头
7　support leg　支腿
8　phase insulation plate　相间绝缘隔板
9　lightning arrester　避雷器
10　push in screw rod　推进螺杆

10.4.10　避雷器　Lightning Arrester

1　flange with gas diverter nozzle　带气体分流嘴的法兰
2　seal　密封
3　pressure relief diaphragm　压力释放膜
4　compressing spring　压缩弹簧
5　metal oxide resistors　金属氧化物电阻
6　composite polymeric housing　合成聚合物外壳

10.4.11 真空接触器 Vacuum Contactor

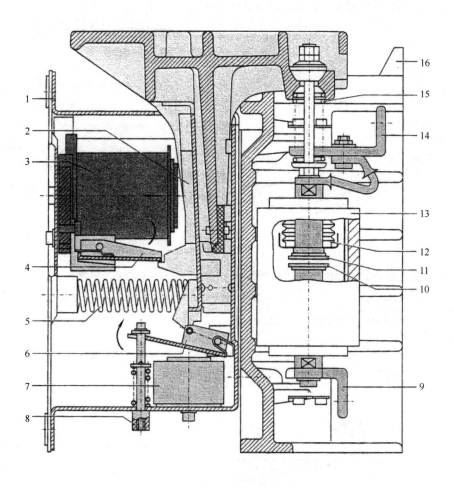

1	steel plate shell 钢板壳体		10	fixed contact 静触头
2	gag bit 衔铁		11	moving contact 动触头
3	electro-magnetic system 电磁系统		12	metal corrugated pipe 金属波纹管
4	mechanic closing lock 机械合闸锁扣		13	vacuum switch tube 真空开关管
5	switch opening spring 分闸弹簧		14	upper wiring terminal 上接线端子
6	trip plate 脱扣板		15	fixed contact compression spring 静触头压力弹簧
7	trip magnet 脱扣磁铁			
8	mechanic trip rod 机械脱扣杆		16	insulation enclosure 绝缘壳体
9	lower wiring terminal 下接线端子			

10.4.12 高压熔断器 HV Fuses

10.4.12.1 高压管式熔断器 HV Fuse Tubes

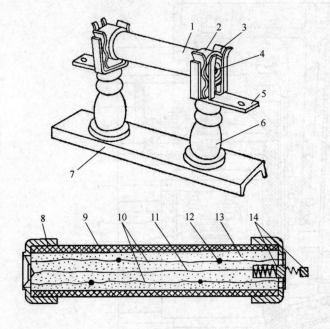

1, 9 ceramic fuse cap   瓷熔管
2, 8 metal cap   金属管帽
3 spring contact holder   弹性触座
4, 14 fuse link indicator   熔断指示器
5 wiring terminal   接线端子
6 porcelain insulator   瓷绝缘子
7 holder   底座
10 function fuse link   熔丝
11 indication fuse link   指示熔丝
12 tin ball   锡球
13 quartz sand filling   石英砂填料

10.4.12.2 高压跌开式熔断器 HV Falling Type Fuse

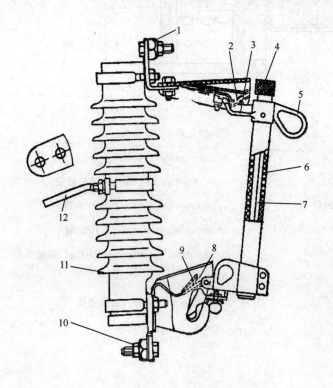

1 upper wiring terminal   上接线端子
2 upper fixed contact   上静触头
3 upper moving contact   上动触头
4 tube cap   管帽
5 operation ring   操作环
6 fuse tube   熔管
7 copper fuse link   铜熔丝
8 lower moving contact   下动触头
9 lower fixed contact   下静触头
10 lower wiring terminal   下接线端子
11 porcelain insulator   绝缘瓷瓶
12 fixed installation plate   固定安装板

10.4.13 补偿电容器成套装置 Compensation Capacitors Cabinet

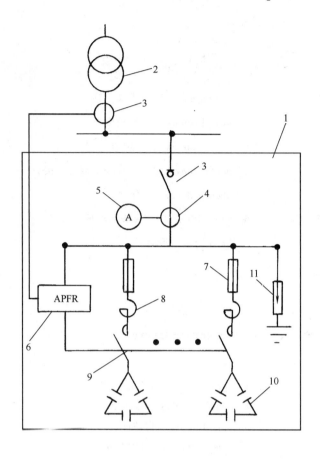

1 compensation capacitor cabinet 补偿电容器成套装置
2 transformer 变压器
3 current transformer 电流互感器
4 load switch 负荷开关
5 ampere meter 电流表
6 automatic power factor regulator 自动功率因数调节器
7 fuse 熔断器
8 reactance 电抗器（消谐线圈）
9 contactor 接触器
10 capacitor 电容器
11 lightning arrester 避雷器

10.4.14 操动机构 Operating Device

10.4.14.1 直动操动机构 Direct Operating Device

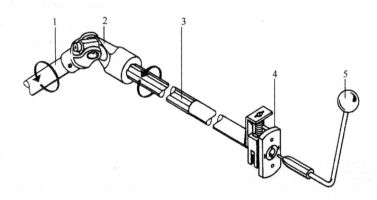

1 switch main axis 开关主轴
2 swivel connector 摆动接头
3 operation axis 操作轴
4 axis with clip 带夹头的轴承
5 hand knob 手柄

10.4.14.2 转向操动机构 Swivel Operating Device

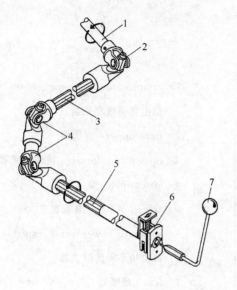

1 switch main axis 开关主轴
2 swivel connector 摆动接头
3 connection axis 连接轴
4 double universal connector 双万向接头
5 operation axis 操作轴
6 axis with clip 带夹头的轴承
7 hand knob 手柄

10.4.14.3 弹簧储能操动机构 Spring-Stored Energy Operating Device

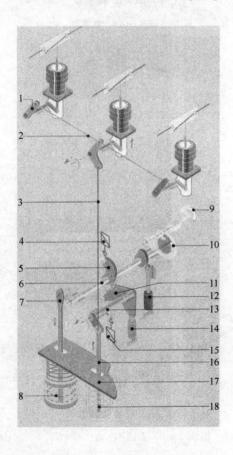

1 corner gears 转角齿轮
2 coupling linkage 耦合器
3 operating rod 操动杆
4 closing release 合闸释放
5 cam plate 凸轮盘
6 charging shaft 储能轴
7 closing spring connecting rod 合闸弹簧连接杆
8 closing spring 合闸弹簧
9 hand-wound mechanism 手摇机构
10 charging mechanism 储能机构
11 roller level 摇动杠杆
12 closing damper 合闸阻尼器
13 operating shaft 操动轴
14 opening damper 开闸阻尼器
15 opening release 开闸释放
16 opening spring connecting rod 开闸弹簧连接杆
17 mechanism house 机构外壳
18 opening spring 开闸弹簧

10.5 低压开关装置 Low Voltage Switchgear

10.5.1 低压开关柜 Low Voltage Switchgear

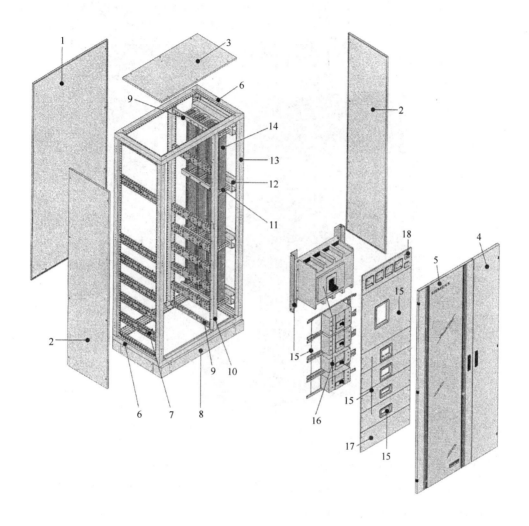

1 rear panel　后盖板
2 side panels　侧封板
3 top　顶板
4 side door　边门
5 transparent door　透明门
6 top frame　顶架
7 double switchgear holder traversal　开关柜双层横梁
8 base　底座
9 single switchgear holder transversal　开关柜单层横梁
10 intermediary upright　中立柱
11 busbar holder　母排支架
12 bar holder transversal　母排支架横梁
13 uprights　立柱
14 vertical busbar　垂直母排
15 masking plates with cut-out/collar/frame or rail supporting the switchgear　开关柜罩板带冲孔/框架或开关柜导轨支架
16 circuit-breaker　断路器
17 solid masking plate　固定罩板
18 masking plates for measuring instruments　用于测量仪表的罩板

10.5.2 框架式空气断路器 Air Circuit Breaker

10.5.2.1 框架式空气断路器前面板 Air Circuit Breaker Front Face

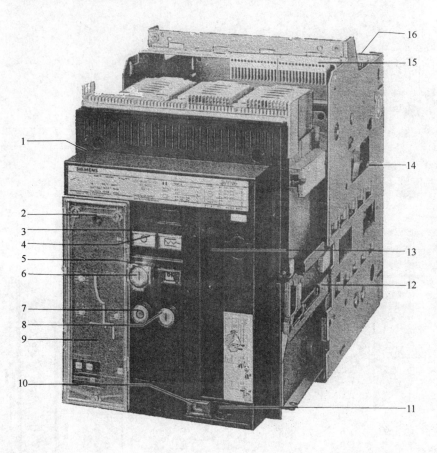

1 draw-out circuit breaker 抽屉式断路器
2 display and reset button for tripped signaling contact and mechanical reclosing lockout 显示和复位按钮，用于脱扣信号发送触头和机械重合闸锁定
3 stored-energy indicator 储能状态显示器
4 switching-state indicator 通断状态显示器
5 ready-to-close indicator 合闸准备就绪状态显示器
6 on button, mechanical 机械合闸按钮
7 off button, mechanical 机械分闸按钮
8 on button, electrical 电气合闸按钮
9 overcurrent release 过电流脱扣器
10 position indicator 位置显示器
11 crank hole 曲柄操作孔
12 guide rails 导轨
13 crank handle 操作手柄
14 guide frame 导向框架
15 auxiliary current connectors 辅助接线端子
16 position signaling switch 位置信号发送开关

10.5.2.2 框架式空气断路器分解图 Air Circuit Breaker Explode Drawing

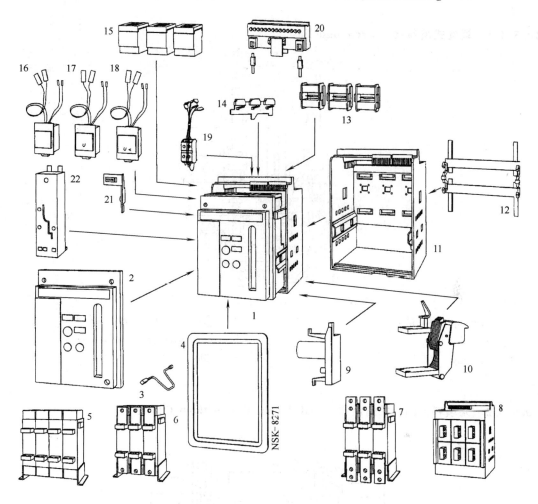

1. basic circuit breaker 断路器本体
2. control panel 控制面板
3. crank handle 曲柄
4. door sealing frame 门密封框
5. main connections at the rear, horizontal 主回路水平后置
6. main connections accessible from the front, top and bottom, single-hole bar ends 主回路连接前置，单孔，顶部和底部
7. main connections accessible from the front, top and bottom, double-hole bar ends 主回路连接前置，双孔，顶部和底部
8. main connections at the rear vertical 主回垂直后置
9. motor for energy storage 储能用电动机
10. main contact sets 主触头
11. guide frame 导向框架
12. shutter 安全挡板
13. current transformer 电流互感器
14. position signaling switches 位置信号发送开关
15. arc chute 灭弧室
16. closing solenoid 电气合闸线圈
17. shunt release 分励线圈
18. undervoltage release 欠电压线圈
19. auxiliary contacts 辅助触头
20. auxiliary connector 辅助接线端子块
21. operating cycle counter 操作计数器
22. solid-state, microprocessor controlled overcurrent release 带微处理器的电子式过电流脱扣器

10.5.3 低压熔断器 LV Fuses

10.5.3.1 螺旋式熔断器 Screw Base Fuse

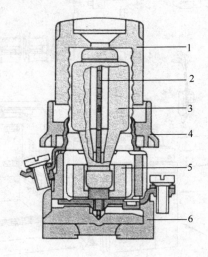

1　screw cap　螺旋帽盖
2　fuse link　熔丝
3　fuse body　熔断体
4　safety protection cover　保护罩
5　compensation ring　配合压环
6　holder　底座

10.5.3.2 无填料封闭管式熔断器 Sealed Tube Fuse without Filling

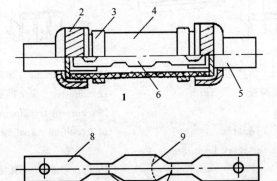

1　fuse tube　熔管
2　copper tube cap　铜管帽
3　tube clamp　管夹
4　fiber fuse tube　纤维熔管
5，8　contact blade　触头片
6　variable zinc fuse link　变截面锌熔片
7　fuse link　熔片
9　overload zone　过载熔断区
10　short circuit zone　短路熔断区

10.5.3.3 有填料快速熔断器 Filled Fast Fuse

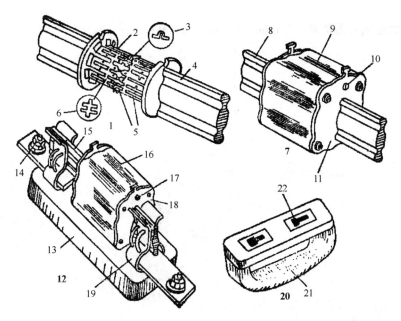

1　fuse　熔体
2，8　copper fuse grid　栅状铜熔体
3　tin bridge　锡桥
4，15　act blade　触刀
5　variable holes　变截面小孔
6　arc grid　灭弧栅
7　fuse tube　熔管
9，16　ceramic fuse tube　瓷熔管
10，17　fuse link indicator　熔断指示器

11，18　end cover plate　端面盖板
12　filled sealed tube fuse　有填料封闭管式熔断器
13　holder　底座
14　wiring terminal　接线端子
19　spring contact holder　弹性触座
20　insulation operation knob　绝缘操作手柄
21　insulation hand pull　绝缘拉手
22　engage hole　扣眼

10.5.3.4 插入式熔断器 Plug-in Fuse

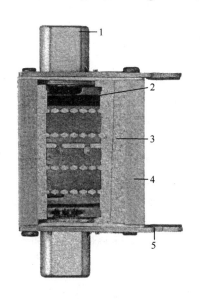

1　contact blade　触头片
2　fuse link　熔丝
3　quartz sand　石英砂
4　ceramic tube　磁熔管
5　cover plate　盖板

10.5.4 微型断路器及附件 Miniature Circuit Breaker and Accessories

10.5.4.1 微型断路器附件 Miniature Circuit Breaker Accessories

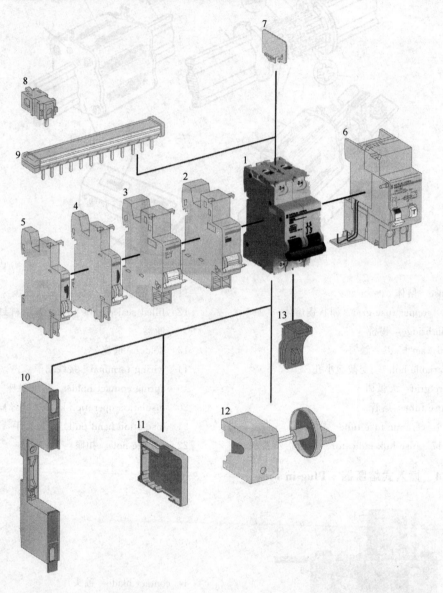

1 miniature circuit breaker 微型断路器
2 undervotage release 欠压脱扣器
3 shunt release 分励脱扣器
4 auxiliary contact 辅助接点
5 alarm contact 报警接点
6 electrical leakage protection module 漏电保护模块
7 interphase barrier 极间隔片
8 insulation connector 绝缘连接头
9 comb bus bar 梳状母排
10 socket base 插拔式底座
11 separation pieces 间隔件
12 rotary hand knob 旋转手柄
13 hand knob lock 手柄锁扣

10.5.4.2 微型断路器 Miniature Circuit Breaker

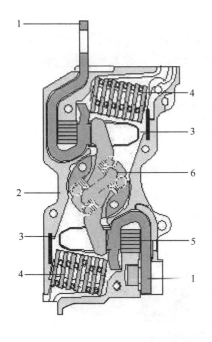

1 fixed contact 静触头
2 closed chamber 封闭小室
3 gas generating 气体产生部分
4 arc suppressing grid 灭弧栅
5 magnetic shield 磁屏蔽
6 moving contact 动触头

10.5.5 塑壳断路器 Moulded Case Circuit Breaker

10.5.5.1 塑壳断路器 Moulded Case Circuit Breaker（MCCB）

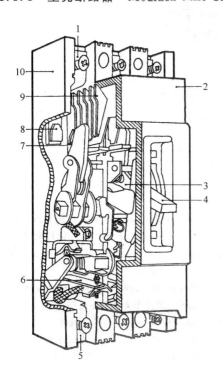

1 line terminal 进线端子
2 moulded cover 塑料盖
3 operation mechanism 操作机构
4 handle 手柄
5 load terminal 负荷端子
6 thermal tripping element 热脱扣元件
7 movable contact 动触头
8 stationary contact 静触头
9 arc quencher 灭弧罩
10 moulded base 塑料底座

10.5.5.2 塑壳断路器模块化系统 MCCB Modularized System

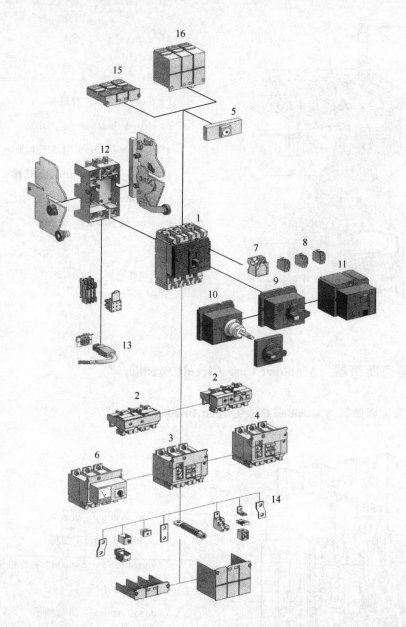

1 switch unit 开关装置
2 release unit 脱扣单元
3 grounding fault protection module 接地故障保护模块
4 insulation monitor module 绝缘监视模块
5 power indication module 带电指示模块
6 ammeter module 电流表模块
7 undervoltage release and shunt release 欠压脱扣线圈及分励线圈
8 versatile auxiliary switch 多功能辅助开关
9 direct rotary hand knob 直接旋转手柄
10 extend rotary hand knob 延伸旋转手柄
11 electrical operating mechanism 电动操作机构
12 plug-in socket 插入式底板
13 plug-in and draw-out auxiliary connection pieces 插入式及抽出式辅助连接件
14 connection pieces 连接附件
15 short terminal cover 短端子罩
16 long terminal cover 长端子罩

10.5.6 真空负荷开关 Vacuum Load Switch

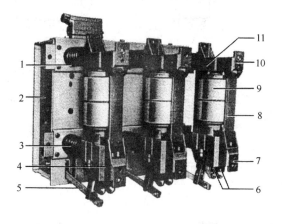

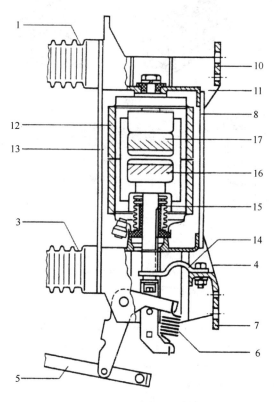

1　upper support insulator　上支持绝缘子
2　operation mechanic box　操作机构箱
3　lower support insulator　下支持绝缘子
4　vacuum switch lower support　真空开关管下支架
5　operation stick　操作杆
6　contact pressure spring and opening spring　接触压力弹簧和分闸弹簧
7　lower wiring terminal　下接线端子
8　front support rod　前支承杆
9　vacuum switch tube　真空开关管
10　upper wiring terminal　上接线端子
11　vacuum switch upper support　真空开关管上支架
12　vacuum switch porcelain shell　真空开关管陶瓷外壳
13　rear support rod　后支承杆
14　soft connection　软连接
15　compensation sheath　波纹管
16　moving contact　动触头
17　fixed contact　静触头

10.5.7 控制开关 Control Switch

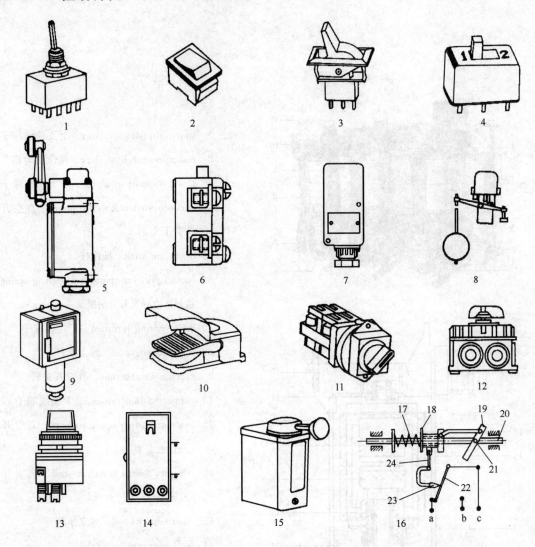

1. toggle switch 拨动开关，钮子开关
2. rocker switch 摇键开关
3. paddle level switch 扳钮开关，扳把开关
4. slide switches with top and side actuating levels 带顶端及侧向板动把的滑动开关
5. limit switch 行程开关，极限开关
6. microswitch 微动开关
7. proximity switch and sensing head 接近开关和检测头
8. float switch 浮子开关
9. pressure switch 压力开关
10. pedal or foot switch 脚踏开关
11. selector switch 选择开关
12. rotary switch, packet switch 转换开关，组合开关
13. voltmeter and ammeter change-over switch 电压表和电流表换相开关
14. time switch 时间开关
15. reversing drum switch 倒顺鼓形开关
16. centrifugal switch 离心开关
17. spring 弹簧
18. fork 叉子
19. fly-iron 飞铁
20. shaft 轴
21. pivot of fly-iron 飞铁支点
22. movable contacting system 动触点系统
23. insulated rod 绝缘杆
24. lever 杆

10.5.7.1 按钮开关 Pushbutton Switches

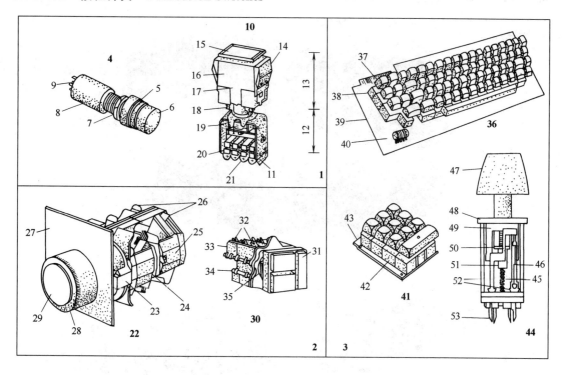

1	panel pushbutton 盘装按钮开关	27	separate label plate 单独标签板
2	industrial pushbutton 工业开关	28	button guard 按钮护套
3	keyboard pushbutton 键盘按钮开关	30	compact size 小型（按钮）
4	integral pushbutton 整体按钮	31	multiple buttons 多按钮
5	button enclosure 按钮外壳	33	contact blocks 开关接点组
6，29	pushbutton 按钮	34	screws assembled blocks 螺钉紧固件栓组装接点组
7	threaded bushing 内外螺纹接头		
8	switch enclosure 开关外壳	35	panel mounting lock nut 盘上安装锁紧螺母
9，21，26，32，53	terminals 端子	36	encoded keyboard for alpha-numeric data entry 字符输入编码键盘
10	built up pushbutton 组合式按钮		
11	insulation between terminals 端子间绝缘板	37	connector 插接片
12	switch assembly 开关组件部分	38，42	individual pushbutton 单按钮
13	actuator 操作机构部分	39，43	printed wiring board 印刷线路板
14	panel mounting spring-clip 装在盘上的卡簧	40	electronic encoding components 电子编码器
15	relegendable button-lens 按钮透镜	41	10-key numeric keyboard 10键数码盘
16	lamp/actuator enclosure 指示灯及操作机构外壳	44	individual pushbutton element 单按钮元件
17	terminal marking 端子标记	45	return spring 复位弹簧
18	lamp terminal 指示灯接线端子	46	reed switch 舌簧开关
19	spring clip 簧片，弹簧线夹	47	truncated button 截头按钮
20	switch contacts 开关接点	48	enclosure 壳体
22	standard size 标准型（按钮）	49	travel stop 行程定位机构
23	mounting collar 安装卡环	50	plunger 活柱
24	first contact block 第一接点组	51	magnet part 磁铁部分
25	second contact block 第二接点组	52	contact leads 接点引线

10.5.7.2 凸轮旋转开关 Cam Switches

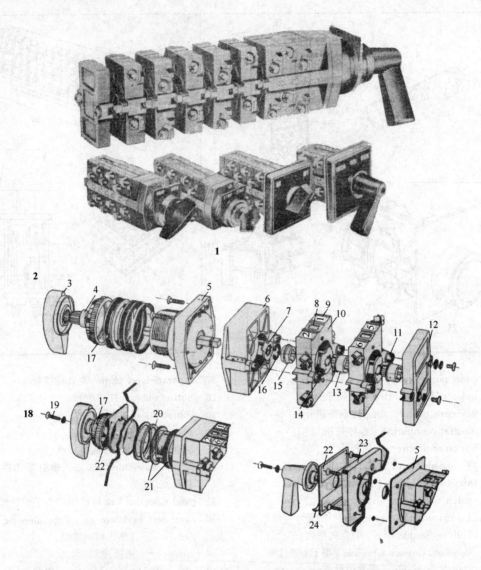

1	cam switches 凸轮旋转开关		13	cam base 凸轮座
2	exploded view of assembly 总成的部件分解图		14	auxiliary cam 辅助凸轮
			15	coupling shaft 联轴器
3	handle 旋钮		16	retaining ring 扣环
4	handle adapter 旋钮插座		17	locking ring 锁紧圈
5	head block unit 前块元件		18	panel mounting 控制盘安装
6	operator block 操作块		19	screw 螺钉
7	stop 止动器		20	ring stoper 止动圈
8	switch indicator 开关指示器		21	rubber washer 橡胶垫圈
9	terminal No. 接线端号		22	nameplate 铭牌
10	contact block 接触块		23	plate 板
11	mounting screw 安装螺钉		24	protection film 保护膜
12	end plate 端板			

10.5.8 变频调速装置 Variable Velocity Variable Frequency Device

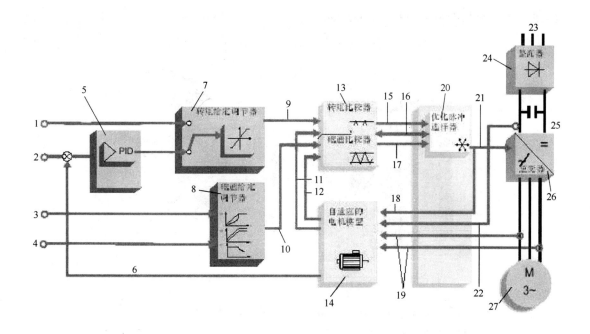

1 torque input 转矩给定
2 speed input 速度给定
3 magnetic flux optimization on/off 磁通优化开关
4 magnetic flux break on/off 磁通制动开关
5 speed regulator, acceleration compensation 速度调节器, 加速补偿器
6 actual speed 实际速度
7 torque input regulator 转矩给定调节器
8 magnetic flux input regulator 磁通给定调节器
9 internal torque input 内部转矩给定
10 internal flux input 内部磁通给定
11 actual torque 实际转矩
12 actual flux 实际磁通
13 torque compare, magnetic flux compare 转矩比较器, 磁通比较器
14 adaptive motor model 自适应电机模型
15 torque status 转矩状态
16 control signal 控制信号
17 magnetic flux status 磁通状态
18 switch position 开关位置
19 motor current 电机电流
20 optimized pulse selector 优化脉冲选择器
21 switch position command 开关位置命令
22 DC busbar voltage 直流母线电压
23 main power supply 主电源
24 rectifier 整流器
25 DC busbar 直流母线
26 inverter 逆变器
27 motor 电机

10.5.9　电动机启动器　Motor Starter

10.5.9.1　手动启动器　Manual Starter

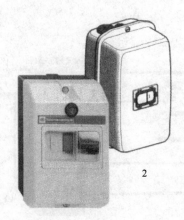

1　thermal-magnetic trip motor circuit breaker　热磁脱扣电动机断路器
2　protection enclosure　防护外壳

10.5.9.2　自动启动器　Automatic Starter

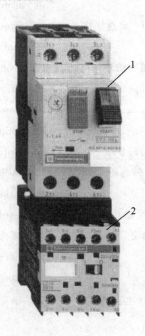

1　thermal-magnetic trip motor circuit breaker　热磁脱扣电动机断路器
2　contactor　接触器

10.6 控制及保护器件 Control and Protection Devices

10.6.1 电磁机构和器件 Electromagnetic Mechanism and Devices

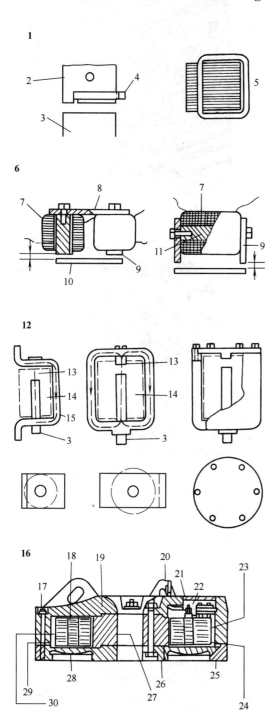

1 shading pole 罩极，屏蔽极
2 stationary pole 静磁铁
3 plunger 插棒式铁芯
4 shading coil 屏蔽线圈，短路环
5 view of pole face 磁极面正视图
6 tractive electromagnet 牵引电磁铁
7 winding section 线圈断面
8 yoke 磁轭
9 pole piece 磁极
10 armature 衔铁
11 core 铁芯
12 iron-clad solenoid actuator 铁壳电磁铁操作机构
13 end stop 端部止挡
14 coil 线圈
15 flux path 磁路
16 circular lifting magnet 圆形起重磁铁
17 counter bolt 沉头螺钉
18 steel spool for coil 线圈钢芯管
19 magnetic shell 磁壳
20 supply cable gland 电源电缆压盖
21 terminal box cover 端子箱盖
22 terminal 端子
23 insulating tubing for flexible lead 引出软线的绝缘管
24 braided flexible lead 纺织引出软线
25 outer pole 外磁极
26 center pole shoe 中间磁极靴
27 center pole 中间磁极
28 manganese steel bumping plate 锰钢冲击板
29 insulating disc 绝缘盘
30 coil disc 线圈盘

10.6.2 电磁继电器 Electromagnetic Relays

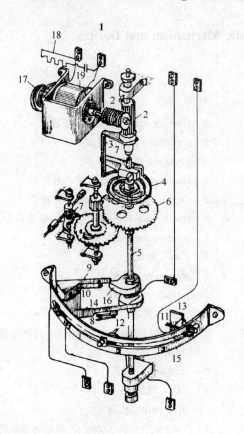
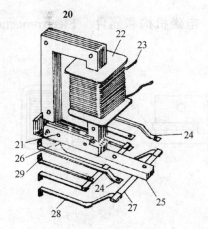
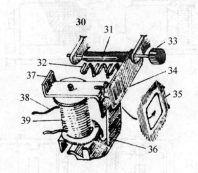

1 time relay 时间继电器
2 worm gear 蜗轮传动机构
3 regulator lever 拨杆
4 hair spring 游丝
5 wheel spindle 轮轴
6 gearing wheel 传动齿轮
7 balance pallets and escape wheel 平衡摆，棘爪和擒纵轮
8 sliding contact 滑动触点
9, 10 stationary contact 固定接点
11 contact pin with fastener 带夹持器的接触针
12 sliding contact pin 滑动接触针
13 backing block 挡块
14 arm for carrying contact 接点支承臂
15 scale for time setting 时间整定刻度盘
16 cam from insulating material 绝缘材料凸轮
17 spring 弹簧
18 resistor and condenser used for spark diminishing 减弱火花用的电阻器和电容器
19 solenoid 螺线管
20 auxiliary relay 辅助继电器，中间继电器
21 armature 衔铁
22 relay coils 继电器线圈
23 coil leads 线圈引线
24 normally opened contacts 常开接点
25 insulation strap 绝缘板
26 armature end 衔铁端
27 bridging contact 桥式触点，短路触点
28 normally closed contacts 常闭触点
29 terminals 端子
30 signal relay 信号继电器
31 moving contact 动触点
32 stationary contacts 静触点
33 reset knob 复位旋钮
34 drop annunciator 掉牌指示器
35 window on a relay cover 继电器罩上的窗孔
36 hinge armature 枢轴衔铁
37 armature core 衔铁铁芯
38 leads of current or potential coil 电流或电压线圈的引线
39 solenoid coil 螺管线圈

10.6.3 固态继电器 Solid State Relays

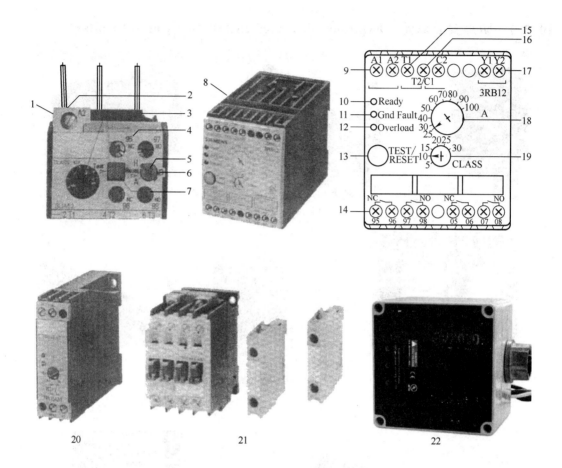

1 overload relay 热过载继电器
2 terminal for contactor coil 接触器线圈接线端子
3 setting current adjusting scale 整定电流旋钮
4 reset button 复位按钮
5 switch position indicator 脱扣指示
6 test button 测试按钮
7 auxiliary contact terminal 辅助触头接线端子
8 solid state overload relay 电子式过载继电器
9 terminals for control supply voltage 控制电源供电端子
10 LED "ready" 就绪指示
11 LED "ground fault" 接地故障指示
12 LED "overload" 过载指示
13 test/reset button 测试/复位按钮
14 overload/thermistor/earth fault tripping 过载/热/接地脱扣
15 thermistor terminal 热敏元件端子
16 terminals for external zero-sequence current transformer 外接零序互感器端子
17 terminals for remote or automatic reset 远动/自动复位端子
18 rotary button for current setting 电流设置旋钮
19 class selector rotary knob 脱扣级别选择开关
20 solid state time relay 电子式时间继电器
21 auxiliary relay with plug-on auxiliary contact blocks 可插装辅助触头座的中间继电器
22 surge protection device (SPD) 浪涌保护器

10.7 电气防爆设备　Explosion Protected Apparatus

10.7.1 防爆安装单元　Explosion Protected Installation Control Units

1　switch　开关
2　junction box　接线盒
3　terminal box　端子箱
4　cable end box　电缆终端箱
5　wall plugs and sockets　墙装插头插座
6　flange plugs and sockets　法兰安装插头插座
7　compacted interlocked receptacles　紧密型带自锁装置的插座
8　transformer module　变压器模块

10.7.2 防爆照明灯具　Explosion Protected Luminaries

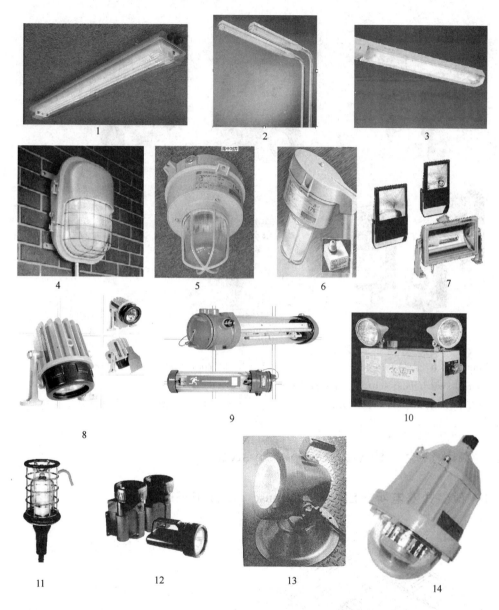

1. fluorescent luminaries with electronic ballast　带电子镇流器的荧光灯
2. pole mounted luminaries for bi-pin fluorescent lamps with electronic ballast　柱装带电子镇流器双脚插头的荧光灯
3. emergency fluorescent luminaries with electronic ballast　带电子镇流器的应急荧光灯
4. wall pack luminaries　壁灯
5. high intensity discharge luminaries (sodium/mercury lamp)　高密度气体放电灯（钠灯/汞灯）
6. pulse start metal halide luminaries　脉冲启动金属卤化物灯
7. flood luminaries　泛光灯
8. tank inspection luminaries　孔视灯
9. exit sign luminaries　出口指示灯
10. halogen sealed beam luminaries (with maintenance free lead acid battery)　密封卤素束灯（带铅酸免维护电池）
11. portable hand lamps　便携式手提灯
12. hand lamps　手提灯
13. portable floodlight　便携式泛光灯
14. LED light fitting　LED 灯

10.7.3 防爆控制设备 Explosion Protected Control Equipment

1　limit switch　限位开关
2　control switch/push button　控制按钮/开关
3　CCTV camera　CCTV 摄像头
4　control station　操作柱
5　horn　喇叭
6　alarm station　火警按钮

10.8 供电系统 Power Supply System

10.8.1 预装式变电所 Electrical House (E-HOUSE)

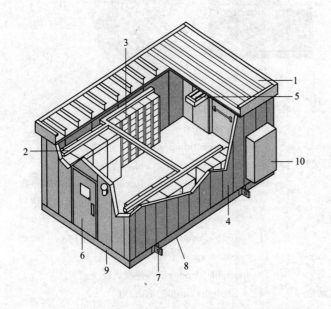

1　ribbed roof　加肋拱顶
2　bus duct　母线槽
3　switchgear　开关设备
4　self framing galvanized steel wall　框架式镀锌钢制墙
5　lighting fixture　照明灯具
6　entrance door　入口门
7　removable lifting lugs　可去除的起重吊耳
8　steel floor　钢板地面
9　welded steel channels and angle supports　焊接槽钢和角钢支承件
10　air condition　空调

10.8.2 热电厂 Thermal Power Plant

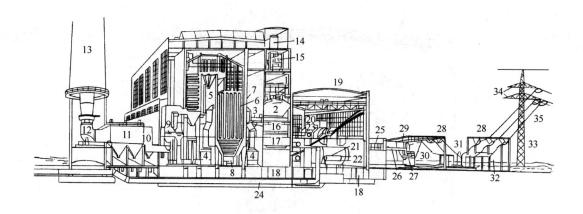

- 1,3 coal conveyor 输煤机
- 2 coal bunker 煤斗
- 4 coal mill 磨煤机
- 5 steam boiler 蒸汽锅炉
- 6 combustion chamber 燃烧室
- 7 water tubes 水管
- 8 clinker pit 灰坑，排渣坑
- 9 air pre-heater 空气预热器
- 10 gas flue 烟道
- 11 electro-static precipitator 静电除尘器
- 12 induced draught fan 引风机
- 13 chimney 烟囱
- 14 de-aerator 除氧器
- 15 feed water tank 给水筒
- 16 switchgear 开关装置
- 17 cable tunnel 电缆隧道
- 18 cable cellar 电缆室
- 19 turbine room 汽轮机间
- 20 steam turbine with alternator 汽轮发电机组
- 21 surface condenser 表面冷凝器
- 22 low-pressure heater 低压加热器
- 23 high-pressure heater 高压加热器
- 24 circulating water pipe 循环水管
- 25 control room 控制室
- 26 busbar 母线
- 27 power transformer 电力变压器
- 28 gantry 门形架
- 29 high-voltage overhead conductor 高压架空导线
- 30 high-voltage conductor 高压导线
- 31 air-blast circuit-breaker 空气吹弧断路器
- 32 surge diverter 避雷器
- 33 transmission line tower or pylon, a lattice-type tower 输电线路、铁塔、钢构架塔
- 34 cross arm 横担
- 35 suspension insulator 悬式绝缘子

10.8.3 变电所屋内配电装置　Indoor Installations of Electric Substation

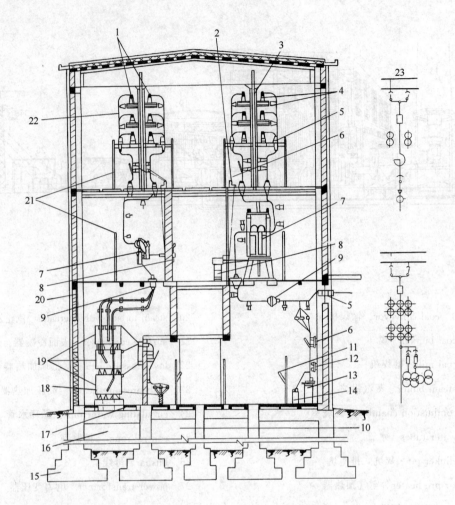

1　horizontal busbar of a double busbar system 双母线系统的水平母线
2　mechanical remote control linkage　机械遥控连杆
3　partition wall　隔墙，间壁
4　supporting insulators　支柱瓷瓶
5　wall bushings　穿墙套管
6　disconnecting switch　隔离开关
7　low-oil type circuit breaker　少油断路器
8　electromagnetic driving mechanism　电磁操动机构
9　through type current transformer　贯穿式电流互感器
10　cable tunnel　电缆隧道
11　manual operating mechanism　手动操动机构
12　current limiting fuses　限流熔断器
13　ground potential transformer　接地电压互感器，零序电压互感器
14　potential transformer　电压互感器
15　foundation　基础
16　ventilating canal　通风沟
17　slab　沟盖板
18　cable head　电缆头
19　current limiting reactor　限流电抗器
20　louver　百叶窗
21　wire fence　铁丝网，栅栏
22　vertical busbar　垂直母线
23　schematic drawing symbols　系统图符号

10.8.4 节能发电厂 Energy Saving Power Plant

1	air discharge pipes from turbo blowers to blast furnace	汽轮鼓风机至高炉的送风管
2	chief engineers office	总工程师办公室
3	locker room	更衣室
4	hot water storage tank	热水储存槽
5	turbine room basement floor	汽轮机室底层
6	overhead travelling crane	桥式吊车
7	turbine generator	汽轮发电机
8	cable pan	电缆盘
9	turbine expansion joint	汽轮机膨胀节
10	condenser	冷凝器
11	hot well pump	凝结水泵
12	circulating water pump	循环水泵
13	heater	加热器
14	boiler feed pump	锅炉给水泵
15	heater drain pump	加热器输水泵
16	conduit tunnel	管线通道
17	discharge treated water tunnel	排出处理水的渠道
18	intake treated water tunnel	进处理水的渠道
19	auxiliary switch room	辅助开关室
20	motor	电动机
21	hydraulic coupling	液压联轴器
22	induced draft fan	引风机
23	forced draft fan	送风机,鼓风机
24	air flow	空气流
25	tempering air duct	调湿风道
26	air inlet to forced draft fan	送风机进风口
27	steam lead	蒸汽导管
28	gas flow	烟气流
29	air preheater	空气预热器
30	economizer	省煤器
31	stack	烟囱
32	superheater	过热器
33	second air duct	二次风管道
34	boiler	锅炉
35	coke oven gas burner	焦炉煤气燃烧器
36	powdered coal burner	煤粉燃烧器
37	blast furnace gas burner	高炉煤气燃烧器
38	ash pit	灰坑
39	coal pulverizer	磨煤机
40	coal scale	煤磅
41	coal bunker	煤斗
42	coal conveyor	输煤机
43	ventilator	通风机
44	blower	鼓风机
45	coke oven gas header	焦炉煤气总管
46	boiler room main floor	锅炉房地面
47	boiler gage board	锅炉仪表盘
48	blast furnace gas regulator	高炉煤气调节器
49	blast furnace gas pipe	高炉煤气管
50	electrical duct run	电线通道
51	treated water to open hearth	至平炉的已处理的水
52	treated service water to coke plant	至焦化厂的已处理的工业用水
53	untreated water to sintering plant	至烧结厂的未处理的水
54	coke oven gas pipe	焦炉煤气管
55	steam pipe	蒸汽管
56	blast furnace gas main	高炉煤气总管

10.8.5 铁塔及电杆 Towers and Poles

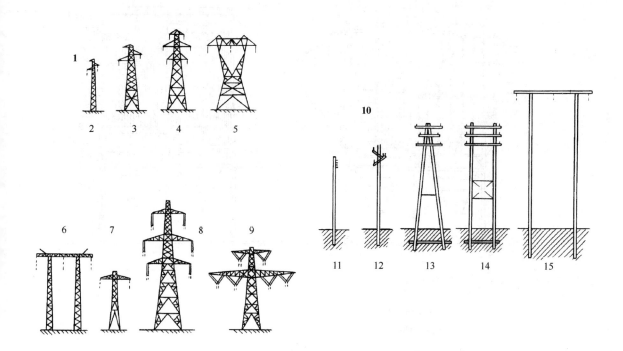

1 lattice-steel tower 钢结构塔
2 narrow-base (single-circuit) 窄基础（单回路）
3 broad-base (single-circuit) 宽基础（单回路）
4，8 double-circuit tower 双回路塔
5 cat's head single-circuit tower 猫头型单回路塔
6 portal single-circuit tower 门形单回路塔架
7 direct-current single-circuit tower 直流单回路塔
9 low-height，double-circuit tower 低高度双回路塔
10 wood pole supports 木电杆
11 3-phase 4-wire distributor 3相4线配电杆
12 3-phase feeder 3相馈线杆
13 A-pole double-circuit feeder A形电杆双回路馈线
14 H-pole double-circuit feeder H形电杆双回路馈线
15 line portal type tower 门形塔架

10.8.6 静电除尘器 Electrostatic Precipitator

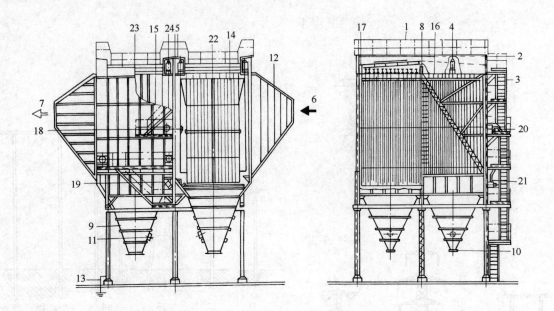

1. rail 栏杆
2. platform 平台
3. stair 梯子
4. supporting insulator 支柱瓷瓶
5. insulator housing 瓷瓶罩
6. dirty gas inlet 含尘气体入口
7. clean gas outlet 清洁气体出口
8. framework 构架
9. dust hopper 集尘斗
10. bottom 底
11. dust-tight cover 尘密盖
12. gas distribution system 气体分配系统
13. ground 接地
14. discharge electrode system 放电极系统
15. collecting electrode system 集尘极系统
16. tubular frames for discharge electrodes 放电极的管架
17. channel frames or collecting electrode 收集极的槽钢架
18. rapping system for discharge electrodes 放电极的捶打系统
19. rapping system for collecting electrodes 集尘极的捶打系统
20. gearmotor drive 齿轮电动机驱动
21. reduction gear 减速齿轮
22. the first section of electric field 电场的第一段
23. the second section of electric field 电场的第二段
24. high voltage DC supply 高压直流电源

10.9 电线、电缆及附件 Wires, Cables and Accessories

10.9.1 电缆、电线结构 Wire and Cables Structure

10.9.1.1 一般结构 General Structure

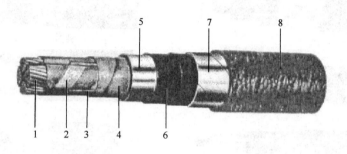

1　conductor　导体
2　conductor screen　导体屏蔽
3　insulation　绝缘材料
4　insulation screen　绝缘屏蔽
5　filler　填料
6　wrapping tape　包带
7　steel tape/wire armored　钢带/丝铠装
8　outer sheath　外护套

10.9.1.2 防火/阻燃电缆结构 Fire-Proof/Flame Retardant Cable Structure

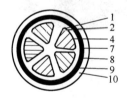 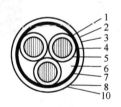 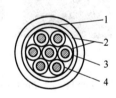

1　conductor　导体
2, 7　common or highly fire-proof layer　普通或高防火层
3　conductor screen　导体屏蔽
4　insulation　绝缘材料
5　insulation screen　绝缘屏蔽
6　filler　填料
8　wrapping tape　包带
9　steel tape/wire armored　钢带/丝铠装
10　outer sheath　外护套

10.9.1.3 铝绞线/钢芯铝绞线结构 Al Twisted Wire/ ACSR Wire

   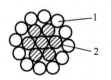

1　Al conductor　铝导体
2　steel conductor　钢导体

10.9.2 电缆、电线产品 Wire and Cables Products

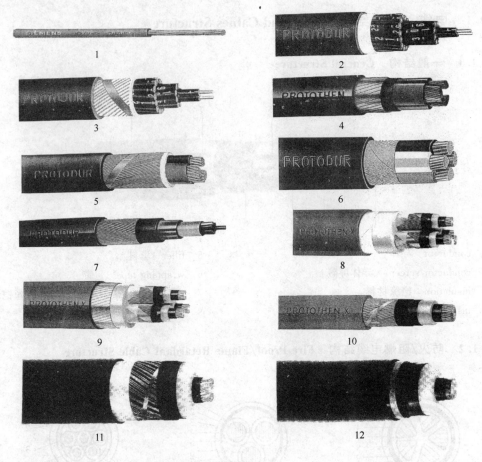

1 PVC insulated copper wire 聚氯乙烯绝缘铜导线
2 control cables with numbered cores 带号芯线的控制电缆
3 control cables with numbered cores and concentric conductor 带号芯线的控制电缆，同心导线
4 0.6/1kV power cables with PVC insulated and sheathed copper conductor 0.6/1kV 聚氯乙烯绝缘及护套铜芯动力电缆
5 power cables with PVC insulated and sheathed, concentric waveform conductor 聚氯乙烯绝缘及护套同心波纹导体动力电缆
6 power cable with XLPE insulation and PVC sheath, steel-wire armoredcopper conductor 交联聚乙烯绝缘，聚氯乙烯护套钢丝铠装动力电缆
7 control cable with PVC insulation and sheath, lead covered copper conductor 聚氯乙烯绝缘及护套，铅包铜芯控制电缆
8 power cables with copper wire screen on each core 每芯带金属屏蔽的动力电缆
9 power cables with copper wire screen and steel tape armoring 带金属屏蔽的钢带铠装动力电缆
10 single core cable with screen 屏蔽单芯电缆
11 64/110kV single core XLPE insulated cable with laminated sheath 64/110kV 交联聚乙烯绝缘综合护套电力电缆
12 64/110kV single core XLPE insulated cable with lead sheath 64/110kV 交联聚乙烯绝缘铅套电力电缆

10.9.3 电缆附件 Accessories for Cables

10.9.3.1 45kV 以下中、低压电缆附件 Up to 45kV Accessories for Low and Medium Voltage Cables

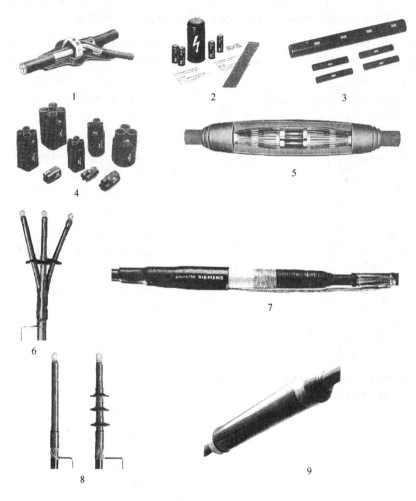

1 branch joint for multi-core cable　多芯电缆分支连接件
2 end joint for multi-core cable　多芯电缆终端头
3 straight joint for multi-core cable　多芯电缆直通连接件
4 termination for multi-core cable　多芯电缆终端头
5 straight joint for 3-core and multi-core cable　3芯以上电缆直通连接件
6 termination for 3-core cable　3芯电缆接线端子
7 straight joint for 1-core cable　单芯电缆直通连接件
8 termination　电缆头
9 straight joint for 1-and 3-core cable　单芯和三芯电缆直通连接件

10.9.3.2 高压电缆附件 Accessories for High Voltage Cables

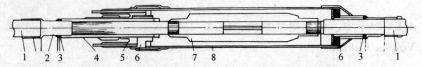

straight joint for high voltage cables 高压电缆中间接头

1. outer sheath 护套
2. metallic sheath 金属护套
3. lead covered 铅封
4. thermal shrinkable conduit 热缩管
5. grounding wire 接地线
6. rubber prefabricated 橡胶预制件
7. straight joint 中间接头
8. enclosure 护套

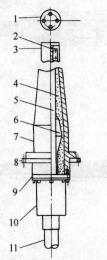

GIS end joint for 110kV cables
110kV 电缆 GIS 终端

1. connection interface between terminal and GIS 终端与 GIS 结合面
2. conductive hardware fittings 导电金具
3. screen sheath 屏蔽层
4. insulation oil 绝缘油
5. cable insulating core 电缆绝缘线芯
6. stress prick 应力锥
7. epoxy conduit 环氧套管
8. binding ring 卡环
9. sealed bottom plate 密封底盘
10. end conduit 尾管
11. XLPE cable 交联聚乙烯电缆

1. leading pole 出线杆
2. porcelain jacket 瓷套
3. insulation oil 绝缘油
4. stress prick 应力锥
5. bottom plate 底板
6. insulators 绝缘子
7. end conduit 尾管

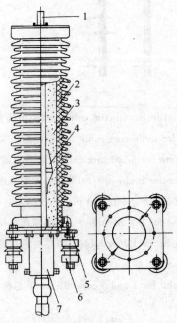

outdoor type sealing end for high voltage cables
高压电缆户外终端

10.9.3.3 电缆梯架 Cable Ladder

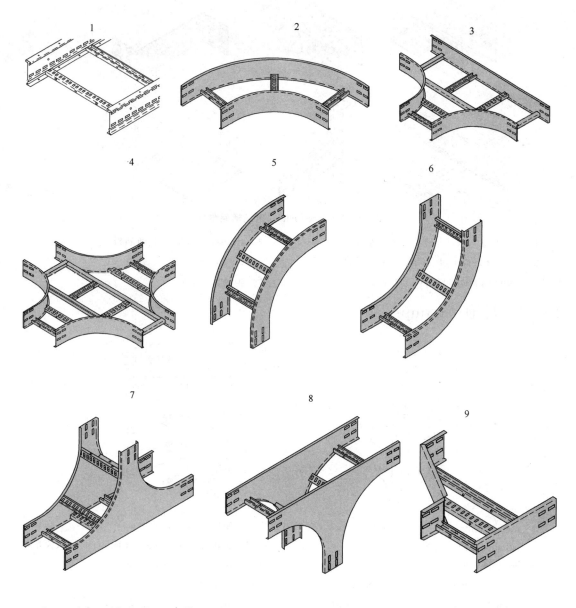

1　straight cable ladder　直通
2　horizontal bend　水平弯通
3　horizontal tee　水平三通
4　horizontal cross　水平四通
5　vertical outside　垂直凸弯通
6　vertical inside　垂直凹弯通
7　vertical tee up　垂直上三通
8　vertical tee down　垂直下三通
9　reducer　变径

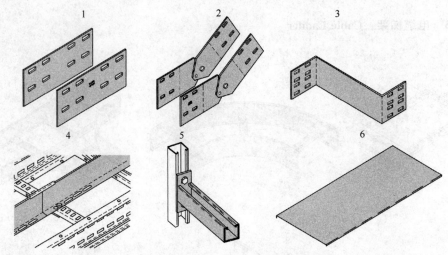

cable ladder accessories 电缆梯架附件

1 standard splice plate 直通连接片
2 vertical adjustable splice plate 垂直调角片
3 reducing coupler plate 调宽片
4 straight divider 直通隔板
5 cantilever bracket 托臂
6 covers 盖板

10.10 灯泡 Lamps

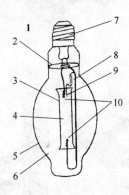
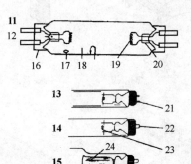

1 mercury-arc lamp 水银荧光灯
2 polished disc 抛光盘
3 arc tube 电弧管
4 argon gas and mercury 氩气和汞
5 outer bulb (glass bulb) 外壳玻璃泡
6 phosphor coating 荧光涂层
7 screw base 螺口灯头
8 starting resistor 启动电阻
9 arcing electrode 引弧电极
10 operating electrode 工作电极
11 fluorescent lamps 荧光灯
12 hot-cathode fluorescent lamp 热阴极荧光灯
13 preheat and rapid start fluorescent lamp
   预热快速启动荧光灯
14 instant start and slimline fluorescent lamp
   瞬时启动细荧光灯
15 cold cathode fluorescent lamp 冷阴极荧光灯
16 bi-pin cap 双脚灯头
17 mercury 水银
18 argon mercury vapour 氩汞蒸气
19 cathode 阴极
20 pinch 夹座
21 bi-pin terminal 双脚插柱
22 single-pin terminal 单脚插柱
23 hot cathode 热阴极
24 large area cylindrical cathode 大面积圆柱形阴极

10.11 电伴热系统 Electrical Heat Tracing System

10.11.1 电伴热产品 Electrical Heat Tracing Products

10.11.1.1 自调控电伴热带 Self-Regulating Heat Tracing Cable

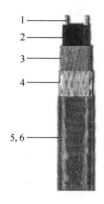

1　nickel-plated copper bus wire　镀镍铜母线
2　radiation cross-linked semi-conductive heating matrix　辐射交联半导体发热矩阵
3　radiation cross-linked dielectric insualtion　辐射交联绝缘材料
4　tinned copper braid　镀锡铜线编织层
5　polylefin over-jacket over tinned plated copper braid　镀锡铜线编织层外的聚烯烃护套
6　fluoropolymer over-jacket over nickel plated copper braid　镀镍铜线编织层外的含氟聚合物外护套

10.11.1.2 限功率伴热带 Power Limiting Heat Tracing Cable

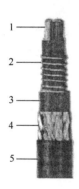

1　nickel-plated copper bus wire　镀镍铜母线
2　composite metal alloy/fiber　合成金属合金与纤维的复合结构
3　fluoropolymer dielectric insulation　含氟聚合物绝缘
4　nickel-plated copper braid　镀镍铜线编织层
5　fluoropolymer over-jacket over nickel-plated copper braid　镀镍铜线编织层外的含氟聚合物绝缘护套

10.11.1.3 并联恒功率伴热带 Parallel Constant Watt Cable

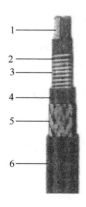

1　nickel-plated copper bus wire　镀镍铜母线
2　nichrome heating element　镍铬合金加热元件
3　fiber glass overlay　玻璃纤维衬层
4　fluoropolymer dielectric insulation　含氟聚合物绝缘材料
5　nickel-plated copper braid　镀镍铜丝编织层
6　fluoropolymer over-jacket over tinned copper　在镀镍铜丝编织层外的含氟聚合物外护套

10.11.1.4 串联恒功率伴热带 Series Constant Watt Cable

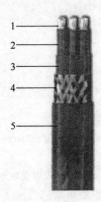

1　nickel-plated copper bus wire　镀镍铜母线
2　fluoropolymer dielectric insulation　含氟聚合物绝缘材料
3　fluoropolymer pairing jacket　含氟聚合物外套
4　nickel-plated copper braid　镀镍铜丝编织层
5　fluoropolymer over-jacket over nickel-plated copper braid　在镀镍铜丝编织层外的含氟聚合物外护套

10.11.1.5 矿物绝缘伴热带 Mineral Insulated Cable

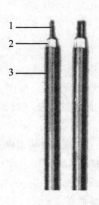

1　solid alloy or copper conductor　固体合金或铜导线
2　compacted magnesium oxide insulation　紧密氧化镁绝缘层
3　alloy 825 sheath　825合金护套

10.11.1.6 罐和容器的加热板 Tank and Vessel Heating Panel

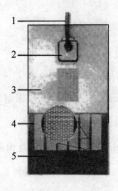

1　high temperature lead wire　高温导线
2　over-temperature safety cutout switch　超温安全断路开关
3　protective metal jacket　金属保护板
4　parallel circuit high temperature alloy heating element　并联型高温合金加热元件
5　heat laminated，high temperature silicon rubber insulation　热挤压耐高温硅橡胶绝缘层

10.11.1.7 仓斗伴热板 Hopper Heating Module

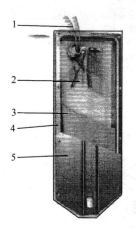

1 fluoropolymer insulated high temperature lead wires 含氟聚合物绝缘耐高温导线
2 parallel circuit high temperature alloy heating elements 并联型耐高温合金加热元件
3 temperature rated insulation 额定温度绝缘层
4 20-gauge aluminized steel protected enclosure 20号镀铝钢外护板
5 20-gauge aluminized steel protected cover with reinforcing attachment channel 带有紧固槽的20号镀铝钢外护板

10.11.2 典型电伴热系统 Typical Heat Tracing System

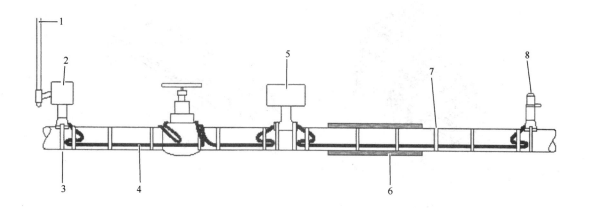

1 to ground-fault protected power 接至带接地故障保护的电源
2 power connection box 电源接线盒
3 pipe strap 管线扎带
4 heating cable 伴热带
5 tee 三通
6 thermal insulation 保温绝缘
7 glass tape 玻璃胶带
8 end seal 终端封头

10.11.3 电伴热控制和监测 Heat Tracing Control and Monitoring

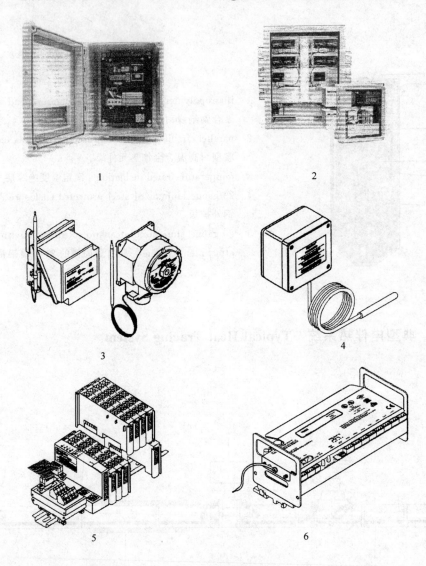

1　single-circuit controller　单回路控制器
2　multi-circuit controller　多回路控制器
3　mechanical thermostats　机械式温控器
4　electronic thermostats　电子式温控器
5　heat tracing remote control module　电伴热远控模块
6　heat tracing remote monitoring module　电伴热远程监测模块

10.12 阴极保护 Cathodic Protection

10.12.1 阴极保护测试站和接线箱 Cathodic Protection Test Stations and Junction Box

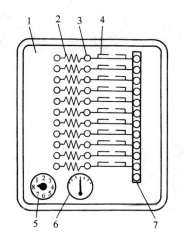
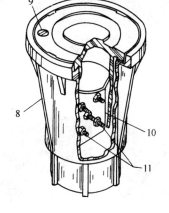

1　junction box　接线箱
2　resistor　电阻
3　terminal lugs　端子头
4　shunt　旁路
5　selector switch　选择开关
6　ammeter　电流表
7　distribution bar　分线排
8　cathodic protection test station　阴极保护测试站
9　locking lid　盖锁
10　terminal board　端子板
11　terminals　端子

10.12.2 长效参比电极 Permanent Reference Cell

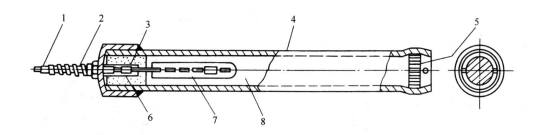

1　lead wire　导线
2　strain relief　应力消除
3　SPL sealant　SPL 密封剂
4　PVC pipe　PVC 管
5　plug　堵头
6　crimped connection　波纹连接
7　silver reference element　银参比元件
8　AgCl mixture　氯化银混合物

10.12.3 阳极 Anodes

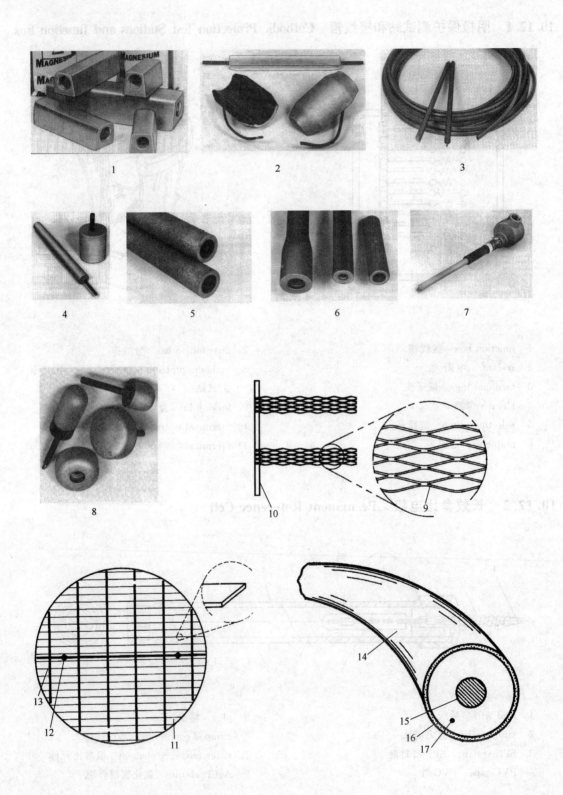

1　cast anodes　块状阳极
2　bracelet nodes　环状阳极
3　ribbon anode　带状阳极
4　cylindrical anode　柱状阳极
5　tubular cast iron anodes　管状铁阳板
6　solid rod cast iron anodes　实心棒铁阳极
7　button and bullet cast iron anodes　纽扣和子弹铁阳极
8　mixed metal oxide probe anode　混合金属氧化物探针阳极
9　mixed metal oxide expanded mesh anodes　混合金属氧化物网状阳极
10　bare titanium cross piece　裸钛连接片
11　mixed metal oxide ribbon anodes　混合金属氧化物带状阳极
12　power feed　供电电源
13　conductor bar　导电板
14　mixed metal oxide wire anodes　混合金属氧化物线状阳极
15　copper core　铜芯
16　mixed metal oxide coating　混合金属氧化物护套
17　titanium substrate　钛衬底

10.12.4　填料　Backfill

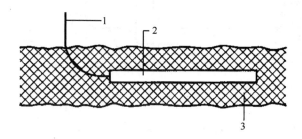

1　coke backfill　焦炭填料
2　anode　阳极
3　lead wire　导线

10.12.5　整流器　Rectifier

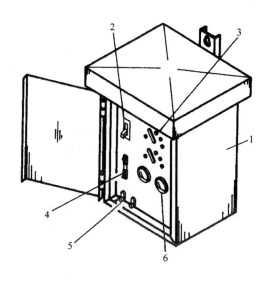

1　rectifiers　整流器
2　circuit breaker　断路器
3　adjustment taps　调整旋钮
4　shunt　旁路
5　output terminals　输出端子
6　output meters　输出电表

10.12.6 绝缘材料 Insulation Material

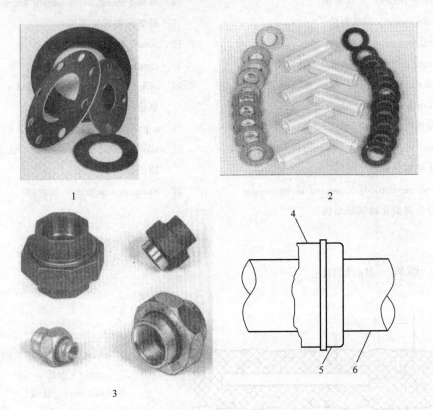

1   flange insulation gaskets   法兰绝缘垫片
2   pipe insulation sleeves and washers   管道绝缘套管和垫片
3   dielectric pipe unions   绝缘管接头
4   casing   外壳
5   end seal   端密封
6   carried pipe   管道

10.12.7 接地电池 Grounding Cells

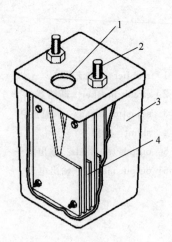

1   fill hole   填充孔
2   cell terminals   电池端子
3   potassium hydroxide solution   氢氧化钾溶液
4   stainless steel plates   不锈钢板

10.12.8 阴极保护系统 Cathodic Protection Systems

10.12.8.1 深阳极系统 Deep Anode System

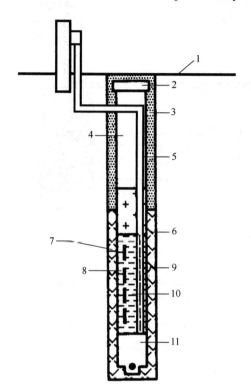

1　earth surface　地面
2　sealed pressure cap　密封压盖
3　coarse bentonite rocks　粗斑脱土石
4　solid plastic casing　塑料外壳
5　sealed conduit　密封管
6　conductive grout　导电水泥浆
7　perforated casing　冲孔外壳
8　anodes　阳极
9　slotted pipe vent　排气小孔
10　carbon backfill　碳回填料
11　steel lead with flow control　流量控制钢芯

10.12.8.2 水罐系统 Water Storage Tank System

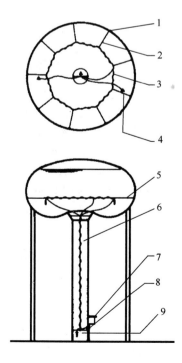

1　steel anchors welded to side wall　焊在罐壁的铁锚
2　polyster rope supports　聚酯绳支承
3，6　precious metal anode　稀有金属阳极
4，9　reference cell　参比电极
5　submerged anode support system　水中电极支承系统
7　rectifier　整流器
8　pressure fitting　压力连接件

10.12.8.3 地上储罐罐底保护 Protection for Above Ground Storage Tank Bottoms

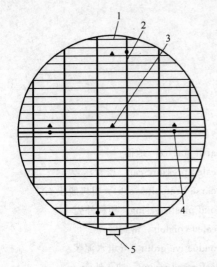
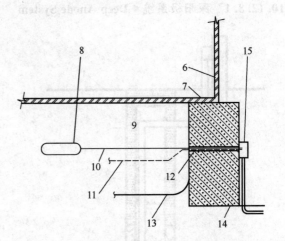

1 titanium anode ribbon　钛带状阳极
2 conductor bar　导电板
3,8 reference electrodes　参比电极
4 power feed point　供电点
5,15 junction box　接线箱
6 tank wall　罐壁
7 tank bottom　罐底
9 compacted sand backfill　压实的沙土回填
10 lead wire　导线
11 titanium anode ribbon　钛阳极带
12 steel conduit　钢管
13 secondary containment　二次密封
14 ring wall　圈梁

10.12.8.4 地下储罐保护 Protection for Underground Storage Tanks

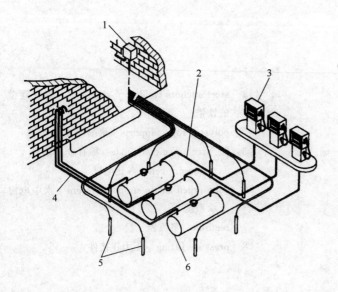

1 rectifier　整流器
2 product piping　工艺配管
3 dispenser　配料器
4 vents　排气管
5 anodes　阳极
6 underground tanks　地下储罐

684

10.13 接地防雷系统 Grounding and Lightning System

10.13.1 接地产品 Grounding Products

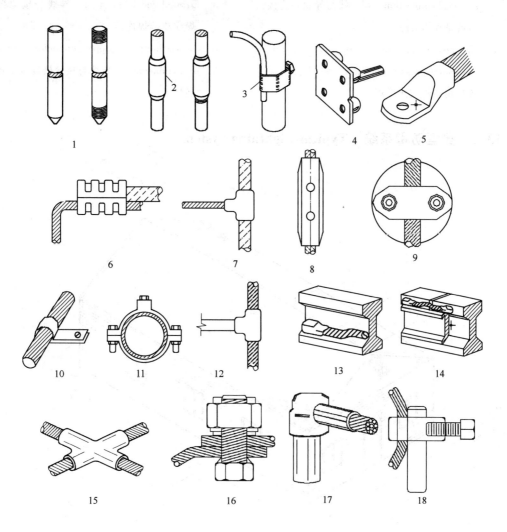

1　grounding rods　接地极
2　grounding rods splice　接地极连接件
3　grounding rod clamp　接地极夹具
4　ground plates　接地板
5　compression lug for conductor　连接导线用压接线鼻子
6　compression tap for conductor　导线压接接头
7　thermoweld or cadweld copper conductor T connection　导线 T 形热焊或热熔焊连接
8　compression splice for conductors　导线压接
9　bronze supporting clamp for conductor　导线用青铜吊环
10　galvanized fastening clip for conductor　导线用镀锌紧固卡
11　galvanized grounding clip for steel conduit　钢管用镀锌接地卡
12　thermoweld or cadweld conductor to galvanized steel flat connection　导线与镀锌钢平面连接用热焊或热熔焊
13　thermoweld or cadweld copper conductor to

	rail foot connection 铜导线与导轨底连接用热焊或热熔焊	16	bronze bolted connection for T copper connection 导线 T 形青铜螺栓连接
14	thermoweld or cadweld copper conductor to rail head connection 铜导线与导轨顶连接用热焊或热熔焊	17	thermoweld or cadweld copper conductor to ground rod connection 导线与接地极热焊或热熔焊连接
15	thermoweld or cadweld copper conductors horizontal cross-connection 铜导线四通连接用热焊或热熔焊	18	brass clamp for copper conductor to ground rod connection 导线与接地极连接用黄铜夹具

10.13.2 典型防雷系统 Typical Lightning System

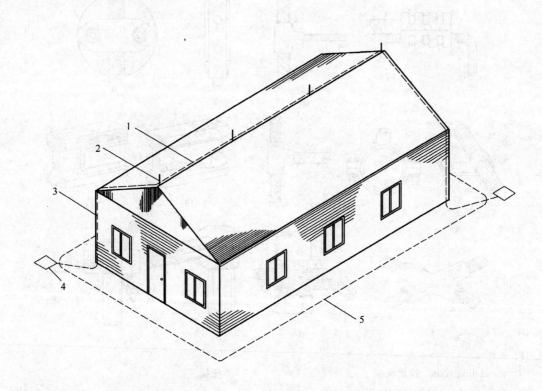

1 lightning protected network 避雷网
2 air terminal 接闪器
3 down lead 引下线
4 grounding rods 接地极
5 grounding network 接地网

10.14 监测控制和数据采集 Supervise Control and Data Acquiring (SCADA)

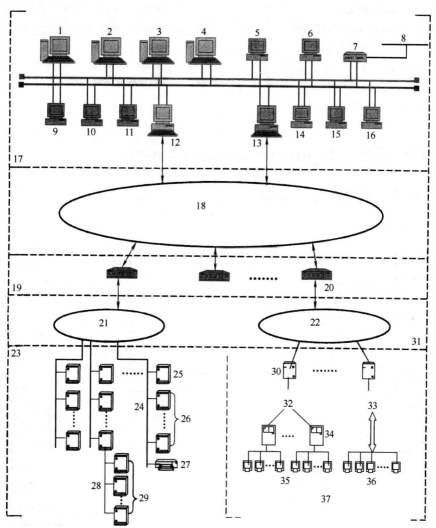

1	main server 主服务器	11	application workstation 应用工作站
2	back-up server 备用服务器	12	gate computer A 前置机 A
3	GIS server 地理信息系统服务器	13	gate computer B 前置机 B
4	WEB server 网络服务器	14	fault analysis workstation 事故分析工作站
5	quantity of electrical recording workstation 电量集抄工作站	15	maintenance workstation 维护工作站
6	quantity of electrical administration workstation 电量管理工作站	16	reading workstation 浏览工作站
		17	main station system 主站系统
7	router 路由器	18	communication 通信
8	management information system network 管理信息系统网	19	sub-station system 子站系统
		20	communication processor 通信系统
9	dispatcher workstation 调度员工作站	21	communication network in station 站内通信网
10	GIS workstation 地理信息系统工作站		

22　exchanging network　交换网
23　equipment　设备
24　fiber　光纤
25　terminal unit　接线单元
26　feeder terminal unit　馈线终端
27　data acquirer　数据采集器
28　communication cable　通信电缆
29　transformer terminal unit　变压器终端
30　concentrator　集中器
31　communication system　通信系统
32　low voltage carrier wave　低压配电载波
33　low voltage carrier wave　低压配电载波
34　remote terminal unit　采集终端
35　impulse watt-hour meter　脉冲电能表
36　carrier wave watt-hour meter　载波电能表
37　remote automatic recording system　远程自动抄表系统

10.15　通信系统　Telecommunication System

10.15.1　用户电话交换机　PABX

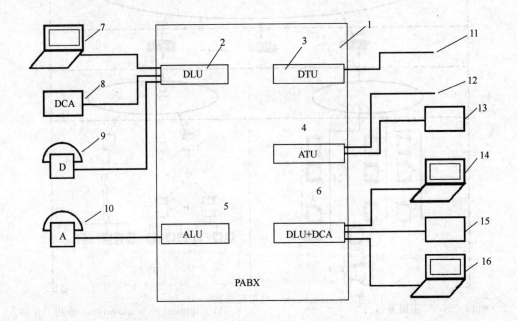

1　PABX（private automatic branch exchanger）用户电话交换机
2　digital line unit　数字用户单元
3　digital trunk unit　数字中继单元
4　analogue trunk unit　模拟中继单元
5　analogue line unit　模拟用户单元
6　digital line unit and digital communication adapter　数字用户单元及数字通信适配器
7　attendant console　话务台
8　digital communication adapter　数字通信适配器
9　digital telephone set　数字电话机
10　analogue telephone set　模拟电话机
11　digital trunk　数字中继线
12　analogue trunk line　模拟中继线
13　voice mail box　语音信箱
14　accounting terminal　计费终端
15　modem　调制解调器
16　maintenance terminal　维护终端

10.15.2 结构化布线系统 Structured Cabling System

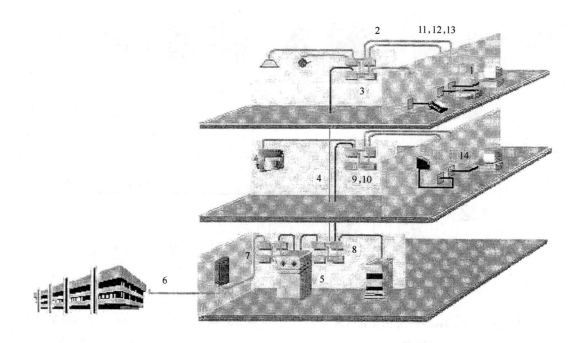

1　work area subsystem　工作区子系统
2　horizontal subsystem　水平布线子系统
3　administration subsystem　管理子系统
4　backbone subsystem　主干子系统
5　equipment subsystem　设备室子系统
6　campus subsystem　建筑群子系统
7　campus distributor　建筑群配线架
8　MDF（main distribution frame）　主配线架
9　intermediate distribution frame　分配线架
10　floor distributor　楼层配线架
11　STP（shielded twisted pair）　屏蔽双绞电缆
12　UTP（unshielded twisted pair）　非屏蔽双绞电缆
13　FTP（foil twisted pair）　金属箔双绞电缆
14　information outlet　信息插座

10.15.3 扩音对讲系统　Paging System

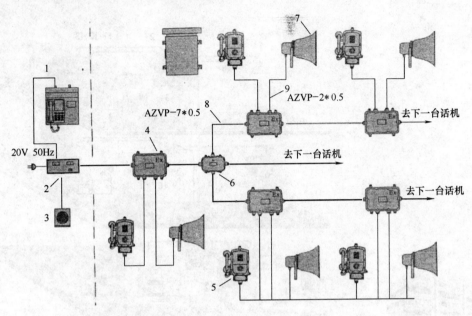

1　indoor desktop station　室内桌式话站
2　indoor amplifier　室内放大器
3　indoor speaker　室内音箱
4　explosion proof amplifier　防爆扬声器放大器
5　explosion proof wall station　防爆墙式话站
6　explosion proof junction box　防爆接线盒
7　explosion proof horn　防爆号筒式扬声器
8　system cable　系统电缆
9　speaker cable　扬声器电缆

10.15.4 有线电视系统　Cable TV System

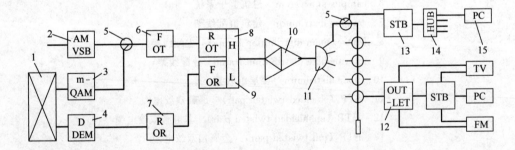

1　exchanger　交换机
2　VSB AM broadcast　残留侧边带调幅广播
3　m-QAM digital broadcast　多值正交调幅数字广播
4　digital demodulation　数字解调器
5　directional coupler　定向耦合器
6　forward optic transmitter　前向光发射器
7　reward optic receiver　后向光接收器
8　high pass filter　高通滤波器
9　low pass filter　低通滤波器
10　amplifier　放大器
11　distributor　分配器
12　outlet　输出口
13　STB　共用机顶盒
14　HUB　集线器
15　personal computer　个人计算机

11 实验室 Laboratory

11.1 实验室常用仪器 General Apparatus and Instruments

11.1.1 玻璃器皿 Glassware

1 beaker 烧杯
2 three-necked flask 三颈瓶
3 washing bottle 洗瓶
4 funnel 漏斗
5 volurnetric flask 容量瓶
6 round-bottomed flask 圆底烧瓶
7 cylinder 量筒
8 separatory funnel 分液漏斗
9 conical flask 锥形烧瓶
10 tube 试管

11.1.2 实验室主要分析仪器 General Laboratory Equipment

1 Abbe refractometer 阿贝折光仪
2 analytical balance 分析天平
3 atomic absorption spectrometer（AAS） 原子吸收光谱仪
4 atomic emission spectrometer（AES） 原子发射光谱仪
5 atomic fluorescence spectrometer（AFS） 原子荧光光谱仪
6 automatic density meter 自动密度计
7 automatic distillation 全自动蒸馏仪
8 automatic potentiometric titrator 自动电位滴定仪
9 automatic vacuum distillation 全自动减压蒸馏仪
10 balance 天平
11 biological oxygen demand analyzer（BOD） BOD 分析仪
12 cetane number test machine 十六烷值机
13 chemical oxygen demand analyzer（COD） COD 分析仪
14 closed cup flash point tester 闭口闪点测定仪
15 conductivity meter 电导率仪
16 cold atomic absorption spectrometer 冷原子吸收光谱仪
17 cold atomic emission spectrometer 冷原子发射光谱仪
18 copper strip corrosion tester 铜片腐蚀测定仪
19 dew point meter 露点仪
20 digital thermostatic waterbath 数显恒温水浴
21 dissolved oxygen tester 溶解氧测定仪
22 drying oven 烘箱
23 electronic balance 电子天平
24 element analyzer 元素分析仪
25 existent gum tester 实际胶质测定仪
26 FT-IR spectrometer（FTIRS） 傅里叶变换红外光谱仪
27 FT-Raman spectrometer 傅里叶变换拉曼光谱仪
28 freezing point tester 冰点测定仪
29 gas chromatograph（GC） 气相色谱仪
30 GC-MS 气相色谱质谱联用仪

31　gel permeation chromatograph（GPC）　凝胶渗透色谱仪
32　heating plate　电热板
33　high pressure/performance liquid chromatograph（HPLC）　高效液相色谱仪
34　HPLC-MS　液相色谱质谱联用仪
35　inductive coupled plasma emission spectrometer（ICP）　电感耦合等离子发射光谱仪
36　infrared spectrophotometer　红外分光光度计
37　ion chromatograph（IC）　离子色谱仪
38　ion meter　离子计
39　Karl-Fischer titrator　卡尔费休水分分析仪
40　kinematic viscometer　运动黏度计
41　laser particle size analyzer　激光粒度仪
42　magnetic stirrer with heater　带磁力搅拌加热器
43　muffle furnace　马弗炉
44　nuclear magnetic resonance spectrometer（NMR）　核磁共振光谱仪
45　octane number test machine　辛烷值机
46　oil bath　油浴
47　open cup flash point tester　开口闪点测定仪
48　pH meter　pH 计
49　platinum cobalt colorimeter　铂钴比色计
50　portable conductivity meter　便携电导仪
51　portable pH meter　便携 pH 计
52　pour point tester　倾点测定仪
53　Saybolt colorimeter　赛波特比色计
54　solidification point tester　凝点测定仪
55　supercritical fluid chromatograph（SFC）　超临界流体色谱仪
56　thermostatic waterbath　恒温水浴
57　total chlorine analyzer（microcoulometry）　总氯分析仪（微库仑法）
58　total nitrogen analyzer（chemiluminescence spectrometry）　总氮分析仪（化学发光法）
59　total organic carbon analyzer（TOC）　TOC 分析仪
60　total sulfur analyzer（ultraviolet fluorescence spectrometry）　总硫分析仪（紫外荧光光谱法）
61　true boiling point distillation　实沸点蒸馏仪
62　turbidimeter　浊度计
63　ultraviolet-visible spectrometer（UV-VIS）　紫外可见光谱仪
64　UP water purification system　超纯水仪
65　vacuum drying oven　真空烘箱
66　vapour pressure tester—Reid method　雷德蒸汽压测定仪
67　visible spectrophotometer　分光光度计
68　water bath　水浴
69　X-ray fluorescence spectrometer（XRF）　X 射线荧光光谱仪

11.2　化学试剂和药品　Chemicals and Reagents

11.2.1　无机试剂　Inorganic Reagent

1　amalgam zinc　锌汞齐
2　ammonium bromide　溴化铵
3　antimony oxychloride　氧氯化锑
4　anhydrous calcium chromate　无水铬酸钙
5　arsenic trioxide　三氧化二砷
6　barium carbonate　碳酸钡
7　beryllium sulfate　硫酸铍
8　bismuth chloride　氯化铋
9　boric acid　硼酸
10　bromine water　溴水
11　cadmium　镉
12　caesium hydroxide　氢氧化铯
13　chromic anhydride　铬酐
14　cobaltic oxide　氧化高钴
15　cupric dichromate　重铬酸铜
16　ferric oxalate　草酸铁
17　fluoboric acid　氟硼酸
18　gallium nitrate　硝酸镓
19　hydrochloric acid　盐酸
20　hydrogen peroxide　过氧化氢
21　indium chloride　氯化铟
22　iodine monobromide　一溴化碘
23　lead sulfate　硫酸铅
24　lithium fluoride　氟化锂
25　magnesium perchlorate　高氯酸镁
26　manganous carbonate　碳酸锰
27　nickel chloride　氯化镍
28　nitric acid　硝酸

29	perchloric acid 高氯酸	36	thallous carbonate 碳酸亚铊
30	phosphorous acid 亚磷酸	37	thorium nitrate 硝酸钍
31	potassium carbonate 碳酸钾	38	tungstic acid 钨酸
32	silicon dioxide 二氧化硅	39	vanadium tetrachloride 四氯化钒
33	sodium hydroxide 氢氧化钠	40	yttrium oxide 氧化钇
34	sodium metaphosphate 偏重亚硝酸钠	41	zirconium phosphate 磷酸锆
35	sodium tetrahydrotellurate 碲酸四氢钠		

11.2.2 有机试剂 Organic Reagent

1	acenaphthenequinone 苊醌	40	diethyl phosphate 磷酸二乙酯
2	acetaldehyde 乙醛	41	dimethyl sulfoxide 二甲基亚砜
3	acetamide 乙酰胺	42	ethyl alcohol 乙醇
4	acetic anhydride 乙酸酐	43	ethyl acetate 乙酸乙酯
5	acetone 丙酮	44	ethylenediamine 乙二胺
6	acetylsalicylic acid 乙酰水杨酸	45	ethyl ether 乙醚
7	acrolein 丙烯醛	46	ethyl salicylate 水杨酸乙酯
8	allylurea 丙烯脲	47	formaldehyde 甲醛
9	p-aminobenzoic acid 对氨基苯甲酸	48	furan 呋喃
10	aniline-p-thiocyanate 对硫氰基苯胺	49	glycine 甘氨酸
11	anisole 茴香醚	50	n-hexane 己烷
12	anthraquinone 蒽醌	51	hydroxylamine hydro chloride 羟胺盐酸盐
13	anthrone 蒽酮	52	indole 吲哚
14	azobenzene 偶氮苯	53	iodoacetic acid 碘乙酸
15	barbituric acid 巴比妥酸	54	laurylamine 月桂胺
16	benzamide 苯甲酰胺	55	malononetrile 丙二腈
17	benzene 苯	56	methyl bromoacetate 溴乙酸甲酯
18	benzenesulfonic acid 苯磺酸	57	morpholine 吗啡啉
19	benzyl alcohol 苄醇	58	naphthalene 萘
20	benzyl cellusolve 苄基溶纤剂	59	naphthol 萘酚
21	m-bromoaniline 间溴苯胺	60	nitrobenzene 硝基苯
22	bromobenzene 溴苯	61	phenanthrene 菲
23	o-bromobenzoic acid 邻溴苯甲酸	62	phenetole 苯乙醚
24	n-butanethiol 丁硫醇	63	phenol 苯酚
25	iso-butyl acetate 乙酸异丁酯	64	potassium citrate 柠檬酸钾
26	carbon tetrachloride 四氯化碳	65	pyrene 芘
27	catechol 邻苯二酚	66	pyridine 吡啶
28	chlorobenzene 氯苯	67	pyrrole 吡咯
29	chlorobutane 氯丁烷	68	ouinoline 喹啉
30	chloroform 氯仿	69	salicylaldehyde 水杨醛
31	chloropicrin 三氯硝基甲烷	70	sorbic acid 山梨酸
32	cinnamaldehyde 肉桂醛	71	sodium succinate 琥珀酸钠
33	creatinephosphoric acid 肌磷酸	72	tetrahydrofuran 四氢呋喃
34	cyanoacetamide 氰乙酰胺	73	thiophene 噻吩
35	cyanuric acid 三聚氰酸	74	thiophenol 苯硫酚
36	cyclohexane 环己烷	75	toluene 甲苯
37	decahydronaphthalene 十氢萘	76	triethylamine 三乙胺
38	diazoaminobenzene 重氮氨基苯	77	urea 脲
39	dichloromethane 二氯甲烷	78	vanillic acid 香草酸

11.2.3 指示剂 Indicator

1. aluminon 铝试剂
2. bromophenol blue 溴酚蓝
3. bromophenol red 溴酚红
4. calmagite 钙镁试剂
5. congo red 刚果红
6. crystal violet 结晶紫
7. eriochrome black T 铬黑 T
8. fluorescein 荧光素
9. indigo 靛蓝
10. methyl orange 甲基橙
11. methyl red 甲基红
12. naphthyl red 萘红
13. neutral red 中性红
14. phenolphthalein 酚酞
15. thionine 硫堇
16. xylenol orange 二甲酚橙

11.3 实验室常用词汇 Laboratory Vocabulary

1. AAS (atomic absorption spectrometry) 原子吸收光谱法
2. absolute deviation 绝对偏差
3. absolute error 绝对误差
4. absolute precision 绝对精度
5. absolute refractive index 绝对折射率
6. absorptiometric method 吸光测定法
7. absorption band 吸收谱带
8. absorption cell 吸收池
9. absorption curve 吸收曲线
10. absorption filter 吸收滤光片
11. absorption flame photometry 火焰吸收光度法
12. absorption maximum 最大吸收
13. abundance 丰度
14. accumulated error 累积误差
15. accuracy 准确度
16. achromatic 消色的，消色差的
17. analytical error 分析误差
18. back titration 反滴定
19. boiling point 沸点
20. calcinate 焙烧
21. catalyzer 催化剂
22. centrifuge 离心机
23. check analysis 对照分析
24. chelate 螯合物
25. chemical pure 化学纯
26. color change interval 变色范围
27. colorimetric analysis 比色分析
28. conductometry 电导分析法
29. constant 恒量
30. constituent 成分
31. correction factor 校正因子
32. coulometry 库仑法
33. decomposition 分解
34. degradation 降解
35. deliquescence 潮解
36. demasking 解蔽
37. distillation 蒸馏
38. electrode 电极
39. electrolysis 电解
40. electron orbit 电子轨道
41. electron cloud 电子云
42. electronegativity 电负性
43. electrophoresis 电泳
44. equivalent 当量
45. explosion limit 爆炸极限
46. extraction 萃取
47. fractional distillation 分馏
48. nonaqueous solvent 非水溶剂
49. number of theoretical plates 理论塔板数
50. particle diameter 粒径
51. passivation 钝化
52. peak area 峰面积
53. peak height 峰高
54. peak width 峰宽
55. percentage error 百分误差
56. plate column 板式塔
57. polarity 极性
58. polarographic analysis 极谱分析法
59. qualitative analysis 定性分析
60. quantification 定量
61. radioactivity 放射性
62. reading accuracy 读出精度
63. recrystallization 重结晶
64. reference peak location 参比峰位
65. reflectivity 反射率
66. resolution 分辨率
67. retention constant 保留常数
68. retention time 保留时间
69. retention volume 保留体积
70. sagging 沉降

11.4 主要分析仪器示例图 Legend for Main Laboratory Equipment

11.4.1 分析天平 Analytical Balance

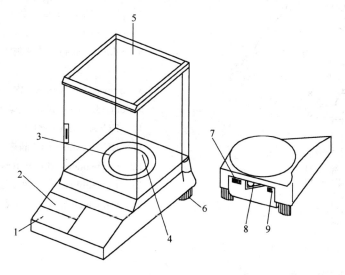

1 keyboard 操作键
2 display 显示屏
3 anti-wind ring 防风圈
4 scale pan 称盘
5 anti-wind enclosure 防风罩
6 leveller 水平调节脚
7 RS-232A cable RS-232A 接口
8 level bubble 水平泡
9 switch 插座

11.4.2 电极 Electrode

11.4.2.1 玻璃电极 Glass Electrode

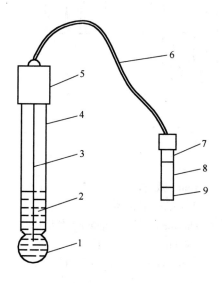

1 pH sensitive ball membrane pH 敏感球膜
2 buffer solution 缓冲溶液
3 Ag-AgCl internal reference electrode Ag-AgCl 内参比电极
4 high resistance glass tube 高阻玻璃管
5 insulated fixed port 绝缘固定端
6 signal cable (with shielding layer) 信号电缆（带屏蔽层）
7 end of output switch 输出插头尾部
8 high insulator 高绝缘物
9 top of output switch 输出插头顶端

11.4.2.2 甘汞电极 Calomel Electrode

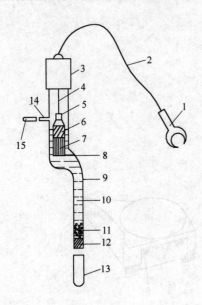

1   metal welding   金属焊片
2   lead circuit   电路引线
3   insulated fixed port   绝缘固定端
4   platinum filament   铂丝
5   pure mercury   纯汞
6   calomel mash   甘汞糊状物
7   cotton   棉花
8   internal glass tube   内玻璃管
9   outside glass tube   外玻璃管
10   KCl saturated solution   饱和 KCl 溶液
11   KCl crystal   KCl 晶体
12   multi-micro core material   多微孔物质
13，15   rubber cap   橡皮帽
14   top connection   上端接口

11.4.3 色谱仪类 Chromatography

11.4.3.1 气相色谱法原理图 Gas Chromatograph Schematic Diagram

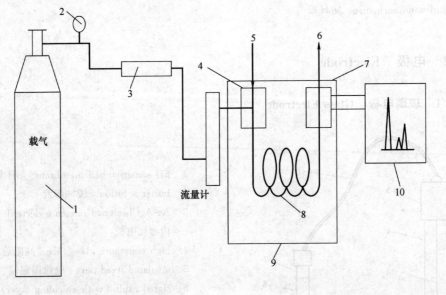

1   carrier gas   载气
2   needle valve   针形阀
3   filter   过滤器
4   vaporizer   气化室
5   inlet   进样口
6   vent   放空口
7   detector   检测器
8   chromatographic column   色谱柱
9   column oven   柱温箱
10   computer workstation   计算机工作站

11.4.3.2 气相色谱仪外观图 Gas Chromatograph Outline

1. inlet 进样口
2. column oven 柱温箱
3. detector 检测器
4. control panel 控制面板
5. LCD display 液晶显示屏

11.4.3.3 气相色谱进样模式 Gas Chromatograph Injection Mode

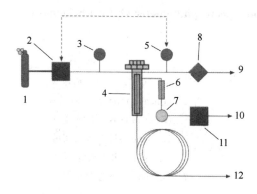

splitless mode 不分流模式

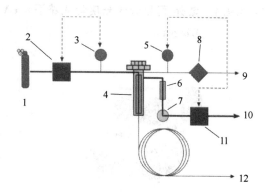

split mode 分流模式

1. gas source 气源
2, 11. proportional valve 比例阀
3. flow sensor 流量传感器
4. capillary inlet 毛细进样口
5. pressure sensor 压力传感器
6. trap 捕集器
7. on/off valve 开/关阀
8. septum purge regulator 隔垫吹扫调节器
9. septum vent 隔垫出口
10. split outlet 分流出口
12. detector 检测器

11.4.3.4 液相色谱示意图 Liquid Chromatograph Schematic Diagram

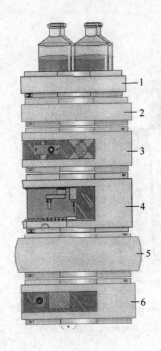

1 solvent cabinet 溶剂储槽
2 degasser 脱气机
3 pump module 泵系统
4 injector/autosampler 进样/自动进样
5 column compartment 柱箱
6 detector 检测器

11.4.4 光谱类仪器 Spectrometric Instruments

11.4.4.1 紫外-可见分光光度计光路图 UV-Visible Spectrophotometer Optical Schematic Diagram

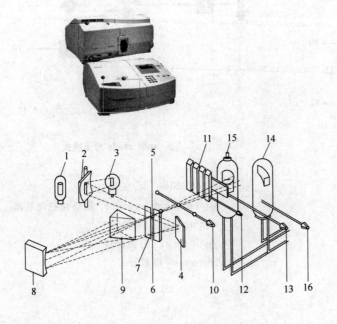

1, 3 light source 光源
2 concave mirror 凹面镜
4 plane mirror 平面镜
5 incidence slit 入射狭缝
6 exit slit 出口狭缝
7 curved slit 弯曲狭缝
8 collimating mirror 准直镜
9 prism 棱镜
10 spectral filter 滤光片
11 cell 比色皿
12 sample support 试样架
13 dark current control strobe 暗电流控制闸门
14, 15 photoelectric tube 光电管
16 sliding support of photoelectric tube 光电管滑动架

698

11.4.4.2 傅立叶变换红外光谱仪原理图 Fourier Transform Infrared Spectrometer Schematic Diagram

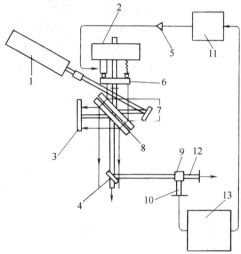

1 helium-neon laser 氦氖激光
2 piezoelectricity component 压电元件
3 float mirror 动镜
4 sample chamber 样品室
5 power amplifier 功率放大器
6 static mirror 固定镜
7 infrared beam 红外光束
8 beamsplitter 分束器
9 polarized light beam separator 偏振光分束器
10 detector A (detection for interference condition) 检测器A（检测干涉条件）
11 digital signal processor 数字式信号处理技术
12 detector B (data collection) 检测器B（数据采集）
13 detection circuit of interference condition 干涉条件测定线路

11.4.4.3 原子吸收分光光度计示意图 Atomic Absorption Spectrophotometer Schematic Diagram

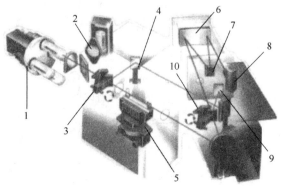

1 hollow cathode lamp 空心阴极灯
2 deuterium lamp 氘灯
3 light cutter 切光器
4, 9, 10 reflector 反射镜
5 burner 燃烧头
6 concave mirror 凹面镜
7 grating 光栅
8 photoelectric multiplier 光电倍增管

11.4.5 电化学类仪器 Electrochemistry Instruments

11.4.5.1 卡尔费休水分滴定仪 Karl Fisher Auto Titrator

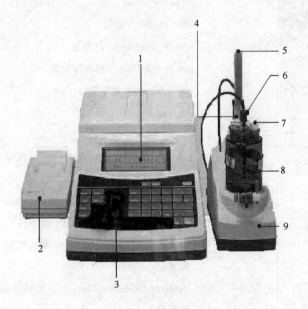

1　display　显示屏
2　printer　打印机
3　keyboard　键盘
4　inner burette　内滴管
5　desiccant tube　干燥试管
6　detection electrode　测量电极
7　sample injection port stopper　样品注射剂端口塞
8　titration cell　滴定池
9　stirrer　搅拌器

11.4.5.2 卡尔费休水分滴定仪原理图 Karl Fisher Auto Titrator Schematic Diagram

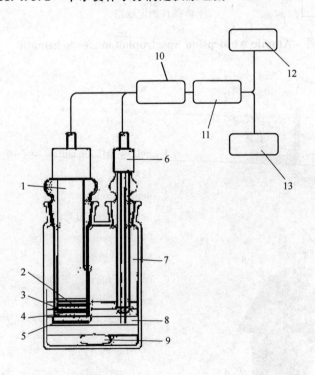

1　generator　发生器
2　catholyte　阴极电解液
3　cathode　阴极
4　membrane　隔膜
5　anode　阳极
6　detection electrode　测量电极
7　titration cell　滴定池
8　analyte　分析物
9　stirrer rotor　转子
10　detection control unit　滴定控制单元
11　microprocessor　微处理机
12　display　显示屏
13　printer　打印机

11.4.5.3 pH计 pH Meter

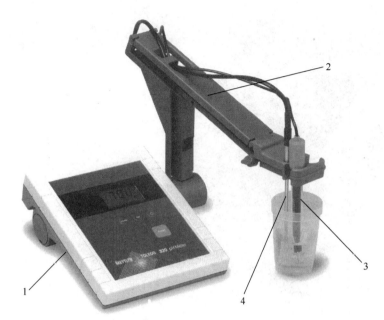

1 pH meter   pH计
2 electrode rack   电极支架
3，4 electrode   电极

11.4.6 水质分析仪器 Apparatus for Water Analysis

11.4.6.1 生物耗氧量测定仪 Apparatus for Biological Oxygen Command Determination

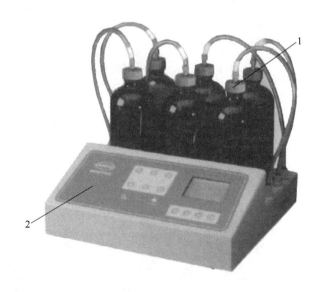

1 BOD bottles   BOD瓶
2 BOD detector   BOD测定仪

11.4.6.2 化学耗氧量测定仪　Apparatus for Chemical Oxygen Command Determination

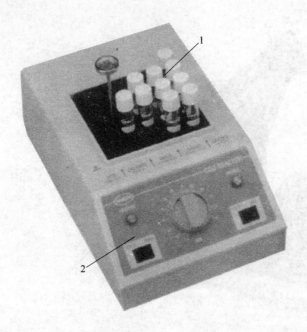

1　COD bottles　COD 瓶
2　COD detector　COD 测定仪

12 土建工程 Civil Engineering

12.1 建筑 Architectural

12.1.1 平面图、立面图、剖面图 Plan, Elevation and Section

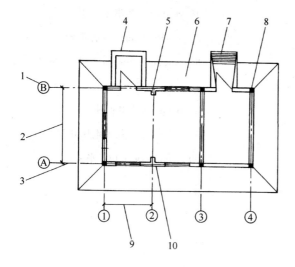

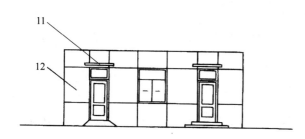

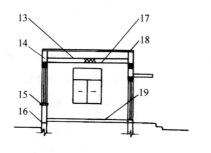

1　number of axis　轴线号
2　depth　进深
3　axis　轴线
4　step　踏步
5　pier　墙垛
6　apron　散水
7　ramp　坡道
8　structural concrete column　构造柱
9　bay　开间
10　pier between windows　窗间墙
11　canopy　雨棚
12　facing　饰面
13　thermal insulation　保温层
14　parapet　女儿墙
15　windowsill　窗台
16　damp-proof course　防潮层
17　roof slab　屋面板
18　coping　压顶
19　floor　地面

12.1.2 楼梯、电梯 Staircase and Lift

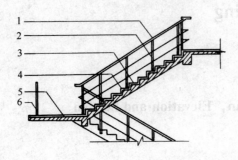

1　handrail　扶手
2　balustrade　栏杆
3　steps　踏步
4　reeding　防滑条
5　stair landing　楼梯平台
6　landing balustrade　平台栏杆
7　car　电梯车厢
8　hoistway　电梯井
9　hoistway door　电梯井门
10　counterweight　平衡锤

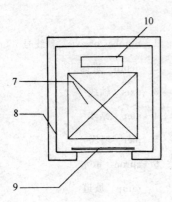

12.1.3 卫生间 Toilet

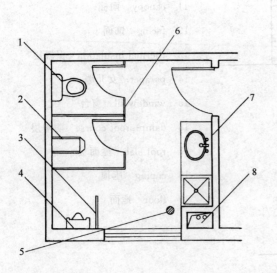

1　WC pan (European type)　马桶
2　WC pan (Turkish type)　大便器
3　partition　隔板
4　urinal　小便器
5　floor drain　地漏
6　toilet partition　厕所隔间
7　wash basin　洗面盆
8　pipe well　管道井

12.1.4 门的形式 Type of Doors

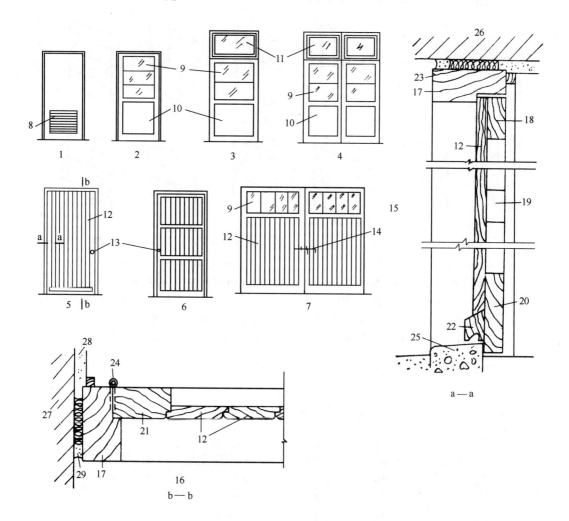

1. flush door with louver for ventilation 带有通风百叶的平板门
2. glazed door for kitchen 厨房用玻璃门
3. glazed door with transom 带顶亮窗的玻璃门
4. glazed double-door with transom 带顶亮窗的双扇玻璃门
5. ledged door with four-railed frame for exterior wall 带四冒头框的拼板外门
6. inner elevation of above mentioned door 上述拼板门内立面
7. ledged door 拼板大门
8. louver for ventilation 通风百叶
9. glass 玻璃
10. panel board 镶板
11. transom 顶亮窗（上亮子）
12. ledge-board 拼板
13. lock 门锁
14. latch bar 门栓
15. section a—a of ledged door 拼板门的剖面 a—a
16. section b—b of ledged door 拼板门的剖面 b—b
17. door frame 门框
18. top rail 上冒头
19. cross rail 中冒头
20. bottom rail 下冒头
21. stile 门挺
22. water stopper 挡水条
23. fiber packing 纤维填料
24. door hinge 门铰链
25. door sill 门槛
26. lintel 过梁
27. wall 墙
28. plaster 抹灰层
29. mastic filling 胶泥填塞

12.1.5 窗的形式 Type of Windows

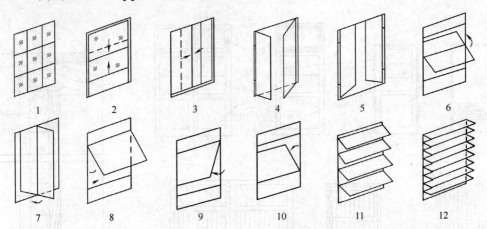

1 fixed window 固定窗
2 double-hung window 双挂拉窗
3 horizontal sliding window 推窗
4 outswinging casement window 向外平开窗
5 inswinging casement window 向内平开窗
6 horizontal pivoted window 平悬转窗
7 vertical pivoted window 立悬转窗
8 top-hinged outswinging window 顶悬外开撑窗
9 top-hinged inswinging window 顶悬内开撑窗
10 bottom-hinged inswinging window 底悬内开翻窗
11 awning window 蓬形窗
12 jalouise window 遮阳窗

12.1.6 窗的组成 Components of Windows

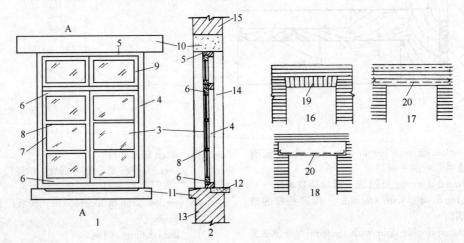

1 elevation of window 窗的立面
2 vertical section of window A—A 窗的竖剖面 A—A
3 casement 窗扇
4 window frame 窗框
5 head 窗楣
6 rail 冒头
7 stile 窗挺
8 muntin 窗棂
9 transom 气窗
10 lintel 过梁
11 window sill 窗台
12 sill plate 窗台板
13 wall 墙
14 side jamb 窗侧壁
15 top jamb 窗顶壁
16 flat brick arch 砖砌平拱
17 reinforced brick lintel 钢筋砖过梁
18 reinforced concrete lintel 钢筋混凝土过梁
19 bricks laid on edge 立砌砖
20 reinforcing bar 钢筋

12.1.7 屋顶的形式 Type of Roofs

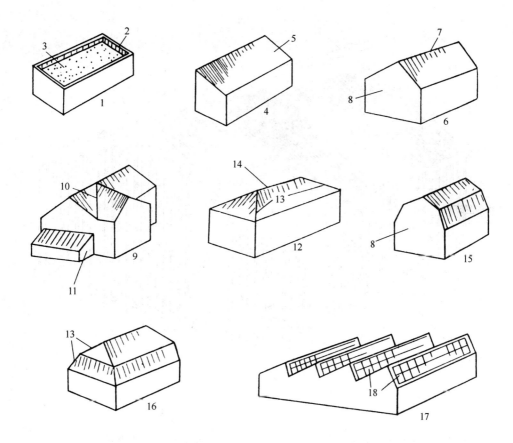

1 flat roof with parapet wall 具有女儿墙的平屋顶
2 parapet well 女儿墙
3 deck 平台
4 shed roof 单坡屋顶
5 shed 棚，坡屋
6 gable roof 双坡屋顶
7 ridge 屋脊
8 gable 山墙
9 shed lean-to 倚墙坡屋
10 valley 斜天沟
11 lean-to 倚墙
12 hip roof 四坡屋顶
13 hip 斜脊
14 ridge 正脊
15 gambrel roof 多折双坡屋顶
16 mansard roof 多折四坡屋顶
17 saw-toothed roof 锯齿形屋顶
18 skylights 天窗

12.1.8 钢结构厂房 Steel Structure House

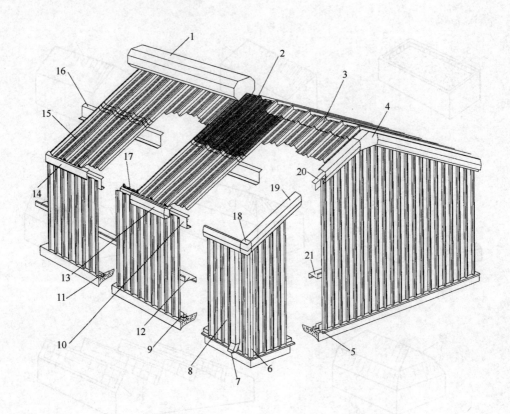

1	continuous ridge vent 屋脊通风器	11	cast-in-place 现浇踢脚
2	light transmitting panel 轻质采光板	12	sidewall girt 边墙檩
3	ridge cap 屋脊盖板	13	sculptured eave gutter 压型檐沟
4	peak box 硬山屋脊包边料	14	eave trim 檐口收边料
5	base trim 踢脚收边料	15	"R" panel "R"形板
6	corner trim 墙角包边料	16	purlin 檩条
7	downspout with 45°elbow 雨水管（带45°弯头）	17	closure strip 带状密封堵头
		18	sculptured corner box 硬山屋角包边料
8	"M" panel "M"形板	19	sculptured rake 压型硬山包边料
9	base angle 角钢踢脚	20	rake angle 硬山顶部角钢墙梁
10	eave strut 檐口边墙梁	21	endwall girt 山墙檩

12.1.9 钢结构厂房节点　Detail of Steel Structure House

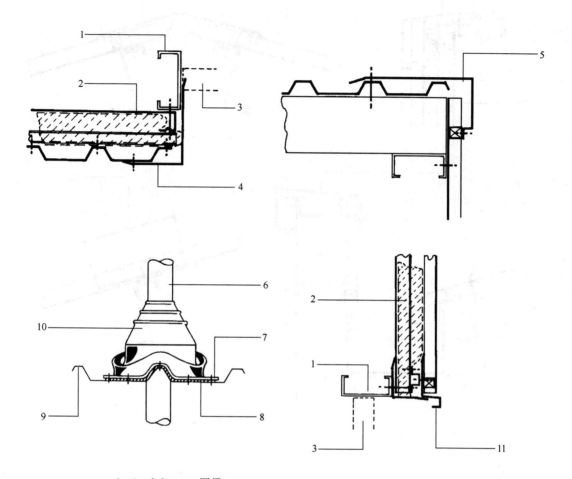

1　jamb beam　围梁
2　thermal insulating wool panel　保温棉
3　window　窗
4　window frame opening flashing　窗边框泛水
5　rake flashing　山墙泛水
6　penetrating pipe dia. 450mm　直径穿管 450mm
7　ring aluminum sheet rivet with roof sheet　环形铝板与屋面板铆接
8　caulking　密封胶层
9　roof sheet　屋面板
10　draped rubber sleeve　褶皱橡胶套筒
11　window head flashing　窗上部泛水

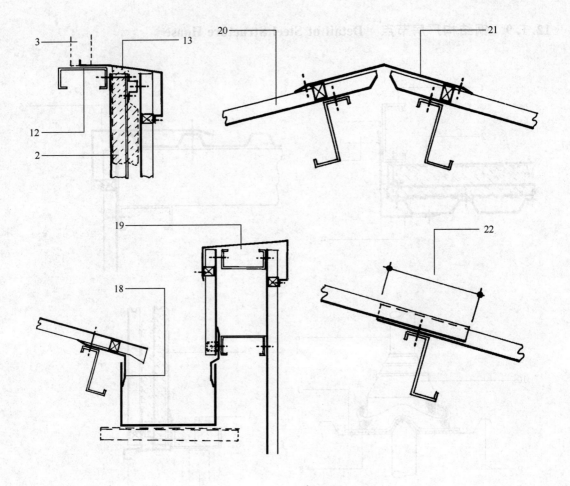

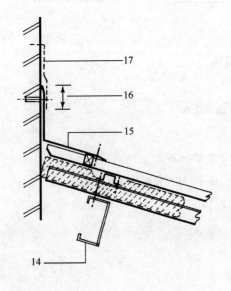

12　sill joist　围檩
13　sill flash　窗台泛水
14　purlin　檩条
15　flashing　泛水
16　overlapping minimum 75mm　至少搭接75mm
17　metal flashing　金属防水板
18　valley gutter flashing　天沟泛水
19　coping　女儿墙压顶泛水
20　corrugating　波纹
21　ridge cap　屋脊盖板
22　provide minimum 230mm　最小边搭接长度230mm

12.1.10 薄壳屋顶 Shell Roofs

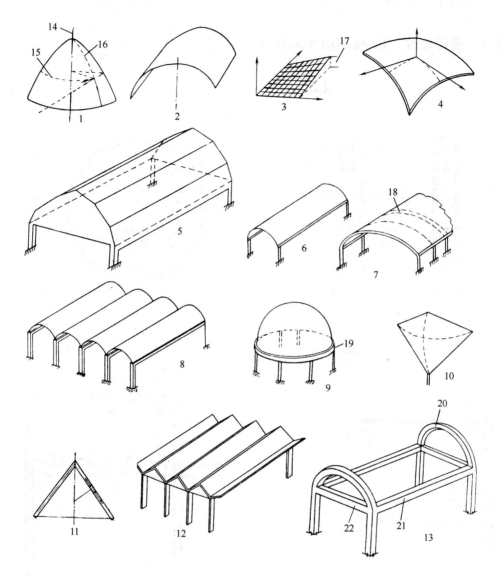

1　shell of revolution　旋转壳
2　cylindrical thin shell　圆柱薄壳
3　hyperbolic-paraboloid thin shell　双曲抛物面薄壳
4　elliptical-paraboloid thin shell　椭圆抛物面薄壳
5　folded-plate roof　折板屋顶
6　long barrel　长筒形壳
7　short barrel　短筒形壳
8　multiple barrel-arch roof　多波筒形拱顶
9　dome　圆屋顶
10　hyperbolic paraboloid　双曲抛物面
11　conical thin shell　圆锥薄壳
12　"V" shape folded-plate structure　V形折板结构
13　stiffening members in thin-shell arch roof　薄壳拱顶的加劲部件
14　shell axis　壳轴
15　parallel　平行线
16　meridian　子午线
17　camber　反挠度
18　rib　肋
19　ring beam　环梁
20　end rib　端肋
21　edge beam　过梁
22　tie　拉杆

12.2 结构 Construction

12.2.1 设计条件 Design Information

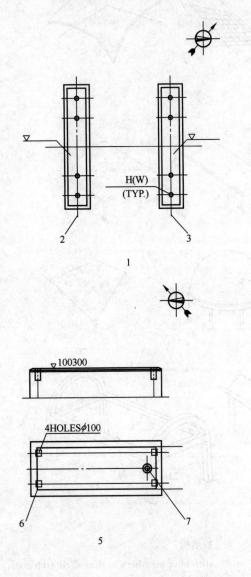
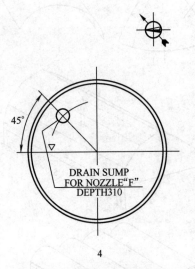
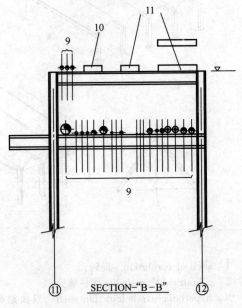

1　foundation information of horizontal equipment　卧式设备基础条件
2　fixed side　固定端
3　sliding side　滑动端
4　foundation information of tank　罐基础条件
5　foundation information of skid　橇式设备基础条件
6　base-plate　设备底板
7　discharge line　出口中心线
8　piping information　管道条件
9　pipe size　管道尺寸
10　instrument cable tray　仪表电缆桥架
11　electrical cable tray　电气电缆桥架

12.2.2 材料 Material

12.2.2.1 混凝土 Concrete

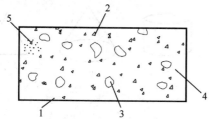

1 concrete 混凝土
2 fine aggregate 细骨料
3 coarse aggregate 粗骨料
4 cement 水泥
5 sand 砂子

12.2.2.2 钢筋 Reinforcement

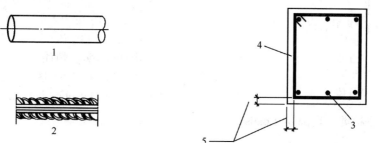

1 plain reinforcing bar 光圆钢筋
2 deformed reinforcing bar 变形钢筋
3 longitudinal bar 纵向钢筋
4 stirrup 箍筋
5 cover 保护层

12.2.2.3 结构钢 Structural Steel

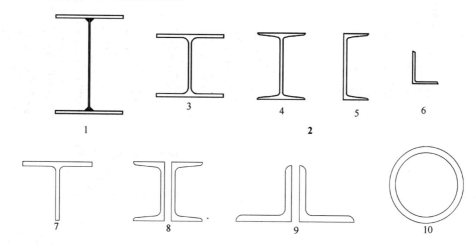

1 welded type 焊接型
2 hot-rolled type 热轧型
3 H-shape H形钢
4 I-shape 工字钢
5 channel 槽钢
6 angle 角钢
7 T-shape T形钢
8 double channel 双槽钢
9 double angle 双角钢
10 tube 钢管

12.2.3 荷载条件 Load Condition

12.2.3.1 几何图形 Geometry

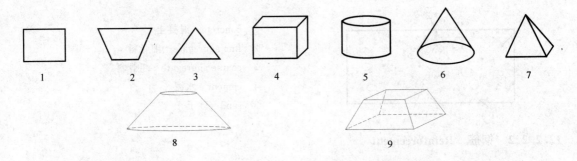

1 rectangular 矩形
2 trapezium 梯形
3 triangle 三角形
4 cube 立方体
5 cylinder 圆柱体
6 cone 圆锥体
7 pyramid 棱锥体
8 frustum cone 圆台
9 frustum pyramid 棱台

12.2.3.2 荷载种类 Type of Loads

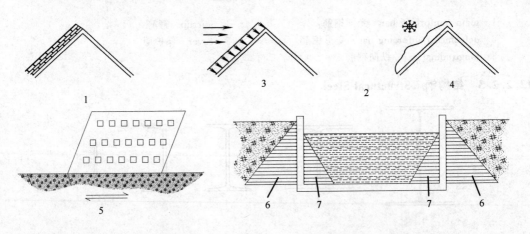

1 dead load 恒荷载
2 live load 活荷载
3 wind load 风载
4 snow load 雪载
5 seismic action 地震作用
6 earth pressure 土压力
7 water pressure 水压力

12.2.3.3 施加载荷 Imposed Loads

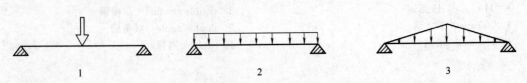

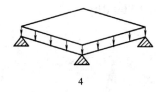
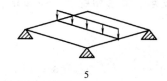

1 concentrated load 集中荷载
2 uniformly distributed load 均布荷载
3 triangular load 三角形荷载

4 uniformly distributed load on slab 板上均布荷载
5 linear load 线荷载

12.2.4 结构计算 Structure Calculation

12.2.4.1 受力及受力条件 Force and Application

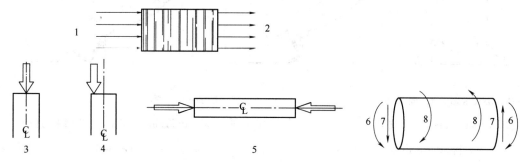

1 compression 压力
2 tension 拉力
3 central line 中心线
4 eccentrically 偏心

5 longitudinal axial force 纵向轴力
6 moment 弯矩
7 shear 剪力
8 torsion 扭矩

12.2.4.2 参数 Parameter

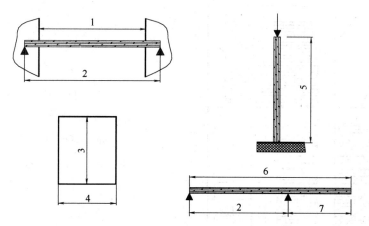

1 effective length 有效长度
2 span 跨度
3 clear height 净高
4 clear width 净宽

5 effective height 有效高度
6 overall length 全长
7 length of cantilever 悬臂长度

715

12.2.4.3 支承、变形和稳定 Support, Deformation and Stability

 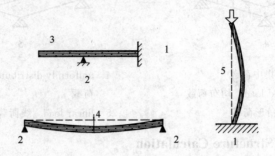

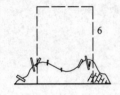 

1　fixed support　固定端
2　simple support　简支端
3　free support　自由端
4　deflection　挠度
5　buckling　弯曲
6　stability against collapse　抗倒塌的稳定
7　stability against overturning　抗倾覆的稳定
8　stability against sliding　抗滑移的稳定

12.2.4.4 框架 Frames

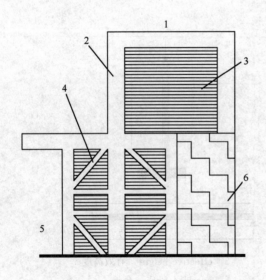

1　framework　框架
2　skeleton　骨架
3　fill-in brickwork　填充砖墙
4　brace　支承
5　lattice frame　格构架
6　wall panel　墙板

12.2.5 土壤与基础 Soil and Foundation

12.2.5.1 土壤符号 Symbols for Soil

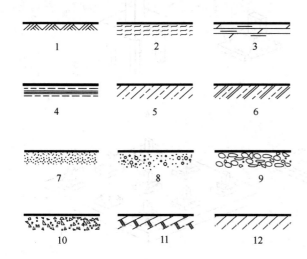

1 field soil 场地土
2 silt 粉土，淤泥
3 peat 泥炭土
4 clay 黏土
5 silty clay 粉质黏土
6 clayey silt 黏质粉土
7 fine sand 细砂
8 coarse sand 粗砂
9 boulder 漂石
10 gravel 砾石
11 hardpan or caliche 泥板岩或石灰岩层
12 weathered rock 风化岩石

12.2.5.2 土层剖面图 Subsoil Profile

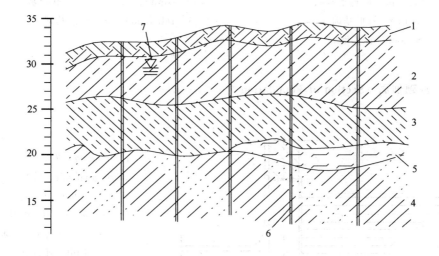

1 humus 腐殖土
2 silty clay 粉质黏土
3 clayey silt 黏质粉土
4 sand with medium density 中密砂
5 silt 淤泥
6 bore hole 钻孔
7 underground water table 地下水位

12.2.5.3 基础类型 Type of Foundation

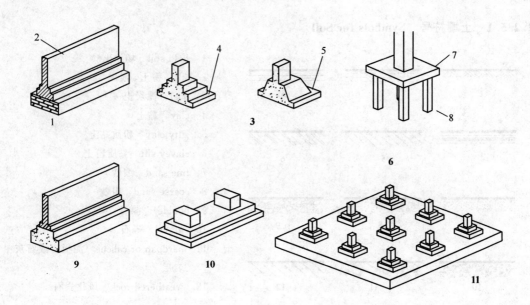

1 rigid footing 刚性基础
2 wall 墙
3 spread footing 扩展基础
4 stepped footing 阶梯形基础
5 sloped footing 锥形基础
6 pile foundation 桩基础
7 pile cap 承台
8 pile 桩
9 strip foundation 条形基础
10 combined foundation 联合基础
11 raft foundation 筏形基础

12.2.5.4 基础剖面 Section of Foundation

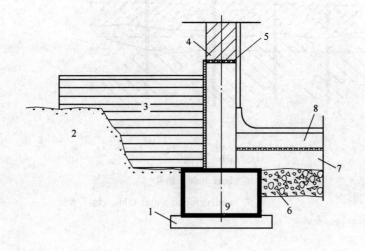

1 blinding 垫层
2 original soil 原土
3 back filling 回填土
4 perimeter wall 外墙
5 damp proof course 防潮层
6 gravel fill 砾石垫层
7 hardcore 夯实砾石层
8 floor slab 地板
9 footing 基础

12.2.5.5 桩的形式 Type of Pile

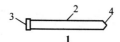

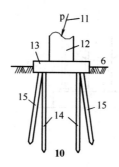
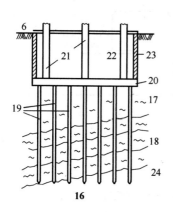
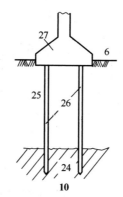
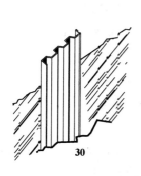
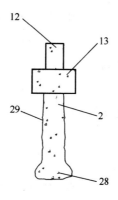

1　precast pile　预制桩
2　pile shaft　桩身
3　pile helmet　桩帽
4　pile tip　桩尖
5　cast-in-situ pile　灌注桩
6　ground　地面
7　pile hole　桩孔
8　reinforcing cage　钢筋笼
9　concrete cast-in-situ　就地灌注的混凝土
10　pile foundation　桩基础
11　inclined loading　斜向荷载
12，21　column　柱
13　pile cap　桩承台
14　vertical pile　直桩
15　batter pile　斜桩
16　box foundation　箱形基础
17　soft clay　软黏土
18　firm clay　坚实黏土
19　friction pile　摩擦桩
20　base plate　底板
22　basement　地下室
23　basement wall　地下室墙
24　bearing stratum　持力层
25　compressible soil　压缩性土
26　end-bearing pile　端承桩
27　column footing　柱基础
28　pile foot　桩足
29　skin friction　表面摩擦力
30　sheet wall piling　板桩

12.2.5.6 设备基础 Foundation for Equipment

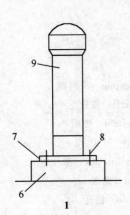
1

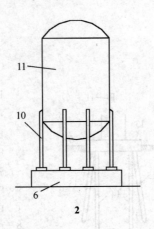
2

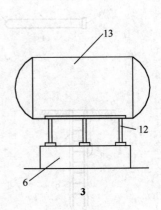
3

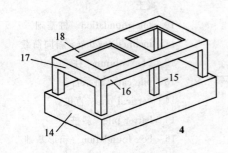
4

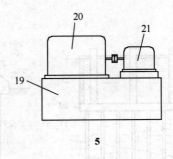
5

1	tower foundation 塔基础	10	legs 支腿
2	8-legged support of vessel 容器的八腿支架	11	vertical vessel 立式容器
		12	supporting saddle 支架座
3	support of horizontal vessel 卧式容器的支架	13	horizontal vessel 卧式容器
		14	concrete base slab 混凝土底板
4	frame foundation of machine 构架式机器基础	15	concrete column 混凝土柱
		16	longitudinal beam 纵梁
5	block foundation of machine 块体式机器基础	17	transverse beam 横梁
		18	top slab 顶板
6	concrete foundation 混凝土基础	19	concrete block foundation 混凝土块体基础
7	base plate of tower 塔底板	20	compressor 压缩机
8	anchor bolt 地脚螺栓	21	motor 电动机
9	tower 塔		

12.2.5.7 大型设备的锚固　Anchorage of Heavy Machine

(1) 大型设备的锚固（1）　Anchorage of Heavy Machine（Ⅰ）

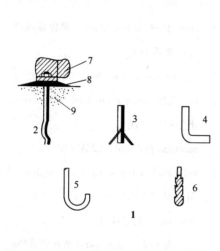
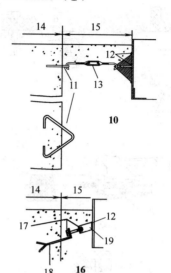
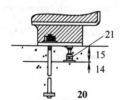
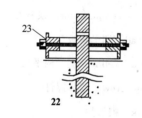
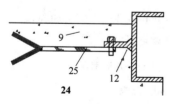
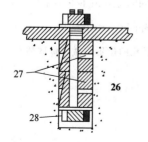

1　anchor bolt　地脚螺栓
2　wave shape　波形
3　saw cut and open to V-shape　锯开扳成 V 形
4　L-shape　L 形
5　hook　钩形
6　rag-bolt end　棘端螺栓
7　machine pedestal　设备支座
8　grout　灌浆
9　cement concrete　水泥混凝土
10　turnbuckle anchor　花篮地脚螺栓
11　embedded round strap　预埋钢环
12　weld　焊缝
13　turn-buckle　花篮螺栓
14　first pour　一次灌浆
15　second pour　二次灌浆
16　adjustment-bolt anchor　调整地脚螺栓
17　slot　间隙
18　flat strap　扁钢
19　channel　槽钢
20　aligning jack and anchor bolt　找正千斤顶和地脚螺栓
21　jack　千斤顶
22　lateral-alignment anchor　横向找正螺栓
23　machine frame　设备框架
24　anchor strap　锚固扁钢
25　flat strap　扁钢
26　expansion sleeves　胀紧套筒
27　expansion sleeves of slotted steel or soft metal　开槽钢环或软金属制的胀紧套筒
28　screw bolt　螺栓

（2）大型设备的锚固（2） Anchorage of Heavy Machine（Ⅱ）

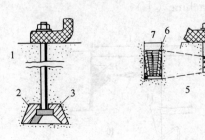

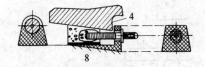

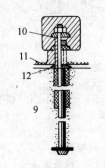

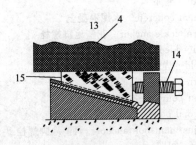

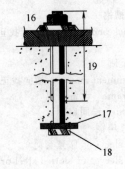

1 embedded anchor block 埋置锚固件，预埋锚固件，埋置锚块
2 square base anchor bolt 方头锚栓
3 cast-iron block embedded in concrete 埋入混凝土中的铸铁块
4 machine pedestal 设备支座
5 split anchor-nut with wedge action 具有楔牢作用的裂张锚固螺母
6 split nut 裂张螺母
7 hole drilled or pre-cast 钻孔或预留孔
8 cylindrical wedge block for uneven or sloping floor 不平楼面或有坡度楼面用的圆柱形楔块
9 leveling screw with anchor bolt 带地脚螺栓的调平螺钉
10 leveling screw 调平螺钉
11 bottom of machine frame 设备底座下缘
12 leveling pad 调平垫板
13 edge-block leveling 楔形调平块
14 regulating screw 调节螺钉
15 edge block 楔形块
16 anchor bolt for location adjustment 可调整位置的地脚螺栓
17 square plate 方形钢板
18 pot weld 点焊
19 elastic deflection of bolt shank (provides small adjustment for hole locations) 螺栓杆的弹性偏移段（对孔的位置可做少量调整）

12.2.6 多层工业厂房 Multistory Industrial Buildings

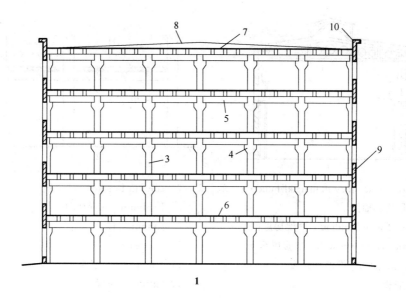

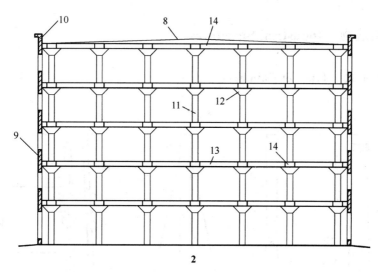

1 section of a prefabricated R.C. multistory industrial building 装配式钢筋混凝土多层厂房的剖面图
2 section of a R.C. flat slab multistory industrial buildings 钢筋混凝土无梁楼盖多层厂房的剖面图
3 column of rectangular cross-section 矩形截面柱
4 bracket of column 柱牛腿
5 pre-cast beam 预制梁
6 pre-cast floor panel 预制楼板
7 pre-cast roof panel 预制屋面板
8 water-proof layer 防水层
9 wall 墙身
10 parapet wall 女儿墙
11 column of circular cross section 圆形截面柱
12 post cap 柱帽
13 flat slab 无梁楼板
14 strip between column 柱间板带

12.2.7 钢筋混凝土结构　Reinforced Concrete Construction

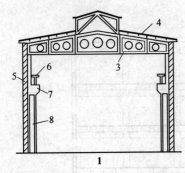

1　section of single story industrial building　单层工业厂房剖面图
2　section of frame building　框架房屋剖面图
3　roof beam with open web　空腹屋面梁
4　channel roof slab　槽形屋面板
5　panel　墙板
6　crane girder　吊车梁
7　column bracket　柱牛腿
8　I-column　工字形柱
9　column　柱
10　secondary beam　次梁
11　floor slab　楼板
12　girder　主梁

12.2.8 工业构筑物　Industrial Structures

12.2.8.1 烟囱的形式　Types of Chimneys

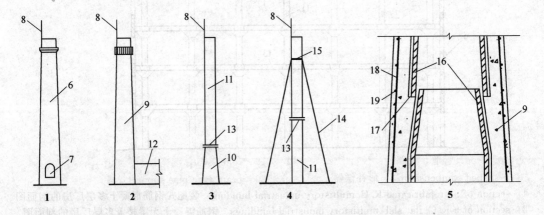

1　brick chimney　砖烟囱
2　reinforced concrete chimney　钢筋混凝土烟囱
3　self-supporting chimney　自承式钢烟囱
4　guyed steel chimney　牵拉式钢烟囱
5　vertical section of reinforced concrete chimney　钢筋混凝土烟囱竖向剖面图
6　shaft of brick chimney　砖烟囱的筒身
7　flue opening　烟道孔
8　lightening arrester　避雷针
9　reinforced concrete shaft　钢筋混凝土筒身
10　tapered steel shaft　锥形钢筒身
11　steel shaft　钢筒身
12　flue　烟道
13　connecting flange　连接法兰
14　guy wire　牵拉绳，拉绳
15　connecting ring　连接环
16　lining of shaft　筒身衬里
17　corbel　挑台
18　vertical reinforcing bar　竖筋
19　annular reinforcing bar　环筋

12.2.8.2 管支架 Pipe Supports

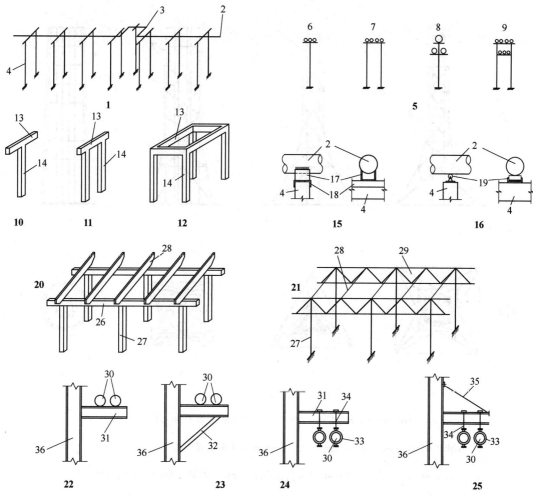

1 sketch for the pipe supporting system 管道支架系统示意图
2,30 pipe 管道,管路
3 π-shaped compensator π形补偿器
4 pipe support 管道支架
5 types of pipe supports 管道支架形式
6 single-post 独柱式
7 double-post 双柱式
8 single-post with double decks 独柱双层式
9 double-post with double decks 双柱双层式
10 single-post pipe support 独柱管架
11 double-posts pipe support 双柱管架
12 space pipe support 空间管架
13 cross beam 井字梁
14,27 post 立柱
15 fixed support 固定支座
16 sliding support 滑动支座
17 steel channel seat 槽钢管座
18 steel angle 角钢
19 roller seat 滚座
20 beam type pipe support 梁式管道支架
21 truss type pipe support 桁架式管道支架
22~25 cantilever pipe support on the column 柱上悬臂式管道支架
26 longitudinal beam 纵梁
28 transversal beam 横梁
29 truss 桁架
31 shape steel beam 型钢梁
32 steel bracing 钢支承
33 pipe stirrup 管套环
34 hanging rod 吊杆
35 tie rod 拉杆
36 column 墙

12.2.8.3 排气筒 Vent Stacks

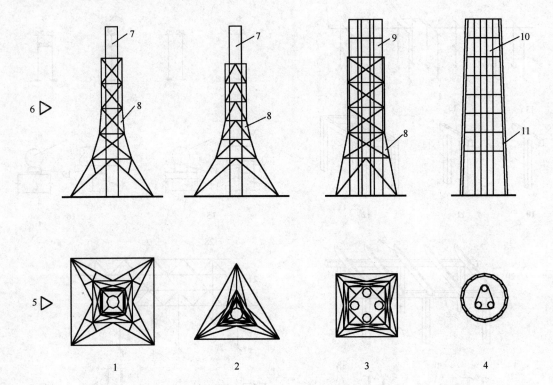

1 quadruped-frame supported vent stack 四腿钢架支承的排气筒

2 tripod-frame supported vent stack 三腿钢架支承的排气筒

3 four-vent assemblage with steel supporting frame 钢架支承的四筒组合体

4 three-vent assemblage with steel supporting frame 钢架支承的三筒组合体

5 plans 平面

6 elevations 立面

7 vent shaft 筒身

8 supporting frame 支架

9 four-vent assemblage 四筒组合体

10 three-vent assemblage 三筒组合体

11 R. C. supporting shaft 钢筋混凝土支架筒

12.2.8.4 水塔的形式 Types of Water Towers

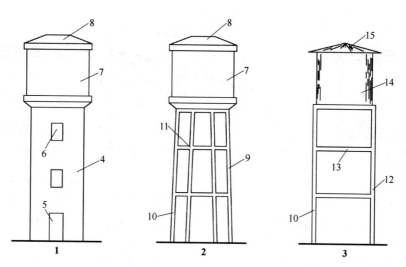

1. water tank on brickwork shaft 砖筒支承水箱
2. water tank on reinforced concrete frame 钢筋混凝土支架水箱
3. steel water tank on R.C. rigid frame 钢筋混凝土刚架支承的钢水箱
4. brickwork shaft 砖砌筒身
5. door opening 门洞
6. window opening 窗洞
7. R.C. water tank 钢筋混凝土水箱
8. R.C. dome 钢筋混凝土圆顶
9. R.C. frame 钢筋混凝土支架
10. R.C. column 钢筋混凝土柱
11. R.C. bracing 钢筋混凝土支承
12. R.C. rigid frame 钢筋混凝土刚架
13. R.C. beam 钢筋混凝土梁
14. steel water tank 钢制水箱
15. conical roof 锥顶

12.2.8.5 冷却塔 Cooling Tower

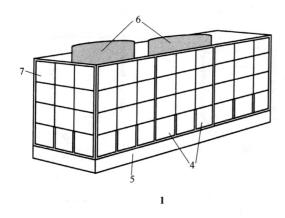
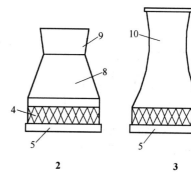

1. cooling tower of multisystem 多系列冷却塔
2. cooling tower with mechanical draft 机械通风冷却塔
3. cooling tower with natural draft 自然通风冷却塔
4. vent opening 进风口
5. semi-underground water reservoir 半地下水池
6. shaft for draft fan 抽风机筒身
7. wall of tower 塔壁
8. conical shaft 圆锥形筒身
9. inverted conical outlet 倒锥形出风口
10. hyperbolic shaft of tower 双曲线塔身

12.2.8.6 通廊和栈桥 Galleries and Trestles

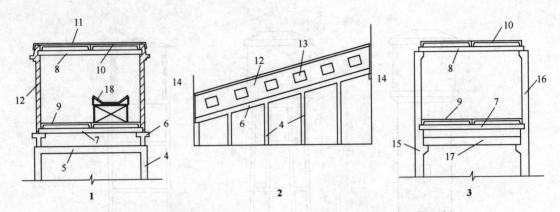

1　section of a gallery　通廊的剖面图
2　elevation of a trestle　栈桥的立面图
3　section of a gallery of alternative design　另一种通廊的剖面图
4　supporting frame　支架
5　frame beam　支架梁
6　longitudinal beam　纵梁
7　transverse beam　横梁
8　head beam　顶横梁
9　flooring panel　楼面板
10　roofing panel　屋面板
11　roofing felt　屋面油毡
12　brick wall　砖墙
13　window opening　窗洞
14　building　建筑物
15　strut of supporting frame　支架立柱
16　strut with wall pillar on top　带扶壁柱的立柱
17　pre-cast beam　预制梁
18　belt conveyer　带输送机

12.2.8.7 筒仓和储斗 Silos and Bunkers

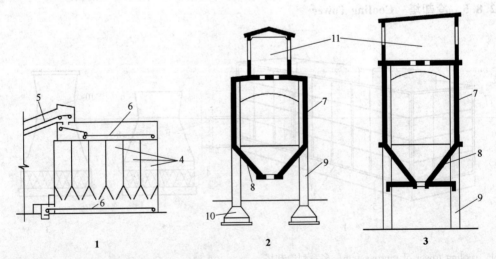

1　group of silos　筒仓群
2　bunker　储斗
3　silo　筒仓
4　cellular silos　分格筒仓
5　gallery　通廊
6　belt conveyor　带输送机
7　wall of silo or bunker　筒仓壁或储斗壁
8　hopper　漏斗
9　column　柱子
10　footing　基础
11　inlet　入口

12.2.8.8 低温液化气体储罐 Tank for the Storage of Refrigerated Liquefied Gases

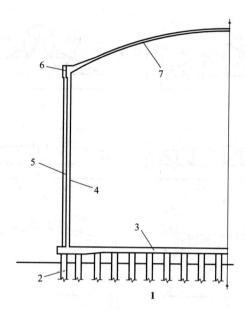

1

1 concrete outer tank vertical section 混凝土外罐竖向剖面图
2 pile 桩
3 elevated slab 架空底板
4 pre-stressing R.C. tank wall 预应力钢筋混凝土罐壁
5 buttress 扶壁柱
6 eaves beam 顶部环梁
7 roof 罐顶

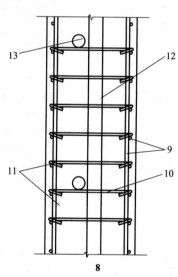

8

8 details of tank wall 罐壁详图
9 cryogenics rebar 低温钢筋
10 cryogenics stirrups 低温箍筋
11 ordinary rebar 普通钢筋
12 vertical pre-stressing tendon 竖向预应力筋束
13 horizontal pre-stressing tendon 水平预应力筋束

12.2.9 钢结构 Steel Construction

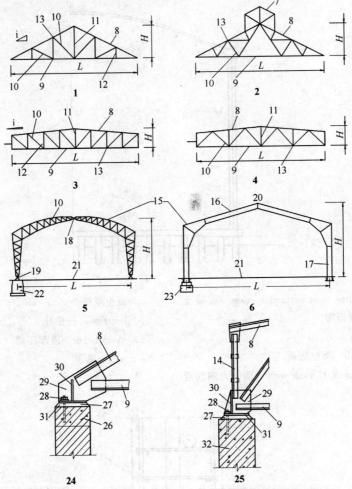

1 triangular truss 三角形屋架
2 Fink truss 芬克式屋架
3 Pratt truss 普勒特式屋架
4 Warren truss 华伦式屋架
5 three-hinged portal frame 三铰门式桁架
6 hingeless portal rigid frame 无铰门式刚架
7 monitor 天窗
8 top chord 上弦
9 bottom chord 下弦
10 inclined web member 斜腹杆
11 central web member 中腹杆
12 vertical web member 竖腹杆
13 panel joint 节点
14 end strut 端柱
15 elbow 弯肘
16 rafter 人字梁
17 leg 支腿
18 crown pin 顶铰
19 base pin 底铰
20 crown 架顶
21 ground 地面
22 pedestal footing 墩式基础
23 stepped footing 台阶式基础
24 support of triangular truss 三角形屋架支点
25 support of Warren truss 华伦屋架支点
26 concrete sleeper 混凝土垫块
27 cement mortar 水泥砂浆
28 anchor bolt 锚固螺栓
29 gusset plate 连接板
30 stiffener 加劲板
31 steel plate 钢板
32 reinforced concrete column 钢筋混凝土柱

12.2.9.1 钢构件的连接　Connection of Steel Members

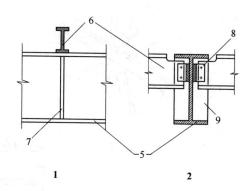

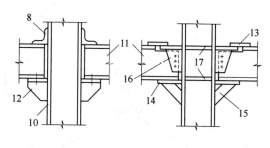

1　beam-on-girder connection　梁搁置在主梁上
2　connection between beam and girder　梁和主梁的连接
3　pinned connection between beam and column　梁和柱的铰接连接
4　rigid connection between beam and column　梁和柱的刚接连接
5　girder　主梁
6　beam　次梁
7　supporting stiffener　支承加筋肋
8　connecting angle　连接角钢
9　web stiffener　腹板加筋肋
10　column　柱
11　beam　梁
12　bracket　支承板
13　connecting cover　连接盖板
14　horizontal connecting plate　水平连接板
15　bracing　支承
16　stiffened connecting plate　加筋连接板
17　horizontal stiffener　水平加筋板

12.2.9.2 钢结构连接详图 Structural Steel Connection Details

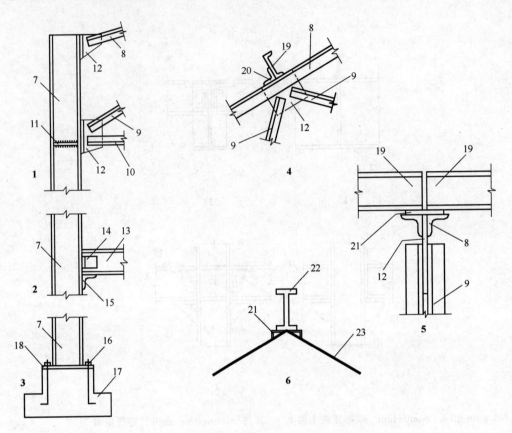

1 connection between roof truss and column   屋架与钢柱的连接
2 connection between beam and column   钢梁和钢柱的连接
3 connection of column to footing   钢柱与柱基的连接
4 connection of purlin to roof slope (section)   钢檩与屋顶坡面的连接（剖面）
5 connection of purlin to roof slope (elevation)   钢檩与屋顶坡面的连接（立面）
6 connection of purlin to roof ridge   钢檩与屋脊的连接
7 steel column   钢柱
8 top chord   上弦
9 web member   腹杆
10 bottom chord   下弦
11 stiffener   加筋肋
12 gusset plate   连接板
13 steel beam   钢梁
14 fastening angle   连系角钢
15 bracket   支托
16 anchor bolt   地脚螺栓
17 column footing   柱基
18 base plate   底板
19 purlin on slope   坡面檩条
20 clip angle   固定角钢
21 cushion plate   垫板
22 ridge purlin   脊檩
23 roof truss   屋架

12.2.9.3 钢栏杆 Steel Balustrade

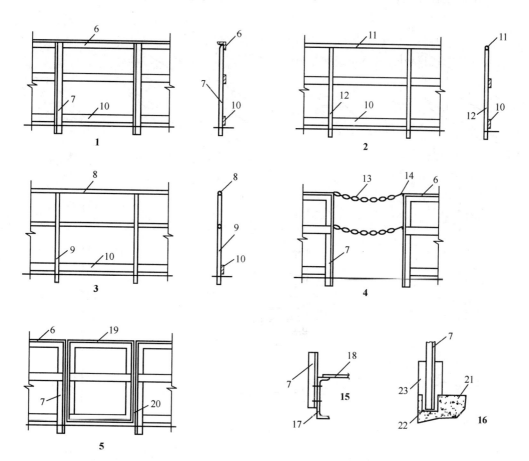

1　balustrade made of steel angles　角钢栏杆
2　balustrade made of steel rods　圆钢栏杆
3　balustrade made of steel tubes　钢管栏杆
4　balustrade with movable chains　带活动链条的栏杆
5　balustrade with hinged door　带活动门的栏杆
6　railing of steel angle　角钢扶手杆
7　baluster of steel angle　角钢栏杆柱
8　railing of steel tube　钢管扶手杆
9　baluster of steel tube　钢管栏杆柱
10　toe plate　踢脚板
11　railing of steel rod　圆钢扶手杆
12　baluster of steel rod　圆钢栏杆柱
13　steel chain　钢制链条
14　steel hook　钢制挂链钩
15　detail of connection of steel baluster with steel platform　钢栏杆柱与钢平台连接细部
16　detail of connection of steel baluster with concrete floor　钢栏杆与混凝土楼板连接细部
17　steel channel　槽钢
18　flooring plate　钢楼板
19　steel-framed door　钢框门
20　hinges　铰链
21　concrete floor　混凝土楼板
22　concrete grouting　混凝土灌浆
23　concrete casting block　混凝土灌注块

733

12.2.9.4 钢扶梯和爬梯 Steel Stairs and Ladders

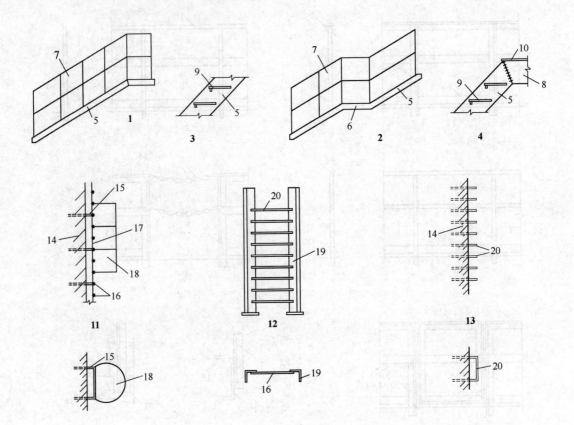

1　one-run stair　单跑扶梯
2　two-run stair　双跑扶梯
3　stair detail　扶梯详图
4　detail of stair with platform　扶梯连平台详图
5　string beam　楼梯斜梁
6　resting platform　休息平台
7　railing　扶手
8　beam for platform　平台梁
9　step plate　踏步板
10　flooring plate　楼面板
11　steel ladder　钢梯
12　straight steel ladder　钢直梯
13　climbing ladder　爬梯
14　wall or shaft　墙或井筒
15　embedded support　预埋支架
16　rungs　梯蹬
17　string　梯帮
18　safety cage　安全笼
19　steel string angle　角钢梯帮
20　rungs with embedded ends　带埋入端的梯蹬

12.3 暖通 Heating Ventilation and Air Conditioning

12.3.1 采暖系统与散热器 Heating System and Radiator

12.3.1.1 散热器 Radiator

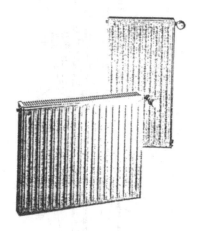

steely panel radiator 钢制板式散热器

bathroom steel tube radiator
卫浴钢管散热器

steely tube column radiator 钢制管柱式散热器

cast iron column radiator 铸铁柱式散热器

copper aluminium combined radiator
铜铝复合散热器

12.3.1.2 暖风机 Unit Heater

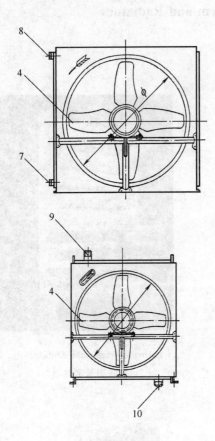

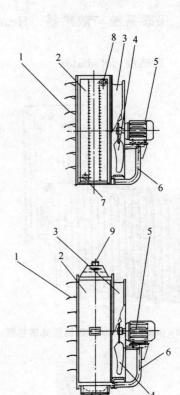

1 gravitational shutter 自垂百叶
2 heating coil 加热盘管
3 air collector 集风器
4 blade wheel of fan 风机叶轮
5 motor 电机
6 support of fan 风机支架
7 hot water inlet 热水进口
8 hot water outlet 热水出口
9 steam inlet 蒸汽进口
10 condensate outlet 凝结水出口

12.3.1.3 自动及手动排气阀 Automatic and Manual Air Vent-Valve

automatic air vent-valve 自动排气阀

  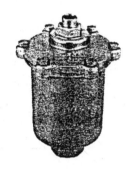

manual air vent-valve for terminal of radiator　　　　**large air displacement automatic air vent-valve**
散热器端头手动排气阀　　　　　　　　　　　　　　　**大排气量自动排气阀**

 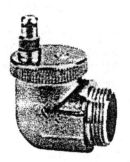

automatic air vent-valve for terminal of radiator
散热器端头自动排气阀

12.3.1.4 热水分/集水器和蒸汽分汽缸　Supply/Return Header for Hot Water and Supply Header for Steam

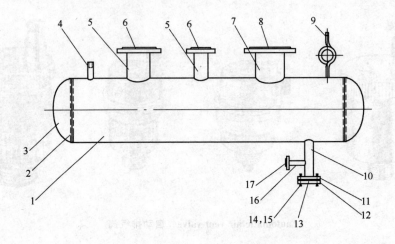

supply header for steam　蒸汽分汽缸

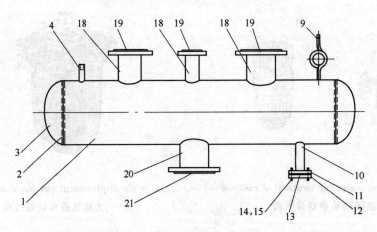

supply/return header for hot water　热水分/集水器

1	body of header　筒体		9	pressure gauge bending pipe　压力表弯管
2	bushing ring　衬环		10	blowdown nipple　排污短管
3	ellipsoidal head　椭圆形封头		11，17	flange　法兰
4	thermometer joint　温度计接口		12	blind flange　法兰盖
5	steam outlet nipple　蒸汽出口短管		13	gasket　垫片
6	steam outlet flange　蒸汽出口法兰		14	bolt　螺栓
7	steam inlet nipple　蒸汽入口短管		15	nut　螺母
8	steam inlet flange　蒸汽入口法兰		16	steam trap nipple　疏水器接口短管
18	outlet nipple of supply header/inlet nipple of return header　分水器出口短管/集水器进口短管
19	outlet flange of supply header/inlet flange of return header　分水器出口法兰/集水器进口法兰
20	inlet nipple of supply header/outlet nipple of return header　分水器进口短管/集水器出口短管
21	inlet flange of supply header/outlet flange of return header　分水器进口法兰/集水器出口法兰

12.3.1.5 开式膨胀水箱 Open Type Expansion Tank

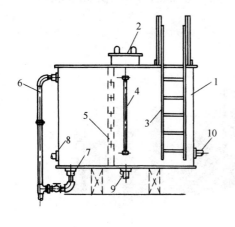
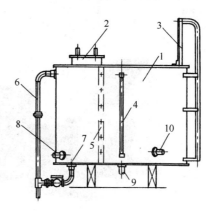

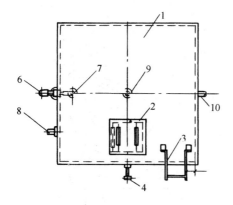

open type square expansion tank
开式方形膨胀水箱

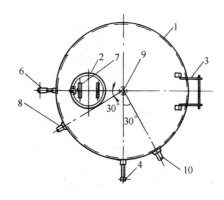

open type round expansion tank
开式圆形膨胀水箱

1　body of tank　箱体
2　access hole　人孔
3　outside ladder　外人梯
4　glass tube water gauge　玻璃管水位计
5　inside ladder　内人梯
6　over flow pipe　溢流管
7　blow off pipe　排水管
8　circulating pipe　循环管
9　expansion pipe　膨胀管
10　signal pipe　信号管

12.3.2 通风与空调设备　Ventilation and Air Conditioning Equipment

12.3.2.1 离心风机及减震台座　Centrifugal Fan and Shock Absorption Support

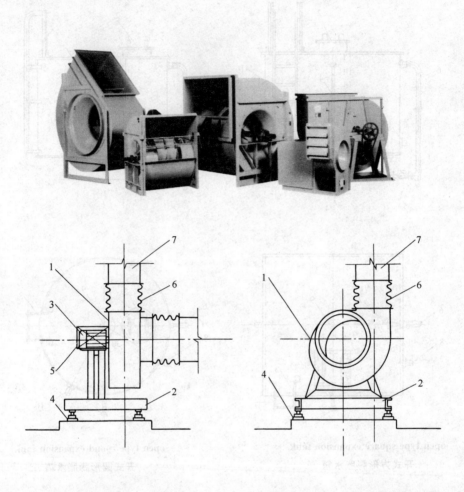

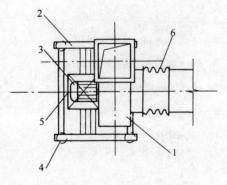

1　centrifugal fan　离心风机
2　shock absorption support　减震钢架
3　motor　电动机
4　spring shock absorber　弹簧减震器
5　motor rainhat　电动机防雨罩
6　flexible duct　柔性软管
7　air duct　风管

12.3.2.2 轴流风机　Axial Fan

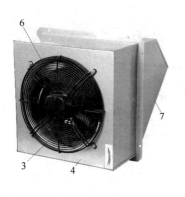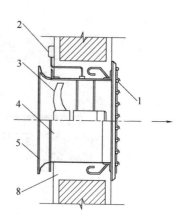

1　gravity shutter　自垂百叶风口
2　terminal block　接线盒
3　wheel　叶轮
4　fan crust　机壳
5　air collector　集风器
6　protective cage　防护网
7　weather hood　防雨罩
8　cement mortar　水泥砂浆

12.3.2.3 混流风机　Mixed Flow Fan

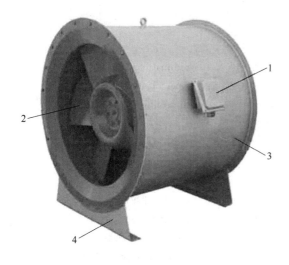

1　terminal block　接线盒
2　wheel　叶轮
3　fan crust　机壳
4　bracket　托架

12.3.2.4 屋顶风机 Roof Fan

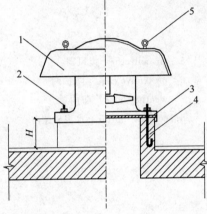

1　case　机壳
2　bolt　螺栓
3　rubber mat　橡胶垫
4　foundation　基础
5　lifting lug　吊环

roof axial fan installation drawing
轴流式屋顶风机安装图

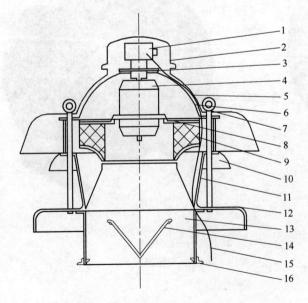

1　terminal block　接线盒
2　upper blast cap　上风帽
3　connecting flange　连接盘
4　motor　电机
5　lower blast cap　下风帽
6　motor bracket　电机支架
7　lifting lug　吊环
8　wheel　叶轮
9　protecting net　护网
10　diffuser　扩压器
11　supporting rod　支杆
12　base plate　机座
13　throttle tube　调节阀筒
14　throttle plate　阀片
15　wire　引线
16　connecting flange　连接法兰

explosion-proof centrifugal type roof fan
防爆离心式屋顶通风机

12.3.2.5 风机箱 Cabinet Fan

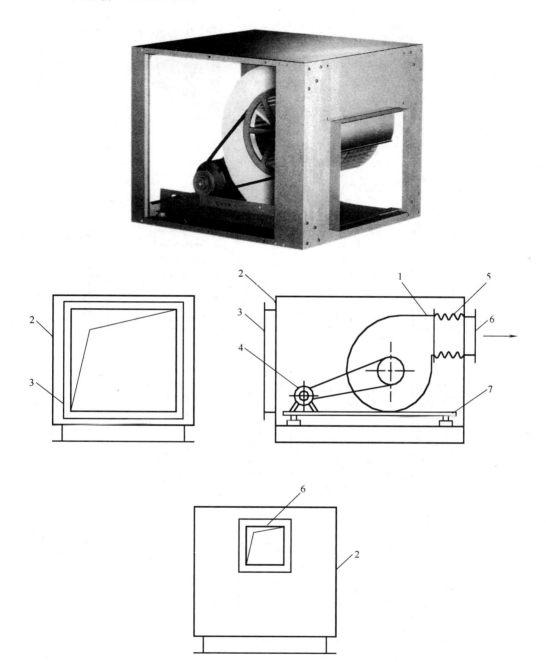

1 centrifugal fan 离心风机
2 cabinet 机箱
3 air inlet flange 进风口法兰
4 motor 电机
5 flexible duct 柔性软管
6 air outlet flange 出风口法兰
7 shock absorption support 减震钢架

12.3.2.6 组合式空气处理机 Modular Air Handling Unit

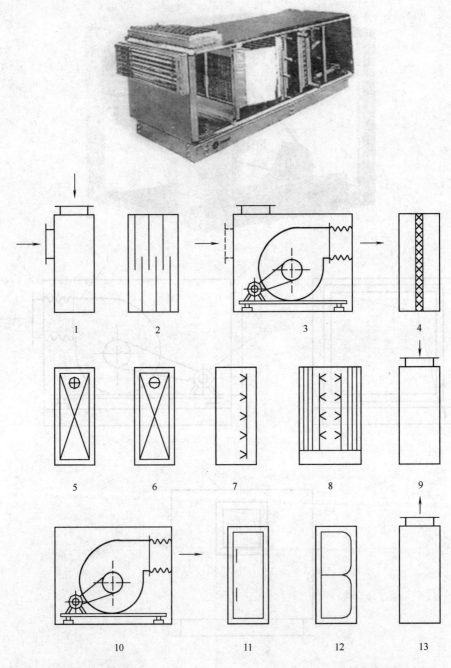

1　mixing box section　混合段
2　muffler section　消声段
3　return fan section　回风机段
4　primary filter section　初效过滤段
5　heating coil section　加热段
6　cooling coil section　冷却段
7　steam humidifier section　蒸汽加湿段
8　spray chamber section　喷水段
9　air return section　回风段
10　supply fan section　送风机段
11　empty section　中间段
12　medium efficiency air filter section　中效过滤段
13　air supply section　送风段

12.3.2.7 整体式水冷空调机　Water-Cooled Packaged Air Conditioner

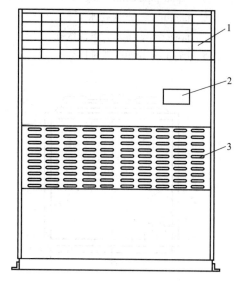
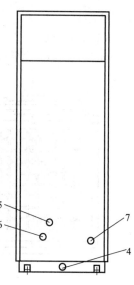

1　air outlet　送风口
2　control panel　操作面板
3　air inlet　回风口
4　pan drain　底盘排水
5　cooling water outlet　冷却水出水
6　cooling water inlet　冷却水进水
7　power wire hole　电源线孔

12.3.2.8 分体式风冷空调机 Air-Cooled Split Air Conditioner

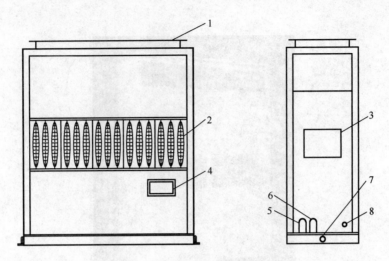

indoor unit 室内机

1 air outlet 送风口
2 air inlet 回风口
3 control panel 操作面板
4 gas pipe connecting port 气管接口
5 liquid pipe connecting port 液管接口
6 pan drain 底盘排水
7 power wire hole 电源线孔
8 fresh air inlet 新风口

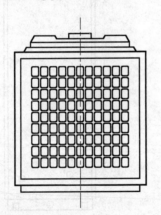
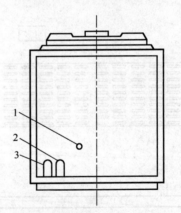
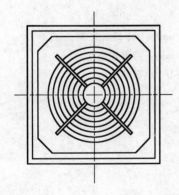

1 outdoor unit power wire hole 室外机电源线孔
2 liquid pipe connecting port 液管接口
3 gas pipe connecting port 气管接口

outdoor unit 室外机

12.3.2.9 屋顶式空调机 Roof Air Conditioning Unit

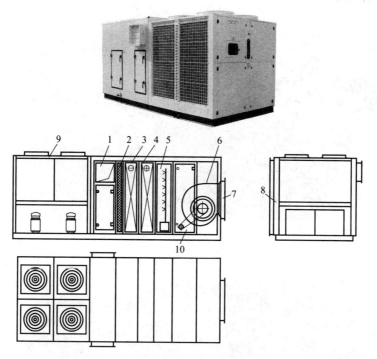

1	fresh air inlet 新风口	6	air supply fan 送风机
2	air filter section 过滤段	7	air outlet 送风口
3	cooling coil section 冷却段	8	air inlet 回风口
4	heating coil section 加热段	9	refrigeration section 制冷段
5	humidifier section 加湿段	10	inspection section 检修段

12.3.3 空调制冷与末端设备 Refrigerating and Terminal Equipment

12.3.3.1 离心式冷水机组 Centrifugal-Type Water Chiller

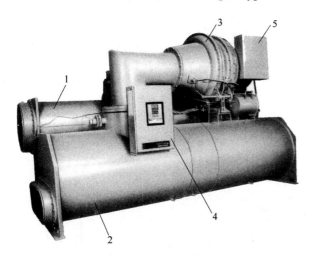

1. condenser 冷凝器
2. evaporator 蒸发器
3. centrifugal compressor 离心式压缩机
4. control cabinet 控制柜
5. starter cabinet 启动器箱

water-cooled centrifugal-type water chiller 水冷离心式冷水机组

12.3.3.2 直燃式溴化锂吸收式冷水机组 Direct-Fired Lithium-Bromide Absorption Water Chiller

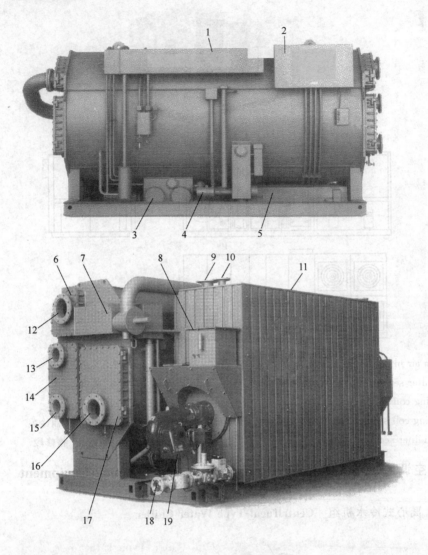

1	automatic exhaust unit	自动抽气装置	11	high-pressure generator room	高压保温房
2	control cabinet	控制柜	12	cooling water outlet	冷却水出口
3	filter	过滤器	13	chilled water outlet	冷冻水出口
4	canned pump	屏蔽泵	14	evaporator	蒸发器
5	heat exchanger	换热器	15	chilled water inlet	冷冻水入口
6	condenser	冷凝器	16	cooling water inlet	冷却水入口
7	low-pressure evaporator	低压蒸发器	17	absorber	吸收器
8	stack vent	排气口	18	gas inlet	燃气入口
9	warm-water inlet	温水入口	19	fire machine	燃烧机
10	sanitary hot water inlet	卫生热水入口			

12.3.3.3 蒸汽型溴化锂吸收式制冷机 Steam Type Lithium-Bromide Absorption Water Chiller

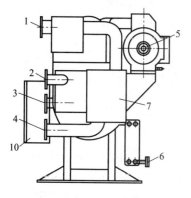
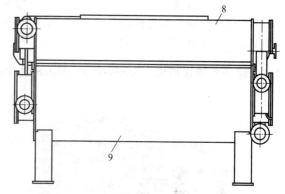

1 cooling water outlet 冷却水出口
2 chilled water outlet 冷冻水出口
3 chilled water inlet 冷冻水入口
4 chilled water inlet 冷却水入口
5 steam inlet 蒸汽入口
6 condensate water outlet 冷凝水出口
7 absorber 吸收器
8 condenser 冷凝器
9 evaporator 蒸发器
10 control cabinet 控制柜

12.3.3.4 热水型溴化锂吸收式制冷机 Hot Water Type Lithium-Bromide Absorption Water Chiller

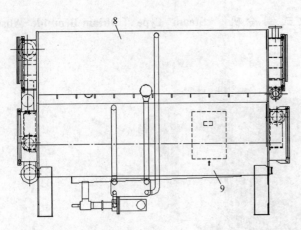
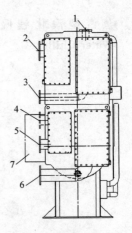

1　hot water outlet　热水出口
2　cooling water outlet　冷却水出口
3　hot water inlet　热水入口
4　chilled water outlet　冷冻水出口
5　chilled water inlet　冷冻水入口
6　cooling water inlet　冷却水入口
7　control cabinet　控制柜
8　condenser　冷凝器
9　evaporator　蒸发器

12.3.3.5　螺杆式冷水机组　Screw-Type Water Chiller

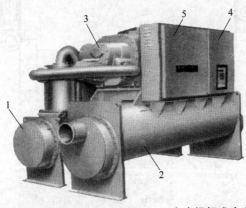

1　condenser　冷凝器
2　evaporator　蒸发器
3　screw compressor　螺杆式压缩机
4　control cabinet　控制柜
5　starter cabinet　启动器箱

water-cooled screw-type water chiller　水冷螺杆式冷水机组

1　condenser　冷凝器
2　chilled water outlet　冷冻水出口
3　chilled water inlet　冷冻水入口
4　control cabinet　控制柜
5　fan of condenser　冷凝器风机

air-cooled screw-type water chiller　风冷螺杆式冷水机组

12.3.3.6 新风净化机组 Fresh Air Unit

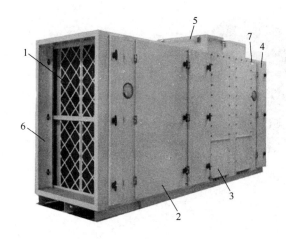

1 pre-filter 初效过滤
2 fan section 风机段
3 discharge opening 卸料口
4 air outlet 出风段
5 chemical filter section 化学过滤段
6 air inlet 进风段
7 secondary filter section 中效过滤段

12.3.3.7 多联机空调 VRV System

1 fan 风扇
2 heat dissipation coil 散热盘管

outdoor unit 室外机

indoor unit 室内机

1 air outlet 送风口
2 refrigerant outlet 冷媒出口
3 refrigerant inlet 冷媒进口
4 drain 排水口

12.3.3.8 新风换气机　Fresh-Air Exchanger

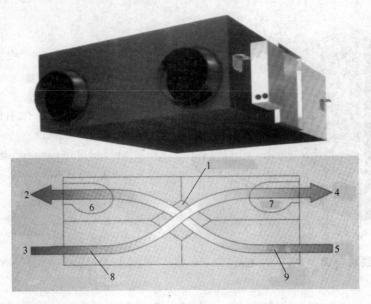

1　thermal exchanger　热交换元件
2　air exhaust　排气
3　fresh air　新鲜空气
4　air supply　供气
5　air return　回气
6　exhaust fan　排气扇
7　supply fan　供气扇
8　outdoor temperature inductor　室外温度感应器
9　indoor temperature inductor　室内温度感应器

12.3.3.9 风机盘管　Fan Coil

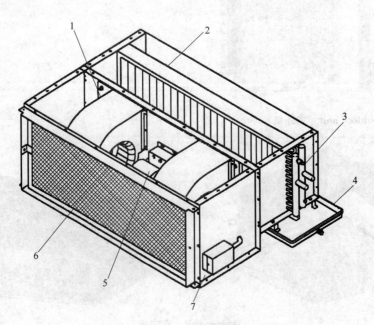

1　fan　风机
2　coil　盘管
3　air vent　跑风
4　drain pan　滴水盘
5　motor　电机
6　air filter　空气过滤器
7　terminal block　接线盒

12.3.4 通风管道及部件 Duct and Fitments

12.3.4.1 风管 Duct

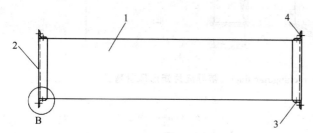

rectangular duct 矩形风管

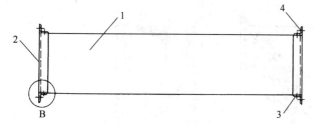

circular duct 圆形风管

1　duct　风管
2　flange　法兰
3　rivet　铆钉
4　bolt hole　螺栓孔
5　welding　焊接

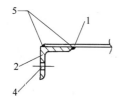 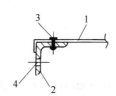

details of B 节点 B 详图

12.3.4.2 风管弯头 Elbow for Duct

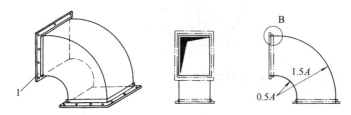

elbow for rectangular duct 矩形风管弯头

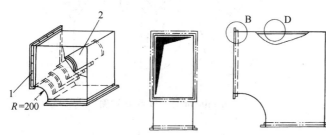

1　flange　法兰
2　guide vanes　导流片

vaned elbow for rectangular duct 带导流片矩形风管弯头

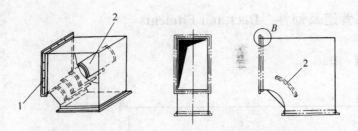

vaned elbow for rectangular duct 带导流片矩形风管弯头

1 flange 法兰　　　　　2 vanes 导流片

12.3.4.3 风管三通 Tee for Duct

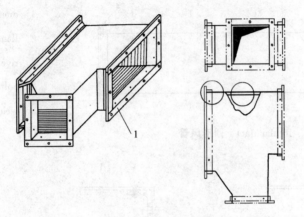

tee for rectangular duct 矩形风管三通

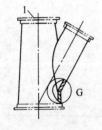 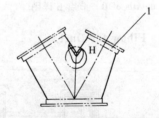

tee for round duct 圆形风管三通

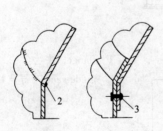 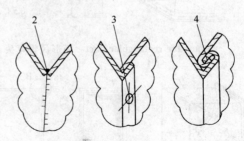

details of G 节点 G 详图　　　　details of H 节点 H 详图

1 flange 法兰　　　　　3 riveting 铆接
2 welding 焊接　　　　4 seaming 咬口

12.3.4.4 风管异径 Reducer for Duct

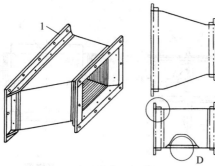
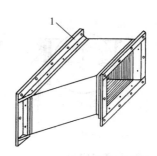
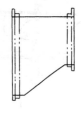

double taper reducer for rectangular duct
双面偏矩形异径管

single taper reducer for rectangular duct
单面偏矩形异径管

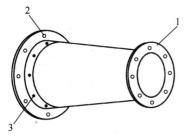
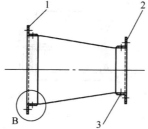

1 flange 法兰
2 bolt hole 螺栓孔
3 rivet 铆钉

reducer for round duct 圆形风管异径

12.3.4.5 手动对开多叶调节阀 Manual Opposed Multi-Blade Regulating Damper

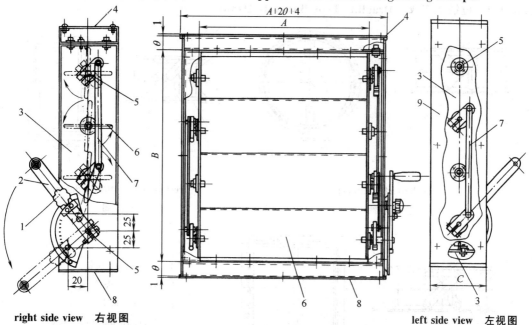

right side view 右视图

1 lock bolt 锁紧螺栓
2 hand grip 手柄
3 body of damper 阀体
4 upper cover plate 上盖板
5 blade shaft 叶片轴

6 blade 叶片
7 linkage tie bar 联动杆
8 lower cover plate 下盖板
9 side cover plate 侧盖板

left side view 左视图

12.3.4.6 手柄式蝶阀 Hand Grip Type Butterfly Damper

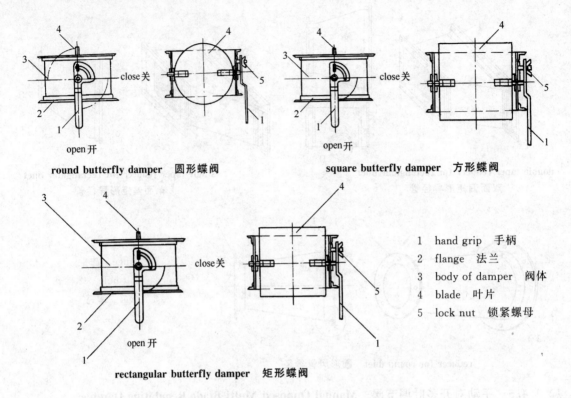

round butterfly damper 圆形蝶阀

square butterfly damper 方形蝶阀

rectangular butterfly damper 矩形蝶阀

1 hand grip 手柄
2 flange 法兰
3 body of damper 阀体
4 blade 叶片
5 lock nut 锁紧螺母

12.3.4.7 拉链式蝶阀 Dragline Type Butterfly Damper

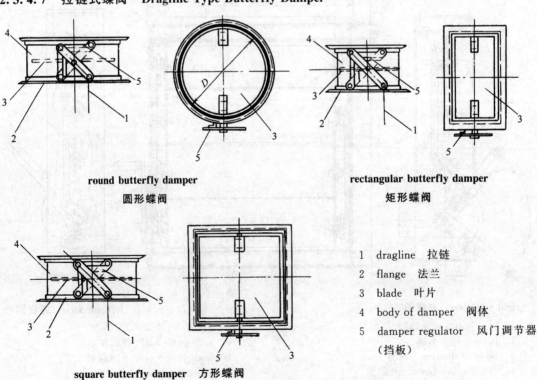

round butterfly damper 圆形蝶阀

rectangular butterfly damper 矩形蝶阀

square butterfly damper 方形蝶阀

1 dragline 拉链
2 flange 法兰
3 blade 叶片
4 body of damper 阀体
5 damper regulator 风门调节器（挡板）

12.3.4.8 风管止回阀 Check Damper

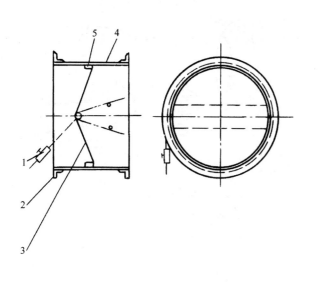
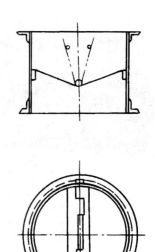

round check damper 圆形风管止回阀

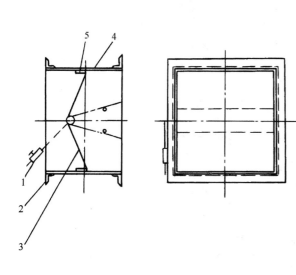
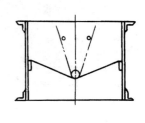
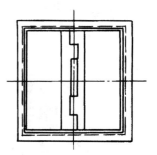

square/rectangular check damper 方形/矩形风管止回阀

1　heavy bob　重锤
2　flange　法兰
3　blade　叶片
4　body of damper　阀体
5　damper regulator　风门调节器（挡板）

12.3.4.9 变风量文丘里阀 Variable Air Volume Venturi Valve

 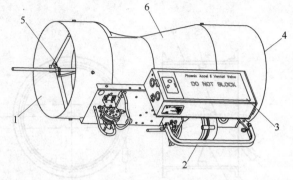

1　air outlet　出风口
2　linkage　连杆
3　control cabinet　控制器
4　air inlet　进风口
5　valve rod　阀杆
6　valve body　阀体

12.3.4.10 抗爆阀 Blast Valve

 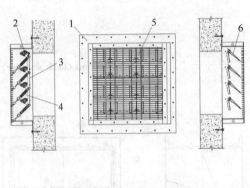

1　frame　框架
2　bearing　轴承
3　spring　弹簧
4　linkage　连杆
5　grid　格栅
6　blade　阀叶

 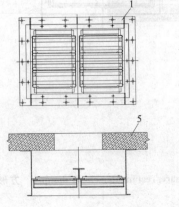

1　frame　框架
2　valve element　阀块
3　clamping mid set　中间连接块
4　clamping end set　端部挡板
5　wall　墙体

12.3.4.11 通风空调系统送回风口 Supply Air Outlet and Return Air Intake for Ventilation and Air Conditioning System

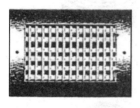

double deflection register
双层百叶风口

square diffuser 方形散流器

single deflection register
单层百叶风口

square grid return air intake 蛋格式回风口

single deflection side-wall grille
侧壁格栅风口

open type single deflection side-wall grille
可开启侧壁格栅风口

return air intake for floor 地板回风口

 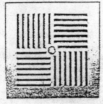

twist outlet 旋流风口　　　　**circular diffuser for floor** 圆形地板散流器

weather proof fixed louver board type air intake
防水固定百叶风口　　　　**globe nozzle type outlet** 球形喷射风口

 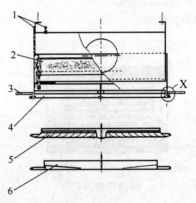

supply air outlet with high efficiency fiter 高效过滤器风口

1　piezometer tube　测压管
2　high efficiency filter　高效过滤器
3　ceiling board　天花板
4　diffusing pore plate outlet　扩散孔板风口
5　diffuser (interchangeable)　散流器风口（可换件）
6　twist outlet (interchangeable)　旋流风口（可换件）

12.3.4.12 防火阀 Fire Damper

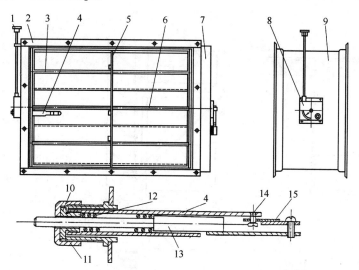

1　hand wheel　手轮
2　flange　法兰
3　blade shaft　叶片轴
4　temperature melting breaker　温度熔断器
5　reinforced linkage bar　加强联杆
6　main shaft　主动轴
7　cover plate of adjustable linkage bar　调节联杆盖板
8　regulator　调节器
9　body of damper　阀体
10　gland bush of spring　弹簧压盖
11　gland nut　压紧螺母
12　pull-out piece spring　压簧
13　draw rod of fusible link bar　易熔片拉杆
14　screw　螺钉
15　fusible link bar　易熔片

12.3.4.13 排气罩 Exhaust Hood

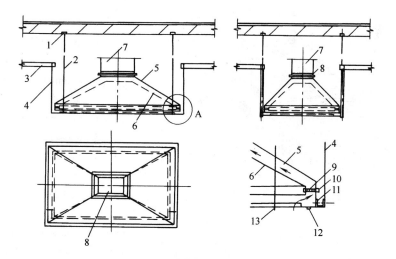

1　expansion bolt　膨胀螺栓
2　hanger rod　吊杆
3　ceiling board　天花板
4　decorative board　装饰板
5　outer sheet of hood　外罩板
6　inner sheet of hood　内罩板
7　exhaust duct　排风管
8　air outlet　出风口
9　sleeve　套管
10　bolt　螺栓
11　flange of inlet port　罩口法兰
12　drain hole　排水口
13　nut　螺母

12.3.4.14 风管支吊架 Bracket Support and Hanger Frame for Duct

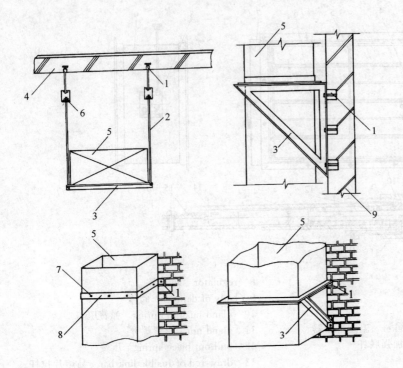

1. expansion bolt 膨胀螺栓
2. hanger rod 吊杆
3. angle steel 角钢
4. slab 楼板
5. duct 风管
6. rubber isolator 防振器
7. tapping screw 自攻螺钉
8. flat steel 扁钢
9. wall 墙

12.3.4.15 风管保温 Insulation of Duct

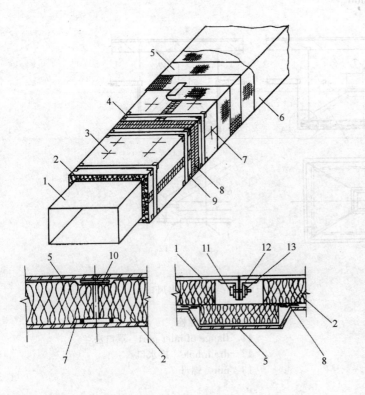

1. duct 风管
2. insulation board 保温板
3. aluminum foil glasscloth 铝箔玻璃布
4. iron wire 铁丝
5. glasscloth 玻璃布
6. fireproof paint 防火涂料
7. insulation nail 保温钉
8. adhesive tape 胶带
9. angle sheet steel 加固薄角铁
10. self-locking packing ring for insulation nail 保温钉自锁垫圈
11. bolt 螺栓
12. flange of duct 风管法兰
13. nut 螺母

12.3.4.16 管式消声器 Cell Type Attenuator

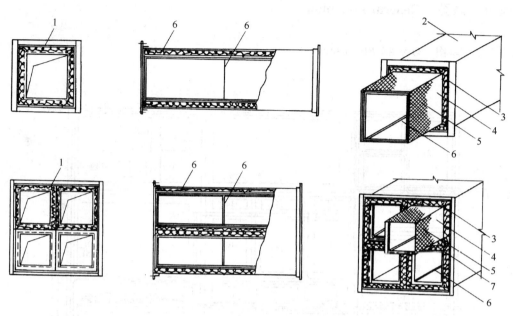

1 flange 法兰
2 galvanized steel sheet outside casing plate 镀锌钢板外壳
3 extra-fine glass-wool felt 超细玻璃棉毡
4 glasscloth 玻璃布
5 galvanizedsteel wire mesh 镀锌钢丝网
6 reinforcing wire 钢筋框
7 plaster slab 石膏板

12.3.4.17 片式消声器 Plate Type Attenuator

 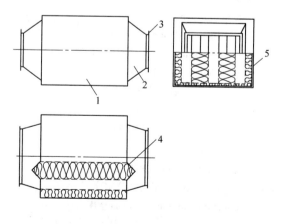

1 shell 外壳
2 reducing pipe 异径管
3 flange 法兰
4 absorption plate 消声片
5 absorption cavity 消声腔

12.4 总图运输 General Plot Plan and Transportation

12.4.1 总图 General Plot plan

12.4.1.1 总图 Overall Plot Plan

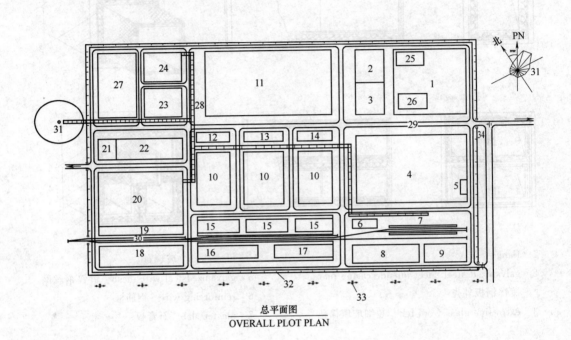

总平面图
OVERALL PLOT PLAN

1	administration area 行政办公区	19	coal storage 煤堆场
2	fire-fighting station 消防站	20	power station 热电站
3	maintenance workshops Center 检维修中心	21	demineralized water and condensate polishing station 脱盐水及凝液精制
4	tank farm 罐区		
5	scale station 地衡站	22	general substation 总变电站
6	acid and alkali station 酸碱站	23	air separation & compression 空分及空压
7	liquid unloading station 液体卸车站	24	water supply and fire water pump station 给水及消防水泵站
8	open storage area 露天堆场		
9	truck parking area 汽车停车场	25	central control building 中央控制楼
10	process area 工艺装置区	26	analytical laboratory center 分析化验中心
11	future process area 预留装置区	27	cooling water system 冷却水系统
12	substation 变电所	28	pipe rack 管廊
13	local control room 现场控制室	29	road 道路
14	workshop office 车间办公室	30	railway 铁路
15	production warehouse 成品仓库	31	flare 火炬
16	chemical warehouse 化学品库	32	main entrance and exit 主进出口
17	spare part warehouse 备品备件库	33	fence 围墙
18	wastewater treatment 污水处理场	34	high voltage corridor 高压线走廊

12.4.1.2 风玫瑰 Wind Rose

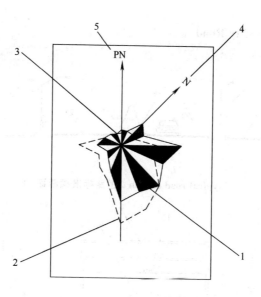

1　yearly prevailing wind　全年主导风向
2　summer prevailing wind　夏季盛行风向
3　yearly minimum frequent wind　全年最小频率风向
4　survey north (true north)　测量北（真北）
5　plant north　工厂北

12.4.1.3 地形图 Topographic Map

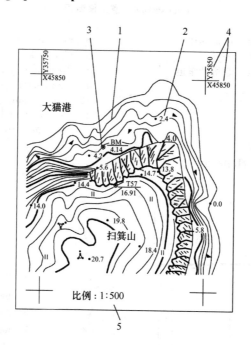

1　contour line　等高线
2　survey elevation　测量标高
3　bench mark (BM)　水准点
4　survey coordination system　测量坐标系统
5　drawing scale　图纸比例

12.4.2 道路 Road

12.4.2.1 城市型道路 City Road

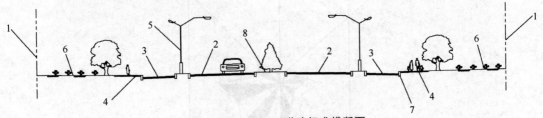

typical road section 道路标准横断面

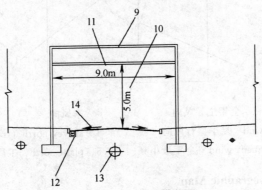

road section 道路横断面

1 road property line 道路征地界线
2 motor vehicle lane 机动车道
3 non-motor vehicle lane 非机动车道
4 sidewalk 人行道
5 road light 路灯
6 green belt 绿化带
7 curb 路缘石
8 median divider 中央分割带
9 pipe rack 管架
10 vertical clearance 垂直净空
11 horizontal clearance 水平净空
12 catch basin 汲水口
13 under ground pipe 地下管线
14 pavement transverse slope 路面横坡

12.4.2.2 公路型道路 Highway

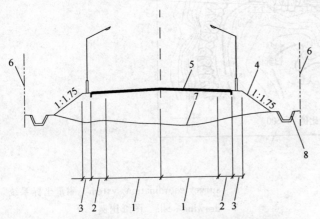

1 carriageway 行车道
2 hard shoulder 硬路基
3 clay shoulder 土路基
4 side slope 边坡
5 pavement 路面
6 road property line 道路征地界线
7 natural ground 自然地形线
8 side ditch 边沟

typical section for highway 公路型道路标准横截面

12.4.2.3 道路结构　Road Structure

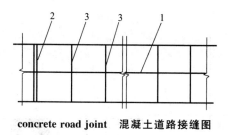

concrete road joint　混凝土道路接缝图

1　longitudinal joint　纵缝
2　expansion joint　胀缝
3　contraction joint　缩缝

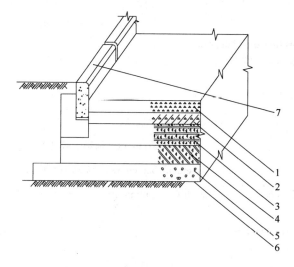

typical structure details for bituminous concrete road
沥青混凝土路面典型结构

1　wearing course of bituminous concrete　沥青混凝土磨耗层
2　binder course of bituminous concrete　沥青混凝土连结层
3　up-base course　上基层
4　down-base course　底基层
5　bed course　垫层
6　compacted subgrade　夯实路基
7　curb　道牙

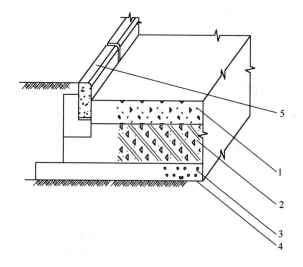

typical structure details for concrete road
混凝土路面典型结构

1　concrete surface course　混凝土面层
2　base course　基层
3　bed course　垫层
4　compacted subgrade　夯实路基
5　curb　道牙

12.4.3 铁路 Railway

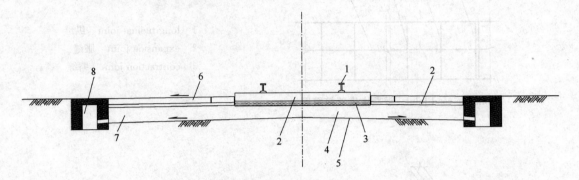

typical section for plant railway
厂内铁路标准横断面

1 rail 轨道
2 sleeper board 轨枕板
3 bed course of sleeper 轨枕板垫层
4 base course (crushed stone) MZ 碎石路基
5 subgrade 路槽底
6 (hard) pavement 铺砌（硬化）地面
7 pavement bed course 铺砌地面垫层
8 ditch (with blind drain) 排水沟（带盲沟）

12.4.4 厂区竖向 Plant Vertical Arrangement

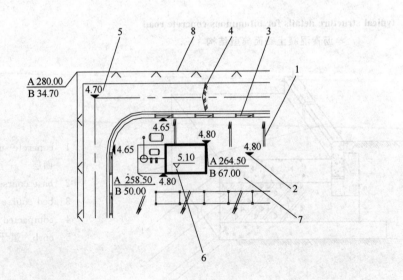

1 drainage direction 排水方向
2 design elevation 设计标高
3 covered ditch 盖板明沟
4 road crown type 路拱形式
5 road center elevation 路中心标高
6 building floor elevation 建筑物室内标高
7 building coordinates 建、构筑物坐标
8 fence 围墙

12.4.5　土方工程　Earth Work

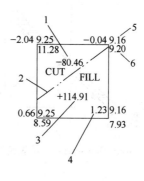

1　excavation quantity　挖方量
2　zero line　填挖零线
3　fill quantity　填方量
4　fill or excavation depth　填挖高度
5　design ground level　设计标高
6　natural ground level　原地面标高

12.4.6　绿化　Greening

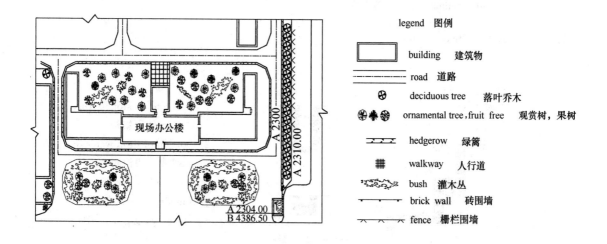

greening plan　绿化平面图

12.4.7 其他构筑物　Other Structure

12.4.7.1 护坡　Slope Protection

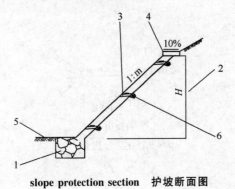

slope protection section　护坡断面图

1　slope protection foundation　护坡基础
2　slope protection height　护坡高度
3　seepage hole　泄水孔
4　top cover　封顶
5　ground　地面
6　crushed stone filter　碎石滤层

12.4.7.2 急流槽　Chute

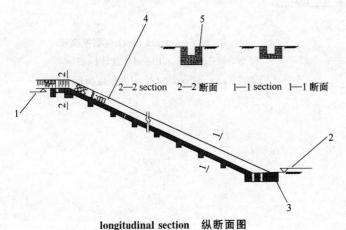

longitudinal section　纵断面图

1　bottom of chute top　（急流槽顶）边沟底
2　low lands　地洼地
3　foundation　基础
4　slope　坡度
5　concrete or stone strcture　混凝土或块石结构

12.4.7.3 跌水　Drop Water

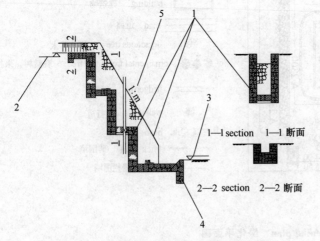

drap water vertical alignment　跌水纵断面图

1　concrete or stone structure　混凝土或块石结构
2　bottom of drop water top　（急流槽顶）边沟底
3　low lands　地洼地
4　foundation　基础
5　slope　坡度

12.4.7.4 挡土墙 Retaining Wall

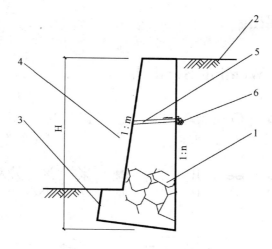

retaining wall section 挡土墙断面

1　stone or concrete structure　块石或混凝土结构
2　filling soil behind the retaining wall　墙后填土
3　toe of retaining wall　挡土墙墙址
4　slope　坡度
5　seepage hole　泄流孔
6　crushed stone filter　碎石滤层

12.4.7.5 涵洞 Culvert

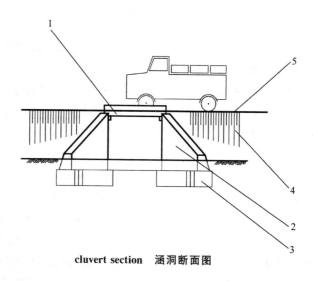

cluvert section 涵洞断面图

1　box culvert cover　箱涵盖板
2　inlet and outlet side-wall　进出口八字翼墙
3　side-wall foundation　翼墙基础
4　side slope　边坡
5　road surface　路面

13 焊接 Welding

13.1 焊接符号 Welding Symbols

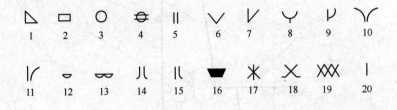

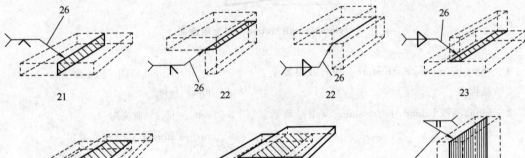

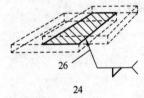

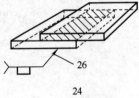

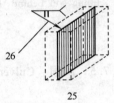

1 fillet 填角焊
2 plug or slot 塞焊或槽焊
3 spot or projection 点焊或凸焊
4 seam 缝焊，滚焊
(5～11 groove 坡口)
5 square 直边坡口
6 V V形坡口
7 bevel 单边V形坡口
8 U U形坡口
9 J J形坡口
10 flare V 喇叭V形坡口
11 flare bevel 喇叭单边V形坡口
12 back or backing 封底焊
13 surfacing 堆焊
(14，15 flange 翻边)
14 edge 双翻边
15 corner 单翻边
16 arc-seam or arc-spot 电弧焊缝或电弧焊点
17 resistance spot 电阻焊焊点
18 projection 凸焊
19 resistance seam 电阻焊焊缝
20 flash or upset 闪光对焊或电阻对焊
21 butt joint 对接接头
22 corner joint 角接接头
23 tee joint T字接头
24 lap joint 搭接接头
25 edge joint 端接接头
26 arrow of welding symbol 焊接符号的箭头

13.2 金属焊接 Welding of Metals

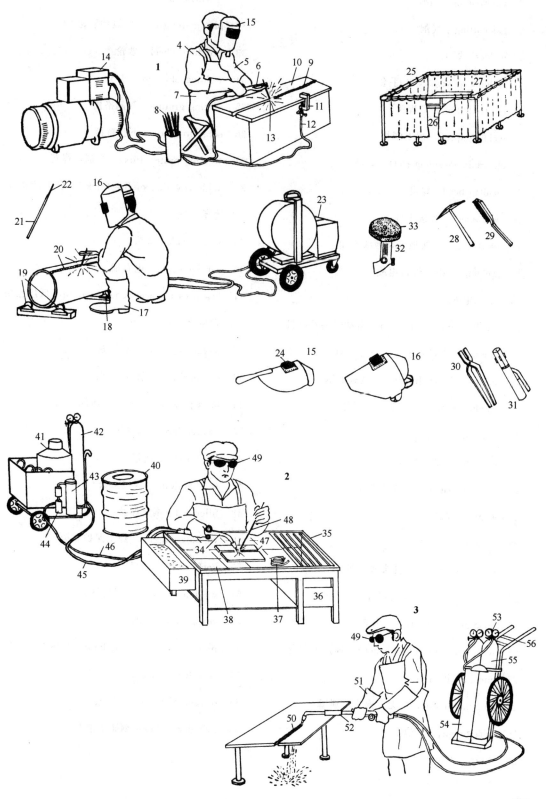

1	metallic are welding	金属电弧焊
2	gas welding	气焊
3	gas cutting	气割
4	welder	焊工
5	welding gloves	焊工手套
6	electrode holder	焊钳
7	welding apron	焊工围裙
8	electrode（welding rod）	焊条
9	welding bead	焊道
10	workpiece	焊接工件
11	earth clamp	接地卡，地线卡钳
12	welding cable	焊接电缆
13	arc	电弧
14	welding motor generator	电动焊接发电机
15	hand shield（helmet）	手持式面罩
16	head shield（helmet）	头戴式面罩
17	welding spats	焊工护脚，鞋罩
18	idler roll	从动滚胎
19	drive roll	驱动滚胎
20	longitudinal seam	纵向焊缝
21	coating	药皮
22	core wire	焊芯
23	welder with taps	抽头式焊机
24	screening glass	护目镜片
25	welding booth	焊接工作间
26	work table（welding bench）	焊接工作台
27	curtain	围帘
28	chipping hammer	尖头榔头，刨锤
29	wire brush	钢丝刷
30	tongs	夹钳
31	electrode-holder with insulated handle	带绝缘手柄的焊钳
32	pneumatic grinder	风动砂轮
33	grinding wheel	砂轮
34	welding blowpipe	焊炬
35	cutting grating	气割格栅
36	scrap-iron box	下脚料箱
37	welding paste（flux）	焊剂，焊药
38	table covering of fire bricks	耐火砖面工作台
39	water tank	水箱
40	carbide drum	电石桶
41	low pressure acetylene generator	低压乙炔发生器
42	oxygen cylinder	氧气瓶
43	gas purifier	气体净化器
44	back pressure valve	背压阀
45	oxygen hose	氧气带
46	gas hose	瓦斯带
47	welding joint	焊接接头
48	filler rod	填充焊丝
49	welding goggles	焊工保护眼镜
50	kerf	切缝，切口
51	welding sleeve	焊工套袖
52	cutting torch	割炬
53	pressure reducing regulator	减压器
54	acetylene cylinder	乙炔瓶
55	oxygen cylinder	氧气瓶
56	cylinder trolley	钢瓶手推车

13.3 保护式电弧焊原理　Principles of Shielded Arc Welding

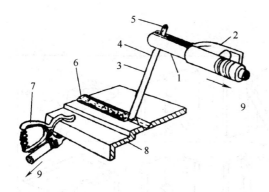
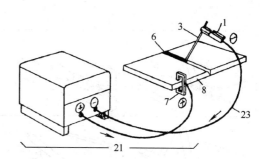
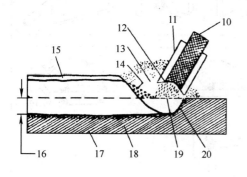
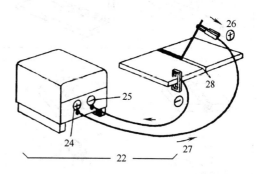

1　electrode holder　电焊钳
2　insulated handle　绝缘手柄
3　electrode　焊条
4　flux-covered portion　药皮部分，涂料覆盖部分
5　bare section for contact　裸露接触部分
6　weld metal　焊缝金属，焊着金属
7　ground clamp　接地卡，地线卡
8　workpiece　焊件
9　to power supply　至电源
10　core wire　焊芯
11　electrode covering, flux　焊条药皮，涂料
12　cup formed on electrode tip　焊条前端形成的杯状
13　gaseous shield　气体保护层，弧焰
14　weld puddle　熔池
15　slag blanket　焊渣覆盖层
16　depth of fusion　熔深，熔透深度
17　base metal　基本金属，母材金属
18　heat affected zone　热影响区
19　arc stream　弧柱
20　weld crater　熔池
21　straight polarity　正极法
22　reverse polarity　反极法
23　welding cable　焊接缆线
24　positive　正极
25　negative　负极
26　direction of welding　焊接方向
27　direction of current flow　电流方向
28　welding joint　焊接接头

13.4 焊接位置、接头形式及焊接形式　Welding Positions, Types of Joints and Welding

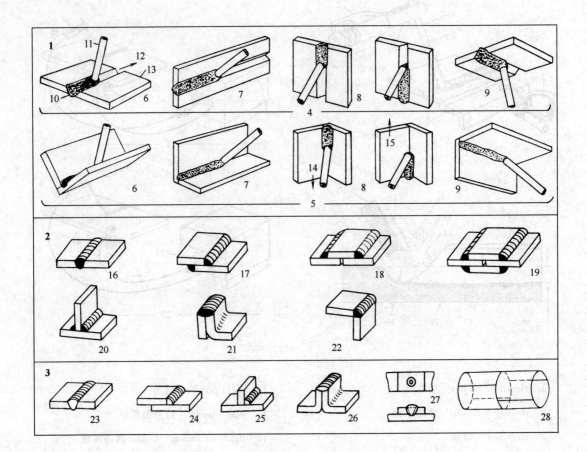

1	welding position 焊接位置	15	up 上向
2	types of joints 接头形式	16	butt joint 对接接头
3	types of welds 焊接形式	17	lap joint 搭接接头
4	groove welds 坡口焊	18	single strapped joint 单面盖板接头
5	fillet welds 填角焊	19	double strapped joint 双面盖板接头
6	flat 平焊	20	tee joint T形接头，T字接头
7	horizontal 横焊	21	edge joint 卷边接头
8	vertical 立焊	22	corner joint 角接接头
9	overhead 仰焊	23	butt welding 对接焊
10	weld metal 焊接金属	24	lap welding 搭接焊
11	electrode 焊条	25	fillet weld 填角焊
12	direction of welding 焊接方向	26	edge weld 角接焊，卷边焊
13	workpiece 工件	27	plug welding 塞焊
14	down 下向	28	circular seam welding 周围焊接，环缝焊接

13.5 坡口 Grooves

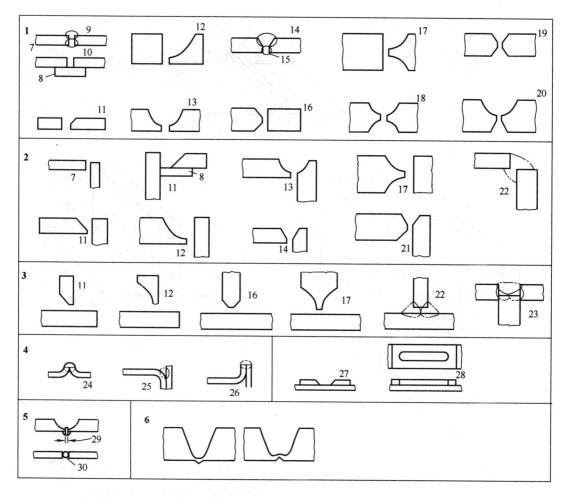

1 grooves for butt joints 对接接头坡口
2 grooves for corner joints 角接接头坡口
3 grooves for T joints T字接头坡口
4 grooves for flange joints 卷边接头坡口
5 grooves for joints using backing filler metal 采用底层填充金属的坡口
6 prepared root joints 锁底接头
7 square groove 直边坡口
8 backing strip 垫板
9 welding from both sides 双面焊
10 welding from one side 单面焊
11 single bevel groove 单面单边V形坡口
12 single J groove 单面J形坡口
13 single U groove 单面U形坡口
14 single V groove 单面V形坡口
15 back run 封底焊
16 double bevel groove K形坡口
17 double J groove 双面J形坡口
18 double U groove 双面U形坡口
19 double V groove X形坡口
20 combination U and V groove U-V组合坡口
21 combination bevel V groove 单边V-V组合坡口
22 double fillet 双面填角
23 three piece T joint 三工件T字接头
24 edge flare 卷边对接
25 single flare bevel 喇叭单边V形，单卷边V形
26 corner flange 卷边端接
27 plug weld 塞焊
28 slot weld 槽焊
29 root opening 根部间隙
30 insert 嵌条

13.6 坡口详图 Detail of Grooves

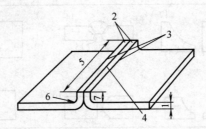
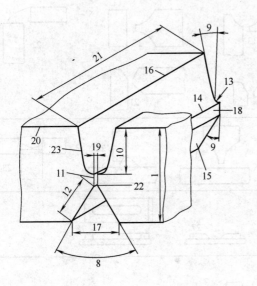
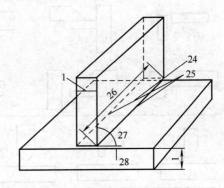

1	thickness of workpiece 工件厚度		15	fusion face (groove face) 坡口面
2	end surface 卷边端面		16	groove long edge 坡口长边
3	flanging long edge 卷边长边		17	groove width 坡口宽度
4	flanging obbutting edge 卷边对边		18	root face (shoulder) 钝边
5	flanging length 卷边长度		19	root gap (root opening) 根部间隙
6	flanging radius 卷边弯曲半径		20	side edge of workpiece 工件侧边
7	flanging height 卷边高度		21	fusion length 焊缝长度
8	groove angle (angle of Vee) 坡口夹角		22	thickness (width) of root face 钝边厚度
9	bevel angle (angle of preparation) 坡口斜角		23	bevel side edge 坡口侧边
10	groove depth 坡口深度		24	front surface 端面
11	root face side edge 钝边侧边		25	front long edge 端面长边
12	bevel length 坡口长度		26	fillet length 角焊缝长度
13	root radius 根部半径		27	fillet angle 角焊缝角度
14	root long edge 钝边长边		28	clearance 端面间隙

13.7 填角焊、角焊 Fillet Welding

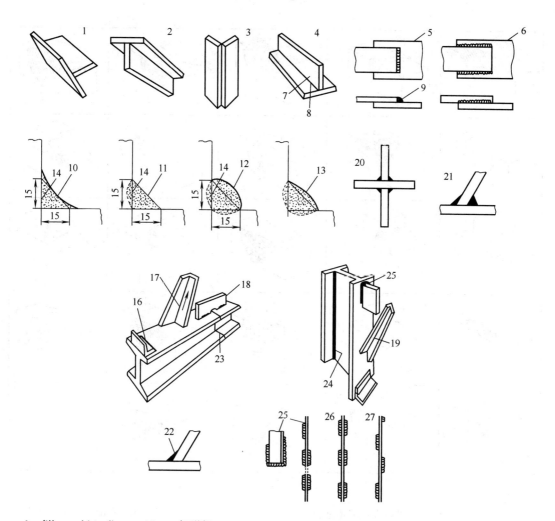

1 fillet weld in flat position 船形焊
2 fillet weld in overhead position 仰角焊
3 fillet weld in vertical position 立角焊
4 fillet weld in horizontal position 横向角焊
5 fillet weld in normal shear 正面角焊缝
6 fillet weld in parallel shear 侧面角焊缝
7 stiffener 加强板
8 weld bone（weld junction） 焊口
9 fillet 角焊缝
（10～13 size of fillet weld 角焊接尺寸）
10 concave fillet weld 凹形角焊缝
11 flat fillet 平顶角焊缝
12 convex fillet weld 凸形角焊缝
13 profile of fillet weld 角焊缝断面形状
14 leg of fillet weld 焊脚
15 leg length 焊脚长

16 light fillet weld 小角缝焊接
17 upward welding in inclined position 上坡焊
18 strongback 加强板
19 overhead position welding with weld inclined 倾斜仰焊
20 cruciform joint 十字接头
21 inclined tee joint 斜交 T 形接头
22 oblique fillet weld 斜交角焊缝
23 melt run（line of fusion） 熔合线
24 continuous fillet weld 连续角焊缝
（25～27 intermittent fillet weld 断续角焊缝）
25 boxing 端部周边焊缝
26 chain intermittent fillet weld 并列断续角焊缝
27 staggered intermittent fillet weld 交错断续角焊缝

13.8 焊接缺陷 Defects of Welding

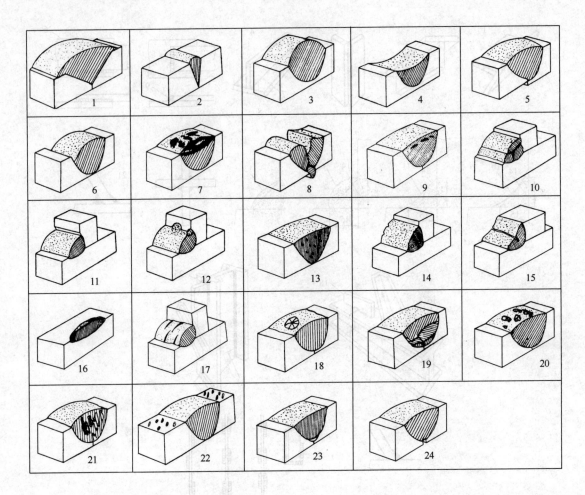

1　excessive width　宽度过大，熔宽过剩
2　insufficient width　宽度太窄，熔宽不足
3　excessive reinforcement　焊缝增强过量
4　insufficient reinforcement　焊缝增强不足
5　mismatch　错位
6　incomplete filled groove　坡口未填满
7　pit　弧坑
8　burn through　焊穿
9　slag inclusion (single-pass weld)　夹渣（单道焊）
10　slag inclusion (multiple-pass weld)　夹渣（多道焊）
11　undercut　咬边，咬肉
12　overlap (lap)　焊瘤
13　porosity (along entire length of weld bead)　气孔（沿整个焊道）
14　wormhole porosity　虫状气孔
15　centerline crack　焊道中心纵向裂纹
16　underbead crack　焊道下裂纹
17　transverse crack　横向裂纹
18　crater crack　弧坑裂纹
19　root crack　根部裂纹
20　fish eye　鱼眼，白点
21　line inclusion　链状夹杂物
22　excessive spatter　严重飞溅
23　incomplete fusion　未熔合
24　incomplete penetration　根部未焊透

13.9 焊接方法及设备 Type of Welding and Welding Machine

13.9.1 自动埋弧焊 Automatic Submerged Arc Welding

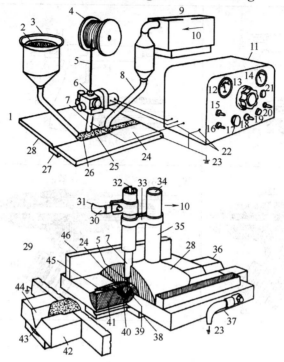

1	typical welding machine 典型焊接机	23	ground 接地
2	flux 熔剂	24	solidified slag 固化焊渣（熔渣）
3	flux hopper 熔剂料斗	25	electrode cable 焊丝导线
4	electrode-wire reel 焊丝盘	26	granular flux 细粒熔剂
5	eletrode-wire 焊丝	27	backing strip 垫板
6	wire feed motor 焊丝送进电动机	28	workpiece 工件
7	contact tube 导电嘴，焊嘴	29	automatic welding 自动焊接
8	unfused-flux recovery tube 未熔熔剂回收管	30	connector 连接头
9	vacuum pick-up system 真空抽吸系统	31	to power supply 至电源
10	direction of travel 焊接走向	32	to automatic wire feed 至焊丝自动给进机构
11	voltage and current control 电压及电流控制装置	33	welding gun 焊枪
12	ammeter 电流表	34	to flux hopper 至熔剂漏斗
13	welding-voltage adjustment 焊接电压调节旋钮	35	flux-delivery tube 熔剂给进管
14	voltmeter 电压表	36	runoff tab 收弧板
15	current adjustment 电流调节旋钮	37	base metal 母材，基体金属焊缝垫板
16	travel control 运行方向控制开关	38	weld backing plate 焊接垫板
17	inch button 微动按钮	39	weld puddle 熔池
18	retract feed 焊丝提件按钮，焊丝缩回按钮	40	arc 电弧
19	weld stop 停焊按钮	41	bead of solidified weld metal 固化金属焊道
20	contactor 开关	42	starting tab 引弧板
21	start 启动	43	joint 接合处
22	voltage-pickup leads 电源插线	44	welding groove 焊接坡口
		45	flux cover 渣壳
		46	flux dam 熔剂挡板

13.9.2 金属极气体保护焊焊枪 Electrode Guns of Gas Metal-Arc Welding

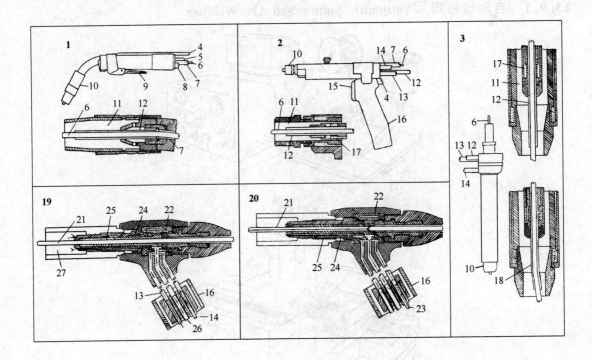

1. air cooled manual electrode holder 空冷式手工焊枪
2. water cooled manual electrode holder 水冷式手工焊枪
3. water cooled automatic electrode holder 水冷式自动焊枪
4. control cable 控制电缆
5. gas duct 气体导管
6. electrode wire 焊丝
7. wire-feed conduit 焊丝导管
8. power inlet 电源馈线
9. lever switch 拨杆开关
10. nozzle 喷嘴
11. shielding gas 保护气体
12. contact tube 导电嘴
13. water inlet 水入口
14. water outlet 水出口
15. trigger switch 启动开关，触发开关
16. handle 手柄
17. water (coolant for contact tube) 水（导电嘴冷却剂）
18. bent contact tube 弯导电嘴
19, 20 torches for manual gas tungsten-arc welding 手工钨极气体保护焊焊炬
21. tungsten electrode 钨极
22. collet nut 连接螺母
23. current cable 电缆
24. collet holder 钨极嘴
25. collet 钨极夹子
26. shielding gas inlet 保护气体入口
27. shielding gas outlet 保护气体出口

13.9.3 气体保护电弧焊 Gas-Shielded Arc Welding

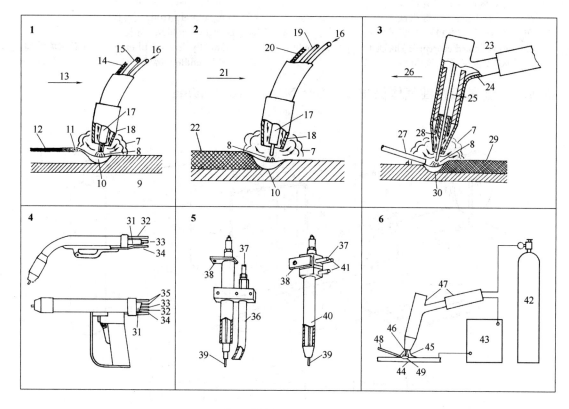

1 gas shielded flux-cored welding 管状焊丝气体保护焊
2 gas-metal arc welding (MIG welding) 金属极惰性气体保护焊,熔化极惰性气体保护焊
3 gas tungsten-arc welding (TIG welding) 钨极惰性气体保护电弧焊
4 guns for semis utomatic welding 半自动焊焊枪
5 nozzle assemblies for full automatic welding 全自动焊焊嘴
6 manual TIG welding 手工钨极惰性气体保护电弧焊
7, 37, 45 shielding gas 保护气体
8 arc 电弧
9, 44 work 工件
10 molten weld metal 熔池
11 molten slag 熔渣
12 solidified slag 焊渣
13, 26 travel 焊接方向

14, 20 current conductor 导线
15 flux-cored electrode 管状焊丝
16 shielding gas in 保护气体入口
17 wire guide and contact tube 送丝管及导电嘴
18 gas nozzle 气体喷嘴
19 solid wire electrode 实心焊丝
21 travel 焊接方向
22 solidified weld metal 焊缝金属
23 TIG torch 钨极保护焊焊炬
24 electrical conductor 导体连线
25 tungsten electrode 钨极
27 filler wire 填充焊丝
28 gas passage 气体通道
29 solidified weld metal 焊缝金属
30 molten weld metal 熔融焊接金属
31 power cable 电缆
32 contactor leads 引线
33, 39 electrode 电极
34 gas line 气体管线

783

35	water lines 水管线	41	circulating water 循环水
36	air cooled side shielded nozzle assembly 空冷式侧面保护气体喷嘴	42	gas supply 气源
		43	welding machine 电焊机
38	power leads 电源引线	46	tungsten electrode 钨极
39	electrode 电极	47	electrode holder 焊枪
40	water cooled concentric shielded nozzle assembly 水冷式同心保护气体喷嘴	48	manually feeding filler rod 手工充填焊丝
		49	weld puddle 熔池

13.9.4 金属极惰性气体保护焊 Gas Metal-Arc Welding

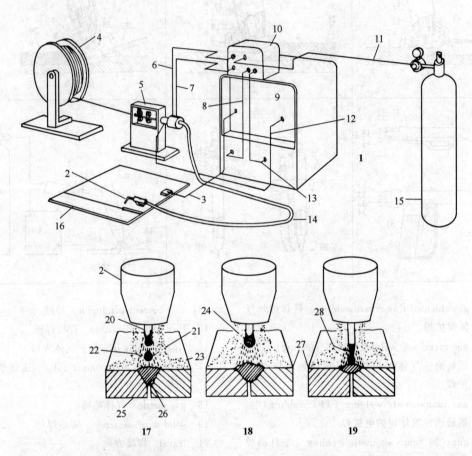

1 MIG welding 金属极惰性气体保护焊
2 electrode holder 焊枪
3 ground 接地，地线
4 electrode-wire supply 焊丝盘
5 feed-roll mechanism 送丝机构
6 gun control 焊枪控制线
7 gas line (out) 气体出口管线
8 contactor control 开关控制线
9 power supply 电源
10 control system 控制系统
11 gas line (in) 气体入口管线
12 welding current input 焊接电源线
13 110 volt supply 110V 电源
14 flexible conduit 软管
15 gas-supply cylinder 气体钢瓶
16 workpiece 工件
(17～19 Modes of metal transfer 熔滴过渡形式)
17 spray transfer 喷射过渡
18 globular transfer 滴状过渡
19 short-circuiting transfer 短路过渡
20 consumable electrode 熔化极
21 droplet 熔滴
22 arc 电弧
23 shielding gas 保护气体
24 single globule 单滴
25 base metal 基体金属，母材
26 weld puddle 熔池
27 oxide film 氧化膜
28 drop touches weld puddle 熔滴接通熔池

13.9.5 普通电渣焊 Conventional Electroslag Welding

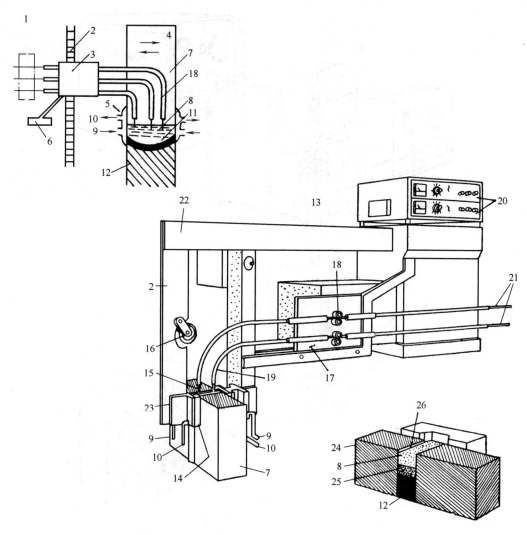

1　principle of electroslag welding　电渣焊原理
2　vertical track (rail)　垂直导轨，立轨
3　welding head　焊接机头
4　oscillated electrode wire　焊丝摆动
5　mold shoes　成型滑套
6　control station　控制台
7　workpiece　焊件
8　molten flux (flux pool)　熔化焊剂，渣池
9　water inlet　进水
10　water outlet　出水
11　metal pool　金属熔池
12　solidified weld metal　焊缝金属
13　equipment of conventional electroplag welding　普通电渣焊设备
14　water-cooled dam　水冷滑块
15　consumable electrode　熔化电极
16　roller guide　导轮
17　wire feed drive　送丝机构
18　electrode-wire feed rolls　焊丝送进轮，送丝滚轮
19　guide tube　送丝导管
20　control panels　控制台
21　electrode wire　焊丝
22　arm　横梁
23　retaining dam　成型滑块
24　base metal　基体金属，母材
25　molten metal　熔化金属
26　surface of workpiece　焊件表面

13.9.6 熔嘴电渣焊 Electroslag Welding by Consumable Guide Tube

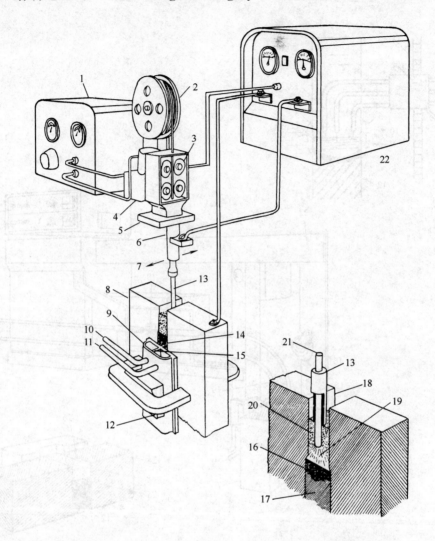

1 control panel 控制盘
2 electrode-wire supply 焊丝盘
3 wire-feed rolls 送丝滚轮
4 wire-feed drive 送丝机构
5 welding head 焊接机头
6 nonconsumable portion of guide tube 导管不熔部分
7 oscillation 摆动方向
8 workpiece 焊件
9 water-cooled dam 水冷滑块
10 water inlet 进水
11 water outlet 出水
12 wedge 固定楔
13 consumable guide tube 熔嘴
14 recess 凹槽
15 baffle 隔板
16 molten metal 熔融金属
17 solidified weld metal 焊缝金属
18 base metal 基体金属, 母材
19 arcless 伪电弧, 弧状体
20 pool of molten flux 渣池
21 electrode wire 焊丝
22 power supply 电源

13.9.7 电气焊 Electrogas Welding

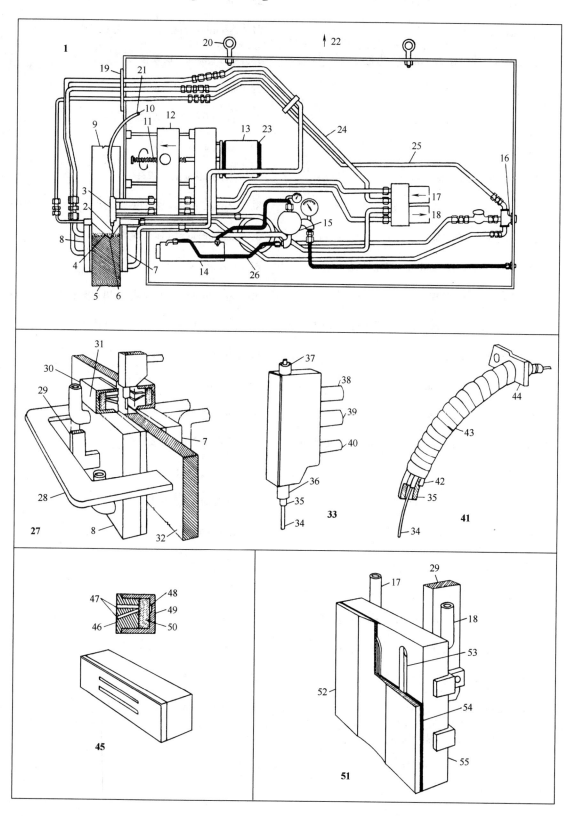

#	English	中文
1	typical unit	典型的设备
2	electrode	焊条
3, 33, 41	electrode holder	焊枪
4	weld puddle	熔池
5	weld metal	焊缝金属
6, 30, 45	gas box	气盒
7	movable dam	活动挡板
8, 51	fixed dam	固定挡板
9	one plate	金属板
10	electrode wire conduit	焊丝导管
11	lead screw	丝杠
12	electrode-holder oscillator	焊条夹持摆动机构
13	motor	电动机
14	air cylinder	气瓶
15	air valve	送气阀
16	shielding-gas manifold assembly	保护气体集合管
17	water inlet	（冷却）水入口
18, 39	water outlet	（冷却）水出口
19	manifold plate	分配板
20	loop	吊环
21	to electrode wire supply	至焊丝送进装置
22	to hoisting mechanism	至悬挂机构
23	to power supply	至电源
24	water	水
25	shielding gas	保护气体
26	air	空气
27	essential components	主要部件
28	strongback	加强板
29	support	支持铁
31	gas inlet	气体入口
32	workpiece	工件
34	electrode wire	焊丝
35	contact tube	导电嘴
36	gas nozzle	气体喷嘴
37	casing	焊丝套管
38	current and water inlet	电流及水入口
40	shielding gas inlet	保护气体入口
42	contact-tube support	导电嘴支持杆
43	insulating tap	绝缘带
44	welding current connection and mounting bracket	通用焊接接头和安装板
46	cover	盖板
47	gas ports	气体口
48	gas channel	气道
49	body	盒体
50	filler	填料
52	face plate	平面
53	water channel	水通道
54	gasket	垫片
55	body (brass)	黄铜基体

13.9.8 管状焊丝电弧焊 Flux-Cored Arc Welding

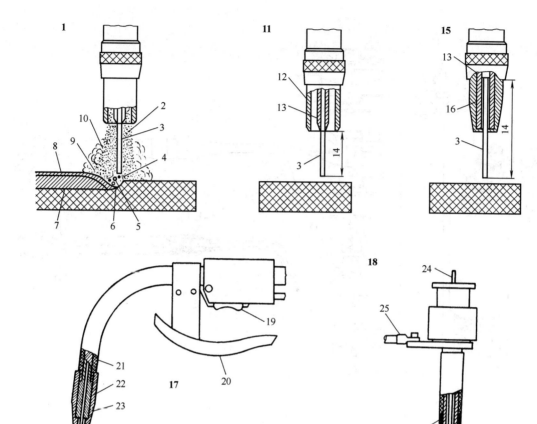

1	flux-cored arc welding with auxiliary gas shielding 管状焊丝辅助气体保护电弧焊	14	electrical stick-out 焊丝伸出长度
2	external gas shielding 外部气体保护层	15	nozzle for use with self shieldng 自保护喷嘴
3	flux-cored electrode 管状焊丝	16	insulated guide wire 绝缘衬套
4	arc 电弧	17	hand-manipulated electrode holder 手工焊焊把
5	metal droplets 熔滴		
6	weld puddle 熔池	18	mechanically manipulated electrode holder 自动焊焊把
7	weld metal 焊缝金属		
8	solidified slag 焊渣	19	switch 开关
9	molten slag 熔渣	20	protective shield 护板
10	shield produced by flux 焊剂保护层	21	flexible guide tube 软芯焊嘴
11	nozzle for use with auxiliary gas shielding 辅助气体保护喷嘴	23	insulated extension guide 绝缘外套
12	gas cup 气室	24	electrode wite 焊丝
13,22	contact tube 导电嘴,焊嘴	25	connection to supply 电源接头

13.9.9 气焊设备 Gas Welding Equipment

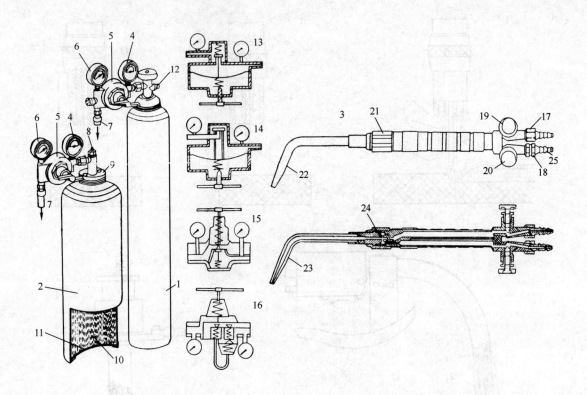

- 1 oxygen cylinder 氧气瓶
- 2 acetylene cylinder 乙炔瓶
- 3 oxyacetylene welding torch 氧炔焊炬
- 4 cylinder-pressure gage 钢瓶压力表
- 5 pressure regulator 减压器
- 6 working-pressure gage 工作压力表
- 7 to torch 至焊炬
- 8 valve stem 阀杆
- 9 fusible plug 易熔塞
- 10 porous material containing 多孔性填料
- 11 safety plug 安全堵盖
- 12 frangible safety plug 易碎安全塞
- 13 single stage regulator 单级式减压器
- 14 two stage regulator 双级式减压器
- 15 direction actuation (nozzle type) regulator 正作用式减压器
- 16 inverse actuation (stem type) regulator 反作用式减压器
- 17 oxygen gland nut 氧气压盖螺母
- 18 acetylene gland nut 乙炔压盖螺母
- 19 oxygen inlet valve 氧气进气阀
- 20 acetylene inlet valve 乙炔进气阀
- 21 sleeve nut 套筒螺母
- 22 welding tip 焊嘴
- 23 equal-pressure mixing chamber 等压式混合管
- 24 injector mixing chamber 射吸式混合管
- 25 hose gland 橡皮管接头

13.9.10 移动式乙炔发生器 Portable Acetylene Generators

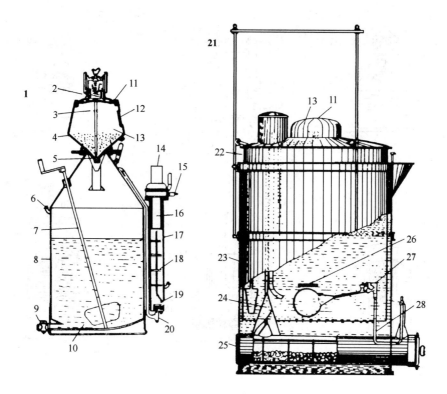

1	medium-pressure generator 中压式发生器	16	filter 滤清器
2	compound spring 压簧	17	flashback arrestor 回火防止器
3	feed rod 加料杆	18	buffle 挡板
4	small granulated carbide 粒状电石	19	hydraulic valve 止回阀
5	feed valve 加料阀	20	drain plug 排放丝堵
6	water inlet 加水口	21	low-pressure generator 低压式发生器
7	agitator 搅拌器	22	rising gas-bell 储气室
8	water tank 水桶	23	outlet pipe 乙炔出口管
9	sludge outlet 出渣口	24	gas collection pipe 集气管
10	bottom 桶底	25	carbide container 电石篮
11	diaphragm 薄膜，隔膜	26	bar for operating valve 溢水阀操纵杆
12	window 看窗	27	float type water control valve 浮球式溢水阀
13	carbide hopper 电石储存室	28	water supply pipe to carbide containers 电石篮供水管
14	safety valve 泄压阀		
15	acetylene outlet 乙炔气出口		

13.9.11 乙炔发生器的基本形式 Basic Types of Acetylene Generators

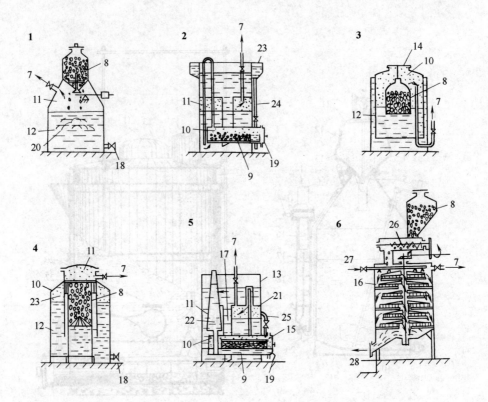

1 carbide to water 电石入水式，投入式
2 water to carbide 水入电石式
3 dipping type 沉浮式
4 recession type 排水式
5 combination of water to carbide and recession type （水入电石和排水）联合式
6 dry-residue type 干渣式
7 acetylene 乙炔
8 carbide holder 电石篮
9 receive carbide tray 电石筐
10 generating chamber 发生室
11 gas holder 储气室
12 water tank 水桶，水箱
13 shell 桶体
14 floating-bell gas holder 浮钟式储气室
15 communicating-vessel gas holder 与锥形罩连通的储气室
16 constant gas holder 固定式储气室
17 circulating pipe 循环管
18 sludge outler 放污口
19 lid 端盖
20 grate 格栅
21 partition 隔板
22 cone-shaped vessel 锥形罩
23 annular reservoir 环形水封室
24 infeed pipe 注水管
25 siphon tube 虹吸管
26 angle guide 斜导轨
27 water 水
28 calcium hydroxide 氢氧化钙

13.10 塑料焊接 Welding of Plastics

(1) 塑料焊接 (1) Welding of Plastics (Ⅰ)

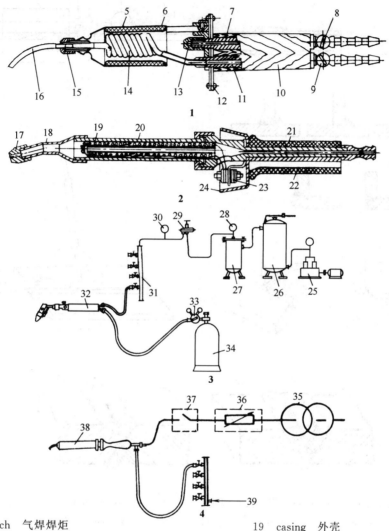

1　gas torch　气焊焊炬
2, 38　electrical torch　电焊焊炬
3　diagram of feeding for gas torch　气体焊炬给料简图
4　diagram of feeding for electrical torch　电气焊炬给料简图
5　heat-insulating layer　隔热层
6　casing for spiral tube　盘管外壳
7　tubing for combustible gas　可燃气管
8　cock　旋塞
9　regulating cock　调整旋塞
10　handle　手把
11, 21　tubing for gas heat-carrier　载热气体管
12　shield plate　护板
13　mixer　混合器
14　spiral tube　盘管
15　connecting nut　连接螺母
16　nozzle　喷嘴
17　nozzle nipple　喷嘴头
18　nozzle tubing　喷嘴管
19　casing　外壳
20　ceramic tubing　陶瓷管
21, 22　handle　手把
23　clamping bolt　夹住螺栓
24　shield cup　杯形护板
25　compressor unit　压缩机组
26　tank for compressed air　压缩空气罐
27　oil-water separator　油水分离器
28　manometer for inlet　入口压力计
29　reductor　减压器
30　manometer for outlet　出口压力计
31　manifold　集管
32　torch　焊炬
33　pressure regulator　减压器
34　heating gas cylinder　加热气体瓶
35　transformer　变压器
36　rheostat　可变电阻器
37　switch　开关
39　compressed air inlet　压缩空气入口

(2) 塑料焊接 (2)　Welding of Plastics (Ⅱ)

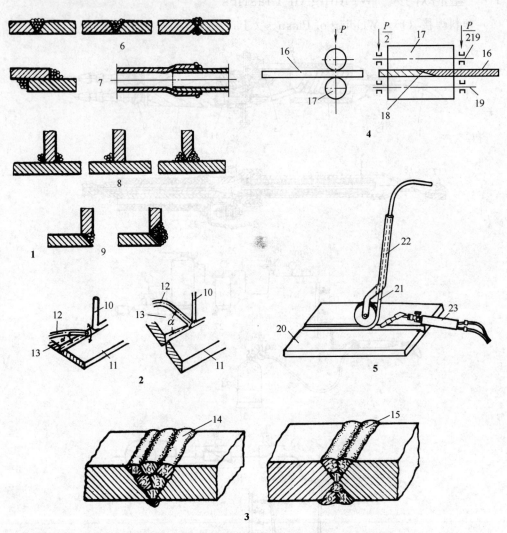

1　types of welded connections　焊接接头型式
2　angle of inclination of the torch nozzle for weld seam　焊炬喷嘴对焊缝的倾斜角
3　sequence of the laying of electrode　焊条施焊顺序
4　welding of thermplastics sheet without electrode　热塑塑料板的无焊条焊接
5　apparatus for welding of polyethylene with electrode　用焊条焊聚氯乙烯的焊接工具
6　butt joint　对接接头
7　lap joint　搭接接头
8　tee joint　T形接头
9　corner joint　角接头
10　electrode　电极
11　workpiece　工件
12，23　torch　焊炬
13　angle of inclination　倾斜角
14　V-shape groove　V形槽
15　X-shape groove　X形槽
16　weldment　焊接件
17　pressing roller　压滚
18　weld seam　焊缝
19　bearing　轴承
20　plates to be welded　待焊的板
21　welding rod　焊条
22　rod-feeder　送条器

13.11 焊接术语 Welding Term

1. base metal, parent metal 母材金属
2. continuous weld 连续焊缝
3. depth of fusion 熔深
4. double groove 双面坡口
5. fillet weld leg 焊脚
6. girth weld, circumferential weld 环缝
7. groove face 坡口面
8. heat-affected zone 热影响区
9. intermittent weld 断续焊缝
10. overheated zone 过热区
11. post heat 后热
12. post weld heat treatment 焊后热处理
13. preheat 预热
14. residual stress 焊接残余应力
15. root of joint 接头根部
16. root opening 根部间隙
17. seal weld 密封焊缝
18. single groove 单面坡口
19. spiral weld, helical weld 螺旋形焊缝
20. weld ability 焊接性
21. welding deformation 焊接变形
22. welding procedure 焊接工艺
23. welding procedure specification 焊接工艺规范
24. welding process 焊接方法
25. welding parameter 焊接参数
26. welding residual deformation 焊接残余变形
27. welding stress 焊接应力
28. weld length 焊缝长度
29. weldment 焊件
30. weld metal 焊缝金属
31. weld reinforcement 余高
32. weld root 焊根
33. weld toe 焊趾
34. weld width 焊缝宽度
35. weld zone 焊缝区

13.12 熔焊术语 Fusion Welding Term

1. arc length 弧长
2. arc stability 电弧稳定性
3. backgouging 清根
4. backing welding 打底焊
5. bead 焊道
6. fusion welding 熔焊
7. gas metal arc welding (GMAW) 气体保护电弧焊
8. gas tungsten arc welding (GTAW) 钨极惰性气体保护焊
9. inclination of weld axis 焊缝倾角
10. polarity 极性
11. pulsed arc 脉冲电弧
12. reversed polarity 反接
13. spatter 飞溅
14. striking 引弧
15. straight polarity 正接
16. surfacing 堆焊
17. welding arc 焊接电弧
18. welding position 焊接位置
19. weld rotation 焊缝转角

13.13 钎焊术语 Soldering Term

1. brazing 硬钎焊
2. soldering 软钎焊

13.14 焊接材料术语 Welding Material Term

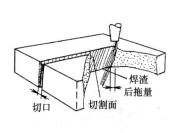
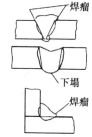

1. backing 焊接衬垫
2. burn-through 烧穿
3. chipping hammer 敲渣锤
4. cold crack 冷裂纹
5. delayed crack 延迟裂纹
6. drag 后拖量
7. excessive penetration 下塌
8. face of cut 切割面
9. fixture 焊接夹具
10. hot crack 热裂纹
11. kerf 切口
12. overlap 焊瘤
13. shielding gas 保护气体
14. size of electrode 焊条规格
15. slag 熔渣
16. solidified slag 焊渣
17. spats 护脚
18. usability of electrode 焊条工艺性
19. weld crack 焊接裂纹
20. welding glove 焊工手套
21. welding material 焊接材料
22. welding wire 焊丝

14 无损检验 Non-destructive Testing

14.1 无损检验方法 Non-destructive Testing Method

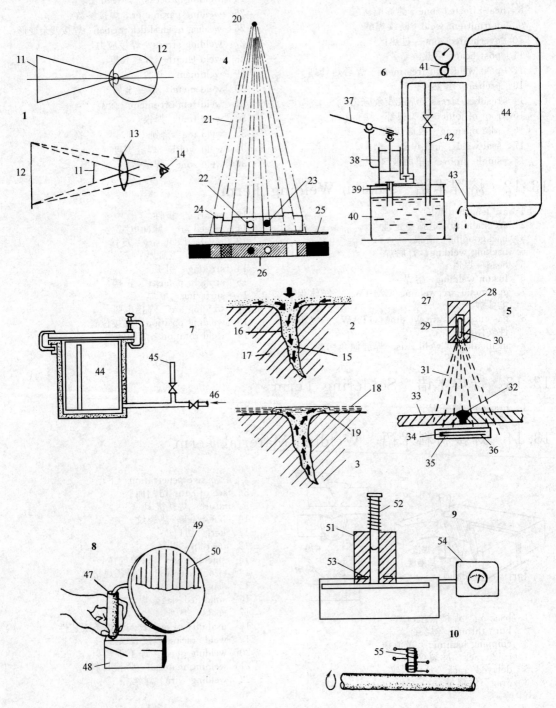

1　visual testing　肉眼检查，外观检验
2　dye penetrant inspection　着色渗透试验
3　fluorescent penetrant inspection　荧光渗透试验
4　X-ray radiography　X射线照相
5　gamma radiography　γ射线照相
6　hydraulic testing　水压试验
7　pneumatic testing　气密试验
8　ultrasonic test　超声波探伤
9　magnetic particle test　磁粉探伤
10　eddy current test　涡流探伤
11　object　物体
12　retinal image　视网膜成像
13　lens　透镜
14　eye　眼睛
15　penetrant　渗透液
16　crack　裂纹
17，24　specimen　工件，试样
18　developer powder　显色粉
19　indication　伤纹
20　X-ray source　X射线源
21　X-rays　X射线
22　porosity　气孔
23　inclusion　夹渣
25　film　底片
26　X-ray image　X射线影像
27　cast iron casing　铸铁外壳
28　lead cylinder　铅室
29　safeguard　保护器
30　radiative cobalt　放射性钴
31　gamma rays　γ射线
32　weld bead　焊道
33，48　workpiece　工件
34　film　底片
35　film cartridge　底片匣
36　intensifying screen　增感屏
37　lever　杠杆
38　hydraulic pump　水压泵
39　check valve　止回阀
40　water tank　水槽
41　pressure gauge　压力表
42　return pipe　回水管
43　leakage　泄漏
44　pressure vessel　压力容器
45　to atmosphere　排入大气
46　from compressor　来自压缩机
47　ultrasonic flaw detector　超声波探伤器
49　ultrasonic monitor　超声波监视器
50　reflected sound　反射波
51　magnet housing　磁铁套
52　spring lifted magnet　弹簧提升磁铁
53　probe head　探头
54　instrument　仪表
55　double probe-coil　双励磁探头线圈

14.2 无损检验设备 Non-destructive Testing Apparatus

14.2.1 探孔镜及显微镜 Borescope and Microscope

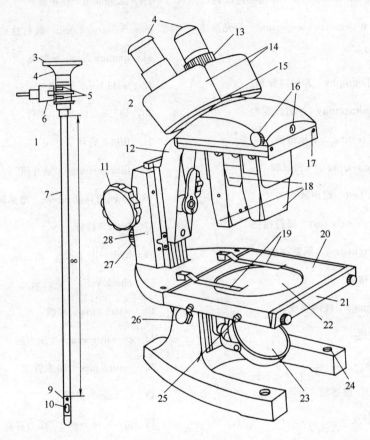

1　right-angle borescope　直角管孔探测镜，直角式探孔镜
2　stereomicroscope　立体显微镜
3　indicating button　基准底线
4　eye piece　目镜
5　contact rings　接触环
6　rotating contact　旋转触点
7　telescope tube　望远镜筒
8　working length　工作长度，观测长度
9　right angle objective lens　直角物镜
10　lamp　灯泡
11　focusing knob　焦距旋钮
12　arm　镜臂
13　acuity balancing adjustment　（双目）平衡微调螺钉
14　prism housings　棱镜外壳
15　eye-spacing adjustment gear　目镜调节齿轮
16　illuminator attachment hole　投光灯安装孔
17　nosepiece　物镜转换器
18　objectives　物镜
19　hold down clips　握持弹簧夹
20　glass stage　载物台
21　removable front strip　活动边框
22　contrast plate　对比板
23　substage mirror　下反光镜
24　mirror attachment hole　反光镜安装孔
25　hand rest attachment screw　手动螺钉
26　substage base　基座
27　thumb screw　指旋螺钉
28　focusing mechanism　聚焦机构

14.2.2　X射线发生器及其线路　X-Ray Tube and Its Circuit

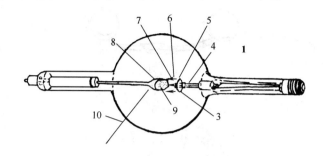

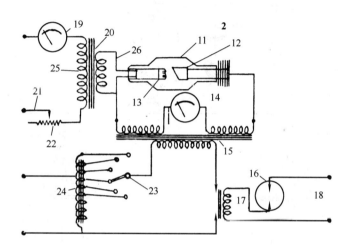

1	coolidge X-ray tube　热阴极X射线发生管	14	milliammeter　毫安表
2	basic X-ray circuit　X射线原理线路	15	H. V. transformer　高压变压器
3	cathode　阴极	16	timer　计时器
4	spiral tungsten filament　螺旋钨丝	17	coil　线圈
5	metal cup　隔电子裙	18	to autotransformer　至自耦变压器
6	electrons　电子	19	ammeter　安培计
7	thermionic emission　热阴极发射	20	filament transformer　灯丝变压器
8	tungsten target　钨极，钨靶	21	230-volt line　230V线路
9	focal spot　焦点	22	rheostat　滑线电阻
10	X-rays　X射线	23	KVP selector switch　高压峰值选择开关
11	line focus X-ray tube　线性聚焦X射线发生管，线焦X射线管	24	autotransformer　自耦变压器
		25	primary circuit　一次回路，初级线圈
12	tungsten filament　钨丝	26	heating current　灯丝加热电流
13	target　阳极靶		

14.2.3 射线照相及电子照相 Photoradiography and Electroradiography

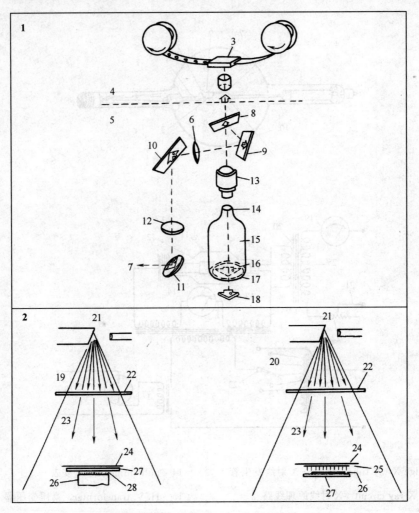

1 photoradiography 射线照相
2 electroradiography 电子照相
3 camera 照相机
4 cine-fluorex 底片显像
5 regular fluorex 正规显像
6 small field lens 小物镜，窄角镜头
7 observer 监测人员
8 1st mirror 第一反光镜
9 2nd mirror 第二反光镜
10 3rd mirror 第三反光镜
11 4th mirror 第四反光镜
12 window 观察孔
13 objective lens 物镜
14 output phosphor 输出荧光
15 fluorex tube 显像管
16 photoelectric surface 光电面
17 fluorescent screen 荧光屏
18 object and X-ray target 物体及X射线目标
19 transmission electron radiography 透射式电子照相
20 back-emission electron radiography 反辐射式电子照相
21 X-ray tube X射线管
22 filter to harden X radiation 硬X射线滤光板
23 X-rays X射线
24 light tight covering 增感屏贴层
25 lead foil 铅箔
26 specimen 工件，试样
27 photographic plate 照相底片
28 photoelectrons 光电子

14.2.4 轻便 X 射线机及透度计　Mobile X-Ray Unit and Penetrometer

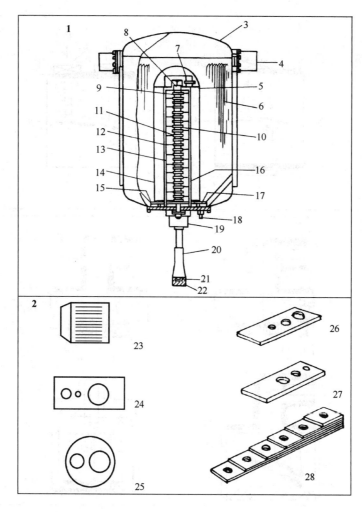

1　mobile X-ray unit　轻便 X 射线机
2　penetrometer　透度计
3　steel tank　钢壳
4　cooler　冷却器
5　end turn filament　端旋灯丝
6　laminated shield　叠层屏蔽
7　variable reactor　可变电抗器
8　cathode assembly　阴极组件
9　first intermediate electrode　第一中间电极
10　glass envelop　玻璃封套
11　shields　护罩
12　tap lead　中接线头
13　glass tie rod　玻璃拉杆
14　secondary coils　二次线圈，次级线圈
15　primary winding　一次线圈，初级线圈
16　insulating filament control shaft　灯丝绝缘调节杆
17　laminated steel bottom　叠层钢制底座
18　filament control motor　灯丝控制电机
19　focusing coil　聚焦线圈
20　lead shield　铅罩
21　tungsten target　对阴极，钨靶
22　lead diaphragm　铅膜，铅隔板
23　wire type penetrometer　金属线型透度计
24　flat type penetrometer　片型透度计，平板式透度计
25　bead type penetrometer　沟槽片型透度计，圆盘式透度计
26　three-hole penetrometer　三孔透度计
27　boiler code penetrometer　锅炉规范透度计
28　step penetrometer　台阶型透度计，阶梯型透度计

14.2.5 超声波探伤方法及探头 Ultrasonic Test Methods and Search Units

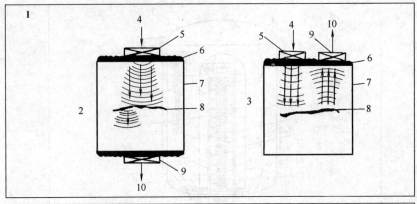

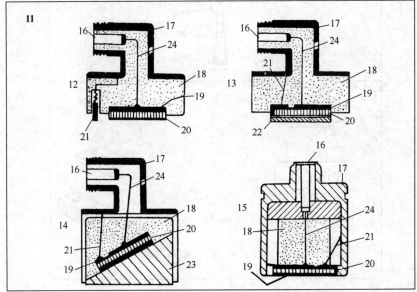

1 basic ultrasonic test methods 超声波探伤的基本方法
2 transmission technique 穿透式超声波探伤法，透射法
3 pulse echo technique 脉冲反射式超声波探伤法，脉冲反射法
4 from electrical generator 自高频电流发生器
5 transmitter transducer 发射探头
6 acoustic couplant 耦合剂，贴合剂
7 specimen 工件
8 internal flaw 内部缺陷
9 to receiver transducer 至接收探头
10 to electrical indicator 至指示器（荧光屏）
11 typical search units 典型的探头
12 straight-beam contact search unit 接触式直探头
13 straight-beam faced unit 表面式直探头
14 angle beam search unit 斜探头
15 immersion search unit 液浸式探头，水浸式探头
16 coaxial connector 同轴接头
17 case 外壳
18 mount and backing 压板
19 electrode 电极
20 crystal （压电）晶片，振子
21 ground connection 接地
22 wear plate 保护片，磨损片
23 wedge 斜楔
24 connected line 接线

14.2.6 超声波发射探头及接收探头 Ultrasonic Transducer and Refraction

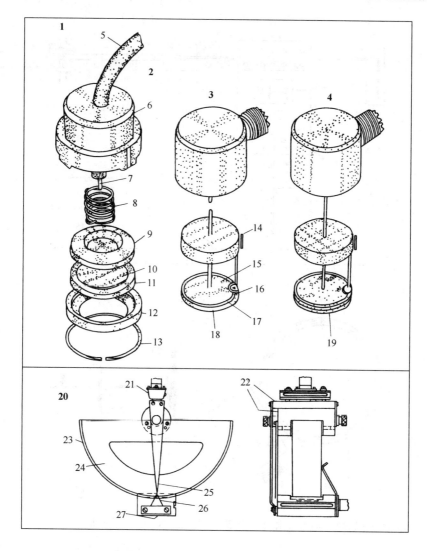

1 ultrasonic transducer 超声波探头
2 edge-damped transducer 角阻尼探头
3 damped transducer 阻尼探头
4 transducer with wear plate 带保护片的探头
5 coaxial cable 同轴电缆
6 holder 把手
7 connection to cable 缆线接头
8 contact spring 压簧
9 contact plate 压板
10 silver coating 银膜套
11 quartz crystal 石英晶片
12 plastic ring 塑料环
13 snap ring 开口环，卡环
14 grounded at holder 把手接地
15 inner silver 内银套膜
16 outer silver 外银套膜
17 piezo electric ceramic 压电陶瓷
18 bonded at assembly 包层
19 wear plate 保护片，磨损片
20 refraction analyzer 折射角测定仪
21 search unit 探头
22 variable angle wedge assembly 活动斜楔组件
23 semi-cylindrical sample 半圆形试样
24 protractor 量角器，半圆规
25 refracted angle indicator 折射角指针
26 incident angle indicator 入射角指针
27 to receiver search unit 至接收探头

14.2.7 配管焊缝的超声波探伤 Ultrasonic Testing of Weld in Piping

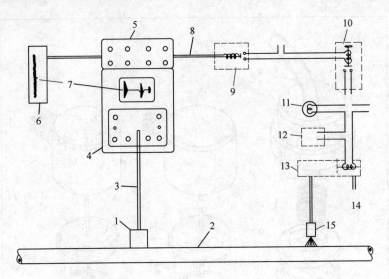

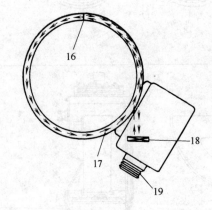

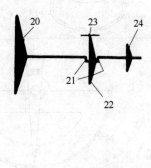

1,18　crystal　（压电）晶片，振子，探头晶片

2　piping　管道

3,19　coaxial cable　同轴电缆

4　reflectoscope　反射式超声波探伤仪

5　RA monitor　反射信号指示器

6　brush recorder　刷式记录仪

7　defect　缺陷波

8　6.3V signal　6.3V 信号

9　6.3V contactor　6.3V 信号接点

10　pneumatic timer　气动定时器

11　signal light　信号灯

12　signal horn　信号笛

13　solenoid operated 4-way valve　电磁四通阀

14　air supply　气源

15　air brush　空气刷，风刷

16　weld zone　焊缝区

17　sound bead　超声波通路

20　initial pulse　始脉冲波

21　weld zone gate　焊缝区截止波

22　reflected sound　反射波

23　alarm level set　报警整定值

24　travel sound　穿透波

14.2.8 成套式渗透探伤仪 Self-Contained Penetrant Inspection Unit

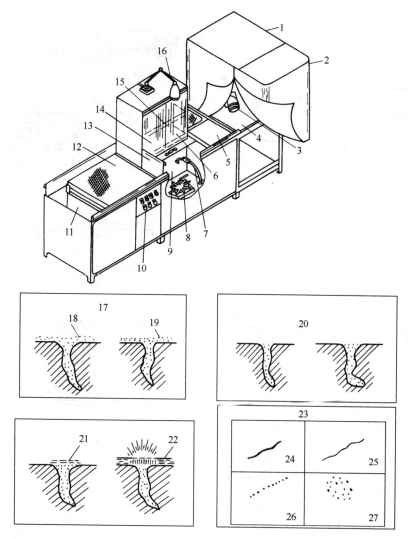

1	inspection hood 试验护罩	15	removable shelf 活动搁板
2	curtain 围帘	16	black light for rinse tank 淋洗池暗室灯
3	black light 暗室灯	17	applying penetrant 涂刷渗透液
4	inspection table 试验台	18	dye penetrant 着色渗透液
5	developer tank 显示罐	19	fluorescent penetrant 荧光渗透液
6	drier curtain 烘箱护罩	20	water spray 水洗
7	handhose 水管	21	dry developer 干粉显示剂,干性显示粉
8	whirlwash 旋转洗盘	22	wet developer 湿粉显示剂,湿性显示粉
9	rinse tank 水池,淋洗池	23	penetrant indications 渗透伤纹
10	control panel 控制盘	24	large crack (opening) 裂缝
11	penetrant tank 渗透液储槽	25	tight crack (cold shut) 裂纹
12	drain pan 脱液盘,搁板	26	partially welded lap 局部焊接皱皮
13	splash shield 防溅护板	27	pits (porosity) 凹痕,气孔
14	drier 干燥箱		

14.3 磁化法 Methods of Magnetization

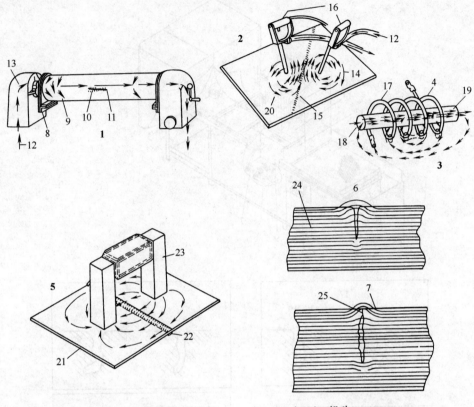

1 circular magnetization 环向磁化
2 transverse magnetization 横向磁化
3 longitudinal magnetization 纵向磁化
4 coil magnetization 线圈磁化
5 yoke magnetization 轭架磁化，永久磁铁磁化
6 surface leakage field 表面泄漏磁场
7 subsurface flaw leakage field 内部缺陷泄漏磁场
8 contact plate 接触板
9 circular magnetic field 环形磁场
10 magnetic particles 磁粉
11 crack 裂纹
12 magnetizing current 磁化电流
13 head 机头
14 magnetic lines of force 磁力线
15 weld 焊缝
16 contact probe 触头
17 wire coil 线圈
18 parts 零件
19 longitudinal magnetic field 纵向磁场
20 localized circular field 局部环形磁场
21 workpiece 工件
22 weld bead 焊道
23 permanent magnet yoke 永久磁铁轭架
24 intensity of magnetizing force 磁力强度
25 depth of defect 缺陷深度

14.4 超声波检测术语 Ultrasonic Testing Term

1 defect 缺陷
2 dynamic range 动态范围
3 refraction 折射
4 sound velocity 声速
5 total reflection 全反射
6 transmission 透射
7 ultrasonic flaw detector 超声探伤仪
8 ultrasonic microscope 超声显微镜
9 ultrasonic testing system 超声检测系统
10 ultrasonic thickness gauge 超声测厚仪
11 ultrasonic wave 超声波
12 wave form 波形
13 wave length 波长

14.5 射线检测术语 Radiation Testing Term

1. absorbed dose 吸收剂量
2. absorption 吸收
3. activity 活度
4. contrast 对比度
5. definition 清晰度
6. development 显影
7. electromagnetic radiation 电磁辐射
8. exposure 曝光
9. fixing 定影
10. fog （底片）灰雾
11. gamma rays γ射线
12. graininess 颗粒性
13. granularity 颗粒度
14. luminance 亮度
15. radiation 辐射
16. X-radiation X射线
17. reflection density 反射密度
18. reticulation 网纹

14.6 渗透检测术语 Penetrant Testing Term

1. bleedout 渗出
2. blotting 吸取
3. capillary action 毛细管作用
4. combined colour contrast and fluorescent penetrant 着色荧光渗透剂
5. comparison test block 比较试块
6. contaminant 污染物
7. contrast 对比度
8. defect 缺陷
9. dye penetrant 着色渗透剂
10. emulsifiable penetrant 可乳化渗透剂
11. evaluation 评定
12. flaw 伤
13. fluorescent dry deposit penetrant 干沉积荧光渗透剂
14. fluorescent penetrant 荧光渗透剂
15. inspection 检查
16. interpretation 解释
17. penetrant 渗透剂
18. penetrant flaw detection 渗透探伤
19. penetrant station 渗透工位
20. post emulsifiable penetrant 后乳化渗透剂
21. relevant indication 相关指示
22. replenishes 补充剂
23. solvent-removal penetrant 溶剂去除型渗透剂
24. spray nozzle 喷嘴
25. test piece 试片
26. thixotropic penetrant 摇溶渗透剂
27. water-washable penetrant 可水洗型渗透剂
28. wetting action 润湿作用

14.7 磁粉表面检测术语 Magnetization Surface Testing Term

1. electrode 电极
2. electromagnet 电磁铁
3. fluorescent magnetic powder 荧光磁粉
4. lines of force 磁力线
5. magnetic circuit 磁路
6. magnetic field 磁场
7. magnetic field distribution 磁场分布
8. magnetic field strength 磁场强度（H）
9. magnetic flux 磁通
10. magnetic force 磁化力
11. magnetic particle 磁粉
12. magnetic pole 磁极
13. magnetic saturation 磁饱和
14. near surface defect 近表面缺陷
15. peak current 峰值电流
16. permanent magnet 永久磁铁
17. prods 触头
18. reluctance 磁阻
19. residual magnetism 剩磁
20. yoke 磁轭

14.8 涡流检测术语 Eddy Current Testing Term

1. absolute 绝对式线圈
2. A.C. magnetic saturation 交流磁饱和
3. comparator coil 比较式线圈
4. D.C. magnetic saturation 直流磁饱和
5. depth of penetration 渗透深度
6. differential coil 差动线圈
7. discrimination 甄别
8. dynamic currents 动态电流
9. eddy current 涡流
10. eddy current flaw detector 涡流探伤仪
11. edge effect 边缘效应
12. ID coil ID线圈（内插式线圈）
13. impedance 阻抗
14. impedance analysis 阻抗分析
15. impedance plane diagram 阻抗平面图
16. lift-off effect 提离效应
17. magnetic flux density 磁通密度
18. magnetic permeability 磁导率
19. primary coil 一次线圈
20. probe coil 点式线圈
21. probe coil clearance 点探头式线圈间隙（提离）
22. recovery time 恢复时间
23. selectivity 选择性
24. skin effect 集肤效应
25. suppression 抑制

15 施工设备和工具 Construction Equipment and Tool

15.1 起重机械 Hoisting Machinery

15.1.1 麻绳、绳结、吊索和吊具 Hemp Ropes, Knots, Sling and Hardware

15.1.1.1 麻绳和绳结 Hemp Ropes and Knots

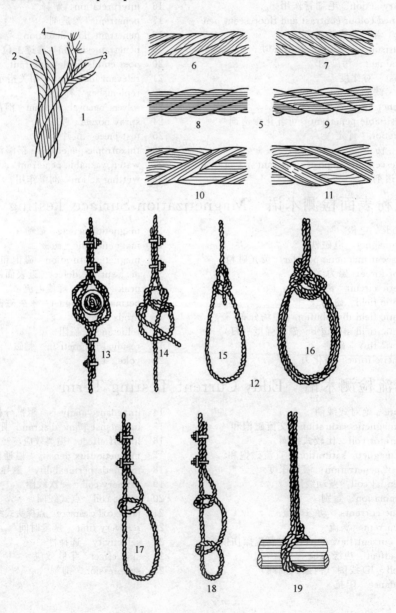

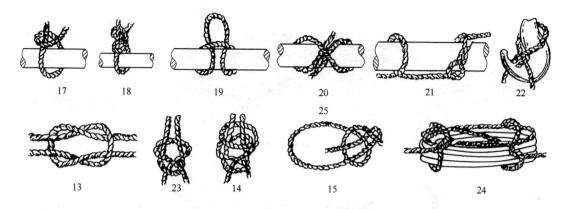

1　hemp rope　麻绳	13　thief knot　平结或直结
2　cord　线	14　double sheet bend　双圈展帆结
3　strand　胶线	15　running bowline　救生结，水手结，航海结
4　hemp yarn　麻纱	16　double running bowline　双航海结
5　twisting　捻绳方式	17　half hitch　单套圈结
6　left-hand lang lay　左向同向捻，左向顺捻	18　two half hitch　双套圈结
7　right-hand lang lay　右向同向捻，右向顺捻	19　builder's knot　死结
8　left-hand cross lay (regular lay universal lay)　左向交互捻，左向逆捻	20　clove, builder's knot　梯绳结，交叉结
9　right-hand cross lay　右向交互捻，右向逆捻	21　timber and half hitch　系木结
10　left-hand mixed lay　左向混合捻	22　blackwall hitch　缠钩结
11　right-hand mixed lay　右向混合捻	23　sheet, becket bend　单圈滑结，绞绳结
12　knot of cable　钢丝绳结，绳扣	24　sheepshank knot　缩绳结
	25　knot of hemp rope　麻绳打结型式

15.1.1.2　吊索　Sling

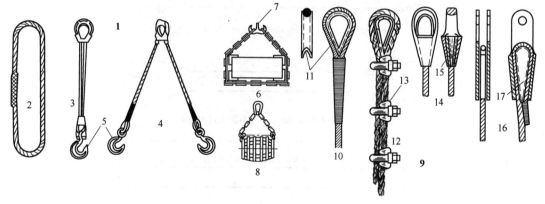

1　sling　吊索	9　eye-splice　索眼
2　endless rope　无接头式吊索	10　wire wound　钢丝缠绕式
3　hook rope　单钩吊索	11　heart-shaped dead eye　钢索套环
4　double hook rope　双钩吊索	12　clip clipped　绳卡夹紧式
5　eye hook　环钩	13　rope clip　绳卡，锚爪子
6　binding chain　捆紧链	14　lead filled type　灌铅式
7　double ramshorn, sister hook　双抱钩，和合钩，双钩	15　lead block　铅块
8　barrel chain　桶链	16　wedge clipped　楔块式紧式
	17　wedge block　楔块

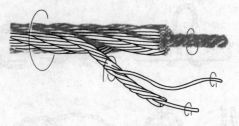
wire rope sling 钢丝绳吊索

typical wire rope sling 典型钢丝绳吊索

metal mesh sling 金属网吊索

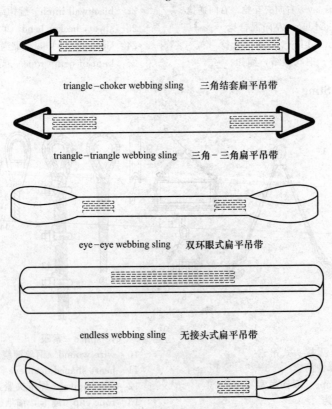

triangle-choker webbing sling 三角结套扁平吊带

triangle-triangle webbing sling 三角-三角扁平吊带

eye-eye webbing sling 双环眼式扁平吊带

endless webbing sling 无接头式扁平吊带

reverse eye webbing sling 反环眼式扁平吊带

webbing sling 扁平吊装带

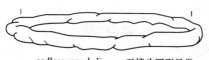

endless round sling 无接头圆形吊带

endless round sling with centre cover (eye-eye)
带有环眼的无接头圆形吊带

round sling 圆形吊装带

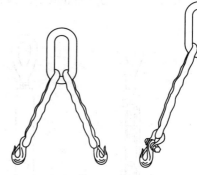

endless round sling with fittings
带有配件的无接头圆形吊带

15.1.1.3 吊具 Hardware

（1）卸扣 Shackles

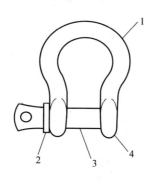

1　bow　扣体
2　shoulder　台肩
3　pin　销轴
4　ear　环眼

screw pin type 螺旋销式

screw pin type 螺旋销式

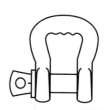

screw pin type 螺旋销式

bolt type 螺栓销式
anchor shackles
弓形卸扣

bolt type 螺栓销式
chain shackles
D形卸扣

bolt type 螺栓销式
synthetic sling shackles
合成吊带卸扣

typical shackle component 典型卸扣形式

811

(2) 可调节式吊具　Adjustment Hardware

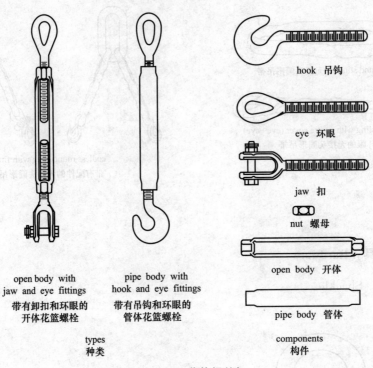

turnbuckle 花篮螺丝扣

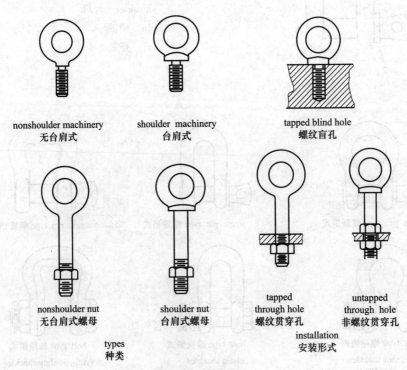

eyebolt 单眼螺栓

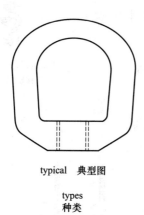

typical 典型图

types
种类

through hole no nut　贯穿孔无螺母

through hole top nut　贯穿孔带顶螺母

through hole bottom nut　贯穿孔带底部螺母

installation 安装形式

eye nuts　单眼螺母

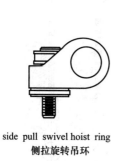
side pull swivel hoist ring
侧拉旋转吊环

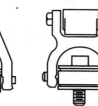
bail swivel hoist ring
直拉旋转吊环

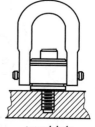
tapped hole
螺纹孔

chain swivel hoist ring
链式旋转吊环

webbing swivel hoist ring
带式旋转吊环

through hole
贯穿孔

types
种类

installation
安装

Swivel hoist ring　旋转吊环螺栓

813

(3) 紧式吊具 Compression Hardware

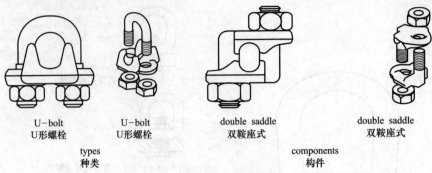

U-bolt
U形螺栓

U-bolt
U形螺栓

double saddle
双鞍座式

double saddle
双鞍座式

types
种类

components
构件

wire rope clip 钢丝绳卡

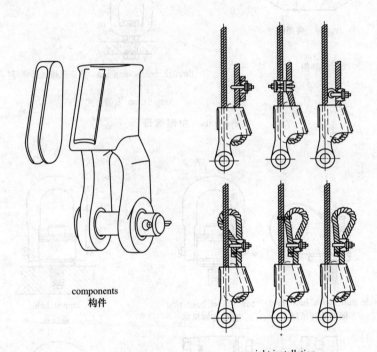

components
构件

right installation
正确安装方式

wedge socket 楔形紧套

(4) 链环和转环 Links, Rings and Swivels

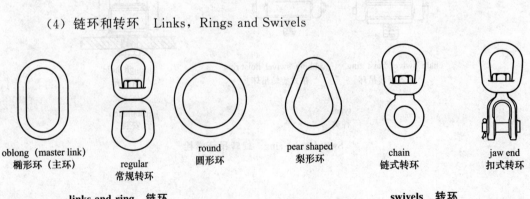

oblong (master link)
椭形环（主环）

regular
常规转环

round
圆形环

pear shaped
梨形环

chain
链式转环

jaw end
扣式转环

links and ring 链环

swivels 转环

(5) 吊索组 Rigging Block

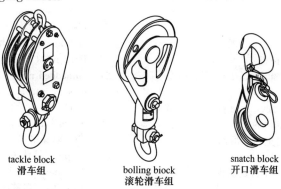

tackle block
滑车组

bolling biock
滚轮滑车组

snatch block
开口滑车组

15.1.2 起重机构造与吊车手语 Crane Structure and Sign Language

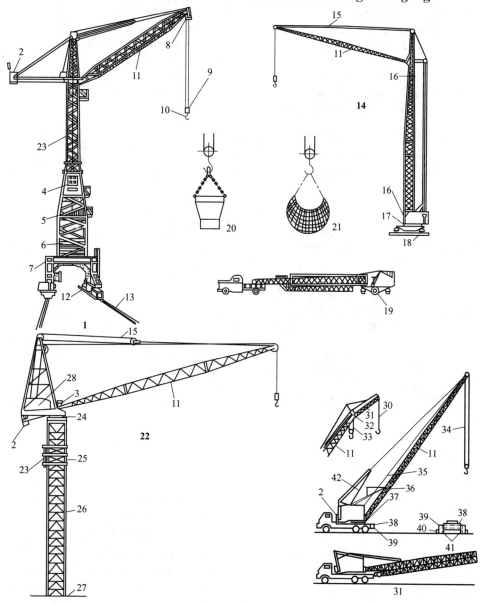

1 column（tower）crane 塔式起重机，塔吊
2 counter weight 平衡重
3 mast 桅杆
4 operator cab 司机室
5 winch（crab） 卷扬机
6 weight 压块，配重
7 gantry 龙门架
8 fast pulley（fixed pulley, dead pulley） 定滑轮
9 movable pulley 动滑轮
10 hoisting hook 起重钩，吊钩
11 boom 动臂，吊杆
12 driving mechanism 行走机构
13 rail 钢轨
14 tower slewing crane 塔式回转起重机
15 topping lift 吊挂钢绳
16 control station 操作室
17 bull wheel 转盘齿轮
18 track 轨道
19 road mounting 公路拖运车轮
20 concrete bucket 混凝土浇筑桶
21 lifting net（cargo net） 吊货网
22 climbing crane 爬升式起重机
23 hydraulic crane 液压千斤顶
24 pivot 枢轴
25 add on section 增高段
26 fixed mast 固定桅杆
27 embedded mast 埋置桅杆
28 diesel controls 柴油机操纵机构
29 truck crane 汽车起重机
30 jib load line 吊臂起重绳
31 jib 鹅头吊臂
32 jib pin 吊臂销
33，34 main load line 主起重绳
35 load line 起重绳
36 safety strut 安全支撑
37 turntable 转台
38 chassis 底盘
39 outriggers 外伸挂架
40 blocks 垫块
41 tires 轮胎
42 topping lift 背绳

1 main hoist 要主钩
2 auxilliary hoist 要副钩
3 hoist load 起钩
4 hoist load slowly 缓缓起钩
5 stop 停止
6 raise boom 起杆
7 raise boom & lower load 起杆下钩
8 lower load 下钩
9 lower load slowly 缓缓下钩
10 emergency stop 紧急停止
11 lower boom 降杆
12 lower boom & raise load 降杆起钩
13 swing boom 转杆
14 swing boom slowly 缓缓转杆
15 travel 吊车移动
16 retract boom 2 hands 收杆（双手）
17 retract boom 1 hand 收杆（单手）
18 extend boom 2 hands 伸杆（双手）
19 extend boom 1 hand 伸杆（单手）
20 dog everything 工作完成

15.1.3 各类起重机 Types of Crane

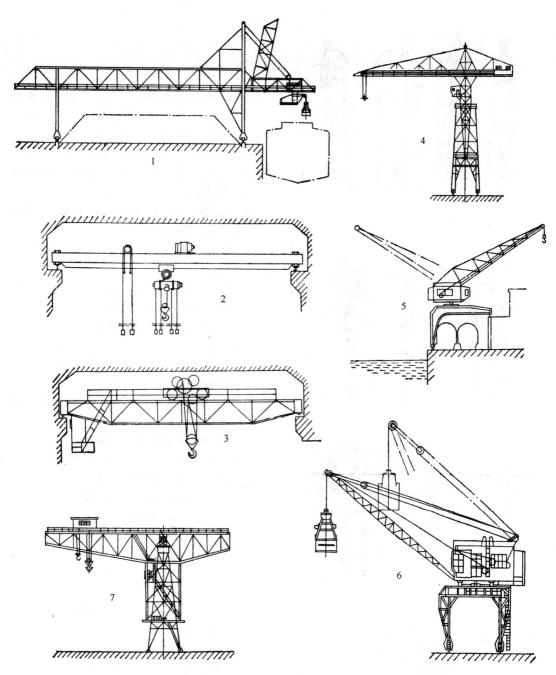

1　gantry crane with slewing man-trolley　带司机小车的旋臂龙门起重机
2　overhead travelling crane with hoist　电动葫芦桥式起重机
3　overhead travelling crane with crab　绞车式桥式起重机
4　hammer-head crane　锤式起重机
5　semi-portal crane　半门座悬臂起重机
6　high pedestal jib crane（portal jib crane）　门座式悬臂起重机
7　hammer-head crane　锤式起重机

15.1.4 手动葫芦和千斤顶 Manual Hoist and Lifting Jack

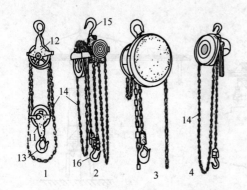
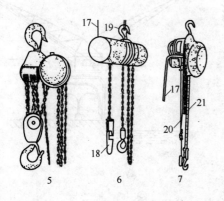
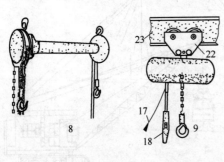
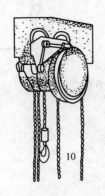

1 differential chain hoist (differential chain block) 差动手拉葫芦，差动倒链
2 screw-geared hoist 蜗轮蜗杆倒链
3 spur-geared chain hoist 正齿轮倒链
4 spur-geared, single-reeved hiost 单穿链正齿轮倒链
5 multiple-reeved, spur-geared hoist 多股穿链正齿轮倒链
6 lightweight electric chain hoist 轻型电动倒链
7 electric chain hoist with pendant rope control 带吊索操纵器的电动倒链
8 twin-hook hoist with link chain 扁节链双钩倒链
9 lightweight electric trolley chain hoist 轻型移动式电动倒链
10 low-headroom trolley hoist 低净空用的小车倒链
11 grooved single sheave 有槽单链轮
12 dual-pocketed upper sheave 双槽上链轮
13 load hook 起重钩，吊钩
14 endless reeved chain 循环链
15 shank hook 挂钩
16 hoisting hook 起重钩
17 flexible cable 电缆
18 pushbutton control 操纵按钮
19 suspension hook 挂钩
20 pendant rope control 吊索操纵器
21 roller chain 滚子链
22 trolley 小车
23 trolley beam 小车梁
24 hoist installation 起重机的配置
25 hoist and trolley clearance 葫芦与行车的净距
26 power supply collectors 电源集电器

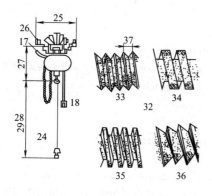

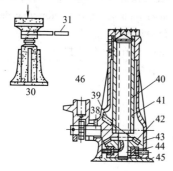

27　head room　净空
28　lift　提升高度
29　travel of hook　钩的升程
30　screw jack　螺旋千斤顶
31　tommy bar　转动用插棒
32　jackscrew　起重螺杆
33　V-section jackscrew　V形起重螺杆
34　square jackscrew　矩形起重螺杆
35　acme jackscrew　梯形起重螺杆
36　buttress jackscrew　锯齿形起重螺杆
37　pitch　螺距
38　ratchet　棘轮组机构
39　drive gear　传动齿轮
40　slip sleeve　升降套筒
41　copper nut　铜螺母
42　large bevel gear　大伞齿轮
43　main support　主架
44　one-direction thrust ball bearing　单向推力球轴承
45　base frame　底座
46　hand screw jack　手动螺旋千斤顶

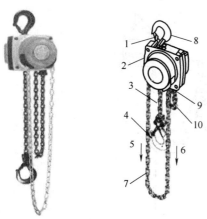

hand chain hoist　手拉链葫芦

1　hook latch　搭钩
2　hoist frame　起重机架
3　load chain　起重链
4　lower hook　挂钩
5　to lower hook pull on this part of chain　拉侧链条降钩
6　to raise lower hook pull on this part of chian　拉侧链条起钩
7　hand chain　手拉链
8　upper hook　上部钩
9　loose end connector　解链端销子
10　loose end of chain　链条解链端

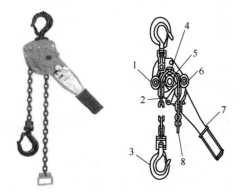

lever hoist　手扳葫芦

1　chain guide roller-lower hook side　链条导轨轮-挂钩侧
2　stripper　棘爪
3　lower hook block　带钩滑车
4　load chain　起重链
5　liftwheel　提升轮
6　chain guide roller-chain stop side　链条导轨轮-链条停止侧
7　handler　手柄
8　chain stop position　链条停止位置

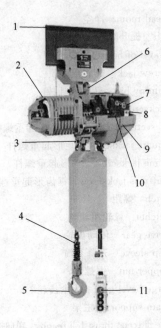

1 structure 结构
2 magnetic backing device 磁力后退装置
3 limit switch 限位开关
4 chains 链
5 hook 钩
6 support frame 机架
7 transformer 变压器
8 electromagnetic contactor 电磁接触器
9 mechanical dual braking 机械双重刹车
10 inverse phase sequence protecting device 逆向保护装置
11 push button 操作按钮

electric chain hoist 电动葫芦

15.1.5 电动吊篮 Electromotion Basket

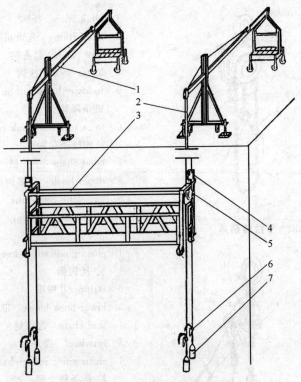

electromotion basket 电动吊篮

1 roof sliding frame 屋面悬挂装置
2 brake of position indicating and control device 限位器止挡
3 basket 工作吊篮
4 safety lock 安全锁定器
5 elevating conveyor 提升机
6 clip for rope plummet 绳坠夹子
7 iron plummet 吊铁

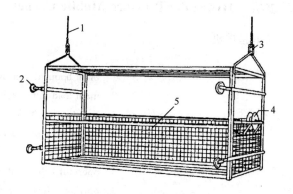

basket (three persons) 吊篮（供三人用）

1 steel cable 钢丝绳
2 pulley 滑轮
3 clamp ring 卡环
4 pulley for hopper 上料滑轮
5 handrail 护栏

15.1.6 手动吊篮 Manual Operation Basket

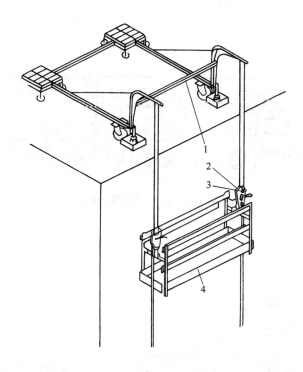

manual operation basket 手动吊篮

1 roof sliding frame 屋面悬挂装置
2 elevating conveyor 提升机
3 safety lock 安全锁定器
4 basket 工作吊篮

15.1.7 液压汽车起重机 Hydraulic Pressure Mobile Cranes

(1) Truck Crane 汽车起重机

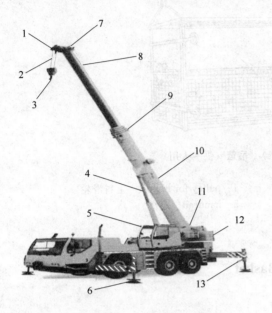

1 main boom tip sheave 主吊臂尖滑轮
2 main load line 主钢丝绳
3 load block 吊钩
4 boom hoist or lift cylinder 起升机构/提升油缸
5 operator cab 司机室
6 outrigger float 支腿浮板
7 main boom head 主吊臂端
8 5th telescoping section 第5节吊臂
9 4th 3rd 2nd 1st telescoping section 第1～4节吊臂
10 base or main section of boom 基础臂
11 boom hinge bin 吊臂销
12 counterweight 配重
13 outrigger 支腿

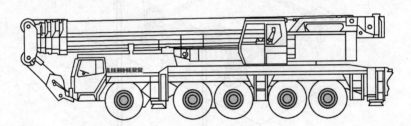

all terrain crane 全路面起重机

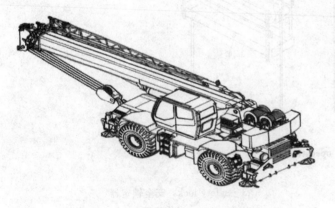

rough terrain crane 越野起重机

boom truck 随车吊

(2) Crawler Crane　履带式起重机

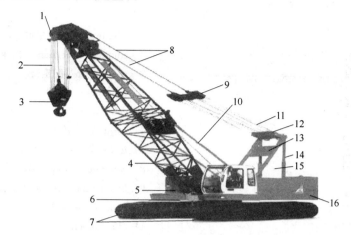

crawler crae　履带式起重机

1　boom tip sheave　主臂端滑车
2　main load line　主钢丝绳
3　load block　吊钩
4　boom heel section　基础臂
5　boom heel pin　吊臂销
6　turntable/swing circle　回转装置
7　crawler assembly　履带
8　boom pendants　吊臂支承拉绳
9　outer bail　背绳滑轮组
10　boom stop　主臂防后倾装置
11　boom hoist reeving　吊臂俯仰拉绳
12　inner bail　门架滑轮组
13　gantry　门架
14　back hitch　后拉钩
15　boom hoist line　吊臂起重绳
16　counterweight　配重

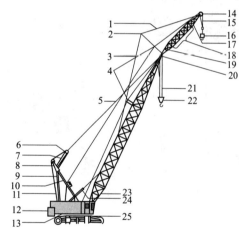

crawler crane with luffing jib　带动臂（附臂）的履带式起重机

1　jib forestay pendant　附臂拉绳
2　jib mast（gantry）　附臂桅杆（门架）
3　backstay pendant　后支索拉绳
4　deflector sheaves　导向轮
5　boom pendants　吊臂支承拉绳
6　outer bail　背绳滑轮组
7　inner bail　门架滑轮组
8　boom hoist reeving　吊臂俯仰拉绳
9　back hitch　后拉钩
10　gantry　门架
11　boom hoist line　吊臂起重绳
12　counterweight　配重
13　machine deck　设备底座
14　jib tip sheave　附臂端滑车
15　auxiliary hoist line　辅助钢绳
16　headache or overhaul ball　吊钩配重球
17　jib tip section　上段附臂
18　jib midsection　中段附臂
19　jib heel section　基础附臂
20　boom tip sheaves　主臂端滑车
21　main hoist line　主钢绳
22　main block　主钩
23　boom heel section　基础主臂
24　boom stop　主臂防后倾装置
25　turntable　回转装置

15.1.8 塔吊 Tower Crane

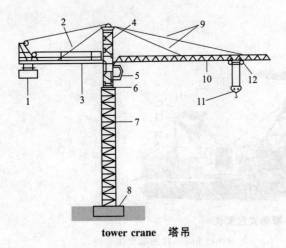

tower crane 塔吊

1　counterweight　配重
2　counter jib pendant　配重臂拉绳
3　counter jib　配重臂
4　top tower　塔帽
5　cab　操作室
6　turntable　转台
7　tower（mast）　塔身
8　concrete footing　基础承台
9　boom pendants　起重臂拉绳
10　boom　起重臂
11　load block　吊钩
12　trolly　小车

15.1.9 桥式起重机 Overhead Crane

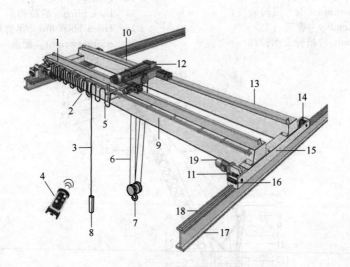

1　panel　接线箱
2　pendant festoon　拉绳滑线
3　pendant cable　电缆
4　radio control　遥控器
5　trolley festoon　小车滑线
6　wire rope　钢丝绳
7　hook block　吊钩
8　pendant　悬垂拉绳
9　conductor track　滑线导轨
10　hoist　起重机
11　bridge drive wheel　桥吊驱动轮
12　trolley frame　小车架
13　bridge girder　桥吊主梁
14　bridge idler wheel　桥吊托轮
15　end truck　端梁
16　end truck bumper　端梁车挡
17　runway beam　行走梁
18　runway rail　行走轨道
19　bridge drive　桥吊驱动

15.1.10 门式起重机 Gantry Crane

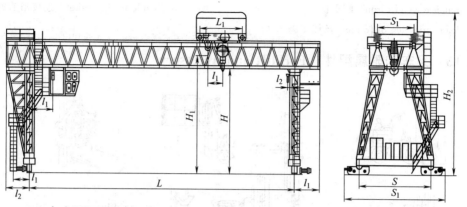

15.1.11 手动式屋面移动吊机 Manual Operating Roof Suspension Hoist Machine

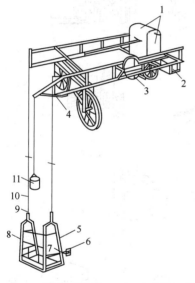

1　gear box　变速箱
2　balance weight case　配重箱
3　frame of steel thread drum for hopper　料斗升降钢丝绳卷轮架
4　movable frame for hopper　摆动送料架
5　basket（one person）　吊篮（供一人用）
6　rubber wheel　橡胶轮
7　handrail of basket　吊篮栏杆
8　safety hook　安全钩
9　steel cable　钢丝绳
10　hopper　料罐
11　pulley for basket　吊篮升降滑轮

manual operating roof suspension hoist machine
手动屋面移动吊机

15.1.12 电动式屋面移动吊机 Electrocution Roof Suspension Hoist Machine

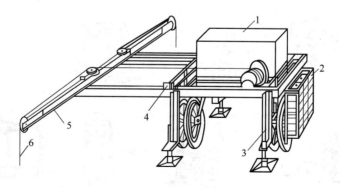

electrocution roof suspension hoist machine　电动屋面移动吊机

1 electromotor, worm gear jacket 电动机，蜗轮箱护罩
2 balance weight iron 配重铁
3 lead screw leg of frame 丝杠支脚
4 switch 开关
5 iron pole 铁扁担
6 steel cable to fix basket 连接吊篮的钢丝绳

15.1.13 叉车和附属配件 Forklift and Accessory

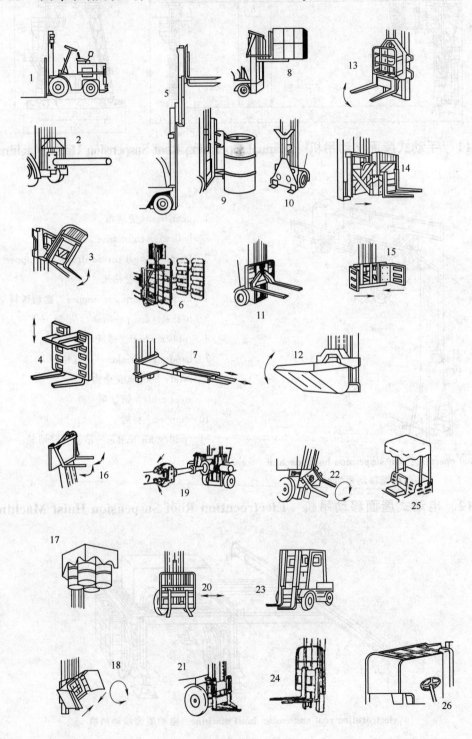

1　fork lift truck　叉车
2　ram　串杆
3　dumping fork　卸货叉
4　load stabilizer　货物稳定叉
5　triple telescoping uprights　三节伸缩式升举器
6　carton clamp　纸盒夹具
7　charger　装货机构
8　pusher device　推货设备
9　drum grab　桶抓
10　vertical crate grab　立式板条箱抓紧器
11　gripping forks　夹紧叉
12　bucket　货斗
13　hinged　铰链连接的叉
14　reach fork　取货叉
15　load clamp　货夹
16　cradle fork　鞍式叉
17　drum carrier　桶架
18　rotating clamp　回转夹具
19　manipulator　机械手
20　side shift　侧向移动
21　swing fork　旋转叉
22　rotating fork　转动叉
23　cab　司机室
24　back rest (load safety rack)　货物安全架
25　weather shade (awning)　遮阳篷
26　overhead guard　护顶

15.2　工具　Tools

15.2.1　木工工具　Wood Working Tools

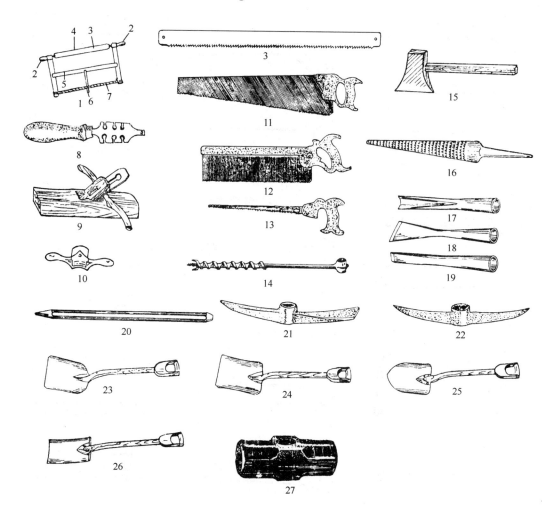

1 frame saw 框锯
2 handle 锯柄
3 saw blade 锯条
4 spring set teeth 锯齿
5 frame (saw bow) 锯架，锯框
6 lever 锯绳杠杆
7 cord 锯绳
8 saw set (saw setting) 整锯器，锯齿整理器
9 carpenter's jack plane 木工粗刨
10 spokeshave 铁弯刨，轴刨，鸟刨
11 hand saw (half-rip saw) 手板锯，木锯
12 back saw 夹背锯，镶边手锯
13 compass saw 鸡尾锯
14 auger (gimlet) 长柄木螺旋钻
15 broad axe 阔斧，宽头斧，木工斧
16 wood rasp 木锉
17 firmer gouge 半圆式木凿
18 beveled firmer chisel 斜口式木凿
19 mortise chisel 平口式木凿
20 drill rod (drill stem, jack hammer drill) 钻杆，钢杆
21 pick-mattock 十字镐，两头锄，扁尖式开山锄
22 pick 鹤嘴锄，尖镐，双尖式开山锄
23 round shoulder shovel 圆平锹
24 square nose shovel 方头锹
25 round nose shove 尖头锹
26 spade 泥锹
27 blacksmith hammer 八角锤，铁工锤

15.2.2 扳手 Wrenches

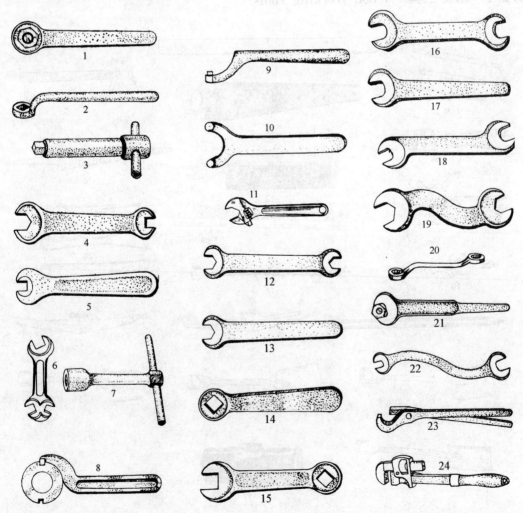

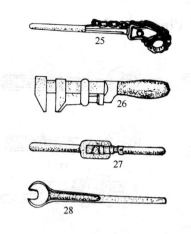
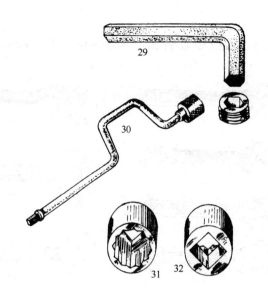

1　box wrench　套筒扳手，中心管扳手
2　chuck wrench　卡盘扳手
3　key wrench (square head wrench)　方头扳手
4　double-end square straight wrench　双手紧定螺钉扳手
5　single-end square straight wrench　单头方形直口扳手
6　triple-end set-screw wrench　三头紧定螺钉扳手
7　socket wrench　套筒扳手
8　hook spanner　钩扳手
9　pin spanner　销子扳手
10　face-pin spanner　端面（销子）扳手
11　adjustable wrench (monkey wrench)　活扳手
12　double-end hexagon straight wrench　双头六角直口扳手
13　single-end hexagon straight wrench　单头六角直口扳手
14　tool-post wrench　刀架扳手
15　double-end tool-post wrench　双头刀架扳手
16　double-end 15-degree angle wrench　双头15°斜口扳手
17　single-end 15-degree angle wrench　单头15°斜口扳手
18　22-degree angle wrench　22°斜口扳手
19　double-s 22-degree angle wrench　S形双头22°斜口扳手
20　double-end hexagon ring wrench　双头梅花扳手
21　construction wrench　大螺母扳手
22　general service wrench　通用扳手
23　pipe tongs wrench　管钳扳手
24　stillson pipe wrench (bulldog wrench)　可调扳手
25　vulcan chain pipe wrench　链扳手，链条扳手
26　screw wrench (shifting wrench)　活动扳手
27　tap wrench　丝锥扳手
28　track wrench　轨道扳手
29　hollow-head screw and wrench　空心螺塞和扳手
30　speed handle　摇杆，快速扳手
31　socket end　套筒端
32　drive end　拧动端

15.2.3 活扳手及管钳 Adjustable Wrenches and Pipe Wrenches

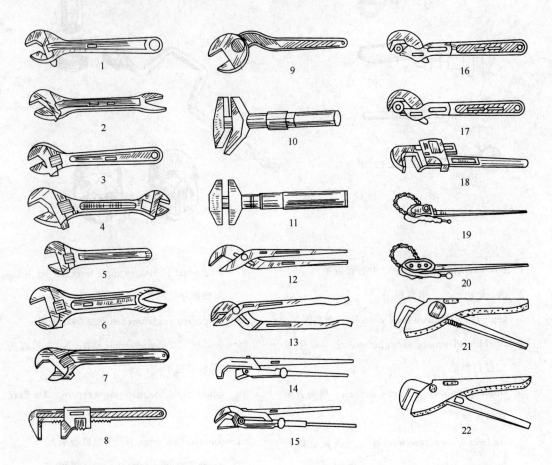

1 double hexagon ring end adjustable wrench 梅花柄端活扳手
2 alligator jaw end adjustable wrench 鳄鱼嘴柄端活扳手
3 standard adjustable wrench 标准扳手
4 double-end adjustable wrench 双头活扳手
5 tyre-lever end adjustable wrench 轮胎撬杠端活扳手
6 alligator jaw end adjustable wrench 鳄鱼嘴柄端活扳手
7 Swedish pattern adjustable wrench 瑞典式活扳手
8, 9 adjustable wrench 活扳手
10 double jaw head wrench 双卡头扳手
11 light type double jaw head wrench 小型双头扳手
12 mechanics service pliers 机修钳
13 universal grip pliers 多用卡钳
14 Swedish pattern tube pliers 瑞典式管钳
15 special tube pliers 专用管钳
16 one hand quick grip wrench 单手快夹紧卡钳
17 one hand quick grip tube pliers 单手快速夹紧管卡钳
18 American pattern pipe pliers 美式管钳
19 chain pipe wrench 链条管钳
20 double jaw pattern wrench 双颚式链条管钳
21 roll nut adjusted quick grip pliers 螺母调节快速夹紧卡钳
22 screw adjusted quick grip pliers 螺钉调节快速夹紧卡钳

15.2.4 钳和剪钳　Pliers and Nippers

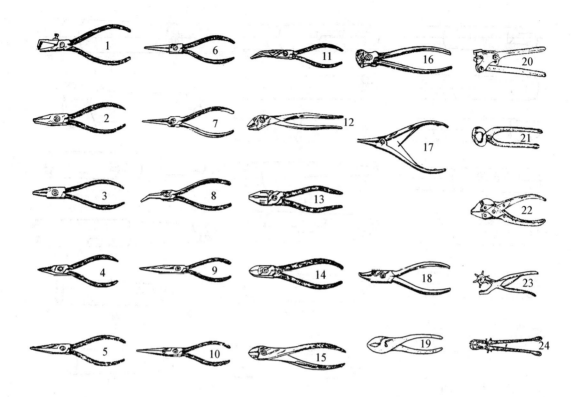

1　wire stripping pliers　剥线钳
2　plastolit insulated wire stripping pliers　塑料绝缘柄剥线钳
3　short round nose pliers　短圆嘴钳
4　chain nose pliers　链用钳
5　long flat nose pliers　长扁嘴钳
6　long round nose pliers　长圆嘴钳
7　needle nose pliers　尖嘴钳
8　bent needle nose pliers　弯尖嘴钳
9　long flat nose side cutting pliers　长扁嘴偏口钳
10　long half round nose side cutting pliers　长半圆嘴偏口钳
11　bent half round nose side cutting pliers　弯半圆嘴偏口钳
12　angle nose pliers　弯颈钳
13　super quality combination pliers　高级多用钳
14　diagonal cutting nippers　斜嘴偏口剪钳
15　heavy duty diagonal cutting nippers　大型斜嘴偏口剪钳
16　end cutting nippers　顶口剪钳
17　straight nose pattern pliers for external circlip　直嘴式外簧环用钳
18　bent nose pattern pliers for external circlip　弯嘴式内簧环用钳
19　slip joint pliers　鳄鱼钳
20　sealing pliers　封印钳
21　tower pinchers　胡桃钳
22　cutting pliers　割线钳
23　revolving punch pliers　活络打眼钳
24　bolt clippers　断线钳

15.2.5 刀具的柄部、套节及套筒 Shakes, Sockets and Sleeves

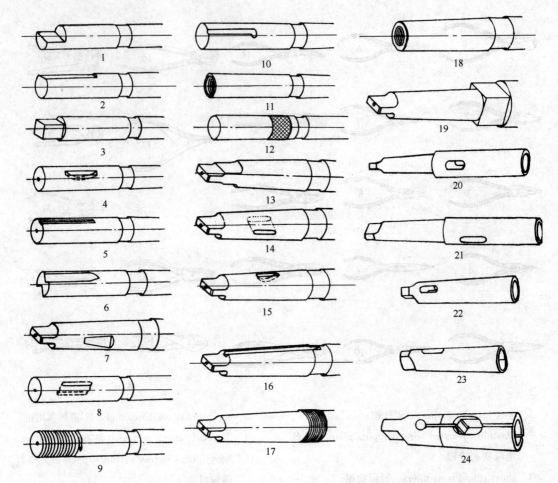

1	shank with half tang 半尾刀柄	15	shank with tang and woodruff keyset 扁尾月牙键槽刀柄
2	shank with special flat 特殊平台刀柄	16	shank with tang and keyway 扁尾键槽刀柄
3	shank with square tang 方尾刀柄	17	threaded on large end of shank 大端带螺纹的锥柄
4	shank with woodruff keyset 月牙键槽刀柄	18	taper shank with threads for draw bar 拉杆螺纹锥柄
5	shank with keyway 键槽刀柄	19	shank with driving square 带驱动方头的刀柄
6	shank with groove 沟槽刀柄	20	socket for taper shank 锥柄套筒
7	shank with whistle notch 哨形凹槽刀柄	21	use-em-up fitted socket 通用套筒
8	shot through shank 长方形透孔刀柄	22	plain sleeve for taper shank 普通锥柄套筒
9	threaded shank 螺纹刀柄	23	use-em-up sleeve for taper shank 通用锥柄用套筒
10	shank with bayonet slot 销钉槽刀柄	24	slotted driver for straight shank 直柄牙槽套筒
11	shank with threads for draw bar 拉杠螺纹刀柄		
12	knurl on shank 刀柄滚花		
13	double tang shank 双尾刀柄		
14	shank with driving slot 带卸刀槽的刀柄		

15.2.6 丝锥 Taps

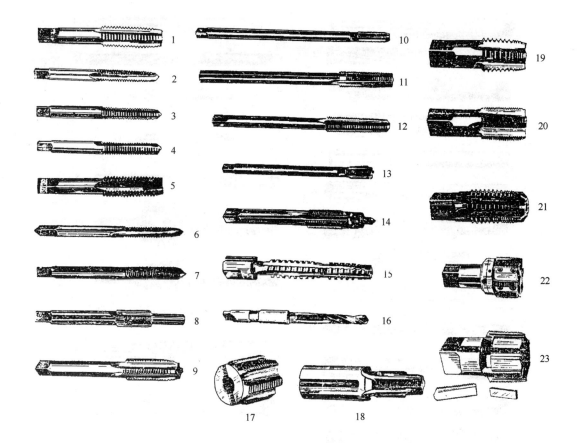

1 plug hand tap 手用精丝锥
2 two fluted hand tap 手用双精丝锥
3 three fluted hand tap 手用三精丝锥
4 machine screw tap 机用丝锥
5 spiral pointed tap 螺尖丝锥
6 spiral pointed machine screw tap 机螺尖丝锥
7 short flute, spiral point tap 短槽螺钉锥
8 tap with plain cylindrical pilot 带普外圆导柱的丝锥
9 extension hand tap 手用长柄丝锥
10 pulley tap 带轮丝锥
11 straight shank taper tap 直柄锥形锥
12 nut tap 螺母丝锥
13 taper pipe tap 锥管丝锥
14 extension tap 长柄丝锥
15 tandem ACME tap 阿克姆梯形丝锥
16 combined pipe tap and drill 管用钻孔攻丝复合刀具
17 shell tap 套式丝锥
18 multiple diameter tap 异径丝锥
19 interrupted thread taper pipe tap 跳牙锥管丝锥
20 straight pipe tap 圆柱管丝锥
21 combined taper pipe tap and core drill 扩孔钻和锥管丝锥复合刀具
22 inserted blade tap 镶齿丝锥
23 inserted blade taper pipe tap 齿锥管丝锥

15.2.7 铰刀 Reamers

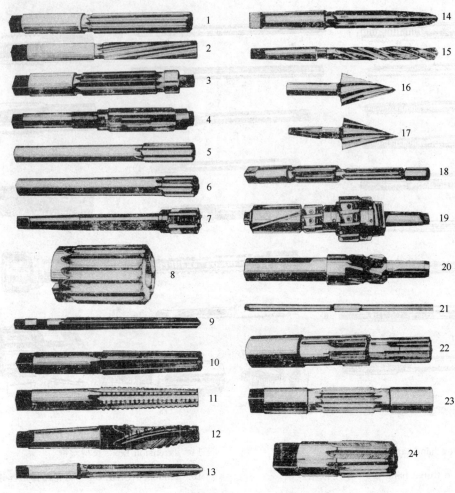

1. straight flute right hand cut reamer 直槽右旋切削铰刀
2. left-hand helical, right-hand cut reamer 左螺旋槽右旋切削铰刀
3. expansion hand reamer 膨胀式（可调）手用铰刀
4. adjustable hand reamer 可调手用铰刀
5. chucking reamer 机用铰刀
6. expansion chucking reamer 膨胀式（可调）机用铰刀
7. taper shank adjustable reamer 锥形柄可调铰刀
8. shell reamer 套式铰刀
9. taper pin finishing reamer 锥形销孔铰刀
10,11. Morse taper roughing reamer 莫氏锥形精加工铰刀
12. spiral-fluted roughing reamer 螺旋槽粗加工铰刀
13. bridge reamer 铆钉控铰刀
14. taper car reamer 锥形短铆钉孔铰刀
15. combined drill and reamer 钻铰复合刀
16. straight shank burring reamer 直柄去毛刺铰刀
17. ratched shank burring reamer 棘柄去毛刺铰刀
18. multiple diameter reamer 异径铰刀
19. built-up line reamer 镶片组合铰刀
20. piloted multi-diameter reamer 带导向柱的异径铰刀
21. piloted expansion line reamer 带导向柱的膨胀式铰刀
22. multi-diameter with pilot 异径膨胀式铰刀
23. hand reamer with pilot 带导向柱的手用铰刀
24. taper pipe reamer 锥管铰刀

15.2.8 麻花钻 Twist Drills

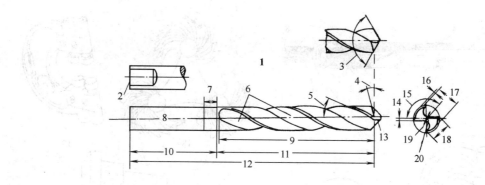

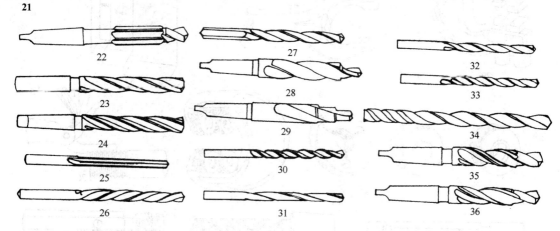

1 twist drills 麻花钻
2 tang 扁尾
3 point angle 顶角
4 lip relief angle 主切削刃后角
5 rake or helix angle 前角或螺旋角
6 flutes 排屑槽
7 neck 颈部
8 axis 轴线
9 flute length 槽长
10 shank 柄部
11 body 工作部分
12 over-all length 全长
13 lip 主切削刀
14 margin 导向刀刃
15 chisel edge angle 横刃斜角
16 body diameter clearance 工作部分径隙
17 clearance diameter land 隙径，留隙直径
18 land 钻棱
19 web 钻心
20 chisel edge 横刃
21 types of drills 各类钻头
22 combination drill and reamer 钻铰复合刀
23 three-groove core drill 三槽扩孔钻
24 four-groove core drill 四槽扩孔钻
25 straight flute drill 直槽钻头
26 left-hand two-flute drill 左旋双槽钻头
27 grooved shank drill 槽柄钻头
28 two-step taper shank drill 锥柄两级钻头
29 three-step core drill 三级扩孔钻
30 high-helix straight shank drill 大螺旋角直柄钻头
31 slow-helix straight shank drill 小螺旋角直柄钻头
32 bolt or cotter-pin drill 螺栓或开尾削孔钻头
33 heavy-web straight shank drill 厚钻心直柄钻头
34 roll-forged straight shank drill 滚轧直柄钻头
35 muti-land drill and reamer 多棱钻铰刀
36 two diameter multi-land drill 双直径多棱钻头

15.2.9 检测规　Inspection Gages

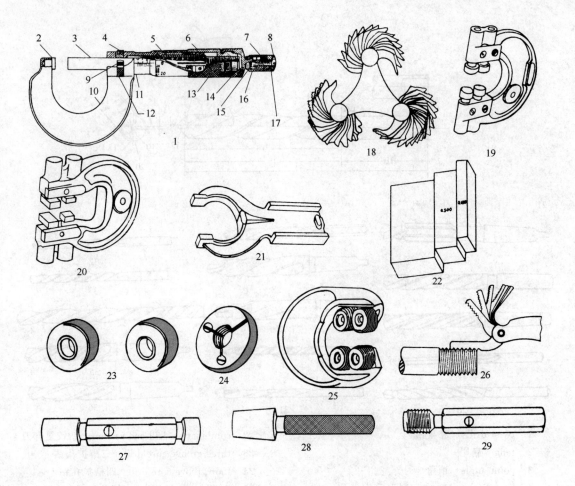

1	micrometer 千分尺	16	ratchet plunger 棘爪柱
2	anvil 测砧	17	ratchet thimble 棘盘
3	spindle 测微螺杆	18	thread pitch gage 螺距规
4	clamp rig 卡环	19	snap gage 卡规
5	screw nut 螺杆螺母	20	solid snap thread gage 整体螺纹卡规
6	compression nut 压紧螺母	21	caliper gage （单头）卡规
7	ratchet spring 棘爪弹簧	22	gage blocks 块规
8	ratchet screw 棘爪螺钉	23	ring gages 环规
9	clamp ring roller 卡环滚子	24	ring thread gage 螺纹环规
10	frame 尺架	25	roller snap thread gage 滚柱螺纹距规
11	sleeve or barrel 固定套管	26	screw thread pitch gage 螺钉螺距规
12	locknut 锁紧螺母	27	plug gage 圆柱塞规
13	thimble 微分筒	28	taper plug gage 锥度塞规
14	cap 调整冒盖	29	plug thread gage 螺纹塞规
15	ratchet ring 棘爪环		

15.2.10 量具 Measuring Tools

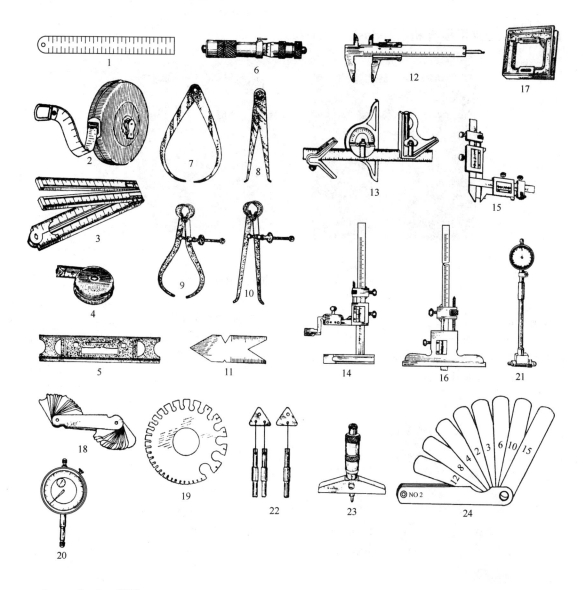

1 steel rule 钢尺
2 tape rule 皮卷尺
3 four-fold rule 四折木尺
4 steel tape 钢卷尺
5 steel spirit level 钢水平尺
6 inside calliper gage 内径千分尺
7 outside calliper 外卡钳
8 inside calliper 内卡钳
9 spring outside calliper 弹簧外卡钳
10 spring inside calliper 弹簧内卡钳
11 center inside calliper 中心规
12 center gage 游标卡尺
13 vernier calliper 万能角度尺
14 height vernier calliper 高度游标卡尺
15 gear tooth vernier gage 齿厚游标卡尺
16 depth gage 深度游标卡尺
17 cross-test level 框式水平仪
18 radius gage 半径规
19 wire gage 线规
20 dial indicator 千分表
21 internal diameter indicator 内径千分表
22 pin gage （螺纹）量尺
23 micrometer depth gage 深度千分尺
24 feeler gage 塞规，塞尺

15.3 混凝土搅拌设备 Concrete Mixer Equipment

15.3.1 工业用搅拌机 Industrial Mixer

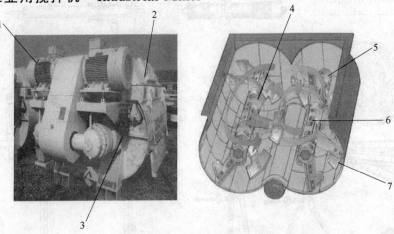

twin shaft mixer 双轴式搅拌机

1 motor 电机
2 housing 外壳
3 lubrication station 润滑油站
4 blade 螺旋叶片
5 shaft 搅拌轴
6 link board 连接板
7 scraper 刮板

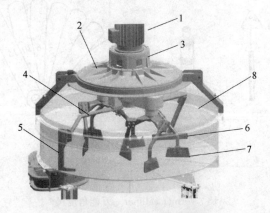

1 motor 电机
2 mixer cover 盖板
3 rotator 转子
4 mixing arm 搅拌臂
5 scraper of drum wall 缸壁刮板
6 scraper of bottom plate 底板刮板
7 mixing blade 搅拌叶片
8 mixing drum 搅拌缸
9 feeding hopper 进料斗
10 planetary concrete mixer 行星式混凝土搅拌机

vertical axis mixer 立轴式混凝土搅拌机

1 feeding hopper 进料斗
2 charge chute 进料槽
3 water pipe 水管
4 drum 筒体
5 mixing blade 搅拌叶片
6 discharge chute 出料槽

reversing drum mixer 反转式搅拌器

15.3.2 现场用搅拌机 On-site Mixer

tilting drum mixer 倾斜式搅拌机

non-tilting drum mixer 非倾斜式搅拌机

15.3.3 混凝土用车 Concrete Mixer Trucks

self loading concrete mixers 自上料混凝土车

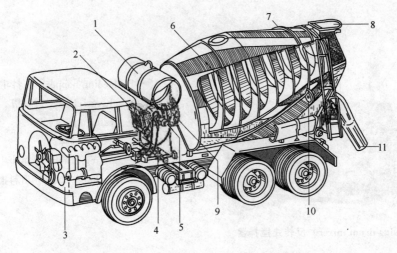

concrete mixer truck 混凝土罐车

1　water clean system　水清洗系统
2　hydraulic oil cooler　液压油冷却器
3　chassis　底盘
4　water dosage unit　注水系统
5　hydraulic engine　液压发动机
6　reversing mixing drum　反转式搅拌罐体
7　cylindrical ring　轮圈
8　charging hopper　进料斗
9　planetary gear　行星齿轮减速器
10　drum supporting rollers　托轮
11　concrete mix discharging chutes　出料槽

15.3.4　预搅拌混凝土站　Ready-mix Concrete Batch Plant

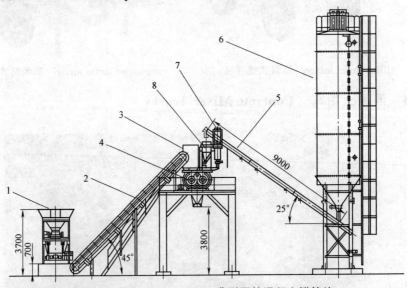

typical ready-mix concrete batch plant　典型预拌混凝土搅拌站

1　aggregate batching machine　骨料配料机
2　conveyor for aggregate feeding　骨料输送带
3　aggregate weighing system　骨料称重系统
4　concrete mixer　混凝土搅拌机
5　screw conveyor　粉料螺旋输送带
6　cement silo　水泥仓
7　water tank　水箱
8　cement weighing system　水泥称重系统

15.3.5 商品混凝土搅拌站 Commodity Concrete Mixing Station

concrete mixing station stereogram　混凝土搅拌站立体图

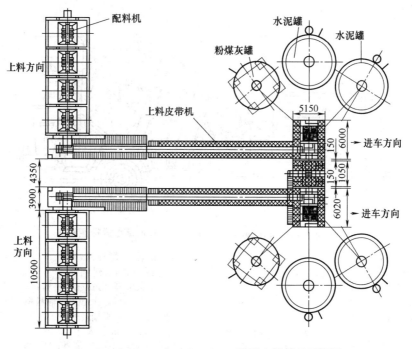

concrete mixing station plot plan　混凝土搅拌站平面图

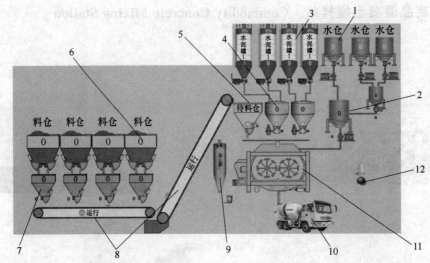

automatic batching system for concrete mixing station 混凝土自动化配料系统

1 water bin 水仓
2 water measure tank 水计量罐
3 cement bin 水泥仓
4 cement measure tank 水泥计量罐
5 sand and aggregate waiting bins 砂石待料仓
6 sand and aggregate bin 砂石料仓
7 sand and aggregate measure system 砂石计量系统
8 belt conveyor for sand and aggregate 砂石皮带输送机
9 fly ash bin 粉煤灰罐
10 concrete-mixer transportation 混凝土搅拌运输车
11 double shaft forced concrete mixer 双轴强制式混凝土搅拌机
12 control buttons in operating room and ringer 操作控制按钮及电铃

15.3.6　汽车式混凝土搅拌运输车　Concrete Truck Mixer

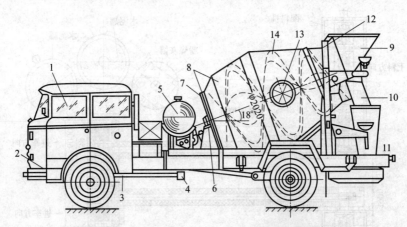

JY1-3 concrete truck mixer JY1-3型汽车混凝土搅拌运输车

1 drivers cabin 驾驶室
2 force dividing mechanism 分力箱
3 driver shaft 传动轴
4 gear box 伞形齿轮箱
5 water tank 水箱
6 axial direction bearing fixer on mixer 拌筒中心轴承座
7 chain wheel 链轮
8 mixer 搅拌筒
9 feed inlet 进料斗
10 discharge outlet 出料口
11 conveying trough 输料槽
12 frame 机架
13 man hole 人孔
14 mixer blade 搅拌叶片

15.4 土方设备　Earthwork Equipment

15.4.1 挖掘机　Excavator

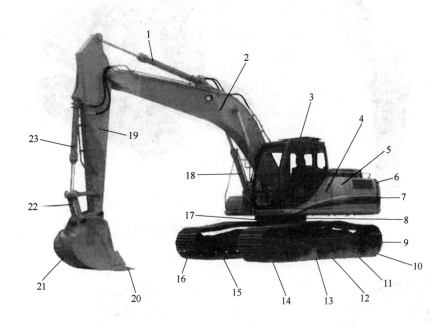

Excavator　挖掘机

1　hydraulic stick cylinder　斗杆液压缸
2　boom　动臂
3　cab　驾驶室
4　main control valve　主控阀
5　engine　发动机
6　counterweight　配重
7　hydraulic pump　液压泵
8　slewing rings　旋转环
9　final drive　末端驱动
10　sprockets　链齿轮
11　top roller　上辊
12　bottom roller　下辊
13　track adjuster　履带张紧器
14　front idler　引导轮
15　track chains　履带链
16　track pads　履带块
17　swing drive　回转驱动
18　hydraulic boom cylinder　动臂液压缸
19　stick　斗杆
20　ground engaging tools　地面接触工具
21　bucket　铲斗
22　linkage　连杆
23　hydraulic bucket cylinder　铲斗液压缸

excavator rock breaker 岩石破碎机

backhoe excavator 反铲挖掘机

1 backhoe excavator 反铲挖掘机　　2 hydraulic impact breaker 液压冲击破碎器（锤）

wheel excavator 轮式挖掘机

1 wheel excavator 反铲挖掘机
2 hydraulic impact breaker 液压冲击破碎器（锤）

15.4.2 推土机 Bulldozer

1 transmission/torque converter 变速器/扭矩转换器
2 engine 发动机
3 lift cylinder 提升臂气缸
4 blade 铲
5 ground engagint tools 地面接触工具
6 push frame 顶推架
7 front ldler 引导轮
8 top roller 上辊
9 bottom roller 下辊
10 track adjuster 履带张紧器
11 final drive 主减速器
12 sprockets 链齿轮
13 track chains 履带链
14 track pads 履带块
15 cab enclosure 驾驶室

15.4.3 装载机 Loader

wheel loader 轮式装载机

1　cab enclosure　驾驶室
2　hydraulic cylinder　液压缸
3　loader bucket　装载斗
4　ground engaging tools　地面接触工具
5　boom　提升臂
6　tires/wheels/rims　轮胎/轮毂
7　axle assembly　车轴总成
8　engine　发动机
9　transmission/torque converter　变速器/扭矩转换器

sketch of track-type tractor 履带式推土机外形

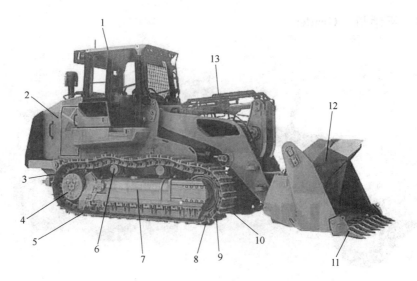

Track loader 履带式装载机

1 cab enclosure 驾驶室
2 engine 发动机
3 sprockets 链齿轮
4 final drive 主减速器
5 bottom rollers 下辊
6 top rollers 上辊
7 track adjuster 履带张紧器
8 track chains 履带链
9 front idler 引导轮
10 track pads 履带块
11 ground engaging tools 地面接触工具
12 loader bucker 装载斗
13 hydraulic cylinder 液压缸

15.4.4 挖掘装载机（两头忙）Backhoe

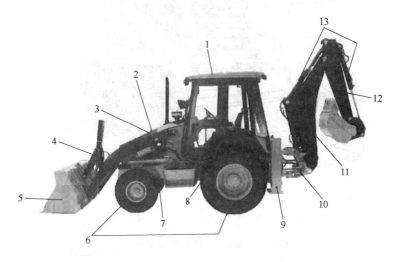

1 cab enclosure 驾驶室
2 engine 发动机
3 transmission 变速器
4 linkage 连杆
5 loader bucket 装载斗
6 tire/rim/wheel 轮胎/轮毂
7 front axle 前轴
8 rear axle 后轴
9 stabilizer 固定器
10 swing tower 摇臂
11 boom 提升臂
12 stick 斗杆
13 hydraulic cylinders 液压缸

847

15.4.5 平地机 Grader

15.4.6 滑移转向装载机（机器猫） Skid Steer Loader（Bobcat）

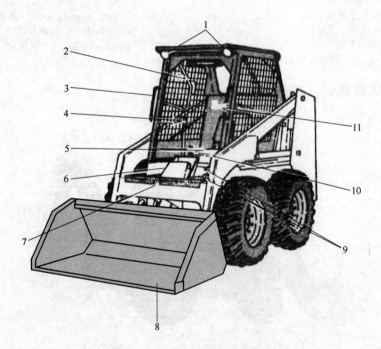

skid steer loader 滑移转向装载机

1　front lights　前车灯
2　seat bar　座位防护罩
3　grab handle　扶手
4　steering lever　操作转向杆
5　seat belts　安全带
6　tilt cylinder　摆动液压缸
7　safety tread　安全踏板
8　bucket　铲斗
9　auxiliary hydraulic lines　辅助液压管
10　operator seat　操作手座位
11　operator handbook　操作手册

15.5 地基处理机械 Foundation Treatment Machinery

15.5.1 步履式打桩机 Walking Piling Rig

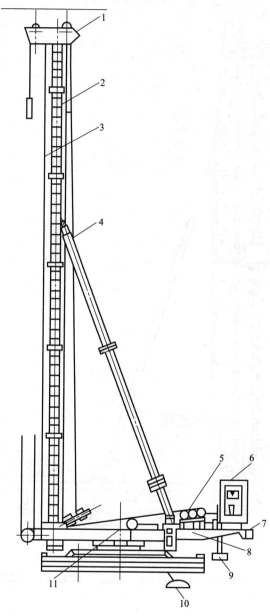

configuration of walking piling rig 步履式打桩架示意图

1 pulley block on top 顶部滑轮组
2 vertical pile 立桩
3 steel cable for rammer and pile 锤、桩钢丝绳
4 inclined brace 斜撑
5 winch for rammer and pile 锤、桩用卷扬机
6 driver cabin 驾驶室
7 balance weight 配重
8 rotating platform 旋转平台
9 hydraulic outrigger 液压支腿
10 walking chassis 步履底盘
11 rotating device 回转机构

15.5.2 螺旋钻打桩机 Auger Piling Rig

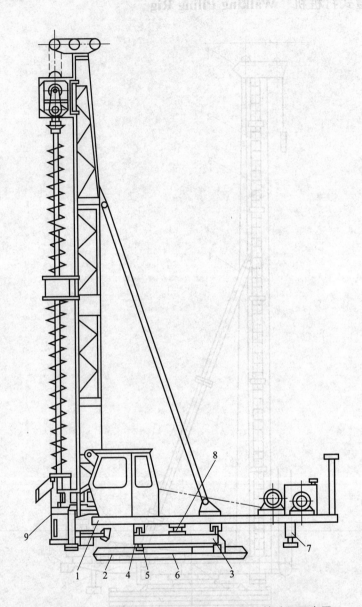

configuration of walking drill machine 步履式钻孔机示意图

1 upside platform 上盘
2 lower platform 下盘
3 middle platform 中盘
4 turning truckle 转向滚轮
5 walking truckle 行走滚轮
6 steel cable pulley 钢丝滑轮
7 outrigger 支腿
8 center axle 回转中心轴
9 soil discharge hole 出土筒

15.5.3 碾压机 Compactor Rollers

(1) 轻质土压实设备 Light Soil Compaction Equipment

rammer compactor 手持压实机　　　　**vibrating plate compactors** 振动板式压实机

duplex road roller/manual 双轮压实机（手动）

(2) 重质土压实设备 Heavy Soil Compaction Equipment

vibratory rollers 振动压路机

1　司机室　driver cabin　　　　4　发动机　engine
2　振动碾　vibrating roller　　　5　驱动轮胎　driving tyre
3　刮板　scraping blade

smooth wheeled rollers 光轮压路机

grid roller 格栅式压路机

pneumatic tyred roller 气胎压路机

sheep foot roller 羊角碾式压路机

pad foot/tamping roller 夯机式碾压机

tamping rammer 冲击夯

mini roller　小型压路机

vibratory roller　振动式压路机

tire roller　轮胎式压路机

15.5.4　水平定向钻孔机　Horizontal Directional Drilling Machinery

sketch of horizontal directional drilling machinery　水平定向钻孔机外形

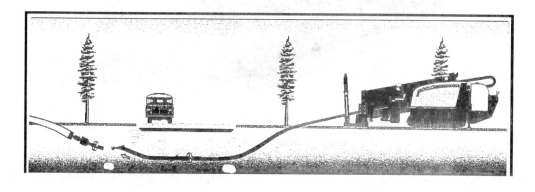
operation of horizontal directional drilling machinery　水平定向钻机工作示意图

15.5.5 钻机 Drilling Machine

rotary drilling rig 旋挖钻机

1 旋挖钻机 rotary drilling rig
2 钻具 drilling tool
3 钻孔灌注桩顶部钢护筒 steel casting on the top of drilling hole filling pile

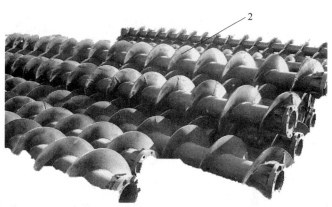

long spiral drilling machine 长螺旋钻机

1 旋挖钻机　rotary drilling rig
2 长螺旋钻杆　long spiral drill pipe
3 混凝土泵管　concrete pump pipe

15.5.6 强夯机械 Rammer Machinery

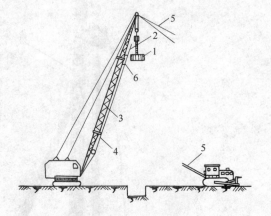

1 rammer head 夯锤
2 leaving hook device 自动脱钩装置
3 boom 起重臂杆
4 pull rope 拉绳
5 anchor 锚绳
6 pneumatic tyre 气胎

track crane rammer 轨道起重式强夯机

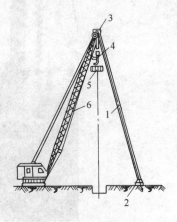

1 φ325×8mm steel ancillary shoring φ325×8mm 钢管辅助桅杆
2 pedestal 底座
3 elbow connector 弯管接头
4 leaving hook device 自动脱钩装置
5 12t rammer head 12 吨夯锤
6 pull rope 拉绳

**15t track crane rammer with ancillary steel shoring
15 吨带辅杆的轨道起重式强夯机**

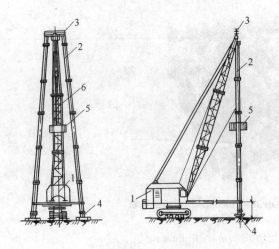

1 15t track crane 15 吨轨道起重机
2 leg of gantry crane 支架
3 gantry crane girder 支架横梁
4 pedestal 底座
5 rammer head 夯锤
6 leaving hook device 自动脱钩装置

**15t track crane rammer with steel gantry frame
15 吨带支架轨道起重式强夯机**

15.6 高处作业平台和工程车辆 Height Platform and Construction Vehicle

15.6.1 叉车和附属配件 Forklift and Attachments

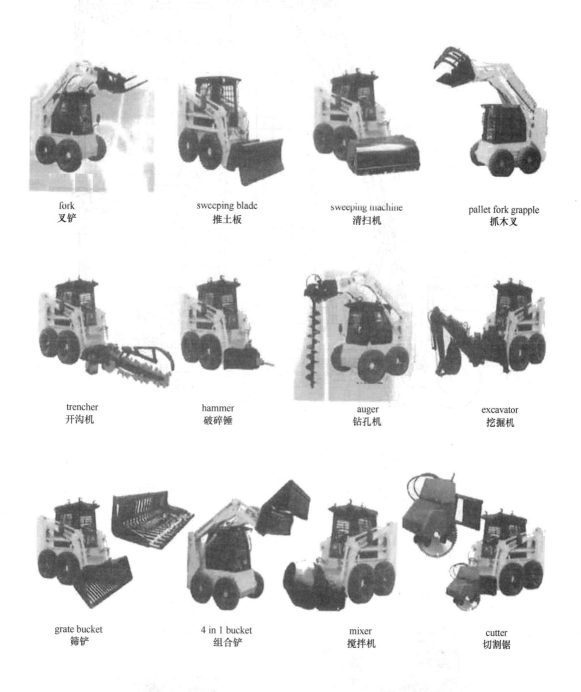

fork 叉铲

sweeping blade 推土板

sweeping machine 清扫机

pallet fork grapple 抓木叉

trencher 开沟机

hammer 破碎锤

auger 钻孔机

excavator 挖掘机

grate bucket 筛铲

4 in 1 bucket 组合铲

mixer 搅拌机

cutter 切割锯

skid steer attachment 各类滑移式工程车

15.6.2 混凝土泵车 Concrete Boom Truck

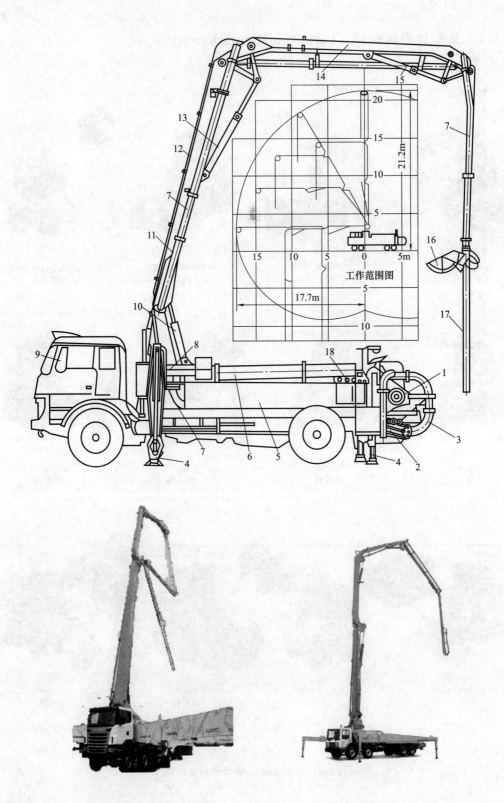

1　hopper and mixer　料斗及搅拌器
2　concrete pump　混凝土泵
3　discharge outlet　Y形出料口
4　hydraulic pressure outrigger　液压外伸支腿
5　water tank　水箱
6　standby pipe for concrete　备用混凝土输送管
7　pipe fixed to rotating platform and along the boom　进入旋转台的和沿臂杆安装的导管
8　rotating platform　旋转台
9　driver cabin　驾驶室
10，13，15　oil tank for folding jib　折叠臂杆用油缸
11，14　boom　臂杆
12　oil pipe　油管
16　support frame of rubber hose　橡胶软管弯曲支管支承架
17　rubber hose　橡胶软管
18　manipulate panel　操纵柜

15.6.3　液压升降平台　Hydraulic Pressure Lift Platform

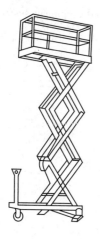

sketch of trailing lift platform
拖式升降平台概示图

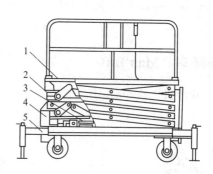

formation and configuration of ZTY hydraulic pressure lift platform　ZTY液压升降平台外形及构造示意图

1　platform　平台
2　vertical boom　叉架
3　hydraulic pressure system　液压系统
4　electronic and equipment system　电气系统
5　chassis　底盘

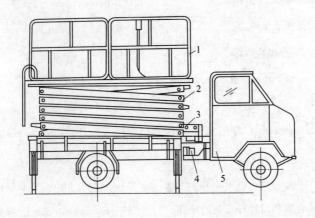

formation and configuration of ZTYC self propelled hydraulic
pressure lift platform   ZTYC 自行式液压升降平台外形及构造示意图

1　platform　平台

2　vertical boom　叉架

3　hydraulic pressure system　液压系统

4　electronic and equipment system　电气系统

5　chassis of BJ130　BJ130 汽车底盘

15.6.4　升降车　Man Lift

（1）悬臂升降机　Boom Lift

articulating booms lift　折叠臂升降机　　　　telescopic booms lift　伸缩臂升降机

(2) 车载悬臂升降机　Trailer Mounted Booms Lift

(3) 剪刀式升降机　Scissor Lift

slab scissor lift　平地型剪刀式升降机　　　　　　**rough terrain scissor lift**　越野型剪刀式升降机

(4) 吊篮车　Bucket Trucks

15.6.5　工程车辆　Construction Vehicle

15.6.5.1　公路型运输车　On-highway Trucks

light truck　轻型卡车

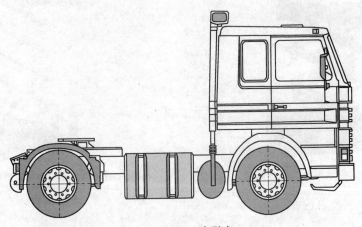

truck tractor　牵引车

tanker truck　罐车

water tanker truck 小型洒水车（属于罐车一种）

dump truck 自卸车

container truck 厢式货车

15.6.5.2　越野型运输车　Off-highway Trucks

mining truck　矿用卡车　　　　off-highway truck　非公路型卡车

15.6.5.3　拖车　Trailer

flatbed trailer　平板拖车

tanker trailer　罐式拖车

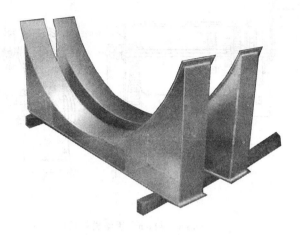

saddle 运输托架

15.7 滑模施工 Slipform Construction

(1) 滑模施工（1） Slipform Construction（Ⅰ）

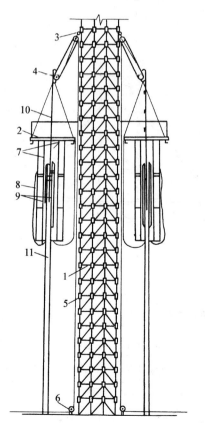

1 frame of shaft 竖井架
2 platform 工作台
3 clamp for steel cable 钢丝绳扣
4 pulley 滑车
5 steel cable 钢丝绳
6 steering pulley 导向滑轮
7 hoisting inside and outside 内外吊装
8 safety net 安全网
9 formwork inside and outside 内外模板
10 chimney steel cable 烟囱钢缆
11 chimney 烟囱筒壁

**sketch of winch elevating process
卷扬机提升工作示意图**

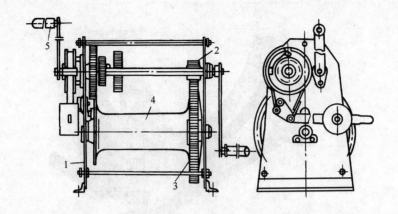

manual winch 手动卷扬机

1　frame　机架
2　small gear　小齿轮
3　large gear　大齿轮
4　roll drum　卷筒
5　control crank　摇把

(2) 滑模施工（2） Slipform Construction（Ⅱ）

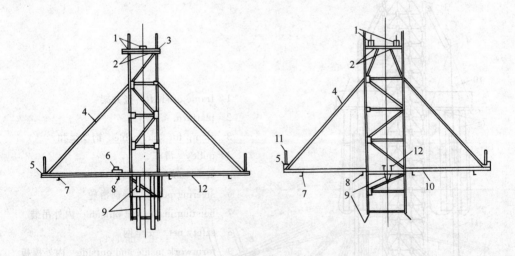

platform headframe 平台井架

1　pulley　滑轮
2　top pulley for guiding rope　导索天滑轮
3　pulley beam on slipform　滑轮模梁
4　steel bevel bracing　钢管斜撑
5　baluster steel ring　栏杆钢圈
6　base of jib　起重臂底座
7　outer steel ring　外钢圈
8　inner steel ring　内钢圈
9　steel shelves　钢挡数
10　radiating beam　辐射梁
11　baluster　栏杆
12　crossbeam　大梁

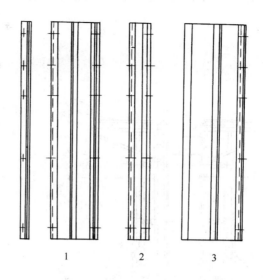

slipform of chimney 烟囱滑升模板

1　fixed formwork　固定模板
2　moving formwork　活动模板
3　tapering formwork　收分模板

（3）滑模施工（3）Slipform Construction（Ⅲ）

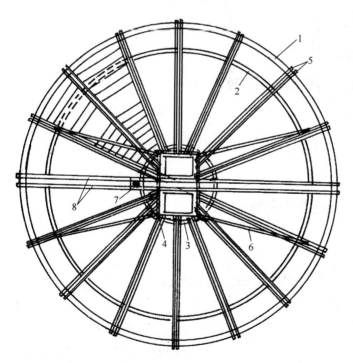

1　baluster steel ring　栏杆钢圈
2　outer steel ring　外钢圈
3　inner steel ring　内钢圈
4　ricker of stationary　平台井架立柱
5　radiating beam　辐射梁
6　bevel bracing　斜撑
7　base of jib　起重机底座
8　crossbeam　大梁

sketch map of operating platform　操作平台示意图

15.8 大模板爬模施工 Large Formwork Climbing Construction

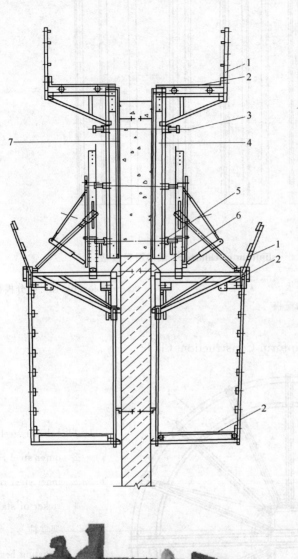

1 protective railing 防护栏杆
2 operating platform 操作平台
3 pull bolt 对拉螺栓
4 wood 木方
5 shaped wood 定型木
6 climbing cone 爬升锥

7 formwork 面板

15.9 钢筋加工机械 Process Machinery for Steel Bar

15.9.1 钢筋切断器 Steel Bar Cutter

（1）钢筋切断器（1） Steel Bar Cutter（Ⅰ）

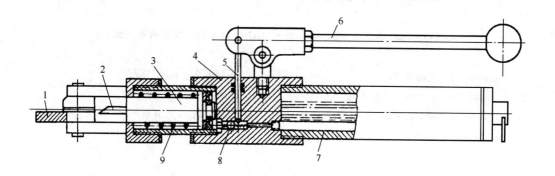

manual operating hydraulic cutter 手动液压切断器

1 slide track 滑轨
2 blade 刀片
3 piston 活塞
4 cylinder body of piston 缸体
5 plugger piston 柱塞
6 compression bar 压杆
7 oil storage column 储油筒
8 oil suction valve 吸油阀
9 releasing spring 回位弹簧

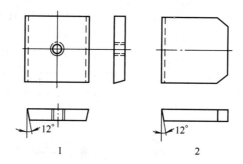

shape of blade in steel bar cutter
钢筋切断机的刀片形状

1 die cutter blade 冲切机刀片
2 fixed blade 固定刀片

(2) 钢筋切断器（2） Steel Bar Cutter（Ⅱ）

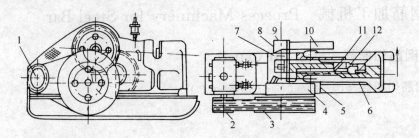

GQ-40 dynamo electric steel bar cutter GQ-40 电动钢筋切断机示意图

1　electromotor　电动机
2，3　pulley conical disk　三角带轮
4，5，9，10　speed reducing gear　减速齿轮
6　fixed blade　固定刀片
7　connecting rob　连杆
8　eccentric axle　偏心轮
11　slide block　滑块
12　movable blade　活动刀片

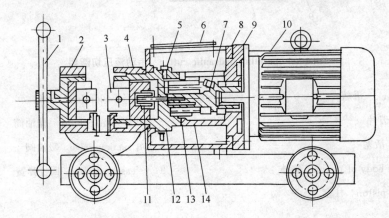

DYJ32 dynamo electric hydraulic steel bar cutter DYJ32 电动液压钢筋切断机示意图

1　stick　手柄
2　frame　支架
3　blade　刀片
4　piston　活塞
5　oil drain valve　放油阀
6　glass window　观察玻璃
7　eccentric axle　偏心轴
8，13　oil tank　油箱
9　link frame　连接架
10　electromotor　电动机
11　rubber cup　皮碗
12　cylinder body of oil tank　油缸体
14　plugger piston　柱塞

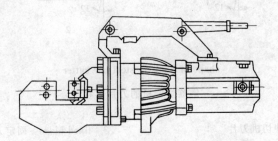

manual steel bar cutter 手动钢筋切断机简图

(3) 钢筋切断器 (3) Steel Bar Cutter (Ⅲ)

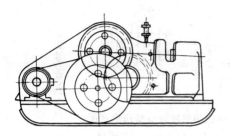

GJ5-40 steel bar cutter
GJ5-40 型钢筋切断机

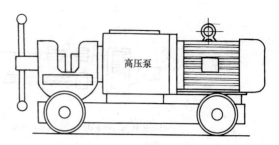

GJ5Y electric hydraulic steel bar cutter
GJ5Y 电动液压切断机

15.9.2 钢筋拔丝机 Steel Bar Stretcher

(1) 钢筋拔丝机 (1) Steel Bar Stretcher (Ⅰ)

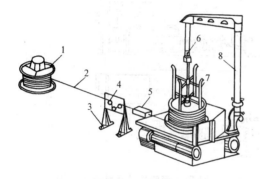

vertical single drum steel bar stretcher
立式单筒拔丝机

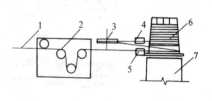

vertical single drum steel bar stretcher with double die 立式单筒双模拔丝机

1　coiled steel bar frame　圆盘架
2　steel bar　钢筋
3　sheave pulley　槽轮
4　steel bar-drawing die　拔丝模
5　pulley　滑轮
6　coiling drum　绕丝筒
7　frame　支架
8　electromotor　电动机

1　steel bar　钢筋
2　sheave pulley　槽轮
3　guiding pulley　导向轮
4, 5　steel bar-drawing die　拔丝模
6　coiling drum　绕丝筒
7　pedestal　机座

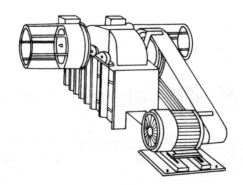

horizontal double drum steel bar stretcher　卧式双筒拔丝机

(2) 钢筋拔丝机 (2) Steel Bar Stretcher (Ⅱ)

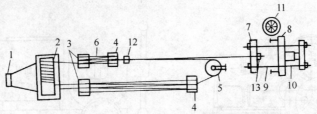

winch type cold-drawing machine 卷扬机式钢筋冷拉机结构示意图

1	earth anchor 地锚	8	fixed beam 固定横梁
2	winch 卷扬机	9	driving shaft 推动杆
3	fixed pulley block 定滑轮组	10	force measuring instrument 测力器
4	movable pulley block 动滑轮组	11	frame for coiled steel bar 放盘架
5	guiding pulley 导向滑轮	12	front-end clamp 前夹具
6	steel cable 钢丝绳	13	back-end clamp 后夹具
7	movable beam 活动横梁		

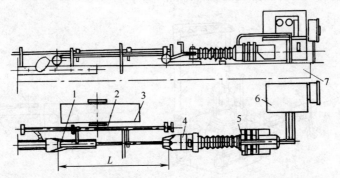

hydraulic pressure type cold-drawing machine 液压式钢筋冷拉机结构示意图

1	towing hook clamp on back-end 尾端挂钩夹具	5	hydraulic tension tank 液压缸
2	feed frame 进料架	6	panel box for pump and valve 泵阀控制器
3	dolly 装料小车	7	concrete base 混凝土基座
4	front-end clamp 前端夹具		

(3) 钢筋拔丝机 (3) Steel Bar Stretcher (Ⅲ)

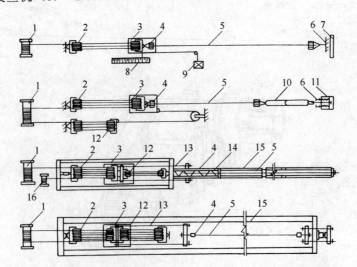

steel bar cold-drawing machine with winch 带卷扬机的冷拉钢筋机

1 winch 卷扬机	9 connecting link 连接杆
2 pulley set 滑轮组	10 spring force measuring instrument 弹簧测力器
3 clamp for steel bar 钢筋夹	11 pulley for reverse direction 回程滑轮组
4 steel bar 钢筋	12 transmission frame 传力架
5 earth anchor 地锚	13 tank style pedestal 槽式台座
6 protecting wall 防护墙	14 winch for reverse direction 回程卷扬机
7 measuring scale 标尺	15 electric scale 电子秤
8 ballast frame for reverse direction 回程荷重架	16 hydraulic pressure jack 液压千斤顶

（4）钢筋拔丝机 (4) Steel Bar Stretcher（Ⅳ）

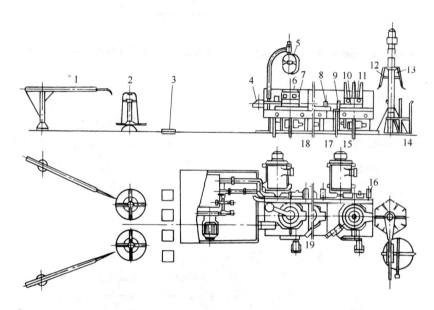

4/550 pulley steel bar cold-drawing machine　4/550型滑轮式钢筋冷拔机构造示意图

1 steel bar collector 集线臂	10 machine halt position for safety 安全停机位置
2 steel bar feeding frame 上线架	11 finished product drum 成品卷筒
3 machine halt position for twisting 乱线停机位置	12 steel bar removing frame 起线架
4 the first straightening die 进料拔丝模盒	13 boom crane 悬臂起重机
5 shelve for guiding pulley 导轮架	14 steel bar receiving frame 落线架
6 straighten drum 拔丝卷筒	15 gear 齿轮
7 hook for steel bar 穿丝钩	16 ventilating device 通风机
8 the second and the third straightening die 中间拔丝模盒	17 electromotor 电动机
9 machine halt position for break steel bar 断线停机位置	18 water-cooling and lubrication system 水冷却及润滑系统
	19 foot switch 脚踏开关

15.9.3 钢筋调直机 Steel Bar Straightening Machines

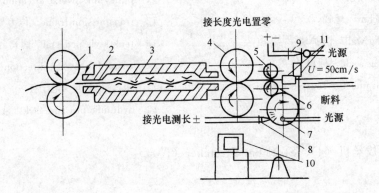

numerical control steel bar straightening machines 数控钢筋调直机结构示意图

1 feeding compressor roller 进料压辊
2 straightening column 调直筒
3 straighten block 调直块
4 drag roller 曳引轮
5 driven roller 从动轮
6 friction drive roller 摩擦轮
7 photo electricity disk 光电盘
8,9 photo electricity cell 光电管
10 electric magnet 电磁铁
11 blade 刀片

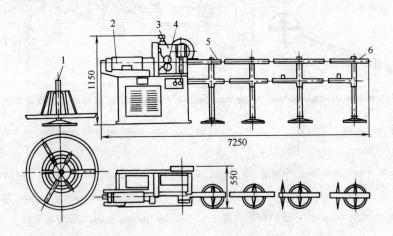

steel bar straightening machines 钢筋调直机结构示意图

1 frame for circled steel bar 放盘架
2 straightening column 调直筒
3 driving box 传动箱
4 pedestal 机座
5 bearing frame 承受架
6 fixing platform for measuring scale 定尺板

15.9.4 钢筋弯曲机 Steel Bar Bender

(1) 钢筋弯曲机 (1) Steel Bar Bender（Ⅰ）

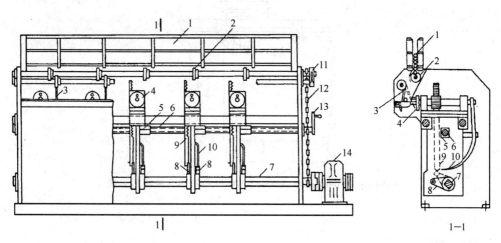

1　steel bar storage vessel　钢筋储槽
2　steel bar feeding device　拨料轮
3　push pole for feed　进料推杆
4　bender device　工作盘
5　nut for opening and close　锁帽
6　lead screw　丝杠
7　driving shaft　传动轴

8, 11　cam　凸轮
9　gear rack　齿条
10　connect bar　拔心轴用的连杆
12　chain　链条
13　manual operation rotating shaft　手摇轮
14　damper　减速器

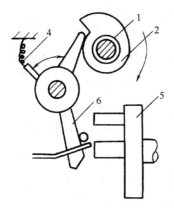 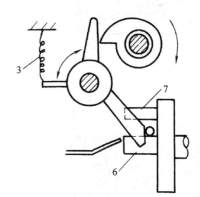

sketch of feeding part of bender　进料机构工作简图

1　axle for feed　进料轴
2　cam　凸轮
3　spring for replacement　复位弹簧
4　bender device　工作盘

5　center axle for bender　心轴
6　shaping axle　成型轴
7　steel bar　钢筋

(2) 钢筋弯曲机（2） Steel Bar Bender（Ⅱ）

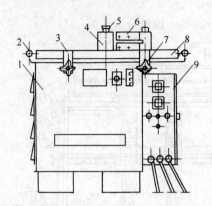

GW40 steel bar power bender　GW40 型钢筋弯曲机示意图

1　frame　机架
2　roller　滚轴
3，7　fasten stick　紧固手柄
4　shaft　转轴
5　adjust hand wheel　调节手轮
6　clamping device　夹持器
8　platform　平台
9　electric cabinet　配电箱

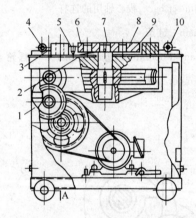

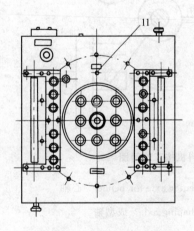

1　frame　机架
2　bender platform　工作台
3　feeding platform　插入座
4　feeding roller　送料辊
5　oil cup　油杯
6　worm gear case　蜗轮箱
7　center axle of bending platform　工作主轴
8　vertical axle　立轴
9　rotating circle platform　工作圆盘
10　worm gear　蜗轮
11　block for lubrication hole　加油孔螺塞

**GJB steel bar power bender
GJB 型钢筋弯曲机示意图**

(3) 钢筋弯曲机 (3) Steel Bar Bender (Ⅲ)

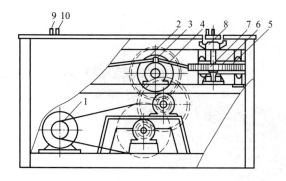

1 electric motor 电动机
2 eccentric wheel 偏心轮
3 eccentric ream 偏心铰
4 connect rob level 连杆
5 chain rack 齿条
6 slide way 滑道
7 spur gear 正齿轮
8 bender platform 工作盘
9 shaping device 成型轴
10 center axle 心轴
11 stop 限位铁

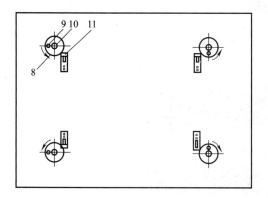

four bender bending machine 四头弯筋机

15.9.5 钢筋笼滚焊机 Rotary Welding Machinery for Steel Cage

15.9.6 数控钢筋切锯套丝机 NC Thread Machinery for Steel Sawing

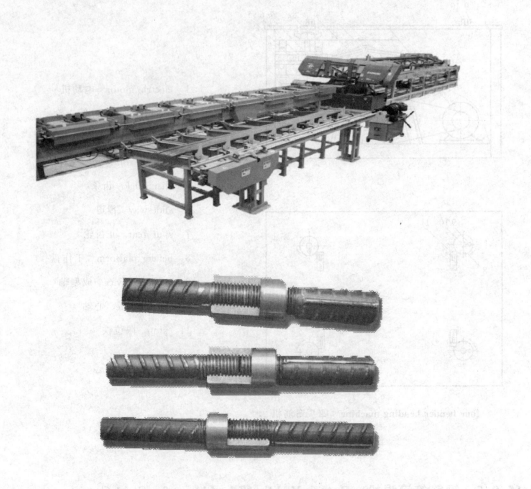

connection of ribbed reinforcing bars 带肋钢筋滚压直螺纹连接

15.9.7 数控钢筋弯曲机 CNC Rebar Bending Machinery

15.10 脚手架 Scaffolding

15.10.1 脚手架构件 Scaffolding Component

base plate 底座

galvanized steel tube 脚手架镀锌管

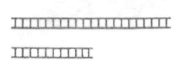

ladder beam 梯梁

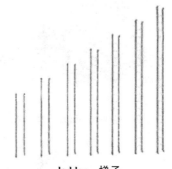

ladder 梯子

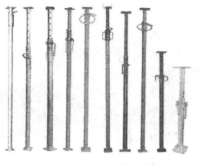

scaffolding props 脚手架支承

scaffolding plank/board 脚手架板

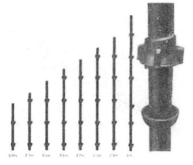

cuplock standard 碗扣式立杆

cuplock horizontal ledger 碗扣式横杆

putlog coupler（Wrap over）
横扣（猪耳扣）

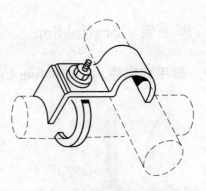
putlog coupler（mills hook）
横扣（钩式）

swivel coupler 旋转扣

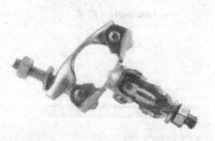
double coupler 十字口（直角扣）

girder clamp 悬梁扣件

ladder clamp 梯子扣

sleeve coupler 外接套管

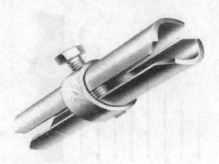
inner joint pin/spigot 内管接

board retaining clip 架板固定卡

brick guard 砖网

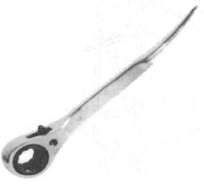
scaffold spanner 脚手架扳手

tube end caps 架杆端盖

guard rail 脚手架护栏

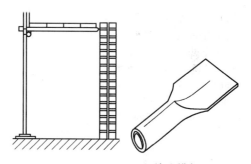
putlog and bridle 入墙式横杆

caster unit 小脚轮

scaffolding safety tags 脚手架安全挂牌

15.10.2 脚手架形式　Types of Scaffolding

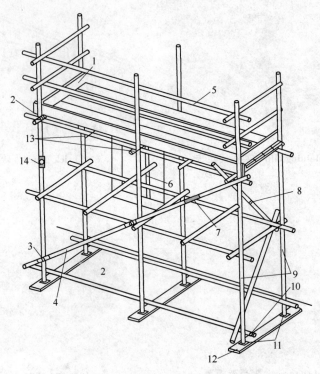

typical scaffolding type　典型脚手架图解

1　toe board　踢脚板
2　double coupler　十字口
3　swivel coupler　旋转扣
4　facade brace　立杆间支承
5　guard rail　护栏
6　window tie　窗口固定
7　sleeve coupler　外接套管
8　ledger to ledger brace　横杆间支承
9　standards　立杆
10　ledgers　横杆
11　base plates　底座
12　sole board　底板
13　transom　横撑
14　joint pin　内接管

steel/tubular scaffolding system　卡扣式钢管脚手架

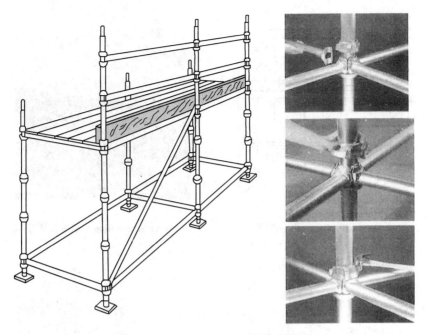

cuplock scaffolding system 碗扣式脚手架

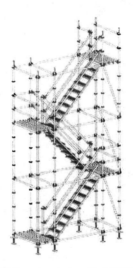

tower scaffolding（stairs） 塔式脚手架（楼梯）

putlog scaffolding 单排/横杆脚手架　　**independent scaffolding** 双排/独立脚手架

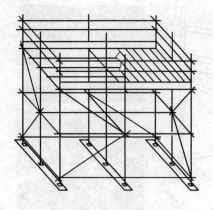
birdcage scaffolding 平台脚手架

cantilever scaffolding 悬挑式脚手架

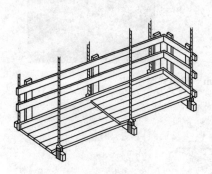
suspended scaffolding 吊式脚手架

mobile scaffolding 移动式脚手架

15.11 测绘 Surveying

15.11.1 经纬仪 Theodolite

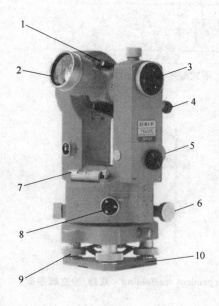

1　optical sight　瞄准镜
2　telescope　望远镜
3　micrometer screw　测微螺旋
4　reading microscope　读数目镜
5　vertical circle tangent screw　垂直微动
6　horizontal circle tangent screw　水平微动
7　alidade level　照准部水准器
8　horizontal clamp knob　水平度盘转换轮
9　leveling screw　脚螺旋
10　base plate　基座

15.11.2 光电测距仪 Photoelectricity Rangefinder

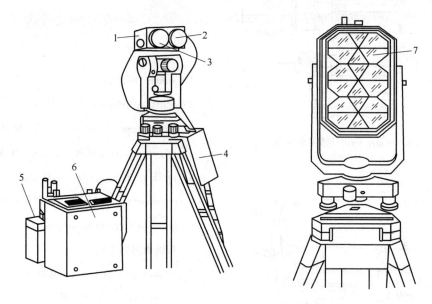

photoelectricity ranger finder 光电测距仪

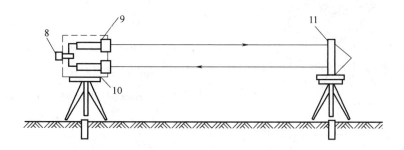

sketch of photoelectricity ranger finder 光电测距仪使用示意

1　aming head　照准头
2　beaming head　发射筒
3　receiving head　接收筒
4　frequency mixer　振荡器，混频器
5　power supply　电源
6　control box　控制箱
7　reflected prism　反射棱镜
8　date display　数字显示器
9　beaming telescope　发射望远镜
10　reflected telescope　接收望远镜
11　reflected mirror　反射镜

15.11.3 水准仪 Optical Level

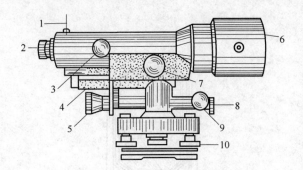

precision water level 精密水准仪

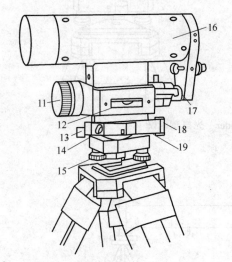

1 aimer 瞄准器
2 ocular of telescope 望远镜目镜
3 focus screw of telescope 望远镜调焦螺旋
4 reflected mirror of water level 水准器反光板
5 precision level screw 微倾螺旋
6 cuneiform glass jacket 楔形保护玻璃
7 horizontal precision measure handwheel 平行玻璃板测微手轮
8 holding on screw 制动螺旋
9 jiggle screw 微动螺旋
10 pedestal screw 脚螺旋
11 object lens 物镜
12 rectangle water level 长水准镜
13 holding on screw 制动螺旋
14 splint 托板部分
15 horizontal adjusting screw 水平螺旋
16 inner steel jacket 钢套内筒
17 ocular 目镜
18 round water level 圆水准器
19 adjusting screw of round water level 圆水准器校正螺旋
20 precision horizontal adjusting screw 微倾螺旋
21 foreside jacket 前护筒套
22 focus adjusting screw 调焦螺旋
23 precision adjusting screw 精调螺旋

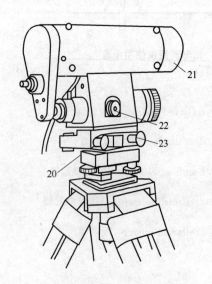

laser water level 激光水准仪

15.11.4 全站仪 Total Station

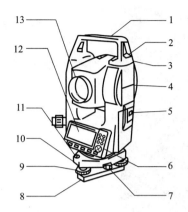

1 handle 提柄
2 handle securing screw 提柄固定螺钉
3 data input/output terminal（remove handle to view） 数据输入/输出终端（拆下手柄可见）
4 instrument height mark 仪器高标记
5 battery cover 电池盒
6 operation panel 操作键盘
7 tribrach clamp 三角基座控制杆
8 base plate 底座
9 levelling foot screw 脚螺旋
10 circular level adjusting screws 圆水准器校正螺旋
11 circular level 圆水准器
12 display 显示器
13 objective lens 物镜
14 tubular compass slot 管状罗盘插
15 optical plummet focussing ring 光学对中器物镜调焦环
16 optical plummet reticle cover 光学对中器分划板护盖
17 optical plummet eyepiece 光学对中器物镜调焦环
18 horizontal clamp 水平制动钮
19 horizontal fine motion screw 水平微动螺旋
20 data input/output connector 数据输入/输出插口
21 external power source connector 外接电源插口
22 plate level 照准部水准器
23 plate level adjusting screw 照准部水准器校正螺旋
24 vertical clamp 垂直制动钮
25 vertical fine motion screw 垂直微动螺旋
26 telescope eyepiece 望远镜目镜
27 telescope focussing ring 望远镜目镜调焦环
28 peep sight 瞄准器
29 instrument center mark 仪器中心标记

15.11.5 其他测绘工具 Other Surveying Tools

RTK GPS base station GPS 测量基站

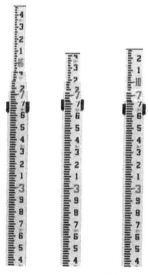

leveling rod 水准尺

tripod 三脚架

索引

A

安全带和安全绳　113
安全填埋场　89
安全栅　593
氨法脱硫　94
氨法烟气脱硫　59
氨合成塔　12
氨合成塔　251
氨氧化炉　24

B

百分误差　694
板式给料机　170
板式换热器　279
板式塔　223
半自动焊焊枪　784
伴管供给管线　449
伴热管　449
保护气体　784
保留常数　694
保留时间　694
保留体积　694
保温层　703
报警阀组　97
苯乙烯反应器　39
苯乙烯/环氧丙烷　38
泵　386
泵体密封环　392
避雷器　638
变电所屋内配电装置　664
变风量文丘里阀　758
变换炉　11，259
变频调速装置　655
变压吸附制氮　42
变压吸附制氮装置　117
表面曝气机　128
丙烷脱氢装置　31

丙烯产品保护床　30
波纹补偿器　494
波纹管密封闸阀　473
玻璃钢设备　309
玻璃器皿　691
泊松比　523
铂钴比色计　691
薄壳屋顶　711
不锈钢管　452

C

采暖系统与散热器　735
操动机构　641
槽钢管座　725
侧式悬臂堆料机　177
测绘　884
测量标高　765
测量坐标系统　765
测温元件　537
柴油发电机组　616
柴油机　442
长螺旋钻机　855
常减压蒸馏装置　8
场地土　717
超声波发射探头　804
超声波探伤　798
超声波探头　804
超声测厚仪　807
超声检测系统　807
超声探伤仪　807
超声显微镜　807
潮解　694
沉降罐、澄清器　325
衬里设备　303
衬砖设备　306
称重系统　596
成套式渗透探伤仪　806
承台　718

齿轮箱 443
翅片 286
冲动式汽轮机 208
冲击夯 853
抽头式焊机 775
出风口 758
出水口闷盖 110
除尘器 92
船形焊 780
窗边框泛水 709
窗洞 727
窗间墙 703
窗台 703
窗台泛水 710
纯水仪 691
磁粉探伤 798
磁化法 807
磁力泵 395
次梁 731
粗砂 717
粗氩塔Ⅰ 41
催化氧化 92
萃取器 319

D

搭接焊 777
搭接接头 777
打桩机 849
大气污染控制 86
带式输送机 133
带式输送机系统 133
袋式过滤器 330
单边V形坡口 772
单级汽轮机 209
单螺杆泵 399
弹簧安全阀 469
弹簧安全泄压阀 480
弹簧架 497
弹簧减震器 740
当量 694
挡土墙 771
道路结构 767
道牙 767
等高线 765

等径三通 453
等径四通 453
等压式混合管 791
低温储罐 302
低温甲醇洗 37
低温冷箱 120
低压开关柜 643
低压熔断器 646
低压式发生器 792
底基层 767
地表水环境质量 86
地基处理机械 849
地脚螺栓 720
地下管道 498
地下管道图例 498
地下管平面布置图 498
地形图 765
第二废热锅炉 383
第一废热锅炉 382
典型的探头 803
典型电伴热系统 677
典型防雷系统 686
点焊 772
电伴热 450
电伴热产品 675
电磁流量计 558
电导仪 575
电动机防雨罩 740
电动机启动器 656
电动执行机构 577
电焊机 785
电弧 775,785
电机支架 742
电极 694
电解 694
电缆、电线 669
电缆附件 671
电气焊 788
电梯 704
电泳 694
电子照相 801
电阻焊焊点 772
垫层 718
垫片 460

吊杆 725
吊挂式链输送机 145
吊篮 821
吊装带 811
跌水 770
蝶阀 478,579
丁二烯抽提装置 10
定量 694
定性分析 694
斗轮堆取料机 176
独柱式 725
锻铁管 452
堆焊 772
对苯二甲酸 38
对苯二甲酸氧化反应器 38
对接焊 777
对接接头 777
对阴极,钨靶 802
钝化 694
多孔性填料 791
多联机空调 751
多效蒸发系统 95

E

颚式破碎机 150
耳罩及耳塞 113
二段转化炉 11
二段转化炉 373

F

发射探头 803
阀杆 791
阀控式密封铅酸蓄电池 629
阀门 461,468,470
阀门操纵机构 482,483
阀门定位器 584
阀塞,阀心 467
筏形基础 718
法兰 458
法兰管件 456
法兰接管 107
反铲挖掘机 844
反辐射式电子照相 801
反渗透膜元件 126

反作用式减压器 791
芳烃联合装置 8
防爆安装单元 660
防爆控制设备 662
防爆离心式屋顶通风机 742
防爆照明灯具 661
防潮层 703,718
防腐 518
防护服 112
防护鞋 114
防护眼镜 114
防渗 96
放射性钴 798
废气污染物 84
废水污染物 84
废液焚烧炉 203
分辨率 694
分级机 163
分解 694
分离器 488
分散控制系统 602
分水罐 37
分体式风冷空调机 746
分析误差 694
分子筛纯化器 41
焚烧厂 90
粉煤加压气化炉 376
粉煤烧嘴 33
粉土 717
风管 740,753
风管弯头 753
风管止回阀 757
风机 432
风机盘管 752
风机叶轮 736
风玫瑰 765
封底焊 772
峰高 694
峰宽 694
峰面积 694
缝焊 772
浮阀塔盘 226
浮筒式液位计 565
腐蚀 521

891

负压气力输送系统 147
腹板加劲肋 731
伽马射线液位计 566

G

干粉罐 99
干粉显示剂,干性显示粉 806
干燥粉磨机 159
干燥器 336
刚度-质点模型 528
钢带造粒机 180
钢焊接管件 453
钢结构 443
钢结构厂房 708
钢筋 713
钢筋滚压直螺纹连接 878
钢筋混凝土梁 727
钢筋混凝土水箱 727
钢筋混凝土圆顶 727
钢筋混凝土支承 727
钢筋混凝土支架 727
钢筋加工机械 869
钢丝刷 775
钢支承 725
钢制板式散热器 735
高密度澄清池 51
高速压缩机 426
高压甲铵冷凝器 14
高压气缸 423
高压熔断器 640
割炬 775
格构架 716
隔离器 594
隔膜电解槽 258
隔膜式压缩机 420
隔热层 794
给料斗 138
工厂空气 43
工程车辆 862
工具 828
工业电视监视系统 600
工艺蒸汽发生器 30
共聚单体干燥器 16
共聚单体脱气塔 16

构造柱 703
骨架 716
鼓风机 430
固定顶储罐 289
固定端 716
固体废物 88
固体混合机械 318
管板 270
管道泵 396
管道编号 5
管道布置 443
管道布置图 444
管道吊架 496
管道动态效应 527
管道工程 443
管道混合器 313
管道机械 493
管道绝热 500
管道盲板阀 469
管道平面布置 444
管道特殊元件 490
管道图例 3
管道应力 524
管道应力分析 522
管道用紧固件 461
管道支吊架 495
管道支架 495
管道组装图 447
管架 365
管架零部件 497
管件 453
管壳式换热器 266
管廊 443
管螺纹 459
管式法低密度聚乙烯 17
管式加热炉 368
管式消声器 763
管箱 271
管牙接口 106,109
管支架 725
管状焊丝 784
管状焊丝电弧焊 790
管状焊丝气体保护焊 784
光电测距仪 885

辊子输送机　144
锅炉　182
锅炉房给水系统　53
锅炉房锅炉给水系统　54
锅炉燃烧系统　55
过滤机　328
过滤器　123，487
过热器　190
过热区　796

H

含盐废水处理　51
涵洞　771
夯实砾石层　718
焊穿　781
焊道　775
焊缝　807
焊工　775
焊件　776
焊脚　796
焊接　772
焊接电缆　775
焊接符号　772
焊接机头　787
焊接接头　776
焊接缺陷　781
焊炬　775
焊钳　775
焊枪　785
焊丝盘　782，785，787
焊条　777
焊芯　775
焊渣　784
合成氨装置　11
合成气制乙二醇装置　23
荷载条件　714
横焊　777
横梁　725
横流冷却塔　122
烘箱　691
干燥箱　805
红外线气体分析仪　573
呼吸器　112
弧坑　781

弧柱　776
户内真空断路器　638
护目镜片　775
护坡　770
护罩　110
滑模施工　865
环管反应器　255
环境空气质量　85
环境质量　85
环网柜　635
环氧乙烷/乙二醇装置　21
换热设备　263
回填土　718
绘图示例氨库装置　5
混合（搅拌）槽　312
混合离子交换器　51，126
混合设备　310
混流泵　397
混凝土　713
混凝土泵车　858
混凝土底板　720
混凝土基础　720
混凝土搅拌运输车　842
混凝土面层　767
活塞　467
活塞式压缩机　418
火车装车机　178
火炬　129
火炬头　131
火炬系统　129
火灾自动报警系统　597

J

基础剖面　718
急救箱　114
急冷油塔　15
急流槽　770
集风器　736
集肤效应　808
集气管　792
计量泵　401
加强板　780
加热盘管　736
加压固定床煤气化炉　378

夹套管　449
甲烷化炉　12
间接干燥器　338
剪切　523
剪应力　523
减温器　191
减压阀　469
减压器　775
减震钢架　740
碱洗塔　15
降解　694
交流不间断电源　630
交流电动机　611
浇注料炉衬结构　362
角钢　725
角焊　780
角接接头　772
角形控制阀　579
脚手架　879
搅拌器　435，792
搅拌设备　839
校正因子　694
接触式直探头　803
接地产品　685
接头形式　777
接线盒　742
结构钢　713
结晶器　51
截止阀　474
金属补偿器　493
金属电弧焊　775
金属封闭式开关装置　633
金属软管　493
金属线型透度计　801
紧急停车系统　605
进深　703
进水口闷盖　110
浸没燃烧式气化器　28，180
浸提设备　319
经纬仪　884
精对苯二甲酸结晶器　38
精氩塔　41
警灯　99
净水场　44

静态荷载　528
静止混合器　314
聚丙烯装置　18
聚合釜　257
聚焦线圈　802
聚结器　40
聚氯乙烯装置　20
聚醚多元醇装置　39
卷边焊　777
卷边接头　777
绝对精度　694
绝对偏差　694
绝对误差　694
绝热　500
绝缘开关装置　632
均匀自动给料机　171

K

卡尔费体水分分析仪　691
开架式气化器　28
开间　703
开式膨胀水箱　739
科氏力质量流量计　549
壳体和封头　345
空分　40
空分单元　36
空分设备　116
空间管架　725
空冷器、空气冷却器　282
空冷式手工焊枪　783
空冷塔　41
空气疏水阀　484
空气预热器　192
空压　40
空压站　43
控制球阀　580
框架　716
框架式空气断路器　644

L

垃圾焚烧炉　205
拉杆　725
拉链式蝶阀　756
拉伸　523

捞渣机　33
雷达液位计　567
累积误差　694
冷分离和冷箱　15
冷媒出口　751
冷却塔　121，727
冷箱　116
离心泵　391
离心风机　740
离心机　332
离心式冷水机组　747
离心压缩机　426
离子色谱仪　691
立焊　777
立角焊　780
立面图　445，703
立柱　725
沥青混凝土联结层　767
沥青混凝土磨耗层　767
砾石　717
砾石垫层　718
连接环　724
连接角钢　731
连续抽提设备　320
连续萃取设备　320
联轴器　392
炼厂　6
炼化一体化　7
量具　838
料仓、料斗与阀门　164
料斗与溜槽　165
裂缝　806
裂解炉　15
檩条　710
流化床分布器　344
流化床煤气化炉　379
流化床吸附　91
流量元件　550
流态化　343
硫化　303
硫化罐硫化　303
硫黄回收单元　36
硫酸装置　10
六氟化硫断路器　636

楼梯　704
炉管、管件及联箱（集合管）　362
炉排　193
卤代烷　112
鲁奇气化炉　378
滤池　51
滤清器　792
履带式起重机　824
绿化　769
轮式挖掘机　845
轮胎式压路机　853
螺杆泵　399
螺杆式冷水机组　750
螺杆压缩机　413
螺纹管件　454
螺旋输送机　143
螺旋钨丝　800

M

马弗炉　692
埋刮板式输送机　143
煤气化工艺　32
煤制甲醇　36
煤制油工艺　34
门洞　727
门式取料机　177
闷盖　107
迷宫式压缩机　419
密封胶层　709
密封圈　109
密相气力输送系统　148
面罩　112，113
膜分离制氮　42
膜分离制氮装置　117
膜组　42
磨机　157
磨煤机　33，66，199
末端试水装置　97
木工工具　828

N

挠度　716
内防护　518
内浮顶罐　295

内人梯 739
泥炭土 717
逆流冷却塔 122
黏土 717
尿素SNCR脱硝系统 61
尿素合成塔 14，253
尿素装置 13
凝结水出口 736
扭转 523
浓盐水结晶系统 51
浓盐水蒸发系统 51
女儿墙 703
暖风机 736
暖通 735

P

爬模施工 868
排放丝堵 792
排气筒 726
排液阀 449
抛煤机 198
泡沫液 99
喷燃器，燃烧器 196
喷射泵 410
喷射过渡 785
喷射装置 411
喷头 97
喷雾干燥器 339
喷油螺杆压缩机 414
膨胀机/再压缩机 441
皮托管泵 413
疲劳曲线 528
片式消声器 763
漂石 717
平焊 777
平面图 703
屏蔽泵 394
坡道 703
坡度 771
坡口 772
破沫器 348
破碎和筛分设备 149
剖面图 703
普通电渣焊 786

Q

气动隔膜泵 406
气动执行机构 577
气割 775
气焊 775
气焊设备 791
气化炉 33
气流（气动）输送干燥器 341
气体保护电弧焊 784
气体报警盘 98
气体净化器 775
气体洗涤器 326
气相腐蚀 521
气相色谱仪 691
汽轮机空冷系统 57
汽轮机循环 207
起重机械 809
千斤顶 819
铅室 798
墙板 716
墙垛 703
桥式起重机 825
轻烃回收装置 8
球阀 476
球磨机 158
驱动机 436
驱动轮 134
取样冷却器 488
全站仪 887

R

燃煤锅炉 182
燃气轮机 219，440
热泵压缩机 31
热导式分析仪 575
热电厂 663
热电站锅炉系统 58
热电站系统 56
热膨胀阀 486
热水伴热管 449
热阴极X射线发生管 800
热载体加热炉 384，385
人孔 739

熔池 776
熔滴 785
熔剂 782
熔渣 784
熔嘴电渣焊 787
润滑油型 6

S

塞焊 777
三层聚乙烯涂层 519
三孔透度计 802
三螺杆泵 400
三通控制阀 579
三维模型 528
散热器 735
散水 703
色谱仪 572
砂轮 775
山墙泛水 709
闪蒸塔 37
伤纹 798
上泵体 440
上基层 767
设备布置 443
射吸式混合管 791
射线照相 801
深冷空分内压缩工艺流程 40
渗透伤纹 806
渗透探伤 808
声环境质量 86
省煤器 60,195
十六烷值机 691
石灰石/石灰-石膏法烟气脱硫 60
石墨设备 307
实心焊丝 784
事故淋浴洗眼器 114
饰面 703
视镜 489
室内机 751
室外机 751
手柄式蝶阀 756
手持式面罩 775
手动葫芦 819
手动矩形蝶阀 469

输送带 133,138
输送机和提升机 133
树脂绝缘干式变压器 623
数控钢筋切锯套丝机 878
双螺杆泵 400
双螺杆连续混合机 315
双柱管架 725
双作用蒸汽往复泵 407
水池 97
水环真空泵 402
水冷式手工焊枪 783
水龙带接口 107
水煤浆气化炉 377
水平加劲板 731
水平连接板 731
水（油）浴炉 384
水浴 692
水准点 765
水准仪 886
伺服液位计 569
送丝滚轮 787
送丝机构 787
送条器 795
塑壳断路器 649
塑料压接管接头 457
酸性水汽提装置 8
碎石滤层 771
缩缝 767

T

塔 720
塔底板 720
塔吊 825
塔基础 720
塔盘型式 226
踏步 703
探孔镜 799
探头 798
羰化反应器 24
搪玻璃设备 304
陶瓷纤维毯结构 358
套管式纵向翅片换热器 278
特种锅炉 202
提离效应 808

897

天平　691
天然气液化　25
填充砖墙　716
填函式伸缩节　493
填角焊　772，777
填料　491
填料塔　237
填料型式　239
挑台　724
条形基础　718
铁塔及电杆　667
通风管道　753
筒身衬里　724
头戴式面罩　775
透度计　802
透平冷凝系统　444
凸焊　772
涂料　520
土层剖面图　717
土方工程　769
土建工程　703
土路基　766
推土机　845
拖车　864
脱气器　51
脱气塔　127

W

挖掘机　843
外浮顶罐　296
外墙　718
外人梯　739
外压缩工艺流程　41
弯曲　523，716
烷基化装置　8
碗扣式　882
碗扣式脚手架　884
往复泵　407
往复式压缩机　416
往复式真空泵　404
微孔曝气器　127
围梁　709
围檩　710
卫生间　704

温度变送器　539
涡流探伤　798
涡流探伤仪　808
污氮气　42
污染控制　86
屋顶风机　742
屋顶式空调机　747
屋脊盖板　710
屋面板　703，709
无机试剂　692
无硫甲醇冷却器　37
无刷同步电动机　614
无损检验　797
无油螺杆压缩机　413
无阻尼振动　528
物料代号　1
雾化喷头、喷雾嘴、雾化器　340

X

吸入筒　392
吸收池　694
吸收剂量　808
吸收谱带　694
吸收曲线　694
稀相气力输送系统　149
洗眼站　451
细砂　717
下泵体　440
下饲炉排　194
显色粉　798
显微镜　799
显影　808
限位开关　586
线性低密度聚乙烯装置　16
线应变　523
箱式变电站　624
橡胶衬里　303
消防泵　97
消防水泵接合器　108
消防水带　Fire Hose　111
消火栓　107
消声片　763
消声器　488
消声腔　763

泄流孔　771
卸荷阀　424
辛烷值机　691
新风换气机　752
新风净化机组　751
型钢梁　725
溴洗涤塔　38
蓄热式氧化器　91
悬挑式脚手架　884
旋片式真空泵　402
旋塞阀　477
旋涡泵，涡流泵　386
旋涡流量计　553
旋涡脱落　528
旋转三通　457
选择阀　98
选择性催化氧化还原　94
选择性催化氧化还原　95
循环流化床锅炉　202

Y

压顶　703
压力变送器　546
压力表　544
压力波　528
压力过滤器　123
压滤机　328
压缩机、鼓风机　413
烟道　724
烟气轮机　441
烟气脱硝 SCR　60
烟羽　89
阳离子交换器　51
仰焊　777
仰角焊　780
氧分析器　573
氧化膜　785
氧氯化反应器　256
氧气混合站　22
氧气瓶　775
氧气压盖螺母　791
摇溶渗透剂　808
药皮　775
叶滤机　329

液氮洗　121
液化天然气接收站　27
液力耦合器　443
液力透平　440
液面控制浮球阀　481
液相色谱　698
液压升降平台　859
液氧泵　41
一段转化炉　11
仪表符号　2
仪表空气　43
仪表空气增压机　43
乙苯过氧化氢反应器　39
乙二醇装置　10
乙炔　793
乙炔瓶　775
乙烯-醋酸乙烯共聚物　17
乙烯装置　15
异径三通　453
异径四通　453
阴极保护系统　683
阴离子交换器　51
应变　522
应力　523
应力腐蚀　521
应力张量　523
荧光磁粉　808
荧光屏　801
荧光渗透试验　798
硬 X 射线滤光板　801
硬路基　766
油浸式变压器　619
有机试剂　693
余热 ORC 发电系统　62
余热锅炉　204
余热回收　380
雨棚　703
预热　796
预装式变电所　662
元素分析仪　691
原土　718
原子发射光谱仪　691
原子吸收光谱仪　691
圆形堆场堆取料机　176

圆振筛 162

Z

再生加热器 41
再生塔/CO_2 汽提塔 250
造粒塔 260
造粒塔扒料机 261
噪声污染控制 88
增感屏贴层 801
增压透平膨胀机 41
闸阀 471，472
张紧装置 133
胀缝 767
真空泵 402
真空抽吸系统 782
真空接触器 639
振动给料机 169
振动筛 161
振动式压路机 853
蒸汽 41
蒸汽伴热管 448
蒸汽动力系统 182
蒸汽轮机 436
蒸汽疏水阀 484
正压气力输送系统 147
正应力 523
正作用式减压器 791
支承加劲肋 731
支架梁 728
直边坡口 772
直流不间断电源 631
直通单座控制阀 578
止回阀 475
指示剂 694
酯化反应塔 24
制冷机 118
制冷装置 118
制氢装置 8
中和器 40
中间罐区 36
中压金属铠装开关装置 634
中压式发生器 792
重力无阀过滤器 123
轴流泵 395

轴流压缩机 430
主冷凝蒸发器 41
主梁 731
柱塞流 528
铸铁管 452
砖炉衬结构 360
砖筒支承水箱 727
砖烟囱 724
转动机器 386
转鼓真空过滤机 331
转化炉 371
转筒造粒机 181
桩 718
装车臂 179
装车撬 179
装卸臂 178
装载机 846
装置 38
锥顶 727
准确度 694
着色渗透剂 808
着色渗透试验 798
紫外可见光谱仪 691
自垂百叶 736
自动放空阀 101
自动埋弧焊 782
自力式调节阀 582
自然地形线 766
自然通风冷却塔 121
纵缝 767
纵梁 720
纵向焊缝 775
阻火器 487
阻尼探头 804
阻尼振动 528
组合电器 631
钻孔机 853

CO_2 吸收塔 249
CO 分离 120
CO 燃烧废热锅炉 381
LNG 储罐 26
MTP 工艺流程 30
pH 分析仪 575

T形接头 777
T字接头 772
V形槽 795
V形控制球阀 580
X射线管 801

X射线照相 798
X形槽 795
γ射线 798
γ射线照相 798
π形补偿器 725